Advances in Intelligent Systems and Computing

Volume 1248

The series "Advances in Intelligent Systems and Computing" contains publications on theory, applications, and design methods of Intelligent Systems and Intelligent Computing. Virtually all disciplines such as engineering, natural sciences, computer and information science, ICT, economics, business, e-commerce, environment, healthcare, life science are covered. The list of topics spans all the areas of modern intelligent systems and computing such as: computational intelligence, soft computing including neural networks, fuzzy systems, evolutionary computing and the fusion of these paradigms, social intelligence, ambient intelligence, computational neuroscience, artificial life, virtual worlds and society, cognitive science and systems, Perception and Vision, DNA and immune based systems, self-organizing and adaptive systems, e-Learning and teaching, human-centered and human-centric computing, recommender systems, intelligent control, robotics and mechatronics including human-machine teaming, knowledge-based paradigms, learning paradigms, machine ethics, intelligent data analysis, knowledge management, intelligent agents, intelligent decision making and support, intelligent network security, trust management, interactive entertainment, Web intelligence and multimedia.

The publications within "Advances in Intelligent Systems and Computing" are primarily proceedings of important conferences, symposia and congresses. They cover significant recent developments in the field, both of a foundational and applicable character. An important characteristic feature of the series is the short publication time and world-wide distribution. This permits a rapid and broad dissemination of research results.

Indexed by SCOPUS, DBLP, EI Compendex, INSPEC, WTI Frankfurt eG, zbMATH, Japanese Science and Technology Agency (JST), SCImago.

All books published in the series are submitted for consideration in Web of Science.

More information about this series at http://www.springer.com/series/11156

Samarjeet Borah · Ratika Pradhan ·
Nilanjan Dey · Phalguni Gupta
Editors

Soft Computing Techniques and Applications

Proceeding of the International Conference on Computing and Communication (IC3 2020)

Springer

Editors
Samarjeet Borah
Sikkim Manipal Institute of Technology
Majhitar, Sikkim, India

Ratika Pradhan
Sikkim Manipal Institute of Technology
Majhitar, Sikkim, India

Nilanjan Dey
Department of Computer Science
and Engineering
JIS University
Kolkata, West Bengal, India

Phalguni Gupta
GLA University
Mathura, Uttar Pradesh, India

ISSN 2194-5357 ISSN 2194-5365 (electronic)
Advances in Intelligent Systems and Computing
ISBN 978-981-15-7393-4 ISBN 978-981-15-7394-1 (eBook)
https://doi.org/10.1007/978-981-15-7394-1

This Springer imprint is published by the registered company Springer Nature Singapore Pte Ltd.
The registered company address is: 152 Beach Road, #21-01/04 Gateway East, Singapore 189721, Singapore

Preface

Soft computing is a subfield of computing that provides vague but practical resolutions to difficult computational problems. This is an emerging research area which tries to acquire the incredible ability of the human mind to dispute and acquire in the situation of ambiguity and cynicism. It is also based on many biologically induced method including evolution, genetics, behaviour of ants, nervous system of humans, particles, etc.

Components of soft computing may include fuzzy-based systems and fuzzy logic, perceptron, evolutionary computing, genetic algorithms, artificial neural networks, metaheuristics, swarm and particle intelligence, expert systems, machine learning, etc. Soft computing provides solutions to many real-life problems where a proper mathematical model is difficult to formulate. Most of the components as well as algorithms are available as tools and software packages, which are being used by researchers to solve wide range of problems.

Looking at its applicability, wide range of domains has adopted soft computing to resolve complex problems. For example, nowadays, fuzzy-based systems and fuzzy logic have been used in many consumer appliances, such as refrigerator, air conditioners, washing machines and microwave ovens. Soft computing also helps in decision support systems, data processing, various machine design, design of electrical circuits, image and video processing, etc.

In view of these, the third International Conference on Computing & Communication (IC3-2020) is cantered upon the theme Soft Computing Techniques and Applications. IC3-2020 is the third edition of the conference. The first and second editions were organized in 2016 and 2018, respectively. The conference is organized by the Department of Computer Applications, Sikkim Manipal Institute of Technology (SMIT), Majhitar, Sikkim. SMIT is a constituent unit of Sikkim Manipal University.

The conference has received papers from diverse engineering domains using soft computing as a tool to solve the complex problems. The papers included in this volume are broadly from several domains of applications of soft computing, such as —fuzzy-based systems, data analytics, biomedical science, image processing, natural language processing, computer network and security and design, circuit and communications. However, this categorization can be considered as a rough one, since research domains always tend to overlap.

The fuzzy-based systems consists of three research works, where the first work presents a numerical study on second-grade fluid flow problems using analysis of fractional derivatives under fuzzy environment. This is followed by a work on optimization of fuzzy inference system through genetic algorithm and an improvised version of on-policy reinforcement learning algorithm.

The volume includes several works on data analytics. Majority of the research works are concentrating on developing a perfect prediction system on various domains of interest, such as air quality, crime rate, fraud rate and suicide cases. Deep learning is being used in solving many real-life problems. Two of such works are included in this volume, where one is working on recognition of CAPTCHA and the other on the medicinal plant.

A handful of research works have been received from biomedical domain. They are basically using soft computing techniques towards classification, prediction and learning process, such as classification of diseases like Alzheimer, cancer, melanoma, etc., detection of Parkinson's disease and prediction of diabetes.

The volume also contains few selective works from the domain of natural language processing. Majority of the works are dealing with speech processing, which include news headline generation, accent classification from speech, word segmentation of noisy speech, etc. The volume is also including works on scene character recognition as well as on predicting the appropriate category of books for online book store.

The recent research in image and video processing issues reveals that effective use of soft computing techniques may lead to better findings. The image processing works included in this volume highlight on rice disease detection and monitoring, object detection, classification of military signs, satellite images, forest fire analysis, mathematical symbols and trigonometric function recognition, etc. A unique work on video processing which deals with Sattriya dance gesture recognition is also a part of the volume.

Soft computing techniques are widely used in security domain. Modern intrusion detection systems are mostly based on various machine learning and soft computing techniques. Large number of works were received in this category, and some have been included in the volume. The works on security domain include DDoS attack in cloud infrastructure, attacks on cognitive radio ad hoc networks, discovery and deterrence of black hole attack, etc. From the computer network point of view, nowadays people mostly talk of sensor network, cognitive radio network and Internet of things (IoT). Specifically, in IoT, which is a fast-growing research area, people are exploring efficient ways of communication and area of applications. This volume is consisting of few works on IoT highlighting the performance of scalable data linked to IoT, ploughing and seeding of agriculture robot using IoT, etc. A work on branch and bound method for maximizing network capacity of cognitive radio networks is also included in this volume.

It has been seen that many simulation software packages are developed based on soft computing algorithms. These simulation software packages are being used in designing a solution for various kinds of hardware-based research issues before going to actual production. Few of such kind of works include induction motor-related issues, minimization of torque ripple of SRM, voltage source inverter-related issues, modelling and analysis of wind turbine, power conditioning unit for wind energy, CMOS technologies, renewable energy system, etc. It is expected that promoting such kind of research work will encourage researchers in developing better hardware products.

This volume includes mostly the ongoing works and research findings, in various research laboratories, universities and institutions from geographically diversified area. It is expected that the findings will lead to the development of demanded and successful products. The volume includes works to show how to apply soft computing methods to solve important and real-life problems as well as how the research has been conducted. The editors of this volume extend heartfelt thanks to all the authors, contributors, reviewers and specially Springer (the publisher) for making this volume possible.

Majhitar, India — Samarjeet Borah
Majhitar, India — Ratika Pradhan
Kolkata, India — Nilanjan Dey
Mathura, India — Phalguni Gupta

Contents

About the Editors

Dr. Samarjeet Borah is currently working as Professor in the Department of Computer Applications, Sikkim Manipal University (SMU), Sikkim, India. Dr. Borah handles various academics, research and administrative activities. He is also involved in curriculum development activities, board of studies, doctoral research committee, IT infrastructure management etc. along with various administrative activities under SMU. Dr. Borah is involved with three funded projects in the capacity of Principal Investigator/Co-principal Investigator. The projects are sponsored by, AICTE (Govt. of India), DSTCSRI (Govt. of India) and Dr. TMA Pai Endowment Fund. Dr. Borah is involved with various journals of repute and book volumes as Editor/Guest Editor.

Dr. Ratika Pradhan is working as Professor in the Department of Computer Science and Engineering, Sikkim Manipal Institute of Technology since July 1999. She received Ph.D. from Sikkim Manipal University (SMU) in 2011 and M.E. (CSE) from Jadavpur University, Jadavpur, in 2004. Her areas of research interest are digital image processing, remote sensing, and GIS. She has published 25 journal papers, and 5 conference papers. Dr. Pradhan also completed few research projects funded by All India Council for Technical Education, Govt. of India.

Nilanjan Dey is an associate professor in the Department of Computer Science and Engineering, JIS University, Kolkata, India. He is a visiting fellow of the University of Reading, UK. Previously, he held an honorary position of a Visiting Scientist at Global Biomedical Technologies Inc., CA, USA (2012–2015). He was awarded his PhD from Jadavpur University in 2015. He has authored/edited more than 70 books with Elsevier, Wiley, CRC Press, and Springer, and published more than 300 papers. He is the Editor-in-Chief of the International Journal of Ambient Computing and Intelligence (IGI Global), Associated Editor of IEEE Access, and International Journal of Information Technology (Springer). He is the Series Co-Editor of Springer Tracts in Nature-Inspired Computing (Springer), Series Co-Editor of Advances in Ubiquitous Sensing Applications for Healthcare (Elsevier), Series Editor of Computational Intelligence in Engineering

Problem-Solving and Intelligent Signal Processing and Data Analysis (CRC). His main research interests include medical imaging, machine learning, computer-aided diagnosis, data mining, etc. He is the Indian Ambassador of the International Federation for Information Processing—Young ICT Group and Senior member of IEEE.

Dr. Phalguni Gupta received his Doctoral degree from Indian Institute of Technology Kharagpur in 1986. He started his carrier in 1983 by joining in Space Applications Centre (ISRO) Ahmedabad, India as a Scientist. In 1987, he joined the Department of Computer Science and Engineering, Indian Institute of Technology Kanpur, India. He is a Professor in the department. Currently, he is the Vice Chancellor of GLA University, Mathura. He has published about 300 papers in International Journals and Conferences. He is also an author of 2 books and 14 book chapters. He has dealt with several sponsored and consultancy projects which are funded by the Government of India. Some of these projects are in the area of Biometrics, System Solver, Grid Computing, Image Processing, Mobile Computing, and Network Flow. He has established the Biometrics Lab at IIT Kanpur.

Scene Character Recognition with Morphological Filtering and HOG Features

Payel Sengupta and Ayatullah Faruk Mollah

Abstract Recognition of scene text is an essential task in the extraction of meaningful information from camera captured images. It is a challenging problem due to heterogeneity in the nature of scene text and complexity in the surrounding background. In this paper, a new recognition method is proposed for recognizing scene characters extracted from scene images. At first, original scene character components are converted into binary character images and then, morphological operations are applied to reduce noises. Subsequently, we extract HOG features from these scene character components and apply multiple classifiers to classify those scene characters components. This method is evaluated on a dataset of 1905 camera captured scene character images and obtained character recognition rates are 56.83% (Naive Bayes), 77.59% (KNN), 75.23% (Random Forest), 78.06% (MLP) and 82.07% (SVM), which is reasonably satisfactory considering scene character complexity.

Keywords Scene character recognition · Morphological opening · HOG features · Pattern classification

1 Introduction

Scene character recognition is an important problem in the field of computer vision. After scene text localization and character segmentation, classification of segmented/dissected characters is the next essential sub-problem in scene text reading. Unlike optically scanned well-segmented binary characters, scene characters are very complex due to numerous factors such as background heterogeneity, non-uniform illumination, text diversity, etc. The sample scene character images are shown in Fig. 1. As a result, traditional character recognition methods cannot

P. Sengupta (✉) · A. F. Mollah
Department of Computer Science and Engineering, Aliah University, IIA/27 New Town, Kolkata 700160, India
e-mail: payel9433@gmail.com

A. F. Mollah
e-mail: afmollah@aliah.ac.in

S. Borah et al. (eds.), *Soft Computing Techniques and Applications*,
Advances in Intelligent Systems and Computing 1248,
https://doi.org/10.1007/978-981-15-7394-1_1

Fig. 1 Sample scene character images from an in-house scene character components dataset

properly recognize scene characters. Thus, scene character recognition stands as a task that requires focused attention. The majority of works in scene text reading [1–4] are found to combine recognition with localized text regions. However, only a few works have been reported with the focus on dissected scene character classification. Some of them are mentioned below. Abdali et al. [5] proposed a CNN model, applied on a combination of EMNIST and Chars74k dataset with augmented data. This CNN model doesn't need a very deep neural network. Recognition accuracy of this proposed method is 92% for both digit and character. Chekol et al. [6] applied curvature-based global image feature on Chars74k and ICDAR 2003 dataset for dissected character recognition. The boundary of the single character image is used to calculate the curvature value of each point. After curvature calculation key points are identified followed by feature descriptor. Then, Support vector machines (SVM) with kernels like quadratic, linear, and cubic are trained to get the recognition accuracy of the proposed method. The accuracy of this method is 65.3%. An effective character recognition using the K-nearest neighbor classifier is proposed by Barnouti et al. [7]. The proposed system is based on character segmentation, feature extraction, and K-nearest classification process. The proposed experiment was evaluated on Chars74K and ICDAR2003 datasets. Jaderberg et al. [8] used CNN-based architecture to introduce a conditional random field (CRF) graphical model. In this graphical model, character at each position of the output is predicted using unary terms provided by CNN.

Network-based models have also been applied for scene character recognition. Luo et al. [9] proposed a multi-object rectified attention network (MORAN) for recognizing scene text. In this framework, rectification and recognition are two processes, using the rectification mechanism regular and irregular scene text can be recognized. Experiment results on different datasets are 91.2% (IIIT5K), 83.3% (SVT), 95.0% (ICDAR2003), 92.4% (ICDAR2013), 68.8% (ICDAR2015), 76.1% (SVT-Perspective), and 77.4% (CUTE80). A Twin Support Vector Mechanism (T-SVM) for scene text classification is proposed by Francis et al. [10]. In this work, during the phase of localization, it can discard the false positives from the localized scene text. This experiment gives a satisfactory result on Low-resolution images with noise. This method gives 84.91% (ICDAR2015), 84.21% (MSRA500) and 86.21% (SVT) accuracy. Liu et al. [11] proposed an innovative method using multi-scales for recognition of scene text. Scale aware feature encoder (SAFE) is designed for scene character recognition. The scale problem by extracting scale-invariant features from the characters is done by SAFE. SAFE is more effective than traditional CNN. This method can handle scene text challenges like distortion, illumination, and poor quality image more efficiently. Results of SAFE are IIIT5K (85.2%), SVT (85.5%), ICDAR2003 (92.9%), ICDAR2013 (91.1%), SVT-Perspective (74.4%) and ICIST-L (65.7%). A Convolutional Recurrent Neural Network (CRNN) framework (combination of Deep Convolutional Neural Network and Recurrent Neural Network) proposed by Shi et al. [12] which is based on sequence-based image recognition. CRNN takes input images of different dimensions and produced predicted images of different lengths. It directly works on word level, without detailed annotations for each individual character in the training phase. These convolutional layers are based on the VGG-Very Deep architectures [13]. As an application of CRNN, a sequence of musical notes are predicted directly from the image. CRNN results on standard datasets are 71.2% on IIIT5K, 82.7% on SVT, 91.9% on ICDAR2003, and 89.6% on ICDAR2013.

In a nutshell, most of the works on scene text reading are reported in connection with text detection. There are relatively fewer attempts on individual dissected scene character recognition, which must be addressed in order to design robust text reading systems from imagery. Performances achieved by the methods reported in this direction are not adequate for practical applications. In this paper, a new character recognition method built with a histogram of oriented gradients and morphological operations is presented. Extracted feature descriptors are passed to multiple classifiers for performance assessment and SVM is found to produce the highest classification accuracy.

2 Proposed Approach

In this work, we considered scene character images obtained from a dataset of scene text images. At first, input character images are converted into grayscale images as outlined in [14]. Such a character image may be defined as $[c(m,n)]_{X\times Y}$ where

Fig. 2 Binarization of scene character images, **a** sample character components, **b** corresponding binary views

$c(m, n) \in [0, 255]$ represents the gray level intensity at (m, n), and X and Y represent the number of rows and columns, respectively. Then, reverse characters that have light text strokes on dark background are detected and converted to normal texts using a method reported in [15]. Binarization is often a prerequisite for many feature extraction methods. In this work, Otsu's global binarization method [16] is applied. It generates a single threshold $\{T : 0 < T < 255\}$ with which binary character image $[b(m, n)]_{X \times Y} \in \{0, 1\}$ is obtained as shown in Eq. 1. [17]. Some character components and their binary versions are shown in Fig. 2.

$$b(m, n) = \begin{cases} 1 \text{ if } c(m, n) \geq T \\ 0 \text{ otherwise} \end{cases} \quad (1)$$

2.1 Morphology Based Filtering

After binarization of character components, it is found that many binary images contain small scattered objects along with the character strokes. Such objects have been reduced by applying a morphological operation called opening [18]. A morphological opening is useful for removing noise/small objects from an image while reasonably preserving the shape and size of the original image. In the current work, we have empirically chosen a kernel size of 7×7 pixels. Subsequently, we apply the bounding box strategy to exclude the background outside the character strokes. After that, we normalize the bounded characters into 32×32 pixels. Figure 3 demonstrates these operations on a sample binary character image.

(a) (b) (c) (d)

Fig. 3 Illustration of morphological filtering, bounding, and normalization on a binary character image, **a** sample image, **b** noise reduction with morphological opening, **c** bounding box, **d** normalized image of dimension 32 × 32 pixels

2.2 Feature Extraction

In this work, we extract histogram of oriented gradients (HOG) [19] from filtered and normalized binary character images. We have taken 324 HOG features to classify 1905 character components with 52 classes. We divide the character images into small connected regions called cells and for each cell, we compute a histogram of gradient directions from the pixels within the cell. Groups of adjacent cells are considered as spatial regions called blocks. The set of these block histograms represents the descriptors. In this experiment, we considered 8 × 8 pixels per cell, 2 × 2 cells per block and orientation is 9 for the HOG descriptor to calculate the features. Visualization of HOG feature for a binary character image is shown in Fig. 4.

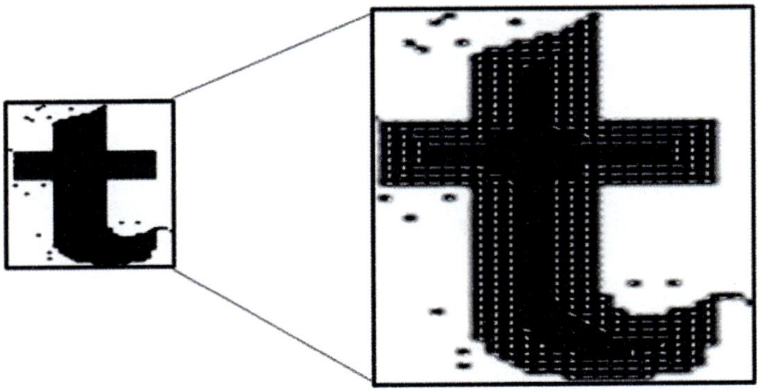

Fig. 4 Visualization of computed HOG features as a plot on a sample binary image

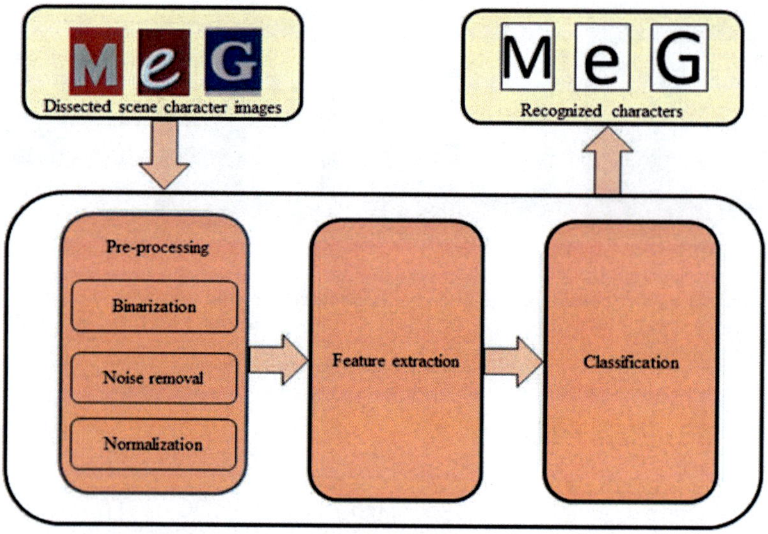

Fig. 5 Pipeline of scene character recognition using HOG feature descriptors

2.3 Scene Character Classification

After feature extraction from binarized character images, features are divided into training and test sets, and subsequently training set is passed to multiple classifiers. In our work, five traditional classifiers viz. Support vector machine (SVM), Multi-layer perceptron (MLP), K-nearest neighbors (KNN), Random Forest, Naive Bayes have been chosen to classify character components. A pipeline of scene character recognition process using HOG feature descriptors and five traditional classifiers is shown in Fig. 5.

3 Experimental Results and Discussion

Experiment has been carried out on 1905 number of camera captured dissected scene character images, while background of some images are nearly homogeneous, many images have non-uniform background. Content of such character images are a collection of the upper case (A–Z) and lower case (a–z) English alphabets. In this experiment, total number of classes is 52 (26 classes for upper case alphabets and rest 26 classes for lower case alphabets). All character images are normalized to 32×32 pixels. We divide the total number of scene images into 80% for training and 20% for testing. Out of 1905 images, 1524 images are obtained for training and 381 images for testing. This is, now, a 52-class classification problem. Results are

Table 1 Recognition results of scene character components using traditional classifiers

Name of classifiers	Precision	Recall	*F* measure	Accuracy
Naive Bayes	0.4577	0.4264	0.4414	0.5683
KNN	0.6050	0.6260	0.6153	0.7759
MLP	0.6805	0.6731	0.6767	0.7806
Random forest	0.6319	0.6027	0.6169	0.7523
SVM	0.6978	0.6946	0.6961	0.8207

obtained in terms of total number of true-positive (TP), true-negative (TN), false-positive (FP), and false-negative (FN) that are measured with the help of input labels and corresponding output labels [20]. Suppose, "a" is a positive class, then "other than alphabet a" is the negative class. True-positive (TP) determines how many characters are correctly predicted. Similarly, true-negative (TN) determines the number of correctly predicted negative samples. False-positive (FP) signifies the number of samples incorrectly predicted as positive class whereas false-negative (FN) signifies the number of samples incorrectly predicted as negative class. After that, F-Measure is computed by Precision (P) and Recall (R) as shown in Eq. 2.

$$F\text{−Measure} = \frac{2 * P * R}{P + R} \tag{2}$$

where Precision(P) $= \frac{TP}{TP+FP}$ and Recall(R) $= \frac{TP}{TP+FN}$. Character recognition accuracy is also calculated as shown in Eq. 3.

$$\text{Accuracy} = \frac{TP + TN}{TP + FP + FN + TN} \tag{3}$$

Using the above statistical parameters, performance is quantitatively measured and obtained results are shown in Table 1.

In the case of KNN classifier, the number of neighbor's, i.e., k is experimentally chosen as 5. In the case of MLP classifier, we considered three hidden layers of 100 neurons each, "relu" activation function and maximum iteration of 8000. Number of neurons and maximum iteration is chosen experimentally. It uses backpropagation algorithm for supervised learning. Random forest classifier constructs decision trees for each character image and get predicted class from each decision tree. After that, it determines the actual class based on the majority of votes of the decision trees. The number of trees in a forest may vary from 10 to 100. In our experiment, we employed 100 trees. For SVM classifiers, we consider tree types of kernels viz. linear, polynomial, and Radial Basis Function (RBF). Out of these three, polynomial kernel yields highest classification accuracy of 82.07%. It is experimentally found that SVM with a polynomial kernel with gamma value 2 gives efficient results than other kernels as well as other classifiers. A classification summary with this model is shown as a confusion matrix in Fig. 6.

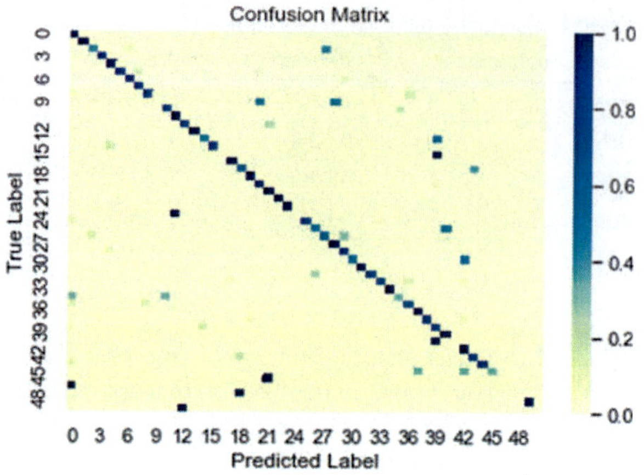

Fig. 6 Confusion matrix using spolynomial kernel for SVM classifier to classify 52 alphabets (26 class for upper case and 26 for lower case character)

It may be noticed that there are some misclassifications in the confusion matrix (see Fig. 6). Some of these misclassifications are found to be due to similar-looking characters like {'*C*', '*c*'}, {'*O*', '*o*'}, {'*X*', '*x*'}, {'*Z*', '*z*'} etc. Moreover, the system is undertrained for some symbols like '*Q*', '*X*', '*Y*', '*Z*', '*q*', '*t*', '*v*', '*x*', '*y*', '*z*' due to very less number of samples present in these classes.

4 Conclusion

In this paper, we proposed a method for recognition of scene character images. The proposed method is evaluated on 1905 number of camera captured scene character images of the English alphabet. It recognized scene character components with 82.07% accuracy using SVM, 78.06% using MLP, 77.59% using KNN, 75.23% with Random Forest, and 56.83% with Naive Bayes classifiers. It may be noted that accuracies may not be as high as conventional character recognition problems, the current work may be considered as an effective approach while taking the complexity of scene characters into account. However, there is scope of improvements. Binarization of scene characters, a really challenging task, should be looked into instead of applying Otsu's global threshold. Some alphabets like '*Q*', '*X*', '*Y*', '*Z*', '*q*', '*t*', '*v*', '*x*', '*y*', '*z*' have an insufficient number of samples for training and testing, which might have undertrained the models. In the near future, we aim to further improve and accelerate the design of effective scene text reading systems.

Acknowledgements The authors are thankful to the Department of Computer Science and Engineering of Aliah University, Kolkata, India for providing every kind of support for carrying out this research work. P. Sengupta is further grateful to the Department of MA and ME, Government of West Bengal for providing Swami Vivekananda Merit cum Means Scholarship.

References

1. Lin, H., Yang, P., Zhang, F.: Review of scene text detection and recognition. Int. J. Arch. Comput. Methods Eng. 1–22 (2019)
2. Zhu, Y., Yao, C., Bai, X.: Scene text detection and recognition: recent advances and future trends. Front. Comput. Sci. **10**(1), 19–36 (2016)
3. Long, S., He, X., Ya, C.: Scene text detection and recognition: the deep learning era. arXiv preprint arXiv:1811.04256 (2018)
4. Liu, X., Meng, G., Pan, C.: Scene text detection and recognition with advances in deep learning: a survey. Int. J. Doc. Anal. Recogn. (IJDAR) **22**(2), 143–162 (2019)
5. Abdali, A.R., Ghani, R.F.: Robust character recognition for optical and natural images using deep learning. In: Proceedings of IEEE Student Conference on Research and Development (SCORD), pp. 152–156 (2019)
6. Chekol, B., Celebi, N., TAŞCI, T.: Segmented character recognition using curvature-based global image feature. Turkish J. Electr. Eng. Comput. Sci. **27**(5), 3804–3814 (2019)
7. Barnouti, N.H., Abomaali, M., Al-Mayyahi, M.H.N.: An efficient character recognition technique using K-nearest neighbor classifier. Int. J. Eng. Technol. **7**(4), 3148–3153 (2018)
8. Jaderberg, M., Simonyan, K., Vedaldi A., Zisserman, A.: Deep structured output learning for unconstrained text recognition. In: Proceeding of International Conference on Learning Representations, pp. 1–10 (2014)
9. Luo, C., Jin, L., Sun, Z.: Moran: a multi-object rectified attention network for scene text recognition. Int. J. Patt. Recogn. **90**, 109–118 (2019)
10. Francis, L. M. and Sreenath, N.: Robust scene text recognition: using manifold regularized twin-support vector machine. J. King Saud Univ. Comput. Inf. Sci. 1319–1578, Elsevier (2019)
11. Liu, W., Chaofeng, C., Wong, K.: SAFE: scale aware feature encoder for scene text recognition. In: Proceedings of Asian Conference on Computer Vision, pp. 196–211, Springer (2018)
12. Shi, B., Bai, X., Yao, C.: An end-to-end trainable neural network for image-based sequence recognition and its application to scene text recognition. IEEE Trans. Patt. Anal. Mach. Intell. **39**(11), 2298–2304 (2016)
13. Simonyan, K., Zisserman, A.: Very deep convolutional networks for large-scale image recognition. arXiv preprint arXiv:1409.1556 (2014)
14. Mollah, A.F., Basu, S., Nasipuri, M.: Computationally efficient and implementation of convolution-based locally adaptive binarization techniques. In: Proceedings of International Conference on Information Processing, pp. 159–168, Springer (2012)
15. Mollah, A.F., Basu, S., Nasipuri, M.: Handheld device-based character recognition system for camera captured images. Int. J. Image Graph. **13**(4), 1350016-1–1350016-21 (2013)
16. Otsu, N.: A threshold selection method from gray level histogram. IEEE Trans. Syst. Man Cybern. **9**(1), 62–66 (1979)
17. Sengupta, P., Mollah, A.F.: Scene text component segmentation using hierarchical distance slicing. Int. J. Comput. Intell. IoT **2**(1), 336–339 (2019)
18. Halder, A., Sarkar, A., Ghose, S.: Adaptive histogram equalization and opening operation-based blood vessel extraction. In: Proceedings of International Conference on Soft Computing in Data Analytics, pp. 557–564, Springer, Singapore (2019)
19. Gogna, A., Majumdar, A.: Discriminative autoencoder for feature extraction: application to character recognition. Int. J. Neural Process. Lett. **49**(3), 1723–1735 (2019)
20. Nayef, N., Patel, Y., Busta, M., Chowdhury, P.N., Karatzas, D., Khlif, W., Matas, J., Pal, U., Burie, J.C., Liu, C.L., Ogier, J.M.: ICDAR2019 robust reading challenge on multi-lingual scene text detection and recognition–RRC-MLT-2019. arXiv preprint arXiv:1907.00945 (2019)

Rice Disease Detection and Monitoring Using CNN and Naive Bayes Classification

Debaniranjan Mohapatra, Jogeswar Tripathy, and Tapas Kumar Patra

Abstract Rice is the most important crop which affects a lot to India's economy. Indian farmer faces many financial problems when their cultivated rice plant is being affected by diseases which lead to the reduction of the Indian economy. Rice leaf blast, bacterial blight, sheath blight, and brown spot are found to be four major diseases that affect a lot in India [1]. In the agricultural industry, the grouping and identification of paddy diseases are most important economic and scientific issues. To recognize the disease, these following activities like quality, outline, and colour can be taken as priority [2]. The objective of this study is to improve an augmented system which is using both machine learning and image processing; the farmer can apply this process easily by taking a photograph of a diseased leaf of rice from his phone, and it will detect the disease so that they can analyse the corresponding precaution quickly. Hence, a farmer can easily detect the disease with a fraction of second and take prevention as soon as possible.

Keywords Max pooling · Convolutional neural network (CNN) · Cloud database · Back propagation algorithm · Naive Bayes classification

1 Introduction

Agriculture is the most important aspect of our global economy. The agricultural system gets pressurized due to vast explosion of population. Precision farming and agri-technology are now termed as digital agriculture, and this is a new scientific

D. Mohapatra
Department of ECE, S'O'A University, Bhubaneswar, Odisha, India
e-mail: mahapatrass1965@gmail.com

J. Tripathy (✉)
Department of CSE, S'O'A University, Bhubaneswar, Odisha, India
e-mail: jogeswar.tripathy@gmail.com

T. K. Patra
Department of I & EE, CET, Bhubaneswar, Odisha, India
e-mail: tkpatra@cet.edu.in

S. Borah et al. (eds.), *Soft Computing Techniques and Applications*,
Advances in Intelligent Systems and Computing 1248,
https://doi.org/10.1007/978-981-15-7394-1_2

field which simultaneously enhances the agricultural production and minimizing its environmental impact [3]. Four major diseases are found in India due to bacteria and fungus. They are brown spot, rice leaf blast, sheath blight, and bacterial blight. These diseases have similar symptoms; hence, manual process of farmers to detect them is not good enough which leads to false detection, and ultimately the farmer faces both financial as well as technical problems. By detecting the wrong disease, they spent money in precaution which is waste of money completely as the detected disease is incorrect. Till now farmers are depending on booklet and some local unauthorized expert advice to detect the disease, for which the exact detection fails, and it affects to Indian food production and economy. They are unable to understand the problems arising for false detection that Indian economy and food production are facing [4, 5]. Hence, an automated and early detection is very much essential so that the exact disease diagnosis can be possible and farmer can take necessary action before the disease makes any damage. Every plant disease has its own growth pattern and period, so whenever a plant is diminished by any disease, the farmers have to focus on only the infection part. This is time-consuming, and it has to take some precaution during the selection of pesticides. Hence, to overcome such problem, an augmented system is developed by us which is enabled by image processing and machine learning.

According to our research, image processing technique helps in diagnosing the disease and classifying it. Now, an exact identification is possible which will increase the predictability and economic growth of all farmers as well as the economy condition of our country. This process not only used for rice disease diagnosis but also it can be used for many applications like identifying corresponding diseases, predict food grading, soil monitoring, weather monitoring, etc. at a cheap rate. Image processing and machine learning procedures are broadly used for different farming applications starting from spotting the leaflet of plant to labelling of different disorders [6–8]. To detect the exact disorder in the good quality of raw image, better feature consideration and best classifier are needed. So an existing system is unable to handle all the tasks at a time. Hence, to get the relative consideration of different task, we carried out a study on different procedures and methodologies used for the revealing of rice plant disorders. To get understanding of domain, we carried out a study on various types of rice plant disorders. We departed through both image processing techniques and machine learning techniques applied in rice disorders for finding the best one.

1.1 Objective

The objective of this study is to improve an augmented system; this can exactly detect the disease of good quality raw image. Our main emphasis is on the image processing and machine learning on which farmer can easily take a photograph of a diseased leaf of rice from his phone, and it will detect the disease of field and also it will analyze the corresponding precaution so that a farmer can easily detect the disease with a fraction of second and take prevention as soon as possible. As our overall process starts with the preprocssing of captured images, and simultaneously the raw images converted into its corresponding matrix form, after then the patterns of input image matrix extracted repeatedly by convolved with number of fillters (kernel filter) [9]. Then, the features are subsampled by the process of pooling. Then, this process proceeds in a loop manner. Finally, the obtained feature pixels are made flatten. This matrix is given to a neural network, and then at the output Naive Bayes classifier [10] works and gives the result.

1.2 Organization

In this research, we introduce a broad analysis of the application of convolutional neural network (CNN) and the Naive Bayes classifier which is utmost preferable classifier in automation in detecting system for diseased image in agriculture. The organization of our deduced analysis is as follows: Sect. 2 describes the motivation of the project. Section 3 introduces the input materials and the employed approach for the assembly and labelling of the presented works. Implementation details and proposed framework are derived in Sect. 4. The investigational results discussed in Sect. 5 and finally conclusion and future work of this paper are discussed in Sect. 6.

2 Motivation

After successfully surveyed all the related papers, we have got an overall idea about the advanced techniques above preprocessing techniques, segmentation approaches, feature extraction mechanism, etc., and also gone through on details about the various classification techniques with using of cloud services. Here also, we go through about the neural network, convolutional neural network, and the process of back propagation algorithm for which we can touch our main motto that is how it works and checks the error and updates it. After all, the use of convolution process has shown us how it is able to detect the features with the operation of pooling for subsampling the input matrix and how it is helping us in the reduction of input image size. Nowadays, supervised is widely used in all authentication purpose and back propagation technique [11] is giving the best performance in system identification as it checks the errors and updates it. It continuously checks the targeted attributes that are given to it, stored in the cloud database during training phase, and then it gives the classified output. After then, it comes under Naive Bayes classifier which gives the output in a probabilistic manner. By giving the identification in percentage manner, it detects the dual diseases also simultaneously.

3 Input Materials and Implemented Methodology

Here, we explained the detailed study of each and individual theories of related topics that are really used and implemented successfully in this paper. First of all, analyse the symptoms of rice disease which are mainly affecting the crop field. Then, it introduces the proposed work to interact with the concern methodologies regarding neural network, convolutional neural network, kernel filtering, deep learning mechanism, cloud computing, and some information about Microsoft Azure.

3.1 Symptoms of Rice Diseases

Rice Bacterial Blight (RBB)

Bacterial blight occurs by Xanthomonas Oryza. In this disease, seed gets damaged, and water soaks to yellowish strips on the leaf. After sometimes, it rises in length and width [12]. Lesions changes from yellow to white as shown in Fig. 1.

Rice Blast (RB)

It is a fungal disease caused by Magnaporthe Oryza. In this disease, wound on leaves and upper leaves becomes dark brown, straight, and progress according to vein. Physically, they are 2–10 mm long and 1–1.5 mm wide as shown in Fig. 2.

Fig. 1 Picture of rice bacterial blight

Fig. 2 Picture of rice blast (RB)

Rice Brown Spot (RBS)

It is spread by fungus. The distinctive injury on leaves and upper leaves converted their colour light to dark brown, progress, and straight parallel to vein. They are normally 2–10 mm long and 1–1.5 mm wide as displayed in Fig. 3.

Rice Sheath Rot (RSR)

It is also a fungal disease spread by two categories of fungus: Sarocladium Oryza and Sacroladium attenuatum [13]. Here, the sheath injury appears at the uppermost leaf sheath encasing the little panicles. It looks oblong or as an unbalanced spot having dark reddish, brown margins and grey centres or brownish grey all through. The image is shown in Fig. 4.

Fig. 3 Picture of rice brown spot

Fig. 4 Picture of rice sheath rot (RSR)

3.2 *Implemented Methodology*

In this research, identifying and classifying rice disease involves convolutional neural network (CNN). A convolutional neural network involves four layers such as

1. Convolution layer
2. ReLu layer
3. Pooling layer
4. Fully connected layer.

After this processing of images, the final fully connected layer is fed as input into the neural network where the back propagation learning algorithm works, and finally we are observing our respective outputs.

The general expression of a convolution is

$$g(x, y) = (w * f)(x, y) = \sum_{s=-a}^{a} \sum_{t=-b}^{b} w(s, t) f(x - s, y - t)$$

where $g(x, y)$ is the filtered image, $f(x, y)$ is the original image, and w is the filter kernel.

Convolution layer

Convolutional layer is introduced just to extract the pattern of input image. Basically here the feature of the total image is being considered. It is a mathematical model that takes two inputs: One is input image matrix, and another one is kernel. Convolution [14] is done between input matrix and filter matrix, shown in Fig. 5. Here, the total number of filter depends upon requirement, which means what amount of features do we require is considered. Filters may be edge detector, object detector, texture

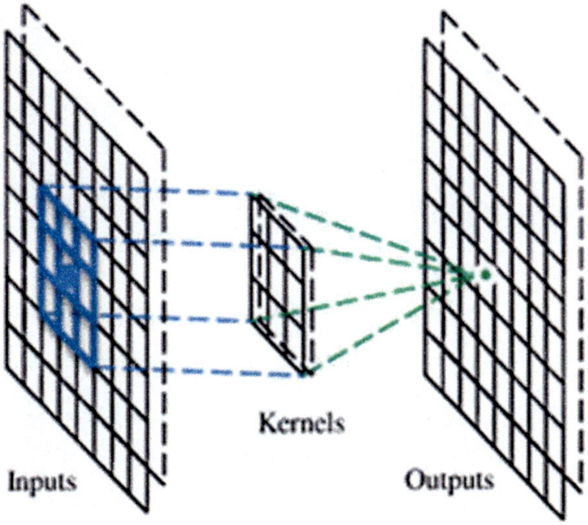

Fig. 5 Picture of a convolution of input matrix with kernel filter

detector, etc; basically, these filters are used to identify the patterns of the input image.

ReLu layer

The full form of ReLu [15] is rectified linear unit and is a type of activation function. We can define it mathematically as:

$$Y = \max(0, x)$$

(where Y is the mapped value in the output and x is the originally calculated value.)

In ANN, a rectifier activation function is used which defines only the positive part of its argument; a part or unit corresponds to rectifier is called as ReLu as shown in Fig. 6.

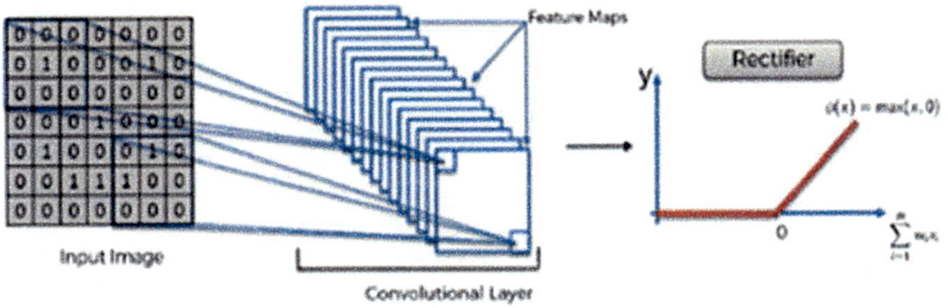

Fig. 6 Picture of a ReLu activation function

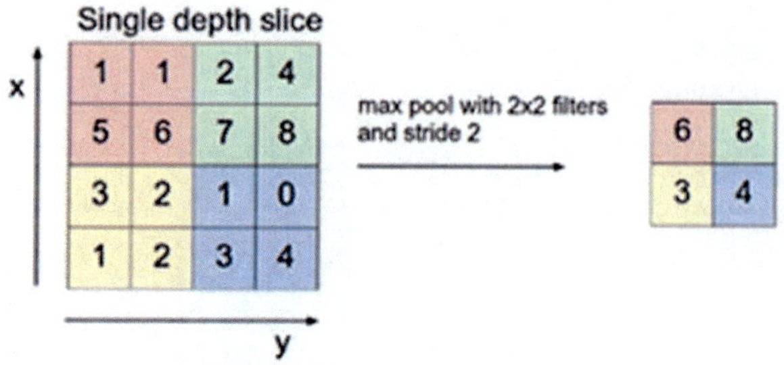

Fig. 7 Picture of a max pooling

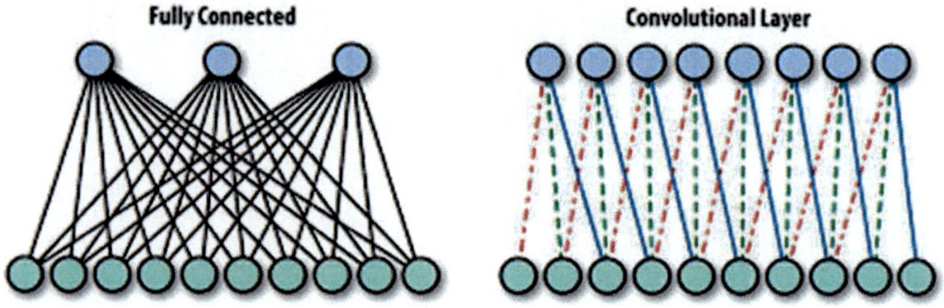

Fig. 8 Picture of a fully connected network

Pooling layer

Pooling layer is also a main part of CNN when time consumption and degree of complexity are considered. It subsamples the image features by reducing its size, and it extracts the important part of the image. Hence, it is called as the process of reduction of size without changing its characteristics. Pooling layer operates on individual feature map independently as shown in Fig. 7.

Fully connected layer

Fully connected layer implies that in neural network all the neurons of a specific layer joined with all neurons in the next layer. At end of output, the last pooling layer acts as input to the neural network, the so-called fully connected layer. In a fully connected layer, each neuron receives input from the previous layer as shown in Fig. 8.

3.3 Neural Network

An ANN is the method of processing of data which was inspired by the biological nervous systems. It has mainly three layers: input layer, hidden layer, and output layer. It is very operational and effective in case of noisy data and untrained pattern as shown in Fig. 9 [16]. There are many applications of neural network, out of which some are given below:

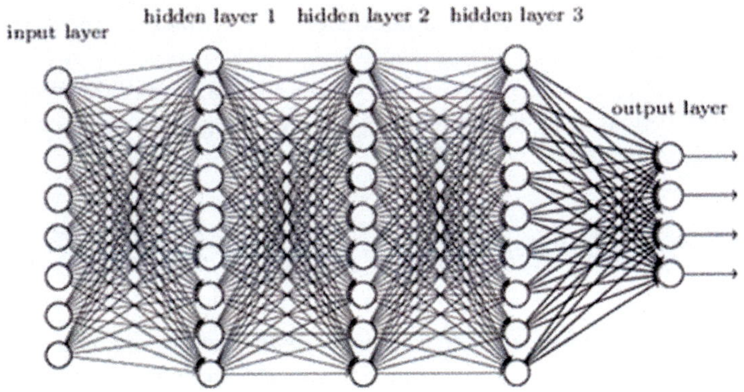

Fig. 9 Basic diagram of neural network

- It allows function approximation or regression analysis where it includes time series prediction and modelling.
- It is very helpful in classification with a great accuracy where it includes pattern and sequence recognition.
- It helps in data processing which includes filtering, clustering, blind signal separation, and compression.

3.4 Back Propagation Algorithm

This is a learning algorithm which follows iterating nature. Here, it learns the data by checking the error and updating it. When the image matrix is made flattened, it is subjected to each neurons and then back propagation algorithm takes place. Back propagation algorithm [15] is a learning algorithm which estimates the gradient for weight and bias updating. It checks the error and updates the weight and bias. It is so called back propagation because it checks the error in a backward manner, which means from the output layer to input layer through the hidden layer.

3.5 Benefits of Machine Learning in the Cloud

There are various advantages of machine learning over cloud network. They are:

- Pay-per-use model of cloud architecture is good for burst AI or machine learning workloads.
- Enterprises are being comfortable to conduct experiment with machine learning capabilities and scale up as projects go into production and demand increases with the help of cloud.

- The cloud makes intelligent capabilities accessible without using advanced skills in artificial intelligence or data science.
- AWS, Microsoft Azure, and Google Cloud Platform provide many machine learning options that do not require heavy knowledge of AI, machine learning theory, or a team of data scientists.

4 Implementation Details and Proposed Framework

Here, the implementation details and future framework for rice disorder identification are introduced. Our future system intended to detect and classify the rice diseases which will help farmer in prevention. The block diagram of the proposed work is shown in Fig. 10. We explain here the dispensation steps of our future work in the following subsections and higher-level steps of our proposed work presented here too. Before arriving at a particular technique for performing a particular operation, e.g. disease segmentation, we have explored various alternative techniques. A detailed discussion of various techniques explored is presented in this section as follows:

4.1 Monitoring Process of the Disease

First of all, the input image is converted into matrix form and then it is convolved with the filter to extract the pattern. Let the input image matrix be (Figs. 11, 12, and

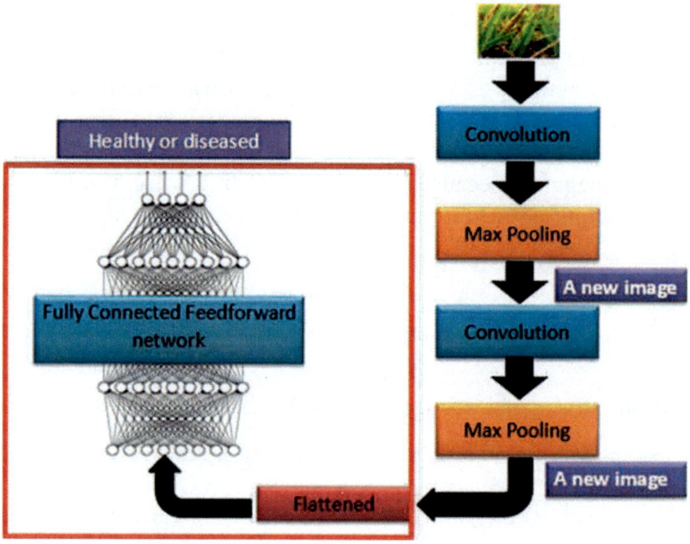

Fig. 10 Input image matrix

Fig. 11 Input image matrix

1	0	0	0	0	1
0	1	0	0	1	0
0	0	1	1	0	0
1	0	0	0	1	0
0	1	0	0	1	0
0	0	1	0	1	0

Fig. 12 Filter 1

-1	1	-1
-1	1	-1
-1	1	-1

13).

Now after convolution of filter 1 with each element of input matrix, the resultant matrix will be as follows (Figs. 14 and 15):

These two 4 * 4 matrices form a matrix of 2 * 4 * 4 matrix, and by continuous convolution between input matrix and filter matrices, we will get a large feature map.

Fig. 13 Filter 2

1	-1	-1
-1	1	-1
-1	-1	1

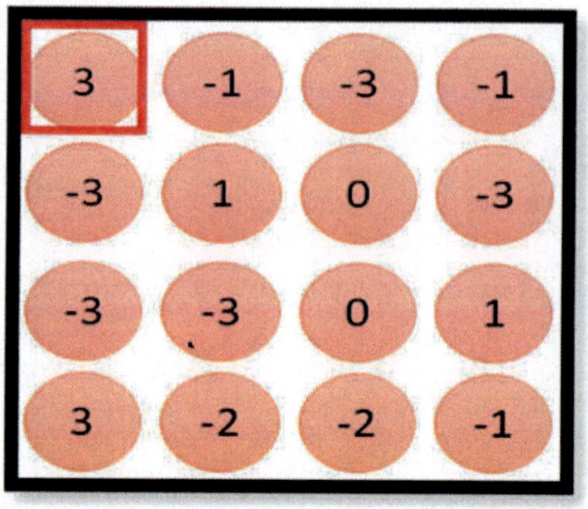

Fig. 14 Convolution with first 3 * 3 strip of input matrix

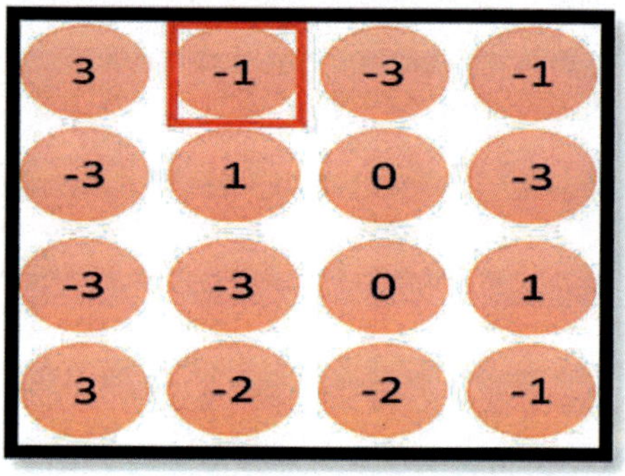

Fig. 15 Convolution with second 3 * 3 strip of input matrix

This feature map matrices will have all the information of the pattern. Now to extract the important features of the image, we have used max pooling.

Max pooling is a method of extraction of important characters of an image without affecting the image characteristics. Subsampling the image will not affect the object, and it also reduces the size of the image which will help in processing the parameters. Here, the 4 * 4 feature maps are subsampled into 2 * 2 matrix first, and thereby the important character (means the maximum value from the 2 * 2 matrix is considered) is extracted.

After this stage, a new matrix is formed which is more convenient and more resized. Hence again, I have convolved and max pooled the image to get the more convenient one. This process processed in a loop till the greatest feature is obtained (Fig. 16).

Finally, this flattened matrix is fed to the neural network (NN) as input. The whole NN is made cloud-based for speeding up of operation and also for the centralized access. Then, the cloud is managed by opening a cloud account in Microsoft Azure. Now inside the network back propagation algorithm works where it continuously checks the error and updates it.

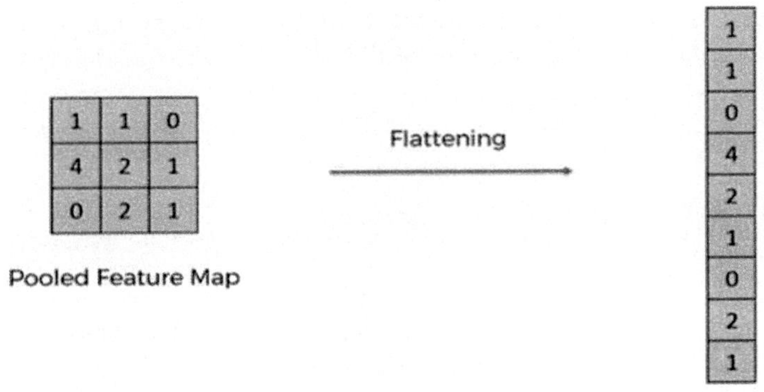

Fig. 16 Flattening of feature from matrix

4.2 *Using of Back Propagation Algorithm*

Let the first layer or the input layer is

$O = I,$ I is the first layer input and O is corresponding output

$M = \left(\sum w1 * O\right) + b1,$ here M is input to hidden layer 1 and w and b are weight and bias

$X1 = {}^{1}/_{1+e^{-M}}$ here $X1$ is the output of hidden layer

Now, repeat the same for all hidden layers.
Now,

$P = \left(\sum w3 * o\right) + b3,$ P is the input of hidden layer 3.
$X3 = 1/e^{(-P)}$ $X3$ is the input of hidden layer 3.

Now the actual algorithm works as

$E3 = O(1 - O)(T - O),$ E is the error of layer 3 and T is the target attribute which is given.

$E1 = O(1 - O) * \sum E3 * w3,$

Now weight will be updated as

$\Delta w = L * E * O,$ where L is the learning rate and $W = w + \Delta w$

Again bias will be updated as

$$\Delta b = L * E \text{and } B = b + \Delta b$$

This process will continue till the terminating condition occurs. Then, the final output is compared with the patterns of the trained datasets of rice and gives the output with a good accuracy.

4.3 Using Naive Bayes Classification

Naive Bayes classification is a class of classifier which works on probabilistic approach. It is a supervised classification in which the system can predict the targeted attribute by analysing the trained feature. In Naïve Bayes classification, each feature is considered and calculated out of hundred. By matching the obtained feature with the trained features, it makes an approximation of probabilities of occurrence of individual disease.

It works with the principle of probability concept. Here, the system collects data that is mainly the features. We have taken four features that is colour, shape, area, and the distance between two consecutive diseased parts. All the features are saved in cloud.

When a farmer is taking a picture of the diseased rice, it has some characteristics. These features are compared with the stored feature in cloud. Then, it compares each feature out of hundred. The feature with maximum probability leads to the exact identification of disease. It is a supervised learning based on Bayes equation:

$$P(A/B) = P(B/A) * \frac{P(A)}{P(B)}$$

It is a probabilistic approach towards the disease diagnosis where the considered features are colour, shape, area, and distance between two diseased spot.

5 The Experimental Result and Discussion

We use Naive Bayes classification as a probabilistic approach to compare the final results with the patterns of the trained dataset features and find out an approximation of probabilities of occurrence of individual disease. We have taken four features that is colour, shape, area, and the distance between two consecutive diseased parts as priority basis. All the features are saved in cloud for easy and quick access with presenting comparison result in 100% accuracy. When a farmer is taking a picture of the diseased rice, it has some characteristics. These features are compared with the stored feature in cloud. Then, it compares each feature out of hundred. The feature with maximum probability leads to the exact identification of disease.

We have taken two target sets to identify whether the rice is healthy or not. Hence, here we used two tags, namely healthy and disease. Finally, we presented some snapshots of our experiment by taking different dataset as input and also getting the good results approx. to 99% accuracy rate as follows:

In the above output of Fig. 17, the system identifies this image as a diseased image with a diseased probability of 100%. We have taken this image of a diseased rice field from a local area, and this network easily identifies it with a good accuracy.

In the above output of Fig. 18, we have processed the healthy image of rice plant which was taken from a reputed website and found 100% of accuracy in this identification process. The system can easily predict the diseased and healthy image with 100% accuracy. Then, rest snapshots are presenting the identification of type of diseases which is shown below.

In the above output of Fig. 19 presenting the user-interface, a sample rice leaf image is given as input and after all processing, system identifying the healthiness of image is approx. 86.27%. Here, the farmer will have to browse the diseased image;

Fig. 17 Detection of diseased rice

Fig. 18 Detection of healthy rice

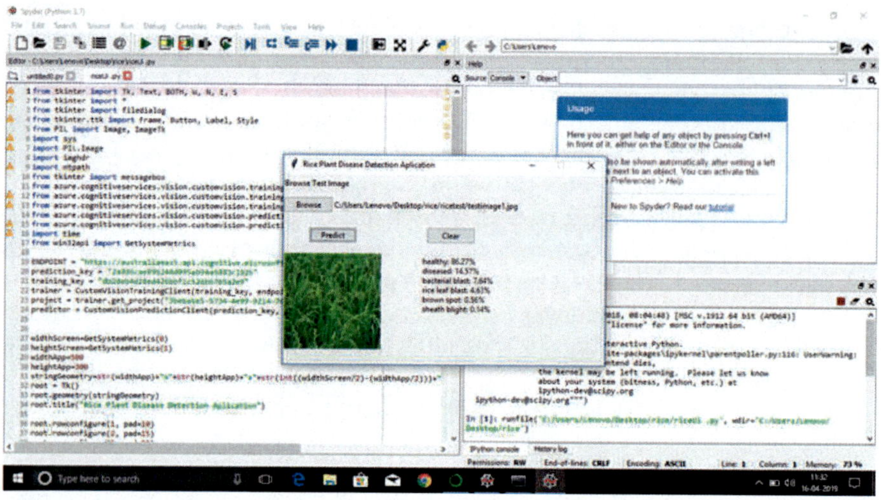

Fig. 19 User interface of a healthy disease

then after clicking on predict button, the display will show the corresponding disease. Some other outputs are placed below:

In the above output of Fig. 20, the identification of sheath blight with a good accuracy is shown. Here, the accuracy is more than 95%. As the features are collected, the input image matches with this image feature which is easily identified.

In the above output of Fig. 21, the identification of brown spot and bacterial blast with a good accuracy is shown. Here, the accuracy is more than 99%. This image

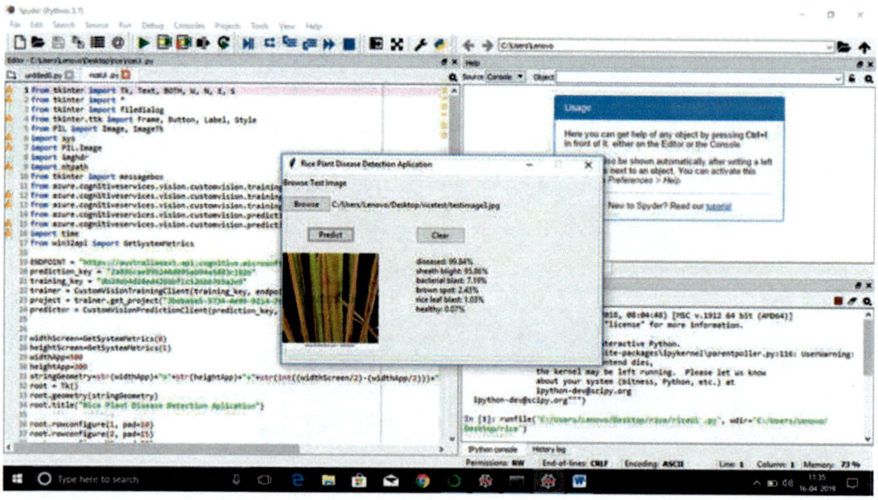

Fig. 20 User interface of a sheath blight disease

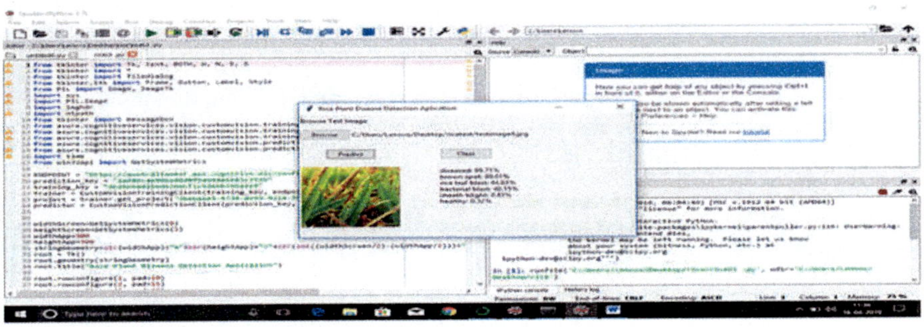

Fig. 21 User interface of a brown spot disease

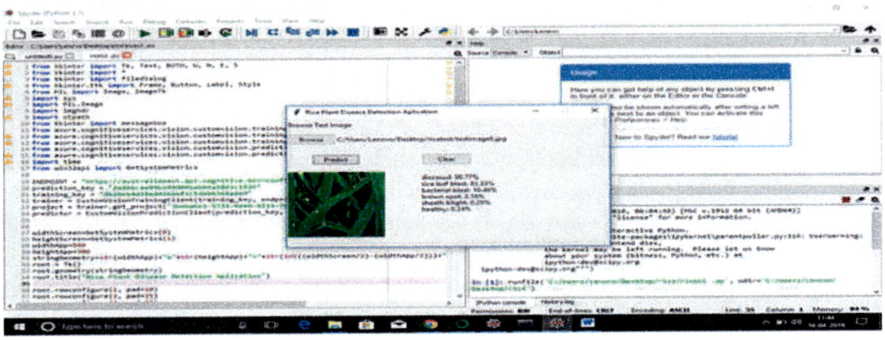

Fig. 22 User interface of a rice leaf blast disease

was obtained from a local field and is tested in this system. We have achieved a good accuracy rate in dual disease detection.

In the above output of Fig. 22, leaf blast disease is identified and its accuracy is over 83%. This image was collected from OUAT, Odisha, and has processed in this network.

In the above output of Fig. 23, brown spot disease is identified and its accuracy is over 97%. This image was collected from NRRI, Odisha, and has processed in this network. It can easily identify any kind of rice disease with a supreme accuracy. According to result of Figs. 17, 18, 19, 20, 21, 22, and 23, we have got the accuracy 100% which is optimum.

6 Conclusion and Future Work

Hence from the above experiment, we have concluded that rice disease identification achieves optimum accuracy by convolutional neural network. It is possible because of max pooling operation in which we have considered all the tinny important samples

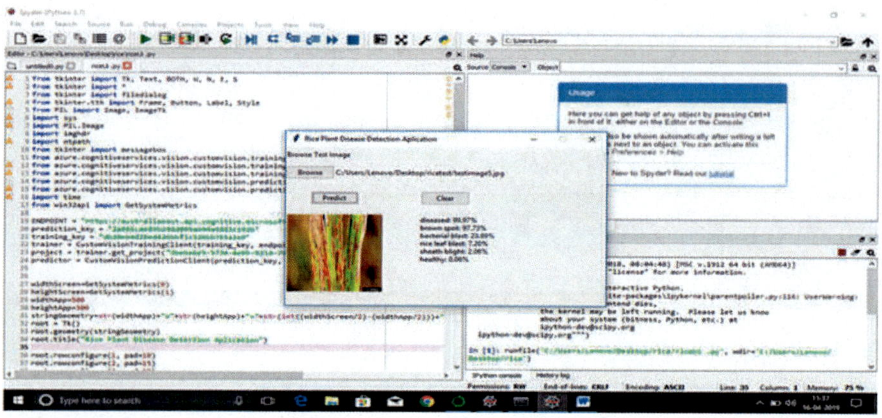

Fig. 23 User interface of a brown spot disease

of the diseased image as we have considered all the captured features and worked over it. From the above, all experiments are also concluded that the Naive Bayes classifier is the most preferable classifier in automation in detecting system for which we have achieved the exact diagnosis and able to find out the multiple diseases in one diseased rice leaf image with probability approach.

So, further research could be done to improve the system prototype into a better system by taking the Augmented Reality extension. The main application of this is that a farmer can easily find the corresponding precaution on his/her mobile in a virtual way.

References

1. Anthonys, G., Wickramarachchi, N.: An image recognition system for crop disease identification of paddy fields in Sri Lanka. In: International Conference on Industrial and Information Systems (ICIIS), pp. 403–407. IEEE, Sri Lanka (2009)
2. Yao, Q., Guan, Z., Zhou, Y.: Application of support vector machine for detecting rice diseases using shape and color texture features. In: International Conference on Engineering Computation, pp. 79–83 (2009)
3. Garci, J., Barbedo, J.A.: Digital image processing techniques for detecting, quantifying and classifying plant diseases. Springer Plus **2**, 1–12 (2013)
4. Charlie Paul, C.K.: Classification of rice plant leaf using feature matching. Int. J. Eng. Technol. Sci. **1**, 290–295 (2014)
5. Suman, T., Dhruva kumar, T.: Classification of paddy leaf diseases using shape and color features. IJEEE **07**, 239–250 (2015)
6. Majid, K., Herdiyeni, Y., Rauf, A.: Mobile application for paddy disease identification using fuzzy entropy and probabilistic neural network. In: Advanced Computer Science and Information Systems (ICACSIS), pp. 403–406. I-PEDIA, Bali (2013)
7. Blumenstein, M., Verma, B., Basli, H.: A novel feature extraction technique for the recognition of segmented handwritten characters. In: Seventh International Conference on Document Analysis and Recognition Proceedings, vol. 1, pp. 137–141 (2003)

8. Golubitsky, Oleg, Stephen, Watt, M.: Distance-based classification of handwritten symbols. In: International Journal on Document Analysis and Recognition (IJDAR), no. 2, pp. 133–146, (2010)
9. Li, J., Li, P.: Human expression recognition based on feature block 2DPCA and Manhattan distance classifier. In: 7th World Congress on Intelligent Control and Automation, pp. 5941–5945. WCICA 2, Chongqing (2008)
10. Joshi, A.A., Jadhav, B.D.: Survey on role of image processing techniques in plants pathology. In: National Conference on Innovation in Engineering Science and Technology, pp. 135–139 (2015)
11. Orillo, J.W, Cruz, J.D., Agapito, L., Satimbre, P.J., Valenzuela, I.: Identification of diseases in rice plant (oryza sativa) using back propagation artificial neural network. In: International Conference on Humanoid, Nanotechnology, Information Technology, Communication and Control, Environment and Management (HNICEM), pp. 1–6, IEEE (2014)
12. Phadikar, S., Sil., J.: Rice disease identification using pattern recognition techniques. In: 11th International Conference on Computer and Information Technology, (ICCIT), pp. 420–423. IEEE (2008)
13. Phadikar, S., Sil, J., Das, A.K.: Rice diseases classification using feature selection and rule generation techniques. Comput. Electron. Agric. **90**, 76–85 (2013)
14. Maharjan, G., Takahashi, T., Zhang, S.H.: Classification methods based on pattern discrimination models for web-based diagnosis of rice diseases. J. Agric. Sci. Technol. **1**, 48–56 (2011)
15. Kurniawati, N.N, Abdullah, S.: Texture analysis for diagnosing paddy disease. In: International Conference on Electrical Engineering and Informatics (ICEEI'09), vol. 1, pp. 23–27, IEEE (2009)
16. Yingyi, Sun, Wei Zhang Hao Gu, Chao Liu, Sheng Hong, Wenhua Xu, Jie Yang, Guan Gui: Convolutional Neural Network Based Models for Improving Super Resolution Imaging. vol. 7, pp. 43042–43051, IEEE (2019)

Air Quality Prediction Using Artificial Neural Network

Limali Sahoo, Bani Bhusan Praharaj, and Manoj Kumar Sahoo

Abstract Pollution is one of the major needs to address in the world's environmental change due to industrialization, urbanization and deforestation. Pollution nature's cause is widely in range and to maintain proper level is a key issue. Here, we are considering industrial and urban-based air quality index predication and this air index is referral to air quality. The air index is a key variable or major factor to establish the relation between the source emission and ambient air concentration. The artificial neural network (ANN) modeling deployment helps environmental management and its planning in a optimize way. Angul-Talcher is one of the industrial zones in eastern region of India. The seasonal variations of air quality indexing parameter like Suspended Particular Matter (SPM) and Respirable Suspended Particular Matter (RSPM), carbon monoxide (CO), sulphur-dioxide (SO_2), nitric-oxide (NO) and nitrogen-dioxide (NO_2) with the environmental effective parameter such as temperature (T_{air}), relative humidity(RH) and air velocity (V_{air}) are considered as inputs for the design and study. Recent development for high speed computing environment helps to analyze this kind of problem. ANN models are shown better result when it is applied to many environmental engineering problems to address for environmental analysis and management. The model, we used to study the air quality, is a feed-forward neural network for predication.

L. Sahoo
S'O'A Deemed to be University, Bhubaneswar, Odisha, India
e-mail: limali25.sahoo@gmail.com

B. B. Praharaj
Shri Rawatpura Sarkar University, Raipur, Chhattisgarh, India
e-mail: praharaj.banibhusan086@gmail.com

M. K. Sahoo (✉)
Biju Patnaik University of Technology, Rourkela, Odisha, India
e-mail: mksahoo@hotmail.com

S. Borah et al. (eds.), *Soft Computing Techniques and Applications*,
Advances in Intelligent Systems and Computing 1248,
https://doi.org/10.1007/978-981-15-7394-1_3

1 Introduction

The pollution in air is a major issue nowadays, which is a direct factor and it influences the living being including human; it affects the health condition in mentally and physically with reflects in activities. The different type or kind of pollutants in the form of chemical substances that contribute to pollution, generate and comes from different sources, natural or artificial. The artificial air pollution sources consider mainly industrial activity and its motorization or surface transport utilities [1]. Most commonly air pollutants are carbon monoxide (CO), carbon-dioxide (CO_2), sulphur-dioxide (SO_2), nitrogen-dioxide (NO_2), nitric-oxide (NO), Suspended Particular Matter (SPM) and Respirable Suspended Particular Matter (RSPM), and after an allowed level, air pollutants show its effect in the population health and ecosystem [1]. Recent report published by World Health Organization (WHO) [2] indicates 4 million individuals loss their lives due to heavily polluted air. The number is quite large as compared to road and transport accidents [3].

Air pollutants are a major threat to the living eco-system as well as for living beings. Effects on living being as well as human health are the primary concern [4]. The pollutant air effect decreases the efficient respiratory flow in the living beings, which leads to impaired capability; oxygen percentage in the blood with a high concentration of air pollutants leads to hazardous respiratory and coronary artery-related disease [5]. Consequently, highly accurate measurement method should be deploying for measurement of air pollution status levels in urban, mining and industrial region or area [4]. The measurement device, system with accuracy need to be used for measurement of air quality in urban, mining and industrial areas [4], which should help for necessary immediate action and controlling strategies to minimize the adverse effects as well as to improve the air quality in long range frame.

Regarding to this, several research and studies on air quality prediction are carried on by using ANN or nonlinear methods [4, 6, 7]. The ANN methods for the model and analysis with no prior assumptions concerning the sample or data distribution and nonlinear relationships nature, allow to trained with highly accurate generalize transfer function, with the data set obtained, against the statistical techniques deployed or used [8]. The ANN method or the algorithm gives a direction on present environment condition and future prediction regarding same, by which a pollution management planning can be done in time frame. This solution is in the direction of good air containing proper amount of oxygen for the eco-system. The pollution in air leads to respiratory system related problem and diseases. It also changes the eco-system which makes other living beings difficult to their existence. It also changes the weather condition which leads to many problem as cultivation in agriculture, livestock development and living being proper natural development. The problem in agriculture and livestock leads to lack of good nutrition food for human and living being.

ANN method or model is used to predict on previous data and work properly on present data or sample set. Here, our research work will attempt to use feed-forward ANN modeling with WHO guidelines (which are refer standard living condition

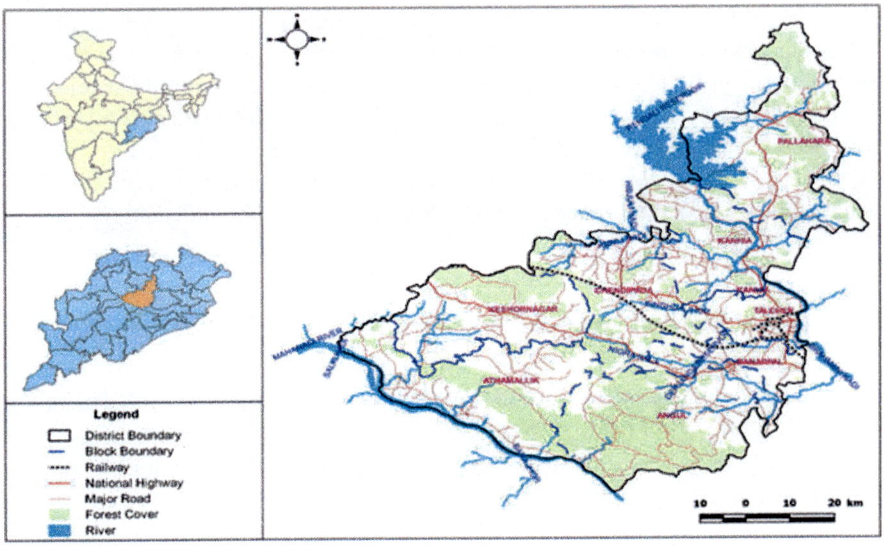

Fig. 1 Geographical map of Angul-Talcher region (district)

in healthy condition) for the prediction need to calculate. The prediction model is developed using NN tool box of MATLAB, simulated and analyzed (Fig. 1).

2 Materials and Methods

2.1 Basic Data Sets

Here, in this case study, we consider Angul-Talcher industrial region form 2015 to 2018 periods with four hours (location listed in Table 1) interval for the data of daily with temperature of ambient air, relative humidity, speed of air and concentration

Table 1 Data collection center for analysis with latitude and longitude

Place	Location	Latitude and longitude	Category of area
Angul	RO, SPCB Building, Angul	20°49′98.3″ N 85°06′25.4″ E	Industrial
Angul	Nalco Nagar, Angul	20°50′77.8″ N 85°09′18.3″ E	Residential
Talcher	TTPS, Talcher	20°54′86″ N 85°12′23.6″ E	Industrial and mining
Talcher	MCL, Talcher	20°50′41.8″ N 85°08′39.9″ E	Residential

of CO, NO, SO_2 and NO_2 as pollutants. Activities in this open cast mines, thermal power plants and aluminum smelter, sponge iron and steel plants become the main sources of air pollutant. In addition to this, the smaller processing plant related to them and the transport system need to be considered. The collected data was grouped into two sets of training datasets for the designed ANN model's training and also for the testing, to investigate effectiveness, efficiency and degree of correctness of the designed ANN structural model in the direction of predication calculation.

2.2 *Software Environment*

For the air quality prediction model development and analysis, we used neural network toolbox of MATLAB computing environment [9]. The neural network toolbox utility in MATLAB helps to solve a quite wide range and flexible parameters for neural network development.

2.3 *ANN Model for Prediction of Air Quality*

The model to predict the air quality represents a three-tier perceptron as shown in Fig. 2. The input tiers which consist of seven neurons in that include four pollutants which is CO, NO, SO_2 and NO_2 and three environmental parameter as variables which is temperature of air (T_{air}), the rate of flow of air (V_{air}) and the humidity RH. The second tier is hidden layer and its value is depending upon the parametric value

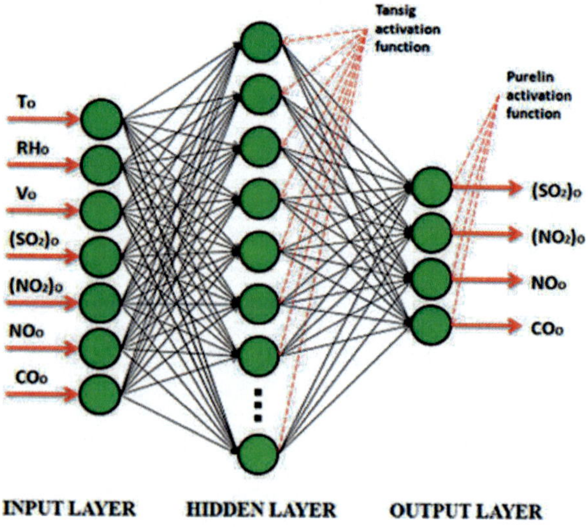

Fig. 2 ANN model for the study

Table 2 Structure and training result for the neural networks model for simulation

Network no.	Network structure	Training function	Learning	Learning rate	Momentum constant	MSE	*R*	R^2
1	7-10-4	Tansig-purelin	Traingdm	0.1	0.6	0.0626	*0.7485*	0.5587
2	7-18-4	Tansig-purelin	Traingdm	0.1	0.6	0.062	*0.7546*	0.5694

for optimization. The final tier is the output tier/layer of model, consisting the target of the prediction. In this model, CO, NO NO_2 and SO_2 were taken as the output variables.

The tangent sigmoid function was used for transfer function of design. The data set suitably segmented into three partitions for faster simulation process. The 65% dataset is used for training, 20% as for validation set, and rest 15% for testing the networks. The performance of model is characterized by mean square error (MSE) for the statistical criteria of the model.

2.4 Results and Discussion

ANN Modeling

The feed-forward method is deployed for the ANN modeling, in this study. The ***tansig*** is used in hidden layer and ***purelin*** functions in output layer neurons. The normalized value range of $[-1, 1]$ is considered for the input and target value in the pre-processing phase. The weight, biases adjustments and values are calculated by using Gradient-descent back-propagation method, while model is in the training phase. The network is put into the simulation and the performance is characterized with mean square error (MSE). The outline of the parameters and their simulated results were summarized in Table 2.

While performing the training with the ANN model, the performance has been found out against the number of epochs. Generally, the network performance initializes with larger value at the first epochs than others, which is changed to diminishing by nature with training as the weights are optimized and minimized by the model.

The best validation performance of the network is plotted with a black color dashed line and the green color line shown in Fig. 3. Figure 3 is a graphical representation of validation training set where the network performance intersects with the black line. The performance curve of the function for the designed model is shown in Fig. 3.

Robustness and regression of the model is analyzed by establishing the correlation of actual value and the predicted one, characterized by correlation coefficient, *R* express in percentage. The perfect fit is a relation between the training data and the simulated results was designated by the value of *R*, when it is equal to 1, i.e., 100%, the ***perfect/best fit***. Figure 4 shows the regression analysis plots of the network structures under validation with data. The regression plot signifies the perfect correlation

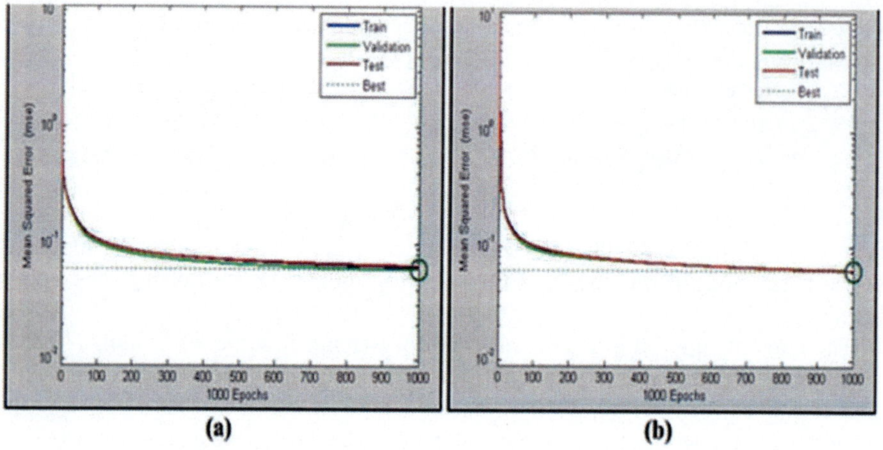

Fig. 3 The performance curve of the function for the network with structure, **a** 7-10-4, **b** 7-18-4

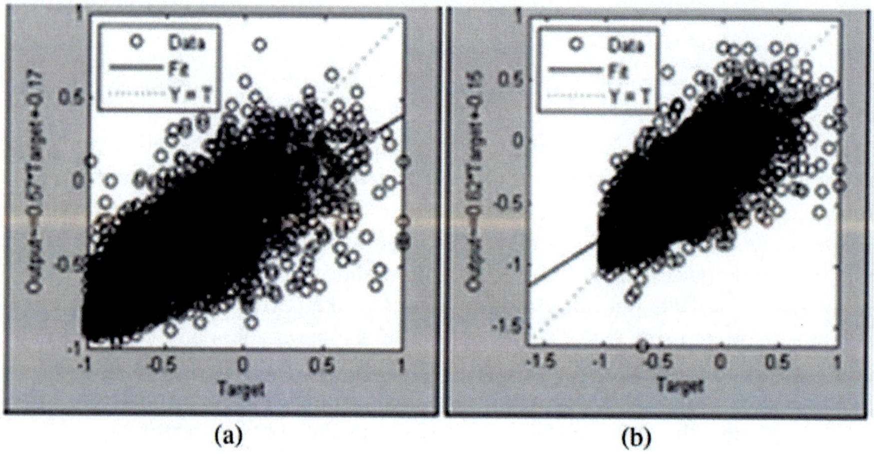

Fig. 4 The regression validation of the networks with structure, **a** 7-10-4, **b** 7-18-4

between the predicted and targets versus epochs are represented by the solid line known as ***best fit***. The dashed line shows the best fit generated by the algorithm.

With the simulation results summarized in Table 2, model with network structure **7-18-4** is the most appropriate, suitable and effective model for the air quality prediction with environmental data, condition and implemented algorithm, which produce the minimized values of MSE, 56.94% coefficient of determination, $\boldsymbol{R^2}$. As shown in Fig. 3, the value of correlation coefficient R for both models did not have much difference between each other. The best model that has been chosen shows a good agreement between predicted and measured values based on the value of correlation coefficient.

3 Conclusions

With simulation results and analysis, conducted with ANN model having structural as 7 (input vector)-18 (hidden layer)-4 (output vector) shows the best result and performance in the prediction of air quality with the values of R and the accuracy for prediction. This model produces R of 0.7546, which indicates better correlation between the targets and predicted outputs. This type of model should be applied for different nature based algorithms before being implemented on a hardware system for real time deployment.

Acknowledgements The author has pleasure to acknowledge **Dr. Subhra Keshari Biswal**, doctoral and post-doctoral in Environmental Science and Engineering, Life member of Indian Association of Environmental management, for his help in this work.

References

1. Dragomir, E.G.: Air quality index prediction using *K*-nearest neighbor technique. Bull. PG Univ. Ploiesti, Series Mathem. Inform. Phys. **LXII**(1), 103–108 (2010)
2. WHO Member State.: Estimated deaths and DALYs attributable to selected environmental risk factors. Accessed 27 Aug 2009
3. Deleawe, S., Kusznir, J., Lamb, B., Cook, D.: Predicting air quality in smart environments. J. Ambient Intell. Smart Environ. **2**(2), 145–154 (2010)
4. Barai, S.V., Dikshit, A.K., Sharma, S.: Neural network models for air quality prediction: a comparative study (2003)
5. Rao, M.N., Rao, H.V.: Air Pollution. Tata McGraw-Hill, New Delhi (2000)
6. Wang, W., Xu, Z., Lu, J.W.: Three improved neural network models for air quality forecasting. Eng. Comput. **20**(2), 192–210 (2003)
7. Li, M., Hassan, M.R.: Urban air pollution forecasting using artificial intelligence-based tools
8. Kurt, A., Gulbagci, B., Karaca, F., Alagha, O.: An online air pollution forecasting system using neural networks. Environ. Int. **34**(5), 592–598 (2008)
9. Demuth, H., Beale, M., Hagan, M.: Neural networks toolbox manual, Math Works Inc. (2009)

Numerical Study on Second-Grade Fluid Flow Problems Using Analysis of Fractional Derivatives Under Fuzzy Environment

Gourangajit Borah, Palash Dutta, and G. C. Hazarika

Abstract A numerical model is presented to study the effects of heat and mass transfer of a conducting magnetohydrodynamic flow of second-grade fluids past a vertical permeable plate, using two analysis of fractional derivatives given by Atangana-Baleanu and Caputo-Fabrizio, respectively. A uniform magnetic field is considered acting perpendicular to the plate. Fuzzy Set Theory (FST) is used to fuzzify the governing partial differential equations, due to the presence of vagueness and uncertainty associated with fluid flow problems. Finite difference scheme is adopted to discretize the equations and suitable computer codes are developed in PYTHON for AB and CF fractional derivatives. Graphical illustrations are made for variation of different rheological parameters on velocity, temperature, and concentration. Comparative results for AB and CF methods is shown in tabular form.

Keywords MHD · Permeability · Second grade fluids · Fuzzification · AB and CF fractional derivatives

Nomenclature

P	pressure,
ρ	fluid density,
ν	inematic viscosity,
μ	viscosity coefficient,

G. Borah (✉) · P. Dutta · G. C. Hazarika
Department of Mathematics, Dibrugarh University, Dibrugarh, Assam 786004, India
e-mail: gjit1993@gmail.com

P. Dutta
e-mail: palash.dtt@gmail.com

G. C. Hazarika
e-mail: gchazarika@gmail.com

S. Borah et al. (eds.), *Soft Computing Techniques and Applications*, Advances in Intelligent Systems and Computing 1248,
https://doi.org/10.1007/978-981-15-7394-1_4

g	gravitational acceleration,
$\overrightarrow{q}$	velocity vector,
$\overrightarrow{u}$	velocity vector $\vec{X}$ direction,
$\overrightarrow{v}$	velocity vector $\vec{Y}$ direction,
$\overrightarrow{w}$	velocity vector $\vec{Z}$ direction,
t	time,
k	thermal conductivity,
$\overrightarrow{\sigma}$	electrical conductivity,
D	molar diffusivity,
β_T	thermal expansion coefficient,
β_C	volumetric expansion coefficient,
B_0	applied magnetic field,
A_0	non-zero parameter,
V_0	non-zero parameter,
$\overline{T}_\infty$	fluid temperature at a considerable distance from the plate,
$\overline{C}_\infty$	ambient concentration when the plate is located far away,
T_w	temperature near the wall of the plate,
C_w	concentration level in the vicinity of the plate,
$\overline{T}$	temperature of the fluid,
$\overline{C}$	species concentration of the fluid,
θ	dimensionless temperature,
$\overline{\phi}$	dimensionless species concentration,
α_1	second-grade fluid viscosity,
α_1	viscoelastic parameter,
α	AB fractional parameter,
β	CF fractional parameter,
C_p	Heat capacity at constant pressure,
M	Magnetic parameter/Hartmann number,
Ec	Eckert number,
Gm	Solutal Grashof number,
Sc	Schmidt number,
Gr	Grashof number for heat transfer,
Pr	Prandtl number,
Γ	Gamma function,
k'	Dimensionless permeability for porous substances,
K	Permeable parameter,
$E_\alpha(-t^\alpha)$	Mittag–Leffler function

1 Introduction

Noll and Coleman were the pioneers of second-grade fluid model [1]. Second-order fluid flow has many practical applications which are more broadly known as transport phenomenon in porous substances. The underlying mechanism of boundary layer flow in viscoelastic materials serve as the base for many manufacturing processes like in fabrication of adhesive tapes, plastic sheets extrusion and also in various blood flow problems. The flow of non-Newtonian fluids together with analysis of heat and mass transfer has special significance because of various engineering applications like food processing and crude oil recovery. The hydromagnetic convection in porous medium has multifarious applications such as in designing MHD generators and accelerations, various groundwater energy storage systems, nuclear power reactors, astrophysics and so on.

Sparrow and Gregg [2], Merkin [3], Loyed and Sparrow [4] are the scientists who were interested in the study of all three modes of convection when in absence of magnetic field, a vicious but incompressible fluid flows over a vertical surface. Somess [5], Soundalgekar and Ganesan [6], Khair and Bejan [7] and Lin and Wu [8] examined heat and mass transfer analysis on a flow past a vertical plate. Chen [9] investigated the combined effects of thermal and solutal transfer analysis in a magnetohydrodynamic flow taking place from a vertical surface, under free convection. The viscous dissipation effects together with ohmic heating were also considered. The ohmic heating effect on the MHD flow of a Newtonian fluid with natural convective heat transfer was studied by Hossain et al. [10]. Asraf et al. [11], studied how mixed convection flow of second-grade fluids behave when we consider the combined effect of chemical reaction along with heat source in a three-dimensional environment. Nayak et al. [12] have taken MHD viscoelastic fluid over a stretching sheet to study the consequences of heat and mass transfer in the presence of chemical reaction. Choudhury and Dhar [13] carefully examined how heat generation giving rise to double-diffusive convection has effects on the MHD flow of a viscoelastic fluid, across a plate which is constantly moving. Choudhury and Dey [14] investigated the manner in which the characteristics of fluid flow is affected when conducting second-grade fluid is allowed to flow across a vertical permeable plate. In order to generate a constant source of energy; understanding diverse cosmic phenomena and technological development, the heat and mass transfer analysis together with the analysis of chemical reaction in MHD viscoelastic fluid flows are very important. Long-term biological problems are finding a solution now.

In fluid dynamics, there exist certain initial value problems, where we are unaware of the initial value precisely and the function may contain an uncertain parameter. Thus, errors might result in measurement, experiment, observation and while applying different operating conditions. Also, the boundary conditions defining a fluid flow problem are uncertain and hence there is nothing wrong to consider them as variables (represented by fuzzy numbers). The solution and behaviour of differential equations or a system of differential equations are fuzzy in nature. The study of fuzzy differential equations will certainly help us to study the fuzzy boundary value problems. So, there is a need to fuzzify the differential equations occurring in various disciplines of science and technology like Fluid Mechanics, Physics and various engineering branches.

Recent studies on the use of Caputo-Fabrizio analysis of fractional derivatives are referred to in [15–18]. But, according to Atangana and Baleanu, there exist two fundamental definitions: first is the Riemann-Liouville definition of fractional derivatives and the other in Caputo sense. One interesting fact about these fractional derivatives is that they involved an anti-derivative of their operators in their definition, *i.e.,* a fractional integral. Atangana-Baleanu (AB) [19], did incorporate the more generalized form of Mittag-Leffler function to solve non-singular kernel problems with non-locality. Nadeem et al. [20] compared the Caputo-Fabrizio and Atangana-Baleanu analysis of fractional derivatives in the generalized fluid model composed of Casson fluid, with the combined effects of chemical reaction and heat generation in it. Fractional derivatives being applied to second-grade fluids are evident in [21, 22].

The specific objective of our study is to numerically analyze the combined effects of heat and mass transfer, of a second-grade fluid flow across a vertical permeable plate under free convection, by employing the AB and CF fractional derivatives. Under the influence of the transverse magnetic field, the flow behaviour is considered and the resulting shear stress is calculated. The partial differential equations and the boundary conditions are made dimensionless by suitable choice of non-dimensional parameters. We then use the Zadeh's Extension Principle of fuzzy set theory to fuzzify the governing equations. The equations are finally discretized by finite difference method and suitable programming codes are developed for AB and CF fractional derivatives in PYTHON. Triangular Fuzzy Numbers are used for the evaluation of codes. The results are depicted graphically for the velocity field, species concentration and temperature distribution, subject to their variation against different rheological parameters. Comparison of the results obtained by both AB and CF methods are shown in tabular form. Graphical illustrations show that AB and CF methods are in excellent agreement.

2 Formulation of the Problem

Here, we consider the unsteady behaviour of two-dimensional boundary layer flow of an incompressible and electrically conducting second-order fluid with heat and mass transfer past a vertical permeable plate taking place under free convection. We then impose a magnetic field B_0 of uniform strength in the tangential direction perpendicular to the plate. The x-axis is considered in the same direction as the vertical infinite plate, whereas y-axis is considered normal to the plate. Let $\overline{u}$ and $\overline{v}$ be the velocity components, respectively, along with the plate (x-direction) and orthogonal to the plate (y-direction). We consider the interaction of induced magnetic field with the flow is much smaller in magnitude or of negligible order than the interaction of fluid flow with imposed magnetic field. Owing to infinite length of the plate, it is clear that the velocity, temperature and concentration are functions of (y, t). The plate is assumed to be accelerating with a velocity of $\overline{u} = V_0$ for $t \geq 0$, in its own plane. The fluid's electrical conductivity parameter is taken to be much smaller in magnitude. Hence the governing flow equations are given by

Equation of continuity:

$$\frac{\partial \overline{v}}{\partial \overline{y}} = 0 \Rightarrow \overline{v} = -V_0(\text{const.}) \tag{1}$$

Equation of conservation of momentum:

$$\frac{\partial \overline{u}}{\partial \overline{t}} + \overline{v}\frac{\partial \overline{u}}{\partial \overline{y}} = \nu_1 \frac{\partial^2 \overline{u}}{\partial \overline{y}^2} + \frac{\nu_2}{\rho}\left[\frac{\partial^2}{\partial \overline{y}^2}\left(\frac{\partial \overline{u}}{\partial \overline{t}}\right) + \overline{v}\frac{\partial^3 \overline{u}}{\partial \overline{y}^3}\right]$$
$$+ g\beta_T\left(\overline{T} - \overline{T}_\infty\right) + g\beta_C\left(\overline{C} - \overline{C}_\infty\right) - \frac{\overline{\sigma} B_0^2 \overline{u}}{\rho} - \frac{\nu_1 \overline{u}}{k'} \tag{2}$$

Equation of energy:

$$\frac{\partial \overline{T}}{\partial \overline{t}} + \overline{v}\frac{\partial \overline{T}}{\partial \overline{y}} = \frac{k}{\rho C_p}\frac{\partial^2 \overline{T}}{\partial \overline{y}^2} - \frac{\nu}{C_p}\left(\frac{\partial \overline{u}}{\partial \overline{y}}\right)^2 - \frac{\nu_2}{\rho C_p}\overline{v}\frac{\partial \overline{u}}{\partial \overline{y}}\frac{\partial^2 \overline{u}}{\partial \overline{y}^2} + \frac{\overline{\sigma} B_0^2 \overline{u}^2}{\rho C_p} \tag{3}$$

Equation of concentration:

$$\frac{\partial \overline{C}}{\partial \overline{t}} + \overline{v}\frac{\partial \overline{C}}{\partial \overline{y}} = D\frac{\partial^2 \overline{C}}{\partial \overline{y}^2} \tag{4}$$

The necessary conditions at the boundary are

$$\overline{y} = 0 : \overline{u} = V_0, \overline{v} = -V_0, \overline{T} = \overline{T}_w, \overline{C} = \overline{C}_w \tag{5}$$

$$\overline{y} \to \infty : \overline{u} \to 0, \overline{T} \to \overline{T}_\infty, \overline{C} \to \overline{C}_\infty \tag{6}$$

where ν_1, ν_2 are kinematic viscosities, k' is the dimensional permeability for porous medium, D denotes the molar diffusivity, k being the thermal conductivity, $\overline{\sigma}$ gives the electrical conductivity, β_C indicates the volumetric expansion coefficient, ρ indicates fluid density, β_T indicates the thermal expansion coefficient, $\overline{C}$ indicates the concentration, $\overline{T}$ gives the temperature of the fluid and g is the measure for gravitational acceleration. When the fluid is situated far away from the plate, $\overline{T}_\infty$ and $\overline{C}_\infty$ are the temperature and concentration of the fluid, respectively, whereas $\overline{T}_w$ and $\overline{C}_w$ gives the value of respective temperature and concentration at the surface of the plate.

Incorporating the following dimensionless parameters we get,

$$u = \frac{\overline{u}}{V_0}, y = \frac{V_0 \overline{y}}{\nu_1}, t = \frac{V_0^2 \overline{t}}{\nu_1}, \theta = \frac{\overline{T} - \overline{T}_\infty}{\overline{T}_w - \overline{T}_\infty}, \phi = \frac{\overline{C} - \overline{C}_\infty}{\overline{C}_w - \overline{C}_\infty},$$
$$\alpha_2 = \frac{\nu_2 V_0^2}{\rho \nu_1^2}, K = \frac{V_0^2 k'}{\nu_1^2}, \text{Pr} = \frac{\mu C_p}{k},$$
$$\text{Sc} = \frac{\nu_1}{D}, \text{Gr} = \frac{g\beta_T\left(\overline{T}_w - \overline{T}_\infty\right)\nu_1}{V_0^3},$$

$$\mathrm{Gm} = \frac{g\beta_C(\overline{C}_w - C_\infty)\nu_1}{V_0^3}, \mathrm{Ec} = \frac{V_0^2}{C_p(\overline{T}_w - \overline{T}_\infty)}, M = \frac{\overline{\sigma} B_0^2 \nu_1}{\rho V_0^2} \tag{7}$$

Here, α_2 is the non-dimensional viscoelastic parameter, K indicates permeable parameter, M denotes Hartmann number, Ec indicates Eckert number, Pr gives the Prandtl number, Gr gives the Grashof number for heat transfer, Sc indicates Schmidt number and Gm gives Solutal Grashof number.

Using the transformations (7), we obtain the non-dimensional forms of (2), (3) and (4) as,

$$\frac{\partial u}{\partial t} - \frac{\partial u}{\partial y} = \frac{\partial^2 u}{\partial y^2} + \alpha_2\left[\frac{\partial^2}{\partial y^2}\left(\frac{\partial u}{\partial t}\right) - \frac{\partial^3 u}{\partial y^3}\right] + \mathrm{Gr}\theta + \mathrm{Gm}\phi - \left(M + \frac{1}{K}\right)u \tag{8}$$

$$\frac{\partial \theta}{\partial t} - \frac{\partial \theta}{\partial y} = \frac{1}{\mathrm{Pr}}\frac{\partial^2 \theta}{\partial y^2} + \mathrm{Ec}\left(\frac{\partial u}{\partial y}\right)^2 + \alpha_2 \mathrm{Ec}\frac{\partial u}{\partial y}\frac{\partial^2 u}{\partial y^2} + M \cdot \mathrm{Ec} \cdot u^2 \tag{9}$$

$$\frac{\partial \phi}{\partial t} - \frac{\partial \phi}{\partial y} = \frac{1}{\mathrm{Sc}}\frac{\partial^2 \phi}{\partial y^2} \tag{10}$$

The corresponding initial and boundary conditions are transformed to:

$$y = 0 : u = 1, \theta = 1, \phi = 1 \tag{11}$$

$$y \to \infty : u \to 0, \theta \to 0, \phi \to 0 \tag{12}$$

3 Fuzzification of the Problem

As we are studying the problem in a fuzzy environment, the governing partial differential equations and the boundary conditions must be transformed into fuzzy form. Here, we fuzzify the functions involved in the problem using the Zadeh's Extension principle. It may be noted that the overhead '~ tilde' sign represents the corresponding fuzzy forms.

For the sake of calculation, we consider an arbitrary function say, $h = f(y, t)$.

(a) $u(y, t) = \frac{\overline{u}}{V_0}$

First method	Alternative Method
Let $u(h) = \frac{\overline{u}}{V_0}$	Let $u(h) = \frac{\overline{u}}{V_0}$
$\Rightarrow u(y) = \frac{\overline{u}}{y\nu/\overline{y}}\left[\text{since } y = \frac{V_0\overline{y}}{\nu}\right]$	$\Rightarrow u(t) = \frac{\overline{u}}{\sqrt{t\nu}/\sqrt{\overline{t}}}\left[\text{since } t = \frac{V_0^2\overline{t}}{\nu}\right]$
$\Rightarrow r = \frac{\overline{u}\cdot\overline{y}}{y\nu}[\text{let } u(y) = r]$	$\Rightarrow r' = \frac{\overline{u}\cdot\sqrt{\overline{t}}}{\sqrt{t}\cdot\sqrt{\nu}}\left[\text{let } u(t) = r'\right]$
$\Rightarrow y = \frac{\overline{u}\cdot\overline{y}}{r\nu}$	$\Rightarrow \sqrt{t} = \frac{\overline{u}\cdot\sqrt{\overline{t}}}{r'\cdot\sqrt{\nu}}$
$\Rightarrow u^{-1}(r) = \frac{\overline{u}\cdot\overline{y}}{r\nu}$	$\Rightarrow t = \frac{\overline{u}^2\overline{t}}{r'^2\nu} \Rightarrow u^{-1}(r') = \frac{\overline{u}^2\overline{t}}{r'^2\nu}$

Now, Zadeh's Extension principle gives,	Now, Zadeh's Extension principle gives,
$\mu_{\widetilde{u}(\widetilde{y})}(r) = \sup_y \mu_{\widetilde{y}}(y)$	$\mu_{\widetilde{u}(\widetilde{t})}(r') = \sup_t \mu_{\widetilde{t}}(t)$
$\Rightarrow \mu_{\widetilde{u}(\widetilde{y})}\left[\frac{\overline{u}.\overline{y}}{y\nu}\right] = \mu_{\widetilde{y}}(y)$	$\Rightarrow \mu_{\widetilde{u}(\widetilde{t})}\left[\frac{\overline{u}.\sqrt{\overline{t}}}{\sqrt{t\nu}}\right] = \mu_{\widetilde{t}}(t)$
$\Rightarrow \widetilde{u}(\widetilde{y}) = \frac{\widetilde{\overline{u}}.\widetilde{\overline{y}}}{\widetilde{y}.\widetilde{\nu}} = \widetilde{r}$	$\Rightarrow \widetilde{u}(\widetilde{t}) = \frac{\widetilde{\overline{u}}.\sqrt{\widetilde{\overline{t}}}}{\sqrt{\widetilde{t}.\widetilde{\nu}}} = \widetilde{r'}$

$$\theta(y, t) = \frac{\overline{T} - \overline{T}_\infty}{\overline{T}_w - \overline{T}_\infty}$$

$$\text{Let } \theta(h) = \frac{\overline{T} - \overline{T}_\infty}{\overline{T}_w - \overline{T}_\infty}$$

$$\Rightarrow s = \frac{\overline{T} - \overline{T}_\infty}{\overline{T}_w - \overline{T}_\infty}, \ [\text{let } \theta(h) = s]$$

(b) Now, Zadeh's Extension principle gives,

$$\mu_{\widetilde{\theta}(\widetilde{h})}(s) = \sup_h \mu_{\widetilde{h}}(h)$$

$$\Rightarrow \mu_{\widetilde{\theta}(h)}\left[\frac{\overline{T} - \overline{T}_\infty}{\overline{T}_w - \overline{T}_\infty}\right] = \mu_{\widetilde{h}}(h)$$

$$\Rightarrow \widetilde{\theta}(\widetilde{h}) = \frac{\widetilde{\overline{T}} - \widetilde{\overline{T}}_\infty}{\widetilde{\overline{T}}_w - \widetilde{\overline{T}}_\infty} = \widetilde{s}$$

$$\phi(y, t) = \frac{\overline{C} - \overline{C}_\infty}{\overline{C}_w - \overline{C}_\infty}$$

(c) Let $\phi(h) = \frac{\overline{C} - \overline{C}_\infty}{\overline{C}_w - \overline{C}_\infty}$

$$\Rightarrow t = \frac{\overline{C} - \overline{C}_\infty}{\overline{C}_w - \overline{C}_\infty}, \ [\text{let } \phi(h) = t]$$

Now, Zadeh's Extension principle gives,

$$\mu_{\widetilde{\phi}(\widetilde{h})}(t) = \sup_{h} \mu_{\widetilde{h}}(h)$$

$$\Rightarrow \mu_{\widetilde{\phi}(h)}\left[\frac{\overline{C} - \overline{C}_{\infty}}{\overline{C}_w - \overline{C}_{\infty}}\right] = \mu_{\widetilde{h}}(h)$$

$$\Rightarrow \widetilde{\phi}(\widetilde{h}) = \frac{\widetilde{\overline{C}} - \widetilde{\overline{C}}_{\infty}}{\widetilde{\overline{C}}_w - \widetilde{\overline{C}}_{\infty}} = \widetilde{t}$$

The fuzzification of the derivatives of the functions is just as similar mathematical calculations. After using the Zadeh's Extension principle to the functions involved in the problem, it is seen that the functions take the same form for both the fuzzy and non-fuzzy forms. The numerical values of the non-dimensional variables are equal in both fuzzy and non-fuzzy equations.

Thus, the fuzzy governing equations and the associated boundary conditions are as follows:

$$\frac{\partial \widetilde{u}}{\partial \widetilde{t}} - \frac{\partial \widetilde{u}}{\partial \widetilde{y}} = \frac{\partial^2 \widetilde{u}}{\partial \widetilde{y}^2} + \widetilde{\alpha}_2\left[\frac{\partial^2}{\partial \widetilde{y}^2}\left(\frac{\partial \widetilde{u}}{\partial \widetilde{t}}\right) - \frac{\partial^3 \widetilde{u}}{\partial \widetilde{y}^3}\right] + \widetilde{\mathrm{Gr}}\widetilde{\theta} + \widetilde{\mathrm{Gm}}\widetilde{\phi} - \left(\widetilde{M} + \frac{1}{\widetilde{\widetilde{K}}}\right)\widetilde{u} \tag{13}$$

$$\frac{\partial \widetilde{\theta}}{\partial \widetilde{t}} - \frac{\partial \widetilde{\theta}}{\partial \widetilde{y}} = \frac{1}{\widetilde{\mathrm{Pr}}}\frac{\partial^2 \widetilde{\theta}}{\partial \widetilde{y}^2} + \widetilde{\mathrm{Ec}}\left(\frac{\partial \widetilde{u}}{\partial \widetilde{y}}\right)^2 + \widetilde{\alpha}_2\widetilde{\mathrm{Ec}}\frac{\partial \widetilde{u}}{\partial \widetilde{y}}\frac{\partial^2 \widetilde{u}}{\partial \widetilde{y}^2} + \widetilde{M} \cdot \widetilde{Ec} \cdot \widetilde{u}^2 \tag{14}$$

$$\frac{\partial \widetilde{\phi}}{\partial \widetilde{t}} - \frac{\partial \widetilde{\phi}}{\partial \widetilde{y}} = \frac{1}{\widetilde{\mathrm{Sc}}}\frac{\partial^2 \widetilde{\phi}}{\partial \widetilde{y}^2} \tag{15}$$

with the boundary conditions,

$$\widetilde{y} = 0 : \widetilde{u} = 1, \widetilde{\theta} = 1, \widetilde{\phi} = 1 \tag{16}$$

$$\widetilde{y} \to \infty : \widetilde{u} \to 0, \widetilde{\theta} \to 0, \widetilde{\phi} \to 0 \tag{17}$$

For simplicity, we drop the overhead tilde (~) notations on the functions and parameters, in the subsequent calculations.

4 Method of Solution

4.1 Solution with Atangana-Baleanu Fractional Derivatives

Simply by replacing the time-dependent governing partial differential equations, with the AB fractional operator of order $0 < \alpha < 1$, we generate the AB fractional model for second-grade fluids. Equations (13)–(15) become,

$$ {}^{AB}\left(\frac{\partial^\alpha u}{\partial t^\alpha}\right) - \left\{1 + \alpha_2{}^{AB}\left(\frac{\partial^\alpha}{\partial t^\alpha}\right)\right\}\frac{\partial^2 u}{\partial y^2} = -\alpha_2\frac{\partial^3 u}{\partial y^3} + \frac{\partial u}{\partial y} + \mathrm{Gr}\theta + \mathrm{Gm}\phi - \left(M + \frac{1}{K}\right)u \tag{18} $$

$$ {}^{AB}\left(\frac{\partial^\alpha \theta}{\partial t^\alpha}\right) = \frac{\partial \theta}{\partial y} + \frac{1}{\Pr}\frac{\partial^2 \theta}{\partial y^2} + Ec\left(\frac{\partial u}{\partial y}\right)^2 + \alpha_2 \mathrm{Ec}\frac{\partial u}{\partial y}\frac{\partial^2 u}{\partial y^2} + \widetilde{M}\cdot\widetilde{\mathrm{Ec}}\cdot\widetilde{u}^2 \tag{19} $$

$$ {}^{AB}\left(\frac{\partial^\alpha \phi}{\partial t^\alpha}\right) = \frac{\partial \phi}{\partial y} + \frac{1}{Sc}\frac{\partial^2 \phi}{\partial y^2} \tag{20} $$

where $\frac{\partial^\alpha u}{\partial t^\alpha}$ is the AB fractional operator of order α defined as,

$$ {}^{AB}\left(\frac{\partial^\alpha u}{\partial t^\alpha}\right) = \frac{1}{1-\alpha}\int\limits_0^t u'(y,t)E_\alpha\left(\frac{-\alpha(z-t)^\alpha}{1-\alpha}\right)\mathrm{d}t $$

Here, $E_\alpha(-t^\alpha) = \sum\limits_{m=0}^{\infty}\frac{(-t)^{\alpha m}}{\Gamma(1+\alpha m)}$ is the Mittag–Leffler function.

4.2 Solution with Caputo-Fabrizio Fractional Derivatives

Just by replacing the time-dependent governing partial differential equations, with the CF fractional operator of order $0 < \beta < 1$, we produce the CF fractional model for second-grade fluids. Equations (13)–(15) become,

$$ {}^{CF}\left(\frac{\partial^\beta u}{\partial t^\beta}\right) - \left\{1 + \alpha_2{}^{CF}\left(\frac{\partial^\beta}{\partial t^\beta}\right)\right\}\frac{\partial^2 u}{\partial y^2} = -\alpha_2\frac{\partial^3 u}{\partial y^3} + \frac{\partial u}{\partial y} + \mathrm{Gr}\theta + \mathrm{Gm}\phi - \left(M + \frac{1}{K}\right)u \tag{21} $$

$$ {}^{CF}\left(\frac{\partial^\beta \theta}{\partial t^\beta}\right) = \frac{\partial \theta}{\partial y} + \frac{1}{\Pr}\frac{\partial^2 \theta}{\partial y^2} + \mathrm{Ec}\left(\frac{\partial u}{\partial y}\right)^2 + \alpha_2 \mathrm{Ec}\frac{\partial u}{\partial y}\frac{\partial^2 u}{\partial y^2} + \widetilde{M}\cdot\widetilde{\mathrm{Ec}}\cdot\widetilde{u}^2 \tag{22} $$

$$ {}^{CF}\left(\frac{\partial^\beta \phi}{\partial t^\beta}\right) = \frac{\partial \phi}{\partial y} + \frac{1}{\mathrm{Sc}}\frac{\partial^2 \phi}{\partial y^2} \tag{23} $$

where $\frac{\partial^\beta u}{\partial t^\beta}$ is the CF fractional operator of order β defined as,

$$ {}^{CF}\left(\frac{\partial^\beta u}{\partial t^\beta}\right) = \frac{1}{1-\beta}\int\limits_0^t u'(y,t)\mathrm{Exp}\left(\frac{-\beta(z-t)}{1-\beta}\right)\mathrm{d}t $$

5 Results and Discussion

The non-dimensional discretized equations along with the non-dimensional boundary conditions are solved with the AB and CF fractional derivative method by developing suitable programming codes in PYTHON. The zero value of the viscoelastic parameter α_2 indicates the Newtonian fluid motion, whereas the non zero values represent the viscoelastic fluid phenomenon. We have considered $\mathrm{Gr} > 0$, which corresponds to the flow structure for an externally cooled plate. $\mathrm{Gm} > 0$ means that the concentration near the boundary surface is greater than the free stream concentration.

This study has been done analysing the impact of various parameters such as α_2, M, Gr, Gm, Ec, Sc, Pr on velocity (u), temperature (θ) and species concentration (ϕ) profiles in presence of time. When we set $\alpha_2 = 0$, we obtain the corresponding results for Newtonian fluids. The numerical results are graphically shown in Figs. 1, 2, 3, 4, 5, 6, 7, 8, 9, 10, 11, 12, 13 and 14. In the following discussion, the mid-values of the parameters are considered as, $\alpha_2 = 0.2$, $M = 0$, $\mathrm{Gr} = 5$, $\mathrm{Gm} = 3$, $\mathrm{Ec} = 0.02$, $\mathrm{Pr} = 3$, $\mathrm{Sc} = 2$, unless otherwise stated.

Figure 1 demonstrates the velocity, subject to the effect of viscoelastic parameter (α_2). In a single graph, it seems very unhandy to indicate the left-value, mid-value and right values, for various values of the parameter. So, to make it more pleading to the eye, in Fig. 2 we depict only the mid-values of the parameters so that we can infer about the behaviour of velocity, for different time slots. Unlike Newtonian fluids, the rise of viscoelastic parameter enhances the fluid velocity at every point of the fluid region. In both the figures, the velocity (u), respectively, increases for increasing values of α_2.

Similar procedure is followed for Fis. 3 and 4 where the velocity profile is drawn for varying magnetic parameter values (M). For weak magnetic fields, preferably ($M < 1$) the velocity may increase with the increase of M. The velocity may initially increase and may, later on, follow a decreasing trend with the further increase of M, unlike the Newtonian fluids.

Fig. 1 Effect of $\alpha_2 (= 0.2)$ on velocity for time $t = 0.1$ s

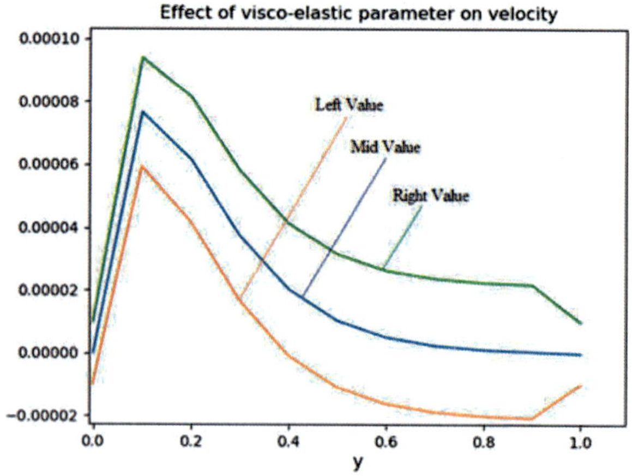

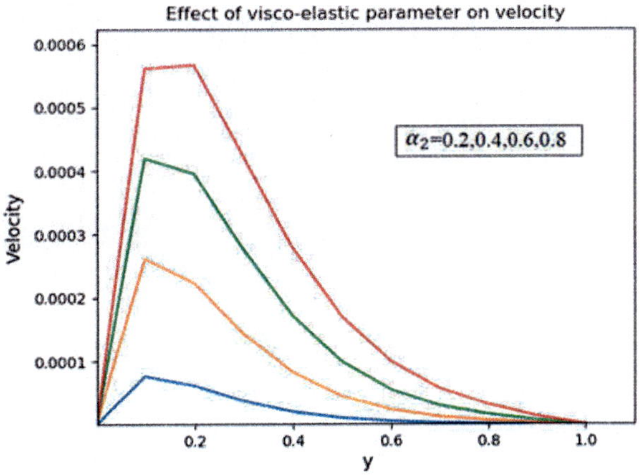

Fig. 2 Velocity profile taking different mid-values of α_2 for time $t = 0.1$ s

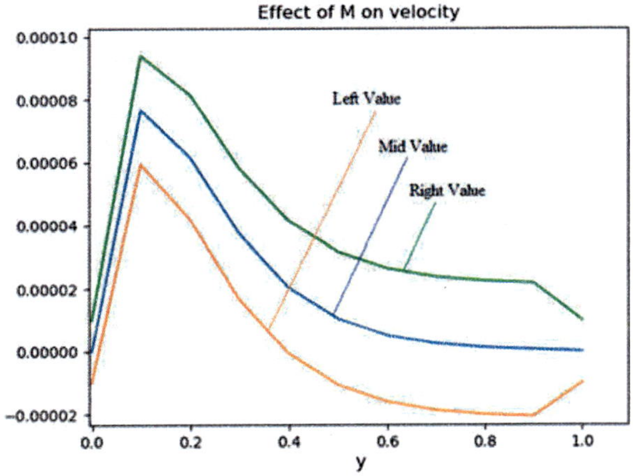

Fig. 3 Effect of M (=0) on velocity for time $t = 0.1$ s

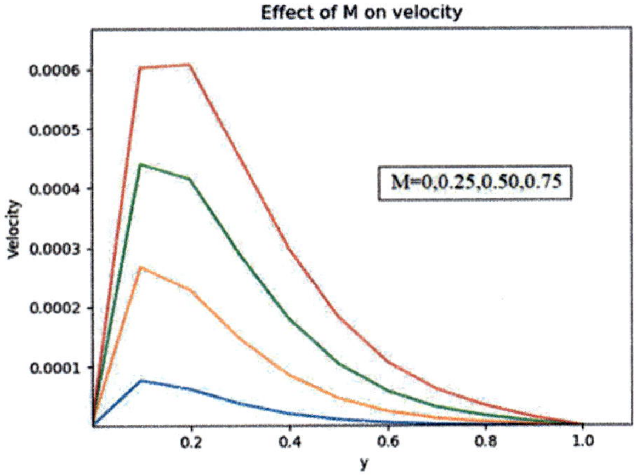

Fig. 4 Velocity profile taking different mid-values of M for time $t = 0.1$ s

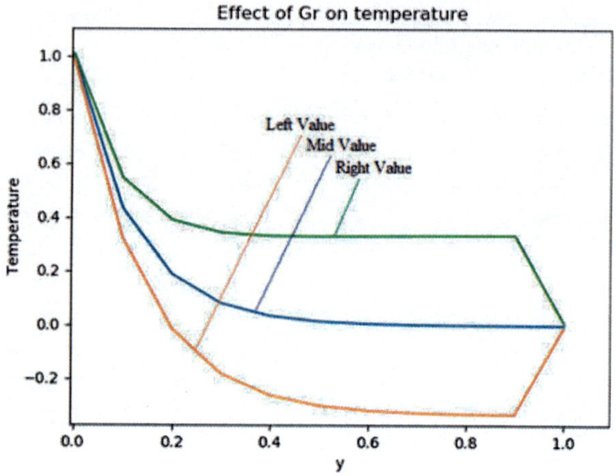

Fig. 5 Effect of Gr (=5) on temperature for time $t = 0.1$ s

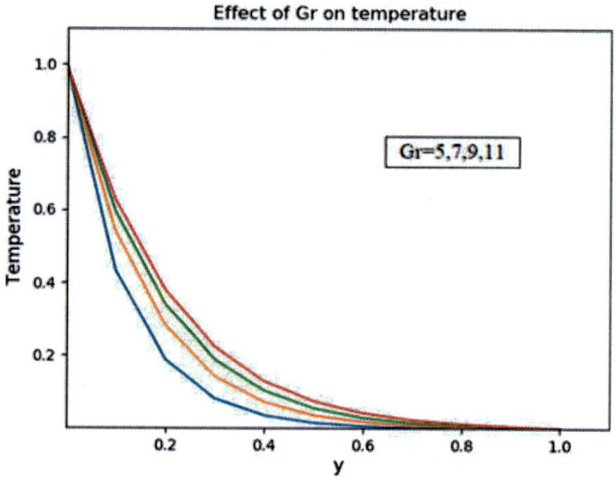

Fig. 6 Temperature profile taking different mid-values of Gr for time $t = 0.1$ s

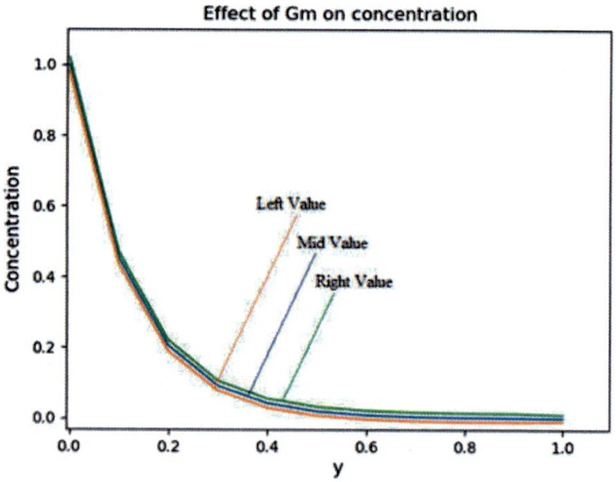

Fig. 7 Effect of Gm (=3) on concentration for time $t = 0.1$ s

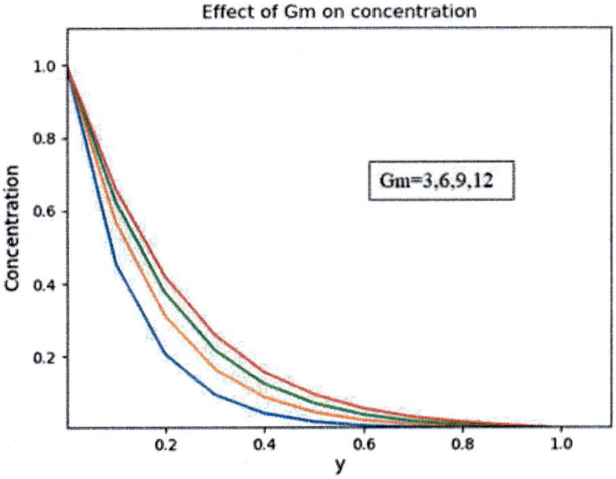

Fig. 8 Concentration profile taking different mid-values of Gm for time $t = 0.1$ s

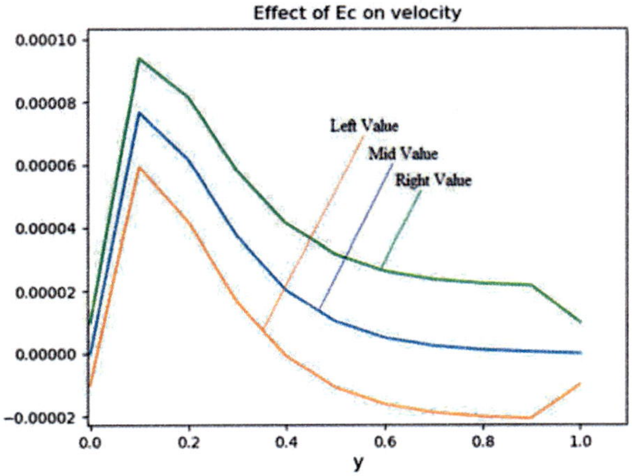

Fig. 9 Effect of Ec (=0.02) on velocity for time $t = 0.1$ s

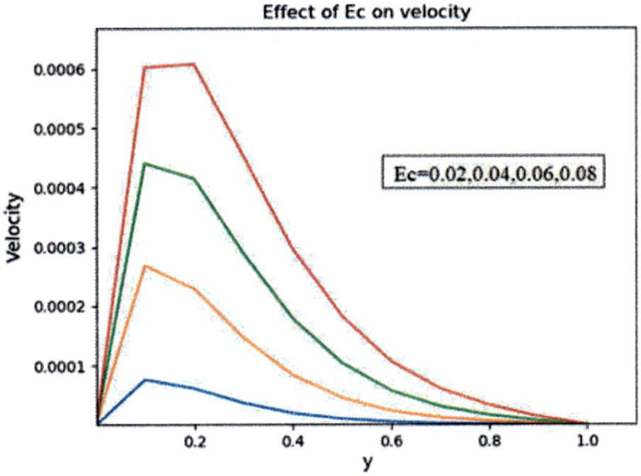

Fig. 10 Velocity profile taking different mid-values of Ec for time $t = 0.1$ s

Fig. 11 Effect of Pr (=3) on temperature for time $t = 0.1$ s

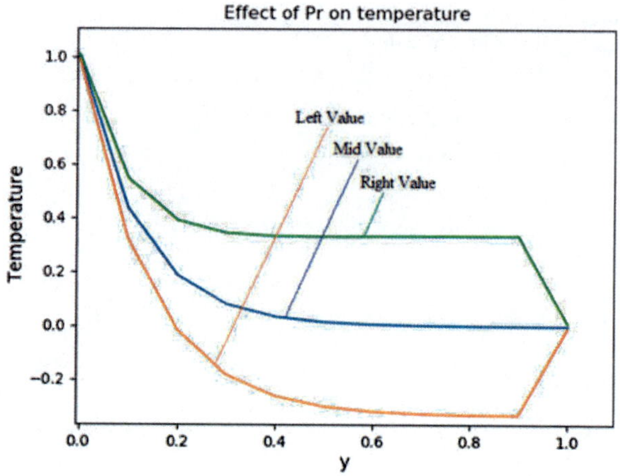

Fig. 12 Temperature profile taking different mid-values of Pr for time $t = 0.1$ s

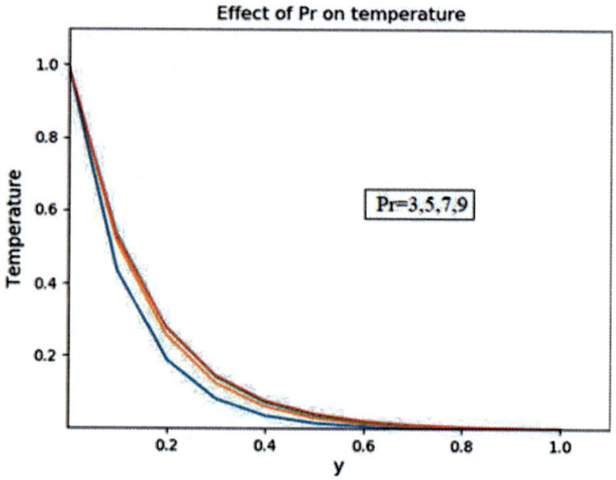

Fig. 13 Effect of Sc (=2) on concentration for time $t = 0.1$ s

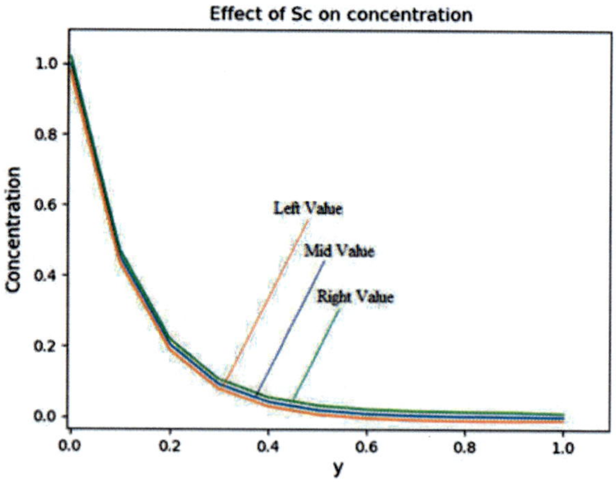

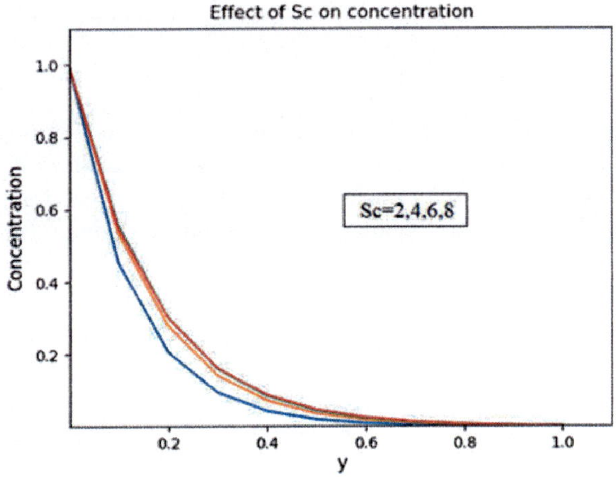

Fig. 14 Concentration profile taking different mid-values of Sc for time $t = 0.1$ s

Figs. 5, 6, 7 and 8 illustrates the temperature and concentration profile for various values of Thermal Grashof number (Gr) and that of Solutal Grashof number (Gm). Results reveal that, in all the four figures of temperature (θ) and concentration (ϕ), they, respectively, increase for sufficiently larger values of Gm and Gr.

Enhancement of Eckert number Ec comprehensively increases the fluid velocity. Ec physically relates the variation of kinetic energy with the enthalpy of the system. A larger Eckert number results in more kinetic energy which implies an increase of the temperature and which accounts for larger buoyancy force. Hence, the velocity increases as a result of significantly larger buoyancy force. Graphical illustrations for this case are evident from Figs. 9 and 10.

Figures 11 and 12 show the variation of temperature (θ) with the parameter Pr. The simultaneous effect of momentum and thermal diffusion in fluid flows is characterized by the Prandtl number and Prandtl number has significant roles in mechanisms involving heat transfer. While a rise of Prandtl number shows an enhancement in fluid temperature.

Unlike for Newtonian fluids, the velocity and temperature increase with the increasing values of Sc for viscoelastic fluids. The Schmidt number characterizes the simultaneous effect of both diffusions of mass and momentum. In mass transfer problems, the presence of this parameter shows comprehensive results. The species concentration is influenced significantly by the Schmidt number (Sc), and it shows an accelerating trend. Here, the corresponding graphs of concentration (ϕ) for the parameter Sc are Figs. 13 and 14.

Further, we provide a comparison of AB and CF Fractional Derivative Methods, providing the data obtained in tabular form.

5.1 Comparison of AB and CF Fractional Derivative Methods in Tabular Form

In Tables 1, 2, 3, 4, 5, 6 and 7, numerical data is provided based on which the graphs are obtained. Results reveal that the AB method has higher rate of convergence than the CF method. The gamma function in the integral operator of AB method is responsible for this outcome, over the exponential function in the CF method.

Table 1 Effect of α_2 on u

$a2$	y	t	Value(s)	u	
				AB	CF
0.2	0.23	0.1	Left value	1.67E-05	1.66E-05
			Mid value	3.75E-05	3.70E-05
			Right value	5.82E-05	5.73E-05
0.4	0.23	0.1	Left value	0.000100997	0.000101103
			Mid value	0.000143322	0.000142232
			Right value	0.000185644	0.000183382

Table 2 Effect of M on u

M	y	t	Value(s)	u	
				AB	CF
0	0.23	0.1	Left value	1.67E-05	1.66E-05
			Mid value	3.75E-05	3.70E-05
			right value	5.82E-05	5.73E-05
0.25	0.23	0.1	left value	0.000104774	0.000104862
			Mid value	0.000147143	0.00014603
			Right value	0.000189508	0.000187219

Table 3 Effect of Gr on θ

Gr	y	t	Value(s)	θ	
				AB	CF
5	0.23	0.1	Left value	0.082177733	0.082177733
			Mid value	0.08218974	0.08201974
			Right value	0.082201886	0.082201886
7	0.23	0.1	Left value	0.143660436	0.141689044
			Mid value	0.144241735	0.134231735
			Right value	0.144411028	0.14422111

Table 4 Effect of Gm on ϕ

Gm	y	t	Value(s)	ϕ	ϕ
				AB	CF
3	0.23	0.1	Left value	0.093902818	0.093903566
			Mid value	0.09391435	0.09391435
			Right value	0.093925883	0.093925134
6	0.23	0.1	Left value	0.164941512	0.164144342
			Mid value	0.165290146	0.164173905
			Right value	0.165550378	0.16420347

Table 5 Effect of Ec on u

Ec	y	t	Value(s)	u	
				AB	CF
0.02	0.23	0.1	Left value	1.67E-05	1.66E-05
			Mid value	3.75E-05	3.70E-05
			Right value	5.82E-05	5.73E-05
0.04	0.23	0.1	Left value	0.000104382	0.000104472
			Mid value	0.000146746	0.000145635
			Right value	0.000189106	0.00018682

Table 6 Effect of Pr on $\boldsymbol{\theta}$

Pr	y	t	Value(s)	$\boldsymbol{\theta}$	
				AB	CF
3	0.23	0.1	Left value	0.082177733	0.080977733
			Mid value	0.08218974	0.08213354
			Right value	0.082201886	0.082200086
5	0.23	0.1	Left value	0.124744233	0.122788223
			Mid value	0.125495664	0.125433766
			Right value	0.125712012	0.12566412

Table 7 Effect of Sc on ϕ

Sc	y	t	Value(s)	ϕ	ϕ
				AB	CF
2	0.23	0.1	Left value	0.093902818	0.093903566
			Mid value	0.09391435	0.09391435
			Right value	0.093925883	0.093925134
4	0.23	0.1	Left value	0.142079286	0.141327334
			Mid value	0.142353043	0.141353554
			Right value	0.142558753	0.141379777

6 Conclusion

The problem of conducting magnetohydrodynamic boundary layer flow of a viscoelastic fluid is studied numerically using fractional derivatives. The combined analysis of heat transfer accompanied by mass transfer is considered. The results of the study are summarized and put pointwise in the following conclusions:

(i) For steady increment in the values of the viscoelastic parameter, there is a positive increment of fluid velocity with time.
(ii) Magnetic parameter/Hartmann number (M) increases the fluid velocity asymptotically to the value 1 (one).
(iii) Grashof number for mass transfer (Gm) and Grashof number for heat transfer (Gr) helps accelerate the fluid motion by, respectively, increasing the temperature and concentration.
(iv) Velocity magnifies for larger values of Ec, in case of viscoelastic fluids. Although temperature and concentration are not significantly affected, yet they show a steady increment with time.
(v) Further enhancement in the values of Pr and Sc, the temperature and the species concentration is noticeably affected, while they show an accelerating trend.
(vi) The presence of gamma function in the integral operator of AB method is responsible for faster convergence rate over the CF method involving the exponential function. Precise results compatible with earlier published works and higher convergence rate of AB method separates it from the CF method, making it a more efficient one.

Disclosure Statement: No potential conflict of interest was reported by the authors.

References

1. Coleman, B.D., Noll, W.: An approximation theorem for functionals, with application in continuum mechanics. Arch. Ration. Mech. Anal. **6**(1), 355–370 (1960). https://doi.org/10.1007/BF00276168
2. Sparrow, E.M., Gregg, J.L.: Buoyancy effects in forced convection flow and heat transfer. Trans. Am. Soc. Mech. Engg. E., J. Appl. Mech. **81**, 133–139 (1959). https://doi.org/10.1016/0017-9310(62)90160-6
3. Merkin, J.H.: The effect of buoyancy forces in the boundary layer over a semi-infinite vertical flat plate in a uniform free stream. J. Fluid Mech. **35**, 439–450 (1969). https://doi.org/10.1017/S0022112069001212
4. Loyed, J.R., Sparrow, E.M.: Combined forced and free convection flow on vertical surfaces. Int. J. Heat Mass Transfer **13**, 434–438 (1970). https://doi.org/10.1016/0017-9310(70)90119-5
5. Somers, E.V.: Theoretical considerations of combined thermal and mass transfer from a vertical flat plate. AMEJ Appl. Mech. **23**, 295–301 (1956)
6. Soundalgekar, V.M., Ganesan, P.: Finite-difference anlaysis of transient free convection with mass transfer on an isothermal vertical flat plate. Int. J. Engg. Sci. **19**, 757–770 (1981). https://doi.org/10.1016/0020-7225(81)90109-9

7. Khair, K.R., Bejan, A.: Mass Transfer to natural convection boundary layer flow driven by heat transfer. Int. J. Heat Mass Transf. **107**, 979–981 (1985). https://doi.org/10.1115/1.3247535
8. Lin, H.T., Wu, C.M.: Combined heat and mass transfer by laminar natural convection from vertical plate. Heat Mass Transf. **30**(6), 369–376 (1995). https://doi.org/10.1007/BF01647440
9. Chen, C.H.: Combined heat and mass transfer in MHD free convection from a vertical surface with ohmic heating and viscous dissipation. Int. J. Engg. Sci. **42**, 669–713 (2004). https://doi.org/10.1016/j.ijengsci.2003.09.002
10. Hossain, M.A., Ahmed, M.: MHD free convection flow from an isothermal plate inclined at a small angle to the horizontal. J. Theor. Appl. Fluid Mech. **1**, 194–202 (1996). https://doi.org/10.1007/BF01170363
11. Ashraf, M., Alsaedi, A., Hayat, T., Shehzad, S.A.: Convective heat and mass transfer in three-dimensional mixed convection flow of viscoelastic fluid in presence of chemical reaction and heat source/sink. Comput. Mathem. Mathem. Phys. **57**(6), 1066–1079 (2017). https://doi.org/10.1134/S0965542517060021
12. Nayak, M.K., Dash, G.C., Singh, L.P.: Heat and mass transfer effects on MHD visco-elastic fluid over a stretching sheet through porous medium in presence of chemical reaction. Propul. Power Res. **5**(1), 70–80 (2017). https://doi.org/10.1016/j.jppr.2016.01.006
13. Choudhury, R., Dhar, P.: Effects of MHD viscoelastic fluid flow past a moving plate with double diffusive convection in presence of heat generation. WSEAS, Trans. Fluid Mech. **9**, 196–205 (2014)
14. Choudhury, R., Dey, B.: Flow features of a conducting viscoelastic fluid past a vertical permeable plate. Global J. Pure Appl. Mathem. Res. India Publ. **13**(9), 5687–5702 (2017)
15. Losada, J., Nieto, J.J.: Properties of a new fractional derivative without singular kernel. Progr. Fract. Differ. Appl. **1**, 87–92 (2015). https://doi.org/10.12785/pfda/010202
16. Baleanu, A.D.: Caputo-Fabrizio derivative applied to groundwater flow within confined aquifer. J. Eng. Mech. **143**, D4016005 [Pub Med] (2016). https://doi.org/10.1061/(asce)em.1943-7889.0001091
17. Hristov, J.: Transient heat diffusion with a non singular fading memory. Therm. Sci. **20**, 757–766 (2016). https://doi.org/10.2298/TSCI160112019H
18. Hristov, J.: Steady-State heat conduction in a medium with spatial non-singular fading memory: Derivation of Caputo-Fabrizio space fractional derivative with Jeffrey's kernel and analytical solutions. Therm. Sci. **21**, 827–839 (2017). https://doi.org/10.2298/TSCI160229115H
19. Baleanu, A.D.: New fractional derivatives with non-local and non-singular kernel: theory and application to heat transfer model. Therm. Sci. 18 (2016). https://doi.org/10.2298/tsci160111018a
20. Nadeem, A.S., Farhad, A., Muhammad, S., Khan, I., Jan, S.A.A., Ali, S.A., Metib, S.A.: Comparison and analysis of the Atangana-Baleanu and Caputo-Fabrizio fractional derivatives for generalized Casson fluid model with heat generation and chemical reaction. Results. Phys. **7**, 789–800 (2017). https://doi.org/10.1016/j.rinp.2017.01.025
21. Khan, M., Nadeem, S., Hayat, T., Siddiqui, A.M.: Unsteady motions of a generalized second grade fluid. Math. Comput. Model. **41**, 629–637 (2005). https://doi.org/10.1016/j.mcm.2005.01.029
22. Khan, M., Ali, S.H., Haitao, Q.: Exact solutions for some oscillating flows of a second grade with a fractional derivative model. Math. Comput. Model. **49**(7–8), 1519–1530 (2009). https://doi.org/10.1016/j.mcm.2008.07.012

Crime Rate Prediction Using Machine Learning and Data Mining

Sakib Mahmud, Musfika Nuha, and Abdus Sattar

Abstract Analysis of crime is a methodological approach to the identification and assessment of criminal patterns and trends. In a number of respects cost our community profoundly. We have to go many places regularly for our daily purposes, and many times in our everyday lives we face numerous safety problems such as hijack, kidnapping, and harassment. In general, we see that when we need to go anywhere at first, we are searching for Google Maps; Google Maps show one, two, or more ways to get to the destination, but we always choose the shortcut route, but we do not understand the path situation correctly. Is it really secure or not that's why we face many unpleasant circumstances; in this job, we use different clustering approaches of data mining to analyze the crime rate of Bangladesh and we also use K-nearest neighbor (KNN) algorithm to train our dataset. For our job, we are using main and secondary data. By analyzing the data, we find out for many places the prediction rate of different crimes and use the algorithm to determine the prediction rate of the path. Finally, to find out our safe route, we use the forecast rate. This job will assist individuals to become aware of the crime area and discover their secure way to the destination.

Keywords Crime · Numerous safety problem · Data mining · KNN (K-Nearest Neighbor) · Safe route

---

S. Mahmud · M. Nuha · A. Sattar (✉)
Department of Computer Science and Engineering, Daffodil International University, 102 Sukrabadh, Mirpur Road, Dhaka 1207, Bangladesh
e-mail: abdus.cse@diu.edu.bd

S. Mahmud
e-mail: sakib15-7962@diu.edu.bd

M. Nuha
e-mail: musfika15-7930@diu.edu.bd

S. Borah et al. (eds.), *Soft Computing Techniques and Applications*, Advances in Intelligent Systems and Computing 1248,
https://doi.org/10.1007/978-981-15-7394-1_5

1 Introduction

In this era of modern world, our popularity is increasing and citification carries enormous general, financial and environmental, while presenting challenges in urban management issues such as traffic resource planning, environment and safe water quality, public policy and public safety services. In addition, represent the most crime rates in larger cities, crime reducing is becoming one of the most important social issues in enormous metropolitan areas as it affects people security issues, youngster growth and person socio-economic status. Crime rate forecast is a scheme that uses different algorithms to determine the crime rate based on prior information. For our daily purposes we have to go many places every day and many times in our daily lives we face numerous security issues such as hijacking, kidnapping, harassment, etc. In general, we see that we are searching for Google Maps when we need to go anywhere at 1st, Google Maps show that one, two or more ways to get to the destination, but we always choose the shortcut route, but we do not comprehend the path condition properly. Is it really safe or not that's why we are faced with many unpleasant circumstances; this research introduces the design and execution of a strategy based on past crime data and analyzes the crime rate in past areas at distinct moments; for this work, we use primary data those are collected from the people based on their previous crime problem. In our train information collection, we used different algorithms to figure out the highest precision between the KNN algorithms that provides the greatest precision. In this paper, we use different models and table to show the different types of crime rate, mostly working data from last 3 years of crime and showing the level of crime prediction in different issues described in Sect. 3. In Sect. 1 gives the introduction about the whole paper, Sect. 2 in literature review section describes the previous work on crime rates, Sect. 3 describes the general paper information in the Methodology section, this chapter discusses the dataset, data processing, crime analysis, crime rate prediction and describes the multiple algorithms that we used in the crime database assessment and lastly demonstrates the greatest precision of the forecast rate and model using python matplotlib and KNN algorithm. In Sect. 5 proves the result and discussion, Sect. 6 dicussed the conclusion about the total work, and finally acknowledgements and references are discussed at the end of this chapter.

2 Literature Review

For this paper we have studied the relationship between crime and different features in the criminology literature. Reduce the crime and detect the techniques of crime and stop the crime before the author uses different techniques. Use Z-Crime Tools and Advanced ID3 algorithm with data mining technology to predict criminal activity.

Identify the appropriate crime pattern and statistical analysis of hidden linked detection algorithm used data. Forensic Tool Kit 4.0 is used to remove research and visualization of data [1]. The uses the K-Means Clustering algorithm for unsupervised learning to determine the crime rate. The model was then analyzed, preprocessed and implemented to taste the set of information and trained the algorithm. K-Means Clustering algorithm provided more than 75% [2].

The author used broken window theory, deep learning algorithm, random forest and naïve Bayes to reduce criminal activity and detect the crime zone. Prepare the data frame to train the model for recognition of images, preprocessing of information and detection of crime hotspot. The model tuned with deep learning provides 0.87% of the best accuracy. Machine learning offers methods of regression and classification used to predict rates of crimes. The author uses multi-linear regression to find the link between dependent and independent variables.

K-Nearest neighbors is used for classification to single and multi-class variable. The Neural Network is used for the prediction's precision. The precision rate by using Neural Network the model accuracy is 60, 93 and 97% [2]. The Author presents a geographical analysis-based and self-regressive approach to automatically identify large danger urban crime areas and to represents crime patterns in each region reliably. Show the result of the algorithm of crime prediction system consisting of a collection of thick crime areas and a set of related crime forecaster. This operate primarily for the large region where the large amounts of individuals live and demonstrates that the suggested strategy achieves excellent precision over rolling time horizons in spatial and temporal crime forecasting. This paper's working process collects raw data the hotspot uses after splitting the data to create the new hotspot model and finally shows the predictive crime rate [3].

Shiju-Sathya devan proposed Apriori algorithm for the identification of criminal trends and patterns. This algorithm is also used to identify association rules in the database that highlight general trends. This paper also suggested the naïve Bayes algorithm by training crime data to create the model. The result showed after testing that the Naive Bayes algorithm gave 90% precision. K. Zakir-Hussain et al. used the methods of information mining to analyze criminal conduct. This paper proposed tool for analyzing criminal investigation (CIA). Within the law enforcement community, this instrument was used to assist resolve violent offenses. This study is about the various type of crime scene. Both from an investigative and a behavioral perspective, the analysis was done. It provided insight into the unknown criminals as well as recommendation for investigation and interview and trial strategies [4].

Classification is one kind unique methods of information mining used to classify each object in an information set into one of the predefined classes or groups, The idea is to define the use of the Criteria for the segmentation of the entire database, once this is done, individual datasets can then fall naturally into one or more groups. By means of classification, existing datasets can be easily understood and it also helps to predict how new individual datasets will behave based on the classification. Data mining generates models of classification by observing classified data and discovering a predictive pattern between those data. Naive Bayes is a classification algorithm used to predict that it works on the principle of Bayesian.

Table 1 Dataset table

Number	Details		
	Name	Type of columns	Descriptions
1	Person_Id	Value type	Person ID in dataset
2	Name	String	Victim person name
3	Year	Numeric	Crime occurs year
4	Ages	Numeric	The ages of the victim
5	Gender	String	Victims neuter
6	Time	String	Time when the crime has occurred
7	Victim area	String	Area where the crime has occurred.
8	Region	String	Region of the victim
9	Home town	String	Home town of the victim
10	Month	String	The year in which the crime has occurred

3 Research Methodology

3.1 Dataset

The crime dataset is extracted from primary data collection based on field work. This dataset consists of about 500 in 10 rows details. The key features such as Name, Years, Months, Crime Type, Crime Areas, Victim Genders, Victim Ages, Victim Areas, and Months are selected from the dataset as the system input features.

The characteristics Perpetrator Ages, Perpetrator Genders and Victims relation are select system's target variables (Table 1).

3.2 Preprocessing

Since unknown is not a value to be considered but an indicator of unfinished papers, we decided to remove unknown values from the information set, Dates and times was recorded as a times window in MM/DD/YY HH/MM format from the documents as direct matching of these dates are complicated by the classification system to match date and time with each other. Classification of the date was considered to classify a particular date into three groups as weekends, weekdays and unaware. This classification is based on the features of the date time windows (Fig. 1).

Explains the system's workflow. The workflow begins by extracting data from data collection, which is a dataset repository on different roles. The primary data will be preprocessed and transformed into a criminal data.

Four target variables are predicted.

i. Linear Regression is used to find the aged based crime accuracy.

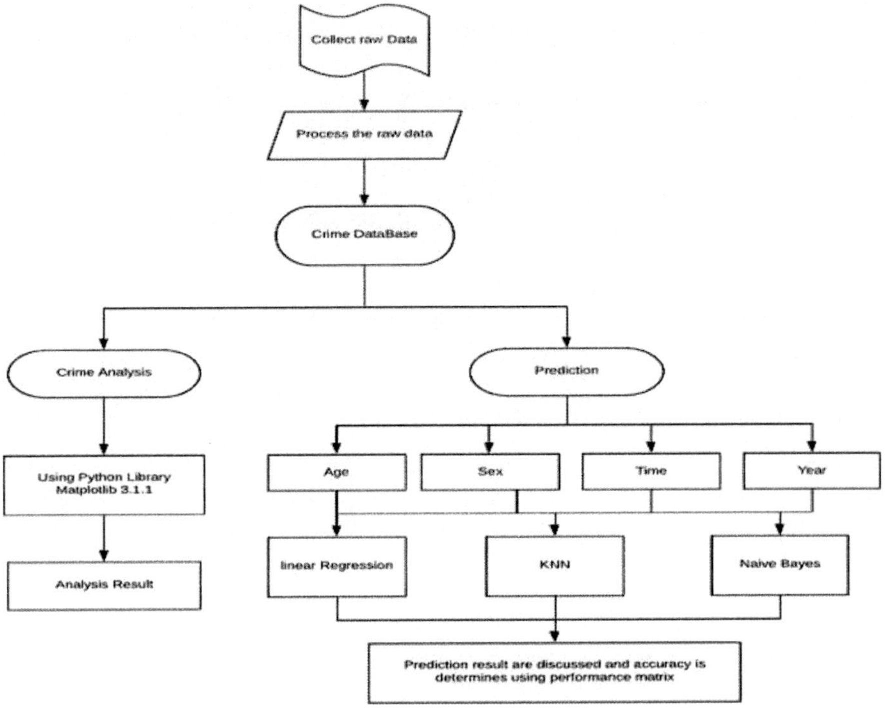

Fig. 1 Work flow diagram

ii. The sex is estimated using the classification of K- Nearest Neighbors.
iii. The gender of the perpetrator is estimated using the classification of K-nearest neighbors.
iv. The Final prediction rate for years based on age, sex, time, and year using K-NN algorithm.

3.3 Crime Dataset

See Table 2.

Few property features of dataset like months, types of crime, victim gender, age, area and the relationship are in the above data set are qualitative from. This qualitative information should be classified as in order 0 or 1 to apply the mathematical models for prediction.

Table 2 Male female identification by binary digit

Gender	Male	Female
	0	1

Table 3 Uses binary value to present the month status

Months	D_A	D_B
Jan–April	1	0
May–August	0	1
Sept–Dec	0	0

As follow the rules male contains 0 and female contains 1 in Table 1. Also adding dummy column by using the crime data in different month of the year in Table 3.

Table 3 adds $N - 1$ dummy columns in the given column for unique N values. The table above explains how data is preprocessed. Three months are selected from the Column state in the data table. In this case, in the column states, the number of unique values is $N = 3$. It is necessary to add dummy columns to categorize the data N s 1. Here are two types of dummy columns those are dummy A and dummy B.

1. 1 is assigned for column A and 0 for column B result between January to April.
2. If 0 is assigned for column A and 1 for column B result between May to August.
3. If both is assigned 0 for column A and column B result between September to December.

4 Algorithm

The domain contains many clustering algorithms. There is widespread use and acceptance of the K-means partitioning method [5]. Apart from the K-means strategy, the Linear regression algorithm [6] is the one we used because it enables consumers to determine the number of clusters based on those values Naïve Bayes is also pretend good result but above two are provides the best accuracy.

4.1 Linear Regression

Multi-linear regression is a sort of mathematical approach to finding a relation between the dependent variables (Victim age) and a set of independent variables those input values gathered from the crime spot. This methodology predicts the Era of the victims age values based on the input characteristics indicate in the metadata column. The multi-linear regression is:

$$Y = \beta_0 + \beta_1 x_1 + \beta_2 x_2 + \cdots + \beta_p x_p \tag{1}$$

Here,

Y performs as the reliant variable
X performs the independent variable
β represent the coefficient formula function of regression.

Linear regression is used in the crime prediction situation to represent the most likely perpetrator age given the crime scenes.

In this figure, the victim rate between male and female is shown, and the past information analyzes show that female victims are quickly increasing between males every day (Fig. 2).

To find out the monthly crime rate we used KNN algorithm at the start we find out each month crime rate for every year. This figure shows the individual month crime rate for three years. We also find out the crime rate which time is more crime occurs in the year, finding that rate divides the years into three dimensions for type 1 from January to April, type 2 may occur to August and type 3 from September to December, and showing that type 1 and type 3 occurred the most crime in those months.

By using the K-NN and Linear regression we find out the aged crime rate for 3 years. To find that rate we divide the age in three types those are Teenager, young and old (Fig. 3).

By using the K-NN and Linear regression we find out the aged crime rate for 3 years. To find that rate we divide the age in three types those are Teenager, young and old (Table 4).

In Dhaka city crimes rate we divide the city in two zone north and south and zone then plot the data by using KNN algorithm X axis for north and Y axis for south, finally show the result in Fig. 4 that is north cities people are more victim then south city in Dhaka (Fig. 5).

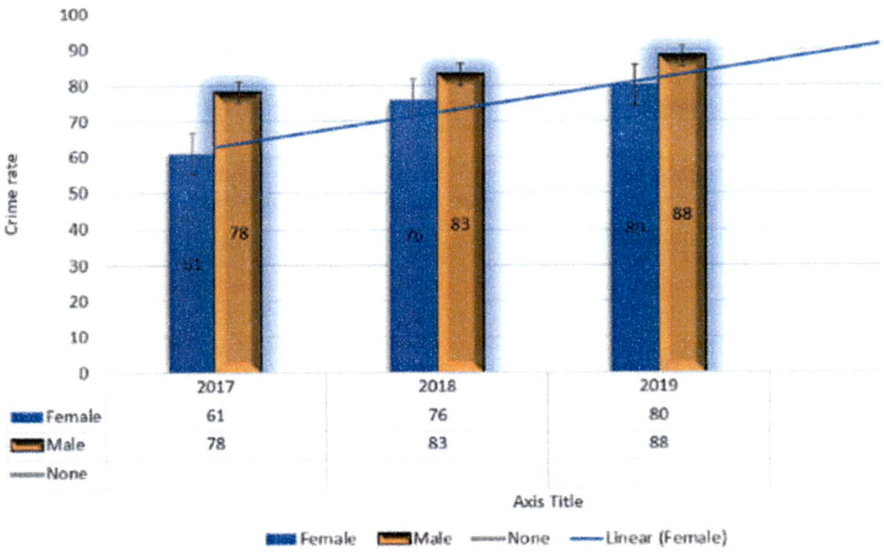

Fig. 2 Male versus female rate

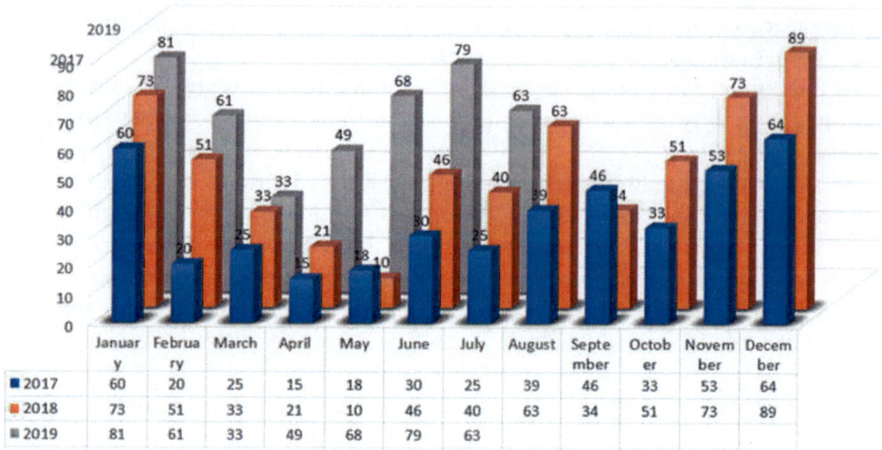

Fig. 3 Male versus female rate

Table 4 Age with range

Age	Range
Teenager	13–19
Young	20–55
Old	56–100
Old	56–100

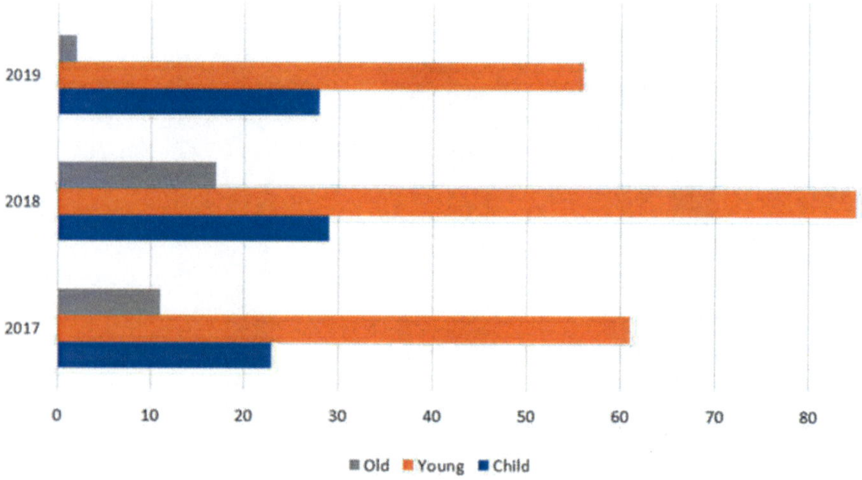

Fig. 4 Age based crime rate

Fig. 5 dhaka city crime rate

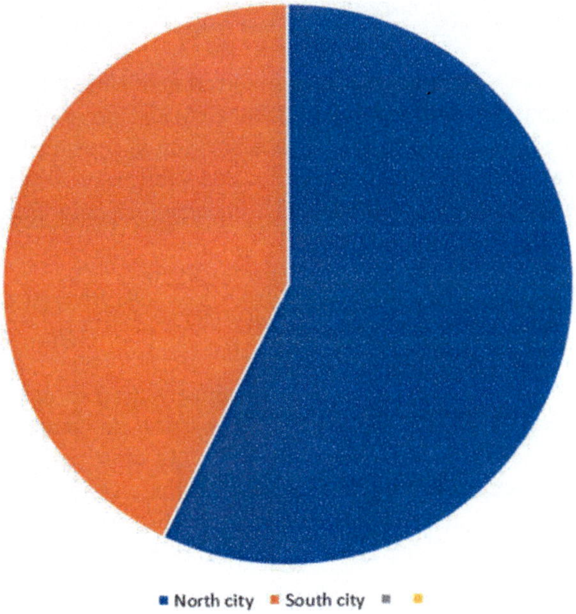

4.2 *K-Nearest Neighbors*

K-nearest neighbors is used when the target variable must be classified in more than two classes. In this dataset, there are three classes of target variable perpetrator sex: male, female, and unknown. Similarly, three categories of young, old and kid are defined in age. To classify these target variables, K-nearest neighbors Classifier is used.

$$D(a, b) = \sqrt{\sum_{i=1}^{n} (a_i - b_i)^2} \tag{2}$$

Pseudo Code:

At first KNN Classifier (Data Entry)

Assign the Number of Cluster in *K*

Choose a set of *K* instances to be cluster centers

Data points for each output

Calculate the Euclidean range

Assign next to the data point the cluster

Perpetually Calculate centroids and reassign the variables of the cluster

Repeat until you reach an appropriate cluster Result, Give back the clusters and their values.

4.3 Naïve Bayes

Naive Bayes classifiers are a collection of Bayes' theorem-based classification algorithms. It is not a single algorithm, but an algorithm family in which all of them share a common principle, each combination of features being classified is separate from each other. Bayes' theorem considers the probability of a case occurring considering the probability of another case occurring already. The theorem of Bayes is indicated as the following equation:

$$p(A|B) = \frac{P(B|A)P(A)}{P(B)} \tag{3}$$

Naive Bayes algorithms are mostly used to determine emotions, delete spam, suggestions, etc. We are fast and easy to implement, but their biggest downside is the need for autonomous predictors. In most real-life situations, the predictors are dependent, hampering the output of the classifier. We do not use that algorithm to find the final result in our work for this problem, but in many previous cases its works are very good and given the best crime rate accuracy.

5 Result and Discussion

In this part summarizes the paper and make aware about the future crime based on algorithms and crime data set we find out the crime rate in various section like age based, male vs female, area based and monthly crime rates. The data sources and methods used to guide forecasting include various type crime statistics, survey of the general people data, literature reviews and statistical models that extrapolate crime trends into the future. Algorithms models that describe the behavior of observe past values can be used to forecast future crime trends by projecting a time series analysis of crime trends into the future. Any predictive model endeavor to show a relationship between certain predictor and a dependent variable. To ensure the greater accuracy those models must identify and predict the scope and nature of a number of factors that will influence crime and victimization in the future. This research paper about future crime rate predictions are much more specific and precise. The accuracy table of different algorithm accuracy (Table 5):

By referring to these 3 algorithms, we demonstrate that K-nearest neighbor provides the crime rate forecast system the greatest precision.

Table 5 Accuracy table

Year	Algorithm	Accuracy
2017, 18, 19	Linear	73.61403
2017, 18, 19	Naïve Bayes	69.5087
2017, 18, 19	KNN	76.9298

6 Conclusion

The sparsity of crime in many areas complicates the application of the prediction rate area-specific modeling. We used the Machine Learning algorithm in that work to create and test age, sex, year, moment, month prediction of crime. In that job we use three types machine learning algorithms Linear regression, Naïve Bayes and K-nearest neighbor among which we discover distinct precision in different instances some linear operates good and provides better precision but the general situation K-nearest neighbor provides the appreciated accuracy other than that's why we use K-nearest neighbor for our Crime Prediction scheme. By using these predict systems we will discover the stronger precision in the future and also by using this precision we will identify and discover the hot zone region in crime rate. In order to finish this job, we would like to use the CNN algorithm to analyze picture information and add the Google API for viewing the hot zone.

Acknowledgements We are very grateful to my Daffodil International University for offering us the chance to be part of the independent research study that contributes to this work being developed. Many thanks to Mr. Abdus Sattar, Assistant Professor, Department of Computer Science and Engineering at Daffodil International University for innumerable debates and feedback that helped me effectively finish the job.

References

1. Lin, Y., Chen, T., Yu, L.: Using machine learning to assist crime prevention. In: 2017 sixth IIAI International Congress on Advanced Applied Science (IIAI-AAI) (2017)
2. Munasinghe, M., Perera, H., Udeshini, S., Weerasinghe, R.: Machine learning based criminal short listing using modus operandi features (2015). https://doi.org/10.1109/icter.2015.7377669
3. Chauhan, C., Sehgal, S.: A review: crime analysis exploitation data processing techniques and algorithms, pp. 21–25 (2017). https://doi.org/10.1109/ccaa.2017.8229823
4. Anon: [online] Available at https://www.researchgate.net/publication/322000460_A_review_Crime_analysis_using_data_mining_techniques_and_algorithms. Accessed 30 Aug. 2019 (2019)
5. Kerr, J.: Vancouver police go high tech to predict and prevent crime before it happens. *Vancouver Courier*, July 23, 2017. [Online] Available https://www.vancourier.com/news/vancouver-police-go-high-tech-topredict-and-prevent-crime-before-it-happens-1.21295288. Accessed 09 Aug 2018
6. Marchant, R., Haan, S., Clancey, G., Cripps, S.: Applying machine learning to criminology: semi parametric spatial demographic Bayesian regression. Security Inform. **7**(1) (2018)

A Case Study and Fraud Rate Prediction in e-Banking Systems Using Machine Learning and Data Mining

Musfika Nuha, Sakib Mahmud, and Abdus Sattar

Abstract Recently banking sector of Bangladesh is undergoing in a revolutionizing change. Over the last few years, Bangladesh's banking industry has achieved remarkable momentum. Especially radical change has come in e-banking and mobile banking sectors. Because of convenience, easy to use, time saving and less complexity, both educated and uneducated people are using those facilities. At the same time, fraudulent activity is also rising rapidly. It is noticed that fraudsters use scary tactics and emotional manipulation to obtain sensitive or confidential customer information instead of coding-based hacking process. As a result, cyber security is the main challenge for the banking sector in Bangladesh. The purpose of the research is to determine the key factors behind increasing fraudulent activities. Concurrently, this study focuses on the relationship between lack of awareness and likeliness to be affected by fraud. In order to acquire the specified purpose of this study, several investigations were conducted on primary and secondary data. Results show that there is a strong correlation between lack of awareness and likeliness to be affected by fraud. 76% people have no idea about e-banking and mobile banking fraud. Furthermore, our findings show that 86.3% of victims of e-banking or mobile banking fraud had no prior knowledge of this type of fraud. Simultaneously, 13.7% of victims in those sectors had prior knowledge of fraud. It is obvious that, behind this type of fraud, lack of knowledge and awareness can be a major fact.

Keywords e-Banking · Mobile baking · Awareness · Challenges · Phishing · Credit card fraud · Banking sector · Bangladesh

M. Nuha · S. Mahmud · A. Sattar (✉)
Department of Computer Science and Engineering, Daffodil International University, 102 Sukrabadh, Mirpur Road, Dhaka 1207, Bangladesh
e-mail: abdus.cse@diu.edu.bd

M. Nuha
e-mail: musfika15-7930@diu.edu.bd

S. Mahmud
e-mail: sakib15-7962@diu.edu.bd

S. Borah et al. (eds.), *Soft Computing Techniques and Applications*, Advances in Intelligent Systems and Computing 1248,
https://doi.org/10.1007/978-981-15-7394-1_6

1 Introduction

According to [3], in 2018 the Internet users are increased day by day the previous users was 3.9 billion around the world and now it's up to 3.65 billion. The e-banking industry has expanded rapidly over the course of time with the growth and widespread Internet. Most businesses and industries have turned their business into online services to deliver e-commerce, easy access and communication to enable better efficiency and accessibility for their customers.

Recently in Bangladesh, the banking sector is undergoing a revolutionizing change. In particular, e-banking and mobile banking are expanding so rapidly. Because of being convenient, easy to use, less time consuming and available, the acceptance of e-banking is increasing in full swing. As a result, people of all classes (both educated and uneducated) are using mobile banking and e-banking. According to [1], 16 banks are currently providing Mobile Financial Services (MFS) in Bangladesh and an average daily transaction (1 crore = 10 million BDT) was 1116.98 crores in January 2019 which clearly indicates that in Bangladesh perspective, this sector is undergoing a revolution. In Bangladesh, mobile banking is more popular to the rural and poor people whom incomes are low and students who generally don't deal with a lot of money. Currently, 53 Banks have transaction card policy in Bangladesh. Simultaneously [2] shows that the total number of Internet subscribers in Bangladesh reached 92.061 million at the end of February, 2019 which is exponential increment than past years. This clearly indicates that activity of e-banking is increasing and concurrently, fraud is also growing rapidly in this sector. Fraud has become a major challenge both for mobile banking and e-banking which is a great concern at present moment. For personal gain, fraud is an illegal and unfair activity by which criminals try to gain something valuable, financial or anything else. Such fraudulent activities have a negative impact on both the individual and organization's reputation. But it is surprisingly true that most fraudsters do not use any type of coding-based hacking process to perform this type of illegal activity. Most of the time, fraudsters are fooling victims and illegally handcuffing their money.

Nowadays in Bangladesh e-banking is considered as a popular and reliable way of transaction money. At the same time, in e-banking sector, clients frequently face various kinds of fraudulent activities in which credit card fraud, identity theft and phishing play the key roles. Credit card is a card that is provided by banks to eligible customers (cardholders) for daily transactions. A cardholder can pay for goods and services using the card at the particular time without having money in their account and can be paid back to banks later. However, credit card is also not safe because fraudster can easily able to clone duplicate cards just using some information of the real card. The research [4] claims that, in Bangladesh, banking services are invulnerable to fraud because of some credit card skimming fraud.

To detect fraudulent activities in e-banking sector, data mining [5, 6] and machine learning [7] techniques are commonly used. However, because of dynamic fraudulent activities exiting systems feel trouble to detect fraudulent transactions. But lack of user's awareness in Bangladesh can be a major factor behind e-banking fraud. The purpose of this paper is to find out the most important factors in both sectors that are responsible for fraudulent activities. Our hypothesis is the lack of awareness and poor knowledge about e-banking and mobile banking fraud is the main factor. Because making people fool and get sensitive information are easier task than coding-based hacking. The number of victims has continued to increase in Bangladesh and most of them have not understood the fraudulent activity until it actually happened to them because they have not any prior idea about these kinds of fraud. At the same time, we will try to provide some reliable and effective solutions against fraud associated with e-banking and mobile banking. In order to conquer the main goal of this paper some specific objectives will be pursued:

Finding out awareness level of those people who are currently using e-banking or mobile banking system.

i. Possibility of becoming a victim for being unconscious of fraud in mobile and e-banking.
ii. Reasons for fraudulent activity occurring.
iii. Effective preventive solutions against fraud in both sectors.

2 Literature Review

Bangladesh is regarded as a developing country. Economy of Bangladesh is now much stronger than ever before. A lot of new industries have emerged with the expansion of information technology, which plays a positive role on the way of growing economy. At the same time, economy of Bangladesh largely depends on non-resident Bangladeshis. According to [10], remittances in Bangladesh had raised to 1202.85 USD Million in December from 1180.44 USD Million in November of 2018. It signifies that banking industry of Bangladesh has achieved remarkable momentum. According to Bangladesh bank [11], at present, there are 59 scheduled banks in Bangladesh in which 41 banks are private. Online banking and mobile banking in Bangladesh are rising rapidly. According to [12], presently a number of mobile phone subscribers have reached 158.438 Million at the end of February, 2019. On the other hand, the total number of Internet subscribers in Bangladesh reached 92.061 million at the end of February, 2019 [2]. It is reported that the number of Debit Card in Bangladesh is 14044338 and number of Credit Card is 1075250 [20]. Online payment services are getting popular day by day in Bangladesh. Internet has revolutionized people's way of shopping. Because of the numerous benefits and advantages, more and more people these days prefer to buy online stuff over the conventional way to go into stores. As a result, e-banking is significantly growing.

At the same time, over 70% of Bangladeshi people lives in urban areas where it is difficult to access formal financial services and fewer than 15% of Bangladeshis are connected to the formal banking system, while over 68% have mobile phones [13]. To solve this problem, mobile banking has opened a new door. As a result, urban and rural people in their daily lives frequently use mobile banking. Nowadays, insidious and fraudulent activities are also increasing rapidly. Credit card fraud, identity theft, various types of phishing such as vishing and SMiShing are increasing at an alarming rate in Bangladesh. According to [14], recently 49 clients of five non-government commercial banks have lost around Tk 2 million when a fraud gang withdrew the money from booths by using duplicate debit and credit cards. However, the studies [15–17] also claimed that many credit card users were victimized by credit card fraud. Another study [18] claimed that approximately 30% of banks are exposed to 'very high' risks of online fraud and security threats. Automated Teller Machine (ATM) and plastic card transactions represent 43% of the highest fraud, followed by 25% of mobile banking, it said. The study pointed out that there is insufficient investment in the professional development of information technology (IT) in the banking sector. This poses a significant threat to digital online fraud and can also trigger serious security threats to the banking sector as a whole.

To detect fraudulent transactions, several machine learning and data mining techniques like Hidden Markov model, Behavior Based Technique, Genetic Algorithm, Deep learning, Logistic Regression, Naïve Bayesian, Support Vector Machine (SVM), Neural Network, Artificial Immune System, K Nearest Neighbor, Decision Tree, Fuzzy logic-based Systems are used by banking organizations. However, still now because of dynamic behavior of the fraudsters, majority of times existing systems become unable to detect. Instead of coding-based hacking, fraudsters are more comfortable in doing fraudulent activities to make fool people and get their sensitive information. Since most people in this sector in Bangladesh do not have sufficient awareness [19], it is easy to be trapped.

3 Research Methodology

The methodological part of this study can be divided into several phases. Problem identification is considered as the first phase where the current crisis of Bangladesh's e-banking and mobile banking sectors is identified. Second phase is data collection. The dataset used in this study was obtained from both primary and secondary resources. In the third phase, collected data was preprocessed to resolve incomplete and inconsistent data. To acquire the purpose of this study, several statistical techniques were followed in the fourth phase (data analysis). According to analysis, in the fifth phase meaningful information was extracted. Based on this information several research findings were discussed in the discussion part (Fig. 1).

3.1 Data Collection

The goal of this research was to recognize the key facts involved behind fraud occurrence in Bangladesh's e-Banking system. By using field survey and online survey, we collected primary data from individuals for the specified purpose. Secondary data from journal articles, conference papers, review articles/documents/chapters, book

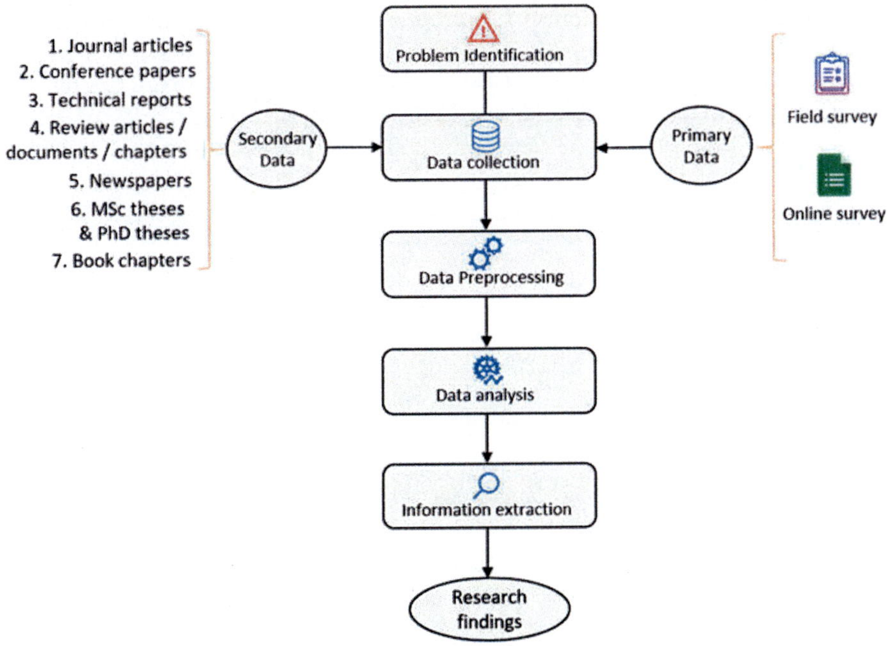

Fig. 1 Represents a methodological framework for this study

chapters, technical reports, M.Sc. theses, and Ph.D. theses were collected. Precise and relevant qualitative and quantitative data were collected in a systematic manner to ensure the best results. We distributed questionnaires to clients from various government and non-governmental banks, students from college and university, general people who are currently enjoying e-banking or mobile banking facilities. We asked them various questions which are related to the purpose of our paper. At the same time, we had to consider carefully the respondent's ethical factors that ensured that their privacy was kept strictly and confidentially secret. We also did not include any question in the questionnaire that could make respondents vulnerable to risk for both online and field surveys; a total of 6000 questionnaires have been distributed and 3720 responses were collected. Both male and female have been answered. Among them, 72.07%.

3.2 *Data Analysis Tools*

Through field survey and online survey, we distributed 6000 questionnaires and collected a total of 3720. For online survey, Google Form was used. In both online and field surveys, the same questionnaires were included. To analyze those collected data, we used python (version: 3.7.2) programming language along with several modules of python and Microsoft excel.

We collected 3720 responses, in which 2681 were male and 1039 were female. The percentage of male respondents was 72.07% and the percentage 27.93 for female respondents. Table 1 and Fig. 2 represent the gender-specific frequency distribution of overall data collected from survey.

Table 1 Dataset table

Number	Details		
	Name	Type of Columns	Descriptions
1	Person_Id	Value Type	Person ID in Dataset
2	Name	String	Victim Person name
3	Year	Numeric	Crime Occurs Year
4	Ages	Numeric	The ages of the victim
5	Gender	String	Victims Neuter
6	Time	String	Time when the crime has occurred
7	Victim Area	String	Area where the crime has occurred.
8	Region	String	Region of the victim.
9	Home Town	String	Home town of the victim
10	Month	String	The year in which the crime has occurred

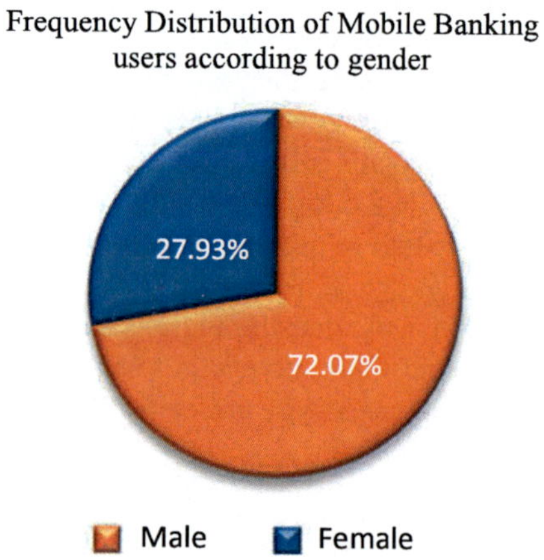

Fig. 2 Gender—based respondents

Table 2 Male Female identification by binary digit

Gender	Male	Female
	0	1

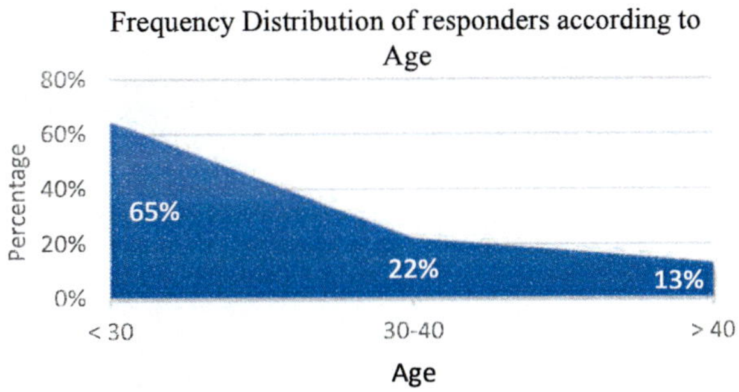

Fig. 3 Frequency distribution of responders according to age

3.3 Data Analysis and Result

Explains the system's workflow. The workflow begins by extracting data from data collection, which is a dataset repository on different roles. The primary data will be preprocessed and transformed into a criminal data.

The respondent's ages are divided into three ranges <30, 30–40, and >40. Table 2 and Fig. 3 represent that the majority of respondents (65%) were less than 30 years, 22% were within 30—40 years and 13% were within or greater >40.

3.4 Mobile Banking or e-Banking Services Used by Respondents

We asked our respondents if they are currently using e-Banking or mobile banking facilities. Out of the 3720 replies, 74% reported that they are using e-Banking or mobile banking facilities, while 26% reported are not using those facilities or mobile banking facilities.

Figure 4 show the frequency distribution of respondents' use of Mobile Banking/e-Banking facilities.

3.5 Victimized by Mobile Banking/e-Banking Fraud

We collected 3720 responses in which 2740 respondents used mobile banking or e- banking facilities. Among them, 1175 respondents were victimized by e-banking/mobile banking fraud and 1565 were non-victims (Fig. 5).

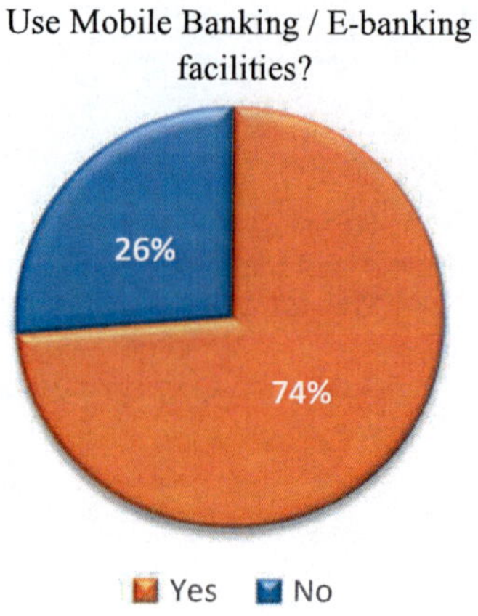

Fig. 4 Frequency distribution of using mobile banking/e-banking facilities

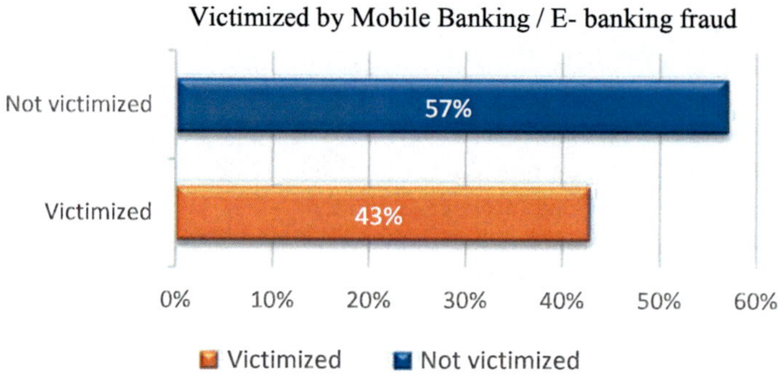

Fig. 5 Victimized by mobile banking/e-banking fraud

3.6 Awareness of e-Banking and Mobile Banking Fraud

We asked the respondents whether they had any idea of e-banking/mobile banking fraud and the responses are demonstrated in Fig. 6. It is evident from the figure that 76% of them said they had no idea of any kind of fraud in this sectors. 14% had idea of phishing by SMS and phone, 12% had idea of phishing, 6% had idea of viruses and Trojans, 8% had idea of spyware and adware, 5% had idea of card skimming, 9% had idea of identity theft and 3% of them had idea of other mobile scams.

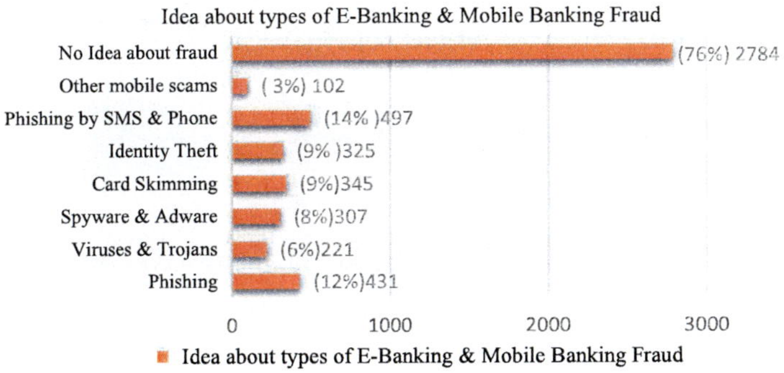

Fig. 6 Idea about types of e-banking and mobile banking fraud

3.7 *Possibility of Becoming a Victim for Being Unconscious of Fraud in Mobile Banking and e-Banking*

We collected 2740 respondents' view who were using e-banking/mobile banking facilities. Among them, 1175 respondent was victimized by e-banking/mobile banking fraud and at the same time, 1565 respondents did not face any kind of fraud in those sectors. We wanted to find out if the nature of e-banking/mobile banking fraud was familiar to them. We asked if they knew about various types of fraud in these sectors. 86.3% of all victims had no prior knowledge of e-banking/mobile banking fraud. At the same instant, 13.7% of the total victims had previous knowledge of fraud in those sectors. It is explicitly visible from Fig. 7 that 36% of total no-victimized people had prior knowledge of e-banking/mobile banking fraud and 64% of non-victimized people had no prior knowledge of e-banking/mobile banking fraud.

In terms of shopping, online purchases, B2B, B2C services and payment services, e-Banking and mobile banking are now one of the most popular medium of making a money transaction in Bangladesh. The volume of e-Banking and mobile banking expanding day by day due to simplicity, convenience, time saving and easy to use. At the same time, insidious activities are also rapidly increasing. Mobile banking's popularity is growing rapidly in Bangladesh. Mobile banking providers are primarily targeting low income earners and rural people who are unable to afford a bank account. But Bangladesh's people are less educated and had no understanding of the fraudulent activities until it actually happened to them. Our aim is to find out the main factors behind frauds in those sectors. The hypothesis was that lack of awareness is the main reason behind fraud victimization in those sectors rather than code-based hacking. We gathered 3720 respondents, 72.07% of whom were male and 27.93% were female. Responses from the people of various age groups are collected by the field survey. Most of them were less than 30 years old (65%), 22% were between 30 and 40 years old, and 13% in over 40 years old. 74% (2740) of total respondents (3720) used e-banking/mobile banking facilities and 26% (980) did not use such

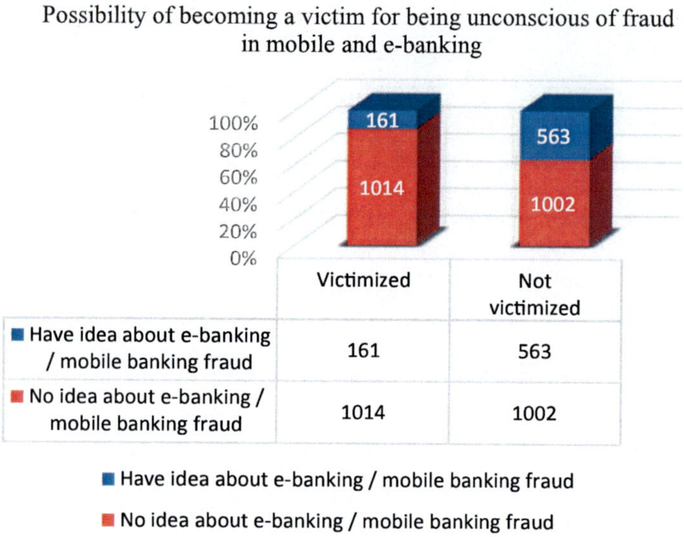

	Victimized	Not victimized
Have idea about e-banking / mobile banking fraud	161	563
No idea about e-banking / mobile banking fraud	1014	1002

Fig. 7 Possibility of becoming a victim for being unconscious of fraud in mobile and

facilities. We asked the reason of not using e-banking and mobile banking facilities to those people who don't use those facilities. Figure 8 shows that, 46.9% people said those kinds of facilities look like inconvenient to them. Most of them complained that service charges on mobile banking is too high. On the other hand, 26.2% people said that they are not interested in mobile banking and e-banking due to security risk. 23.7% people said they don't prefer those kinds of facilities and 3.1% didn't make any comment.

We noticed in Fig. 6 that majority (76%) of the people had no idea about the types of e-Banking/Mobile Banking Fraud which is clear indication of lack of awareness.

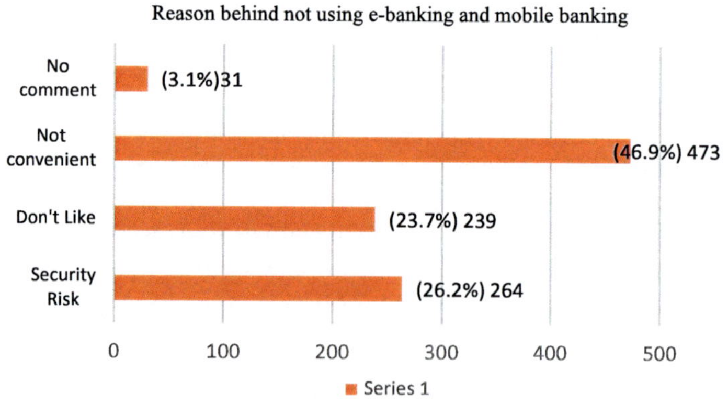

Fig. 8 Reasons behind not using e-banking and mobile banking

We found that 1175 respondents were victimized by e-banking/mobile banking fraud and at the same time, 1565 respondents did not face any kind of fraud in those sectors. However, it is demonstrated in Fig. 7 that 86.3% of all victims had no prior knowledge of e-banking/mobile banking fraud. Also, 13.7% of the total victims had prior knowledge of fraud in those sectors. It indicates that the people having an idea of e-banking and mobile banking fraud, are very unlikely to be victimized than those of unaware people. Fraudsters in Bangladesh are now turning to alternative methods of scamming, such as vishing and SMiShing. They try to trick people into giving up their sensitive information by using association of scare tactics and emotional manipulation. These criminals even create fake Caller ID profiles making phone numbers appear legitimate. Vishing's goal is simple: steal money from victims account. Most users have no idea that legitimate business does not make unsolicited requests for personal, sensitive, or financial information. As a result, they easily provide their personal information to the fraudsters.

Several studies reported that Bangladesh is also experiencing with rising credit card fraud and identity theft [8, 9]. Still now in Bangladesh, EMV card technology is not fully used. EMV is the most recent technology for secure payment which was invented jointly by Europay, MasterCard and Visa and was later adopted by other payments card brands. Transactions from EMV chip card improves security against fraud compared to magnetic stripe card operations that rely on the account owner signature and visual inspection of the card to check for features such as hologram. But still now people use magnetic stripe cards in Bangladesh, which is very easy to clone. A magnetic stripe card is a card containing information embedded in a plastic film strip consisting of iron particles. The front of the card contains information about identification, such as the name of the cardholder and the name of the issuing company.

However, in Bangladesh, vishing and SMiShing have become the most popular way of committing fraud. But question is why vishing and SMiShing? The possible is much population in Bangladesh using use e-banking or mobile banking are not sufficiently aware of fraud. At the same time, they don't know the key facts about e-banking or mobile banking. As a result, they become the fraudsters' easy target. Because of this type of fraudulent activity, general users are unable to trust the existing systems and blame Bangladesh's overall banking system. We asked people whether e-banking system or mobile banking system does provide good customer services; 61% of them agreed, 14% strongly agreed, 11% disagreed and 14% of them did not comment (Fig. 9).

4 Discussion and Conclusion

In Bangladesh, e-banking and mobile banking system are growing rapidly. The banking sector is undergoing a revolutionizing change with Bangladesh's economic growth. There are a significant number of national, private commercial banks and specialized banks operating in Bangladesh for a long time. Fraudulent activities are

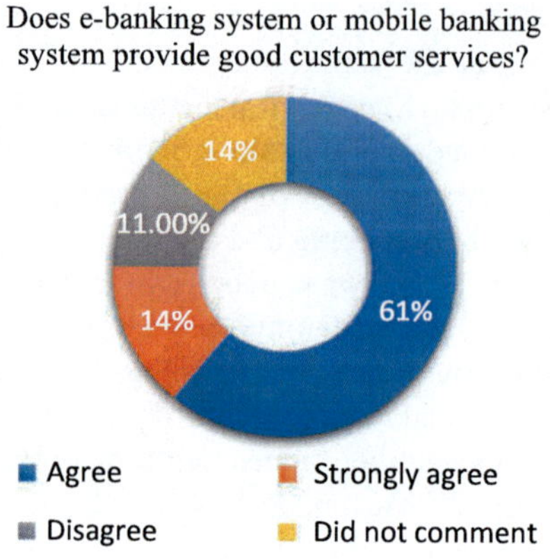

Fig. 9 Does e-banking system or mobile banking system provide good customer services?

also rapidly increasing. This study revealed that the popularity of e-banking and mobile banking in Bangladesh have become more popular than ever before. Around 74% people are using mobile banking or e-banking facilities. But still now, most of the users are not fully aware of fraud which could likely to be occurred while enjoying facilities. Customer-driven fraud and agent-driven fraud are mostly occurring in mobile banking. Mobile banking is predominantly popular among young people, aged and rural people. As the people in rural areas are less, they do not understand the fraudulent activity until it actually happened to them. 86.3% victims of fraud in this sector had no prior knowledge of e-banking mobile banking fraud. At the same time, 13.7% of the total victims had prior knowledge of fraud which could likely to happen in those sectors. Our investigation clarifies that the lack of awareness is regarded as the main reason for fraud associated with mobile banking and e-banking. Therefore, banking sectors should focus on their consumers ‘ awareness. It can be done through either an emotive or a logical message to raise awareness of this need or problem. However Collective incentives, can more effectively reduce fraud in mobile banking and e-banking rather than individual effort.

References

1. Mobile financial services data. Bb.org.bd, 2019 [Online]. Available https://www.bb.org.bd/fnansys/paymentsys/mfsdata.php. Accessed: 12 Mar 2019
2. Internet Subscribers in Bangladesh January, 2019. BTRC. Btrc.gov.bd, 2019 [Online]. Available http://www.btrc.gov.bd/content/internet-subscribers-bangladesh-January-2019. Accessed 12 Mar 2019
3. Number of internet users worldwide 2005–2018 Statista, Statista, 2019 [Online]. Available https://www.statista.com/statistics/273018/number-of-internet-users-worldwide/. Accessed 12 Mar 2019
4. Banking services vulnerable to fraud, The Daily Star, 2017 [Online]. Available https://www.thedailystar.net/editorial/banking-services-vulnerable-fraud-1448149. Accessed 12 Mar 2019

5. Jeeva, S., Rajsingh, E.: Intelligent phishing url detection using association rule mining. Human-Centr. Comput. Inf. Sci. **6**(1), 9 (2016)
6. Pouramirarsalani, A., Khalilian, M., Nikravanshalmani, A.: Fraud detection in e-banking by using the hybrid feature selection and evolutionary algorithms. Int. J. Comput. Sci. Netw. Secur. **17**(8), 271 (2017)
7. Ritu, S.N.: Credit card fraud detection using GSA. Int. J. Sci. Res. Comput. Sci. Eng. Inf. Technol. **2**, 3 (2017)
8. City Bank to start repaying credit card fraud victims Tuesday, Bdnews24.com, 2018 [Online]. Available https://bdnews24.com/business/2018/03/20/city-bank-to-start-repaying-credit-card-fraud-victims-Tuesday. Accessed: 20 Mar 2019
9. Rahman. S. Credit, debit cards: swindling on the rise. The Daily Star, 2017 [Online]. Available https://www.thedailystar.net/frontpage/credit-debit-cards-swindling-the-rise-1447729. Accessed 20 Mar 2019
10. Bangladesh Remittances 2019 Data Chart Calendar Forecast News. Tradingeconomics.com, 2019 [Online]. Available https://tradingeconomics.com/bangladesh/remittances. Accessed 25 Mar 2019
11. Bb.org.bd, 2019 [Online]. Available https://www.bb.org.bd/fnansys/bankfi.php. Accessed 25 Mar 2019
12. Mobile Phone Subscribers in Bangladesh February, 2019. BTRC, Btrc.gov.bd, 2019 [Online]. Available http://www.btrc.gov.bd/content/mobile-phone-subscribers-bangladesh-february-2019. Accessed 25 Mar 2019
13. Company Profile bKash, Bkash.com [Online]. Available https://www.bkash.com/about/company-profile. Accessed 25 Mar 2019
14. Sakib, S.: Tk 2 m stolen thru card forgery, Prothom Alo, 2018 [Online]. Available https://en.prothomalo.com/bangladesh/news/172860/Tk-2m-stolen-thru-card-forgery. Accessed 25 Mar 2019
15. Rahman, S.: Credit, debit cards: swindling on the rise. The Daily Star, 2017 [Online]. Available https://www.thedailystar.net/frontpage/credit-debit-cards-swindling-the-rise-1447729. Accessed 25 Mar 2019
16. City Bank to start repaying credit card fraud victims Tuesday. Bdnews24.com, 2018 [Online]. Available https://bdnews24.com/business/2018/03/20/city-bank-to-start-repaying-credit-card-fraud-victims-tuesday. Accessed 25 Mar 2019
17. Rabbi, A.: Bank card cloning mastermind arrested with 1,400 fake cards, Dhaka Tribune, 2018 [Online]. Available https://www.dhakatribune.com/bangladesh/crime/2018/04/25/bank-card-cloning-mastermind-held-1500-fake-cards. Accessed 25 Mar 2019
18. Banking services vulnerable to fraud, The Daily Star, 2017 [Online]. Available https://www.thedailystar.net/editorial/banking-services-vulnerable-fraud-1448149. Accessed 25 Mar 2019
19. Rahman, M., Saha, N.K., Sarker, M.N., Sultana, A., Prodhan, A.Z.: Problems and prospects of electronic banking in Bangladesh: a case study on Dutch-Bangla Bank Limited. Am. J. Oper. Manag. Inf. Syst. **2**(1), 42–53 (2017). https://doi.org/10.11648/j.ajomis.20170201.17
20. Payment Systems, Bb.org.bd, 2019 [Online]. Available https://www.bb.org.bd/fnansys/paymentsys/natpayswitch.php. Accessed 26 Mar 2019

Novel User Preference Recommender System Based on Twitter Profile Analysis

Narasimha Rao Vajjhala, Sandip Rakshit, Michael Oshogbunu, and Shafiu Salisu

Abstract Recommender systems can help provide preference-based personalized services to consumers and help them make informed decisions. However, a key shortcoming of the recommender systems is the lack of interactive methods to dynamically change the weights of recommendation algorithms. Our proposed system uses the Twitter profile and tweets to identify the interests of a user and then recommends the relevant products and services to that user. Our recommendation system is built to predict and personalize products and services based on the result of mining and analyzing the user's Twitter timeline. The proposed recommender system is built upon an artificial intelligence platform called IBM Watson. The experimental result from the platform displayed the category of goods and services the user is most likely to consume. Our recommender system also showed a strong correlation between the category of products and services a user consumes and his/her tweets.

Keywords Personalization · Collaborative · Recommender · E-commerce · Twitter · Mining · Hybrid · Retrieval

1 Introduction

Recommender systems can be used to provide personalized services to individuals based on their preferences helping customers make smart, informed decisions [1, 2].

N. R. Vajjhala (✉) · S. Rakshit · M. Oshogbunu · S. Salisu
American University of Nigeria, Yola 2250, Nigeria
e-mail: narasimha.vajjhala@aun.edu.ng

S. Rakshit
e-mail: sandip.rakshit@aun.edu.ng

M. Oshogbunu
e-mail: michael.oshogbunu@aun.edu.ng

S. Salisu
e-mail: shafiu.salisu@aun.edu.ng

S. Borah et al. (eds.), *Soft Computing Techniques and Applications*, Advances in Intelligent Systems and Computing 1248,
https://doi.org/10.1007/978-981-15-7394-1_7

Online consumers have access to huge volumes of data presenting unique opportunities for price transparency as well as a wider choice of products for consumers. However, consumers are also exposed to information overload as consumers often end up browsing a lot of irrelevant information [3–6]. The use of recommender systems can address the issue of information overload to a certain extent [5]. Twitter is considered as a major platform for peer interaction allowing classification of latent user attributes mainly demographic information. This information provides key informal content that is essential for recommendation and personalization [7].

Researchers have developed computational models predicting user traits based on their online footprints in social networking sites, such as Twitter [8]. These models can be useful for personalized search and recommendations. Twitter, a micro-blogging platform enables users to share information in the form of tweets and create a social relationship has been recognized as a potentially powerful analytical and marketing platform. In the context of Twitter, Wang, Shu [9] categorize the dimensions, into two categories, namely, interior and exterior. The exterior dimensions include tweet volume and the number of users, while the interior dimension could include the extent to which the tweet goes viral. One of the shortcomings of the recommender systems that are currently in use is the lack of interactive methods to dynamically change the weights of recommendation algorithms [3]. Our proposed system uses the Twitter profile and tweets to identify the interests of a user and then recommends the relevant products and services to that user.

Analyzing the user and predicting their preferred product relies on the use of statistical and knowledge technique [10]. Classification and prediction techniques are now used to personalize email marketing campaigns [11]. Hence, campaigns are now modeled with customer behaviors and preferences in mind in a bid to increase their effectiveness [11]. Our recommendation system is built to predict and personalize products and services based on the result of mining and analyzing the user's Twitter timeline. The proposed recommender system is built upon an Artificial Intelligence platform called IBM Watson. The experimental result from the platform displayed the category of goods and services the user is most likely to consume. Our recommender system also showed a strong correlation between the category of products and services a user consumes and his/her tweets.

2 Review of Literature

2.1 Recommender Systems

Recommender systems are used in several areas, including music, news, online auctions, business-to-commerce sites, and social networks [1, 2]. Recommender systems can be categorized as personalized and non-personalized systems [5]. The focus of non-personalized recommendation systems is on facets, such as top-selling books or most read news articles. In such systems, the emphasis is not on personalized recommendations. However, in the case of personalized recommendation systems, attributes including the behavior of the user and likes are taken into consideration [5]. Recommender systems facilitate the personalization of the system leading to higher levels of customer satisfaction with the system. Especially, in the case of online shopping, recommender systems can provide a sophisticated level of personalization through product suggestions which could influence and help consumers in an online purchasing decision.

Recommender systems are now commercial success and are becoming increasingly popular in various industries [12]. Hence, the ability to recommend products and services based on a user's preference has evinced the interest of companies as well as researchers. Recommender systems have contributed to a higher number of transactions since the items displayed are in alignment with the user's interest. The higher transactions are bound to occur when merchants display items that are frequently bought together or by bundling several products/items for sale as one product at a lower price [13]. Although, other researchers have gone further to argue that it is not just enough to display a product but also take into consideration the price of the product during recommendation [14]. This is because if the price is too high, the customer may not buy. Hence, recommender systems need to also factor in additional attributes, such as cost apart from recommending items that a user might be interested in.

E-commerce companies, such as Wal-Mart and Amazon display products and services based on the consumer's prior shopping history, demographic attributes, or social media profile [14]. They do so by leveraging social media platforms, such as Twitter or Facebook to understand their users. This is because, on such platforms, customers share their opinions about products they recently purchased or they are familiar with [15]. Data about the user is obtained by first allowing the user to connect with their preferred social networking site [16]. Once access is granted, third-party services obtain certain information such as demographics, interests, posts, and tweets about him/her via Open Standards for Authorization, such as the OAuth2. Recommendation systems can be considered as decision-support systems that are based on the user's preferences [3, 4]. The recommendation algorithms are categorized into three key categories, namely, content-based filtering, collaborative filtering, and hybrid filtering approach [2, 17]. Collaborative filtering and content-based filtering techniques are extensively used by recommender systems [1, 2, 18].

2.2 *Content-Based Filtering Algorithms*

In content-based filtering, the selection of the items is based on the correlation between the user and the preferences of the user [1]. Content-based recommender systems analyze and map the past preferences of the user to predict the future ratings of the user [19]. The predictions are based on the user's information in the case of content-based filtering techniques. The main difference between content-based filtering techniques and collaborative filtering techniques is that the contribution from other users is ignored in the case of the former [20]. In the case of collaborative algorithms the focus is on finding correlations between users while in the case of content-based systems the focus is on finding correlation between the content of the items [19]. One of the weaknesses of content-based algorithms is that the users may only receive recommendations similar to their past experiences. Content-based systems are also weak in analyzing multimedia, images, and music [19]. Content-based systems are effective in items that are in textual format because in such documents, each of the item is described through a set of keywords [18]. Content-based systems also suffer from the cold start problem similar to collaborative filtering techniques where a user is new and does not have adequate past experiences to reference [6]. One of the examples of content-based systems is the Vector Space Model (VSM) [18].

2.3 Collaborative Filtering Algorithms

Collaborative filtering techniques provide recommendations based on the preferences of similar users in the recommender system [21]. Collaborative filtering techniques were initially used for the document recommendation system within a newsgroup but were later extensively used in e-commerce recommender systems [9]. Collaborative filtering techniques rely on similarity calculation methods to identify users having similar tastes [17]. In collaborative filtering, the selection of the items is based on the correlation between active users and other users of the system [1]. The neighborhood formation in collaborative filtering refers to users who have rated items similar to what the active user has and recommendations are made based on the active user's rating in conjunction with the neighbor's rating [21]. Collaborative filtering results are considered to be more accurate as compared to content-based filtering techniques because they rely on actual user ranking as compared to pure machine made predictions [1, 2]. Some of the examples of recommender systems using content-based filtering include StumbleUpon and Last.fm [1].

There are two main collaborative filtering models, namely, the neighborhood model (memory-based) and the latent factor model (model-based) [2, 9]. Memory-based approaches provide the recommendation by examining the similarities among the items, for instance, products. An example of this approach would be the case when the retailers recommend a product to a customer because similar customers had also purchased the product [22]. Memory-based algorithms make use of the user data ratings of the items while model-based algorithms build a model between the items for generating recommendations using models, such as Bayesian networks and fuzzy systems [23]. Neighborhood models can be subcategorized to user-based and item-based models [9]. The key difference between user-based and item-based collaborative filtering algorithms is that while the former technique uses user ratings most similar to target user for the unrated item predictions, the latter uses the ratings that are most similar to the unrated user's rating predictions [17].

Some of the shortcomings of the collaborative filtering techniques include sparse datasets, lower accuracy, and efficiency problems with large volumes of data [4, 17, 24, 25]. The collaborative filtering technique also does not necessarily reflect the customers' purchasing behavior [17, 22]. For instance, a customer might be interested in buying gifts because of an occasion or festival, but on another day he might shop for another item [22]. Data sparsity errors in collaborative filtering techniques result in the rating of only a few items in the recommender systems [17, 25]. The drawback of user-based collaborative filtering is mainly scalability [3, 4]. The computation time of these algorithms can increase significantly with an increase in the number of products and customers. Many researchers have proposed using collaborative filtering techniques in a distributed computing environment, such as Hadoop or Spark to address the data sparsity issues [4, 5].

2.4 Hybrid Filtering Algorithms

Both collaborative and content-based algorithms have a set of advantages and disadvantages. Hybrid techniques adopt a combination of algorithms to leverage the advantages of various categories of algorithms [6, 18, 20]. The intention of hybrid systems is to leverage the strengths of the recommender algorithms while avoiding their weaknesses [19]. Some of the recommender systems, such as Google News, Pandora, and Facebook use hybrid techniques which are a combination of both collaborative filtering as well as content-based filtering techniques [1]. Several hybrid recommender systems have been built and tested by researchers in the past. Lu et al. [15] developed a hybrid system, named Customer-Social Identification (CSI) to identify customers by using their basic information. Park et al. [10] developed another hybrid system named Dominance-Influence-Steadiness-Conscientiousness (DISC) to understand the behavioral pattern of buyers. Another example of a hybrid system, named Bundle Purchase with Motives (BPM) developed by Liu et al. [13] to generate product bundles based on a user's preferences. The mix of two or more filtering techniques can help in increasing the performance and accuracy of recommender systems [20]. There are several hybrid techniques, including feature-hybrid and meta-level hybrid techniques [20].

3 Experimental Procedures and Results

Our proposed recommender application named WYDL (What Do You Like) mines a user's tweet, pass it on to IBM Watson to be analyzed and then outputs the category of product and services that the customer would most likely consume. The communication flow is illustrated in Fig. 1.

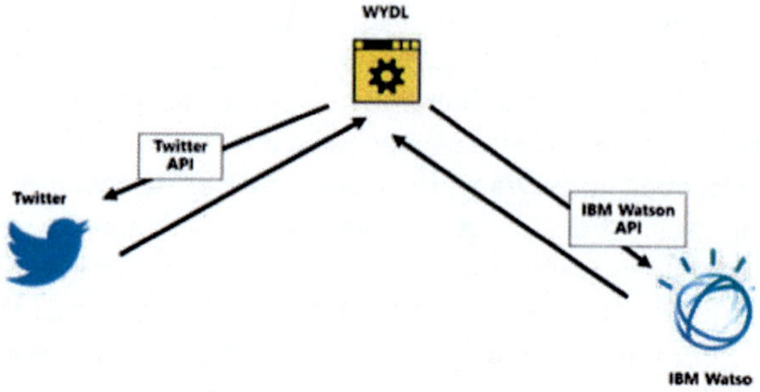

Fig. 1 Communication flow between various services

Our proposed recommender system leverages the computing power of IBM Watson's Natural Language Understanding (NLU) service, and outputs multiple preferences of a user. Our proposed system provides a quantitative measure by scoring each preference as shown in Fig. 1. The proposed recommendation mechanism is a practical method that works well in real life situations. To begin, we requested for API access to Twitter's service. After that was granted, we built our very own WYDL application which would make an HTTP Get request to Twitter's API endpoints and return all tweets by that user. The returned tweets as our data source were later analyzed. The number of tweets mined was limited to the most recent 100 tweets. This resulted in 4756 characters in total. To clean the data, emojis, hashtags and URLs were removed from each tweet. The compiled tweets were then sent to IBM Watson via an HTTP Get request. This request also contains all the necessary authentication tokens. The IBM Watson platform uses Natural language processing for advanced text analysis. The resulting analyzed data as shown in Fig. 2 shows the category of products/services and their respective scores on a scale of 0–1.

The compiled tweets were sent over to IBM Watson via an HTTP Get request. This request also contains all the necessary authentication tokens. The IBM Watson platform uses natural language processing for advanced text analysis. The APIs available over the platform can help analyze text. We could create a custom model


```
[
  {
    "score": 0.676907,
    "label": "/technology and computing/internet technology"
  },
  {
    "score": 0.666597,
    "label": "/technology and computing/hardware"
  },
  {
    "score": 0.621132,
    "label": "/technology and computing/tech news"
  }
]
```


Fig. 2 JSON response from IBM Watson

using some of the available APIs to get specific results tailored for the requirement in this project. We could make use of the default model and worked to understand only the categories of the text. The user's interest plotted in a bar chart is shown in Fig. 3.

In our paper, we propose that recommender systems should focus on the granular level of each category and should not recommend products from the broad top categories. Recommending products from the broad top categories reduces the effectiveness of a system. For Example, as seen in Fig. 1, we identified that a user is interested in a subcategory titled "tech news" under the top category titled "technology and computing". If a system should recommend items using only the top category, then items such as servers, keyboard, routers, and internet cables would be displayed. Meanwhile, the user is more interested in "tech news" and not other items from the top category. Another example is when the system predicts that a user is interested in a subcategory titled "Sea Travel" under "Transport" and under "Travel". Hence the structure of the response is \ Travel\Transport\Sea Travel. Here, recommending items from the top category titled "Travel" would display items such as space travel, pilgrimage, and railroad travel instead of recommending items specifically to what the user truly prefers which is "Sea Travel". Therefore, recommending items using the granular level of a category increases the effectiveness of recommender systems and our system aides such by providing the categories and subcategories.

We made use of content-based filtering method because we are focused on a user and not recommending items based on similar users. Also, we are mining a user's tweet because people often share their opinions about products and experiences on Twitter. A user's tweet could read "The camera of my iPhone is very clear". This tweet serves as a form of product review and a hint that the user is interested in "iPhones" under the category of "Smartphones". Also, our system provides scoring for each preference. The proposed system can sort items from the user's most preferred to the least preferred items. If a user is found to like football as the first, cooking as the

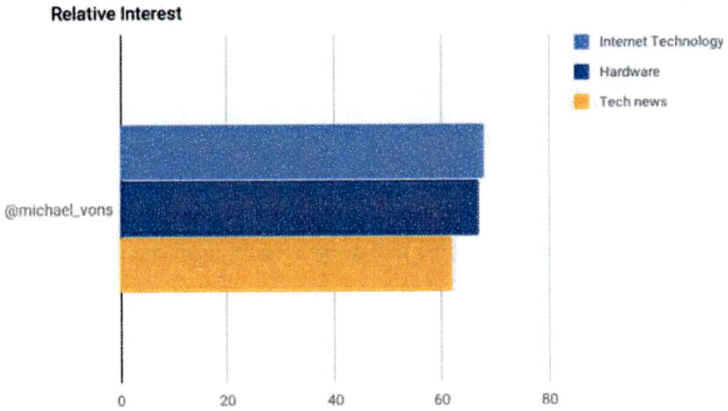

Fig. 3 User's interest plotted in bar chart

second, and movies as the third, then an e-commerce website can display the most preferred items accordingly. This improves the overall customer experience.

4 Conclusion

In this paper, we proposed a system that can predict the category of products and services that the user is interested in by using the Twitter timeline and profile of the user. The system was tested and the results suggested the personalization of the products could be reached at a category level. While extensive studies have been done on recommender systems, there is limited literature focused on the effectiveness of recommender systems along with the ability to predict multiple preferences. In our study, we have scored each preference based on the intensity of interest which is useful for sorting items from most preferred to least preferred. The system needs to be further expanded into a subcategory level personalization. For instance, the personalization level reached in the system is limited to a broader category, such as Electronics. Further research and testing are required to product purchases at a subcategory and product level, for instance Televisions, which is a subcategory within the Electronic category. A complete recommender system would be a logical extension for our system.

References

1. Madadipouya, K., Chelliah, S.: A literature review on recommender systems algorithms techniques and evaluations. BRAIN: Broad Res. Artif. Intell. Neurosci. **8**(2), 109–124 (2017)
2. Feng, J., et al.: An improved collaborative filtering method based on similarity. PLoS ONE **13**(9), 1–18 (2018)
3. Guo, Y., Wang, M., Li, X.: An interactive personalized recommendation system using the hybrid algorithm model. Symmetry **9**(1), 216–233 (2017)
4. Tao, J., Gan, J., Wen, B.: Collaborative filtering recommendation algorithm based on spark. Int J Perform. Eng. **15**(3), 930–938 (2019)
5. Gautam, A., Bedi, P.: Developing content-based recommender system using Hadoop Map Reduce. J. Intell. Fuzzy Syst. **32**(4), 2997–3008 (2017)
6. Çano, E., Morisio, M.: *Hybrid recommender systems: a systematic literature review*. Intell. Data Anal. **21**(1), 1487–1524 (2017)
7. Rao, D., et al.: Classifying latent user attributes in twitter. In: Proceedings of the 2nd International Workshop on Search and Mining User-Generated Contents, pp. 37–44. ACM, Toronto, ON, Canada (2010)
8. Samani, Z.R., et al.: Crossplatform and cross-interaction study of user personality based on images on Twitter and Flickr. PLoS ONE **13**(7), 1–19 (2018)
9. Wang, C., et al.: Behavior-interior-aware user preference analysis based on social networks. Complexity **2018**(1), 1–18 (2018)
10. Park, W., Kang, S., Kim, Y.-K.: Personalized mobile e-commerce system using DISC psychological model. pp. 245–248 (2011)
11. Abakouy, R., En-Naimi, E.M., Haddadi, A.E.: Classification and prediction based data mining algorithms to predict email marketing campaigns. In: Proceedings of the 2nd International

Conference on Computing and Wireless Communication Systems, pp. 1–5. ACM, Larache, Morocco (2017)
12. Zhao, Q.: E-commerce product recommendation by personalized promotion and total surplus maximization. In: Proceedings of the Ninth ACM International Conference on Web Search and Data Mining. pp. 709–709. ACM, San Francisco, California, USA (2016)
13. Liu, G., et al.: Modeling buying motives for personalized product bundle recommendation. ACM Trans. Knowl. Discov. Data **11**(3), 1–26 (2017)
14. Zhao, Q., et al.: E-commerce recommendation with personalized promotion. In: Proceedings of the 9th ACM Conference on Recommender Systems, pp. 219–226. ACM, Vienna, Austria (2015)
15. Lu, C.-T., Shuai, H.-H., Yu, P.S.: Identifying your customers in social networks. In: Proceedings of the 23rd ACM International Conference on Information and Knowledge Management, pp. 391–400. ACM, Shanghai, China (2014)
16. Zhang, Y., Pennacchiotti, M.: Recommending branded products from social media. In: Proceedings of the 7th ACM Conference on Recommender Systems, pp. 77–84. ACM, Hong Kong, China (2013)
17. Wu, Y., et al.: Collaborative filtering recommendation algorithm based on user fuzzy similarity. Intell. Data Anal. **21**(2), 311–327 (2017)
18. Albatayneh, N.A., Ghauth, K.I., Chua, F.-F.: Utilizing learners' negative ratings in semantic content-based recommender system for e-learning forum. Educ. Technol. Soc. **21**(1), 112–125 (2018)
19. Alyari, F., Navimipour, N.J.: Recommender systems: a systematic review of the state of the art literature and suggestions for future research. Kybernetes **47**(5), 985–1017 (2018)
20. Isinkaye, F.O., Folajimi, Y.O., Ojokoh, B.A.: Recommendation systems: principles, methods and evaluation. Egyptian Inf. J. **16**(1), 261–273 (2015)
21. Najafabadi, M.K., Mahrin, M.N.: A systematic literature review on the state of research and practice of collaborative filtering technique and implicit feedback. Artif. Intell. Rev. **45**(2), 167–201 (2015)
22. Wang, P., Chen, J., Niu, S.: CFSH: factorizing sequential and historical purchase data for basket recommendation. PLoS ONE **13**(10), 1–16 (2018)
23. Lee, H.-M., Um, J.-S.: A study on the context-aware hybrid bayesian recommender system on the mobile devices. IAENG Int. J. Comput. Sci. **45**(1), 1–7 (2017)
24. Li, Z.: collaborative filtering recommendation algorithm based on cluster. Int. J. Performability Eng. **14**(5), 927–936 (2018)
25. Khan, M.M., Ibrahim, R., Ghani, I.: Cross domain recommender systems: a systematic literature review. ACM Comput. Surv. **50**(3), 1–34 (2017)

Cycloconverter Fed Capacitor Start Capacitor Run Induction Motor Drive: Simulation Analysis

Pragada Niharika, Vinnakota Vineetha, and K. Durgendra Kumar

Abstract This paper provides a detailed explanation of control principle of cycloconverter-fed capacitor-start capacitor-run induction motor. Analog circuitry scheme for gate pulse generation scheme of cycloconverter, and different industrial applications have been discussed in this study. For motor control application, cycloconverter-fed capacitor-start capacitor-run induction motor has been considered in this study. Fuzzy-PD+I based feedback control approach has been used for control which provides better performance than classical PID control approach. MATLAB based simulation analysis has been provided to validate the fuzzy-PD+I controller design approach.

Keywords Cycloconverter · Induction motor · Electric drive · Fuzzy-PD+I · Feedback control · Capacitor-start capacitor-run

1 Introduction

Cycloconverter is a naturally commutated converter with bi-directional power flow capability. There are variety of applications of cycloconverter such as gearless motor drive in cement industry, centrifugal pump, electric traction, rolling mills, variable speed constant frequency (VSCF) system, ship propellers etc. Evaluation of cycloconverter fed semi-autogenous grinding (SAG) mills have been extensively discussed in [7, 8]. PWM based control technique for forced commutated cycloconverter is discussed in [2]. 3-ϕ to 3-ϕ cycloconverter has been discussed in [3, 5]. Power factor

P. Niharika · V. Vineetha · K. Durgendra Kumar (✉)
Department of Electrical and Electronics Engineering,
Aditya Engineering College, Surampalem, Andhra Pradesh, India
e-mail: durgendrakumar@gmail.com

P. Niharika
e-mail: pragadaniharika@gmail.com

V. Vineetha
e-mail: vineethavinnakota@gmail.com

S. Borah et al. (eds.), *Soft Computing Techniques and Applications*, Advances in Intelligent Systems and Computing 1248,
https://doi.org/10.1007/978-981-15-7394-1_8

improvement in cycloconverter has been discussed in [13]. Harmonic analysis of cycloconverter and suppression of harmonics have been discussed in [14]. Cycloconverter fed gearless mill drive has been discussed in [9] and cycloconverter fed drives for mining applications have been discussed in [11]. The power quality aspects of cycloconverter based control system has been discussed in [6].

This paper provides a discussion of cycloconverter-fed capacitor-start capacitor-run induction motor. Mathematical analysis of induction motor has been discussed in [4, 10, 12]. Soft-computing based feedback control principle has been used to control the motor. Fuzzy-PID controller has been used for controlling the induction motor speed and torque [1]. MATLAB based simulation analysis has been carried out and shown in the paper.

2 Industrial Applications of Cycloconverter

Cycloconverter is used to drive electrical motor and subsequently the mechanical load. The basic drive system (Fig. 1) where fuzzy-PD+I controller has been used.

Figure 2a illustrates the cycloconverter fed gearless drive. Figure 2b illustrates cycloconverter fed synchronous motor drive and Fig. 2c presents cycloconverter based traction system.

3 Operation of Cycloconverter

Figure 3a, b presents the circuit diagram and related waveform for 1-ϕ to 1-ϕ cycloconverter. In this circuit, the output frequency is half or one-third of input frequency.

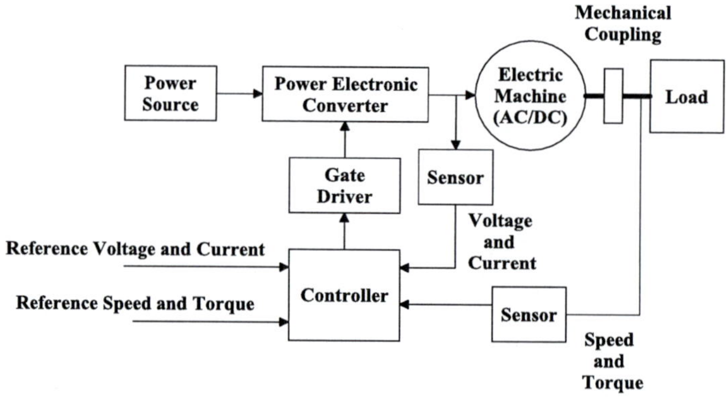

Fig. 1 Generalized block diagram of electric drive

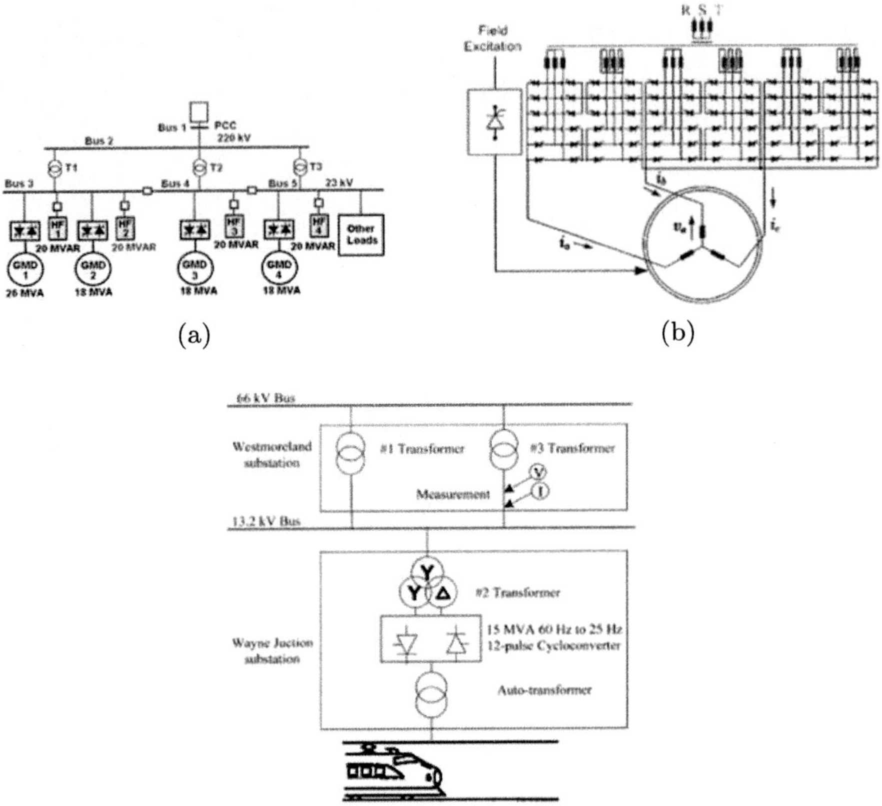

Fig. 2 Industrial applications **a** Cyclcoconverter fed gearless mill drive (GMD), **b** Cycloconverter-fed synchronous motor drive, **c** Cycloconverter based traction system

The RMS value of output voltage at $0 \leq \alpha \leq \pi$ can be represented as

$$V_o = V_s\sqrt{\frac{1}{\pi}\left(\pi - \alpha + \frac{\sin 2\alpha}{2}\right)} \tag{1}$$

The input power factor is

$$\text{p.f.} = \cos\theta\sqrt{\frac{1}{\pi}\left(\pi - \alpha + \frac{\sin 2\alpha}{2}\right)} \tag{2}$$

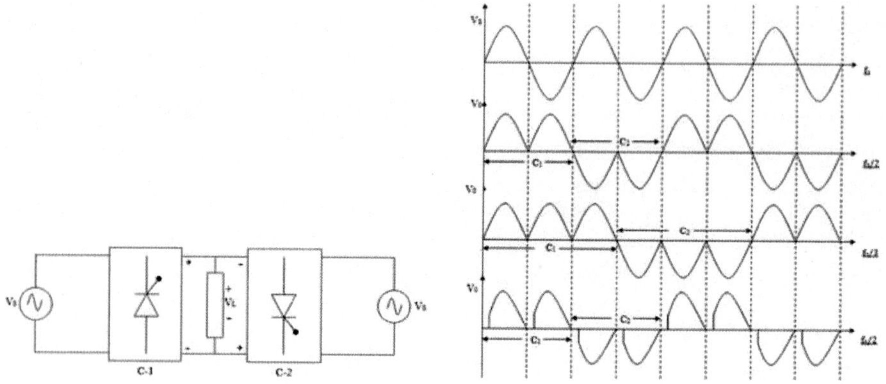

(a) (b)

Fig. 3 Step-down cycloconverter **a** Circuit diagram, **b** Related waveform

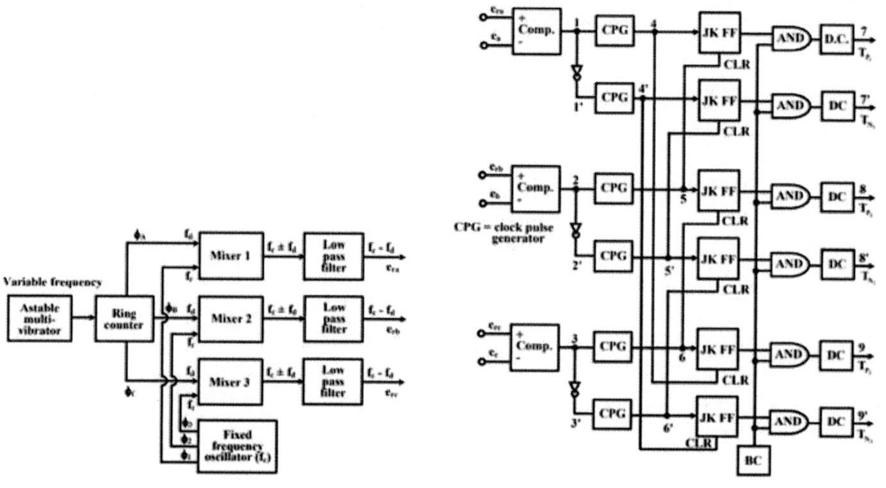

(a) (b)

Fig. 4 Pulse generation for cycloconverter (**a, b**) Block diagram of gate pulse generation scheme

Figure 4a, b illustrates the gate pulse generation scheme for cycloconverter circuit. There are a large number of thyristors in cycloconverter therefore, the gate pulse generation scheme is also fairly complicated.

3.1 Fuzzy-PD+I Controller

Fuzzy-PD+I controller can be represented as

$$K_{PD}u_{PD}(nT_s) = K_p^1 \frac{e(nT_s)}{T_s} + K_d^1 \Delta e(nT_s) \tag{3}$$

$$d(nT_s) = K_{PD}u_{PD}(nT_s) + u_I(nT_s) \tag{4}$$

T_s is sampling time (sec); d is duty cycle

4 Simulation Results

In this section, Fuzzy-PD+I controller is used to control the capacitor-state capacitor-run induction motor with the help of cycloconverter which as a power converter. Simulated waveform of 1-ϕ to 1-ϕ step-down cycloconverter has been illustrated in Fig. 5. As it can be shown the output frequency is half of the input frequency $f_o = \frac{f_{in}}{2}$.

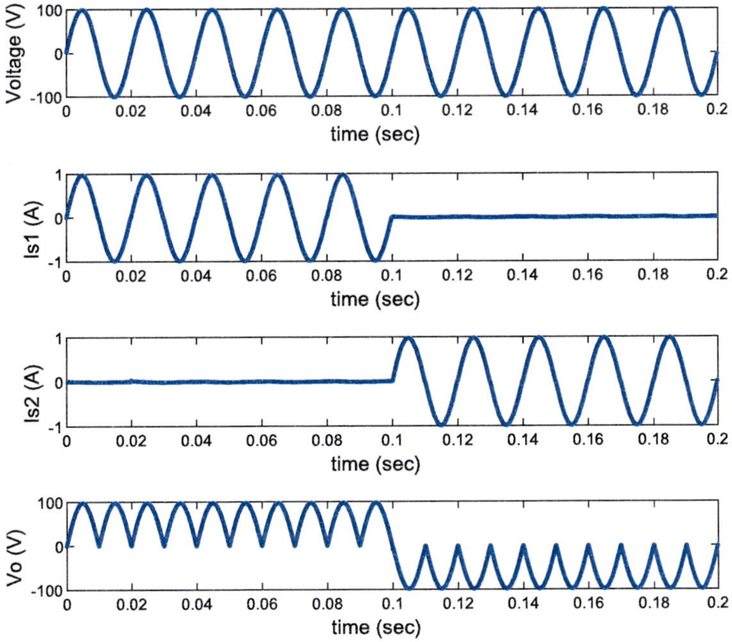

Fig. 5 Simulation result for single-phase to single-phase Step-down cycloconverter

Table 1 Simulation parameters

Parameters	Values
Type of machine	Capacitor-start capacitor-run
Nominal power	0.25 h.p. (186.425 W)
Voltage	240 V (RMS)
Frequency	50 Hz
Stator winding	$R_s = 2.02\Omega$, $L_s = 7.4$ mH
Rotor winding	$R_r = 4.12\Omega$, $L_s = 5.6$ mH
Capacitor start	$R_s = 2\Omega$, $C_s = 254.7$ μF
Capacitor run	$R_r = 18\ \Omega$, $C_r = 21.1$ μF

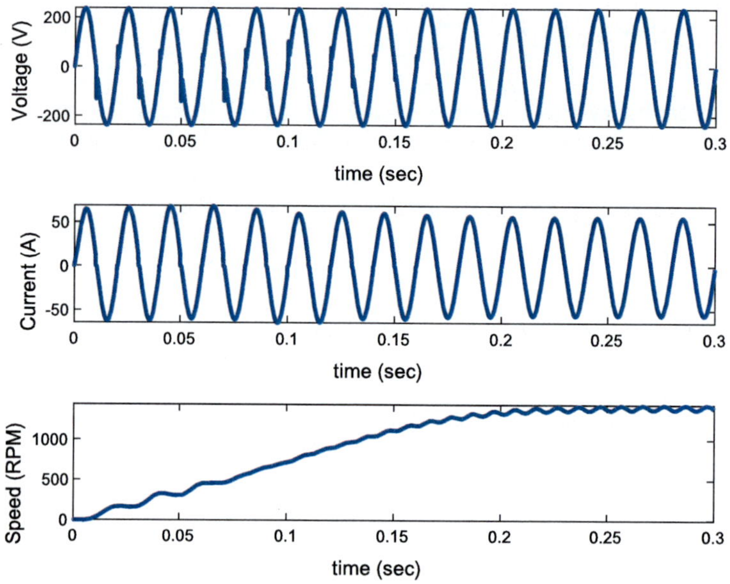

Fig. 6 Simulation result for cycloconverter-fed CSCR induction motor

Table 1 provides simulation parameters used for CSCR induction motor and Fig. 6 illustrates the simulation results for voltage, current and speed (RPM) of cycloconverter-fed CSCR induction motor controlled using Fuzzy-PD+I controller. Settling time t_s is 0.2 s which is very satisfactory in nature.

5 Conclusion

This paper provides a detailed simulation result for cycloconverter-fed capacitor-start capacitor-run induction motor using Fuzzy-PD+I controller. Different application of cycloconverter and gate pulse generation scheme of cycloconverter has been discussed in this paper. Using Fuzzy-PD+I controller, the settling time is very less which is encouraging in nature.

Acknowledgements The authors are thankful to Accendere Knowledge Management Services Pvt. Ltd, CL Educate Ltd for their assistance during the preparation of the manuscript.

References

1. Al-Greer, M., Armstrong, M., Pickert, V.: Selecting appropriate fuzzy PID control structure for power electronic applications. J. Eng. **2019**(17), 4457–4460 (2019)
2. Babaei, E., Heris, A.A.: PWM-based control strategy for forced commutated cycloconverters. In: 2009 IEEE Symposium on Industrial Electronics & Applications, vol. 2, pp. 669–674. IEEE, New York (2009)
3. Basirifar, M., Shoulaie, A.: A comparative study of circulating current free and circulating current cycloconverters. In: 2010 First Power Quality Conference, pp. 1–4. IEEE, New York (2010)
4. Fuchs, E., Vandenput, A., Holl, J., White, J.: Design analysis of capacitor-start, capacitor-run single-phase induction motors. IEEE Trans. Energy Convers. **5**(2), 327–336 (1990)
5. Heris, A.A., Sadeghi, M., Babaei, E.: A new topology for the three-phase to three-phase cycloconverters. In: 2011 2nd Power Electronics, Drive Systems and Technologies Conference, pp. 489–494. IEEE, New York (2011)
6. Liu, Y., Heydt, G.T., Chu, R.F.: The power quality impact of cycloconverter control strategies. IEEE Trans. Power Delivery **20**(2), 1711–1718 (2005)
7. Palavicino, P.C., Valenzuela, M.A.: Modeling and evaluation of cycloconverter-fed two-stator-winding sag mill drivepart I: modeling options. IEEE Trans. Ind. Appl. **51**(3), 2574–2581 (2014)
8. Palavicino, P.C., Valenzuela, M.A.: Modeling and evaluation of cycloconverter-fed two-stator-winding sag mill drivepart II: Starting evaluation. IEEE Trans. Ind. Appl. **51**(3), 2582–2589 (2014)
9. Pontt, J.O., Rodriguez, J.P., Rebolledo, J.C., Tischler, K., Becker, N.: Operation of high-power cycloconverter-fed gearless drives under abnormal conditions. IEEE Trans. Ind. Appl. **43**(3), 814–820 (2007)
10. Popescu, M., Miller, T., McGilp, M., Strappazzon, G., Trivillin, N., Santarossa, R.: Asynchronous performance analysis of a single-phase capacitor-start, capacitor-run permanent magnet motor. IEEE Trans. Energy Convers. **20**(1), 142–150 (2005)
11. Silva, G.F., Morán, T.L., Torres, T.M., Weishaupt, V.C.: A method to evaluate cycloconverters commutation robustness under voltage and frequency variations in mining distribution systems. IEEE Trans. Ind. Appl. **54**(1), 858–865 (2017)
12. Sorrentino, E., Fernandez, S.: Comparison of six steady-state models for single-phase induction motors. IET Electr. Power Appl. **5**(8), 611–617 (2011)
13. Takahashi, I., Akagi, H., Miyairi, S.: Improvement of cycloconverter power factor via unsymmetric triggering method. Electr. Eng. Jpn **96**(1) (1976)
14. Wang, M., Li, Y., Tan, B., Wei, B.: Harmonic analysis and suppression methods study of cycloconverter-feed synchronous motor drive system. In: 2006 International Conference on Power System Technology, pp. 1–6. IEEE, New York (2006)

Control Scheme to Minimize Torque Ripple of SRM

M. Venkatesh, Vijayasri Varshikha Joshi, K. L. Mounika, and B. Veeranarayana

Abstract This paper discusses the DTC and DITC scheme for switched reluctance motor. Modeling, controller and simulation have been discussed in this paper. TSK-based fuzzy controller design has been incorporated in switched reluctance motor to control the speed. MATLAB-based simulation has been provided in this paper. From simulation analysis, TSK fuzzy controller provides better performance.

Keywords SRM · DTC · DITC · TSK fuzzy rules · Fuzzy controller design

1 Introduction

"Switched reluctance motor" is a special motor which has simple construction, low maintenance cost and high torque-to-mass ratio. SRM finds its application in hybrid vehicles where it provides superior performance than other motors. One of the main limitations of SRM is ripple content in torque which causes vibration and noise. Therefore, ripple content of the torque should be minimized so as to use the motor in an effective manner.

A comprehensive design technique and modeling scheme for switched reluctance machine has been discussed in [1, 4]. Speed control with optimization technique for SRM has been discussed in [8, 9]. FEM-based modeling approach of SRM has been discussed in [6]. "DTC" and "DITC" for minimization of torque have been discussed

M. Venkatesh (✉) · V. V. Joshi · K. L. Mounika · B. Veeranarayana (✉)
Department of Electrical and Electronics Engineering,
Aditya Engineering College, Surampalem, Andhra Pradesh, India
e-mail: venkateshmeka1999@gmail.com

B. Veeranarayana
e-mail: b.veeranarayana@aec.edu.in

V. V. Joshi
e-mail: vijayasri.j23@gmail.com

K. L. Mounika
e-mail: kondapallimounika1@gmail.com

S. Borah et al. (eds.), *Soft Computing Techniques and Applications*, Advances in Intelligent Systems and Computing 1248,
https://doi.org/10.1007/978-981-15-7394-1_9

in [3, 7, 10]. Adaptive fuzzy control scheme for control of SRM has been discussed in [2, 5]. TSK-based speed control is discussed in [11].

This study provides mathematical modeling, direct torque control method and direct instantaneous torque control method which have been discussed for SRM. Takagi 'Sugeno' Kang (TSK) speed control scheme has been discussed in this paper.

2 Modeling of SRM

Phase voltage of SRM

$$v = R_s i + \frac{d\psi(\theta, i)}{dt} \quad (1)$$

$$\psi = L(\theta, i)\, i \quad (2)$$

Eq.1 can be re-written as

$$v = R_s i + L(\theta, i)\frac{di}{dt} + \frac{dL(\theta, i)}{d\theta} i\omega_m \quad (3)$$

Power

$$P = R_s \frac{I^2}{2} + \frac{I}{\sqrt{2}} L(\theta, I)\frac{dI}{dt} + \frac{I^2}{2}\frac{dL(\theta, I)}{d\theta}\omega_m \quad (4)$$

where I represents peak value of phase current

$$T(\theta, i) = \sum_{phase} \frac{I^2}{2}\frac{dL(\theta, I)}{d\theta} \quad (5)$$

The efficiency of SRM can be represented as

$$\eta(\%) = \frac{P_{out}}{P_{out} + P_{loss}} \times 100\% \quad (6)$$

Torque ripple

$$\Delta T(\%) = \frac{T_{\max} - T_{\min}}{T_{mean}} \times 100\% \quad (7)$$

3 Control Scheme of SRM

Generalized control scheme for switched reluctance motor is discussed in Fig. 1 and 2.

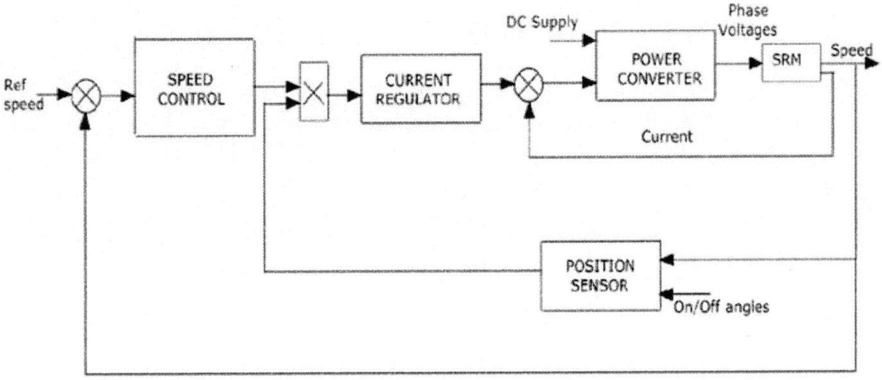

Fig. 1 Control scheme of SRM

Equation 1,

$$vi = R_s i^2 + i\frac{\partial\psi(\theta, i)}{\partial i}\frac{di}{dt} + i\frac{\partial\psi(\theta, i)}{\partial\theta}\frac{d\theta}{dt} \tag{8}$$

Field energy can be represented as

$$dW_f = \frac{\partial W_f}{\partial i}di + \frac{\partial W_f}{\partial\theta}d\theta \tag{9}$$

$$dW_m = i\frac{\partial\psi(\theta, i)}{\partial\theta}d\theta - \frac{\partial W_f}{\partial\theta}d\theta \tag{10}$$

$$T = \frac{dW_m}{dt} = i\frac{\partial\psi(\theta, i)}{\partial\theta} - \frac{\partial W_f}{\partial\theta} \tag{11}$$

3.1 *TSK Fuzzy System*

TSK fuzzy IF–THEN rules can be represented as

$$\mathbf{R}^{(i)} : \text{ IF } x_1 = F_1^i \text{ and } x_2 = F_2^i \text{ THEN } y^i = q_0^i + q_1^i x_1 + q_2^i x_2 \tag{12}$$

4 Simulation Results

Table 1 shows the simulation parameters used for SRM.

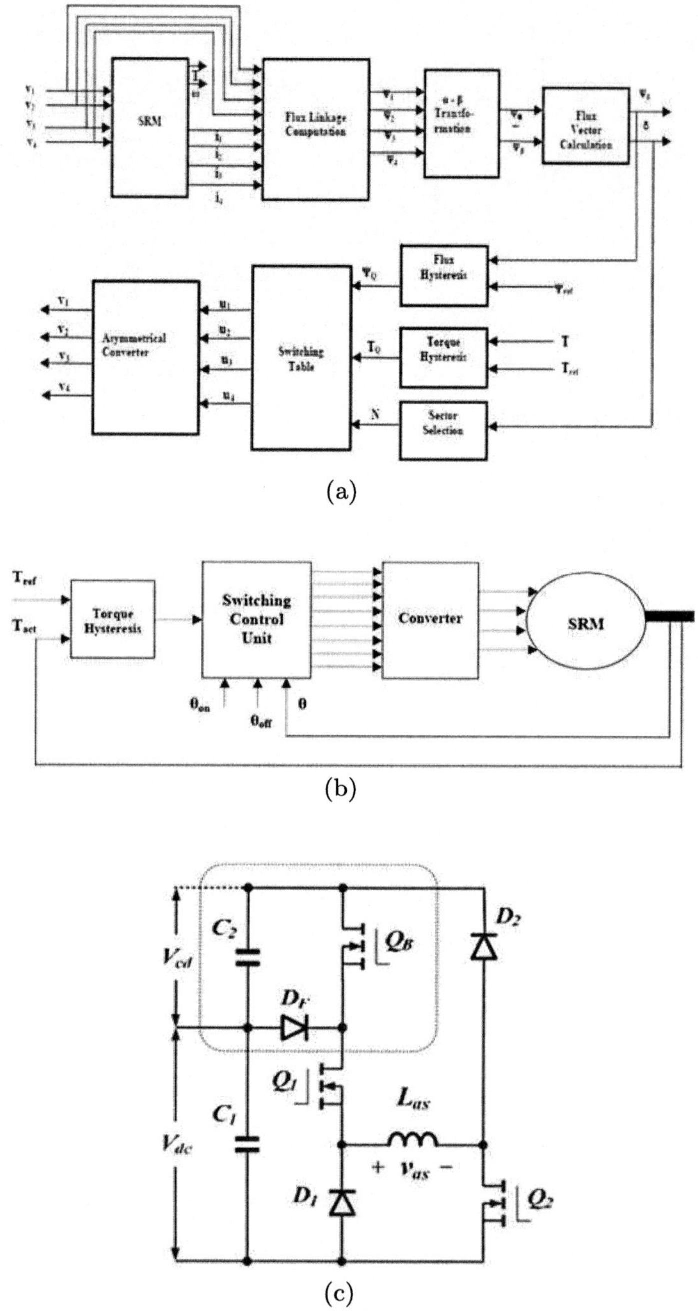

Fig. 2 **a** Direct toque control of SRM, **b** direct instantaneous torque control scheme of SRM, **c** asymmetrical power converter for switched reluctance motor

Table 1 Simulation parameters

Type of motor	6/4 SRM
Stator resistance	0.05 Ω
Inertia	0.5 kgm^2
Friction	0.2 Nms
Inductance	0.67 mH (Unaligned), 23.6 mH (Aligned)
Saturated aligned inductance	0.15 mH

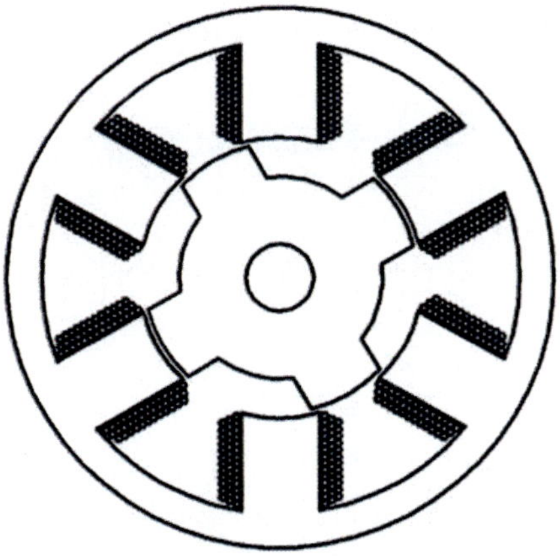

(a)

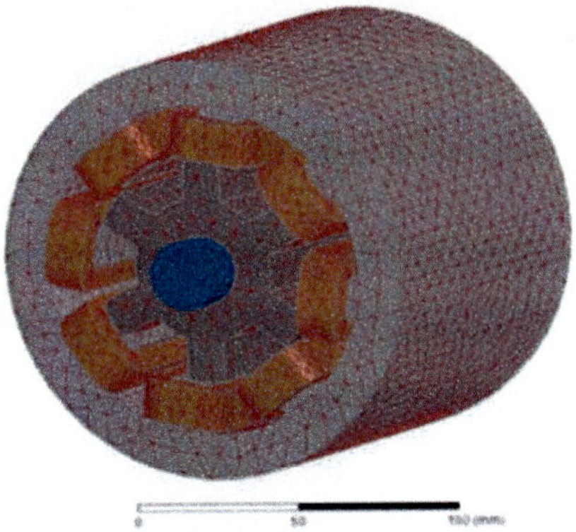

(b)

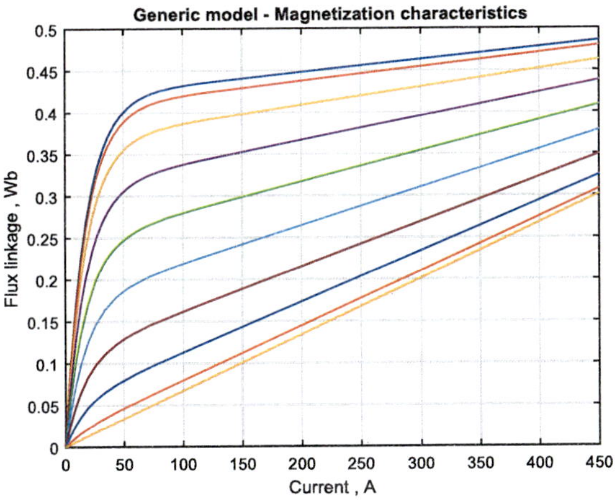

(c)

Fig. 3 **a** Construction of 6/4 SRM, **b** finite element analysis of SRM, **c** magnetization characteristics of specific model of SRM

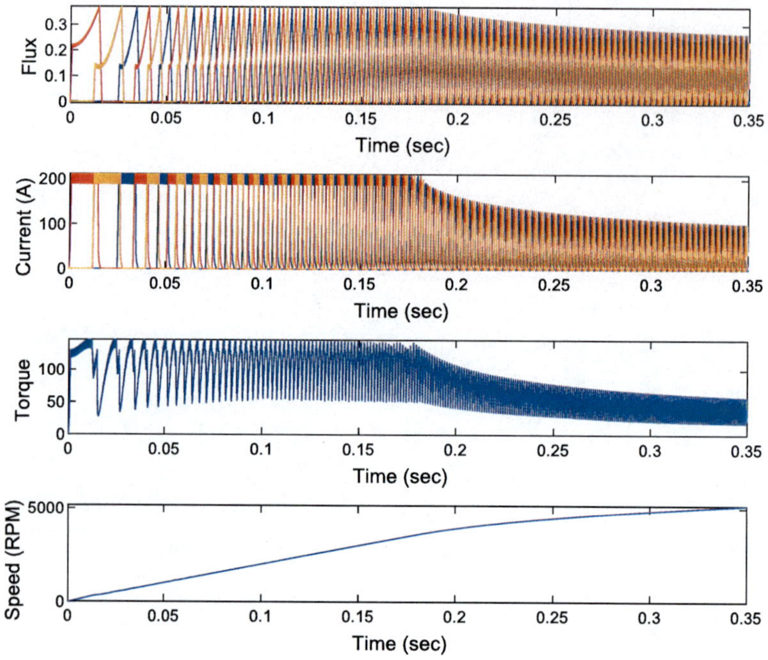

Fig. 4 Flux, current, torque and speed of SRM

Figure 3a shows the construction of 6/4 SRM, Fig. 3b shows the finite element analysis of SRM and Fig. 3c illustrates the magnetizing characteristics of 6/4 SRM. Fig. 4 shows the flux, current, torque and speed characteristics of SRM.

5 Conclusion

This paper provides mathematical model of SRM along with different control schemes. DTC and DITC have been analyzed in this paper. Simulation results have been provided in this paper. TSK fuzzy control has been highlighted in this paper.

Acknowledgements The authors are thankful to "Accendere Knowledge Management Services Pvt. Ltd, CL Educate Ltd" for their assistance during the preparation of the manuscript.

References

1. Anwar, M., Husain, I., Radun, A.V.: A comprehensive design methodology for switched reluctance machines. IEEE Trans. Ind. Appl. **37**(6), 1684–1692 (2001)
2. Bolognani, S., Zigliotto, M.: Fuzzy logic control of a switched reluctance motor drive. IEEE Trans. Ind. Appl. **32**(5), 1063–1068 (1996)
3. Chaple, M., Bodkhe, S.B., Daigavane, P.: Four phase (8/6) SRM with DTC for minimization of torque ripple. Int. J. Electr. Eng. Educ., p. 0020720919841686 (2019)
4. Hwu, K.: Applying powersys and simulink to modeling switched reluctance motor. Tamkang J. Sci. Eng. **12**(4), 429–438 (2009)
5. Masoudi, S., Soltanpour, M.R., Abdollahi, H.: Adaptive fuzzy control method for a linear switched reluctance motor. IET Electr. Power Appl. **12**(9), 1328–1336 (2018)
6. Prasad, N., Jain, S., Gupta, S.: Measurement and optimization of performance parameters of linear switched reluctance motor using finite element method. MAPAN, pp. 1–9 (2019)
7. Pratapgiri, S., Narsimha, P.P.V.: Direct torque control of 4 phase 8/6 switched reluctance motor drive for constant torque load. World J. Modell. Simul **8**(3), 185–195 (2012)
8. Saha, N., Panda, A., Panda, S.: Speed control with torque ripple reduction of switched reluctance motor by many optimizing liaison technique. J. Electr. Syst. Inf. Technol. **5**(3), 829–842 (2018)
9. Saha, N., Panda, S.: Speed control with torque ripple reduction of switched reluctance motor by hybrid many optimizing liaison gravitational search technique. Int. J. Eng. Sci. Technol. **20**(3), 909–921 (2017)
10. Srinivas, P., Prasad, P.: Torque ripple minimization of 4 phase 8/6 switched reluctance motor drive with direct instantaneous torque control. Int. J. Electr. Eng. Inform. **3**(4), 488 (2011)
11. Tseng, C.L., Wang, S.Y., Chien, S.C., Chang, C.Y.: Development of a self-tuning TSK-fuzzy speed control strategy for switched reluctance motor. IEEE Trans. Power Electron. **27**(4), 2141–2152 (2011)

Simulation and Analysis of Seven-Level Voltage Source Inverter

L. Sri Hansitha Priya, K. Rajesh, U. Satya Sai Polaraju, and N. Rajesh

Abstract A seven-level inverter topology with the minimum component count is presented in this paper. The presented topology has low switching stress and fundamental frequency operating switches that enhance the efficiency of the configuration. The operating modes of the proposed inverter are analyzed in detail during zero, positive, and negative levels. The proposed topology is gated using fuzzy based sinusoidal Pulse Width Modulation in MATLAB/Simulink environment.

Keywords Multilevel inverter · Seven-level · Pulse width modulation · FFT · Fuzzy

1 Introduction

Multilevel inverter (MLI) is spreading to several areas FACTS, renewable-energy sources [1, 2]. Although "Neutral point clamped" (NPC), "Flying capacitor clamped" (FCC) and "Cascaded H-bridge" (CHB) converters [3–6] are standard MLI, the component count in these topologies increases drastically w.r.t the number of levels in the output voltage. In order to alleviate the limitations in standard topologies, new MLIs with a reduced number of components are developed and are suggested in many applications [7–9]. From the past few decades, many such Voltage Source Inverters

L. S. H. Priya · K. Rajesh · U. Satya Sai Polaraju · N. Rajesh (✉)
Department of Electrical and Electronics Engineering, Aditya Engineering College, Surampalem, Andhra Pradesh, India
e-mail: rajesh.n@aec.edu.in

L. S. H. Priya
e-mail: srihansi860@gmail.com

K. Rajesh
e-mail: kapavarapurajesh048@gmail.com

U. Satya Sai Polaraju
e-mail: polaraju8499@gmail.com

S. Borah et al. (eds.), *Soft Computing Techniques and Applications*, Advances in Intelligent Systems and Computing 1248,
https://doi.org/10.1007/978-981-15-7394-1_10

(VSIs) have been reported using several combination of power semiconductor devices, isolated dc supplies and other devices [10–12].

The proposed converter offers the benefits of a reduced number of switching devices, a reduced number of dc power supplies, and lower off-state voltage stress across the switching devices. This script is detailed as follows: Section 2 explores the operation, Sect. 3, modulation scheme with improved spectral performance is shown, in Sect. 4, the simulation results at different values of modulation indices are discussed and Sect. 5, discusses the conclusion.

2 System Configuration

Figure 1 shows the proposed single-phase seven-level inverter using eight switching devices and two independent dc sources. All the switching devices used in the proposed configuration are of bi-directional conducting devices with uni-directioucnal voltage sustaining ability. The dc supplies can be obtained either from rectifier circuits, battery banks, or pv arrays.

Figure 2a gives $V_0 = 0^+$, this switching state is used during positive zero-crossing of the output. Figure 2b gives $V_{A0} = 0^-$, this switching state is used during negative zero-crossing of the output. Both the switching states are effectively utilized to ensure the thermal equilibrium of the system. Figure 3 shows the positive-level operating modes of the converter. Figure 3a gives the output voltage $V_0 = V_{dc}/2$, during this state the switching devices S_2, S_4, S_7, and S_8 are triggered. Figure 3b gives the output voltage $V_0 = V_{dc}$, during this state the switching devices S_2, S_3, S_6, and S_8 are triggered. Figure 3c gives the output voltage $V_0 = 3V_{dc}/2$, during this state the switching devices S_2, S_3, S_5, S_7, and S_8 are triggered. Figure 4 shows the negative-level operating modes of the configuration. Figure 4a gives the output voltage $V_0 = -V_{dc}/2$, during this state the switching devices S_1, S_3, S_6, and S_9 are triggered.

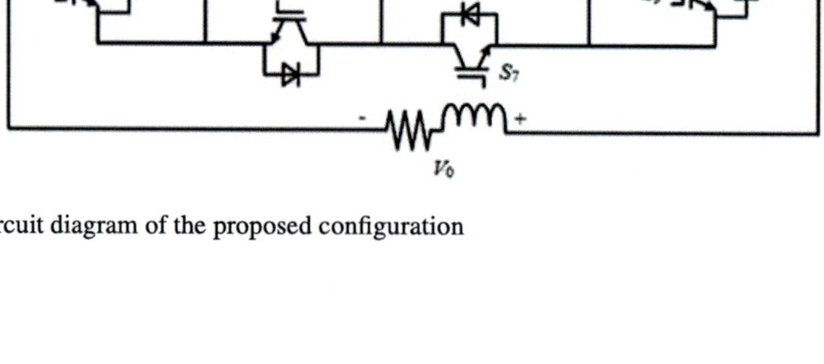

Fig. 1 Circuit diagram of the proposed configuration

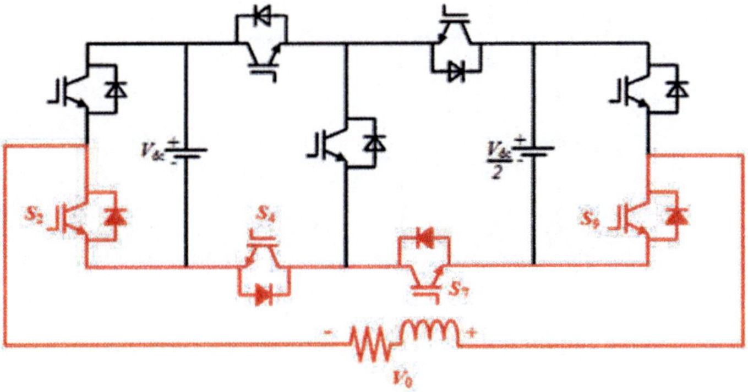

(a)

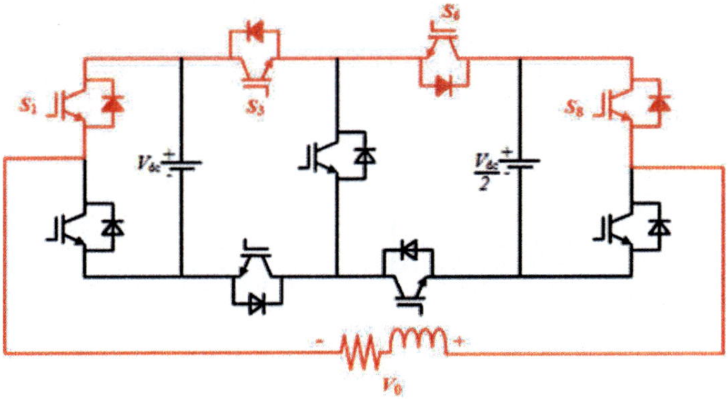

(b)

Fig. 2 Zero-level operating modes: **a** $V_0 = 0^{+}$ and **b** $V_0 = 0^{-}$

Figure 4b gives the output voltage $V_0 = -V_{dc}$, during this state the switching devices S_1, S_4, S_7, and S_9 are triggered. Figure 4c gives the output voltage $V_0 = -3V_{dc}/2$, during this state the switching devices S_1, S_4, S_5, S_6, and S_9 are triggered.

3 Sinusoidal PWM Scheme

Table 1 lists the various switching states of the proposed converter operating in seven-level mode. The digital numbers 1 and 0 is the indication for switch-on and switch-off respectively for the switches. Figure 5 shows the modulation method [13, 14] to generate pulses to the switches in proposed converter. A regular sinusoid is cross-compared with six level-shifted triangular waves to produce the seven-level PWM output voltage (Table 2).

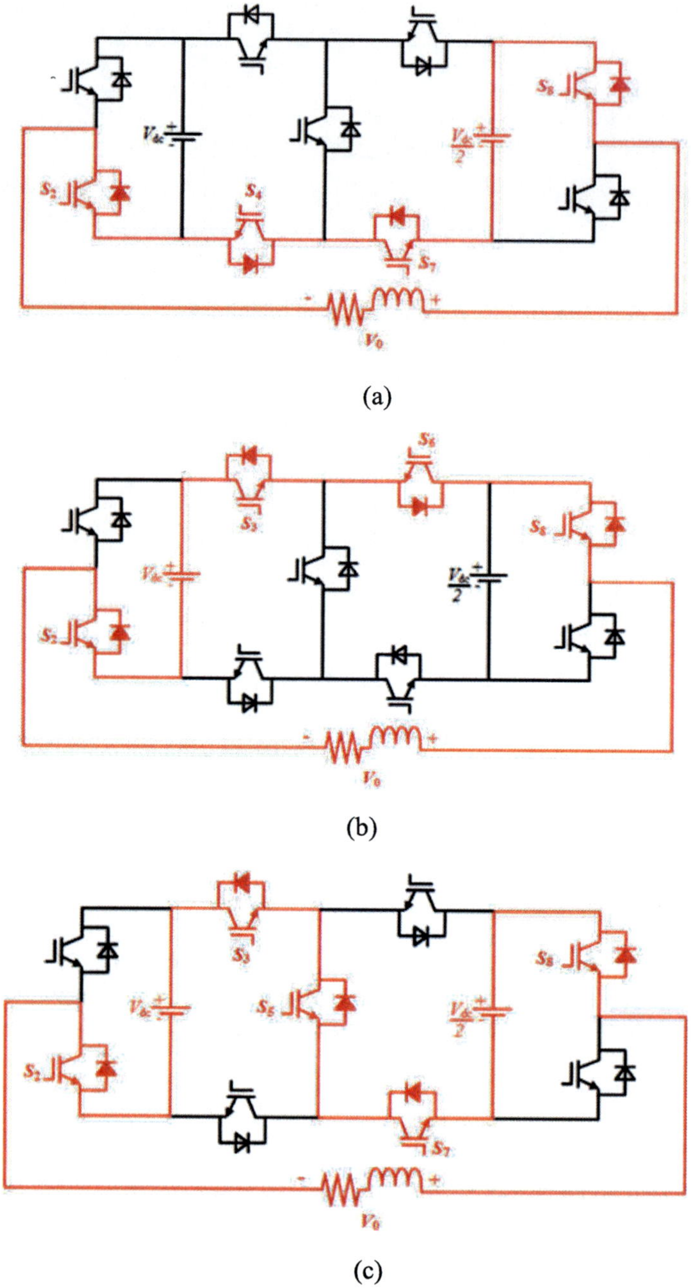

Fig. 3 Positive-level operating modes: **a** $V_0 = V_{dc}/2$, **b** $V_0 = V_{dc}$, and **c** $V_0 = 3V_{dc}/2$

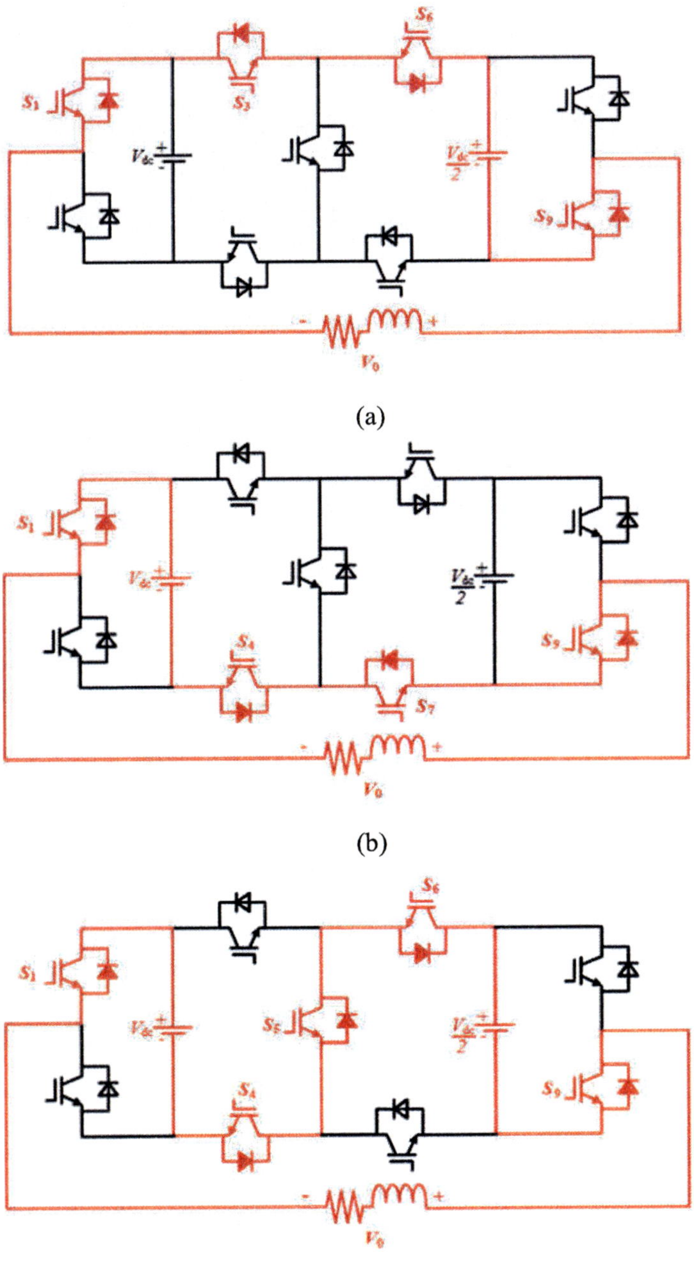

(a)

(b)

(c)

Fig. 4 Negative-level operating modes: **a** $V_0 = -V_{dc}/2$, **b** $V_0 = -V_{dc}$, and **c** $V_0 = -3V_{dc}/2$

Table 1 Switching sequence of the inverter

V_0	S_1	S_2	S_3	S_4	S_5	S_6	S_7	S_8	S_9
$3V_{dc}/2$	0	1	1	0	1	0	1	1	0
V_{dc}	0	1	1	0	0	1	0	1	0
$V_{dc/2}$	0	1	0	1	0	0	1	1	0
0^+	0	1	0	1	0	0	1	0	1
0^-	1	0	1	0	0	1	0	1	0
$-V_{dc}/2$	1	0	1	0	0	1	0	0	1
$-V_{dc}$	1	0	0	1	0	0	1	0	1
$-3V_{dc}/2$	1	0	0	1	1	1	0	0	1

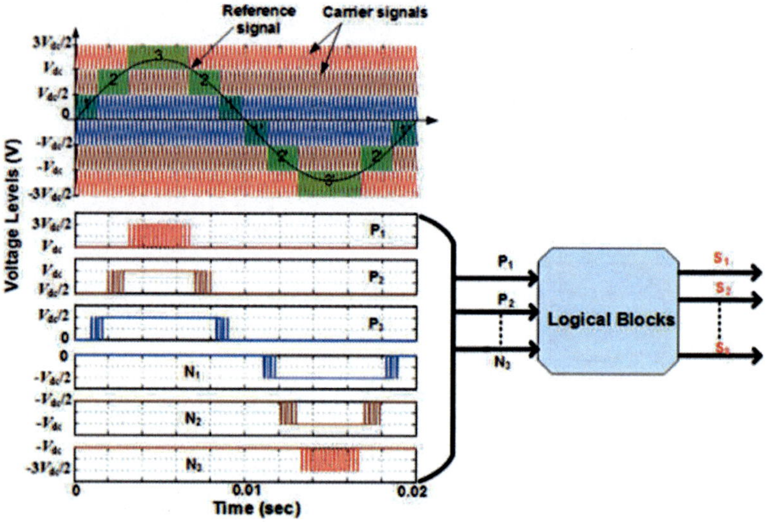

Fig. 5 Sine-triangle comparison PWM scheme

Table 2 Configuration parameters

Parameters	Values
V_{dc}	240 V
P_{output}	730 W, 335 W
V_0	230 V
I_0	3.5 A
Switching frequency (f_{sw})	4 kHz
Fundamental frequency (f_{sw})	50 Hz

$$\text{M.I.} = \frac{V_{0\,\text{peak}}}{3 \times V_{\text{dc}}} \quad (1)$$

4 Simulation Analysis

Figures 6a and 7a show the inverter output results for an M.I. of 0.9 and 0.6 respectively. The inverter is producing 7-level output at M.I. = 0.9 and five-level output at M.I. = 0.6. Due to the reduction in the value of M.I., the peak value of the output also decreases. Figures 6b and 7b illustrate the harmonic analysis of the inverter output voltage. It can be observed that when the inverter is operating at M.I. = 0.9, the magnitude of the peak value of the fundamental component of the output voltage is 323 V and the %Total Harmonic Distortion (%THD) is 22.4. It is also to be noted that the decrease in the value of M.I. increases %THD. Figures 6c and 7c illustrate the harmonic analysis of the inverter output current. It is observed that when the inverter is operating at M.I. = 0.9, the magnitude of the peak value of the fundamental component is 4.9 A and the %THD is 0.38.

5 Conclusion

This paper provides several operating modes of proposed reduced component count MLI. The proposed topology has low frequency operating switches that result in the overall improved efficiency of the system. The modulation technique to generate the firing pulses to the inverter switches has been elaborated. The simulation results at different values of M.I. and at different power ratings are presented. FFT analysis of the output parameters has been carried out.

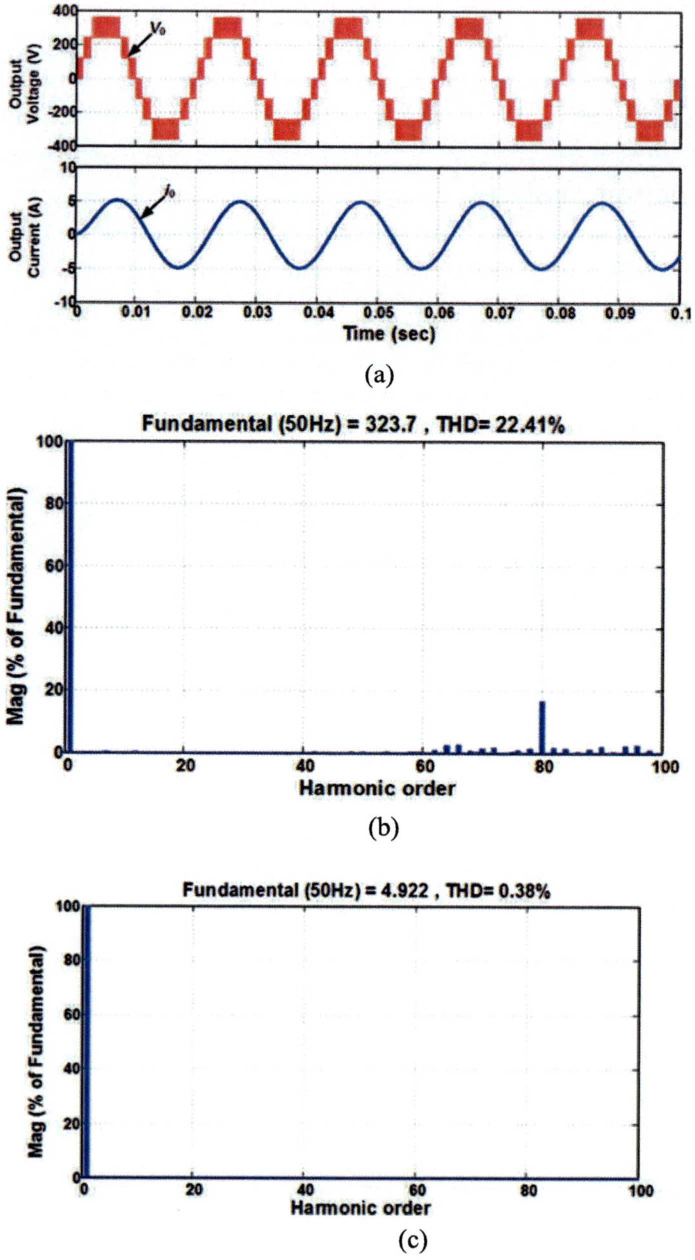

Fig. 6 Simulation outcomes at M.I. = 0.9: **a** System voltage and current waveforms at the output, **b** FFT analysis of V_0, and **c** FFT analysis of I_0

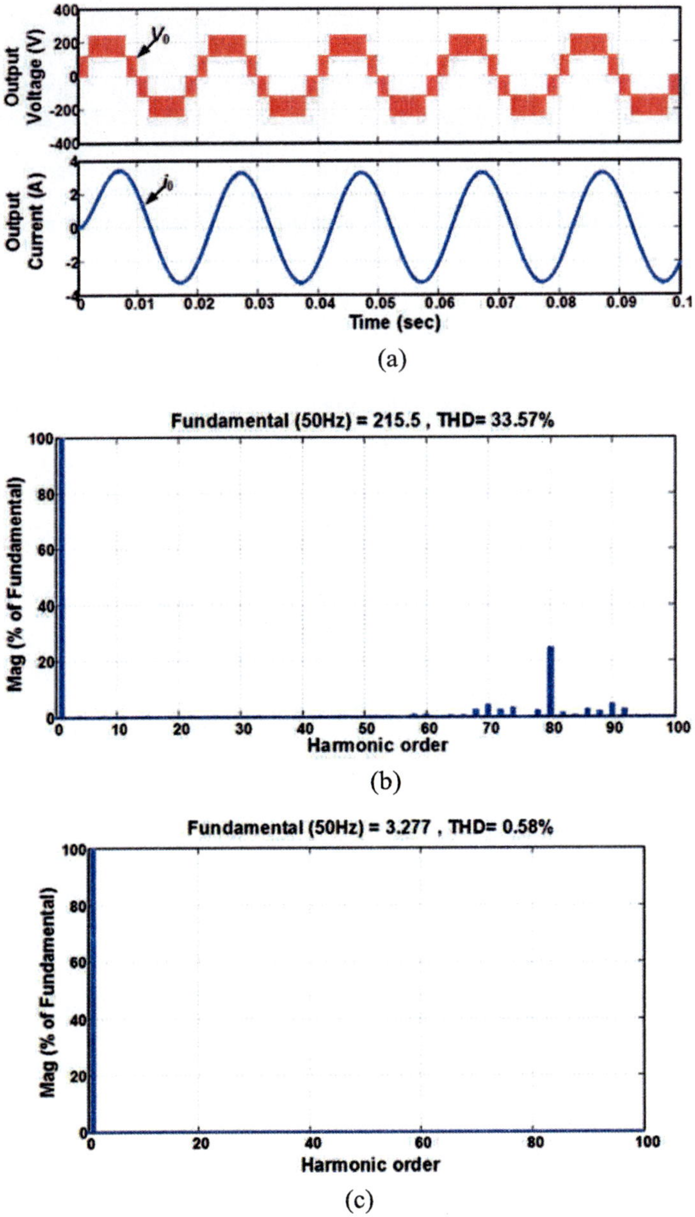

Fig. 7 Simulation outcomes at M.I.= *0.6*: **a** System voltage and current waveforms at the output, **b** FFT analysis of V_0, and **c** FFT analysis of I_0

References

1. Neugebauer, T.C., Perreault, D.J., Lang, J.H., Livermore, C.: A six-phase multilevel inverter for MEMS electrostatic induction micro motors. IEEE Trans. Circuits Syst. II Exp. Briefs **51**(2), 49–56 (2004)
2. Chan, M.S.W., Chau, K.T.: A new switched-capacitor boost multilevel inverter using partial charging. IEEE Trans. Circuits Syst. II. Exp. Briefs **54**(12), 1145–1149 (2007)
3. Nabae, A., Takahashi, I., Akagi, H.: A new neutral-point clamped PWM inverter. IEEE Trans Ind Appl **IA-17**, 518–523 (1981)
4. Lai, J.S., Peng, F.Z.: Multilevel converters—a new breed of power converters. IEEE Trans. Ind. Appl. **32**, 509–517 (1996)
5. Meynard, T.A., Foch, H.: Multi-level choppers for high voltage applications. Eur. Power Electron. Drives J. **2**(1), 45–50 (1992)
6. Abhilash, T., Kirubakaran, A., Somasekhar, V.T.: A seven-level VSI with a front-end cascaded three-level inverter and flying capacitor fed H-bridge. IEEE Trans. Ind. Appl. **55**(6), 6073–6088 (2019)
7. Ounejjar, Y., Al-Haddad, K., Dessaint, L.A.: A novel six-band hysteresis control for the packed U cells seven-level converter: experimental validation. IEEE Trans. Ind. Electron. **59**(10), 3808–3816 (2012)
8. Saeedifard, M., Barbosa, P.M., Steimer, P.K.: Operation and control of a hybrid seven-level converter. IEEE Trans. Power Electron. **27**(2), 652–660 (2012)
9. Vahedi, H., Al-Haddad, K.: Real-time implementation of a seven level packed U-cell inverter with a low-switching-frequency voltage regulator. IEEE Trans. Power Electron. **31**(8), 5967–5973 (2016)
10. Sheng, W., Ge, Q.: A novel seven-level ANPC converter topology and its commutating strategies. IEEE Trans. Power Electron. **33**(9), 7496–7509 (2018)
11. Tsunoda, A., Hinago, Y., Koizumi, H.: Level and phase shifted PWM for 7-level switched-capacitor inverter using series/parallel conversion. IEEE Trans. Ind. Electron. **61**(8), 4011–4021 (2014)
12. Wu, J.-C., Chou, C.-W.: A solar power generation system with a seven level inverter. IEEE Trans. Power Electron. **29**(7), 3454–3462 (2014)
13. Abhilash, T., Kirubakaran, A., Somasekhar, V.T.: A new structure of three-phase five-level inverter with nested two-level cells. Int. J. Circ. Theor. Appl. **47**(9), 1435–1445 (2019)
14. Abhilash, T., Kirubakaran, A., Somasekhar, V.T.: A new hybrid flying capacitor based single phase nine-level inverter. Int. Trans. Electr. Energy Syst. **29**(12), 1–15 (2019)

Study of Protection, Fault and Condition Monitoring of Induction Motor

Ch. Seshu Mohan, E. V. Madhav Kumar, K. Harikrishna, and Petta Sridhar

Abstract This paper provides an overview of different protection schemes for induction motor. The paper also analyzes the rise in temperature on the performance of motor. A systematic classification of faults of induction motor has been discussed in this paper. Fuzzy-based inference mechanism has been developed for condition monitoring of the motor considering the stator current of the motor.

Keywords Condition monitoring · Protection · Induction motor · Fuzzy inference · Stator current

1 Introduction

There are different methods of condition monitoring of an induction motor including monitoring acoustic emission and current signature analysis, etc. These monitoring techniques except motor current signature analysis technique require high precision sensors, DAQ cards and sophisticated signal processing analysis procedures. Different signal processing techniques are implemented on the raw data to extract vital clues and estimate the probable faults in the machine. To automate the diagnostic process, recently, a number of soft computing techniques have been proposed.

The condition monitoring system for induction machine has been reviewed by [6–8]. Online protection of induction machine has been reviewed by [1, 4]. Nowa-

Ch. Seshu Mohan · E. V. Madhav Kumar · K. Harikrishna · P. Sridhar (✉)
Department of Electrical and Electronics Engineering, Aditya College of Engineering and Technology, Surampalem, Andhra Pradesh, India
e-mail: pmahima.sri@gmail.com

Ch. Seshu Mohan
e-mail: seshumohan205@gmail.com

E. V. Madhav Kumar
e-mail: vigneshmadhavkumar2000@gmail.com

K. Harikrishna
e-mail: kalaharikrishna@gmail.com

S. Borah et al. (eds.), *Soft Computing Techniques and Applications*, Advances in Intelligent Systems and Computing 1248,
https://doi.org/10.1007/978-981-15-7394-1_11

days, thermal imaging-based system is widely used for fault detection and condition monitoring [2, 3]. In [5], fuzzy entropy has been used to detect multiple faults. This paper provides an overview of different faults and related protection scheme of induction motor. Different condition monitoring techniques have been discussed in this paper. Section 2 provides a basic idea of induction motor, Sect. 3 provides details of different protection of induction motor, Sect. 4 shows the faults of induction motor and Sect. 5 provides maintenance and condition monitoring system of induction motor. Section 6 concludes the paper.

2 Induction Motor: Preliminary Study

Figure 1 provides a per-phase equivalent circuit diagram of induction motor.

Slip of induction motor can be represented as

$$s = \frac{N_s - N}{N_s} \tag{1}$$

Consider a motor with the following parameters:

- Type of motor: Wound rotor induction motor
- Number of phases: Three-phase connection Y connected
- Power rating: 15 hp
- Number of poles: 2
- R_1: 0.2 Ω
- R_2: 0.12 Ω
- X_m: 15 Ω
- X_l: 0.41 Ω
- Mechanical power: 250 W
- Core loss: 180 W
- Slip: 0.05
- Voltage: 208 V.

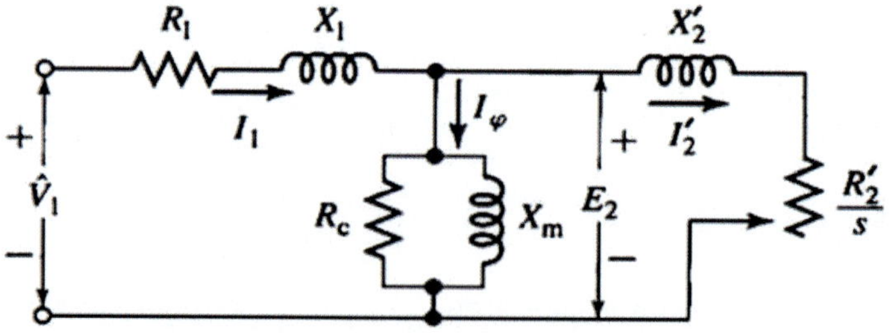

Fig. 1 Per-phase circuit of IM [9]

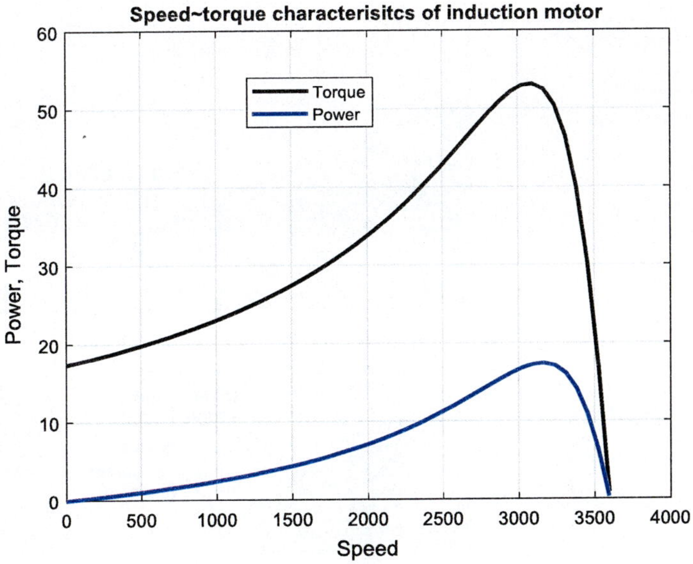

Fig. 2 Speed-torque characteristics of induction motor

The torque and power of IM are illustrated in Fig. 2.

$$Z_f = \frac{1}{\frac{1}{jX_m} + \frac{1}{Z_2}} \tag{2}$$

$$I_L = I_a = \frac{V}{R_1 + jX_1 + R_f + jX_f} \tag{3}$$

2.1 Temperature Effect

With change in ambient temperature, the performance of induction motor also changes. Figure 3 shows the efficiency of motor with change in temperature.

Figure 4 shows different torque region and speed variations in motor.

3 Protection of Induction Motor

There are different electrical protections for electrical motor. Some of the important protection are

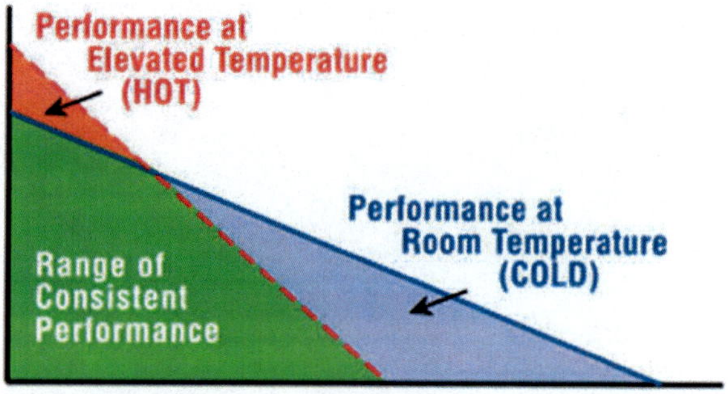

Fig. 3 Motor performance at room temperature and elevated temperature

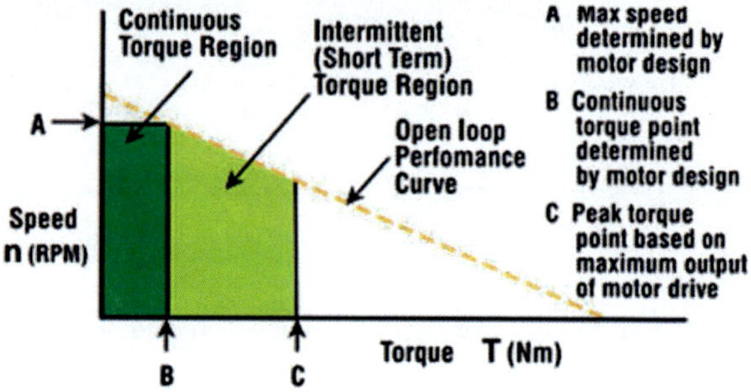

Fig. 4 Different torque region and speed variations

- Thermal overload

(a) Excessive load
(b) High ambient condition
(c) Voltage and current harmonics

- Phase fault
- Ground fault
- Abnormal operating principle

(a) Over and under voltage
(b) Under frequency
(c) Voltage and current unbalance
(d) Load loss
(e) Jamming
(f) Jogging.

3.1 Ground Fault Protection

Figure 5 shows the ground fault protection of induction motor where all the phase conductors are passed through the window of the same current transformer (CT) which is referred to a zero sequence CT. If any of the phases is shorted to ground,

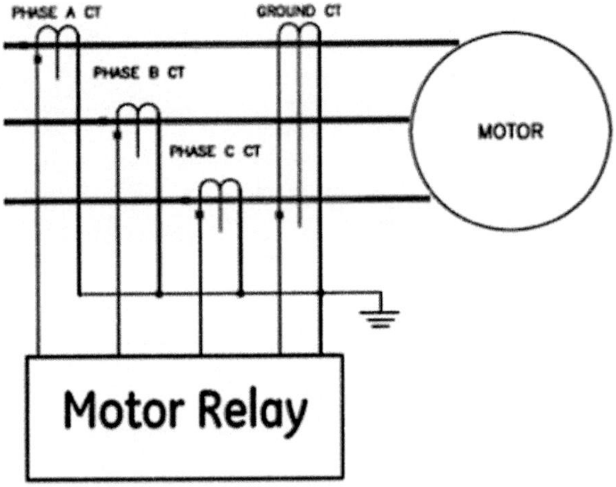

Fig. 5 Ground fault protection of motor

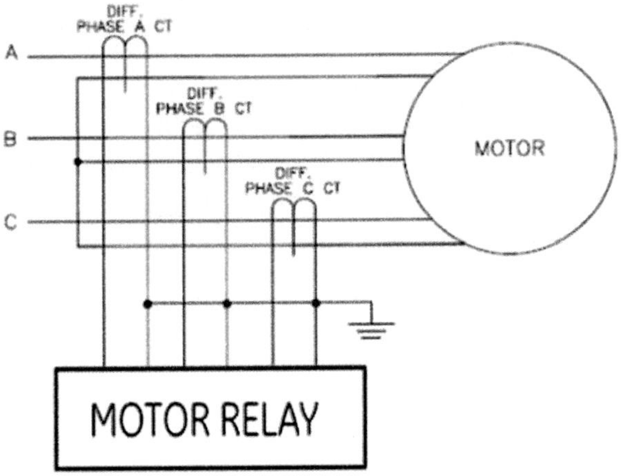

Fig. 6 Differential protection of induction motor

then sum of phase current will not be zero which causes flow of current in the secondary of the zero sequence CT. The current will actuate the motor relay.

3.2 Differential Protection

Differential protection scheme is one of the most widely used techniques for induction motor protection scheme. In Fig. 6, differential protection of induction motor has been shown. The differential protection eliminates the phase-to-phase fault or phase-to-ground fault. In this case, three current transformers (CTs) are used.

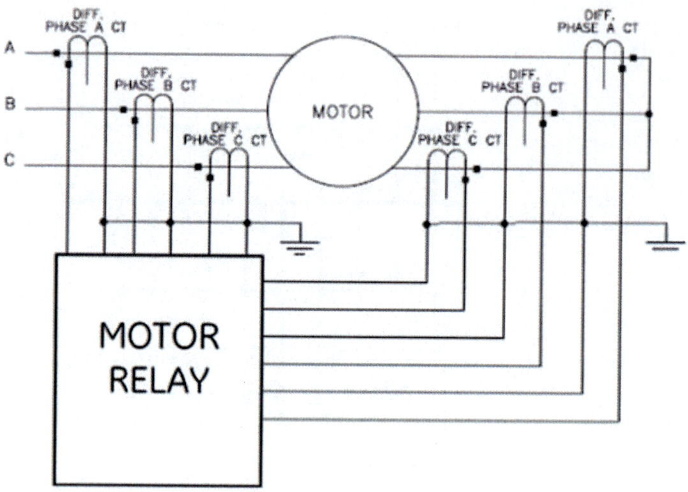

Fig. 7 Biased differential protection of induction motor

3.3 *Biased Differential Protection*

Figure 7 shows the biased differential protection of induction motor. In this case, six current transformers are used unlinke three current transformers in differential protection. Figure 8 shows the protection scheme for three-phase induction motor where overload relay, start and stop button are provided. Fuse is provided in three-phase line.

4 Faults in Induction Motor

Basically, the faults occurring in induction machine are divided in to two parts, i.e., faults due to internal factors and faults due to external factors. Some of the internal faults are eccentricity fault, broken rotor bar fault, bearing fault, dielectric failures, etc. Faults can occur in induction machines due to some external factors like high temperature, high humidity, overload, unbalanced voltage, transient voltage, voltage fluctuations, etc. According to IEEE and EPRI (Electric Power Research Institute) survey results, the faults and losses of induction motor are summarized below (Figs. 9 and 10).

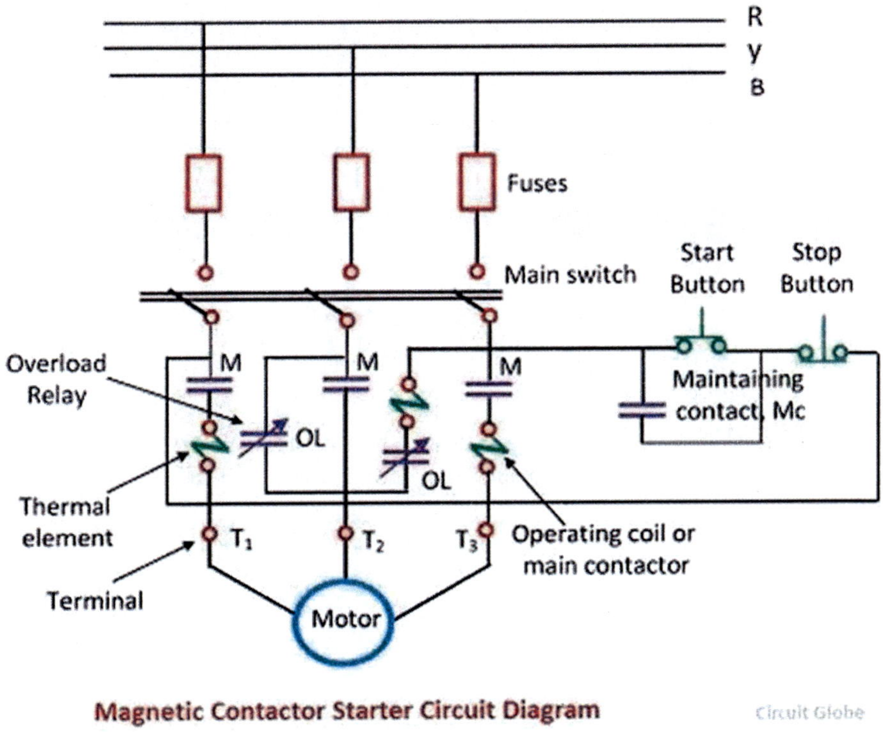

Fig. 8 Protection scheme for three-phase induction motor

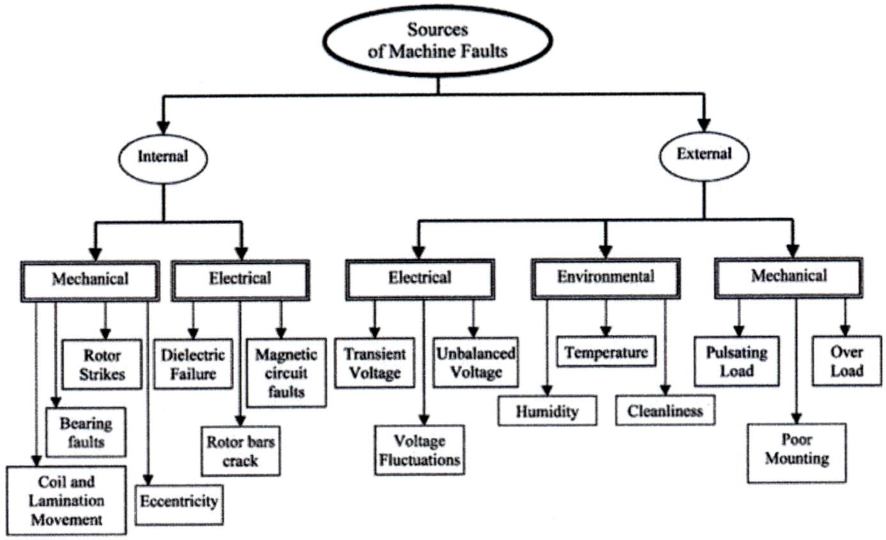

Fig. 9 Classification of faults in induction machine

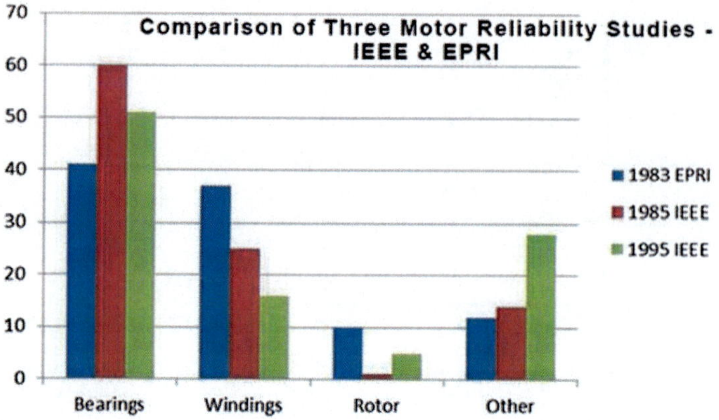

Fig. 10 IEEE and EPRI study about motor failures

5 Condition Monitoring of IM

Traditionally, maintenance procedures in industry follow two approaches

1. Preventive maintenance
2. Breakdown maintenance

In preventive maintenance, maintenance operation is carried out in a regular interval of time. In breakdown maintenance, maintenance is carried out when the breakdown occurs. However, nowadays, with implementation of different scientific approaches, predictive maintenance is possible. In predictive maintenance, different signal processing approaches are used to detect the early signs of defect, and suitable actions are taken before any kind of catastrophic failure occurs.

Figure 11 shows the overall system diagram of condition monitoring system of induction motor where three-phase voltage and three-phase current are measured. Hall effect sensors are used for current and voltage measurement. Signal conditioning circuit is used for interfacing the measurement with PC. Vibration signal is measured using accelerometer. Signal conditioning circuit and data acquisition (DAQ) unit are used to interface the signal from the motor to the personal computer.

5.1 Fault Diagnosis Method

Different fault diagnosis methods are used to detect different faults of induction motor. These techniques are summarized in Fig. 12.

There are basically three techniques for predictive maintenance, namely model-based technique, signal processing techniques and soft computing technique. Signal processing technique is widely used to detect faults in induction machine. Different

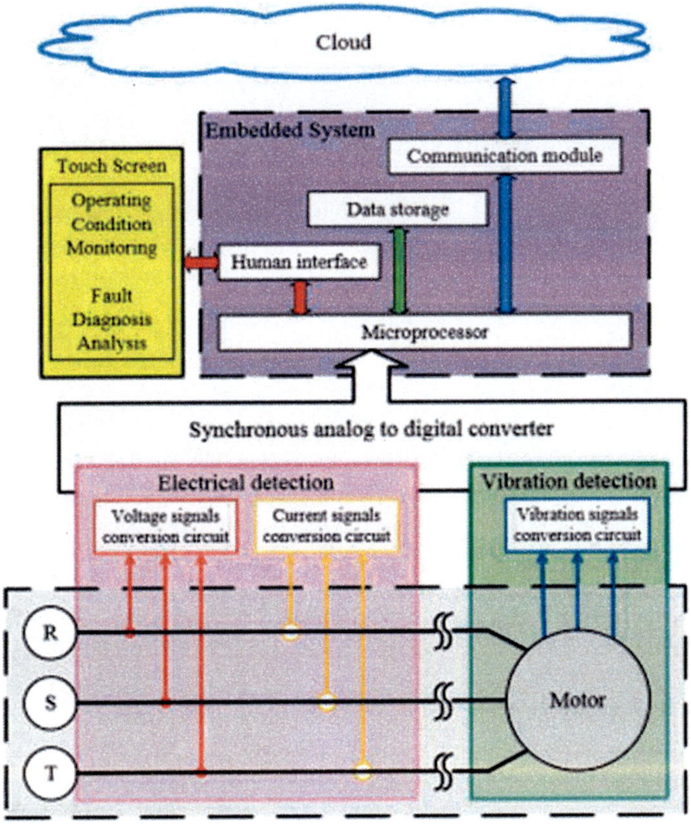

Fig. 11 Overall system diagram for condition monitoring system of induction motor

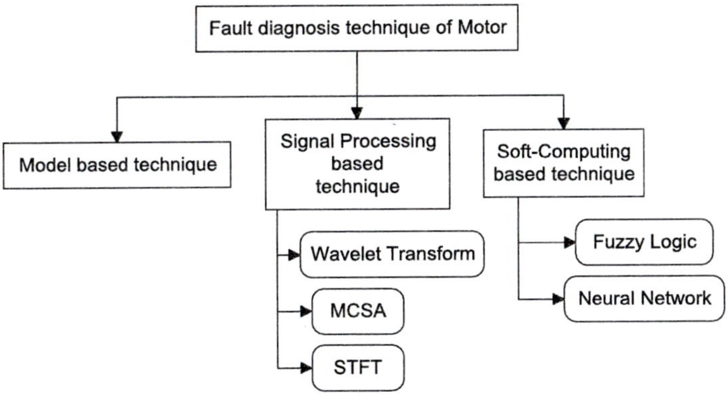

Fig. 12 Induction motor fault diagnosis technique

signal processing algorithms are applied in raw sensor data to extract valuable information, which then provides vital clue regarding the health of the machine. Some of the signal processing techniques widely used for fault detections are wavelet transform, Fourier transform, short-time Fourier transform, motor current signature analysis, power spectral density, etc. Soft computing techniques like neural network and fuzzy logic can also be used for fault detection in induction motor. Different supervised and unsupervised classification techniques like support vector machines and self-organizing maps, respectively, can be implemented in raw sensor data to ascertain the health of the machine.

6 Conclusion

This paper provides an overview of different fault and protection schemes of induction motor and also provides details of different techniques by which condition monitoring of the motor can be achieved. Different signal processing-based feature extraction methods have been discussed.

Acknowledgements The authors are thankful to Accendere Knowledge Management Services Pvt. Ltd, CL Educate Ltd for their assistance during the preparation of the manuscript.

References

1. Colak, I., Celik, H., Sefa, I., Demirbaş, Ş.: On line protection systems for induction motors. Energy Convers. Manage. **46**(17), 2773–2786 (2005)
2. Glowacz, A., Glowacz, Z.: Diagnosis of the three-phase induction motor using thermal imaging. Infrared Phys. Technol. **81**, 7–16 (2017)
3. Janssens, O., Loccufier, M., Van Hoecke, S.: Thermal imaging and vibration-based multisensor fault detection for rotating machinery. IEEE Trans. Industr. Inf. **15**(1), 434–444 (2018)
4. Patel, S.P., Tseng, S., Weeks, K.: Protection of motors: examples of setting with and without complete data. IEEE Ind. Appl. Mag. **20**(6), 64–78 (2014)
5. Romero-Troncoso, R.J., Saucedo-Gallaga, R., Cabal-Yepez, E., Garcia-Perez, A., Osornio-Rios, R.A., Alvarez-Salas, R., Miranda-Vidales, H., Huber, N.: FPGA-based online detection of multiple combined faults in induction motors through information entropy and fuzzy inference. IEEE Trans. Industr. Electron. **58**(11), 5263–5270 (2011)
6. Singh, G.: Multi-phase induction machine drive research: a survey. Electr. Power Syst. Res. **61**(2), 139–147 (2002)
7. Singh, G., et al.: Experimental investigations on induction machine condition monitoring and fault diagnosis using digital signal processing techniques. Electr. Power Syst. Res. **65**(3), 197–221 (2003)
8. Singh, G., et al.: Induction machine drive condition monitoring and diagnostic researcha survey. Electr. Power Syst. Res. **64**(2), 145–158 (2003)
9. Syal, A., Gaurav, K., Moger, T.: Virtual laboratory platform for enhancing undergraduate level induction motor course using matlab/simulink. In: 2012 IEEE International Conference on Engineering Education: Innovative Practices and Future Trends (AICERA), pp. 1–6. IEEE, New York (2012)

Analysis of Inverter Topologies and Controller Schemes in Grid-Connected Photovoltaic Module

Dammala Naveena, A. S. S. V. Lakshmi, and S. Reddy Ramesh

Abstract The present study provides modeling and simulation of grid-connected PV-fed voltage source inverter and analyzes the working principle of the grid-connected PV-fed inverter along with H5 inverter. A detailed circuit analysis along with simulation results has been provided. Fuzzy-based control for grid-connected inverter has been discussed. The fuzzy-based control provides proper control and outperforms other classical controller scheme.

Keywords PV system · H5 inverter · Grid synchronization · Fuzzy-based control · Inverter

1 Introduction

Among different renewable energy systems, photovoltaic is one of the best because of several reasons and features. There is plenty of sunlight all around during daytime; therefore, PV system can generate electricity from the sunlight directly. Solar panel is made up of semiconductor material, and it works fine but it has low efficiency. The solar panel provides DC output which needs to be stepped up and converted to AC so utility load can be connected to the inverter. The circuit analysis and modeling of PV panel has been illustrated in [2, 7, 10].

Low-power photovoltaic application which requires grid connectivity does not usually use transformer at the grid side as it is bulky and costly and it changes the overall dynamics of power system. Due to non-availability of transformer, passive fil-

D. Naveena · A. S. S. V. Lakshmi · S. Reddy Ramesh (✉)
Department of Electrical and Electronics Engineering, Aditya College of Engineering and Technology, Surampalem, Andhra Pradesh, India
e-mail: ramesh516269@gmail.com

D. Naveena
e-mail: naveenadammala@gmail.com

A. S. S. V. Lakshmi
e-mail: sathyasvl236@gmail.com

S. Borah et al. (eds.), *Soft Computing Techniques and Applications*, Advances in Intelligent Systems and Computing 1248,
https://doi.org/10.1007/978-981-15-7394-1_12

ters are used which eliminate the inverter harmonics. In [4, 9], high-speed maximum power point tracking (MPPT) scheme has been elaborated. A different converter topology for PV panel is provided in [6]. DC microgrid for PV panel is discussed in [3]. Battery-connected PV panel experimental setup has been shown in [5]. There are different full-bridge inverter topologies which are used in PV-based application. Standard VSI topology is H4, and modified VSI topology is H5. Current control of grid-connected inverter scheme has been provided in [1]. A comparative analysis of a different inverter topology has been shown in [8].

This paper provides a current controller scheme of full-bridge VSI. Circuit analysis of H5 inverter also has been provided.

2 Grid-Integrated Photovoltaic System

Figure 1 illustrates the grid-integrated photovoltaic system.

2.1 Modeling of PV System

The equivalent circuit diagram of an ideal single-diode PV cell is illustrated in Fig. 2.

$$I = I_{sc} - I_o\left[e^{\frac{qV_d}{kT}} - 1\right] \tag{1}$$

$$I = I_{pv} - I_o\left[e^{\frac{q(V+IR_s)}{nkT}} - 1\right] - \frac{V + R_s I}{R_p} \tag{2}$$

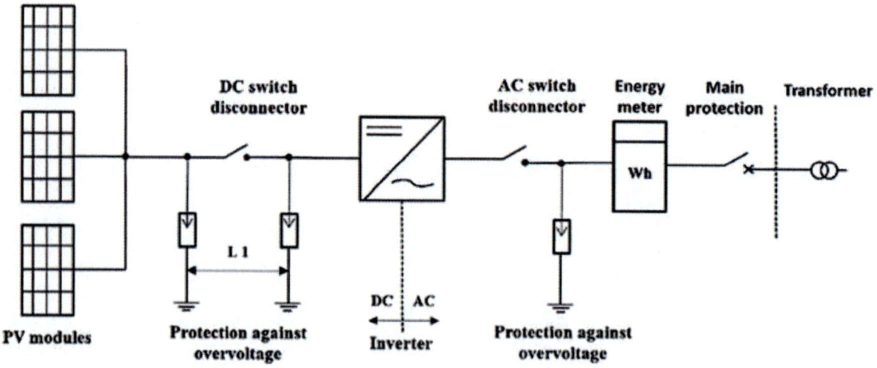

Fig. 1 Grid-integrated photovoltaic system

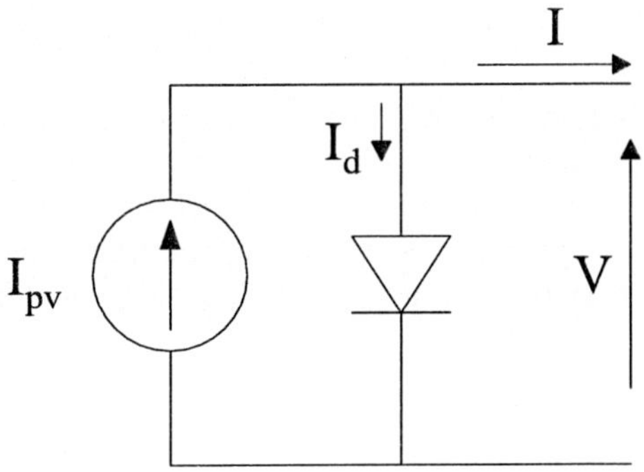

Fig. 2 Equivalent circuit diagram of PV cell

At any given temperature, the short-circuit current can be represented as

$$I_{sc} = I_{sc}\left[1 + K_i\left(T - T_{ref}\right)\right] \tag{3}$$

The photon-generated current also varies based on solar irradiance (G) represented by

$$I_{sc,G} = I_{sc,G_{ref}}\left(\frac{G}{G_{ref}}\right) \tag{4}$$

$$I_{pv} = (I_{sc} + K\Delta T)\left(\frac{G}{G_n}\right) \tag{5}$$

The final expression for PV-based system can be represented as

$$I_o = I_{o,T_{ref}}\left(\frac{T}{T_{ref}}\right)^{\frac{3}{n}} e^{-\left[\frac{qV}{nk}\left(\frac{1}{T} - \frac{1}{T_{ref}}\right)\right]} \tag{6}$$

2.2 Grid Connection Compliance

See Table 1.

Table 1 IEEE standards for grid connection

IEEE 1547		IEC 6127		VDE 0126-1-1	
Voltage range	Disconnection time	Voltage range	Disconnection time	Voltage range	Disconnection time
$V < 50$	0.16 s	$V < 50$	0.1 s	$110 < V < 135$	2 s
$50 < V < 88$	1 s	$50 < V < 85$	2 s		
$110 < V < 120$	1 s	$110 < V < 135$	2 s		

3 Inverter Topologies

In Fig. 3, a PV source is connected to conventional full-bridge inverter, namely H4 inverter.

In Fig. 4, a modified inverter topology, namely H5 inverter, is shown.

Figure 5a–d shows the operation mode of PV-fed full-bridge grid-connected H5 inverter. The current direction is shown in the operating mode.

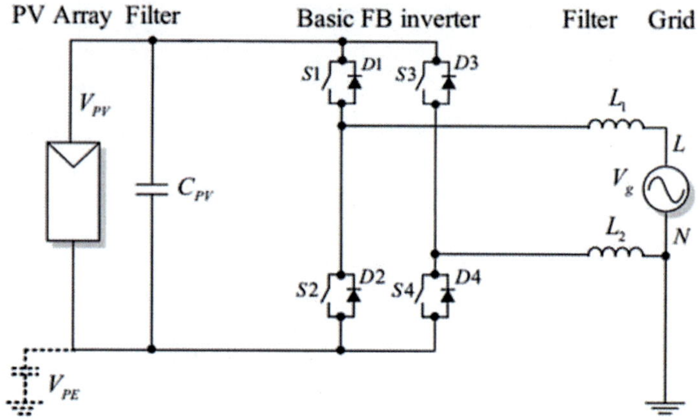

Fig. 3 Full-bridge VSI connected to PV

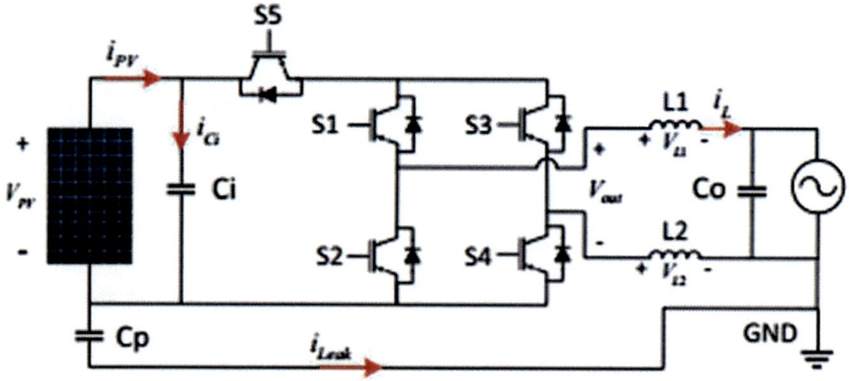

Fig. 4 H5 voltage source inverter connected to PV

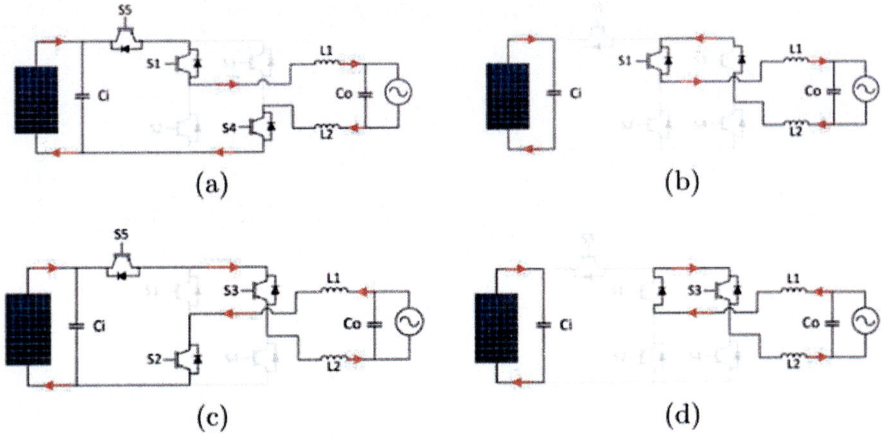

Fig. 5 Operating modes of H5 inverter

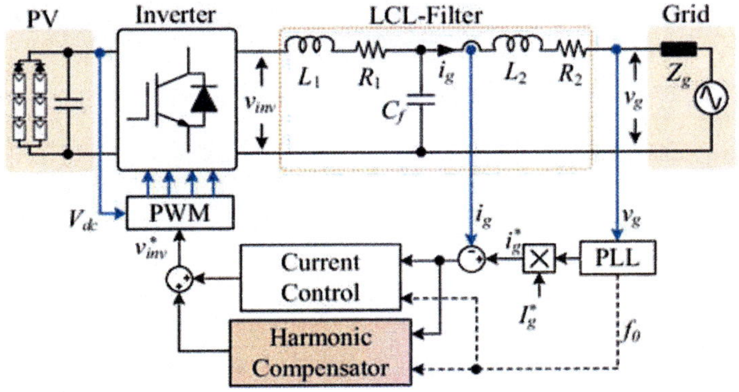

Fig. 6 Controller scheme of grid-connected PV inverter

3.1 *Fuzzy Control of Grid-Connected PV Inverter*

Figure 6 illustrates the generalized controller scheme for PV-based inverter. Fuzzy-based controller has been developed, and controller scheme is provided.

4 Simulation Results

BP solar SX series PV panel (150 W) has been considered for the simulation. Figure 7 illustrates the P–V and V–I characteristics of a single PV module at different ambient temperatures. For simulation, multiple PV modules are used. Table 2 presents

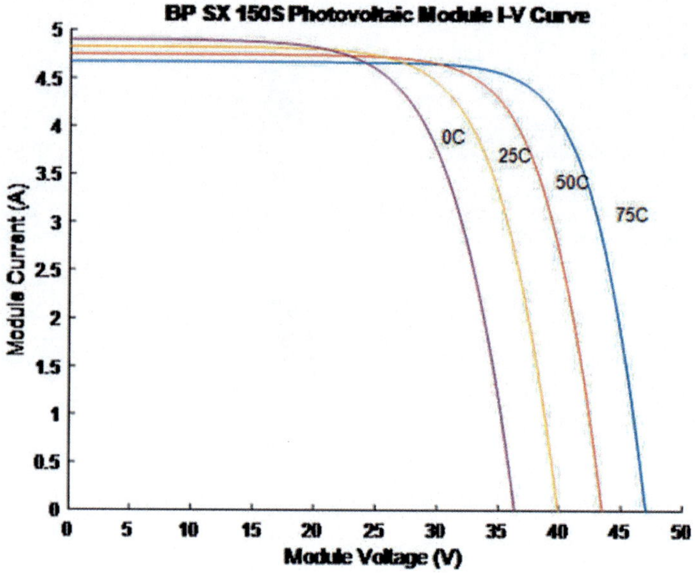

Fig. 7 *P–V* and *V–I* characteristics of 150 *W* PV panel at a different *T*

Table 2 Parameters of inverter

Parameters	Values
V_{dc}	300 V
f_{sw}	20 kHz
L_f	10 mH
C_f	20 μF

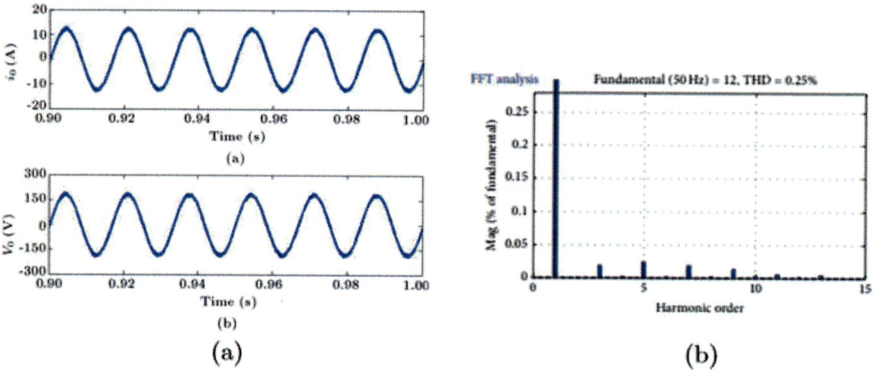

Fig. 8 Inverter results in steady state **a** V_{inv} and I_{inv}, **b** %THD

the inverter parameters. Figure 8a presents the voltage and current profile of grid-connected voltage source inverter, whereas Fig. 8b presents the total harmonic distortion of current waveform.

5 Conclusion

This paper provides a detailed analysis of PV-fed grid-connected inverter. Operation of different inverter topologies such as H4 and H5 inverters has been discussed. The grid-connected inverter is controlled using d-q transform and deregulated fuzzy-based controller. Simulation results have been provided which validate the theoretical analysis, and it is shown that fuzzy-based controller outperforms classical PI controller.

Acknowledgements The authors are thankful to Accendere Knowledge Management Services Pvt. Ltd and CL Educate Ltd for their assistance during the preparation of the manuscript.

References

1. Algaddafi, A., Altuwayjiri, S.A., Ahmed, O.A., Daho, I.: An optimal current controller design for a grid connected inverter to improve power quality and test commercial PV inverters. Sci. World J. **2017** (2017)
2. Bhuvaneswari, G., Annamalai, R.: Development of a solar cell model in Matlab for PV based generation system. In: 2011 Annual IEEE India Conference, pp. 1–5. IEEE, New York (2011)
3. Ingale, G.B., Padhee, S., Pati, U.C.: Design of stand alone PV system for dc-micro grid. In: 2016 International Conference on Energy Efficient Technologies for Sustainability (ICEETS), pp. 775–780. IEEE, New York (2016)
4. Mohanty, M., Selvakumar, S., Koodalsamy, C., Simon, S.P.: Global maximum operating point tracking for PV system using fast convergence firefly algorithm. Turkish J. Electr. Eng. Comput. Sci. **27**(6), 4640–4658 (2019)
5. Mohapatra, D., Padhee, S., Jena, J.: Design of solar powered battery charger: an experimental verification. In: 2018 IEEE International Students' Conference on Electrical, Electronics and Computer Science (SCEECS), pp. 1–5. IEEE, New York (2018)
6. Padhee, S., Pati, U.C., Mahapatra, K.: Overview of high-step-up dc-dc converters for renewable energy sources. IETE Tech. Rev. **35**(1), 99–115 (2018)
7. Rahman, S.A., Varma, R.K., Vanderheide, T.: Generalised model of a photovoltaic panel. IET Renew. Power Gener. **8**(3), 217–229 (2014)
8. Rizzoli, G., Mengoni, M., Zarri, L., Tani, A., Serra, G., Casadei, D.: Comparison of single-phase H4, H5, H6 inverters for transformerless photovoltaic applications. In: IECON 2016-42nd Annual Conference of the IEEE Industrial Electronics Society, pp. 3038–3045. IEEE, New York (2016)
9. Selvakumar, S., Madhusmita, M., Koodalsamy, C., Simon, S.P., Sood, Y.R.: High-speed maximum power point tracking module for PV systems. IEEE Trans. Industr. Electron. **66**(2), 1119–1129 (2018)
10. Villalva, M.G., Gazoli, J.R., Ruppert Filho, E.: Comprehensive approach to modeling and simulation of photovoltaic arrays. IEEE Trans. Power Electron. **24**(5), 1198–1208 (2009)

Aerodynamic Modelling and Analysis of Wind Turbine

K. Giridhar, S. J. Venkata Aravind, and Sana Vani

Abstract This paper provides an fuzzy-based aerodynamic modelling approach for wind turbine. A comparative analysis of different wind turbine and aerodynamic simulation results has been provided in this paper. Different fuzzy-based modeling schemes have been highlighted which provides major performance boost in fault detection, health monitoring, and various other related applications.

Keywords Aerodynamic model · Wind turbine · Fuzzy modeling · Fault diagnosis · Health monitoring

1 Introduction

Accurate mathematical modeling of different parts of WECS is one of the most interesting research area. Modeling of wind turbine helps us to understand the utility of WECS. Review of wind turbine is available at [9]. Performance analysis of vertical axis wind turbine has been discussed in [4, 6]. Modeling and analysis of direct-driven PMSG-based wind turbine system have been discussed in [1]. Energy efficiency analysis of small wind turbine has been discussed in [5]. Design-oriented aerodynamic modeling of wind turbine has been discussed in [2]. Aerodynamic modeling and aerodynamic measurement of wind turbine have been discussed in [3, 7], respectively. Aerodynamic-based model of wind turbine and related testing has been discussed in [8].

K. Giridhar · S. J. Venkata Aravind · S. Vani (✉)
Department of Electrical and Electronics Engineering, Aditya College of Engineering and Technology, Surampalem, Andhra Pradesh, India
e-mail: sanavanimtech@gmail.com

K. Giridhar
e-mail: giridhar791@gmail.com

S. J. Venkata Aravind
e-mail: sjvaravind5247@gmail.com

S. Borah et al. (eds.), *Soft Computing Techniques and Applications*, Advances in Intelligent Systems and Computing 1248,
https://doi.org/10.1007/978-981-15-7394-1_13

Fig. 1 Blade spatial distribution

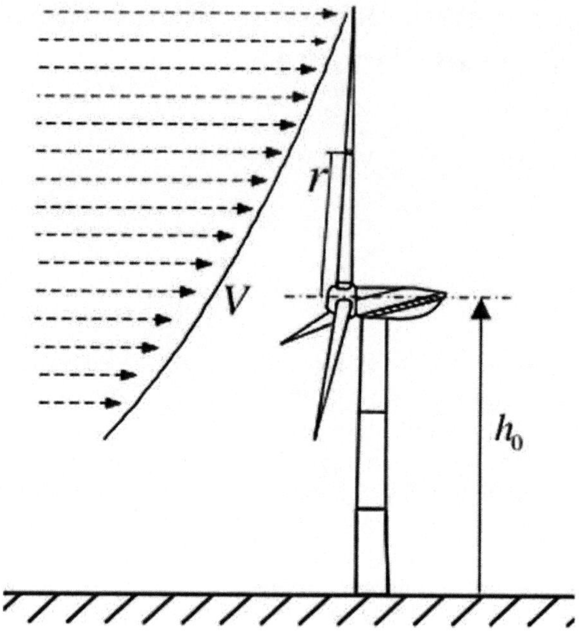

This paper provides fuzzy-based aerodynamic modelling of wind turbine system which is useful for wind energy conversion system.

2 Wind Rotor Aerodynamic Model

The power extracted from turbine is

$$P = 0.5C_p\rho\pi R^2V^3 \tag{1}$$

where

$$C_p(\lambda, \beta) = 0.22\left(\frac{116}{\lambda_i} - 0.4\beta - 5\right)e^{-\frac{12.5}{\lambda_i}} \tag{2}$$

$$\frac{1}{\lambda_i} = \frac{1}{\lambda + 0.08\beta} - \frac{0.035}{\beta^3 + 1} \tag{3}$$

$$\begin{cases} dL = 0.5\rho C_l V_1^2 c dr \\ dD = 0.5\rho C_d V_1^2 c dr \end{cases} \tag{4}$$

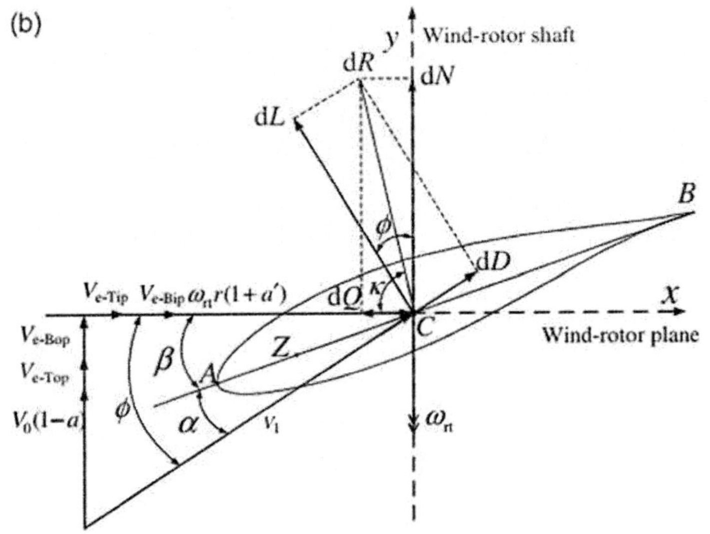

Fig. 2 Airfoil dynamics

$$\begin{cases} dN = 0.5\rho c V_1^2 \left(C_l \cos\phi + C_d \sin\phi\right) dr \\ dQ = 0.5\rho C_d V_1^2 \left(C_l \sin\phi - C_d \cos\phi\right) dr \end{cases} \tag{5}$$

Total blade rotor bending moment can be represented as

$$M_y = \int_0^R dM_y = \int_0^R f_1(r)\, dr \tag{6}$$

Bending moment of blade rotor in flap-wise direction

$$M_y = \int_0^R dM_y = \int_0^R f_1(r)\, dr \tag{7}$$

The aerodynamic force pitch load can be written as

$$M_z = \int_0^R dM_z \tag{8}$$

$$\dot{m} = \rho S_1 V_1 = \rho S_2 V_2 = \text{constant} \tag{9}$$

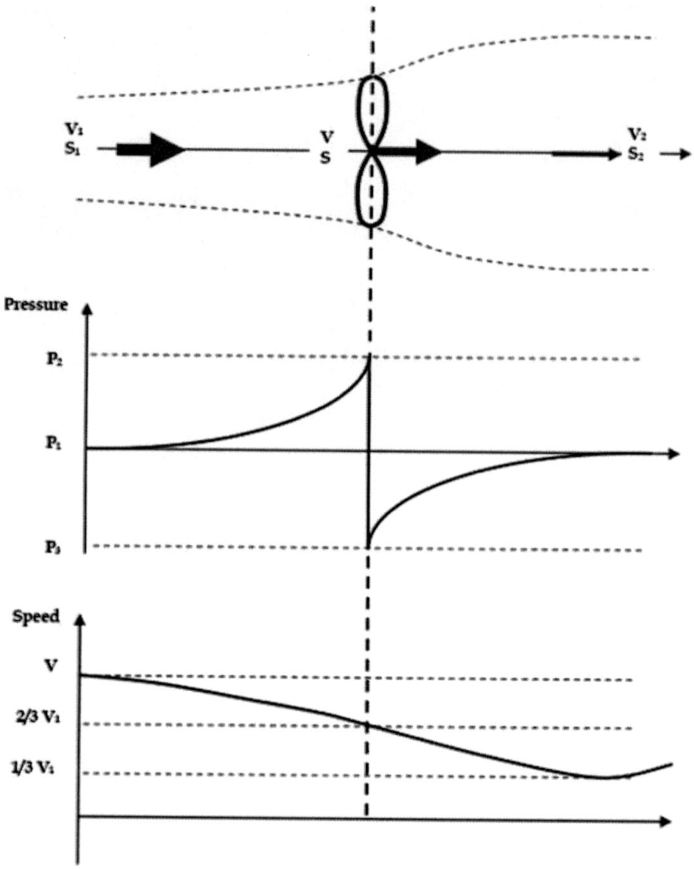

Fig. 3 Variation of pressure and speed in wind turbine

Force exserted is

$$F = m\frac{dV}{dt} = \rho SV\,(V_1 - V_2) \tag{10}$$

The incremental work done in the wind turbine can be represented as

$$dE = Fdx \tag{11}$$

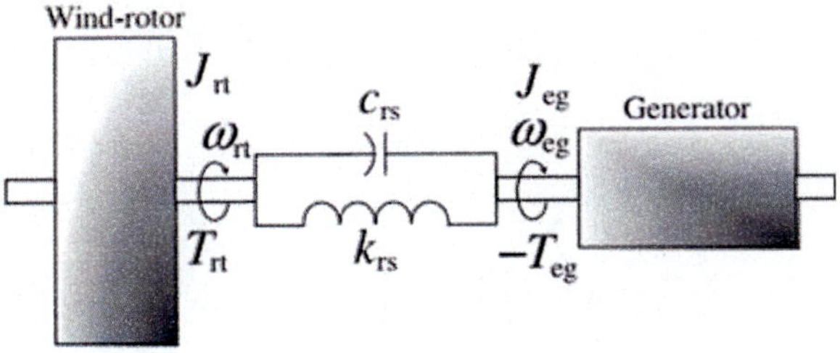

Fig. 4 Model of wind rotor and generator

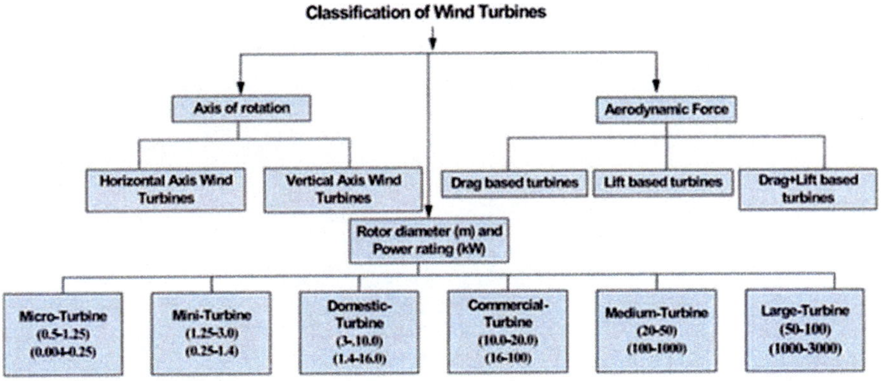

Fig. 5 Classifications of wind turbines

Table 1 Performance classification of wind turbine

Diameter	Height	kWh per year
1–13 m	18–37 m	20,000
13–30 m	35–60 m	600,000
47–90 m	50–80 m	4,000,000

The power content of the wind turbine can be represented as

$$P = \frac{dE}{dt} = F\frac{dx}{dt} = FV \tag{12}$$

$$\begin{aligned} T_{rt} &= J_{rt}\frac{d\omega_{rt}}{dt} + c_{rs}\left(\omega_{rt} - \omega_{eg}\right) + k_{rs}\left(\theta_{rt} - \theta_{eg}\right) \\ T_{eg} &= J_{eg}\frac{d\omega_{eg}}{dt} + c_{rs}\left(\omega_{eg} - \omega_{rt}\right) + k_{rs}\left(\theta_{eg} - \theta_{rt}\right) \end{aligned} \tag{13}$$

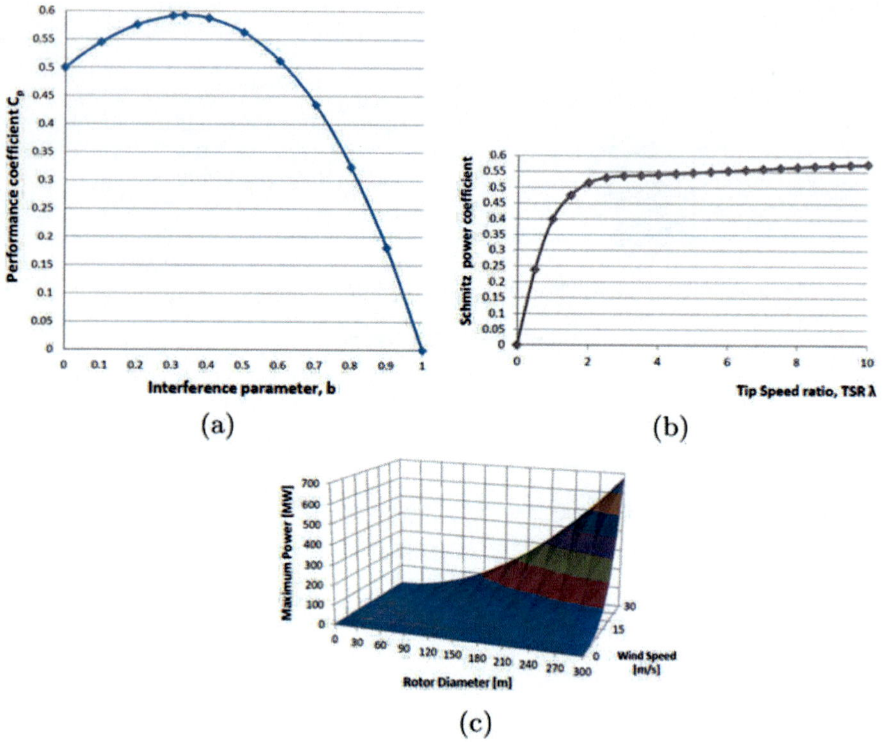

Fig. 6 **a** Graph between interference parameter and performance coefficient, **b** Graph between schimitz power coefficient and tip speed ratio, **c** 3D graph between rotor diameter, wind speed, and maximum power

3 Simulation Results

Figure 6a illustrates the graph between interference parameter and performance coefficient.

4 Conclusion

This paper provides an aerodynamic modelling and simulation studies for wind turbine system. A comparative analysis of different wind turbines have been discussed in this paper. Simulation results have been provided in this paper.

Acknowledgements The authors are thankful to "Accendere Knowledge Management Services Pvt. Ltd, CL Educate Ltd" for assistance during the preparation of the manuscript.

References

1. Dai, J., Hu, Y., Liu, D., Wei, J.: Modelling and analysis of direct-driven permanent magnet synchronous generator wind turbine based on wind-rotor neural network model. Proc. Inst. Mech. Eng. Part A: J. Power Energy **226**(1), 62–72 (2012)
2. Greco, L., Testa, C., Salvatore, F.: Design oriented aerodynamic modelling of wind turbine performance. J. Phys.: Conf. Ser. **75**, p. 012011 IOP Publishing (2007)
3. Hand, B., Cashman, A.: Aerodynamic modeling methods for a large-scale vertical axis wind turbine: a comparative study. Renew. Energy **129**, 12–31 (2018)
4. Jain, P., Abhishek, A.: Performance prediction and fundamental understanding of small scale vertical axis wind turbine with variable amplitude blade pitching. Renew. Energy **97**, 97–113 (2016)
5. Mirecki, A., Roboam, X., Richardeau, F.: Architecture complexity and energy efficiency of small wind turbines. IEEE Trans. Industr. Electron. **54**(1), 660–670 (2007)
6. Refan, M., Hangan, H.: Aerodynamic performance of a small horizontal axis wind turbine. J. Solar Energy Eng. **134**(2) (2012)
7. Snel, H.: Review of aerodynamics for wind turbines. Wind Energy: Int. J. Progress Appl. Wind Power Conversion Technol. **6**(3), 203–211 (2003)
8. Sunny, K.A., Kumar, N.M.: Vertical axis wind turbine: aerodynamic modelling and its testing in wind tunnel. Proc. Comput. Sci. **93**, 1017–1023 (2016)
9. Tummala, A., Velamati, R.K., Sinha, D.K., Indraja, V., Krishna, V.H.: A review on small scale wind turbines. Renew. Sustain. Energy Rev. **56**, 1351–1371 (2016)

Suicidal Intent Prediction Using Natural Language Processing (Bag of Words) Approach

Ononuju Adaihuoma Chidinma, Samarjeet Borah, and Ranjit Panigrahi

Abstract The Bag of Words (BOW) technique is a Natural Language Processing technique that is applied to learn from the text. The basic idea of the Bag of Words is to take any text and count the frequency of words and place those unique words in a dictionary. This research is carried out to predict the causes of suicide in Nigeria by using the Bag of Words (BOW) technique together with three machine learning algorithms. The logistic regression algorithm, Random Forest classifier, and XGBoost classifier are the machine learning algorithms considered for this dataset. For feature extraction, the TfidfVectorizer has been used. The result gotten shows that the logistic regression classifier is suitable for the dataset used in this study, therefore, for similar datasets, the logistic regression together with TfidfVectorizer can be used. However, the logistic regression gave an accuracy score of 89% leaving an 11% chance of misclassification.

Keywords Suicide · Nigeria · XGBoost · Bag of Words · Machine learning · Logistic regression · Random forest · Prediction

1 Introduction

Creating a model using machine learning algorithms require that the features to be used are represented in numerical values. When a column has values that are sentences or phrases, the machine learning algorithm would not be able to interpret

O. A. Chidinma
Department of Computer Science, School of Information and Communications Technology, Federal University of Technology, Owerri, Imo State, Nigeria
e-mail: chidinmaononuju@rocketmail.com

S. Borah (✉) · R. Panigrahi
Department of Computer Applications, Sikkim Manipal Institute of Technology, Sikkim Manipal University, Sikkim, India
e-mail: samarjeetborah@gmail.com

R. Panigrahi
e-mail: ranjit.panigrahi@gmail.com

S. Borah et al. (eds.), *Soft Computing Techniques and Applications*, Advances in Intelligent Systems and Computing 1248,
https://doi.org/10.1007/978-981-15-7394-1_14

these sentences or phrases and that creates a need to represent these words or strings using numbers so that we can fit the data on an algorithm.

The Bag of Words (BOW) technique is applied to learn from the text. The basic idea of the Bag of Words is to take any text and count the frequency of words and place those unique words in a dictionary. Each word is known as a token. For each token, we will have a feature column, and this is called Text Vectorization.

This study is aimed at analyzing posts and comments written in natural language and developing a model that can predict to some extent future and/or similar cases in Nigeria by using Bag of Words, Logistic regression, Random Forest, and XGBoost machine learning algorithms.

This research study could provide valuable insight into the use of Natural Language Processing to investigate cases like the issue of suicide in Nigeria. This study will also be beneficial to the government, including the department of guidance and counseling in schools and workplaces in Nigeria as the study will give the necessary information on the triggers of suicidal thoughts.

Further, this research would be used as a review to anyone looking to create prediction models in this area. To future researchers, this study can provide baseline information in this area. This research work contains a dataset of suicide cases in Nigeria from the year 2016 to the year 2019 collected from articles, newspapers, and questionnaires. The quality and option of data available on suicide and suicide attempts are very poor. The following were limitations faced during this research:

- Lack of enough data for training the models.
- Difficulty in gathering data for this research.

2 Review of Related Works

In this section, some of the research and models built by some researchers in the field were discussed.

Researches have been carried out in some areas of the world with the aid of many machine learning algorithms, some of which include: Support Vector Machine, Artificial Neural Networks, Decision Trees, etc.

Concerning this case study, research was carried out in India which aimed at predicting the causes of suicide in India by using machine learning algorithms and data mining techniques to identify the causes of suicide [1].

For this research, two approaches were taken: Descriptive and statistical approach for finding out the patterns of suicide with age, gender, marital status, social status, education, and professional occupation as features. The predictive approach, a model was generated to predict the future causes of suicide by making use of the information present in the existing data. The features considered in this research work were: State, Year, Gender, Age Group, Total Number of Suicides, Type/cause, Marital Status, Professional Occupation, and Educational Level. The dataset contained approximately 109,200 records [1].

The machine learning algorithm used in this research was the Support Vector Machine (SVM) algorithm and an Artificial Neural Network algorithm. Using the Neural Network model, 77.5% accuracy was achieved with 17% misclassified cases. The SVM model performed much better achieving 81.5% accuracy. The performance of this model, however impressive, still has an 18.5% chance of misclassifying new data which poses a serious problem since a person at risk of suicide could be misclassified.

This work, however, is limited to predicting the causes of suicide in general and does not focus on predicting based on age group or gender factor [1].

In recent years, data mining techniques have been used in aiding the medical decision-making process and they have already proven to be useful in medicine [2–4]. In psychiatry, data mining has been used also in estimating suicide risks by using the information provided by electronic medical records but has not been able to predict individual risks at a given moment [5].

Research to propose a predictive model for suicide risk in a clinical sample of patients with a mood disorder was carried out in the Metropolitan Region of Santiago, Chile. Three hundred and forty-three (343) variables were studied using five questionnaires. The individuals who participated in this study were divided into two groups: (1) those with suicidal behavior who sought treatment for suicide attempts; and (2) those without suicidal behavior who sought treatment for other reasons, without suicidal attempts. The sample dataset was made up of all ages greater than or equal to 14 years [5].

For this research, six data mining techniques were explored: Classification and Regression Tree (CART), Support Vector Machine (SVM), K-Nearest Neighbours (KNN), Random Forest, AdaBoost, and Neural Network Multilayer Perception (MLP) [5].

The SVM model had the greatest accuracy with a mean accuracy of 0.78, a sensitivity of 0.77, and a specificity of 0.79.

This research, however, only focused on individuals who suffered from a mood disorder and not the general public.

Another researcher offers a new approach to the assessment of suicide risk that uses machine learning detection of neural signatures of concepts that have been altered in suicidal individuals. This approach identifies individual concepts from their functional magnetic resonance imaging (fMRI) signatures. This research aimed at determining whether the neural representation of positive, negative, and suicide-related concepts was altered in a group of participants with suicidal ideation relative to a control group. The research made use of the Gaussian Naïve Bayes machine learning algorithm to identify such individuals [6].

Beyond detecting neural signatures, the study also aimed at detecting the emotional component of the neural signatures. These emotional components included nine different types of emotions but mainly focused on four which were: sadness, shame, anger, pride [6].

This research, however, focused on people with suicidal ideation and not psychiatric control participants who are affected by psychopathology in general [6].

In 2018, researchers from Christ University, Bangalore, Karnataka, India used six different data mining classification algorithms namely: Regression algorithm, Logistic Regression algorithm, Random Forest algorithm, Decision Table algorithm, and Sequential Minimal Optimization (SMO) algorithm. These different classifiers were trained and tested on a dataset gotten from the UK data archive [7].

The prediction was done based on some major risk factors which are depression, anxiety, hopelessness, stress, or substance misuse. The results obtained from the various data mining classification algorithms were compared for prediction [7].

Using Weka, these researchers were able to identify the classifier with the highest prediction accuracy which was the Regression classifier.

3 Dataset and Performance Matrices

To analyze the performance of the selected Natural Language Processing technique, data was collected from newspapers, articles, and through questionnaires in natural language (English). The data consisted of two hundred and ninety-one rows (291) and two columns (2). The first column "Reason" contained individual comments, thoughts, and suicidal statements in natural language, and the second column "suicidal" contained binary values to identify the negative and positive statements.

The problems of the BOW technique, however, are as follows:

- We lose word order and
- Counters are not normalized.

To achieve a better Bag of Words, we replace counters with term frequency-inverse document frequency (Tf-Idf) and normalize the result row-wise (divide by L2 norm). This technique was applied to the "Reason" column for further processing and understanding of the dataset.

For performance evaluation, precision, recall, F measure, accuracy, and AUC score is used. The dataset gathered was imbalanced therefore, the performance measure to be used as a benchmark will be the precision/recall curve since it is a useful measure when the classes are very imbalanced.

Table 1 Characteristics of suicide cases in Nigeria Dataset

Dataset	Class type	Sample size	No. of attributes	No. of Distinct Classes
Suicide_Nig	Binary	291	2	2

4 Results and Analysis

Table 1 shows the performance of the three classifiers used with the Bag of Words technique.

The figure below depicts the precision/recall curve plot showing the precision/recall for each threshold.

From Fig. 1, it is interesting to note that the logistic regression classifier emerged with a better precision/recall score. In Table 2, there was also a better performance of the logistic regression model. It has stable performance in all the performance matrices.

For this dataset, the logistic regression is most suitable since it has the highest precision/recall score and outperformed the other two models in all the performance matrices.

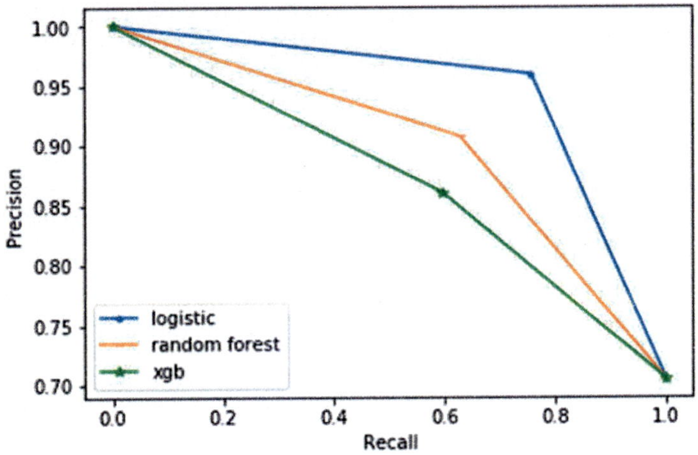

Fig. 1 Precision/Recall plot for logistic regression, random forest, and XGBoost classification algorithm

Table 2 Experimental result

Classifiers	Mean cross_val_score	Roc_auc_score	F measure (f1-score)	Accuracy score
Logistic regression	0.89	0.841	0.847	0.807
Random forest classifier	0.81	0.738	0.743	0.693
XGBoost classifier	0.83	0.683	0.705	0.648

5 Conclusion

Natural Language Processing is an effective approach to dealing with sociological issues. In this work, five performance measures were considered on three different machine learning algorithms. The TfidfVectorizer was used in converting text to numerical values. The dataset used was collected from successful suicide attempts recorded in newspapers, online articles, and questionnaires. It has been discovered that the logistic regression model outperformed the random forest model and XGBoost model based on the dataset used.

Creating a suicide prevention model that performs exceptionally well requires continuous monitoring of the data required for the development of such a model and this data needs to be updated periodically to increase the accuracy of the model. The quality and option of data available on suicide and suicide attempts, however, are very poor. It is important to implement an improved surveillance and monitoring system for suicide and suicide attempts for creating an effective suicide avoidance strategy. Also, with the availability of concrete data, data prediction accuracy will increase, and thus more suicide prevention strategies can be made and implemented.

6 Future Work

For future work, a larger size of the dataset can be used to further investigate the performance of the algorithms used and more performance evaluation methods can be introduced. Introducing more classifiers is also an area to further this study.

References

1. Imran, A., Sobia, S.: Prediction of suicide cases in india using machine learning. J. Independent Stud. Res., pp. 1–6 (2017)
2. Kausar, N., Palaniappan, S., Samir, B.B., Abdullah, A., Dey, N.: Systematic analysis of applied data mining based optimization algorithms in clinical attribute extraction and classification for diagnosis of cardiac patients. In: Applications of intelligent optimization in biology and medicine. Springer, Cham, pp. 217–231 (2016)
3. Varlamis, I., Apostolakis, I., Sifaki-Pistolla, D., Dey, N., Georgoulias, V., Lionis, C.: Application of data mining techniques and data analysis methods to measure cancer morbidity and mortality data in a regional cancer registry: the case of the island of Crete, Greece. Comput Methods Programs Biomed **145**, 73–83 (2017)
4. Kamal, M.S., Dey, N., Ashour, A.S.: Large scale medical data mining for accurate diagnosis: A blueprint. In: Handbook of large-scale distributed computing in smart healthcare. Springer, Cham, pp 157–176 (2017)
5. Jorge, B., Susana, M., Orietta, E., Arnol, G., Jaime, O., Takeshi, A., Catalina, N.: Suicide detection in Chile: proposing a predictive model for suicide risk in a clinical sample of patients with mood disorders. Revista Brasileira de Psiquiatria, pp. 1–11 (2017)

6. Just, Marcel, A., Lisa, P., Vladimir, L.C., Dana, L.M., Christine, C., David, B.: Machine learning of neural representations of suicide and emotion concepts identifies suicidal youth. Nat. Hum. Behav. (October 30) 2–8. https://doi.org/10.1038/s41562-017-0234-y (2017)
7. Alina, J., & Ramamurthy, B.: Suicidal behaviour prediction using data mining technique. Int. J. Mech. Eng. Technol., pp. 293–301 (2018)

Analyzing Performance of Virtualization and Green Computation Applying Firefly Approach

Jyoti Prakash Mishra, Zdzislaw Polkowski, and Sambit Kumar Mishra

Abstract The concept of green computing in general innovates the measures to overcome the issues linked towards computation. In fact, now it is a general consideration the greening concept enhances the publication relations as well as minimizes the cost of processing. The green computation not only maximizes the efficiency but also minimizes the computing resources. Deployment of internet of things or cloud service sometimes is complex while linked to its effectiveness as well as cost. In such a case, it is required to ascertain the challenges towards computation in cloud with suitability and design mechanism towards decision making. While analyzing the mechanisms linked to virtualization, the feature of portability with information exchange is obtained in heterogeneous environments. In such cases, the virtualization technology is dynamically linked to the distributed systems rather than remote servers. Accordingly, the utilization of virtualized environment increases boosting the integration of earlier computing technologies. The virtualization of servers in general consolidates the applications to multiple servers.

Keywords Green computing · Virtualization · Metaheuristic · Cloudlet · Query plan

---

J. P. Mishra (✉) · S. K. Mishra
Gandhi Institute for Education and Technology, Bhubaneswar, Affiliated to Biju Patnaik University of Technology, Rourkela, Odisha, India
e-mail: jpmishra@gietbbsr.com

S. K. Mishra
e-mail: sambitmishra@gietbbsr.com

Z. Polkowski
Wroclaw University of Economics and Business, Wroclaw, Poland
e-mail: zdzislaw.polkowski@ue.wroc.pl

S. Borah et al. (eds.), *Soft Computing Techniques and Applications*,
Advances in Intelligent Systems and Computing 1248,
https://doi.org/10.1007/978-981-15-7394-1_15

1 Introduction

To assess the suitability of adoption of cloud, it is required to deploy the public cloud. It is essential to estimate the risk as well as cost of using public clouds. The tools associated with the system support efficiently while making decisions. The decision in this case enables the system towards deploying public clouds and measuring the risk factors. The sustainability linked to the design approach of components in the cloud optimizes the virtualization up to a minimum of 20%. While observing the processors of multi-core servers, it is clear that the processors maintain cent percent of usage with the available instances. As all the instances usually complete in time, the maintenance of servers may be achieved up to 92%. The consumption of energy, in this case, is the maximum 10% less linked with the physical server. As such, the estimated energy linked to virtualized servers is expensive and the difference within the primitives and virtualized servers increases. So accordingly, the throughput is optimized. In general, the servers in a virtual environment utilize more energy as compared to physical servers. To promote green computation as well as sustainability, it is essential to focus on the design requirements of data centers along with the consumption of electrical power supply. As the physical servers are associated with high operating costs, it leads to significant emission of carbon dioxide in the environment. In the idle condition of virtualized servers, the maximum one percent energy is incurred and ten percent extra energy cost can be contributed. The mechanisms associated with energy in virtual platforms can exhibit within the range 2.9–67% towards the complete workload. To analyze large-sized heterogeneous data, it is essential to keep backup of data as well as maintain a stream track to manage a huge amount of power. As such the power automatically has to be obtained costing more. In connection with the large set of data the application may be responsible for problems linked to hardware. It has been seen that to obtain sustainability towards green computation, the influence of cloud computing must be measured. In such a connection, the green characteristics linked with the information communication system relates to ecological informatics as well as sustainable computing. It is observed that information communication technology, in general, is responsible towards two percent of carbon dioxide. Green cloud aims to decrease data center power consumption. Again a large amount of data are accumulated in Data Centers which can consume more power towards servers to run the servers, large amount of heat can be released. Accordingly, it is very much essential to minimize energy consumption. It is understood that virtualization is intended and linked green computing which provides a hypervisor to focus on logical view of computing only. Basically, it is linked with a processor, storage as well as primary memory. In fact, algorithms discharge the main role in green computing provided minimum energy is consumed with minimum emission of carbon dioxide. While focusing on various approaches of virtualization as well as green computing, it is essential to concentrate on design of data centers as well as additional energy design opportunities. To improve the efficiency as well as performance, it is also essential to estimate the intensification of the algorithm, implementation of servers, and concept of virtualization.

2 Review of Literature

Bruneton et al. [1] during their work have concentrated on viability of component-based design linked towards the security of complex systems with heterogeneity. They opted for cloud security to project unified security architecture.

Smith et al. [2] during their study have focused on management of sharing of physical resources within the concurrent virtual machines. During monitoring and strengthening the activity, they adopted the concept of hypervisor similar to the virtual machines.

Thomas et al. [3] in their research have focused on security vulnerabilities linked to virtualization. The major drawback in this case links to the lack of configuration, malicious device drivers, and the association among virtual machines and the hardware issues. Hypervisor security is such a case is only part of the problem since virtual machines are responsible to bring their own set of vulnerabilities.

Barham et al. [4] during their research have adopted techniques linked to hyper calls to substitute the instruction set architecture in the host system. Practically, it is associated with the communication between the operating system and the hypervisor to enhance efficiency.

According to the specification [5], the user requirements linked to the host machine must be conferred while its association with the software layer in virtualization which acts in between operating system and hardware as an intermediate layer. Accordingly, the classification of hypervisor is done and can run directly on the hardware. But the host operating system may be inclined towards the host hypervisor.

Siddhisena et al. [6] in their work have focused on mechanisms associated with virtualization. Initially, they focused on the server hosted or remote virtualization on the server machine and operated by the client across a network. Subsequently, during their progress, they have taken local or client-hosted virtualization where the secured and virtualized operating environment runs on local machine into consideration.

Alameri I et al. [7] in their study observed that as the cloud service providers as well as users are aware of security concerns, it must be essential to make provision of various cloud computing service models towards security services.

Jeena et al. [8] in their research focused on the shared responsibility cloud model. In fact, the cloud service user knows the impact of privacy, security as well as compliance. Also, the users should retrieve the secure solution from the cloud service provider.

Kazim et al. [9] during their research have discussed the methodologies to protect the virtualization environment in the cloud. In such a case, they have focused on some security measures associated with secure cloud execution, protected network, patch management as well as back up of virtual machine data.

Xiong et al. [10] during their work have thought of mechanism to maximize resource utilization, a number of virtual machines are usually linked to the cloud environment on the same physical server. The reason behind the same is the malicious virtual machine penetrates the isolation between virtual machines and extracts the relevant information.

Jansen et al. [11] during their study have observed that a hypervisor is usually framed to run numerous guest virtual machines with applications simultaneously to provide separation among the guest virtual machines along with single host machine.

According to services and cost linked to the data centers [12], it is seen that servers use 45% of the cloud service, a minimum of 25% is required for infrastructure and moreover 25–30% may be used towards power and networks. So to minimize the cost and improve agility, the policies and mechanisms should be adopted to enhance reliability and performance.

The mechanisms linked to cloud storage are mainly adopted in a different manner [13]. In the early days, hardly any facility was there towards mechanisms for cloud storage. Gradually with the advancement of the mechanisms, Cloud storage models comprise initial position to trade an API to get to the capacity. In customary capacity frameworks, this API is the SCSI convention. Behind the front end is a layer of middleware which is a rationale.

Itani et al. [14] in their work have focused on the main usage of cloud computing. They observed the significance of data linked to servers with heterogeneity instead of specific and dedicated servers in the cloud. During storage of data, the virtual server may be observed comprising virtual space in the cloud.

Jakimoski et al. [15] in their work have focused on distributed loosely coupled environments comprising the services associated with cloud in the form of patterns. Accordingly, the processes are reformed in the cloud considering the data storage and cloud resources.

Sarddar et al. [16] during their research considered many cloud security issues associated with scheduling in virtualization, controlling the concurrency as well as storage management in cloud and designed as computing utility. The resources linked to cloud are usually heterogeneous in nature and having provision to update the clusters with new generations of machines.

Gao et al. [17] in their work have emphasized on secure mechanisms in virtual environment. In fact, to measure the effectiveness of the computing paradigm along with restructuring and designing within the cloud, the secure mechanism linked to cloud sometimes faces new business and management problems arising from virtualization. As the organizations are struggling to identify the features of cloud security issues, it is highly essential to chalk out suitable decisions towards the successful implementation of cloud computing technology.

Bindu et al. [18] during their research have discussed the mechanisms to enhance the efficiency of data centers. They observed that many techniques with green computations are to be developed to maximize server efficiency. They focused on various aspects of cloud computing towards managing cloud resources in an energy-efficient, reliable, and sustainable manner.

Zhou et al. [19] in their study have addressed the problem of energy consumption by the heuristic scheduling algorithm. As per their consideration, first of all, the initial order of tasks is obtained by implementing the specific algorithm. Also, the inefficient processors linked with energy utilization can be combined towards redistribution of tasks and reclaim the slack time.

3 Implementation of Firefly Algorithm in Virtualization

The firefly algorithm is basically a metaheuristic approach with flashing activities towards pre-defined routine. In this approach, the brighter fireflies attract the less bright ones to move randomly. The objective function linked with the flashing light characteristics enables measuring the fitness parameters of the fireflies. Certainly, the present load can be calculated considering all resources in the virtual environments. As soon as the index towards all the resources is achieved, the resources can dynamically be assigned to suitable nodes to obtain optimality. In such a case, to optimize load balancing, an effective optimization procedure can be adopted. Therefore, the firefly algorithm is more suitable towards load balancing and indexing in virtualization as well as green computation.

3.1 Algorithm (Implementation in Firefly Algorithm)

3.1. Algorithm (Implementation in Firefly algorithm)

Step 1 : Estimate the cost function, csf

Step 2 : Define the decision variables and size,ds

Step 3 : Assisn the lower bound and upper bound of decision variables vl=5, vu=10

Step 4 : Specify maximum iterations, itm=100

Step 5 : Define maximum cloudlet size=25

Step 6 : Define the absorption coefficient

Step 7 : Define the base value of the coefficient

Step 8 : Define the mutation coefficient

Step 9 : Determine the mutation coefficient ratio= 0.05*(vu-vl)

Step 10 : if(vu>vl)

```
   maxlimit = (vu-vl)*sqrt(ds);
else
   maxlimit = mutation coefficient ratio *(vu-vl);
```

Step 11 : Define the position of cloudlets and estimate the cost

Step 12 : Create initial query plans linked to cloudlets

```
for i=1:itm
   pop(i).loc= csf(vl,vu,ds);
```

Step 13 : Iterate to achieve new positions of cloudlets consisting query plans

```
for i=1:itm
   newpop= csf(cldqp,itm,1);
   for i=1:itm
      newpop(i).Cost = inf;
      for j=1:nPop
         if pop(j).Cost < pop(i).Cost
            rij= mutation coefficient ratio* (pop(i).Position -
pop(j).Position)/maxlimit;
```

Step 14 : Obtain the positions of cloudlets consisting query plans

Step 15 : if pos cldqp.Cost <= newpop(i).Cost

```
            newpop(i) = pos cldqp;
```

See Tables 1 and 2.

3.2 *Experimental Analysis*

It is practically seen that the server in the standstill position also consumes roughly more than half of the energy even if its computing resource is fully occupied. In such a case, it will be essential to make an association of multiple servers and unifying as well as visualizing as virtualized server. Considering the virtual machine applications and referring to Fig. 1, it is clear that the response time is directly proportional to the size of the virtual machine. Similarly, considering the implementation of tasks in servers and referring to Fig. 2, it is understood that the execution of tasks definitely

Table 1 Virtual machines w.r.t response request

S. No.	Size of virtual machine	Response time (μs)
1	15	0.19
2	29	0.37
3	34	0.64
4	37	0.73

Table 2 Tasks per server w.r.t execution time

S. No.	No. of tasks per server	Estimated execution time (μs)
1	190	0.46
2	370	0.55
3	460	0.82
4	550	0.91

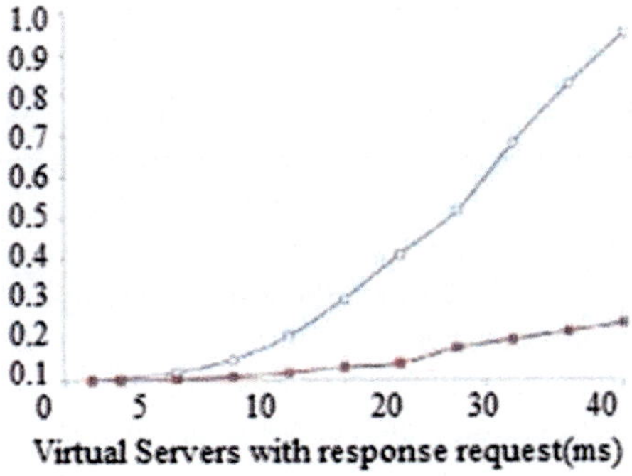

Fig. 1 Virtual servers with response request

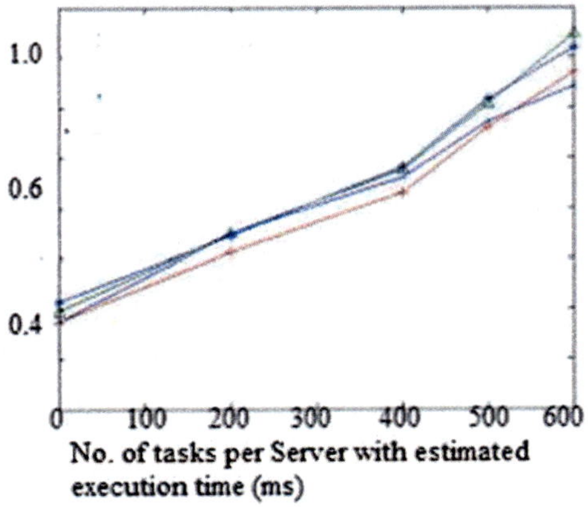

Fig. 2 No. of tasks per server with estimated execution time (μs)

links to the position and size of the servers.

4 Discussion and Future Direction

In the general case, in the idle and normal state of virtualized servers, around ten percent extra energy cost can be incurred. As such, the consolidation of energy towards virtualized servers increases with the increment of resource utilization. Also, the virtual firewall router, virtual network interface, etc. also include energy overhead ranging from a minimum 3–50% for throughput linked to other heterogeneous systems. Sometimes it is seen that partly the virtualized servers contribute around 10% extra energy cost. In order to estimate the impact of virtualization, it is required to focus on virtual network bridge as well as energy overhead and make it proportionate. So deployment of the data centers near renewable energy sources is very much essential along with focusing towards green data centers.

5 Conclusion

Focusing towards green computing, it is required to measure the impact of virtualization on environmental sustainability. The technology liked virtualization minimizes the usage of energy to form a green data center. It will help the environment minimizing the amount of gas remitted into the atmosphere. The importance of virtualization is very much there to minimize the usage of data centers and saving a huge amount of electrical power. In addition to continuous emissions of carbon dioxide will be responsible towards climate change and global warming. If it will be difficult to manage the energy-efficient techniques and algorithms linked to computing resources, then the consumption of energy as well as emission of carbon dioxide will be rapidly increased. Therefore, it is clear that virtualization has the main role towards consideration of green computing as well as the sustainability of environmental conservation.

References

1. Bruneton, E., Coupaye, T., Leclercq, M., Quéma, V., Stefani, J.: The fractal component model and its support in java: Experiences with auto-adaptive and reconfigurable systems. Softw. Pract. Exper. **36**(11–12), 1257–1284 ISSN 0038-0644. https://doi.org/10.1002/spe.v36:11/12. http://dx.doi.org/10.1002/spe.v36:11/12, (September 2006)
2. Smith, I, Nair, R.: Virtual Machines: Versatile Platforms for Systems and Processes (The Morgan Kaufmann Series in Computer Architecture and Design). Morgan Kaufmann Publishers Inc., San Francisco, CA, ISBN 1558609105 (2005)
3. Ristenpart, T., Tromer, E., Shacham, H., Savage, S:Hey, you, get off of my cloud! Exploring information leakage in third-party compute clouds. In: ACM Conference on Computer and Communications Security(CCS) (2009)
4. Barham, P., Dragovic, B., Fraser, K., Hand, S., Harris, T., Ho, A., Neugebauer, R., Pratt, I., Warfield, A.: Xen and the art of virtualization. In: SOSP '03: Proceedings of the Nineteenth ACM Symposium on Operating Systems Principles, pp. 164–177. ACM Press, New York, NY (2003)
5. IBM: IBM systems virtualization, version 2 release 1. http://publib.boulder.ibm.com/infocenter/eserver/v1r2/topic/eicay/eicay.pdf (2005)
6. Siddhisena, B., Wruasawithana, L., Mendis, M.I: Next generation muti tenant virtualization cloud computing platform. In: Proceedings of 13th International Conference on Advanced Communication Technology (ICACT), vol. 12, no. 3; pp. 405–10 (2011)
7. Alameri, I.: Radchenko G Development of student information management system based on cloud computing platform. J. Applied Compu. Sci. Math. **11**, 9–14 (2017). https://doi.org/10.4316/JACSM.201702001
8. Jeena, R., Kumar, S.S., Sudhan, S.K.H.H: Efficient and secure techniques for protecting data in the cloud. In: International Conference on Information Communication and Embedded Systems (ICICES2014), Chennai, pp. 1–5 (2014). https://doi.org/10.1109/icices.2014.7033771 (2014)
9. Kazim, M., Zhu, S.Y.: Virtualization security in cloud computing. In: Zhu, S., Hill, R., Trovati, M. (eds.) Guide to Security Assurance for Cloud Computing. Computer Communications and Networks. Springer, Cham. https://doi.org/10.1007/978-3-319-25988-8 (2015)
10. Xiong, H., Zheng, Q., Zhang, X., Yao, D.: CloudSafe: securing data processing within vulnerable virtualization environments in the cloud. In: 2013 IEEE Conference on Communications

and Network Security (CNS), National Harbor, MD, 2013, pp. 172–180. https://doi.org/10.1109/cns.2013.6682705 (2013)
11. Jansen, W.A.: Cloud hooks: security and privacy issues in cloud computing. In: 2011 44th Hawaii International Conference on System Sciences, Kauai, HI, pp. 1–10. https://doi.org/10.1109/hicss.2011.103 (2011)
12. FalconStor Software, Inc., Demystifying Data Reduplication: Choosing the Best Solution. http://www.ipexpo.co.uk/content/download/20646/353747/file/DemystifyingDataDedupe_WP.pdf. White Paper (2009)
13. Storage Networking Industry Association. Cloud Storage Reference Model (June 2009)
14. Chehab, W.I.A.K.A.: Privacy as a service: privacy-aware data storage and processing in cloud computing architectures. In: Eighth IEEE International Conference on Dependable (2009)
15. Jakimoski, K.: Security techniques for protecting data in cloud computing. Int. J. Grid Distrib. Comput. **9**(1), 49–56 (2016)
16. Sarddar, D., Sen, P., Sanyal, M.K.: Central controller framework for mobile cloud computing. Int. J. Grid Distrib. Comput. **9**(4), 233–240 (2016)
17. Gao, Z., Li, Y., Tang, H., Zhu, Z.: Management process based cloud service. In: International Conference on Cyberspace Technology (CCT 2013), pp. 278–281 (2013)
18. Bindu, G.B.H., Janet, J.: A statistical survey on VM scheduling in cloud workstation for reducing energy consumption by balancing load in cloud. In: 2017 international conference on networks & advances in computational technologies (NetACT). IEEE, New York (2017)
19. Zhou, Z., et al.: Minimizing SLA violation and power consumption in cloud data centers using adaptive energy-aware algorithms. Future Gener. Comput. Syst. (2017)

Model-Order Reduction and Reduced Controller Design Using Routh Approximation and Factor Division Method

Karri Sudheswari, Rayadu Veera Ganesh, Ch. Manoj, Adireddy Ramesh, and K. Manoz Kumar Reddy

Abstract This paper studies model order reduction method and controller design technique using routh-approximation and factor division method. The said technique is simple and computationally efficient. Different simulation examples are considered to validate the above mentioned methods. This method has been compared with a soft-computing based method and its found that soft-computing based method provide better results with increased time complexity. For time saving purposes, this method should be adopted.

Keywords Model order reduction · Controller design · Soft-computing · Time complexity · Reduction scheme

1 Introduction

Stability equation method of model order reduction has been discussed in [3]. Truncation method have been discussed in [5, 10]. Stability equation method, Roth stability criteria, Mihailov criterion and Pades approximation are useful technique to reduce

---

K. Sudheswari · R. V. Ganesh · Ch. Manoj · A. Ramesh (✉) · K. Manoz Kumar Reddy
Department of Electrical and Electronics Engineering, Aditya College of Engineering, Surampalem, Andhra Pradesh, India
e-mail: rameshadireddy007@gmail.com

K. Sudheswari
e-mail: sudheswari.karri@gmail.com

R. V. Ganesh
e-mail: rvganesh366@gmail.com

Ch. Manoj
e-mail: chithaluru.manoj@gmail.com

K. Manoz Kumar Reddy
e-mail: manozreddy111@gmail.com

S. Borah et al. (eds.), *Soft Computing Techniques and Applications*, Advances in Intelligent Systems and Computing 1248,
https://doi.org/10.1007/978-981-15-7394-1_16

model order [4, 7–9, 11]. Reduced controller has been designed in [1, 2, 12] respectively. Direct and indirect method of controller design has been discussed in [6].

This paper provides model order reduction scheme using routh and factor division method. Reduced controller design has been discussed and simulation results using different examples have been provided in this paper.

2 Problem Formulation

Let us consider a higher order model which can be represented as

$$G_n(s) = \frac{\sum_{i=0}^{n-1} d_i s^i}{\sum_{i=0}^{n} e_i s^i} \tag{1}$$

The reduced order model can be represented as

$$G_r(s) = \frac{\sum_{i=0}^{r-1} a_i s^i}{\sum_{i=0}^{r} b_i s^i} \tag{2}$$

where a_i, b_i, d_i and e_i are scalar constant

$$G_{OCL}(s) = \frac{C_f(s)\, G_n(s)}{1 + C_f(s)\, G_n(s)\, H(s)} = G_{ref}(s) \tag{3}$$

On simplification of Eq. 3 yields

$$C_r(s) = \frac{G_{ref}(s)}{G_2(s)\left(1 - G_{ref}(s)\, H(s)\right)} \tag{4}$$

Let us consider the reduced order model

$$G(s) = \frac{N(s)}{D(s)} = \frac{d_{n-1}s^{n-1} + d_{n-2}s^{n-2} + \cdots + d_1 s + d_o}{e_n s^n + e_{n-1}s^{n-1} + \cdots + e_1 s + e_o} \tag{5}$$

1.

$$D_r(s) = \sum_{i=0}^{r} b_i s^i \tag{6}$$

2.

$$N_{r-1}(s) = \frac{N(s)}{D(s)} \times D_r(s) \tag{7}$$

3.

$$G_r(s) = \frac{N(s)}{\frac{D(s)}{D_r(s)}} = \frac{\sum_{i=0}^{n-1} b_i s^i}{\sum_{i=0}^{n-r} e_i s^i} \tag{8}$$

$$\begin{aligned}
c_0 &= \frac{b_0}{e_0}\begin{cases} b_0 & b_1 & b_2 & \cdots & b_{r-1} \\ e_0 & e_1 & e_2 & \cdots & e_{r-1} \end{cases} \\
c_1 &= \frac{q_0}{e_0}\begin{cases} q_0 & q_1 & \cdots & q_{r-2} \\ e_0 & e_1 & \cdots & e_{r-2} \end{cases} \\
c_2 &= \frac{r_0}{e_0}\begin{cases} r_0 & \cdots & r_{r-3} \\ e_0 & \cdots & e_{r-3} \end{cases} \\
c_{r-2} &= \frac{u_0}{e_0}\begin{cases} u_0 & \cdots & u_{r-4} \\ e_0 & \cdots & e_{r-4} \end{cases} \\
c_{r-2} &= \frac{v_0}{e_0}\begin{cases} v_0 \\ e_0 \end{cases}
\end{aligned} \tag{9}$$

where

$$\begin{aligned}
q_i &= b_{i+1} - c_0 e_{i+1}, i = 0, 1, \ldots, r-2 \\
r_i &= q_{i+1} - c_1 e_{i+1}, i = 0, 1, \ldots, r-3 \\
&\vdots \ldots\ldots \vdots \\
v_0 &= u_1 - c_{r-2} e_1
\end{aligned} \tag{10}$$

4.

$$N_{r-1}(s) = c_{r-1}s^{r-1} + c_{r-2}s^{r-2} + \ldots + c_1 s^1 + c_0 = \sum_{i=0}^{r-1} c_i s^i \tag{11}$$

3 Simulation Results

Consider the fourth-order LTI model

$$\begin{aligned}
G_4(s) &= \frac{18.439s^3 + 14.446s^2 + 11.454s + 1.3765}{35.83s^4 + 40.388s^3 + 32.541s^2 + 13.94s + 1} \\
D^*(s) &= s^4 + 1.127s^3 + 0.908s^2 + 0.389s + 0.028
\end{aligned} \tag{12}$$

1. Calculation for denominator polynomial Normalizing G(s) gives

$$\begin{aligned}
&G_{ref}(s) = \frac{2.25}{s^2 + 2.1s + 2.25} \\
&D_2^*(s) = s^2 + 0.4769s + 0.0338 \\
&0.028 \\
&0.389 \\
&c_0 = 0.0465 \\
&c_1 = 0.3954
\end{aligned} \tag{13}$$

$$D(s) = s^4 + 1.127s^3 + 0.908s^2 + 0.389s + 0.028 \tag{14}$$

Routh table can be represented as

$$\begin{matrix} 0.028 & 0.908 & 1 \\ 0.389 & 1.127 & \\ 0.826 & 1 & \\ 0.656 & & \\ 1 & & \end{matrix} \tag{15}$$

Using the first three entries as the first column, form a new array for the reduced system as:

$$\begin{matrix} 0.028 \\ 0.389 \\ 0.826 \end{matrix} \tag{16}$$

$$D_2(s) = 0.826s^2 + 0.389s + 0.028 \tag{17}$$

$$D_2^*(s) = s^2 + 0.4769s + 0.0338 \tag{18}$$

2. Calculation for numerator polynomial $c_o = 0.04651$ and $c_1 = 0.3954$

$$G_2(s) = \frac{0.3954s + 0.0465}{s^2 + 0.0.4769s + 0.0338} \tag{19}$$

$$G_{orr}(s) = \frac{1.56635s^3+1.63157s^2+0.31131\,s+0.01583}{0.69611s^5 + 2.18697s^4 + 3.2239s^3 + 1.9298s^2 + 0.32617\,s + 0.01583} \tag{20}$$

So, $C_r(s)$ is represented as

$$C_r(s) = \frac{2.25s^2 + 1.073s + 0.076}{0.3954s^3 + 0.8763s^2 + 0.0966s} \tag{21}$$

$$C_f(s) = \frac{80.6175s^4 + 90.873s^3 + 73.2172s^2 + 31.365s + 2.25}{18.439s^5 + 53.1679s^4 + 41.7906s^3 + 25.4294s^2 + 2.8896s} \tag{22}$$

$$G_{orr}(s) = \frac{1.56635s^3+1.63157s^2+0.31131\,s+0.01583}{0.69611s^5 + 2.18697s^4 + 3.2239s^3 + 1.9298s^2 + 0.32617\,s + 0.01583} \tag{23}$$

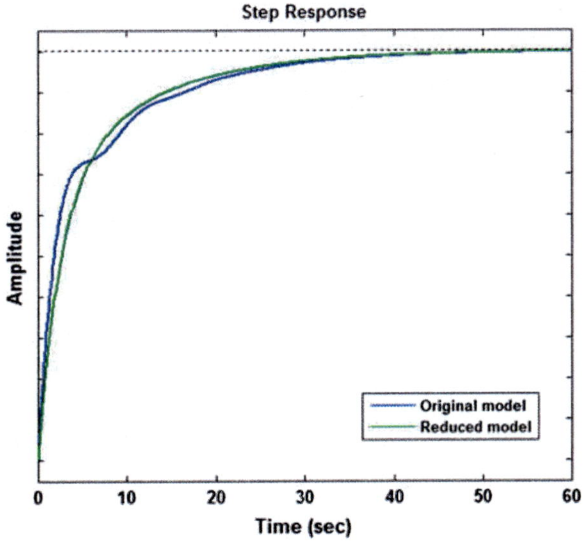

Fig. 1 Comparative analysis

If the original plant $G_4(s)$ along with reduced-order controller $C_r(s)$, then overall closed loop transfer function turns out to be

$$G_{oro}(s) = \frac{41.49s^5 + 52.29s^4 + 42.67s^3 + 16.49s^2 + 2.347s + 0.1046}{14.17s^7 + 47.37s^6 + 93.21s^5 + 90.22s^4 + 58.43s^3 + 18.71s^2 + 2.444s + 0.1046} \quad (24)$$

$$G_{ooo}(s) = \frac{1486.5061s^7 + 2840.2077s^6 + 3586.1962s^5 + 2787.8642s^4 + 1458.303s^3 + 492.5417s^2 + 68.9454s + 3.0971}{660.6694s^9 + 2649.7202s^8 + 5731.23s^7 + 7426.358s^6 + 6836.28s^5 + 4367.796s^4 + 1948.61s^3 + 557.252s^2 + 71.835s + 3.097} \quad (25)$$

See Figs. 1 and 2.

4 Conclusion

This paper provides model-order-reduction technique using factor division method. Different examples are considered to validate the method. The controller design and model order reduction provides satisfactory results. The paper has been compared with soft-computing based reduction technique and analysis has been done.

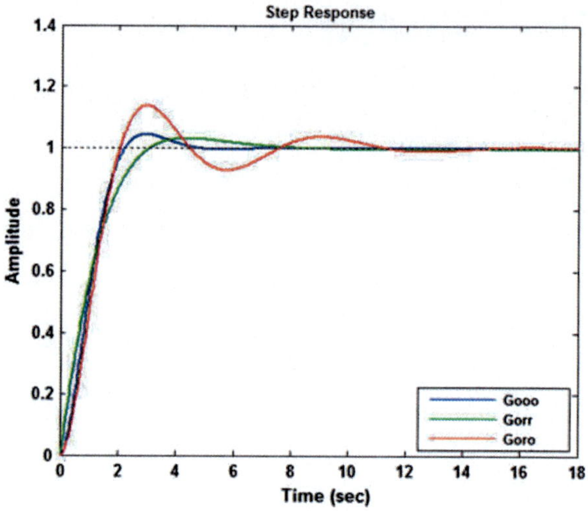

Fig. 2 Comparison of responses of closed loop system

Acknowledgements The authors are thankful to Accendere Knowledge Management Services Pvt. Ltd, CL Educate Ltd for their assistance during the preparation of the manuscript.

References

1. Aguirre, L.: Designing controllers by means of model reduction techniques. Electron. Lett. **29**(4), 389–390 (1993)
2. Anderson, B.D., Liu, Y.: Controller reduction: concepts and approaches. In: 1987 American Control Conference, pp. 1–9. IEEE, New York (1987)
3. Chen, T., Chang, C., Han, K.: Reduction of transfer functions by the stability-equation method. J. Franklin Inst. **308**(4), 389–404 (1979)
4. Chen, T., Chang, C., Han, K.: Model reduction using the stability-equation method and the Padé approximation method. J. Franklin Inst. **309**(6), 473–490 (1980)
5. Gutman, P., Mannerfelt, C., Molander, P.: Contributions to the model reduction problem. IEEE Trans. Autom. Control **27**(2), 454–455 (1982)
6. Hyland, D.C., Richter, S.: On direct versus indirect methods for reduced-order controller design. IEEE Trans. Autom. Control **35**(3), 377–379 (1990)
7. Krishnamurthy, V., Seshadri, V.: Model reduction using the routh stability criterion. IEEE Trans. Autom. Control **23**(4), 729–731 (1978)
8. Shamash, Y.: Model reduction using the routh stability criterion and the Padé approximation technique. Int. J. Control **21**(3), 475–484 (1975)
9. Singh, V., Chandra, D., Kar, H.: Improved routh-Pade/spl acute/approximants: a computer-aided approach. IEEE Trans. Autom. Control **49**(2), 292–296 (2004)
10. Smamash, Y.: Truncation method of reduction: a viable alternative. Electron. Lett. **17**(2), 97–99 (1981)
11. Wan, B.W.: Linear model reduction using Mihailov criterion and Pade approximation technique. Int. J. Control **33**(6), 1073–1089 (1981)
12. Yousuff, A., Skelton, R.: A note on balanced controller reduction. IEEE Trans. Autom. Control **29**(3), 254–257 (1984)

Community Detection in a Patient-Centric Social Network

Swarupananda Bissoyi and Manas Ranjan Patra

Abstract Detecting communities in complex networks and particularly in social networks have been gaining traction in recent years. Not much work has been undertaken in the area of Community Detection for patient-centric social networks. In this work, we have tried to detect significant communities in a patient-centric social network by computing similarity among the unstructured physician notes which are in the form of long texts. Experiments have been conducted using the MIMIC III dataset, which is a large open database containing de-identified health-related patient data. Three modularity based community detection algorithms, viz., FastGreedy, Walktrap, and Louvain have been applied to identify the potential communities in the network. Further, the significance of such communities has been tested using the Wilcoxon rank-total test. It is found that Louvain algorithm fared better in terms of modularity and significance compared to the other two algorithms.

Keywords Community detection · Social network analysis · Patient similarity

1 Introduction

Social Networks have emerged as media for connecting people. There has been growing interest among individuals towards the use of social networks as a tool to discover people who are alike in some aspects and share information with each other. This has impacted the lives of individuals with useful data that are being generated out of such as information sharing. Sharing is now a new form of caring. As social networks thrive, an ever-increasing number of patients are socializing

S. Bissoyi (✉)
Department of Computer Application, North Orissa University, Baripada, India
e-mail: swarupananda.bissoyi@nou.nic.in

M. R. Patra
Department of Computer Science, Berhampur University, Berhampur, India
e-mail: mrpatra.cs@buodisha.edu.in

S. Borah et al. (eds.), *Soft Computing Techniques and Applications*, Advances in Intelligent Systems and Computing 1248,
https://doi.org/10.1007/978-981-15-7394-1_17

online to discover and share their well-being information related to their clinical symptoms, diagnosis, and treatment plans. They find it as an open door for solace and an opportunity to seek information regarding their health [1]. The increased use of social networks by patients solves many purposes; Patients get emotional support, discover self-esteem, and gather information. It also helps them to compare their own conditions with other patients who are in the network [2]. As patients spend more and more time discovering others, patients having similar disease profiles are most likely to form communities within a social network.

Detecting Communities in a Social Network is one significant aspect of Social Network Analysis which deals with identifying people who share some sort of similarity or taste with a possibility of connecting and forming a group and then discovering the boundaries of it. The connection among the group members is dictated by Similarity—"Similarity breeds connection" phenomenon [3]. Social networks inherently exhibit community structures. A group of individuals are said to be in a community when the interaction among themselves is more frequent than those outside of the group [4]. Community Detection can be seen as an extension of the discovery of clusters or modules. Community Detection algorithms are based on Clustering, Graph partitioning, Label propagation, Genetic algorithms, Clique and non-clique based overlap detection, etc. which are reviewed by Bedi and Sharma in detail [5]. In this paper, we focus on Clustering-based Community Detection which is based on an important Social Network parameter called "Modularity".

This organization of the paper is as follows: we first study the state-of-the-art related to community detection, and then we focus on modularity based community detection in detail, especially three algorithms, viz., FastGreedy, Walktrap, and Louvain. A publicly available popular patient dataset has been used to simulate a patient-centric social network by computing similarities among patients. Thereafter, we apply the community detection algorithms on this network with varying weight/similarity thresholding in order to evaluate the communities so evolved by calculating their significance scores.

2 Related Work

Papadopoulos et al. detected communities in Social Media and distinguished between Explicit Communities and Implicit Communities. Explicit communities are the ones created out of the consented acquisition of a connection. On the other hand, implicit communities are the ones that inherently exist in the system waiting to be discovered [6]. Nguyen et al. developed algorithms for detecting overlapped communities and applied on real-world networks like Enron, Facebook, Foursquare, etc. They found that the occurrence of overlapped communities is natural and frequent [7]. Du et al. developed a Community Detector algorithm which first derives the Giant Component of the Social Network Graph and then identifies all of the maximal cliques in it [8]. They applied it in various large-scale networks such as Telecommunication call networks. Clara Pizzuti employed a Genetic Algorithm for detecting communities with a graph partitioning approach. He introduced a parameter called Community Score representing the quality of partitioning which is optimized by the genetic algorithm [9]. Zhao et al. developed a topic-oriented community detection algorithm based on social entities clustering using Entropy Weighting k-Means. They partitioned the cluster entities into topic-based clusters and performed link analysis between them to detect topical communities [10]. Chen et al. applied Game Theory to consider community detection as a community formation game wherein the dynamics of community building are treated as strategic games in which each actor in the social network is selfish; and their selection to join or leave the community is based on a utility function. The community structure is interpreted as the equilibrium of the game [11].

In terms of the type of social networks, on which community detection is applied: Chan et al. applied it on time-varying mobile networks [12], Liu et al. applied on Signed Social Networks containing both positive and negative connections [13]. Xin et al. developed recommendation strategies on the communities detected with similar interests with the selection of influential parameters [14]. Xu et al. combined community detection with sentiment analysis to discover "sentiment communities" in social networks where sentiment community is a group of users sharing some common sentiment towards a topic, or a product or a service [15].

As we focus on a patient-centric social network, we found to the best of our insight, that Community Detection is never applied to such a network that we have achieved in this work.

3 Modularity-Based Community Detection

Community structure in a network relates to arc-arrangement which can be quantified by a measure called 'Modularity' [16]. It's a multiplicative constant i.e., the number of arcs falling inside a community minus the expected number in an identical network with arcs put randomly. Positive modularity shows the conceivable nearness of network structure. Community structure can be discovered in a network by looking at portions of network that have large positive values of modularity. Modularity for a weighted graph is expressed as:

$$Q = \frac{1}{2e} \sum_{ij} \left[W_{ij} - \frac{s_i s_j}{2e} \right] \delta\left(c_i, c_j\right)$$

where

W_{ij} is the weight of the arc between node i and node j
s_i is the sum of weights of arcs connected to node i and s_j for node j
e is the sum of all the weights of all the arcs
$\delta\left(c_i, c_j\right)$ is the Kronecker delta function between two communities c_i and c_j which is defined as: $\delta(p, q) = 1, \quad \text{if} \quad p = q, 0 \quad \text{if} \quad p \neq q.$

There are many community detection algorithms viz. Girvan-Newman, Fast-Greedy, Walktrap, Leading Eigen Vector, Louvain, etc. which use modularity.

3.1 Girvan–Newman Algorithm

Girvan–Newman community detection algorithm is a hierarchical method [17] where edges are removed one by one in the decreasing order of "edge betweenness", i.e., the number of shortest paths going through a given edge. Edges associating various groups are bound to be constrained in multiple shortest paths because of the fact that they are the only alternatives to connect between different groups. However, Community detection using edge betweenness builds a full dendogram and there's no clear cut approach about where to slice the dendogram to obtain the final grouping. In such a case, the modularity score of each of the cluster hierarchy can be used to decide that.

3.2 Fast-Greedy Algorithm

It is a bottom-up hierarchical based algorithm, which uses a greedy technique to optimize the modularity score [18]. The algorithm begins with each node belonging to a community and then these communities are iteratively consolidated in a manner so that it maximizes the modularity score. This process is repeated until there's no possibility of a further increase of modularity.

3.3 Walktrap Algorithm

This is a hierarchical clustering algorithm that works on the idea that short distance random walks are more likely to be bound inside the same community because there are just a few of the edges that lead outside of a given community [19]. This algorithm makes a random walk of 3–5 steps and uses the results for merging separate communities in a bottom-up manner. Again one can use the modularity score to figure out where to cut the dendogram the way it's done in FastGreedy algorithm. This algorithm is a bit slower than the fast-greedy algorithm but accuracy is more.

3.4 Louvain Algorithm

The Louvain algorithm is a greedy optimization method to detect communities by optimizing modularity [20]. In this method, initially, small communities are formed with local-modularity optimization on all the nodes, then these small communities are grouped to a single node and the previous step is repeated. Like FastGreedy algorithm, this process is repeated several times until there's no possibility of further increase of modularity.

4 Methodology

In this work, we simulate a Social Network of patients using the data taken from a large patient dataset. Here, the network is represented as a graph with the nodes representing patients. An edge between any two nodes is possible if there is some degree of similarity between the patients representing the nodes. Though there can be a maximum of ${}^nC_2$ possible connections among n patients, the non-similarity between the patients will result in no edge between the corresponding nodes. We treat the similarity score between the patients as the weights on the edges connecting the corresponding nodes. Thus, we build a Similarity matrix of the patients which is used to build the network. We then, apply the three algorithms, viz., FastGreedy, Walktrap, and Louvain on this network to detect communities and then evaluate them.

4.1 Dataset Used

The MIMIC-III dataset has been used which is an open and free clinical database containing de-identified records of patients from CCU of "Beth Israel Deaconess Medical Center" for the period 2001–2012 [21]. This dataset is of our interest because it contains discharge summaries of patients in the form of long unstructured texts containing patient profiles. The same has been used to build a patient-centric social network for our experimentation.

4.2 Building the Similarity Matrix

Figure 1 depicts the process of building the Similarity Matrix. We use the Unified Medical Language System (UMLS) [22] as a knowledge base to create concept vectors out of each of the unstructured patient profile by applying the Matcher algorithm from one of our earlier work [23]. The similarity is computed among the concept vectors, using the Jaccard similarity coefficient measure [24] resulting in a similarity matrix which we used as a weighted adjacency matrix to build our social network graph.

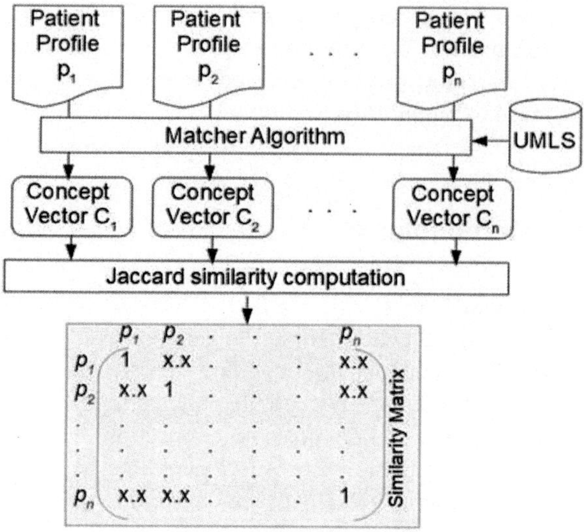

Fig. 1 Building the similarity matrix

4.3 *Building the Network*

We use the Similarity Score as the weight for our social network graph. However, if we consider all the non-zero weights, the resultant graph becomes a graph with high degree distribution; therefore the network created out of it will not be a scale-free network [25]. This can be figured out by checking whether the degree distribution follows power law [26]. If not, we can apply weight thresholding, i.e., by discarding the insignificant weights (weight $\leq \{0.2, 0.3, 04\}$) to see if we get a scale-free network exhibiting behavior of a true social network for applying the community detection algorithms.

5 Experiment

MIMIC III was installed locally in a PostgreSQL database. The 'NOTEEVENTS' table of MIMIC III database was used which is huge table with more than 2 million rows containing physician notes for the patients. We used the first 1000 rows from that table to have 1000 patient profiles. We installed UMLS v2017AA in a MySQL database to run our Matcher algorithm based on our earlier work [23] in order to get the concept vectors. We then applied Jaccard coefficient on these 1000 concept vectors to obtain the similarity matrix of size 1000 $\times$ 1000. We treat the similarity matrix as the weight adjacency matrix to create the graph. Since the Louvain community detection algorithm treats the edge weights as distance to compute the shortest paths, we normalize the weights by subtracting each weight from 1. In this way, the high similarity score between two patients will result in low distance value between

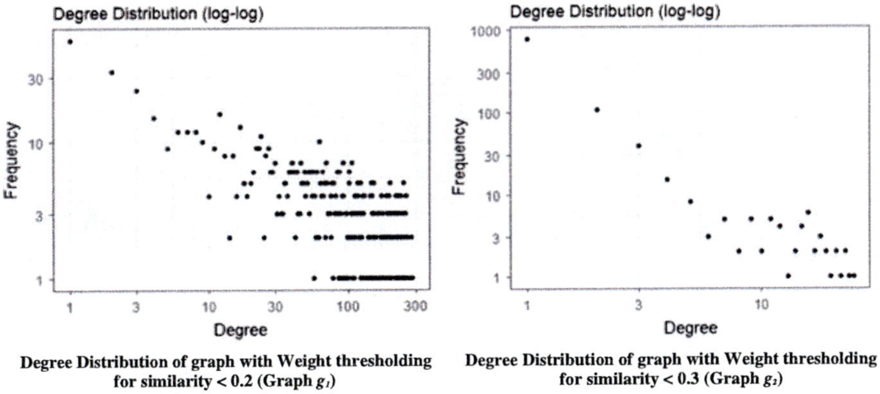

Fig. 2 Degree distribution exhibiting power law

the corresponding nodes. However, for the FastGreedy and Walktrap algorithms, the weights are used as it is since those algorithms don't treat weights as distance. We then applied different thresholding to our graph g with threshold similarity score $\lambda = 0.2, 0.3, 0.4$, and so on, i.e., we delete the edges having weight $\leq \lambda$. Therefore, we obtain subgraph g_1 of the original graph g by applying $\lambda = 0.2$, g_2 by applying $\lambda = 0.3$ and g_3 by applying $\lambda = 0.4$.

Figure 2 shows the degree distribution of g_1 and g_2. The distribution clearly follows a power law, thereby making the network scale-free. We then apply the FastGreedy, Walktrap, and Louvain algorithms on graphs g_1, g_2, and g_3 to obtain the Modularity score, number of communities, and community size for each of the cases. Figure 3a–c shows the result and the community structure. Note that we have ignored very small size communities, i.e., those having community size ≤ 2.

6 Results and Discussion

As shown in Fig. 3a, in the graph g_1, FastGreedy and Louvain algorithms could generate 6 communities with similar modularity scores $Q = 0.2877172$ and $Q = 0.2851899$, respectively. However, Walktrap algorithm generated lesser modularity than the other two but more number of communities. The low modularity ($Q = 0.2655586$) in this case means, there are relatively more external connections than internal connections within the community. On the graph g_2, as shown in Fig. 3b, we got very distinct communities with each algorithm resulting in different numbers of communities. However, in this case, we got very high modularity with the Louvain algorithm compared to FastGreedy and Walktrap. Finally, on graph g_3, as shown in Fig. 3c, all algorithms fare equally, with almost same modularity and same number of communities.

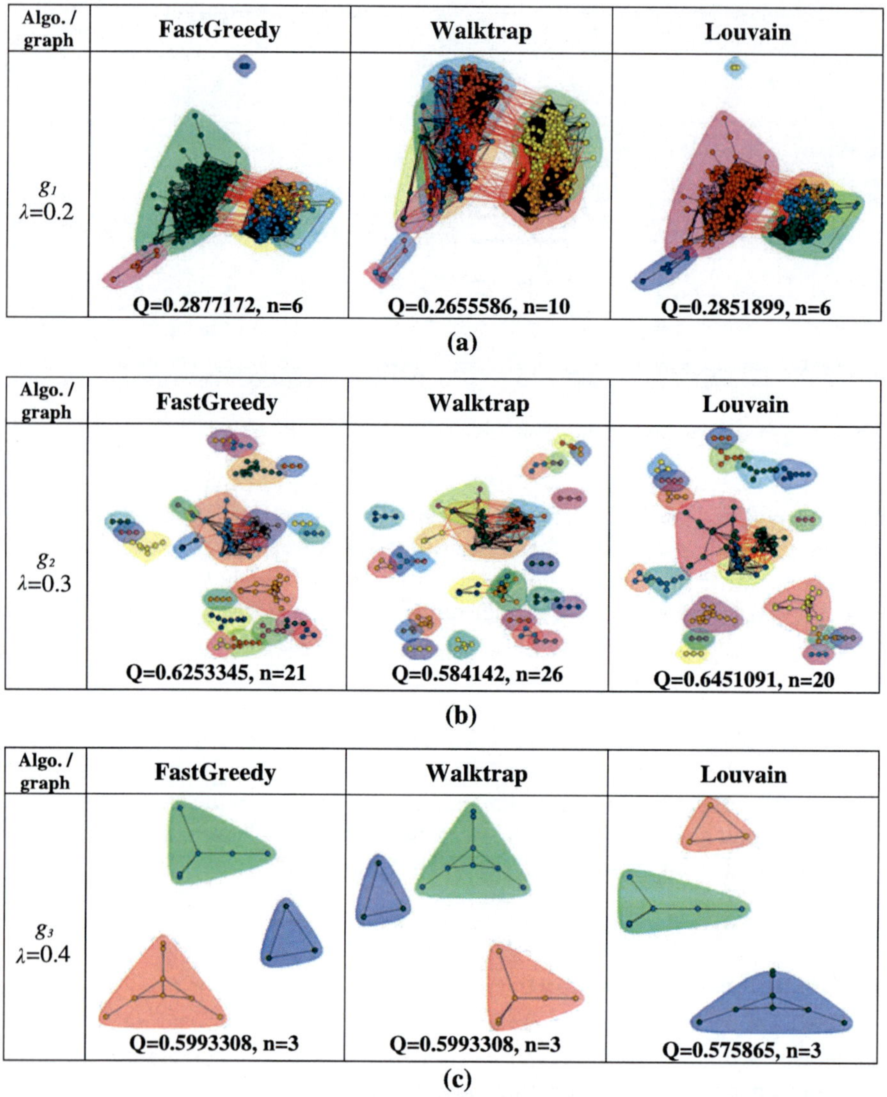

Fig. 3 **a** Community detection algorithms (FastGreedy, Walktrap, and Louvain) applied on subgraphs g_1 showing modularity Q and number of communities n. **b** Community detection algorithms (FastGreedy, Walktrap, and Louvain) applied on subgraphs g_2 showing modularity Q and number of communities n. **c** Community detection algorithms (FastGreedy, Walktrap, and Louvain) applied on subgraphs g_3 showing modularity Q and number of communities n

7 Evaluation

Community detection can be evaluated on the premise of whether the number of internal connections within communities outplays the number of external connections with other communities. We applied the Wilcoxon rank-total test [27] on the internal community degrees and the external community degrees for the communities to find their significance. The null hypothesis here is that the internal community degree and the external community degree are the same. Now, the community is significant when internal community degree is more than the external community degree, otherwise, it is insignificant or "anti-community". Figure 4 shows the outcome of Wilcoxon rank-total test applied to the community detection algorithm results on g_1. In the figure, we show the number of communities along with their p-values, i.e., the level of significance.

It's seen that, with FastGreedy, out of 6 communities detected, only one community was insignificant (p-value $= 0.241$). With Waltkrap, out of 10 communities detected, four communities were insignificant with p-values 0.554, 0.278, 0.642, and 0.447. Finally, with Louvain, only one community was insignificant (p-value $=$ 0.418) out of 6 communities detected.

Similarly, for g_2 subgraph, as shown in Fig. 5, with FastGreedy algorithm, out of 21 communities detected, as many as 13 communities found to be significant (p-value < 0.05). With Walktrap algorithm, out of 26 communities detected, 11 were significant. And, with Louvain algorithm, out of 26 communities, 20 were found to be significant.

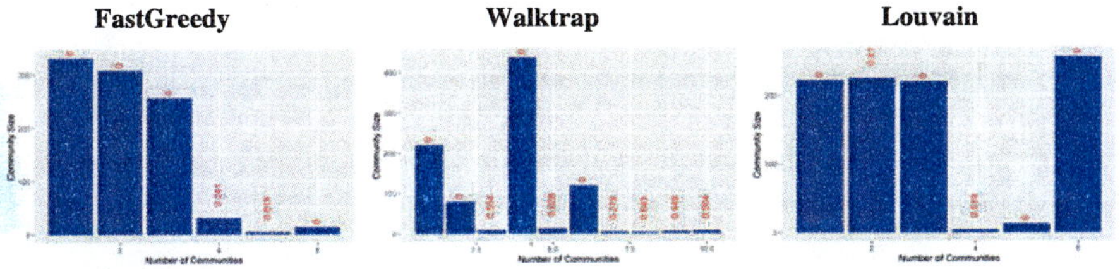

Fig. 4 Number of communities detected along with level of significance for different algorithms applied on g_1

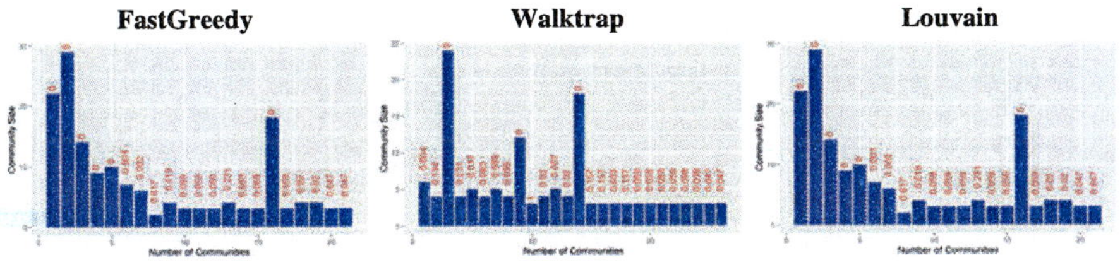

Fig. 5 Number of communities detected along with level of significance for different algorithms applied on g_2

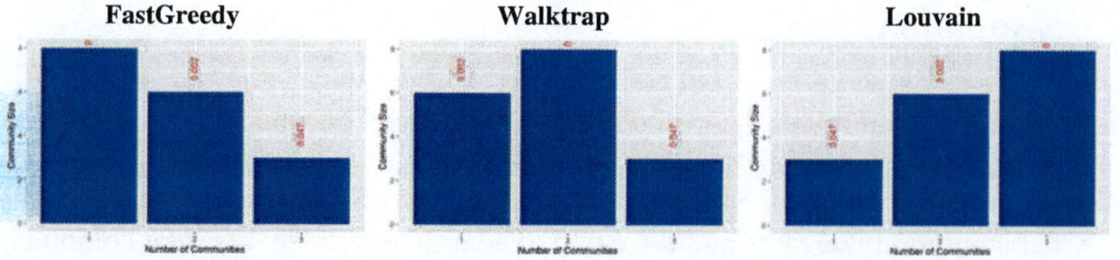

Fig. 6 Number of communities detected along with level of significance for different algorithms applied on g_3

Lastly, for g_3 subgraph as shown in Fig. 6, all the communities detected by applying the three algorithms were found to be significant. Overall, we found that Louvain algorithm fared better in comparison to FastGreedy and Walktrap algorithms with higher percentage of significant community detection.

8 Conclusion and Future Work

This research was intended to identify communities of similar patients in a social network of patients using community detection algorithms. The publicly available MIMIC III dataset was used to create a social network of patients by computing the similarities between them. The weight thresholding method was applied to show that the resulted network is scale-free. Among the three algorithms, viz., FastGreedy, Walktrap, and Louvain, we found that Louvain fares better in terms of modularity as well as the significance of the community.

The communities detected in our work are solely based on the similarity computed using physician notes for the patients. In the future, we will accommodate other parameters of patients for community detection and creation. We will also extend our work to a larger real-life patient-centric network.

References

1. Klenk, S., Dippon, J., Fritz, P., Heidemann, G.: Determining patient similarity in medical social networks. In: Proceedings of the First International Workshop on Web Science and Information Exchange in the Medical Web, pp. 6–14 (2010)
2. Smailhodzic, E., Hooijsma, W., Boonstra, A., Langley, D.J.: Social media use in healthcare: a systematic review of effects on patients and on their relationship with healthcare professionals. BMC Health Services Res. **16**(1), 442 (2016). http://bmchealthservres.biomedcentral.com/articles/10.1186/s12913-016-1691-0
3. McPherson, M., Smith-Lovin, L., Cook, J.M.: Birds of a feather: Homophily in social networks. Ann. Rev. Sociol. **27**(1), 415–444 (2001)
4. Tang, L., Liu, H.: Community detection and mining in social media. Synthesis Lect. Data Mining Knowl. Discovery **2**(1), 1–137 (2010)

5. Bedi, P., Sharma, C.: Community detection in social networks: Community detection in social networks. Wiley Interdiscip. Rev.: Data Mining Knowl. Discovery **6**(3), 115–135 (2016). http://doi.wiley.com/10.1002/widm.1178
6. Papadopoulos, S., Kompatsiaris, Y., Vakali, A., Spyridonos, P.: Community detection in social media: performance and application considerations. Data Mining Knowl. Discovery **24**(3), 515–554 (2012). http://link.springer.com/10.1007/s10618-011-0224-z
7. Nguyen, N.P., Dinh, T.N., Nguyen, D.T., Thai, M.T.: Overlapping community structures and their detection on social networks. In: 2011 IEEE Third International Conference on Privacy, Security, Risk and Trust and 2011 IEEE Third International Conference on Social Computing, pp. 35–40. IEEE, New York (2011)
8. Du, N., Wu, B., Pei, X., Wang, B., Xu, L.: Community detection in large-scale social networks. In: Proceedings of the 9th WebKDD and 1st SNA-KDD 2007 Workshop on Web Mining and Social Network Analysis, pp. 16–25. ACM, New York (2007)
9. Pizzuti, C.: Ga-net: A genetic algorithm for community detection in social networks. In: International Conference on Parallel Problem Solving from Nature, pp. 1081–1090. Springer, Berlin (2008)
10. Zhao, Z., Feng, S., Wang, Q., Huang, J.Z., Williams, G.J., Fan, J.: Topic oriented community detection through social objects and link analysis in social networks. Knowl.-Based Syst. **26**, 164–173 (2012)
11. Chen, W., Liu, Z., Sun, X., Wang, Y.: Community detection in social networks through community formation games. In: Twenty-Second International Joint Conference on Artificial Intelligence (2011)
12. Chan, S.Y., Hui, P., Xu, K.: Community detection of time-varying mobile social networks. In: International Conference on Complex Sciences, pp. 1154–1159. Springer, Berlin (2009)
13. Liu, C., Liu, J., Jiang, Z.: A multiobjective evolutionary algorithm based on similarity for community detection from signed social networks. IEEE Trans. Cybernet. **44**(12), 2274–2287 (2014)
14. Xin, L., Haihong, E., Song, J., Song, M., Tong, J.: Book recommendation based on community detection. In: Joint International Conference on Pervasive Computing and the Networked World, pp. 364–373. Springer, Berlin (2013)
15. Xu, K., Li, J., Liao, S.S.: Sentiment community detection in social networks. In: Proceedings of the 2011 iConference, pp. 804–805 (2011)
16. Newman, M.E., Girvan, M.: Finding and evaluating community structure in networks. Phys. Rev. E **69**(2), 026113 (2004)
17. Girvan, M., Newman, M.E.: Community structure in social and biological networks. Proc. Natl. Acad. Sci. **99**(12), 7821–7826 (2002)
18. Clauset, A., Newman, M.E., Moore, C.: Finding community structure in very large networks. Phys. Rev. E **70**(6), 066111 (2004)
19. Pons, P., Latapy, M.: Computing communities in large networks using random walks. In: International Symposium on Computer and Information Sciences, pp. 284–293. Springer, Berlin (2005)
20. Blondel, V.D., Guillaume, J.L., Lambiotte, R., Lefebvre, E.: Fast unfolding of communities in large networks. J. Stat. Mech: Theory Exp. **2008**(10), P10008 (2008)
21. Johnson, A.E., Pollard, T.J., Shen, L., Li-wei, H.L., Feng, M., Ghassemi, M., Moody, B., Szolovits, P., Celi, L.A., Mark, R.G.: MIMIC-III, a freely accessible critical care database. Sci. Data **3**, 160035 (2016)
22. Bodenreider, O.: The unified medical language system (UMLS): integrating biomedical terminology. Nucl. Acids Res. **32**(suppl_1), D267–D270 (2004)
23. Bissoyi, S., Patra, M.R.: Mapping clinical narrative texts of patient discharge summaries to UMLS concepts. In: Advanced Computing and Intelligent Engineering, pp. 605–616. Springer, Berlin (2020)
24. Jaccard, P.: The distribution of the flora in the alpine zone. 1. New Phytol. **11**(2), 37–50 (1912)
25. Barabási, A.L., Albert, R.: Emergence of scaling in random networks. Science **286**(5439), 509–512 (1999)

26. Albert, R., Jeong, H., Barabási, A.L.: Internet: diameter of the world-wide web. Nature **401**(6749), 130 (1999)
27. Wilcoxon, F.: Individual comparisons by ranking methods. Biometrics Bullet. **1**(6), 80–83 (1945). http://www.jstor.org/stable/3001968

Performance Analysis of Single-Phase VSI Using Variable and Multi-pulse-Width Modulation Techniques

Kurumalla Saithulasi, Panniru Raj kumar, Koppisetti Chandra Mukesh kumar, and K. RamBabu

Abstract This paper demonstrates the variable pulse-width modulation (PWM) methods for a 1-Φ full-bridge DC-AC converter. The performance of the circuit is estimated from the waveforms of output voltage and current. The harmonic spectral behavior of the modulation methods is also taken into consideration to highlight the merits and demerits of each method. A MATLAB/Simulink platform has been employed to obtain the output waveforms and fast Fourier transform (FFT) analysis.

Keywords Multi-level inverter · Fixed pulse-width modulation · Harmonic spectra · FFT · Fuzzy control

1 Introduction

Power electronic technology is finding several applications in industries, the power sector, as well as academia [1–4]. With the advent of high power rated power electronic devices like MOSFETs and IGBTs, several types of converters and inverters are begun to develop [5–8]. The evolution of power electronic technology was started with the power diodes. Later, in order to control the output variables and to obtain the desired output, thyristor-based technology has been introduced [9, 10]. The thyristor-based circuits have the limitations of lower switching frequency, lower

K. Saithulasi · P. Raj kumar · K. C. Mukesh kumar · K. RamBabu (✉)
Department of Electrical and Electronics Engineering, Aditya Engineering College, Surampalem, Andhra Pradesh, India
e-mail: rambabu.k@aec.edu.in

K. Saithulasi
e-mail: saithulasi1920@gmail.com

P. Raj kumar
e-mail: pannirurajkumar@gmail.com

K. C. Mukesh kumar
e-mail: mukeshkoppisetti@gmail.com

S. Borah et al. (eds.), *Soft Computing Techniques and Applications*, Advances in Intelligent Systems and Computing 1248,
https://doi.org/10.1007/978-981-15-7394-1_18

power handling capacity, and lack of self-commutating ability. Due to the aforesaid reasons, self-commutating devices such as MOSFETs, IGBTs, and IGCTs are getting wide attention throughout the world [11–13]. The self-commutating devices are replacing the diodes and thyristors in all power electronic applications.

In this manuscript, the application of IGBTs in DC-AC conversion circuit or inverter is studied. An inverter is basically defined as a power conversion circuit that converts DC power to AC power. The behavior of a single-phase full-bridge (FB) inverter supplied from a DC source is analyzed in this paper using various variable multi-pulse-width modulation techniques. This script is detailed as follows: Sect. 2 describes the operation of single-phase FB inverter, and the operating modes; in Sect. 3, the variable multiple PWM techniques are comprehensively presented; in Sect. 4, the simulation outcomes are presented and discussed; Sect. 5 discusses the conclusion.

2 System Configuration

Figure 1 shows the proposed 1-Φ full-bridge inverter using the self-commutating devices called IGBTs. Four IGBTs along with their body diodes are connected in two legs across a DC source. All the switching devices used in the proposed configuration are of bidirectional conducting devices with a unidirectional voltage blocking capability. The DC sources can be obtained either from rectifier circuits, batteries, or pv arrays.

Figure 2 shows the zero-level operating mode during positive zero-crossings. The switches S_2 and S_4 are triggered during this state. However, the conduction of either the IGBT or the anti-parallel diode is determined by the direction of the output current [14–16]. In Fig. 2a, $V_0 = 0^+$ and i_0 is positive, hence, the conducting devices are S_2 and D_4. In Fig. 2b, $V_0 = 0^+$, but i_0 is negative, hence, the conducting devices are S_4 and D_2.

Figure 3 gives $V_0 = 0^-$, this switching state is used during negative zero-crossing of the output. The switches S_1 and S_3 are triggered during this state. However, the conduction of either the IGBT or the anti-parallel diode is determined by the direction of the output current. In Fig. 3a, $V_0 = 0^-$ and i_0 is positive, hence, the conducting devices are S_1 and D_3. In Fig. 3b, $V_0 = 0^-$, but i_0 is negative, hence, the conducting

Fig. 1 Schematic diagram of the circuit configuration

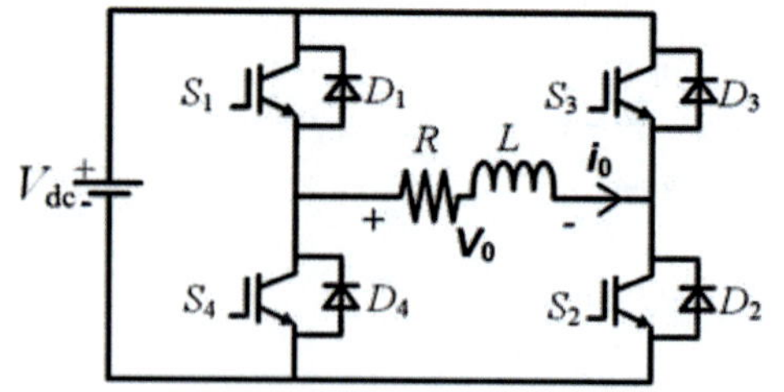

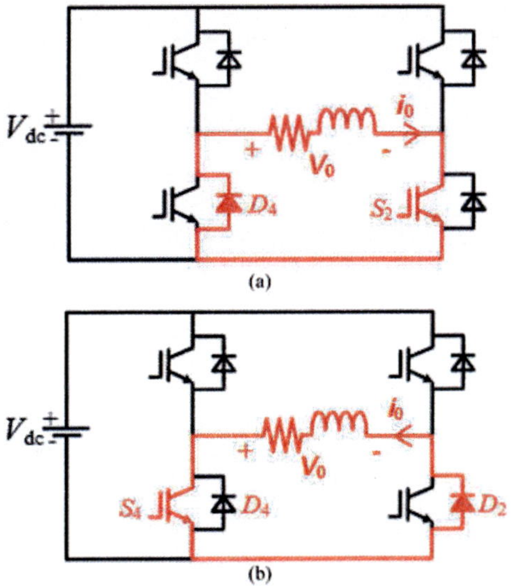

Fig. 2 Positive zero-crossing operating modes: **a** I_0 = positive and **b** I_0 = negative

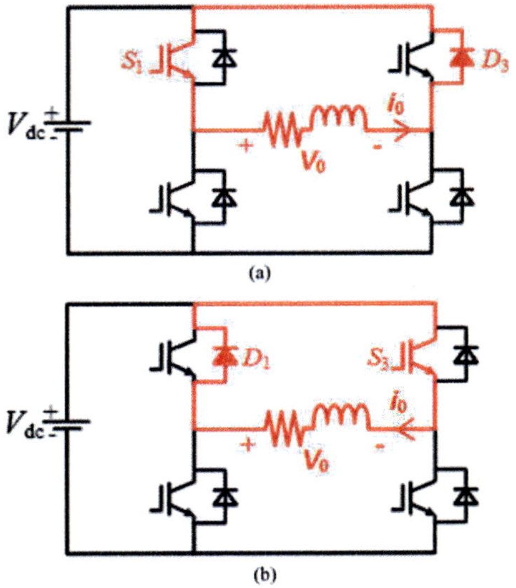

Fig. 3 Negative zero-crossing operating modes: **a** I_0 = positive and **b** I_0 = negative

devices are S_3 and D_1. Both the switching states shown in Figs. 2 and 3 are effectively utilized to maintain the thermal stability of the system.

Figure 4 shows the positive-level operating modes of the converter. Figure 4a gives the output voltage $V_0 = V_{dc}$, during this state the switching devices S_1 and S_2 are triggered. When i_0 is positive, S_1 and S_2 conduct. If i_0 is negative D_1 and D_2 conduct. Figure 5 gives the output voltage $V_0 = -V_{dc}$, during this state the switching devices S_3 and S_4 are triggered. When i_0 is positive, D_3 and D_4 conduct. If i_0 is negative S_3

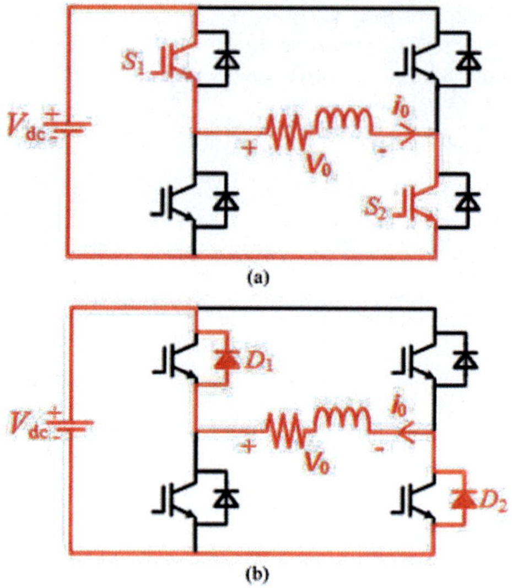

Fig. 4 V_{dc}-level operating modes: **a** I_0 = positive and **b** I_0 = negative

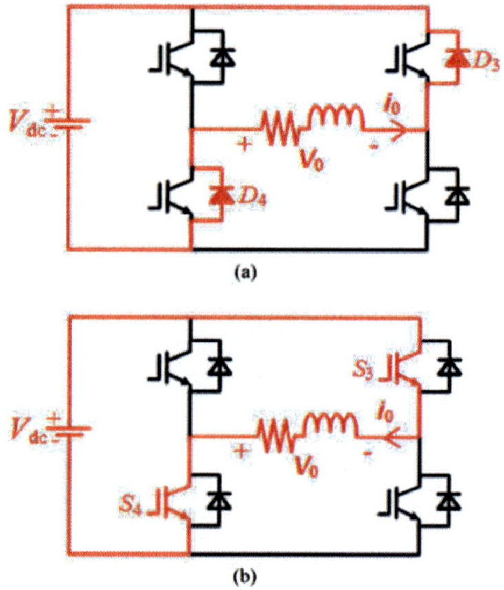

Fig. 5 V_{dc}-level operating modes: **a** I_0 = positive and **b** I_0 = negative

and S_4 conduct. The duration of conduction of the anti-parallel diodes and IGBTs is decided by the values of resistance (R) and inductance (L) in the load.

3 Modulation Techniques

In this section, several methods of triggering the inverter switches using multiple variable PWM techniques are explored. In order to generate the variable and multiple pulses per half-cycle, a sine-triangle comparison PWM method is adopted in the present study. The complete control logic circuit to trigger the inverter is presented in separate figures.

3.1 Unipolar Pulse-Width Modulation

In this technique, during positive half-cycle, the switching device S_2 is clamped to on-state, whereas the switching devices S_1 and S_4 in the first leg are gated with PWM pulses to produce V_{dc} and 0 output voltages, respectively. Similarly, during negative half-cycle, the switching device S_1 is clamped to on-state, whereas the switching devices S_2 and S_3 in the second leg are gated with PWM pulses to produce 0 and $-V_{dc}$ output voltages, respectively. Due to the availability of unipolar voltage in a half-cycle, i.e., V_{dc} during positive half-cycle and $-V_{dc}$ during negative half-cycle, this method is observed as unipolar PWM. Figure 6 shows unipolar PWM control strategy, in which two fundamental sinusoidal waves of equal amplitude (A) and phase-shifted by 180° are compared with a triangular function. The resultant pulses are utilized through logic gates to generate the switching pulses to the inverter. The sinusoidal amplitude (A) is also called the modulation index (M. I), which is defined as follows:

$$\text{M.I.} = \frac{V_{0\text{peak}}}{3 \times V_{dc}} \tag{1}$$

where $V_{0\text{peak}}$ is the peak value of the fundamental component of the output voltage and V_{dc} is the magnitude of the input DC voltage.

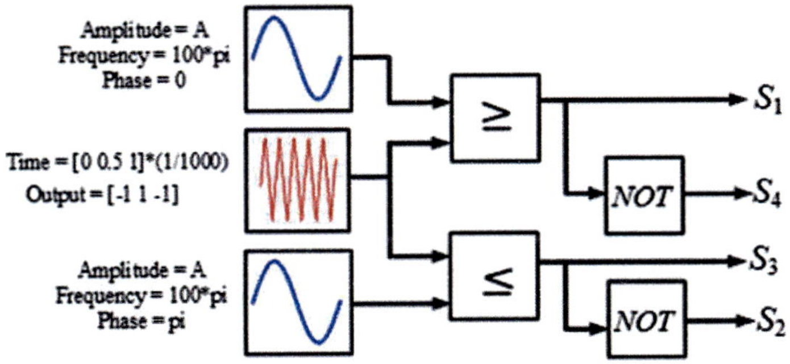

Fig. 6 Unipolar pulse-width modulation scheme

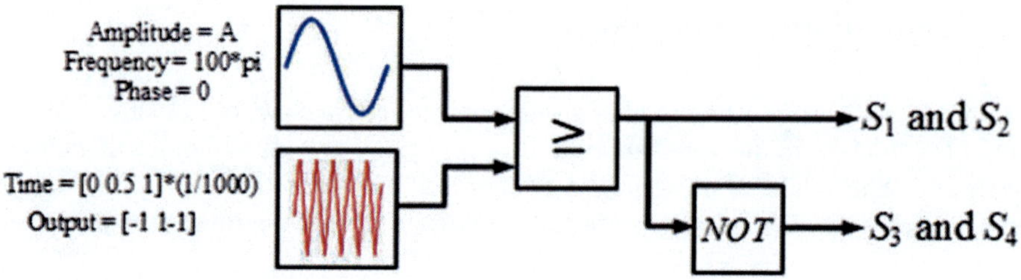

Fig. 7 Bipolar pulse-width modulation scheme

Table 1 Switching sequence of the inverter

V_0	G_1	G_2	i3	G_4	Conducting devices	
					$I_0 > 0$	$I_0 < 0$
V_{dc}	1	1	0	0	S_1 and S_2	D_1 and D_2
0^+	0	1	0	1	S_2 and D_4	D_2 and S_4
0^-	1	0	1	0	S_1 and D_3	D_1 and S_3
$-V_{dc}$	0	0	1	1	D_3 and D_4	S_3 and S_4

3.2 Bipolar Pulse-Width Modulation

Figure 7 shows bipolar PWM control strategy, in which a single fundamental sinusoidal wave of amplitude (A) with 0° phase shift is compared with a triangular function. The resultant pulse is given to the diagonal switches in the inverter circuit, i.e., S_1 and S_2. The complementary of this pulse is obtained through a *NOT* gate and is given to the other group of diagonal switches in the inverter circuit, i.e., S_3 and S_4.

Table 1 lists the various switching states of the converter operating in fixed PWM mode. The gate pulses given to the inverter switches are represented as G_1–G_4. The ON and OFF states of the gate pulses are indicated as 1 and 0, respectively. The device that conducts in any state depends on the direction of the current I_0. The conducting devices for different directions of the load current are also listed in Table 1. It can be observed that only two devices either the switches or diodes or the combination of a switch and diode conduct for a given voltage level.

4 Simulation Results

Table 2 lists the parameters considered in the simulation. Figures 8a and 9a show the inverter output results for unipolar and bipolar modes, respectively. It can be observed that the inverter is producing a three-level output in unipolar pwm due to the presence of V_{dc}, 0, and $-V_{dc}$ levels. The same inverter circuit operates in two-level mode by gating with the bipolar pwm pulses. It can be understood from

Table 2 Configuration parameters

Parameters	Values	
	Unipolar PWM	Bipolar PWM
Switching device	IGBT	IGBT
V_{dc}	300 V	300 V
P_{output}	705 W	936 W
V_0	227 V	300 V
I_0	3.4 A	3.46 A
I_{dc}	1.9 A	2.0 A
Fundamental frequency (f_m)	50 Hz	50 Hz

Table 2 that the root mean square (RMS) value of the output voltage in a bipolar mode is greater than that of unipolar mode. However, the RMS value of currents in both the modes is almost the same. The harmonic spectra of output voltage and currents in unipolar pwm mode are shown in Fig. 8b, c, respectively. The harmonic spectra of output voltage and currents in bipolar pwm mode are shown in Fig. 9b, c, respectively. The %total harmonic distortion (%THD) of the inverter output voltage is almost doubled from unipolar to bipolar pwm mode, whereas the current THD is increased by five times. Finally, it can be concluded that the %THD of current harmonics is less in unipolar pwm, which is below the IEEE 1547 grid standard.

Figures 8d and 9d provide the input DC bus current of the inverter in unipolar and bipolar pwm modes, respectively. It is noted that the average value of DC bus current is approximately equal in both the modes of operation. The magnitude of DC bus currents is also listed in Table 2.

5 Conclusion

This paper provides a detailed analysis and explanation of variable multi-pulse-width modulation strategies existing for a single-phase full-bridge inverter. The control strategy of both the unipolar and bipolar PWM methods is presented in detail. The simulation model is evaluated in MATLAB/Simulink to show the inverter performance and outputs for different modulation techniques. Finally, the output results and spectral performance are presented for each method.

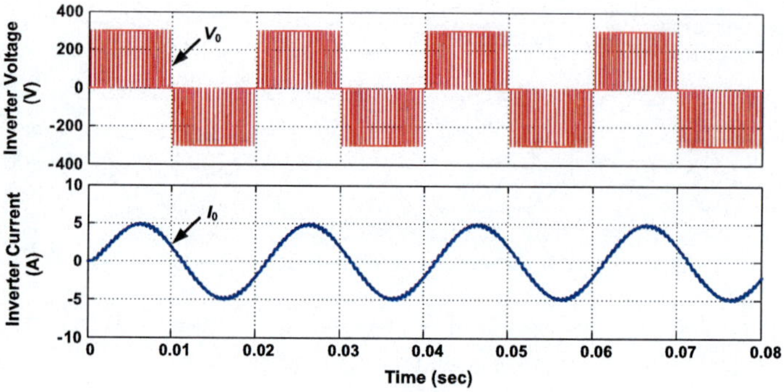

(a)

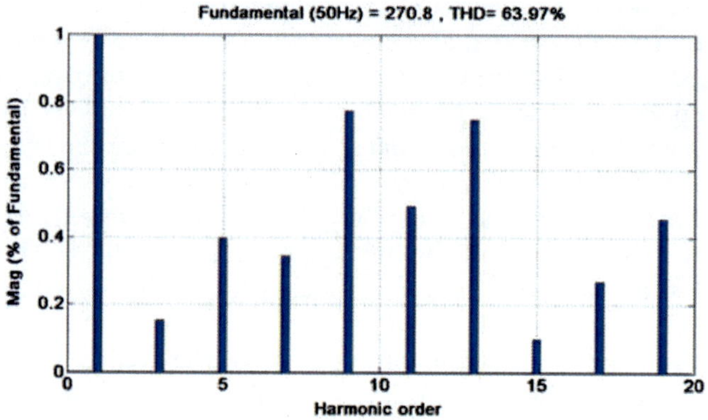

(b)

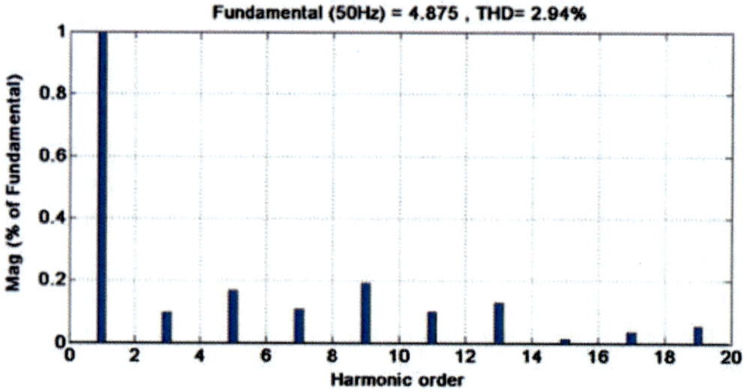

(c)

Fig. 8 Simulation outcomes for unipolar PWM: **a** inverter voltage and currents, **b** harmonic analysis of inverter voltage, and **c** harmonic analysis of inverter current, **d** input DC current

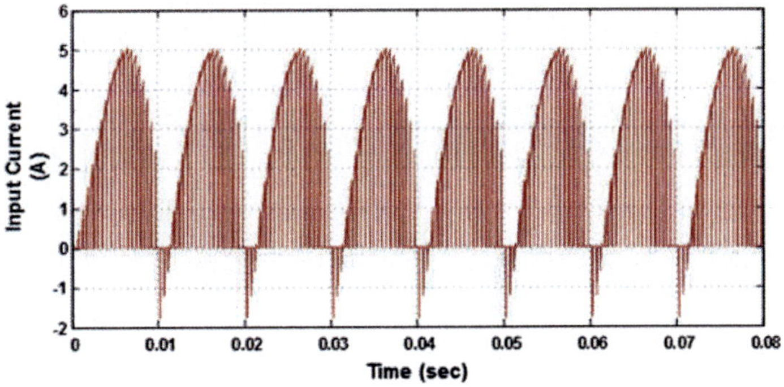

(d)

Fig. 8 (continued)

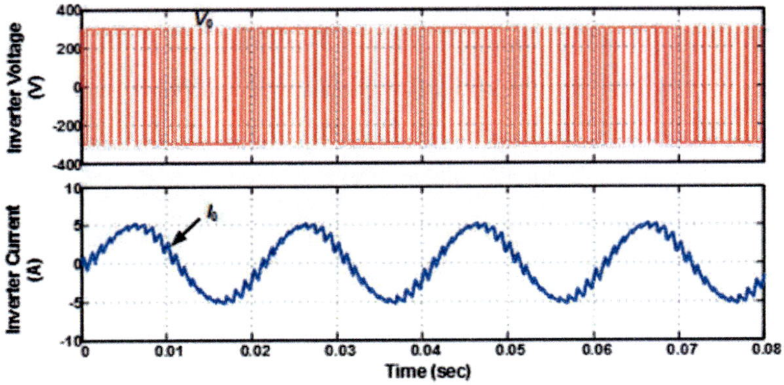

(a)

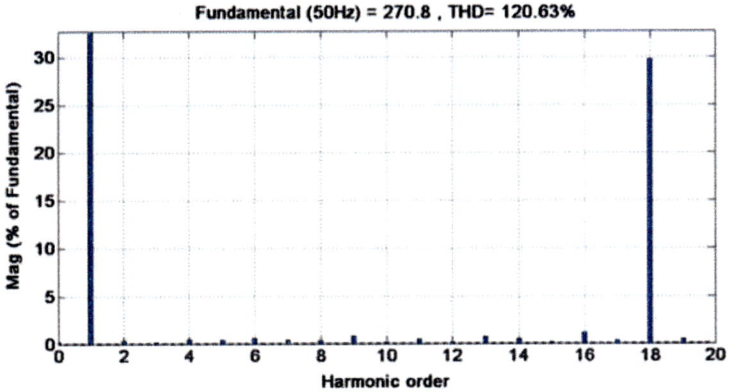

(b)

Fig. 9 Simulation outcomes for bipolar PWM: **a** inverter voltage and currents, **b** harmonic analysis of inverter voltage, and **c** harmonic analysis of inverter current, **d** input DC current

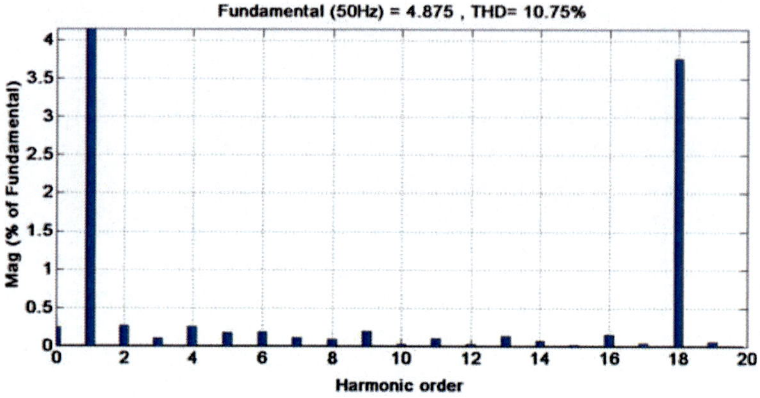

(c)

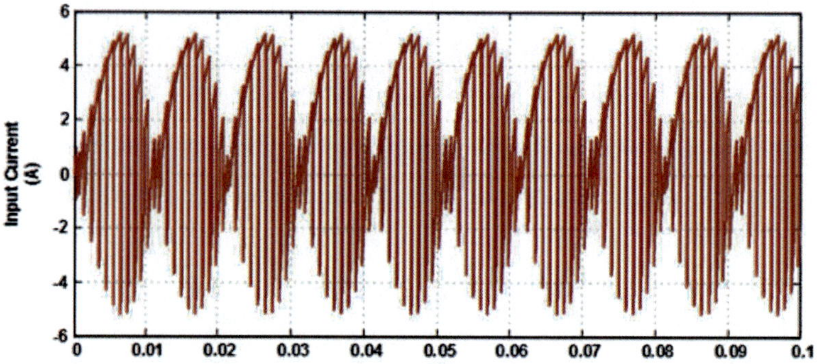

(d)

Fig. 9 (continued)

References

1. Carroll, E.I.: Power electronics: where next? Power Eng. J., pp. 242–243 (1996)
2. Chan, M.S.W., Chau, K.T.: A new switched-capacitor boost multilevel inverter using partial charging. IEEE Trans. Circuits Syst. II, Exp. Briefs 54(12), 1145–1149 (2007)
3. Nabae, A., Takahashi, I., Akagi, H.: A new neutral-point clamped PWM inverter. IEEE Trans. Industry Appl. **IA-17**, 518–523 (1981)
4. Bernet, S.: Recent developments of high power converters for industry and traction applications. IEEE Trans. Power Electron. **15**(6), 1102–1117 (2000)
5. Meynard, T.A., Foch, H.: Multi-level choppers for high voltage applications. Euro. Power Electron. Drives J. **2**(1), 45–50 (1992)
6. Hoft, R.G.: Semiconductor Power Electronics. Van Nostrand Reinhold, New York (1986)
7. Ounejjar, Y., Al-Haddad, K., Dessaint, L.A.: A novel six-band hysteresis control for the packed U cells seven-level converter: Experimental validation. IEEE Trans. Ind. Electron. **59**(10), 3808–3816 (2012)
8. Saeedifard, M., Barbosa, P.M., Steimer, P.K.: Operation and control of a hybrid seven-level converter. IEEE Trans. Power Electron. **27**(2), 652–660 (2012)
9. Bedford, B.D., Hoft, R.G.: Principle of Inverter Circuits. Wiley, New York (1964)

10. Sheng, W., Ge, Q.: A novel seven-level ANPC converter topology and its commutating strategies. IEEE Trans. Power Electron. **33**(9), 7496–7509 (2018)
11. Tsunoda, A., Hinago, Y., Koizumi, H.: Level and phase shifted PWM for 7-level switched-capacitor inverter using series/parallel conversion. IEEE Trans. Ind. Electron. **61**(8), 4011–4021 (2014)
12. Thorborg, K., Nystorm, A.: Staircase PWM: an uncomplicated and efficient modulation technique for ac motor drives. IEEE Trans. Power Electron. **PE3**(4), 391–398 (1988)
13. Taniguchi, K., Irie, H.: PWM technique for power MOSFET inverter. IEEE Trans. Power Electron. **PE3**(3), 328–334 (1988)
14. Abhilash, T., Kirubakaran, A., Somasekhar, V.T.: A new hybrid flying capacitor based single phase nine-level inverter. Int. Trans. Electr. Energy Syst. **29**(12), 1–15 (2019)
15. Abhilash, T., Kirubakaran, A., Somasekhar, V.T.: A new structure of three-phase five-level inverter with nested two-level cells. Int. J. Circuit Theory Appl. **47**(9), 1435–1445 (2019)
16. Abhilash, T., Kirubakaran, A., Somasekhar, V.T.: A seven-level VSI with a front-end cascaded three-level inverter and flying capacitor fed H-bridge. IEEE Trans. Ind. Appl. **55**(6), 6073–6088 (2019)

Design of Power Conditioning Unit for Wind Energy Conversion System Using Resonant Converter

K. Ganesh Sai Reddy, K. Sai Babu, D. V. L. N. Murthy, and K. Prabharani

Abstract This paper provides design details of power conditioning unit for wind energy conversion system. Power conditioning unit using DC-DC resonant converter are discussed in this paper. Different resonant DC-DC converter topologies have been investigated with respect to the wind energy conversion system. Fuzzy based control scheme has been discussed for DC-DC converter. Simulation results and analysis have been provided.

Keywords Resonant converter · Wind energy conversion system · Simulation · Fuzzy control · Modeling

1 Introduction

To effectively satisfy the increase in electricity demand, wind energy have been used extensively. As wind speed variation is difficult to predict, weibull probability distribution function is used. For extracting electrical energy from wind turbine, power conditioning unit (PCU) is used [6, 7]. Resonant converter are special type DC-DC converter which uses soft-switching concept using resonant tank of L and C [3, 4, 8].

K. Ganesh Sai Reddy · K. Sai Babu · D. V. L. N. Murthy · K. Prabharani (✉)
Department of Electrical and Electronics Engineering, Aditya College of Engineering and Technology, Surampalem, Andhra Pradesh, India
e-mail: prabharani3212@gmail.com

K. Ganesh Sai Reddy
e-mail: ganeshsaireddy70@gmail.com

K. Sai Babu
e-mail: kalimisai0@gmail.com

D. V. L. N. Murthy
e-mail: nareshdhoni864@gmail.com

S. Borah et al. (eds.), *Soft Computing Techniques and Applications*, Advances in Intelligent Systems and Computing 1248,
https://doi.org/10.1007/978-981-15-7394-1_19

Generator capacity for wind turbine has been studied in [11]. Different techniques for high power wind conversion system has been discussed in [10]. Discrete modelling technique has been discussed in [2, 9]. Grid code for wind power generation has been discussed in [1]. In [5], optimal parameters for wind turbine has been determined and in [11], generator capacity under different heights have been discussed.

This paper provides a detailed modelling and simulation aspect of wind energy conversion system where stochastic nature of wind is modelled using Weibull distribution and resonant converter is used. Comparative analysis of different resonant converter has been provided. Fuzzy-controller has been designed.

2 Wind Energy Conversion System

Figure 1 illustrates PCU for wind energy conversion system. The unregulated AC provided by permanent magnet synchronous generator (PMSG) is converted to DC voltage using PCU.

Power of wind at any time instance can be represented as

$$P_w = \frac{1}{2}\rho A v^3 \tag{1}$$

Cost of energy (COE) can be represented as

$$\text{COE} = \frac{C_c FCR}{E_a} + \bar{C}_{OM} \tag{2}$$

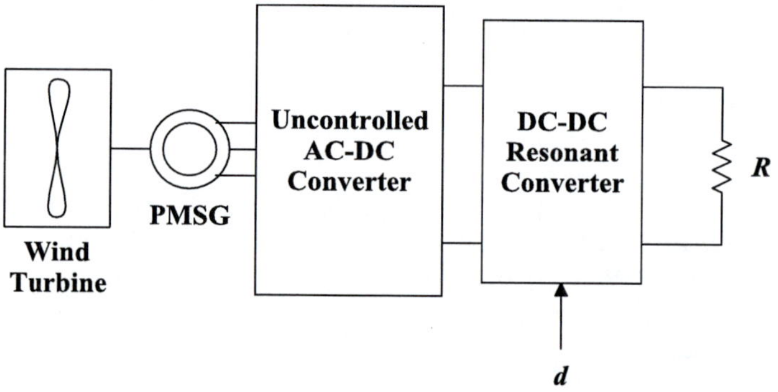

Fig. 1 Block diagram of power conditioning unit

2.1 Modeling of PMSG

Dynamics of PMSG in d–q reference frame (Fig. 2) is

$$\begin{aligned} \frac{di_{sd}}{dt} &= -\frac{R_{sa}}{L_{sd}} i_{sd} + \omega_s \frac{L_{sq}}{L_{sd}} i_{sq} + \frac{1}{L_{sd}} u_{sd} \\ \frac{di_{sq}}{dt} &= -\frac{R_{sa}}{L_{sq}} i_{sq} - \omega_s \left(\frac{L_{sd}}{L_{sq}} i_{sd} + \frac{1}{L_{sq}} \psi_p \right) + \frac{1}{L_{sq}} u_{sq} \end{aligned} \tag{3}$$

$$T_e = 1.5 \frac{P}{2} \left(\psi_p i_{sq} + i_{sd} i_{sq} \left(L_{sd} - L_{sq} \right) \right) \tag{4}$$

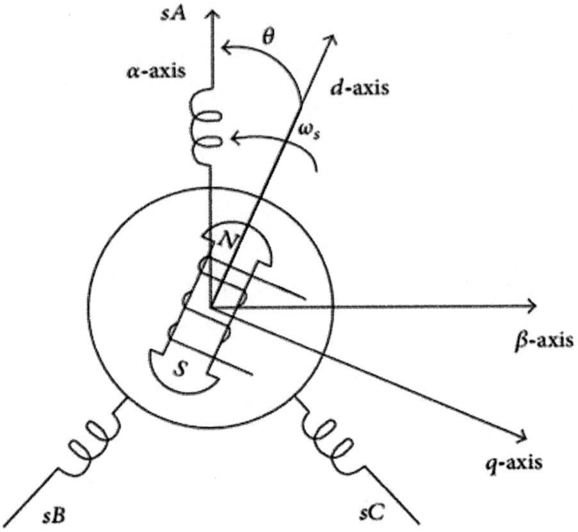

Fig. 2 d-q model of permanent magnet synchronous generator (PMSG)

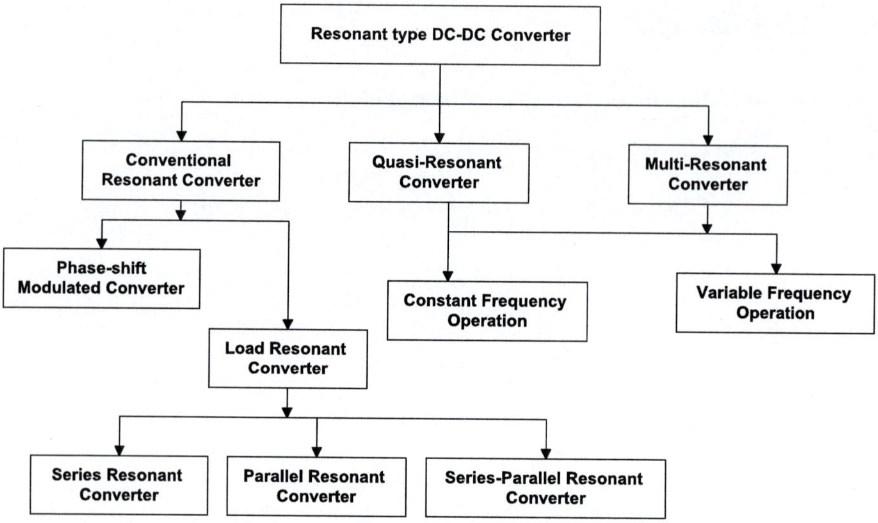

Fig. 3 Classification of resonant converter

3 Resonant Converter

Classification taxonomy of resonant converter is represented (Figs. 3 and 4).

Simulation parameter and voltage gain have been provided in Table 1.

Figures 5a, b provides probability density function of wind. Figure 6 illustrates the power wind turbine characteristics.

Figure 7a and subsequent figure shows the gain plot of resonant converter.

4 Conclusion

This paper provide details of power conditioning unit for wind energy conversion system. In the PCU, resonant DC-DC converter is used. Different characteristics of resonant converter is described in this paper. Different topologies of resonant DC-DC converter has been discussed in this paper. A comparative review of different topologies have been carried out. Fuzzy based control scheme and implementation scheme can also be provided.

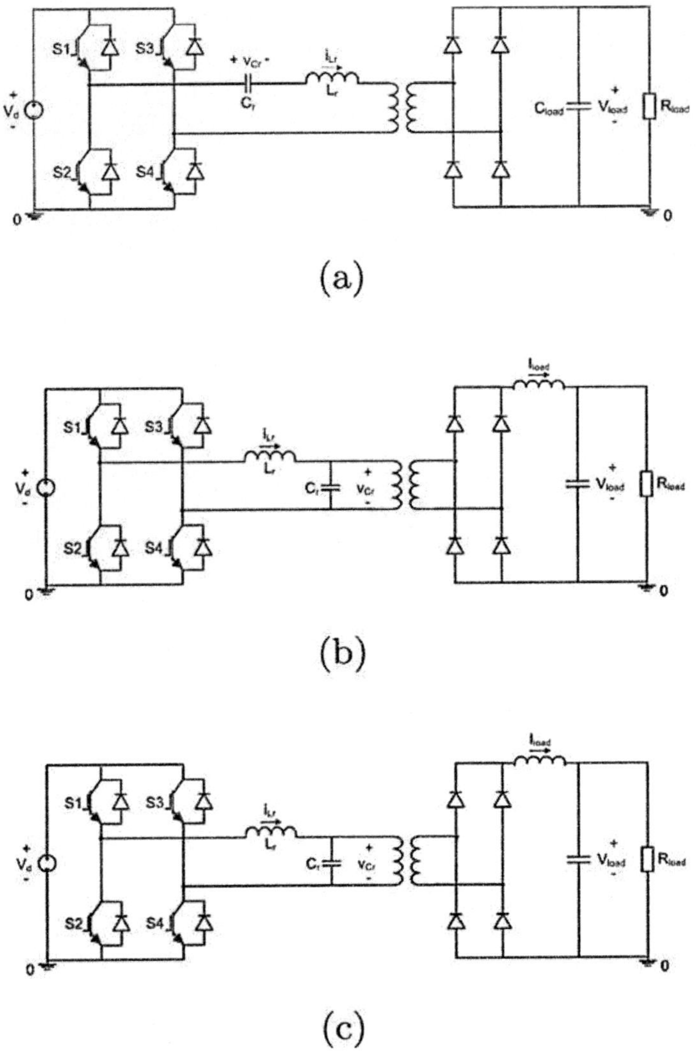

Fig. 4 Resonant converter **a** Series, **b** Parallel, **c** Series-Parallel

Table 1 Simulation parameters

Parameters	SRC	PRC	SPRC
Transformer turns ratio	9:1	6:1	6:1
L_r	58 μH	58 μH	72 μH
C_{sr}	11.7 nF	11.7 nF	17.7 nF
C_{pr}			17.7 nF
Gain	$\frac{V_O}{V_{in}} = \frac{1}{1+j\left(\frac{X_L}{R_{ac}}-\frac{X_C}{R_{ac}}\right)}$	$\frac{V_O}{V_{in}} = \frac{1}{1-\frac{X_L}{X_C}+j\frac{X_L}{R_{ac}}}$	$\frac{V_O}{V_{in}} = \frac{1}{1+\frac{X_{CS}}{X_{CP}}-\frac{X_L}{X_{CP}}+j\left[\frac{X_L}{R_{ac}}-\frac{X_{CS}}{R_{ac}}\right]}$

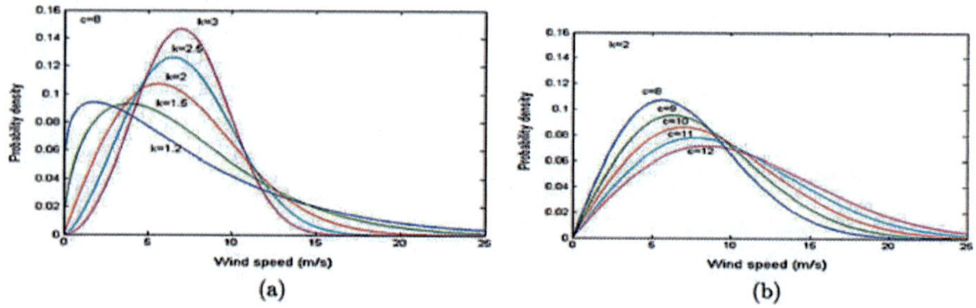

Fig. 5 Probability density function of wind speed

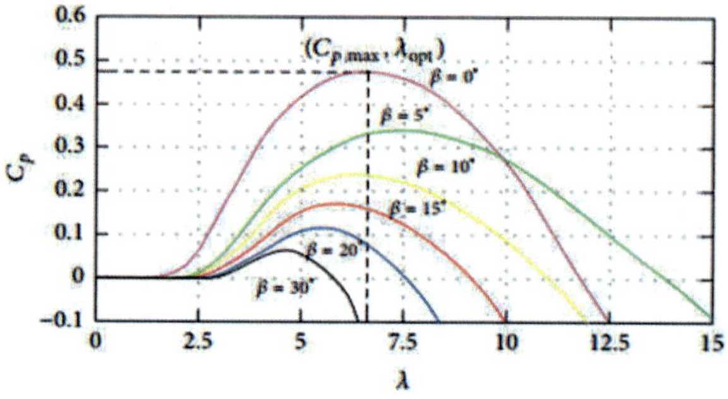

Fig. 6 Power wind turbine coefficient

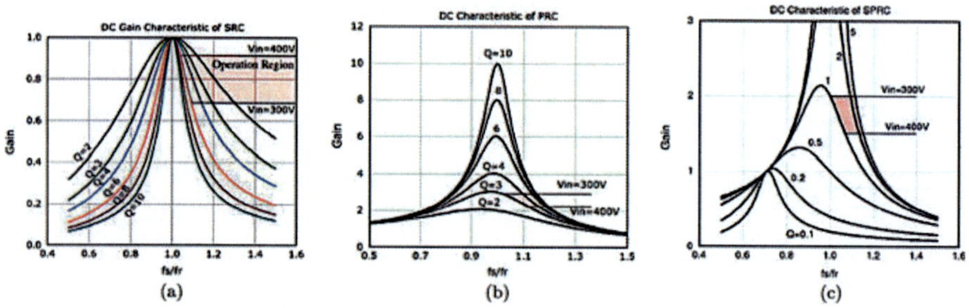

Fig. 7 Gain plot of resonant converter **a** Series, **b** Parallel, **c** Series-parallel

Acknowledgements The authors are thankful to "Accendere Knowledge Management Services Pvt. Ltd", "CL Educate Ltd" for their assistance during the preparation of the manuscript.

References

1. Bhawalkar, M., Gopalakrishnan, N., Nerkar, Y.: A review of Indian grid codes for wind power generation. J. Inst. Eng. (India): Ser. B, pp. 1–9 (2019)
2. Chen, Y.H., Dincan, C.G., Kjær, P., Bak, C.L., Wang, X., Imbaquingo, C.E., Sarrà, E., Isernia, N., Tonellotto, A.: Model-based control design of series resonant converter based on the discrete time domain modelling approach for dc wind turbine. J. Renew. Energy **2018** (2018)
3. Dincan, C., Kjaer, P., Chen, Y.H., Munk-Nielsen, S., Bak, C.L.: Analysis of a high-power, resonant dc–dc converter for dc wind turbines. IEEE Trans. Power Electron. **33**(9), 7438–7454 (2017)
4. Dincan, C.G., Kjaer, P., Chen, Y.H., Sarrá-Macia, E., Munk-Nielsen, S., Bak, C.L., Vaisambhayana, S.: Design of a high-power resonant converter for dc wind turbines. IEEE Trans. Power Electron. **34**(7), 6136–6154 (2018)
5. Jangamshetti, S.H., Rau, V.G.: Normalized power curves as a tool for identification of optimum wind turbine generator parameters. IEEE Trans. Energy Convers. **16**(3), 283–288 (2001)
6. Max, L., Lundberg, S.: System efficiency of a dc/dc converter-based wind farm. Wind Energy: Int. J. Progress Appl. Wind Power Convers. Technol. **11**(1), 109–120 (2008)
7. Padhee, S., Pati, U.C., Mahapatra, K.: Overview of high-step-up dc-dc converters for renewable energy sources. IETE Tech. Rev. **35**(1), 99–115 (2018)
8. Steigerwald, R.L.: A comparison of half-bridge resonant converter topologies. IEEE Trans. Power Electron. **3**(2), 174–182 (1988)
9. Tobías-González, A., Peña-Gallardo, R., Morales-Saldaña, J., Medina-Ríos, A., Anaya-Lara, O.: A state-space model and control of a full-range pmsg wind turbine for real-time simulations. Electr. Eng. **100**(4), 2177–2191 (2018)
10. Yaramasu, V., Wu, B., Sen, P.C., Kouro, S., Narimani, M.: High-power wind energy conversion systems: state-of-the-art and emerging technologies. Proc. IEEE **103**(5), 740–788 (2015)
11. Yeh, T.H., Wang, L.: A study on generator capacity for wind turbines under various tower heights and rated wind speeds using Weibull distribution. IEEE Trans. Energy Convers. **23**(2), 592–602 (2008)

A Numerical Study on Atangana–Baleanu and Caputo–Fabrizio Analysis of Fractional Derivatives for the Effects of Variable Viscosity and Thermal Conductivity on an MHD Flow Over a Vertical Hot Stretching Sheet with Radiation and Viscous Dissipation

Dipen Saikia, Utpal Kumar Saha, and G. C. Hazarika

Abstract A numerical analysis has been made to study the effects of variable viscosity and thermal conductivity over a vertical hot stretching sheet by using Atangana–Baleanu (AB) and Caputo–Fabrizio (CF) fractional derivatives under the application of magnetic field in two-dimensional steady flow. As the viscosity and thermal conductivity of a fluid are dependent on temperature, these properties are taken as a variable. The effects of radiation and viscous dissipation are also taken into account. The governing partial differential equations along with the boundary conditions are transformed into ordinary form by similarity transformations so that physical parameters appear in the equations and interpretations on these parameters can be done suitably. The equations so obtained are discritized using ordinary finite difference scheme and the discritized equations are solved numerically using a method based on the Gauss–Seidel iteration scheme. Numerical techniques are used to find the values from AB and CF formulae for fractional derivatives on time. The effects of various parameters involved in the problem viz., parameter of viscosity, parameter of thermal conductivity, parameter of magnetic field, radiation parameter, heat source parameter, Schmidt number, chemical reaction parameter, etc., on velocity, temperature, and concentration distribution are represented graphically. The

D. Saikia (✉) · U. K. Saha
Department of Basic & Applied Science, NIT Arunachal Pradesh, Yupia 791112, India
e-mail: deepsand2009@gmail.com

U. K. Saha
e-mail: utpal@nitap.ac.in

G. C. Hazarika
Department of Mathematics, Dibrugarh University, Dibrugarh, Assam 786004, India
e-mail: gchazarika@gmail.com

S. Borah et al. (eds.), *Soft Computing Techniques and Applications*, Advances in Intelligent Systems and Computing 1248,
https://doi.org/10.1007/978-981-15-7394-1_20

effects of each parameter are prominent. A comparison has been given on AB and CF methods in tabular form. It is observed that both the methods agreed well.

Keywords Variable viscosity · Thermal conductivity · Hot stretching sheet · Radiation · Viscous dissipation

1 Introduction

The study of fluid flow in the boundary layer and heat transfer of a viscous fluid over a hot stretching sheet has so many important roles in fluid dynamics such as blowing of glass, processing of polymer, and metal extrusion. So the study of variable viscosity and thermal conductivity on an MHD flow over a vertical hot stretching sheet has gained more significance.

Saikiadis [1] carried out pioneering work on stretching flow problem. Different aspects of the problem have been investigated on the flow caused by the stretching sheet by many authors such as Crane [2], Sharidan [3], Carraagher et al. [4], Cortell [5], Xu and Liao [20], Hazarika [6], and Hayat and Sajid [7]. Magyari and Keller [8] studied heat and mass transfer on boundary layer flow due to an exponentially continuous stretching surface. The researchers Gupta and Gupta [9], Dutta et al. [1] analyzed the work of Crane [2] by including the effect of heat and mass transfer analysis under different circumstances.

The temperature distribution is changed by viscous dissipation and it does so by playing a role like an energy source that leads to affected heat transfer rates. The value of the impact of viscous dissipation relies upon whether the plate is being cooled or heated. Again, the physical properties of the liquid may change fundamentally with temperature. The growth in temperature ends in the growth inside the transport phenomena through lowering the viscosity throughout the momentum boundary layer and due to which the heat transfer rate at the wall is likewise affected. Many authors in [3, 10–17] are discussed about the effect of temperature-dependent viscosity in different situations.

The main aim of this paper to investigate the effects of variable viscosity and thermal conductivity on a vertical stretching sheet in present of heat and mass transfer. The effects of radiation and viscous dissipation have been also in use into account. The main partial differential equations along with the boundary conditions are made dimensionless by using some suitable non-dimensional parameters. The non-dimensional main equations with the non-dimensional boundary conditions are discretized with ordinary finite difference kernel solved numerically with the help of AB fractional derivative and CF fractional derivative method by developing suitable programming code in MATLAB. Comparisons of the results obtained by both the methods are shown in tabular form.

2 Formulation of the Problem

Here, we consider a steady two-dimensional incompressible viscous fluid bounded by a stretching sheet, where $\overline{x}$-axis is taken along the stretching sheet in the direction of the motion and $\overline{y}$-axis is perpendicular to it. At $t = 0^{+}$, the double diffusion from the plate to the fluid has gained the temperature $\overline{T}_w$, and concentration level near the plate is $\overline{C}_w$. A variable magnetic field $B(x) = B_0 e^{x/2L}$ is applied normally to the sheet surface while the induced magnetic field is negligible, where B_0 is a constant magnetic field.

Under the usual boundary layer approximation, the flow and heat transfer in the presence of radiation effects with variable viscosity and thermal conductivity are governed by the following equations:

Continuity equation:

$$\frac{\partial \overline{u}}{\partial \overline{x}} + \frac{\partial \overline{v}}{\partial \overline{y}} = 0 \tag{1}$$

Conservation of momentum equation:

$$\begin{aligned} \frac{\partial \overline{u}}{\partial \overline{t}} + \overline{u}\frac{\partial \overline{u}}{\partial \overline{x}} + \overline{v}\frac{\partial \overline{u}}{\partial \overline{y}} &= \vartheta \frac{\partial^2 \overline{u}}{\partial \overline{y}^2} + \frac{1}{\rho}\frac{\partial \mu}{\partial \overline{y}}\frac{\partial \overline{u}}{\partial \overline{y}} - \frac{\sigma B_0^2 \overline{u}}{\rho} \\ &+ g\beta_T(\overline{T} - \overline{T}_\infty) + g\beta_C(\overline{C} - \overline{C}_\infty) \end{aligned} \tag{2}$$

Conservation of energy equation:

$$\rho C_p\left(\frac{\partial \overline{T}}{\partial \overline{t}} + \overline{u}\frac{\partial \overline{T}}{\partial \overline{x}} + \overline{v}\frac{\partial \overline{T}}{\partial \overline{y}}\right) = \frac{\partial \lambda}{\partial \overline{y}} \cdot \frac{\partial \overline{T}}{\partial \overline{y}} + \lambda \frac{\partial^2 \overline{T}}{\partial \overline{y}^2} + \mu\left(\frac{\partial \overline{u}}{\partial \overline{y}}\right)^2 + \frac{\overline{J}^2}{\sigma} - \frac{\partial \overline{q}_r}{\partial \overline{y}} \tag{3}$$

Concentration equation:

$$\frac{\partial \overline{C}}{\partial \overline{t}} + \overline{u}\frac{\partial \overline{C}}{\partial \overline{x}} + \overline{v}\frac{\partial \overline{C}}{\partial \overline{y}} = \frac{\partial D_m}{\partial \overline{y}}\frac{\partial \overline{C}}{\partial \overline{y}} + D_m \frac{\partial^2 \overline{C}}{\partial \overline{y}^2} + D_t \frac{\partial^2 \overline{T}}{\partial \overline{y}^2} \tag{4}$$

The initial and boundary conditions are:

$$\begin{aligned} &\overline{t} \le 0 : \overline{u} = 0, \overline{T} = \overline{T}_\infty, \overline{C} = \overline{C}_\infty \quad \forall y \\ &\overline{t} \succ 0 : \overline{u} = U_0 \cos(\overline{\omega t}), \overline{T} = \overline{T}_\infty + \left(\overline{T}_w - \overline{T}_\infty\right)A\overline{t}, \\ &\quad \overline{C} = \overline{C}_\infty + \left(\overline{C}_w - \overline{C}_\infty\right)A\overline{t} \quad \text{at} \quad \overline{y} = 0 \\ &\overline{t} \succ 0 : \overline{u} \to 0, \overline{T} \to \overline{T}_\infty, \overline{C} \to \overline{C}_\infty \quad \text{at} \quad \overline{y} \to \infty \end{aligned} \tag{5}$$

where $\overline{u}$ and $\overline{v}$ are the fluid velocities in $\overline{x}$ and $\overline{y}$ directions, respectively, ϑ is the kinematic viscosity, ρ is the fluid density, μ is the viscosity of the fluid, σ is the electrical conductivity, g is the acceleration due to gravity, β_T is the coefficient of volume expansion for heat transfer, β_C is the coefficient of volume expansion for mass transfer, $\overline{T}$ is the temperature and $\overline{T}_\infty$ is the temperature at free stream of the fluid, $\overline{C}$ is the concentration and $\overline{C}_\infty$ is the concentration at free stream of the fluid, λ is the thermal conductivity of the fluid, C_p is the specific heat at constant pressure, $\overline{q}$ is the radiative heat flux, D_m is the mass diffusivity, D_t is the heat diffusivity, U_0 is the velocity of the plate, $\overline{C}_w$ is the species concentration at the surface of the plate and A is a constant, $\overline{J}$ is the electric current density. By Ohm's law $\overline{J} = \sigma\left[\vec{E} + \vec{q} \times \vec{B}\right]$, $\vec{q}$ be fluid velocity at a particular point, $\vec{B} = B_0\hat{j}$ be the applied magnetic field. Here, no electric field is applied for which$\vec{E} = 0$.

The last two terms of Eq. (2) represent the thermal and concentration buoyancy effects, respectively. In Eq. (3), the last two terms denote the heat absorption and thermal radiation effects, respectively. Also, the last term of Eq. (4) represents the chemical reaction effect.

Using the Rosseland approximation the radiative heat flux is given by

$$\overline{q}_r = -\frac{4\overline{\sigma}}{3\overline{a}}\frac{\partial \overline{T}^4}{\partial \overline{y}} \tag{6}$$

where$\overline{a}$ is the mean absorption coefficient and $\overline{\sigma}$ is the Stefan–Boltzmann constant.

Assuming that the temperature differences within the flow are such that the term $\overline{T}^4$ may be expressed as a linear function of the temperature, we expand $\overline{T}^4$ in Taylor's series about$\overline{T}_\infty$ as follows:

$\overline{T}^4 = \overline{T}_\infty^4 + 4\overline{T}_\infty^3(\overline{T} - \overline{T}_\infty) + 6\overline{T}_\infty^2(\overline{T} - \overline{T}_\infty)^2 + \cdots$

By neglecting the higher-order terms beyond First degree in $\overline{T} - \overline{T}_\infty$ we have

$$\overline{T}^4 = 4\overline{T}_\infty^3\overline{T} - 3\overline{T}_\infty^4 \tag{7}$$

Using Eqs. (6) and (7), Eq. (3) reduces to

$$\rho C_p\left(\frac{\partial \overline{T}}{\partial \overline{t}} + \overline{u}\frac{\partial \overline{T}}{\partial \overline{x}} + \overline{v}\frac{\partial \overline{T}}{\partial \overline{y}}\right) = \frac{\partial \lambda}{\partial \overline{y}}.\frac{\partial \overline{T}}{\partial \overline{y}} + \lambda\frac{\partial^2 \overline{T}}{\partial \overline{y}^2} + \mu\left(\frac{\partial \overline{u}}{\partial \overline{y}}\right)^2 + \frac{\overline{J}^2}{\sigma} - \frac{16\overline{\sigma}\overline{T}_\infty^3}{3\overline{a}}\frac{\partial^2 \overline{T}^4}{\partial \overline{y}^2} \tag{8}$$

Applying the following non-dimensional quantities:

$$x = \frac{U_0\overline{x}}{\vartheta_\infty}, y = \frac{U_0\overline{y}}{\vartheta_\infty}, u = \frac{\overline{u}}{U_0}, \theta = \frac{\overline{T} - \overline{T}_\infty}{\overline{T}_w - \overline{T}_\infty}$$
$$\varphi = \frac{\overline{C} - \overline{C}_\infty}{\overline{C}_w - \overline{C}_\infty}, t = \frac{U_0^2\overline{t}}{\vartheta_\infty}, \omega = \frac{\overline{\omega}\vartheta_\infty}{U_0^2}, A = \frac{U_0^2}{\vartheta_\infty} \tag{9}$$

The viscosity and thermal conductivity of the fluid are assumed to be inverse linear function of temperature [18] as follows:

$$\frac{1}{\mu} = \frac{1}{\mu_\infty}\{1 + \gamma(\overline{T} - \overline{T}_\infty)\} \tag{10}$$

$$\frac{1}{\lambda} = \frac{1}{\lambda_\infty}\{1 + \delta(\overline{T} - \overline{T}_\infty)\} \tag{11}$$

where γ and δ are constants.

We define two parameters as $\theta_r = \frac{\overline{T}_r - \overline{T}_\infty}{\overline{T}_w - \overline{T}_\infty}$ called viscosity parameter and$\theta_c = \frac{\overline{T}_c - \overline{T}_\infty}{\overline{T}_w - \overline{T}_\infty}$called thermal conductivity parameter.

Using these two parameters in (10) and (11), we have the viscosity and thermal conductivity, respectively, as

$$\mu = -\frac{\mu_\infty \theta_r}{\theta - \theta_r}, \lambda = -\frac{\lambda_\infty \theta_c}{\theta - \theta_c} \tag{12}$$

Using the transformations (9) and (12), the non-dimensional forms of (2), (3), and (4) are

$$\frac{\partial u}{\partial t} + u\frac{\partial u}{\partial x} + v\frac{\partial u}{\partial y} = -\frac{\theta_r}{\theta - \theta_r}\frac{\partial^2 u}{\partial y^2} + \frac{\theta_r}{(\theta - \theta_r)^2}\frac{\partial \theta}{\partial y}\frac{\partial u}{\partial y} - \text{Mu} + \text{Gr}\theta + \text{Gr}_m\varphi \tag{13}$$

$$\Pr\left(\frac{\partial \theta}{\partial t} + u\frac{\partial \theta}{\partial x} + v\frac{\partial \theta}{\partial y}\right) = \left(\frac{4}{3}\text{Kr} - \frac{\theta_c}{\theta - \theta_c}\right)\frac{\partial^2 \theta}{\partial y^2} + \frac{\theta_c}{(\theta - \theta_c)^2}\left(\frac{\partial \theta}{\partial y}\right)^2 - \left(\frac{\theta_r}{\theta - \theta_r}\right)\Pr .\text{Ec}\left(\frac{\partial u}{\partial y}\right)^2 + \text{Ec}.u^2 \tag{14}$$

$$\frac{\partial \varphi}{\partial t} + u\frac{\partial \varphi}{\partial x} + v\frac{\partial \varphi}{\partial y} = \frac{\theta_r}{(\theta - \theta_r)^2}.\frac{1}{\text{Sc}}\frac{\partial \theta}{\partial y}.\frac{\partial \varphi}{\partial y} - \frac{\theta_r}{\text{Sc}(\theta - \theta_r)}\frac{\partial^2 \varphi}{\partial y^2} + \text{So}\frac{\partial^2 \theta}{\partial y^2} \tag{15}$$

The given initial and boundary conditions are transformed to:

$$\begin{aligned} t \le 0 &: u = 0, v = 0, \theta = 0, \varphi = 0 \quad \forall y \\ t \succ 0 &: u = \cos(\omega t), v = 0, \theta = t, \varphi = t \quad \text{at} \quad y = 0 \\ t \succ 0 &: u \to 0, v \to 0, \varphi \to 0 \quad \text{at} \quad y \to \infty \end{aligned} \tag{16}$$

where$\text{Ec} = \frac{U_o^2}{\overline{T}_w C_p}$ is the Eckert number, $\text{So} = D_t\frac{\overline{T}_w - \overline{T}_\infty}{\vartheta_\infty(C_w - C_\infty)}$ is the Soret number, $\text{Sc} = \frac{\vartheta_\infty}{D_m}$ is the Schmidt number of fluid,$M = \frac{\sigma B_0^2 \vartheta_\infty}{\rho U_0^2}$ is the magnetic field parameter,$\text{Kr} = \frac{16\overline{a}\vartheta_\infty \overline{\sigma}\overline{T}_\infty^2}{3\lambda_\infty U_0^2}$ is the radiation parameter, $\text{Gr} = \frac{\vartheta_\infty g\beta_T(\overline{T}_w - \overline{T}_\infty)}{U_0^2}$ is the

Grashof number, $\mathrm{Gr}_m = \frac{\vartheta_\infty g \beta_C (\overline{C}_w - \overline{C}_\infty)}{U_0^2}$ is the concentration buoyancy parameter andPr $= \frac{\rho \vartheta_\infty C_p}{\lambda_\infty}$ is the Prandtl number.

3 Atangana–Baleanu Fractional Derivatives

In order to generate the AB fractional model [19], we replace governing partial differential equations with respect to time by the AB fractional operator of the order$0 \prec \alpha \prec 1$, Eqs. (13)–(15) become

$$\mathrm{AB}\left(\frac{\partial^\alpha u(y,t)}{\partial t^\alpha}\right) = -u\frac{\partial u}{\partial x} - v\frac{\partial u}{\partial y} - \frac{\theta_r}{\theta - \theta_r}\frac{\partial^2 u}{\partial y^2} + \frac{\theta_r}{(\theta - \theta_r)^2}\frac{\partial \theta}{\partial y}\frac{\partial u}{\partial y} - \mathrm{Mu} + \mathrm{Gr}\theta + \mathrm{Gr}_m\varphi \tag{17}$$

$$\mathrm{AB}\left(\frac{\partial^\alpha \theta(y,t)}{\partial t^\alpha}\right) = -u\frac{\partial \theta}{\partial x} - v\frac{\partial \theta}{\partial y} + \frac{1}{\mathrm{Pr}}\left(\frac{4}{3}\mathrm{Kr} - \frac{\theta_c}{\theta - \theta_c}\right)\frac{\partial^2 \theta}{\partial y^2} + \frac{1}{\mathrm{Pr}}\frac{\theta_c}{(\theta - \theta_c)^2}\left(\frac{\partial \theta}{\partial y}\right)^2 - \left(\frac{\theta_r}{\theta - \theta_r}\right).\mathrm{Ec}\left(\frac{\partial u}{\partial y}\right)^2 + \frac{\mathrm{Ec}}{\mathrm{Pr}}u^2 \tag{18}$$

$$\mathrm{AB}\left(\frac{\partial^\alpha \varphi(y,t)}{\partial t^\alpha}\right) = -u\frac{\partial \varphi}{\partial x} - v\frac{\partial \varphi}{\partial y} + \frac{\theta_r}{(\theta - \theta_r)^2}.\frac{1}{\mathrm{Sc}}\frac{\partial \theta}{\partial y}.\frac{\partial \varphi}{\partial y} - \frac{\theta_r}{\mathrm{Sc}(\theta - \theta_r)}\frac{\partial^2 \varphi}{\partial y^2} + \mathrm{So}\frac{\partial^2 \theta}{\partial y^2} \tag{19}$$

where $\frac{\partial^\alpha u(y,t)}{\partial t^\alpha}$ is the AB fractional operator of orderα defined as

$$\mathrm{AB}\left(\frac{\partial^\alpha u(y,t)}{\partial t^\alpha}\right) = \frac{1}{1-\alpha}\int_0^t u'(y,t)E_\alpha\left(\frac{-\alpha(z-t)}{1-\alpha}\right)\mathrm{d}t \tag{20}$$

where $E_\alpha(-t^\alpha) = \sum_{m=0}^{\infty}\frac{(-t)^{\alpha m}}{\Gamma(1+\alpha m)}$ is the Mittage–Leffler function.

4 Caputo–Fabrizio Fractional Derivatives

In order to generate the CF fractional model [19], we replace governing partial differential equations with respect to time by the CF fractional operator of the order $0 \prec \beta \prec 1$, Eqs. (13)–(15) become

$$\mathrm{CF}\left(\frac{\partial^\beta u(y,t)}{\partial t^\beta}\right) = -u\frac{\partial u}{\partial x} - v\frac{\partial u}{\partial y} - \frac{\theta_r}{\theta - \theta_r}\frac{\partial^2 u}{\partial y^2} + \frac{\theta_r}{(\theta - \theta_r)^2}\frac{\partial \theta}{\partial y}\frac{\partial u}{\partial y} - \mathrm{Mu} + \mathrm{Gr}\theta + \mathrm{Gr}_m\varphi \tag{21}$$

$$\mathrm{CF}\left(\frac{\partial^\beta \theta(y,t)}{\partial t^\beta}\right) = -u\frac{\partial \theta}{\partial x} - v\frac{\partial \theta}{\partial y} + \frac{1}{\mathrm{Pr}}\left(\frac{4}{3}Kr - \frac{\theta_c}{\theta - \theta_c}\right)\frac{\partial^2 \theta}{\partial y^2} + \frac{1}{\mathrm{Pr}}\frac{\theta_c}{(\theta - \theta_c)^2}\left(\frac{\partial \theta}{\partial y}\right)^2 - \left(\frac{\theta_r}{\theta - \theta_r}\right).\mathrm{Ec}\left(\frac{\partial u}{\partial y}\right)^2 + \frac{\mathrm{Ec}}{\mathrm{Pr}}u^2 \tag{22}$$

$$\mathrm{CF}\left(\frac{\partial^\beta \varphi(y,t)}{\partial t^\beta}\right) = -u\frac{\partial \varphi}{\partial x} - v\frac{\partial \varphi}{\partial y} + \frac{\theta_r}{(\theta - \theta_r)^2}.\frac{1}{\mathrm{Sc}}\frac{\partial \theta}{\partial y}.\frac{\partial \varphi}{\partial y} - \frac{\theta_r}{\mathrm{Sc}(\theta - \theta_r)}\frac{\partial^2 \varphi}{\partial y^2} + \mathrm{So}\frac{\partial^2 \theta}{\partial y^2} \tag{23}$$

where $\frac{\partial^\beta u(y,t)}{\partial t^\beta}$ is the CF fractional operator of order β defined as

$$\mathrm{CF}\left(\frac{\partial^\beta u(y,t)}{\partial t^\beta}\right) = \frac{1}{1-\beta}\int_0^t u^{/}(y,t)\mathrm{Exp}\left(\frac{-\beta(z-t)}{1-\beta}\right)\mathrm{d}t \tag{24}$$

5 Numerical Method for Solution

Solutions of Eqs. (17)–(20) or (21)–(24) are obtained by using ordinary finite difference scheme. Discritization is performed using the following formulae:
$\frac{\partial u}{\partial x} = \frac{u_{l,m+1,n} - u_{l,m,n}}{\Delta x}$, $\frac{\partial u}{\partial y} = \frac{u_{l,m+1,n} - u_{l,m,n}}{\Delta y}$, $\frac{\partial^2 u}{\partial y^2} = \frac{u_{l,m,n+1} - 2u_{l,m,n} + u_{l,m,n-1}}{\Delta y^2}$ etc.

The fractional derivatives given by (20) or (24) are calculated using numerical integration. Finally, the set of Eq. (17)–(19) or (21)–(23) together with boundary condition (16) completely discritized and the discritized equations are solved by using an iterative method based on Gauss–Seidel scheme.

6 Results and Discussion

The non-dimensional discretized governing equations along with the non-dimensional boundary conditions are solved with the help of AB and CF fractional derivative method by developing suitable programming code in MATLAB using the method described in Sect. 7. This analysis has been done to study the effects of various parameters such as θ_r,θ_c, M, α, and β, Kr, Sc, Ec, Pr, etc., on velocity

(u), temperature(θ) and species concentration(φ) profiles in presence of time. The numerical results are shown graphically in Figs. 1, 2, 3, 4, 5, 6, 7, 8, 9, 10, 11, 12, and 13. In the following discussion, the initial values of the parameters are considered as $\theta_r = -20$, $\theta_c = -12$, $M = 1, \alpha = 0.25$, $\mathrm{Gr}_m = 0.1$ $\mathrm{Gr} = 0.1$, $\beta = 0.25$, $\mathrm{Kr} = 0.05$, $\mathrm{Sc} = 0.5$, $\mathrm{Ec} = 0.1$, $\mathrm{Pr} = 0.25$, $\mathrm{So} = 0.5$unless otherwise stated.

Figures 1 and 3 demonstrate the velocity, temperature, and concentration profile for various values of the AB fractional operator (α)/CF fractional operator (β). It is observed in all the three figures that the velocity, temperature, and concentration u, θ and φ, respectively, increases for increasing values of α/β.

In Fig. 4, it is seen that with the increasing value of the Hartmann number, velocity decreases. The presence of magnetic field in the normal direction of the flow in an

Fig. 1 α/β on velocity profile

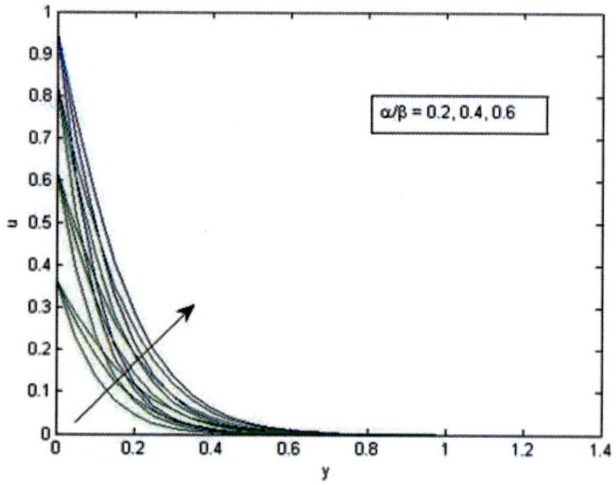

Fig. 2 α/β on temperature profile

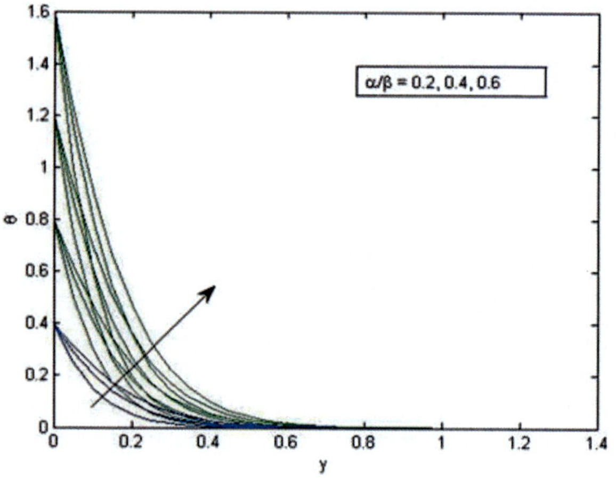

Fig. 3 α/β on concentration profile

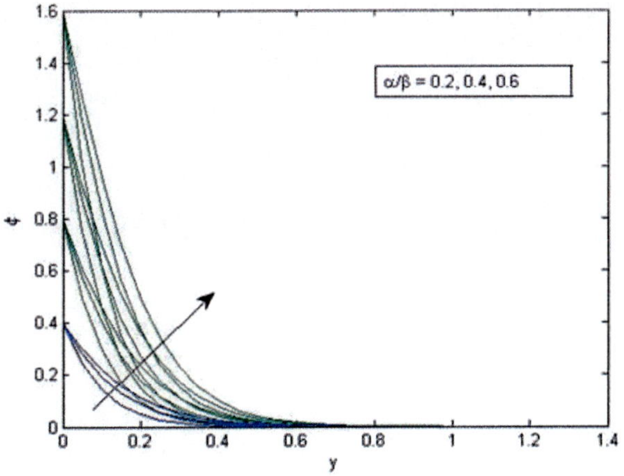

Fig. 4 *M* on velocity profile

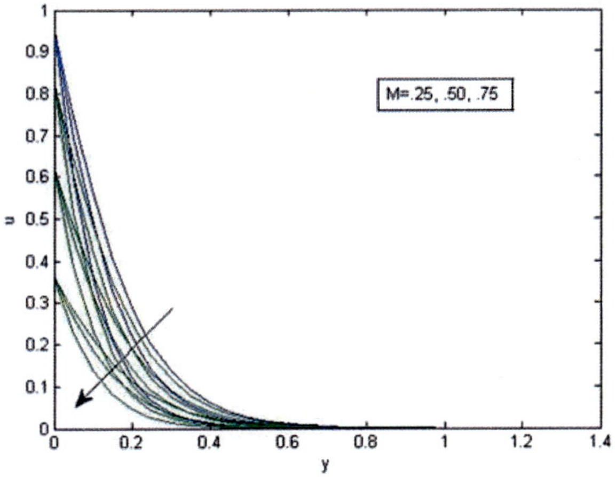

Fig. 5 Effects of *M* on temperature

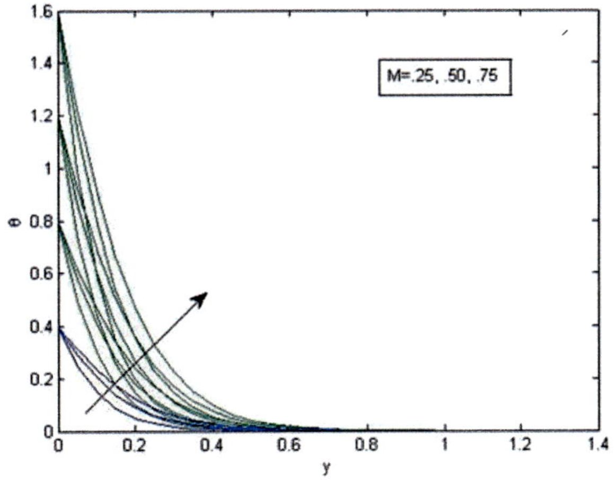

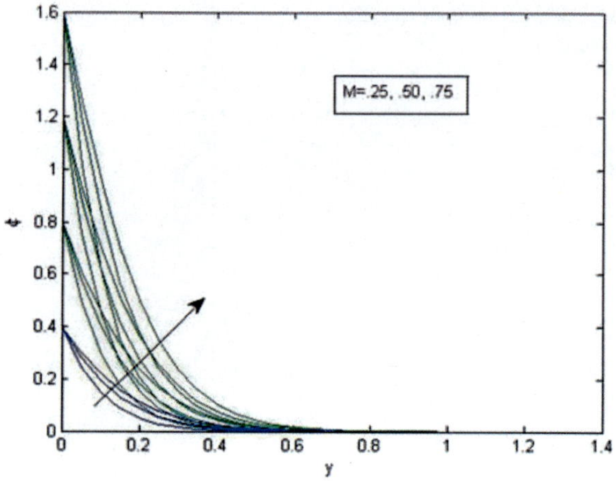

Fig. 6 Effects of M on concentration

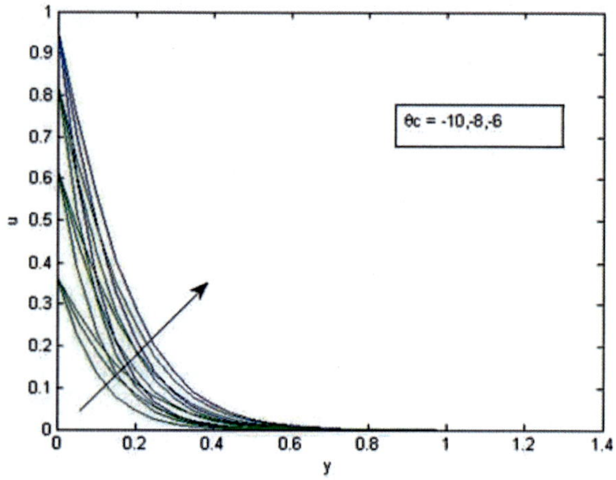

Fig. 7 θ_c on velocity profile

electrically conducting fluid produces Lorentz force which opposes the flow. To overcome this opposing force, some extra work should be done which is transformed to heat energy, as a result temperature increases (Fig. 5). With the increase of M species concentration also increases (Fig. 6). Figure 7 depicts the distribution of velocity with the variation of the thermal conductivity parameter θ_c. Velocity increases with the increasing value of θ_c. In Fig. 8 temperature decreases with the increasing value of θ_c which implies decreasing of viscosity and so velocity increases. Again species concentration is increases with the increasing value of θ_c (Fig. 9). The effects of viscosity parameter θ_r on velocity, temperature, and species concentration distribution are plotted in Figs. 10, 11, and 12. Figure 10 displays that dimensionless velocity decreases with the increases of θ_r. This is due to the fact that with the increase of the viscosity parameter the thickness of the velocity boundary layer decreases. Physically, this is because of that a larger θ_r implies higher temperature difference between

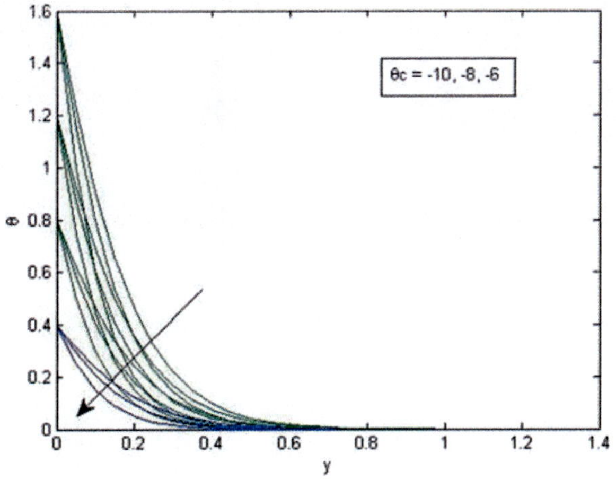

Fig. 8 θ_c on temperature profile

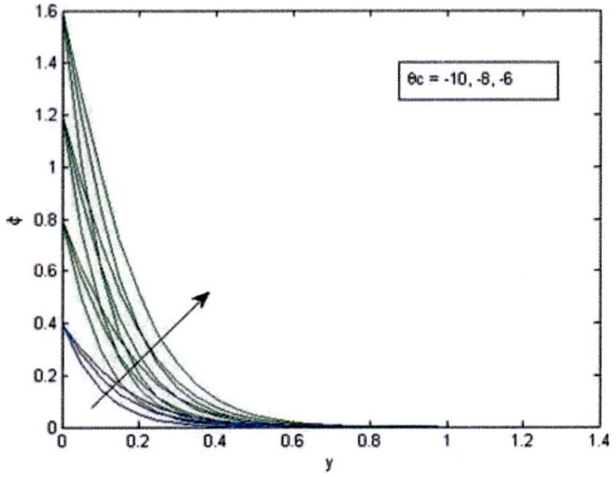

Fig. 9 Effects of θ_c on concentration

the fluid and the surface. Figure 11 shows that temperature is increases with the increasing value of θ_r. The species concentration decreases for increasing value of θ_r (Fig. 12). With the increasing value of Ec, the thickness of both velocity boundary layer and thermal boundary layer increase. Hence, both the velocity and the temperature increase with increasing Ec (Figs. 13 and 14). Increasing the value of radiation parameter Kr enhances the velocity (Fig. 15). In Fig. 16, it is noticed that with the increase of Kr temperature increases. This is due to the fact that the thermal boundary layer thickness increases with the increase of Kr and hence temperature. Velocity decreases with the increasing value of Prandtl number Pr (Fig. 17). This is due to the fact that with the increase of Pr, viscosity increases, so velocity decreases. In Fig. 18, it is noticed that with the increasing value of Pr temperature of the fluid decreases.

Fig. 10 Effects of θ_r on velocity profile

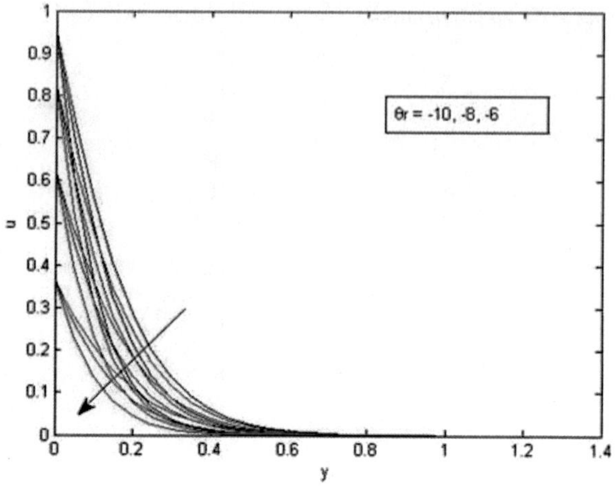

Fig. 11 θ_r on temperature profile

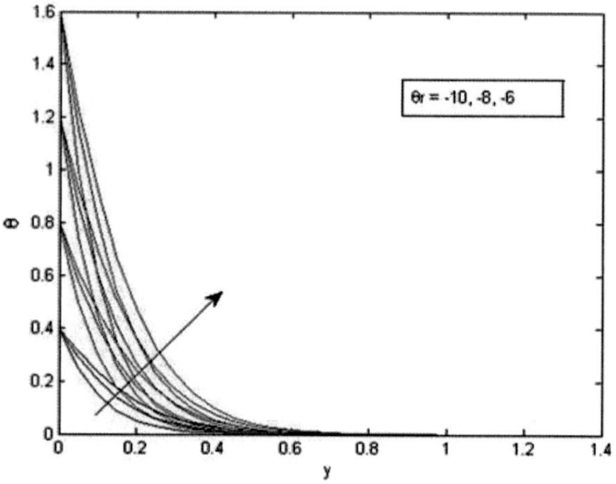

Fig. 12 θ_r on concentration profile

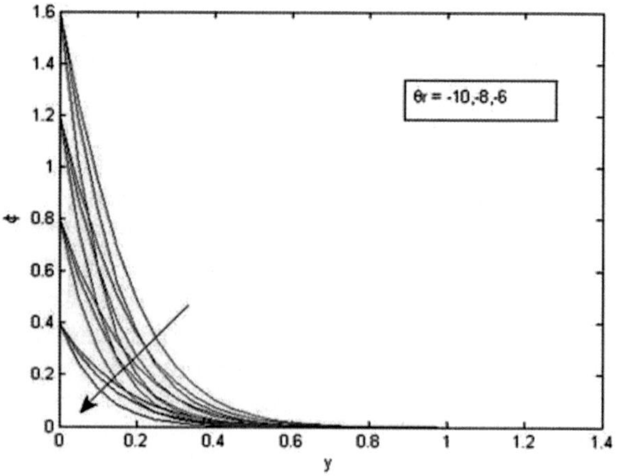

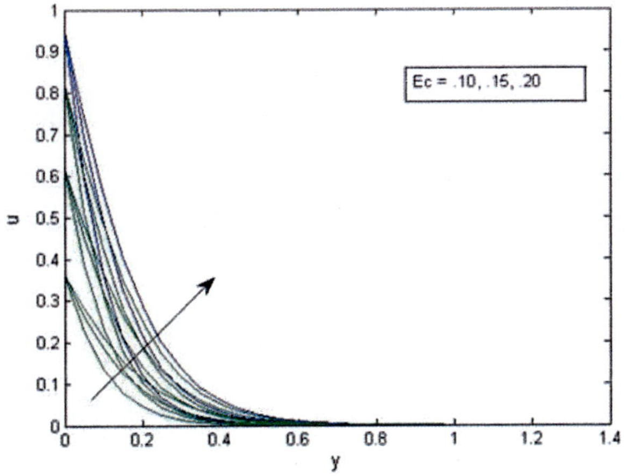

Fig. 13 Ec on velocity profile

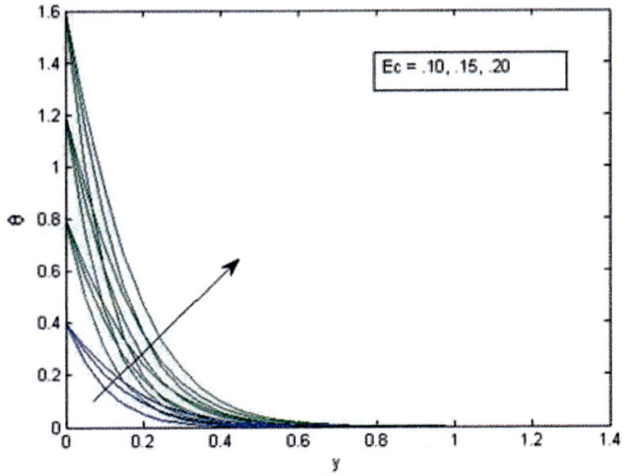

Fig. 14 Ec on temperature profile

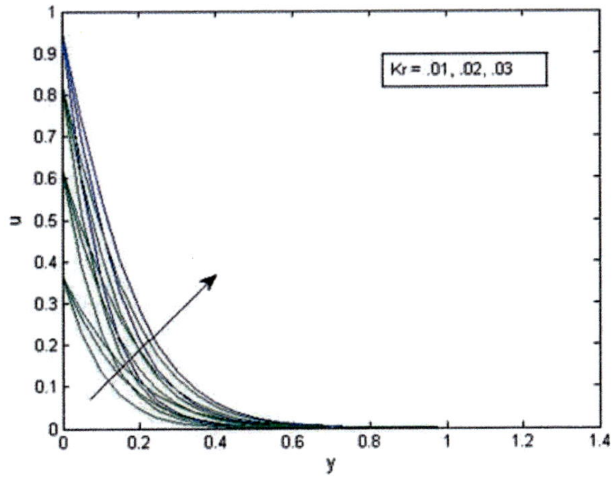

Fig. 15 Kr on velocity profile

Fig. 16 Kr on temperature profile

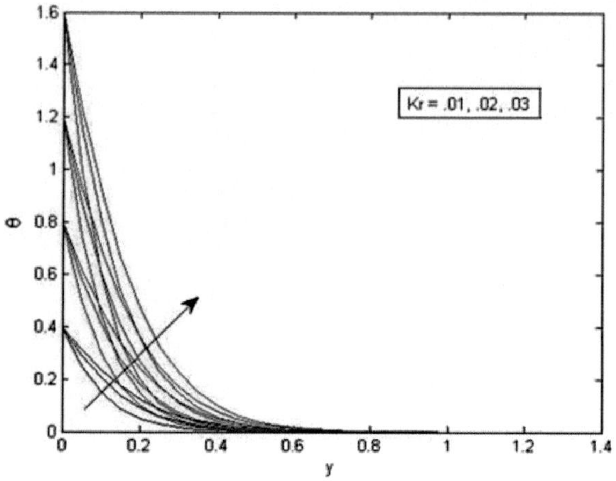

Fig. 17 Effects of Pr on velocity

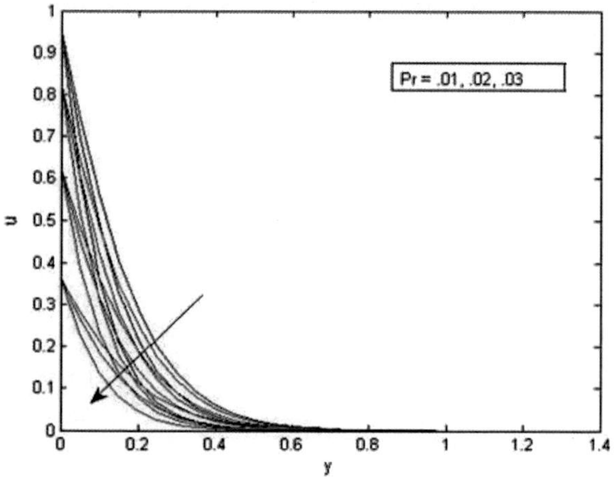

Fig. 18 Effects of Pr on temperature

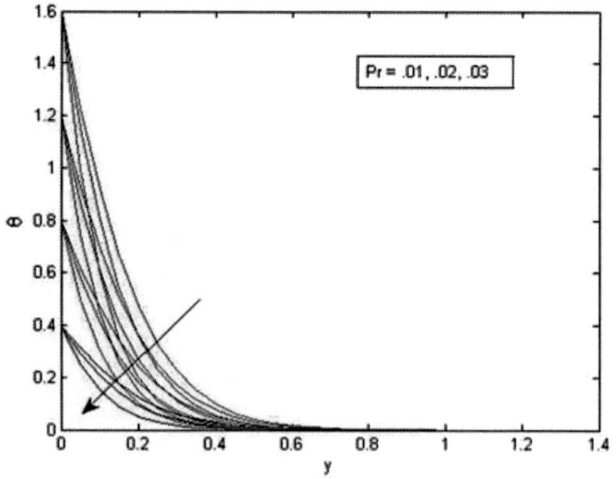

For higher Prandtl number, the fluid has a relatively high thermal conductivity which decreases the temperature.

Figure 19 shows the variation of velocity profile with the variation of Schmidt number Sc. As Sc is increased the concentration boundary layer becomes thinner than the viscous boundary layer, as a result of which velocity reduces. With thinner concentration boundary layer, the concentration gradients are enhanced causing a decrease in concentration of species in the boundary layer (Fig. 20).

Fig. 19 Sc on concentration profile

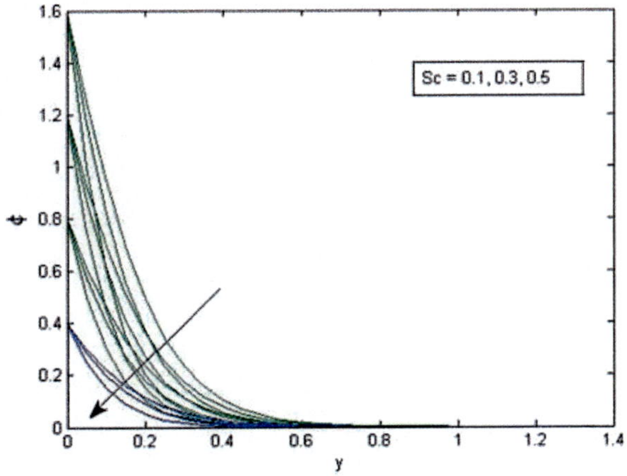

Fig. 20 Sc on velocity profile

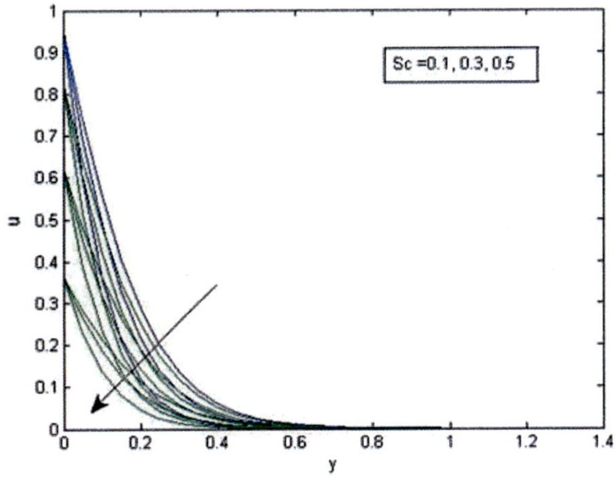

7 Comparison of AB and CF Fractional Derivative Methods for Various Values of the Parameters

See Tables 1, 2, 3, 4, 5, 6, 7, and 8.

Table 1 Effect of α/β on and φ

α/β	y	t	u		θ		φ	
			AB	CF	AB	CF	AB	CF
0.2	0.4	0.4	0.020391	0.020404	0.008651	0.008655	0.008436	0.008436
		0.8	0.017675	0.017699	0.017189	0.017206	0.017133	0.017133
		1.2	0.013412	0.013441	0.025687	0.025726	0.025913	0.025913
		1.6	0.008001	0.008026	0.034148	0.034222	0.034762	0.034761
0.4	0.4	0.4	0.054469	0.054519	0.023390	0.023387	0.021828	0.021813
		0.8	0.047164	0.047269	0.046056	0.046079	0.045348	0.045296
		1.2	0.035754	0.035916	0.068466	0.068555	0.069381	0.069261
		1.6	0.021375	0.021599	0.090224	0.090874	0.094404	0.094380
0.6	0.4	0.4	0.091599	0.091687	0.039660	0.039611	0.035661	0.035620
		0.8	0.079359	0.079601	0.077699	0.077637	0.075159	0.074978
		1.2	0.060825	0.061336	0.115090	0.115121	0.115950	0.115464
		1.6	0.037842	0.037842	0.152881	0.155850	0.153696	0.153435

Table 2 Effect of Mon and φ

M	y	t	u		θ		φ	
			AB	CF	AB	CF	AB	CF
0.25	0.4	0.4	0.036310	0.036588	0.008650	0.008654	0.008439	0.008439
		0.8	0.060579	0.060758	0.017186	0.017204	0.017140	0.017140
		1.2	0.079846	0.079959	0.025679	0.025723	0.025928	0.025927
		1.6	0.092230	0.092280	0.034133	0.034218	0.034789	0.034789
0.50	0.4	0.4	0.021552	0.021691	0.023380	0.023378	0.021848	0.021838
		0.8	0.036048	0.036159	0.046038	0.046054	0.045382	0.045347
		1.2	0.047537	0.047615	0.068443	0.068500	0.069422	0.069347
		1.6	0.054904	0.054941	0.090305	0.090710	0.094434	0.094428
0.75	0.4	0.4	0.008057	0.008086	0.039595	0.039575	0.035782	0.035764
		0.8	0.013510	0.013544	0.077513	0.077494	0.075483	0.075427
		1.2	0.017808	0.017835	0.114838	0.114844	0.116372	0.116254
		1.6	0.020547	0.020562	0.150642	0.151433	0.159960	0.159948

Table 3 Effect of θ_c on and φ

θ_c	y	t	u		θ		φ	
			AB	CF	AB	CF	AB	CF
−10	0.4	0.4	0.020389	0.020403	0.008648	0.008652	0.008443	0.008443
		0.8	0.017670	0.017698	0.017177	0.017195	0.017163	0.017163
		1.2	0.013406	0.013440	0.025657	0.025702	0.025982	0.025981
		1.6	0.007996	0.008025	0.034096	0.034181	0.034885	0.034885
−8	0.4	0.4	0.054477	0.054514	0.023351	0.023349	0.021920	0.021909
		0.8	0.047169	0.047246	0.045914	0.045929	0.045696	0.045661
		1.2	0.035772	0.035882	0.068166	0.068223	0.070127	0.070052
		1.6	0.021391	0.021530	0.089819	0.090224	0.095677	0.095672
−6	0.4	0.4	0.091609	0.091658	0.039482	0.039462	0.036065	0.036048
		0.8	0.079313	0.079426	0.077050	0.077030	0.076657	0.076602
		1.2	0.060184	0.060362	0.113818	0.113824	0.118971	0.118853
		1.6	0.036088	0.036364	0.148869	0.149660	0.164496	0.164490

Table 4 Effect of θ_r on and φ

θ_r	y	t	u		θ		φ	
			AB	CF	AB	CF	AB	CF
−10	0.4	0.4	0.090885	0.035411	0.008650	0.008654	0.035576	0.035559
		0.8	0.078121	0.059070	0.017186	0.017204	0.074725	0.074669
		1.2	0.058894	0.078234	0.025678	0.025723	0.114768	0.114648
		1.6	0.035137	0.090934	0.034133	0.034218	0.157360	0.157348
−8	0.4	0.4	0.054212	0.054249	0.023378	0.023376	0.021777	0.021766
		0.8	0.046728	0.046805	0.046032	0.046048	0.045128	0.045093
		1.2	0.035292	0.035400	0.068437	0.068494	0.068887	0.068812
		1.6	0.021033	0.021170	0.090301	0.090706	0.093557	0.093551
−6	0.4	0.4	0.020353	0.007974	0.039582	0.039562	0.008430	0.008430
		0.8	0.017610	0.013373	0.077489	0.077469	0.017110	0.017110
		1.2	0.013339	0.017637	0.114813	0.114820	0.025867	0.025867
		1.6	0.007946	0.020367	0.150628	0.151418	0.034688	0.034687

8 Conclusions

From the above study, we may conclude that:

(i) Velocity, temperature, and species concentration are increases with the increasing value of AB fractional (α) parameter and CF fractional parameter (β).

Table 5 Effect of *Sc* on and φ

ScSc	*y*	*t*	*u*		θ		φ	
			AB	CF	AB	CF	AB	CF
0.1	0.4	0.4	0.091613	0.091662	0.008650	0.008655	0.036463	0.036446
		0.8	0.079308	0.079421	0.017186	0.017205	0.075666	0.075610
		1.2	0.060167	0.060344	0.025679	0.025724	0.115655	0.115537
		1.6	0.036053	0.036329	0.034133	0.034219	0.157770	0.157758
0.3	0.4	0.4	0.054479	0.054516	0.023383	0.023381	0.022380	0.022374
		0.8	0.047168	0.047245	0.046040	0.046056	0.045518	0.045497
		1.2	0.035766	0.035876	0.068444	0.068501	0.068846	0.068801
		1.6	0.021378	0.021516	0.090305	0.090710	0.092658	0.092654
0.5	0.4	0.4	0.020389	0.020404	0.039600	0.039581	0.008556	0.008556
		0.8	0.017670	0.017698	0.077517	0.077497	0.017153	0.017153
		1.2	0.013406	0.013439	0.114839	0.114845	0.025756	0.025756
		1.6	0.007994	0.008023	0.150642	0.151432	0.034361	0.034361

Table 6 Effect of Kr on and φ

Kr	*y*	*t*	*u*		θ		φ	
			AB	CF	AB	CF	AB	CF
0.01	0.4	0.4	0.020389	0.020403	0.008650	0.008654	0.008439	0.008439
		0.8	0.017670	0.017698	0.017186	0.017205	0.017139	0.017139
		1.2	0.013406	0.013440	0.025680	0.025724	0.025924	0.025924
		1.6	0.007996	0.008025	0.034136	0.034220	0.034783	0.034782
0.02	0.4	0.4	0.054477	0.054514	0.023375	0.023373	0.021862	0.021851
		0.8	0.047167	0.047244	0.046037	0.046052	0.045384	0.045350
		1.2	0.047167	0.047244	0.046037	0.046052	0.045384	0.045350
		1.6	0.021386	0.021525	0.090334	0.090730	0.094359	0.094353
0.03	0.4	0.4	0.091608	0.091657	0.039570	0.039551	0.035842	0.035825
		0.8	0.079307	0.079420	0.077501	0.077482	0.075512	0.075459
		1.2	0.060172	0.060350	0.114854	0.114860	0.116330	0.116216
		1.6	0.036069	0.036345	0.150733	0.151496	0.159727	0.159715

(ii) Increasing value of magnetic field parameter decreases the value of velocity but increases the values of temperature and species concentration.

(iii) When the viscosity parameter increases, the velocity and the species concentration decreases whereas temperature increases.

(iv) With the increasing thermal conductivity parameter, the velocity and the species concentration increases but the temperature decreases.

Table 7 Effect of Pr on and φ

Pr	y	t	u		θ		φ	
			AB	CF	AB	CF	AB	CF
0.01	0.4	0.4	0.091603	0.091657	0.039570	0.039551	0.008439	0.008439
		0.8	0.079309	0.079420	0.077501	0.077482	0.017139	0.017139
		1.2	0.060172	0.060350	0.114854	0.114860	0.025924	0.025924
		1.6	0.036069	0.036345	0.150733	0.151496	0.034783	0.034782
0.02	0.4	0.4	0.054474	0.054514	0.023375	0.023373	0.021862	0.021851
		0.8	0.047187	0.047244	0.046037	0.046052	0.045384	0.045350
		1.2	0.047194	0.047244	0.046037	0.046052	0.045384	0.045350
		1.6	0.021386	0.021525	0.090334	0.090730	0.094359	0.094353
0.03	0.4	0.4	0.020387	0.020403	0.008650	0.008654	0.035842	0.035825
		0.8	0.017670	0.017698	0.017186	0.017205	0.075512	0.075459
		1.2	0.013406	0.013440	0.025680	0.025724	0.116330	0.116216
		1.6	0.007992	0.008025	0.034136	0.034220	0.159727	0.159715

Table 8 Effect of Ec on and φ

Ec	y	t	u		θ		φ	
			AB	CF	AB	CF	AB	CF
0.10	0.4	0.4	0.020389	0.020403	0.008650	0.008655	0.008437	0.008437
		0.8	0.017670	0.017698	0.017186	0.017205	0.017139	0.017139
		1.2	0.013406	0.013440	0.025679	0.025724	0.025927	0.025927
		1.6	0.007986	0.008025	0.034133	0.034219	0.034789	0.034788
0.15	0.4	0.4	0.054477	0.054514	0.023529	0.023527	0.021473	0.021463
		0.8	0.047167	0.047244	0.046149	0.046165	0.045100	0.045065
		1.2	0.047167	0.047244	0.068506	0.068563	0.069261	0.069186
		1.6	0.021386	0.021525	0.090326	0.090731	0.094379	0.094373
0.20	0.4	0.4	0.091608	0.091657	0.040223	0.040203	0.034198	0.034181
		0.8	0.079307	0.079420	0.077981	0.077962	0.074296	0.074240
		1.2	0.060172	0.060350	0.115102	0.115109	0.115697	0.115578
		1.6	0.036069	0.036345	0.150732	0.151523	0.159729	0.159717

(v) Velocity and temperature increases with the increasing value of Eckert number and radiation parameter.

(vi) Velocity and temperature increases with the increasing value of Prandtl number.

(vii) Temperature and species concentration decrease with the increasing value of the Schmidt number.

(viii) The values of the velocity, temperature, and concentration profiles for various parameters are almost the same for both the methods—AB and CF fractional derivative. As gamma function is present inside the exponential function in AB fractional derivative method, so the result obtained by it is more accurate over the CF fractional derivative method.

References

1. Sakiadis, B.C.: Boundary layer behaviors on continuous solid surface. AIChE J. **7**(2), 221–225 (1961)
2. Crane, L.J.: Flow past a stretching plate. Z Angew Math. Phys. **21**, 645–647 (1970)
3. Soid, S.K., Ishak, A., Pop, I.: MHD flow and heat transfer over a radially stretching/shrinking disk. Chin. J. Phys. **56**(1), 58–66 (2018)
4. Carragher, P., Crane, L.J.: Heat transfer on continuous stretching surface. ZAMM **62**, 564–565 (1982)
5. Cortell, R.: Effectts of viscous dissipation and work done by deformation on the MHD flow and heat transfer of a viscoelastic fluid ever a stretching sheet. Phys. Lett. A **357**, 298–305 (2006)
6. Hazarika, G.C.: Effects of variable viscosity and thermal conductivity on MHD flow over a radially stretching disk. J. Comput. Math. Sci. **9**(9), 1282–1291 (2018)
7. Hayat, T., Sajid, M.: Influence of thermal radiation on Yhe boundary layer flow due to an exponentially stretching sheet. Int. Comm. Heat Mass Transf. **35**, 347–356 (2008)
8. Magyari, E., Keller, B.: Heat and mass transfer in the boundary layers on an exponentially stretching continuous surface. J. Phys. D: Appl. Phys. **32**, 577–585 (2000)
9. Gupta, P.S., Gupta, A.S.: Heat and mass transfer on a stretching sheet with suction or blowing. Can. J. Chem. Eng. **55**, 744–746 (2009)
10. Prasad, K.V., Vajravelu, K., Datti, P.S.: The effects of variable fluid properties on the hydromagnetic flow and heat transfer over a non-linearly stretching sheet. Int. J. Therm. Sci. **49**(3), 603–610 (2010)
11. Alam, M.S., Rahman, M.M., Sattar, M.A.: Transient magnetohydrodynamic free convective heat and mass transfer flow with thermophoresis past a radiate inclined permeable plate in the presence of variable chemical reaction and temperature dependent viscosity. NonlinearAnal. Modell. Control **14**(1), 3–20 (2009)
12. Ali, M.E.: The effect of variable viscosity on mixed convection heat transfer along a vertical moving surface. Int. J. Therm. Sci. **45**(1), 60–69 (2006)
13. Makinde, O.D.: Laminar falling liquid film with variable viscosity along an inclined heated plate. Appl. Math. Comput. **175**(1), 80–88 (2006)
14. Mukhopadhyay, S., Layek, G.C., Samad, S.A.: Study of MHD boundary layer flow over a heated stretching sheet with variable viscosity. Int. J. Heat Mass Transf. **48**, 4460–4466 (2005)
15. Mukhopadhyay, S., Layek, G.C.: Effects of thermal radiation and variable fluid viscosity on free convective flow and heat transfer past a porous stretching surface. Int J. Heat Mass Transf. **51**(9–10), 2167–2178 (2008)
16. Mukhopadhyay, S., Layek, G.C., Samad, S.A.: Effects of variable fluid viscosity on flow past a heated stretching sheet embedded in porous medium in presence of heat source/sink. Meccanica **47**, 863–876 (2011)
17. Anjali Devil, S.P., Ganga, B.: Effects of viscous and Joules dissipation on MHD flow, heat and mass transfer past a stretching porous media. Nonlinear Anal. Model Control, **14**(3), 303–314 (2009)

18. Lai, F.C., Kulacki, F.A.: The effect of variable viscosity on convective heat and mass transfer along a vertical surface in standard porous media. Int. J. Heat Mass Transf. **33**, 1028–1031 (1990)
19. Khan, A., Abro, K.A., Tassaddiq, A., Khan, I.: Atangana–Baleanu and Caputo Fabrizio analysis of fractional derivatives for heat and mass transfer of second grade fluids over a vertical plate: a comparative study. Entropy **19**(279), 1–12 (2017)

Optimization of Fuzzy Inference System Using Genetic Algorithm

Seela Naga Veerababu, Konna Roja, M. V. Kumar Reddy, K. Manoz Kumar Reddy, and Adireddy Ramesh

Abstract With increase in number of fuzzy membership function, fuzzy rules also get increased. The increased number of fuzzy rules creates computational complexity, and it becomes computationally more expensive. To optimize the fuzzy inference system, this paper provides a genetic algorithm-based approach which reduces the number of rules as well as reduces the computational time and complexity. A detailed analysis has been provided, and simulation results have also been provided. Mamdani-based FIS has been optimized using GA for a control system application.

Keywords Fuzzy inference system · MAMDANI · Fuzzy controller · Optimization · Genetic algorithm

1 Introduction

Fuzzy systems are rule-based system which is capable of dealing with imprecise information. Fuzzy logic works on the basis of heuristic and human interpretation. As the real-world system gets more complicated, the mathematical modelling of the real-system gets more and more difficult. Successful design of fuzzy inference system relies on several factors such as (a) membership function, (b) rule-base, (c)

S. N. Veerababu · K. Roja · M. V. Kumar Reddy · K. Manoz Kumar Reddy · A. Ramesh (✉)
Department of Electrical and Electronics Engineering, Aditya College of Engineering, Surampalem, Andhra Pradesh, India
e-mail: rameshadireddy007@gmail.com

S. N. Veerababu
e-mail: veerababuseela1999@gmail.com

K. Roja
e-mail: konnasharon15@gmail.com

M. V. Kumar Reddy
e-mail: kumarreddyrajanew@gmail.com

K. Manoz Kumar Reddy
e-mail: manozreddy111@gmail.com

S. Borah et al. (eds.), *Soft Computing Techniques and Applications*, Advances in Intelligent Systems and Computing 1248,
https://doi.org/10.1007/978-981-15-7394-1_21

inference mechanism and (d) defuzzification technique. As number of membership function increases to accurately model the input parameter, the rule base also exponentially increases due to combination of several inputs. To execute a particular rule for a particular input combination at a particular sample time, the processor takes considerable computational burden. Due to the complexity of the problem, it can be treated as an optimization problem.

Different researchers have looked this problem in different manner. A reliable method to identify a complete rule-based fuzzy system has been discussed in [10]. In [9], the authors have used SOGARG technique. Its three-stage hierarchical mechanism has been tested by the authors in different examples. Extended Kalman filter has been used to optimize the membership function for modelling of system in [4]. Ant colony optimization algorithm is used for fuzzy MF [2, 13]. H-inf filtering optimization has been discussed in [3]. PSO-based method to optimize fuzzy rule based has been discussed in [1, 5], and GA-based method to optimize fuzzy rule base has been discussed in [11, 12].

2 Problem Formulation

Consider a fuzzy-based control system (Fig. 1). A fuzzy inference mechanism is used. The fuzzy inference mechanism can be optimized as shown in Fig. 2.

3 Fuzzy Inference System

FIS is intelligent technique which computes outputs based on the fuzzy inference rules and present inputs. FIS methods use fuzzification, inference and defuzzification processes. The mapping from the presented inputs to fuzzy sets defined in the corresponding universe is the process of fuzzification that results in fuzzy inputs. In order to generate the corresponding fuzzy outputs of these inputs, the decision-

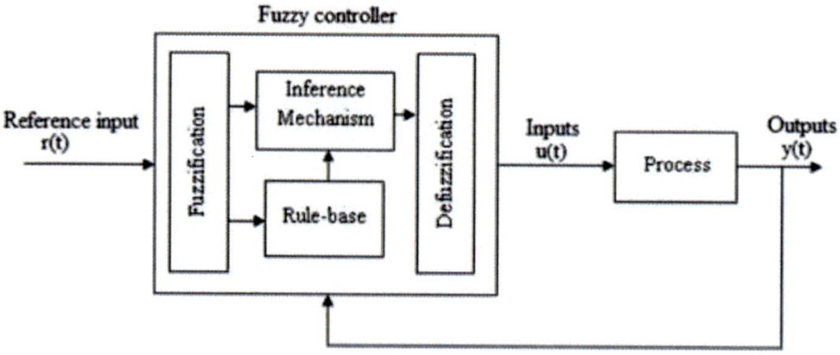

Fig. 1 Generalized fuzzy-based control

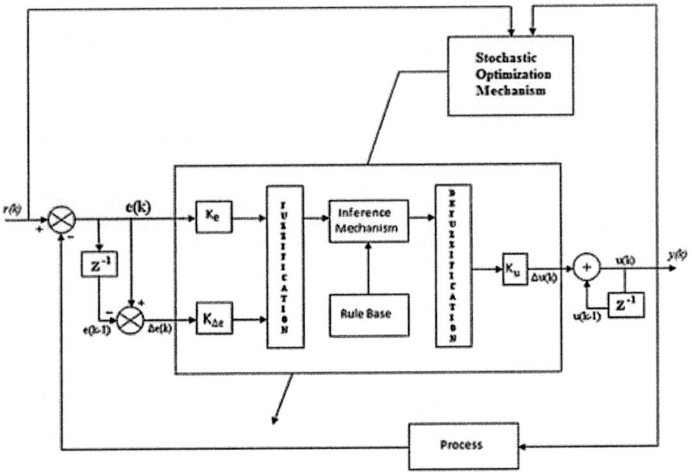

Fig. 2 Optimized fuzzy-based control

making inference process employs fuzzy inference rules. The defuzzification process produces nonfuzzy outputs.

There are three types of fuzzy inference system

1. Mamdani FIS
2. Takagi, Sugeno and Kang FIS.

Mamdani FIS is a IF-Else rule-based FIS where input and output is fuzzy set, whereas in TSK FIS, the output is a mathematical function (Fig. 3).

4 Optimization of FIS

FIS can be optimized using three different ways

1. Optimization of membership function
2. Optimization of fuzzy rule base
3. Optimization of membership function and rule base.

4.1 Genetic Algorithm-Based Optimization of FIS

First of all, two rule bases are created and single rule from each rule base are treated as parent chromosomes, and they are encoded accordingly. After encoding

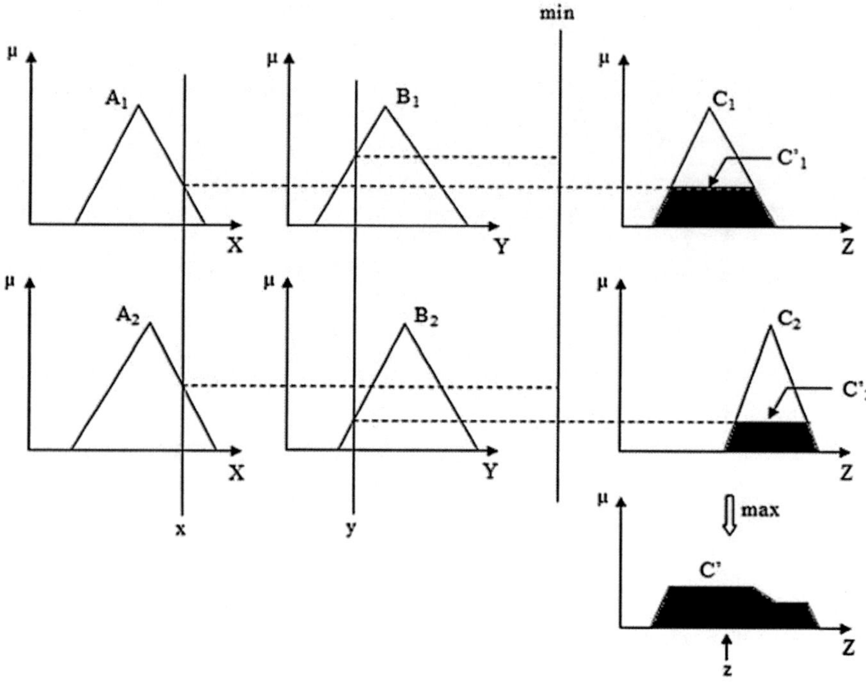

Fig. 3 Mamdani fuzzy inference

the performance index, fitness of each chromosome is calculated. The performance index for the GA is defined as the ITAE (Integral Time Absolute Error) and defined as

$$\text{ITAE} = \int_{0}^{\infty} t\,|e(t)|\,dt \tag{1}$$

$$\text{fitness} = \frac{1}{\int_{0}^{\infty} t\,|e(t)|\,dt} \tag{2}$$

After fitness value is calculated, the fit chromosomes are allowed for two-point crossover. The two-point crossover results in production of offsprings. These offsprings are the first-generation offsprings. So after the first generation, the mating pool is increased, the least fit chromosomes are eliminated, and the termination criterion (ITAE) is checked. If the criterion is not satisfied, again crossover operation is performed, till the termination criterion is fulfilled (Fig. 4).

After the termination criterion is fulfilled, the chromosomes are mutated and the optimal rule is generated. This procedure is done for each rule, and after that the rule base is created.

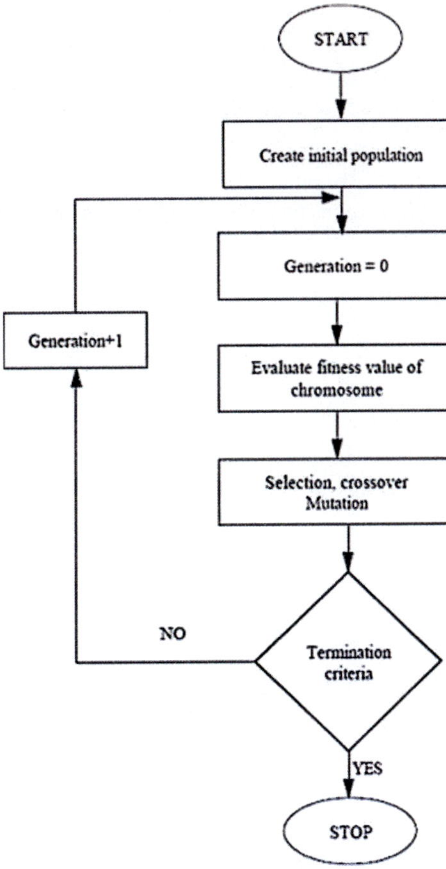

Fig. 4 Flow chart of genetic algorithm

5 Simulation Results

Consider a transfer function [6–8]

$$G(s) = \frac{5}{90s^2 + 33s + 1} e^{-2s} \tag{3}$$

For developing fuzzy controller, we have considered the following input and output variables (Fig. 5).

For each inputs and outputs, a seven-membership function is defined, i.e. NB refers to negative big, NM refers to negative medium, NS refers to negative small, ZO refers to zero, PS is positive small, PM is positive medium, and PB is positive big (Figs. 6 and 7).

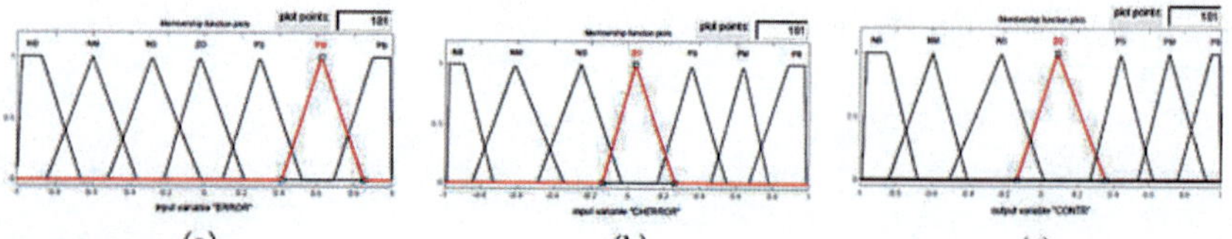

(a) (b) (c)

Fig. 5 Fuzzification of crisp inputs and output **a** Input variable1, **b** Input variable2 , **c** Output variable

Fig. 6 Fuzzy rule base

Δe_N / e_N	NB	NM	NS	ZE	PS	PM	PB
NB	NB	NB	NB	NB	NM	NS	ZE
NM	NB	NB	NB	NM	NS	ZE	PS
NS	NB	NB	NM	NS	ZE	PS	PM
ZE	NB	NM	NS	ZE	PS	PM	PB
PS	NM	NS	ZE	PS	PM	PB	PB
PM	NS	ZE	PS	PM	PB	PB	PB
PB	ZE	PS	PM	PB	PB	PB	PB

Fig. 7 Convergence analysis

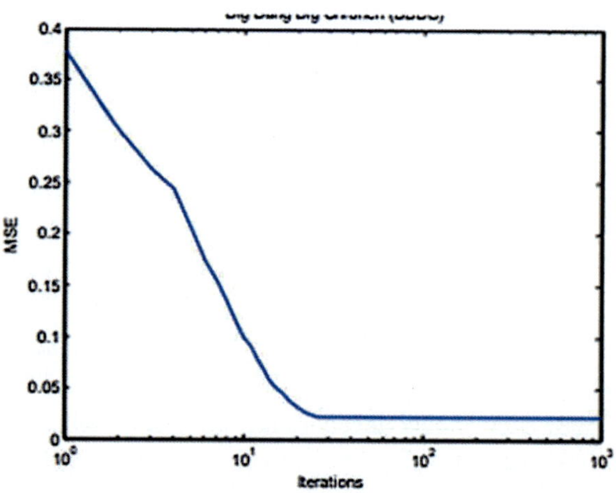

6 Conclusion

This paper gives a brief overview of implementation of genetic algorithm to optimize the fuzzy IF-THEN rule bases. It also gives an overview of identification and estimation of membership functions using H-inf filtering. It takes a sample fuzzy IF-THEN rule base and proposes genetic algorithm-based optimization method to optimally place the rules so that the computation power and search speed are minimized. In future scope of this research, a comparative study of different optimization algorithms can be studied, and time complexity of the fuzzy rule base can also be investigated.

Acknowledgements The authors are thankful to Accendere Knowledge Management Services Pvt. Ltd and CL Educate Ltd for their assistance during the preparation of the manuscript.

References

1. Fang, G., Kwok, N.M., Ha, Q.: Automatic fuzzy membership function tuning using the particle swarm optimization. In: 2008 IEEE Pacific-Asia Workshop on Computational Intelligence and Industrial Application, vol. 2, pp. 324–328. IEEE, New York (2008)
2. Juang, C.F., Chang, P.H.: Designing fuzzy-rule-based systems using continuous ant-colony optimization. IEEE Trans. Fuzzy Syst. **18**(1), 138–149 (2009)
3. Kharrati, H., Khanmohammadi, S.: Genetic algorithm combined with h-inf filtering for optimizing fuzzy rules and membership functions. J. Appl. Sci. **8**(19), 3439–3445 (2008)
4. Kosanam, S., Simon, D.: Fuzzy membership function optimization for system identification using an extended Kalman filter. In: NAFIPS 2006-2006 Annual Meeting of the North American Fuzzy Information Processing Society, pp. 459–462. IEEE, New York (2006)
5. Kumar, M., Jangra, A., Diwaker, C.: Genetic optimization of fuzzy rule-base system. Int. J. Inf. Technol. **2**(2), 287–293 (2010)
6. Padhee, S.: Controller design for temperature control of heat exchanger system: simulation studies. WSEAS Trans. Syst. Control **9**, 485–491 (2014)
7. Padhee, S., Khare, Y.B., Singh, Y.: Internal model based PID control of shell and tube heat exchanger system. In: IEEE Technology Students' Symposium, pp. 297–302. IEEE, New York (2011)
8. Padhee, S., Singh, Y.: A comparative analysis of various control strategies implemented on heat exchanger system: a case study. Proc. World Congress Eng. **2**, p873–877 (2010)
9. Pal, T., Pal, N.R.: Sogarg: a self-organized genetic algorithm-based rule generation scheme for fuzzy controllers. IEEE Trans. Evol. Comput. **7**(4), 397–415 (2003)
10. Pomares, H., Rojas, I., González, J., Prieto, A.: Structure identification in complete rule-based fuzzy systems. IEEE Trans. Fuzzy Syst. **10**(3), 349–359 (2002)
11. Prado, R., Exposito, J.M., Yuste, A., et al.: Knowledge acquisition in fuzzy-rule-based systems with particle-swarm optimization. IEEE Trans. Fuzzy Syst. **18**(6), 1083–1097 (2010)
12. Saleh, C., Avianti, V., Hasan, A.: Optimization of fuzzy membership function using genetic algorithm to minimize the mean square error of credit status prediction. In: The 11th Asia Pacific Industrial Engineering and Management Systems Conference The 14th Asia Pacific Regional Meeting of International Foundation for Production Research (2010)
13. Zhao, Y., Li, B.: A new method for optimizing fuzzy membership function. In: 2007 International Conference on Mechatronics and Automation, pp. 674–678. IEEE, New York (2007)

Representation of Moving Object in Two-Dimensional Plane Through Object Detection Using Yolov3

Bipal Khanal, Chiranjeevi Chowdary Yanamadala, Nitesh Rai, Chinmoy Kar, and Debanjan Konar

Abstract Path detection is an emerging field which is challenging in itself. Here path detection is achieved through the coordinate values of anchor box which is used to bound the object throughout the video. The proposed work is implemented using YOLOv3 (You Only Look Once) network which is used mostly for real-time object detection. The proposed tracking and path detection model have been tested and implemented in real-time videos. The coordinate values are used further to generate a path of the moving object. The proposed algorithm here is tested on 20 number of videos of different resolution and frames successfully.

Keywords Yolo · Object detection · Yolov3 · Path detection

1 Introduction

Object detection and tracking is a challenging task and it has a wide range of applications in the modern world. Few of them are video surveillance and security systems, traffic monitoring, human computer interaction, autonomous vehicles etc. The task of object detection in a video can be summarized as the task of finding the position of the object in every frame. There has been a substantial improvement in object detection and tracking techniques like 'City Tracker: multiple object tracking in urban

B. Khanal (✉) · C. C. Yanamadala · N. Rai · C. Kar · D. Konar
Sikkim Manipal Institute of Technology, Manipal, India
e-mail: khanalbipal98@gmail.com

C. C. Yanamadala
e-mail: chiruychowdary@gmail.com

N. Rai
e-mail: niteshrai182@gmail.com

C. Kar
e-mail: info.chinmoy@gmail.com

D. Konar
e-mail: konar.debanjan@gmail.com

S. Borah et al. (eds.), *Soft Computing Techniques and Applications*, Advances in Intelligent Systems and Computing 1248,
https://doi.org/10.1007/978-981-15-7394-1_22

mixed tracking scenes' [1] which uses Yolov3 [2] and deep SORT [3] algorithms for object detection and tracking, Deep affinity network for multiple object tracking [4] etc. In this paper we present object detection with Yolov3 [2] in a video and collect the detected points of each frame and represent those points to track the movement of object in the entire video.

YOLO means 'you only look once' because the it predicts class probabilities and bounding boxes using a single neural network directly from full image in one evaluation [5]. It is one of the latest techniques for object detection and classification. Faster real-time object detection can be achieved by Yolov3 [2] providing space for video surveillance and other real-time systems. It uses darknet-53 architecture [2] for feature extraction and provides better result for smaller objects detection as compare to its previous versions Yolo9000 [5]. The basic motivation behind this topic is to be able to detect object and track its movement in a video by just using Yolov3 [2]. This paper explains how to track movement of an object and represent it in a 2D plane using Yolov3 [2].

1.1 Architecture of Yolov3

As per the original paper 'Yolov3: An incremental improvement' [2] Yolov3 is based on darknet-53 instead of darknet-19 used in Yolov2 or Yolo9000. Darknet-53 means that the network has 53 convolutional layers which is used as a feature extractor which has some shortcut connections as well as shown in the Fig. 1. Some of the features of the network are as follows.

The architecture consists of skip connections and upsampling layer. It upscales the image in 3 scales as shown on the above figure. The '+' does addition and '*' does concatenation. The boxes and '.' represents different layers. To generate output,

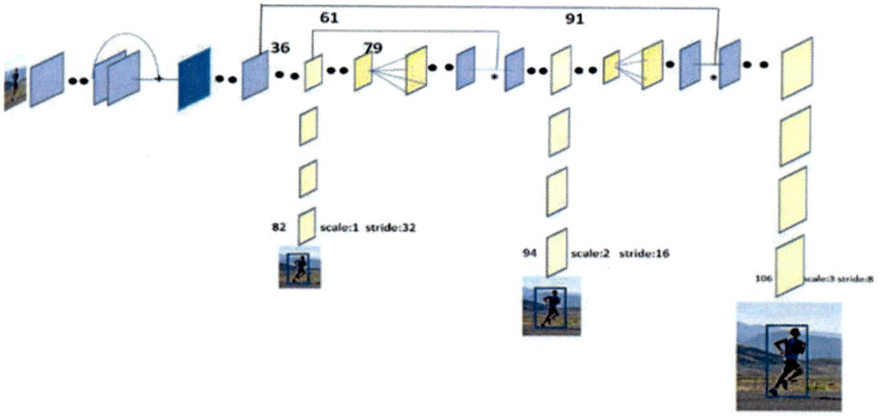

Fig. 1 Yolov3 architecture

we apply 1×1 detection kernels on a feature maps of three different sizes at three different locations in the network. [2]

1.1.1 Batch Normalization

Batch normalization [5] is used for regularization. To Improve mAP (mean Average Precision) and to provide stability of the neural network batch normalization is added to the convolutional layers and overfitting is also prevented [5].

1.1.2 Anchor Boxes

Yolo uses fully connected layers on top of feature extractor to predict coordinates of bounding boxes. These anchor boxes are used to predict the bounding box. These anchor boxes are designed according to the dataset instead of keeping a fixed anchor boxes these anchor boxes are obtained by using k-means clustering [5] in the training dataset object points.

1.1.3 Bounding Boxes Prediction

The bounding box comprise of four coordinate points which gives score for the object in each bounding box. Here t_x, t_y, t_w and t_h are four coordinates of bounding box which are calculated as follows,

$$b_x = \sigma(t_x) + C_x \tag{1}$$

$$b_y = \sigma\left(t_y\right) + C_y \tag{2}$$

$$b_w = p_w e_w^t \tag{3}$$

$$b_h = p_h e_h^t \tag{4}$$

where C_x and C_y are the offset from the origin and p_w and p_h are prior width and height.

The prediction rule of Yolov3 [2] is similar to FPN. Here for every location of input image is predicted based on three number of prediction and features are extracted from each prediction. By doing so it has better ability to detect object at different scales. Each prediction contains boundary box, objects and class scores.

2 Related Work

This section deals with prior papers which are used for implementing YOLOv3. The work done by Joseph [5] gives a picture about YOLO and its overall performance. For class prediction they have used independent logistic classifier instead of softmax in order to achieve better performance. They have used Darknet neural network framework for training and testing of YOLOv3 [2]. Prediction of bounding boxes using dimension clusters is defined as anchor boxes. Object score for each bounding box uses logistic regression. This model of YOLO involves a part of YOLOv2 and Darknet-19. Yolov3 network is more powerful than Darknet-19 and efficient than ResNet-101 or ResNet-152. As stated in YOLOv2 [5] which uses batch normalization on all convolutional layers which helps to get 2% of improvement in mAP(mean Average Precision). Anchor boxes is used to predict bounding boxes instead of fully connected layers, and it down samples the image by a factor of 32. K-means clustering has been implemented which leads to a good score. YOLOv2 bounds the location using logistic activation σ function, which makes the value fall between 0 and 1.

3 Experiments

In this section, we discuss about the experiments carried out in order to represent the detected points of an object in a two-dimensional plain using Yolov3 object detector. There are three phases in this process. Preparing a dataset, Training and testing the model and Visualization of the acquired results.

3.1 Preparing a Dataset

There are a lot of existing datasets freely available for any deep learning or machine learning approach like COCO dataset [6] which has 80 classes and more than 330,000 images, Pascal VOC 2012 [7] which has 20 classes and more than 11,500 images etc., but our objective was to prepare a custom dataset consisting of only one class. This dataset consists of images of persons from various sources and consists of 1000 images. These images are divided into 60/40 ratio for training and testing respectively. These images are annotated manually using LabelImg [8] a free image labelling tool and the corresponding annotation files for each of those images were generated in Yolo format and saved in .txt extension, the annotation file has one row consisting the values class which is 0 and is labelled as Person in this case, x coordinates of the object in the image, y coordinates of the object in the image, width of the object, height of the object each separated by single space. For example, an annotation file of image 1 .jpg would be image 1 .txt as shown in Fig. 2.

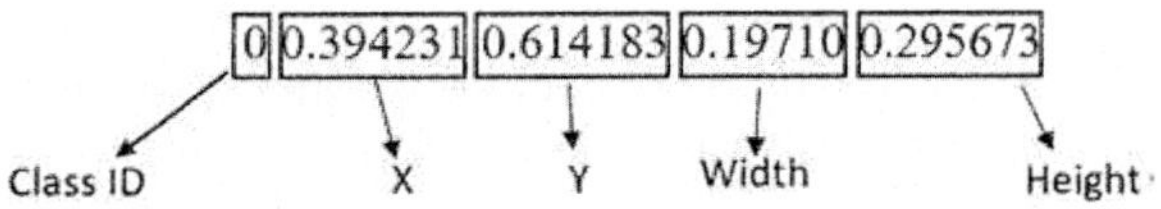

Fig. 2 Contents of annotation file generated for training

Here in annotation files the coordinates and size are as a proportion of the entire image size. For example, an image of 600 × 600px and coordinate (x, y) as (300,400) would be represented by (x, y) as (0.5, 0.66666). After getting all the annotation files of all the images, a text file should be created which consists of locations of all the training images or the path of all the training images one per row. Also, to note that the images and the annotation file should be in the same directory and with the same name just with different extensions.

3.2 Training and Testing the Model

Before starting the training, process there are few things that are to be done like making changes to the configuration (.cfg) file and creating some files that is used for training. Firstly the .names file is created which has the names of object classes that are to be trained, in our case we have only one class Person that is in the 0th index of the .names file. Now .data file is created where few things about our training data are specified like number of classes, path of list of training data, path of list of testing data, location of .names file and backup which is optional is a path to back up the trained weights at any stage of training.

After creating all this files, we are ready to train our model but not until we modify the configuration file i.e. .cfg file of the Yolov3. In the .cfg file there are few parameters we need to modify for training our custom dataset, in the .cfg there are many sections the first section is the net section which gives information about the network where the batch value and subdivisions should be modified according to your GPU memory. The more the batch size the better and faster the training but it will take more memory. Also, in this section it is possible to change the learning rate, the max batch size etc. There are different sections in the .cfg file like convolutional, shortcut, Yolo etc. which are layers of the model. In Yolo section set the classes to the number of classes to be trained, in our case it is 1, there are 3 Yolo layers so the changes are to be made in each of the Yolo sections and change the filters value of convolution section just above the Yolo section. The value of filters will be equal to:

$$\text{filters} = (\text{classes} + 5) * 3 \tag{6}$$

So, for 1 class the filters value would be 18.

Now we start the training using all the files mentioned above, the implementation of the network has been done in python using Pytorch and the weights generated after each 100 epocs is being saved so that we can pick any checkpoint and start training again for fine tuning the results or detections, and we can even try to detect some objects loading those checkpoint weights and whichever checkpoint weights gives the suitable result can be picked. The training process can go on for 4–5 h or days depending on your machine. We trained our model on a system having NVIDIA Geforce RTX 2070 GPU with 8gb of GPU memory and it took around 4–8 h to complete the training. The predefined darknet53 weights are used to initialize the network before starting the training process. Once the training is done, we follow up with testing the trained weights with the images separated while preparing the dataset. Once the test results are satisfactory, we can conclude the training and testing process.

3.3 Visualization of the Acquired Results

Once the mode is trained and we have the trained and tested weights now we visualize the outputs of the network in the form of bounding boxes around the object in the image as shown below. As shown in Fig. 3 the input image and the corresponding

Fig. 3 Input image

Fig. 4 Output image with bounding box

output in Fig. 4 the model is detecting the class, we trained it for.

4 Proposed Work

Now we have a model that is trained to detect one class, same model can be used to detect single class of objects in videos too. Videos are a sequence of frames and each frame can be considered as an image then we can perform object detection in videos using this model. Let us consider a 30 fps video of 3 s, we have 90 frames, each is like an image so the detection is needed to be made in each frame, which means there will be a total of 90 detections of an object in the entire video. Now if the person is moving from point A to point B in that video then we will have different points for detection of that object in the video. Here point A is the starting point and point B is the ending point for that moving object. We take all those points of detections and try to plot it in a two-dimensional graph and find the path followed by the object in the image which basically means tracking an object in an image.

Object tracking is not done by Yolov3 as it is just an object detector, to perform object tracking an object detection algorithm with object tracking algorithms are used

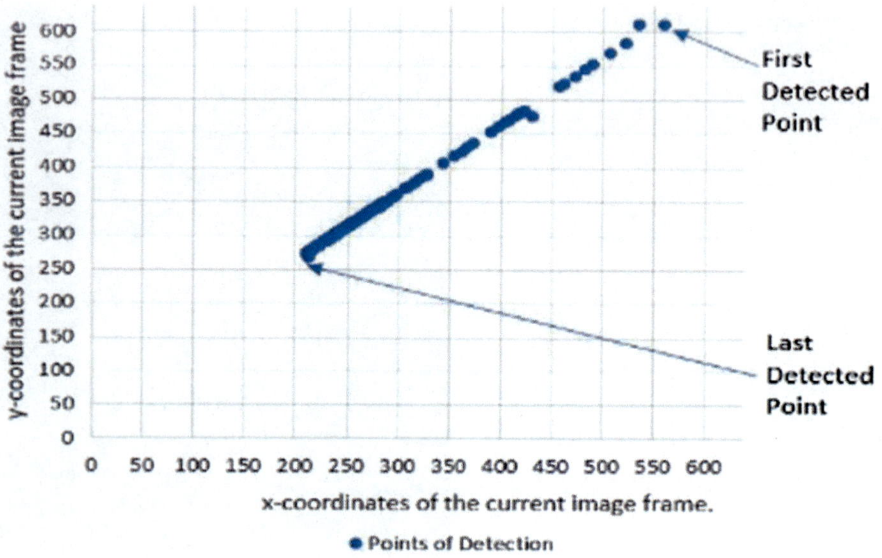

Fig. 5 Points of detection

like with Yolov3 which is a real-time object detector an object tracking algorithm named SORT (simple online and real-time tracking) [9] is used. Here instead of using a tracking algorithm to track an object in video we are storing the bounding boxes values which are *x*, *y*, height and width which is being calculated by the model while detection. The values that are stored is further plotted in a 2-d graph to generate a path of that moving object by just curve fitting as shown below.

The points shown in Fig. 5 are obtained from detecting an object from each frame of a video, here the object first detected on the frame was on point (*x*, *y*) (256, 230) and we can see that the object is being moving from that point let say A towards point B (*x*, *y*) (520, 610) following a linear pattern, thus applying a linear curve fitting we get the path as shown in Fig. 6. Few more examples are provided below (Figs. 7, 8 and 9).

As shown in Fig. 10 the detector has detected two objects both as person but there is no other person in that frame, this video was shot on low light and as the blue bounding box shows a person to a shadow of a person. This result leads to an important finding in this entire process of object detection and representation of these detected objects in a two-dimensional frame that, if there are multiple objects in a same frame then the points generated in that frame would also be multiple and if the objects are close to each other then there is an overlapping points as well, as we can verify this result in Fig. 11 at the end of the straight line there are multiple points almost overlapped. These multiple points are nothing but the other object detected near to the prime object.

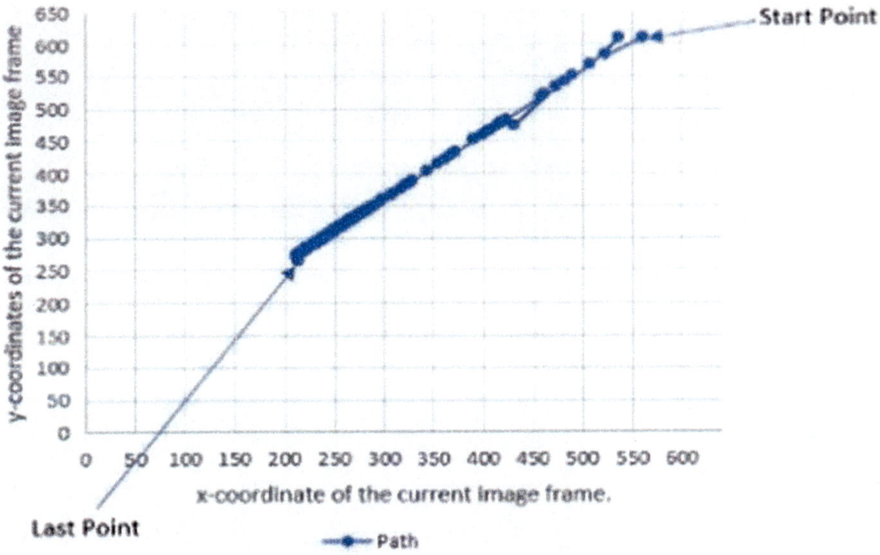

Fig. 6 Applying curve fitting to the points

Fig. 7 First input frame

Fig. 8 Last input frame

Fig. 9 First detected frame

Fig. 10 Last detected frame

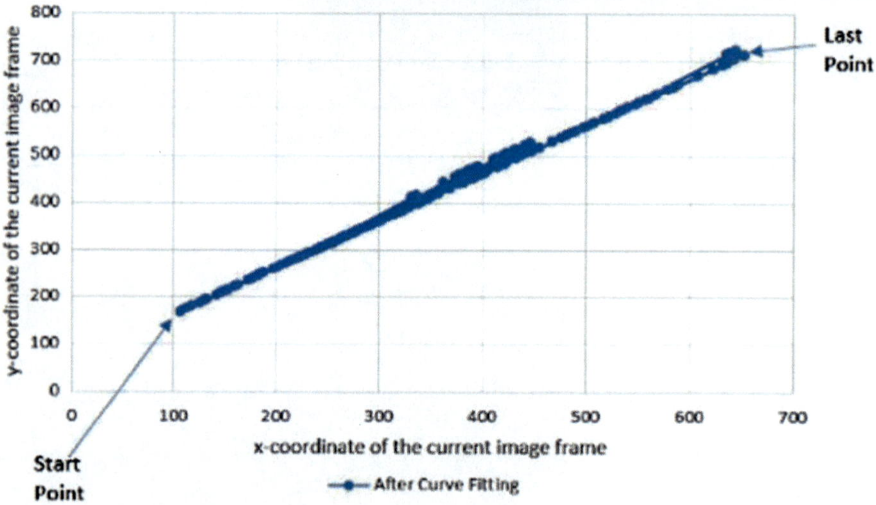

Fig. 11 Points after curve fitting which is the path followed by the object in a video

5 Conclusion

This paper proposes a method of tracking an object from a video using the points of bounding boxes generated by Yolov3 object detector. It is a kind of object tracking but without using any object tracking algorithms. Thus, Yolov3 performs excellent in real-time object detection; generating a path directly using the proposed method will make the tracking process even faster. In our experiment it is shown that training and testing of Yolov3 can be done for any arbitary classes of objects which may belong to medical science, defence, surveillance systems etc., which may enhance the process of involvement of the field of computer vision in every aspect of human life. This method performs best when detecting a single object but can be modified to work for multiple objects. While detecting multiple objects, if the objects are of different classes then the points can be separated and a separate path can be found for each object but if the objects are of same class then the classification of points for different object of a same class is a challenging task to accomplish.

References

1. Chan, Z.Y., Suandi, S.A.: City tracker: multiple object tracking in urban mixed traffic scenes. In: 2019 IEEE International Conference on Signal and Image Processing Applications (ICSIPA), Kuala Lumpur, Malaysia, 2019, pp. 335–339. https://doi.org/10.1109/icsipa45851.2019.8977783
2. Redmon, D.J., Farhadi, A.: Yolov3: An incremental improvement. arXiv preprint arXiv:1804.02767 (2018)
3. Wojke, N., Bewley, A., Paulus, D.: Simple online and realtime tracking with a deep association metric. In: 2017 IEEE International Conference on Image Processing (ICIP), Beijing, 2017, pp. 3645–3649. https://doi.org/10.1109/icip.2017.8296962
4. Sun, S., AKHTAR, N., Song, H., Mian, A.S., Shah, M.: Deep affinity network for multiple object tracking. In: IEEE Transactions on Pattern Analysis and Machine Intelligence
5. Redmon, J., Farhadi, A.: Yolo9000: Better, faster, stronger. In: 2017 IEEE Conference on Computer Vision and Pattern Recognition (CVPR), pp. 6517–6525. IEEE, New York (2017)
6. Lin, T.Y., et al.: Microsoft COCO: common objects in context. In: Fleet, D., Pajdla, T., Schiele, B., Tuytelaars, T. (eds.) Computer Vision—ECCV 2014. ECCV 2014. Lecture Notes in Computer Science, vol. 8693. Springer, Cham (2014)
7. Everingham, M., Van Gool, L., Williams, C.K.I., Winn, J., Zisserman, A.: The PASCAL visual object classes challenge 2012 (VOC2012) Results (2012)
8. Tzutalin. LabelImg. Git Code (2015). https://github.com/tzutalin/labelImg
9. Bathija, A., Sharma, G.: Visual object detection and tracking using Yolo and sort. Int. J. Eng. Res. Technol. (Ijert) **08**(11) (2019)

Analysis of DDoS Attack in Cloud Infrastructure

Anurag Sharma, Md Ruhul Islam, and Dhruba Ningombam

Abstract The computing world was revolutionized when the concept of Cloud Computing emerged. Cloud Computing became a solution to various problems of organizations from small scale to large scale organizations thanks to the benefits associated with it which includes Service on demand, Scalability, Flexibility, Availability, Ability to access anywhere only with an internet connection and other various added advantages. However, with the advancement of any technology, there have always been security issues related to it. One of the major security issues that was there previously and is continuing to emerge in Cloud Computing is the Denial of Service attack (DoS). The DoS attack in cloud is more harmful than ever and their effect in cloud is huge as it affects the scalability and the pay-per-use model of Cloud Computing. The paper will focus on the analysis of the DDoS attack on cloud infrastructure through simulation in CloudSim and flooding of ICMP packets using Ping of Death.

Keywords Cloud computing · Scalability · Flexibility · Availability · Service on demand

1 Introduction

The basic concept of Cloud Computing is providing various computational facilities over the internet, where the user might want some storage services, computing power, and the ability to access it anywhere only with the help of an internet connection. With the advancement of Cloud Computing the process of On-demand Infrastructure

A. Sharma · Md R. Islam (✉) · D. Ningombam
Department of Computer Science and Engineering, Sikkim Manipal Institute of Technology, Sikkim Manipal University, Rangpo, India
e-mail: ruhulislam786@gmail.com

A. Sharma
e-mail: anuragamrahs@gmail.com

D. Ningombam
e-mail: dningombam@gmail.com

S. Borah et al. (eds.), *Soft Computing Techniques and Applications*, Advances in Intelligent Systems and Computing 1248,
https://doi.org/10.1007/978-981-15-7394-1_23

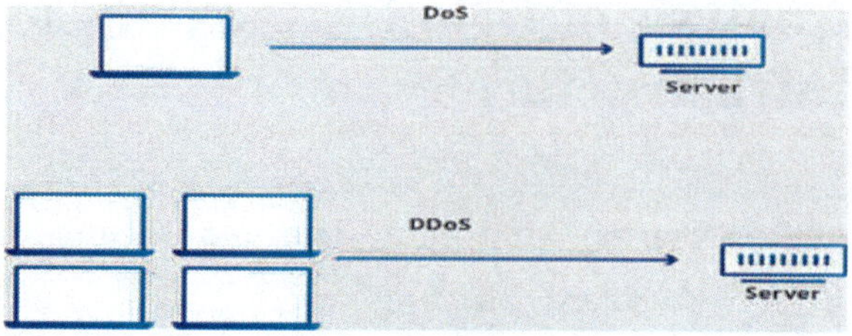

Fig. 1 DoS attack and DDoS attack

services has become simpler and has benefited various organizations since the Cloud works on pay-per-use model and it also has the ability to increase or decrease the need for resources. The most widely used services in cloud include Gmail, Facebook, and Dropbox, etc. these services can be accessed anywhere in any device like mobile phones, laptops, tablets provided an internet connection [1].

Security is and always has been a major concern for technology like cloud computing. The cloud services can be threatened by various types of risks which might cause a great loss to the user or the Cloud Service Provider. The availability of the resources is harmed by one major threat which is known as Denial of Service where the computation power or the resources are used up by the intruder so that the authentic users do-not get any computational power and if in case they get it then it will incur a huge computational cost according to pay-per-use model of Cloud Computing. Such an attack can have a great effect on the cloud in terms of resources or financially, short term for customers as their work might not be done and long term for Cloud Service Providers as they might lose their customers.

Types of Denial of Service attacks:-

(i) Denial of Service (DoS): Denial of Service (DoS) attack is one of the most growing types of attack in cloud, one of the most problematic and affecting the normal functioning and workflow of various types of organizations. The attacker floods the system with a huge number of requests thereby denying legitimate users of the resource. Attacking through a single source machine for the disruption of the service is known as DoS attack.
(ii) Distributed Denial of Service (DDoS): The Distributed Denial of Service (DDoS) is an extension of DoS attack, here the attack is initiated from multiple un-protected devices which increase the level of attack on the target machine exponentially. The ultimate goal of DoS and DDoS attacks is to crash the service by flooding the network with multiple requests and taking up all the resources (Fig. 1).

2 Information Security Principles

In order to have a cloud communication that is secure, there are certain principles that we need to follow; those principles are known as Information Security Principles. One of the known security models which deal with the aspect of Information Security

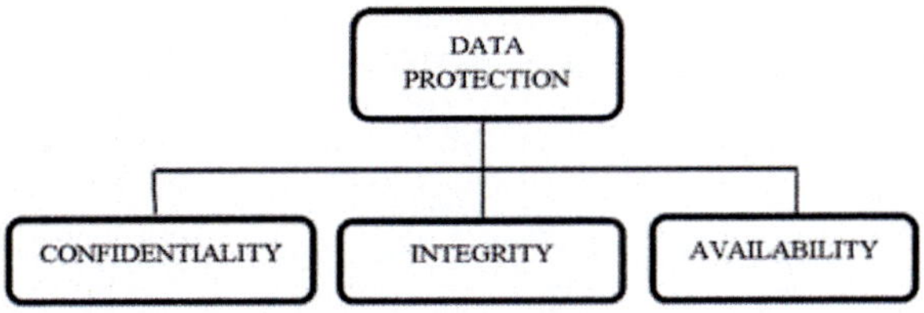

Fig. 2 Division of data protection into CIA

is CIA. The C stands for Confidentiality; I stand for Integrity and A stands for Availability.

(I) Confidentiality: Confidentiality refers to the protection of information for users who are not authorized. The main aim is to hide the information from others so that they may not access it. The increase in users in the cloud has led to an increase in threats in clouds so the confidential data needs to be saved.
(II) Integrity: Integrity refers to the data being consistent and seeing how accurate the data is. The data should not be manipulated by any unauthorized users. The data should not be altered while being transmitted to other systems.
(III) Availability: Availability refers that the data should be available every time the user wants to access it, however, and whenever the user needs it. It basically refers to that the data should be always available for the authorized users to access [2] (Fig. 2).

3 Related Works

In [3] an attack in cloud environment has been proposed by the authors. The authors have proposed an attack in the application layer using XML and HTTP DDoS attack and Amazon EC2 has been used. Here the request is made by the users using XML and then the request is sent using HTTP protocol. Attacker then implements the attack which will be hard for the intrusion detection system to detect.

In [4] the authors have proposed a model for DoS attack by mixing filtering, memory depletion, and bandwidth exhaustion, to demonstrate the attack. The testing shows the influence of the attacker and the properties of the victim on the possibility of success of DoS attack. This DoS attack model has a very high accuracy of estimation of attack.

In [5] the authors in their paper describe various types of DDOS attacks which they have described in two types one which causes the exhaustion of resources and the other is exhaustion of bandwidth. The bandwidth exhaustion attacks consist of flooding attacks like UDP flooding and ICMP. The exhaustion of resources is mainly due to exploitation of protocol and the attacks based on them. The attacks like TCP-SYN flood attack. The authors have also shown various classic tools like Trinoo, TFN, mstream, etc. that can be used for DoS attacks.

In [6] the authors have described the package for the creation of virtual network and simulation of DDoS attacks and defence against those attacks. The proposed method for defence method was Distack. This package was built using system OMNeT++, which included libraries of INET and ReaSE.

4 Proposed Design Strategy

The experimental model and the flowchart given below depict the various actions taking place in the system and the conditions responsible for the actions. It also serves the purpose of giving a clear picture of the working of the system. In this paper, two Denial of Service attacks are performed and it is then analysed how the system reacts normally when comparing to the system when DoS attack is performed.

(i) Ping of Death [7]: The command ping is used for testing the availability of the network. The Ping of Death uses this as an advantage of sending the data packets which are more than the limit (65,536 bytes) which is in the case of TCP. Now since the data packets which are sent are of larger size than that of what size server can handle, so thus the server might reboot or crash, or in case of cloud the server might be scaled larger thereby causing an additional cost to the organization. The general command for the ping of death is **(Ping victim's_ip–t|packets)**

In this experiment we are going to flood the victim with infinite data packets of 65,500, the victim and the attacker are connected with an Ethernet (Fig. 3).

(ii) Simulation of DDoS attack using CloudSim [8]: In the second part, a simulation of DDoS attack is being performed over instances of Virtual Machines. As the services under Cloud are provided as pay-per-use model so if any DDoS attack is being done then the increase in computation will incur huge losses on the organization.

a. CloudSim: Cloudsim is a java framework that was developed by CLOUDS Lab, "University of Melbourne" Australia headed by Dr. Buyya. The Version 3.03 has been used. This software provides the framework for modeling the cloud environment and simulating cloud services and infrastructures. Under CloudSim, simulation of virtualized datacentre will be done and creation of virtual machines under datacentre will be done. The creation and simulation of jobs are done under CloudSim.

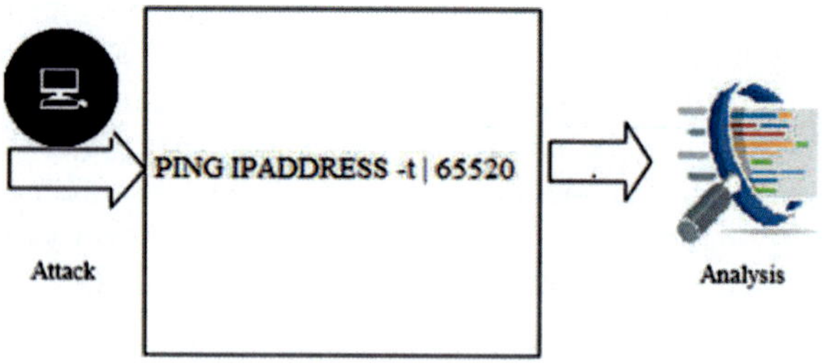

Fig. 3 Experimental model of ping of death

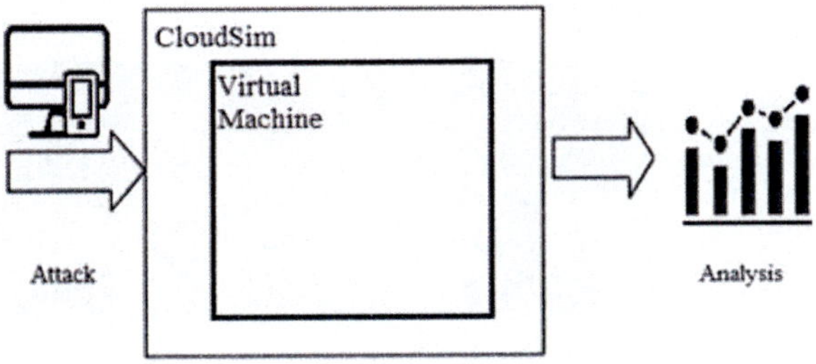

Fig. 4 Experimental model of DDoS in cloudSim

b. Programming Tools: The programming language used will be Java. Eclipse will be used for the CloudSim framework installation and as well for programming. The prebuilt libraries are called using Java programming language for modelling and simulation. There will be the use of JDK for development of Java program in Netbeans (Fig. 4).

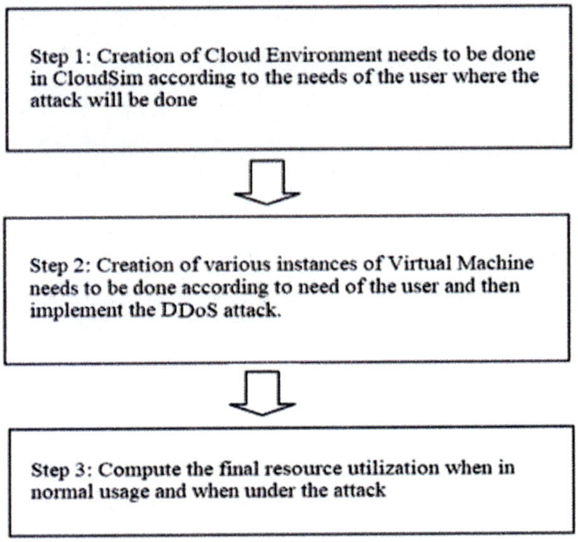

Flow Chart for DDoS attack using CloudSim.

5 Experimental Result

(i) Ping of Death (see Figs. 5, 6, 7)
(i) *Simulating DDoS using CloudSim* (see Figs. 8, 9, 10)

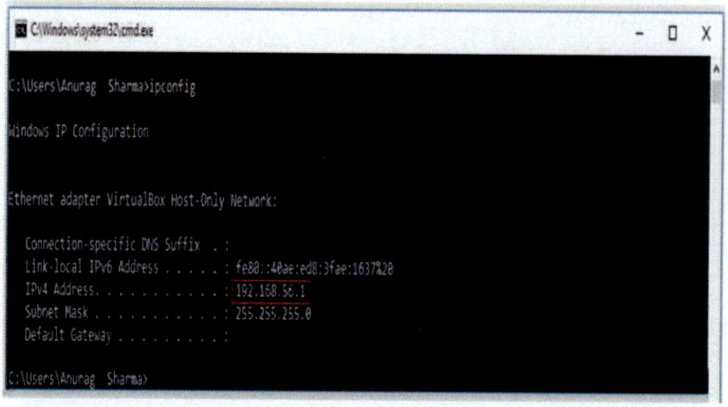

Fig. 5 Checking the victim's IP address

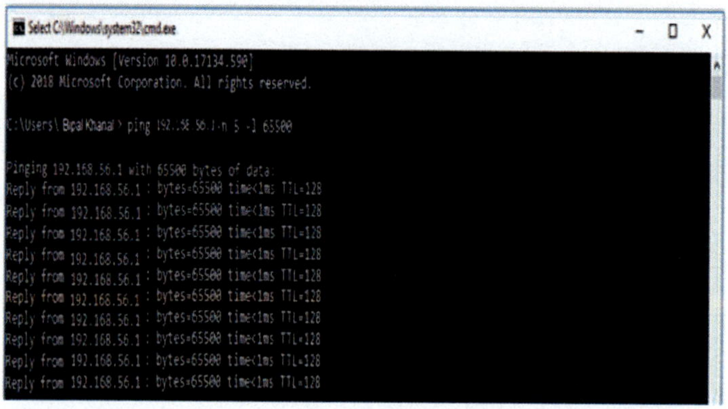

Fig. 6 Flooding of packets from attacker to victim

6 Conclusion

The resulting graph of Ping of Death clearly shows the increase in traffic and how it can be dangerous in causing harm to the system. The results of the simulation in Fig. 10 show that the resource required by Virtual Machine under attack is huge as compared when the Virtual Machine is under normal work. When in high octane attack the normal working of Virtual Machine might be disturbed and the original user might be denied from service if the resources are not enough. Since Cloud services use pay-per-use utility the organization will need to pay a huge sum for the service as well as there would be customer dissatisfaction with the unavailability of the resource.

There is no any doubt that the current technological development in cloud computing has proven to be of huge benefit to most of the industries, start-up, and consumers in terms of availability of different types of services with the help of

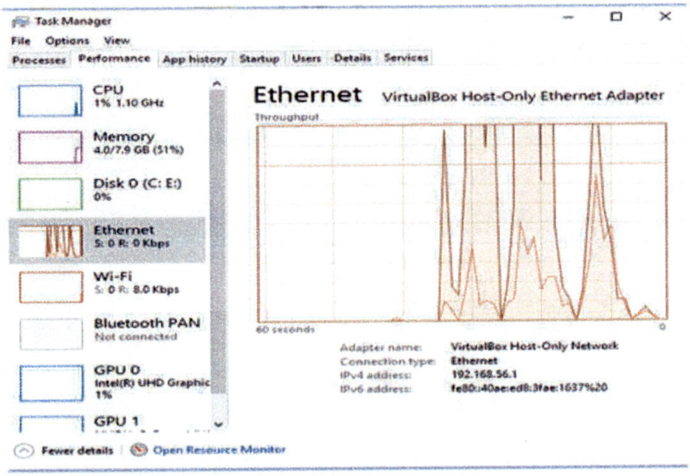

Fig. 7 Increase in network activity determines the success of ping of death

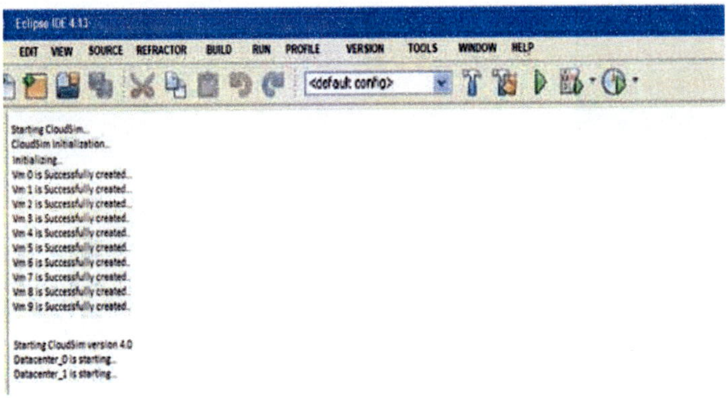

Fig. 8 Creation of virtual machine and data center

internet across the world but the threat in Cloud Security still is a huge problem which might be a roadblock for organizations which are trying to fully migrate on to Cloud. Denial of Service is hugely problematic by looking at the outcome that might cause a big loss in terms of resources or finance. The cost and time to recover from these issues might not be easy, which is sufficient for organizations to back out. Collective measures of protection need to be done which would reduce the influence and reduce the scale of such attacks. The administrator should detect the issues in networks in an active manner to reduce the threats of those attacks and protect the devices and overall protect themselves from such attacks.

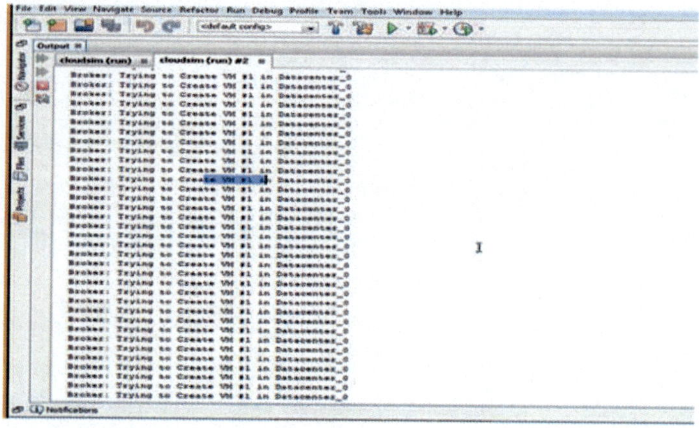

Fig. 9 Attack on virtual machine by sending multiple requests

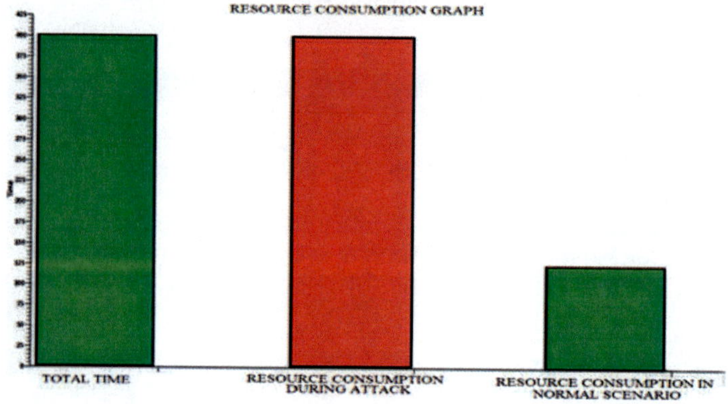

Fig. 10 Ren consumption with attack and without attack

7 Future Works

Denial of Service has been a major problem since the inception of Cloud Computing so in future, we want to see out ways in which DoS can be prevented or we can decrease the frequency of this attack by creating various countermeasures.

References

1. Huang, V.S.-M., Huang, R., Chiang, M.: A DDoS mitigation system with multi-stage detection and text-based turingtesting in cloud computing. In: 27th International Conference on Advanced Information Networking and Applications Workshops, IEEE (2013)

2. Georgescu, M., Suicirnezov, N.: Issues regarding security principles in cloud computing. USV Ann. Econ. Public Adm **12**(2)(16), (2012)
3. Karnwal, T., Sivakumar, T., Aghila, G.: A Comber approach to protect cloud computing against XML DDOS and HTTP DDOS attack. In: IEEE Student Conference on Electrical, Electronics and Computer Science (2012)
4. Ramanauskaite, S., Cenys, A.: Composite DoS Attack Model. ISSN 2029–2341 print/ISSN 2029-2252 Online (2012)
5. Kumar, A.R., Selvakumar, S.: Distributed denial of service threat in collaborative environment—a survey on DDOS tools and traceback mechanism. In: IEEE International Advance Computing Conference (2009)
6. Gamer T., Mayer C.: In: Large-scale Evaluation of Distributed Attack Detection II 2nd International Workshop on OMNeT++ (2009)
7. Yihunie, F., Abdelfattah, E., Odeh, A.: Analysis of Ping of Death DoS and DDoS Attacks, Sacred Heart University, Al Maarefa Colleges for Science and Technology
8. Karthik, S., Shah, J.J.: Analysis of simulation of DDOS attack in cloud. ICICES2014- S.A. Engineering College, Chennai, Tamil Nadu, India, IEEE. ISBN No. 978–1-4799-3834-6/14. (2014)

Prospective SD–WAN Shift: Newfangled Indispensable Industry Driver

Sudip Sinha, Rajdeep Chowdhury, Anirban Das, and Amitava Ghosh

Abstract Software-Defined Wide Area Network (SD-WAN) has been developed to introduce Next Gen ready WAN infrastructure designed for easy cloud-centric application access in a secured and cost-effective manner. SD-WAN abridges maneuver and supervision of WAN by disengaging the network tool from its regulation modus operandi. The aim of SD-WAN is to replace existing inflexible WAN technologies with much advanced features to enhance ROI for the Enterprise customers. The scope of the study is to determine the drivers for the Enterprise customer, to move from Legacy WAN environment to SD-WAN for long term business benefit through a set of questionnaires to justify their investment for technology migration. It will also help IT managers to structure standard questions that they will use to select the best SD-WAN solution providers to meet their business requirements.

Keywords Wide area network (WAN) · MPLS · Internet-based VPN service · Cloud service · Private cloud · Public cloud · Split tunnel · SD–WAN · NBFW · Hybrid network

S. Sinha (✉)
Seacom Skills University, Kolkata, West Bengal, India
e-mail: sudipepi@gmail.com

R. Chowdhury
Life Member, Computer Society of India, West Bengal, India
e-mail: dujon18rc@gmail.com

A. Das
Department of Computer Science, University of Engineering and Management, Kolkata, India
e-mail: anirban-das@live.com

A. Ghosh
School of Management Studies, Seacom Skills University, Kolkata, West Bengal, India
e-mail: amitava67@gmail.com

S. Borah et al. (eds.), *Soft Computing Techniques and Applications*,
Advances in Intelligent Systems and Computing 1248,
https://doi.org/10.1007/978-981-15-7394-1_24

1 Introduction

In the industry version 5.0, organizations are focusing on expanding their network connectivity across the globe such that customers can reach them anytime anywhere by any means when they essentially need it [1–4]. Demand for IoT is expanding rapidly and focus has been given tremendously on Artificial Intelligence–Machine Learning, Big Data Analytics, Robotics Automation, etc. to achieve next-generation Business Intelligence for getting a competitive advantage. While connectivity demand has accentuated, safekeeping has become a critical challenge for making sure that customer's data and information is well protected, while providing a focus on end customer's easy application admittance.

Data Privacy, GDPR, PCI-DSS, ISO Compliance requirements are becoming stiffer for the Enterprise customers who operate globally. WAN safekeeping has become a mandatory necessity to support the impending business necessity. Quicker Network enablement, On-Demand scaling up or scaling down, and Cost Optimization is the critical factors for business expansion within limited resources—beget and time maintaining the same quality of service delivery.

SD-WAN is claimed to be apposite for purpose of future industry demand, eradicating inflexibility, longer lead time, high investment, and path optimization with enhanced security features and expanded application throughput. It also focuses on Virtualization technologies and On-Demand feature enablement, which potentially adds worth towards flexibility and dynamism of infrastructure components [1–8].

SD-WAN Solution focuses on increasing end user's gratification index for application admittance, response time, application performance, mainly for Public Cloud-hosted applications. It also provides end to end visibility on traffic flow across the network along with granular details of link health. In summary, SD-WAN Solution will endow with effortless deployment, effortless life cycle management, effortless environment management through a solitary window platform (Figs. 1, 2).

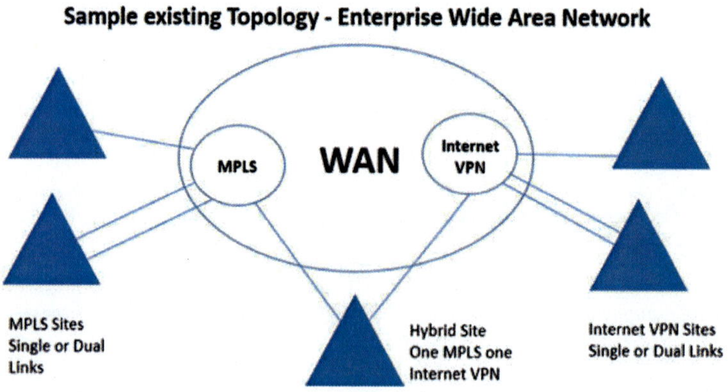

Fig. 1 Prevailing WAN architecture for enterprise network

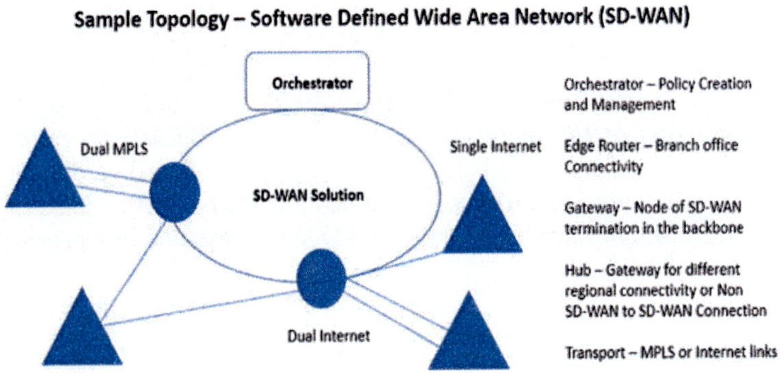

Fig. 2 SD-WAN architecture for enterprise network

2 Literature Review

In [5], probable risk and mitigation approach regarding Quality of Service, Scalability, Security, and Load Balancing areas have been extensively highlighted, but not focused on customer's investment protection by technology optimization.

In [1], features and benefits of Cisco Based SD-WAN Solution and security enhancement has been described meticulously, but less emphasis has been given on how other SD-WAN Solutions are designed.

SD-WAN survey was conducted by Packet Pushers Interactive LLC and the outcome was published, which was regarding existing infrastructure categorization and preference of SD-WAN migration approach. No Validation questions were framed in the survey to select the best SD-WAN vendor [2].

SD-WAN Solution comparison has been described in an assortment of contemporary articles, where the uniqueness of diverse SD-WAN Solutions has been highlighted. It helps organizations to make their SD-WAN requirement based on technical solution offerings, amid the existence of pros and cons [3, 4, 6–10].

3 Objective

The aim of the contemporary study is to identify the common feature sets of SD-WAN solutions from leading vendors and highlighting the business drivers to move from Legacy WAN Solution to SD-WAN. Significant focus should be given in preparing the top 20 questions that IT managers should get answers, before adhering to purchase decisions for SD-WAN Solution as part of WAN technology migration.

- What is the common SD-WAN feature/key enabler for businesses to grow differently?
- What exactly should be asked to SD-WAN Solution providers to validate their solution?

4 Methodology

Detailed study has been conducted for various SD-WAN Solution features to document the common benefits among them. In-depth analysis has been performed for IT and business risks and probable mitigation approaches. Based on the risk assessment, best practice questionnaire has been constructed, which will enable Enterprise customers in conscious decision making for SD-WAN Solution procurement.

5 Obtained Outcome

1. For Legacy WAN environment, MPLS Solution deployment takes longer lead time but overload lead time for SD-WAN Solution deployment is quicker.
2. SD-WAN Solution can work on any transport. Country can take local internet line from the preferred vendor which can be used for SD-WAN and may be enabled in weeks rather than months' time.
3. For MPLS Router deployment, it requires complex CLI-based deployment script.

SD-WAN provides configuration at central server and plug and play push for Edge router.

4. For MPLS—Quality of Service is static configuration. In SD-WAN, dynamic application response has been focused on.
5. For MPLS, Load Sharing across multiple links are complex. In SD-WAN, the same has been designed as standard practice to send traffic across multiple links.
6. In MPLS Primary/Backup situation, backup link will not be activated until primary link has attained complete outage. SD-WAN can use a re-transmission mechanism with error correction so that with high jitter, latency link, traffic will be sent; which was usually dropped in case of MPLS. Since SD-WAN provides Active–Active mode of communication by design, overall aggregated WAN bandwidth will be higher compared to Legacy MPLS, for getting better ROI.
7. Legacy MPLS was having high latency path for Cloud-centric application, which was either via Express Route connection or regionally broken via VPN based Cloud breakout solution. SD-WAN core infrastructure has been strategically designed such that every region has got proximity to the Cloud service provider via Gateway Router placement and each site will have lower latency reaching the Cloud-based applications.
8. MPLS or Internet VPN based Solutions were having a static application flow mechanism with a dynamic path optimization technique of SD-WAN. Real-time traffic like voice, video can be flipped to other links with good quality (Low latency/error/jitter/packet drops) to provide enhanced application throughput.
9. In Legacy WAN environment, it's difficult to manage policy framework across the organization, SD-WAN by design is a single window managed solution

where each site has got configuration complied with organization policy. Illustrating the same, viz; payment related transaction entails traversal via MPLS private link for enhanced security requirement; Cloud specific application access is preferred via Internet link rather than MPLS; Business-specific applications entail being Load Shared across the links, etc.

10. In Legacy WAN Solution, Routers are mainly focused on routing function. The base of SD-WAN has been developed on Virtualization. Network Function Virtualization has been added as a unique creative feature where single Edge device can function as Router, Layer 3 Switch, DHCP function, Firewall, Anti X device, WAN Acceleration device, etc. thereby reducing rack space and power requirement in the server room (Fig. 3).

Considering the aforementioned techno-commercial benefit, SD-WAN seems to be the change-maker for imminent IT infrastructure but Enterprise customers should wisely select SD-WAN considering short-mind and long-term strategy for getting competitive advantage to make their business expanding with more future-readiness and closer reach with supplier and customers.

Following top 20 germane questions (Q01–Q20) entail to be answered for Enterprise customers as best practice guideline, enabling them to intelligently select SD-WAN Partner for technology enabler in sustainable business progress, namely;

Q01. How many customers are already employing SD-WAN solutions across the globe?

Q02. What is the redundancy level of SD-WAN Backbone?

Is DR Drill being performed periodically to ensure business continuity?

Q03. Is there any dependency on transport options? Customer Managed versus Provider Managed?

Is mixed-mode supported for single site?

Q04. What is the overall SLA? Time to restore and Edge router replacement timer?

What is the general Edge Router delivery lead time (Compared to MPLS Router delivery lead time of 4-6 weeks)?

Q05. Is there Site-to-Site latency commitment?

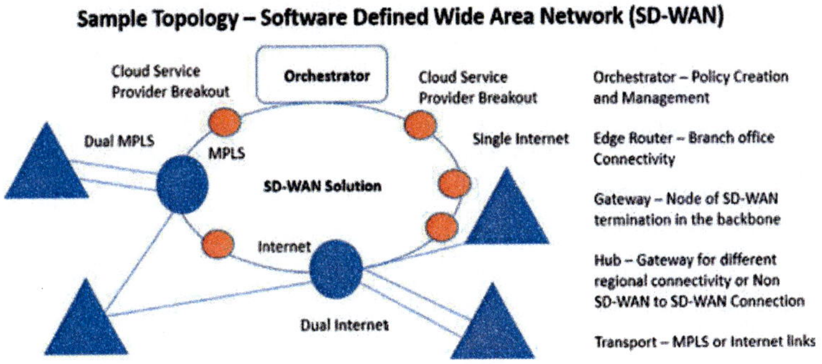

Fig. 3 SD-WAN architecture with CSP breakout

Q06. What is Hub deployment strategy for other customers, minimizing latency across regions?

Q07. Will SNMP and Net Flow-based information be available for capacity planning?

Q08. How Network segments of Enterprise can be mapped with SD-WAN Solution?

[Specimen 3rd Party VPN, Joint Venture, Production Network, Engineering Network, etc.]

Q09. How SD-WAN Solution is meeting safekeeping level as per industry standards?

Q10. What compliance are SD-WAN Solution meeting?

Q11. If any security violation happens, what is the SLA for SD-WAN vendor to fix it?

Q12. Is SD-WAN vendor ready to share penalty cost due to compliance issues, if found related to SD-WAN Infra?

Q13. If any Gateway towards Cloud service provider node fails (Viz; O365 node) in specific country/region, how resiliency can be ensured? Zero Packet drop?

Q14. What are the expected skills at the branch level for SD-WAN Solution deployment?

Q15. How is Inbound Load Sharing addressed for Bandwidth intensive applications?

Q16. What is the SLA for Cloud-based Proxy Service? Are custom Ports supported?

Q17. Is there any limitation for number of SD-WAN Sites supported per region?

What is the maximum limit that has been supported by the vendor until now?

Q18. What is the cost component for On-Premises Orchestrator versus Provider Managed Orchestrator? What are the pricing components related to License?

[Details of Site-specific license-number of concurrent user/flow/volumes of data/number links]

Q19. How End-to-End QOS will be mapped with SD-WAN Solution?

Will DSCP Marking be trusted by the Edge Router?

Is there any major issue reported for VOICE/Video Integration with SD-WAN Solution for Enterprise customers?

Q20. Overlay IPSEC or SD-WAN Tunnel will work on top of Underlay Routing function.

How will the migration strategy be framed for SD-WAN Sites?

How Non-SD-WAN Sites will communicate with migrated SD-WAN Sites?

What is the latency consideration?

6 Conclusion

SD-WAN will be the game changer in prospective times and 75% of Enterprise customers are eagerly waiting for adapting to the solution. 3 years back, only 5%

of customers showed their interest to grab the new technology. Since SD-WAN technology is relatively new for numerous global customers, architecture design will play an indispensable role and key for the success of the migration project.

The aforementioned generic questions entail to be answered for greater visibility of the vendor-specific SD-WAN Solution, thereby selecting right partner with proper risk assessment before finalizing the procurement.

Hub design will be the key for taking control over the latency and license cost and will be a relatively newer component for SD-WAN. Cost versus benefit justification will be relatively easy to prepare business case, once the generic questions are answered and clarified as typical best practice.

References

1. Cisco White Paper.: In Search of the right SD-WAN solution. Cisco SD-WAN Security. Cisco Systems, San Jose, California, United States of America, pp. 1–21 (2019)
2. Packet Pushers Interactive.: Packet pushers SD-WAN survey. Packet Pushers Interactive, LLC, Northfield, NH, United States of America, pp. 1–13 (2018)
3. CenturyLink White Paper: Benefits of SD–WAN to the distributed enterprise, pp. 1–6. CenturyLink, Monroe, Louisiana, United States of America (2016)
4. Aryaka White Paper.: SD-WAN Versus MPLS: Key Considerations For Your Global Enterprise Network. Aryaka Networks Inc., San Mateo, California, United States of America, pp. 1–12 (2015)
5. Govindarajan, K., Chee Meng, K., Ong, H.: A literature review on software-defined networking (SDN) research topics, challenges and solutions. In: Proceedings of 2013 5th International Conference on Advanced Computing (ICoAC), Chennai, pp. 293–299. ISBN–978–1-4799-3447-8
6. T–Systems White Paper.: SD-WAN. T-Systems International GmbH, Frankfurt, Germany, pp. 1–16 (2019)
7. Riverbed White Paper: Riverbed SD–WAN, pp. 1–13. Riverbed Technology, San Francisco, California, United States of America (2015)
8. ONUG White Paper.: ONUG Software–Defined WAN Use Case. ONUG SD–WAN Working Group, Open Networking User Group (ONUG), New York, United States of America, pp. 1–10 (2014)
9. https://www.velocloud.com/sd-wan
10. SD-WAN Solution Comparison. https://www.cisco.com/c/en/us/products/routers/sd-wan-vendor-comparison.html, https://www.netmanias.com/en/post/oneshot/12481/sd-wan-sdn-nfv/comparison-of-the-sd-wan-vendor-solutions, https://www.thenetworkunion.com/content/sd-wan-comparion-vendor-capabilities, http://cdn2.hubspot.net/hubfs/488500/Comparison_of_SD-WAN_Solutions_TechBrief.pdf, https://searchnetworking.techtarget.com/tip/Infographic-Compare-the-leading-SD-WAN-vendors-before-you-buy, https://www.eweek.com/networking/top-10-sd-wan-vendors, https://www.featuredcustomers.com/software/sd-wan/all, https://www.ctctechnologies.com/the-top-5-sd-wan-vendors-to-watch-in-2019

Comparative Analysis of Adder for Various CMOS Technologies

M Amrith Vishnu, Bansal Deepika, and Garg Peeyush

Abstract The current technologies are moving towards small size, high speed, and cost-effective computing systems. The demand of efficient devices such as operating at high speed and low power is always increasing. Therefore, the work compares various design configuration to check their performance. Adders are the key element of the arithmetic functions such as addition, multiplication, subtraction, and division. Adder has been simulated using spice code to check their functionality. Synopsis HSPICE simulator is used to analyze the circuits for various CMOS technologies. The proposed pseudo domino techniques are proffered to economize power dissipation up to 61% in half adder circuit and 32% in full adder circuit. The 14 transistor based design is economized the power dissipation up to 60% in 1-bit adder circuit.

Keywords CMOS technology · Dynamic power consumption · Pass transistor logic · Propagation delay · Adder

1 Introduction

Currently, nano technology enhances the semiconductor industries to a huge profitable market by its several advantages [1]. Its era has started in 1970s and faced a rapid growth in many electronic applications. With the help of deep sub-micron devices [2], the computer can have multiple cores on a single chip and perform

M. A. Vishnu · B. Deepika (✉)
Department of Electronics and Communication Engineering, Manipal University, Jaipur, Rajasthan, India
e-mail: deepika.bansal@jaipur.manipal.edu

M. A. Vishnu
e-mail: vishnuaarcha@gmail.com

G. Peeyush
Department of Electrical Engineering, Manipal University, Jaipur, Rajasthan, India
e-mail: peeyush01garg@gmail.com

S. Borah et al. (eds.), *Soft Computing Techniques and Applications*, Advances in Intelligent Systems and Computing 1248,
https://doi.org/10.1007/978-981-15-7394-1_25

multiple functions and threads parallelly [3]. The large number of applications in integrated circuits have been rising at a very fast pace with the advent of its designs for high performance computing [4], telecommunication [5], consumer electronics and low bit rate video.

Full adder [6, 7] is a prominent circuit for various module in an integrated circuit as it is used for many arithmetic circuits [8] such as addition, subtraction, multiplication and division. It is also used for various calculating logical addresses. So being a major block, adder must be very efficient in all the way such as speed [9], size, power consumption [10] etc. The designers are highly concerned about area, speed and testability of the deep sub-micron devices [11]. However, when the technology size scaled down to nanometers then the real problem is to be taken care of power consumption. In this paper, few adder circuits have been analyzed and compared in a layered way to track their performance [12]. The power consumption is an unavoidable peak factor, so various techniques have been applied to reduce it.

In this paper, various types of full adder circuits are designed and simulated with 45, 32, 22 and 16 nm CMOS technologies with the help of HSPICE. After that, the circuit performance has been analyzed in terms of dynamic and static power dissipation.

In the modern days technologies static power dissipation dominates all other power dissipations in a circuit. Scaling down in size technology leads to various types leakage current in a circuit even though the circuit got no inputs. So, reduction of the static power is a challenging task in the current VLSI industries. Therefore, few configurations are considered from literature to understand and verify their results.

2 Static CMOS Logic

Standard static CMOS circuits which comprise both pull up and pull-down networks have the capability of providing full swing output voltage. High stability is achieved by having equal number of nMOS and pMOS transistors.

The output function SUM is depending on the other output function carry (COUT). The SUM is carried out by giving the COUT signal to it as shown in Fig. 1. Though it creates some delay and reduces the chip area. The greater number of pMOS further retard the circuits speed. Larger number of transistor count is a primary issue. The major disadvantages of the static CMOS logic are weak output driving capability, glitch and short circuit power dissipation.

Static CMOS circuits are not widely used because of its inefficiency in the area. However, designers include it on their designs due to its robustness circuit. The designing part is very simple as it could be designed from the Boolean expression of SUM and COUT. The circuit module can be splitted into COUT and SUM. In the Fig. 1, transistors number 15 and 16 are feded with the static CMOS logic, which generates the SUM function.

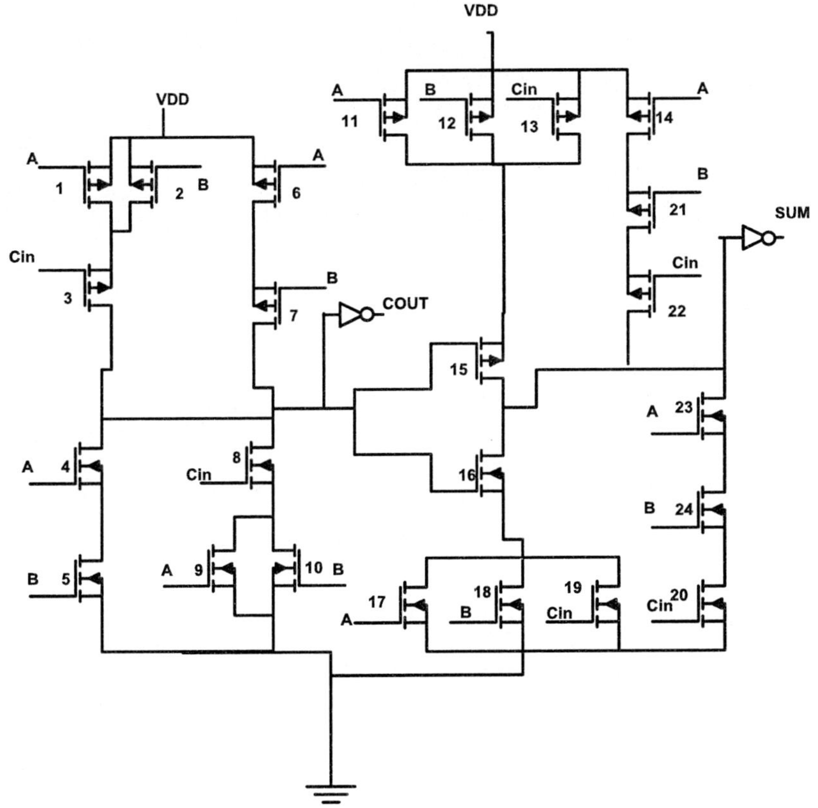

Fig. 1 Static CMOS logic based full adder

Hereby SUM is generated through an inverter, which makes an additional delay. The static CMOS cann't generate COUT and SUM parallel which makes the circuit further slower.

3 Modified Complementary Pass-Transistor Logic

Complementary pass transistor logic (CPL) circuits consist high transistor count compare to static CMOS logic because it gives two output functions (desired and compliment of the desired function). Ultimately, it will consume very large area and dissipates more power.

Threshold voltage drop is major concern of the CPL circuits, which can be resolved to get full swing using modified CPL. In Fig. 2, keeper circuit has been applied at the output node to resolve the threshold voltage drop and produce the full voltage

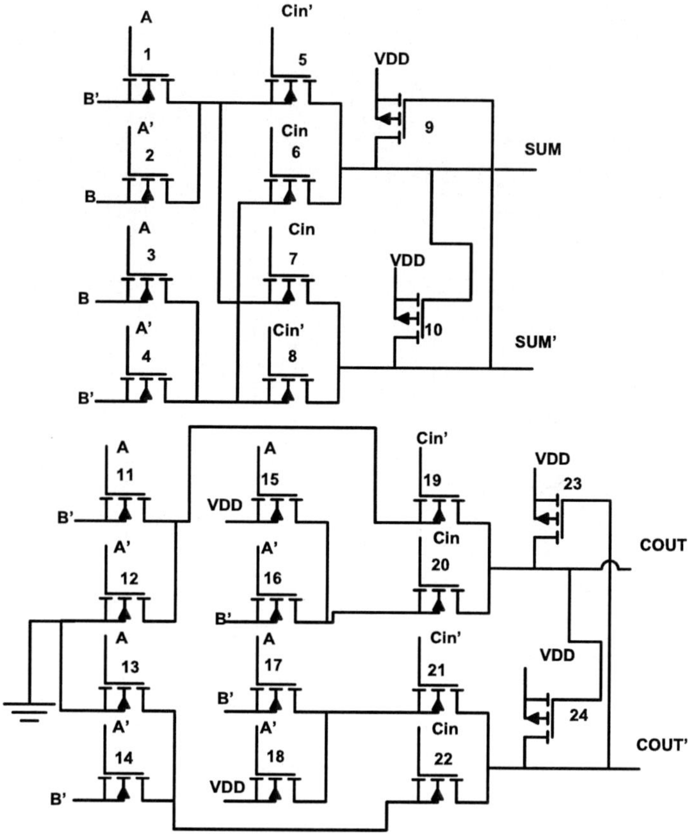

Fig. 2 Modified complementary pass-transistor logic using full adder

swing. On applying keeper transistor, this technique improves the performance and produce high speed than static CMOS circuit due to a smaller number of pMOS.

Since the transistor count and wiring complexity are very high. It may produce high glitches because of high input load. The worst-case propagation delay tends to create glitches. In the CPL full adder circuit, both the sum and carry modules generate their appropriate outputs independently so glitches don't show up in this model. It contains all the primary inputs in complemented form as well. These desired and complemented inputs are used to drive the transistors between the output and input transistors. RC delay is very high due to high transistor count. These circuits are very useful for XOR/XNOR gate because both circuits can be designed in a single circuit.

CPL is not at all recommended because of high power consumption. However, modified CPL produces better voltage swing than other techniques as discussed further in paper. Separate module for generating SUM and COUT reduces glitch but it doesn't help to reduce circuit delay due to high transistor count.

4 14T Configuration

Low power circuit have been designed using only 14 transistors. This design methodology is preferable and efficient due to less transistor count and low power. Although, the design method gives better performance than static CMOS and CPL full adder circuits. This circuit can be preferred because of its simplicity. It is a combination of both static CMOS and transmission gate (TG). The circuit implementation is derived from XOR and XNOR circuits, those are performed simultaneously to generate sum and carry thereby the speed of circuit is improved. Short circuit power dissipation is very low as compared to static CMOS and CPL model. The main drawbacks in this model are glitches and partial voltage swing. But it is easier to design and test the circuit because of the lesser circuit complexity.

It is clearly seen in the Fig. 3 that the circuit is a combination of three modules using 14 transistors. It uses a simple XOR gate followed by an inverter to generate the complement of the XOR function mean XNOR gate. The module which generates XOR and XNOR is composed of three transistors. The output of transistor no. 3 is named as T, which is obtained XOR function and connected with an inverter. The output of the inverter is treated as XNOR gate, which is denoted by T'. T is given into a module with two transmission gates to generate SUM whereas T' is given into another module with two transmission gates to generate carry (COUT). Thus, SUM and COUT are generated independently, which helps in reducing circuit delay. Both

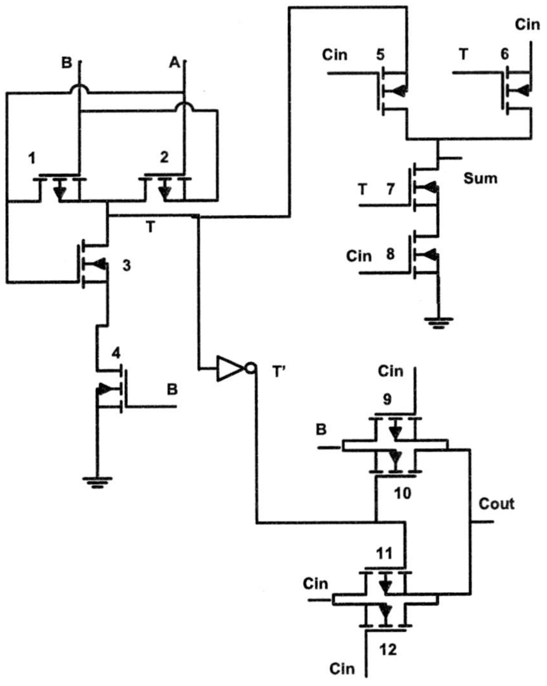

Fig. 3 Full adder circuit using 14 transistors

the outputs (SUM and COUT) are generated by the TG modules, those are having four transistors. So, performance wise both the modules have approximately same speed.

5 Transmission Full Adder

It is a perfect model for good output driving capability. The output is generated by transmission gate circuits. The implementation and functioning of the circuit are entirely based on transmission gate concepts so it obtained the full swing.

In the previous circuit, full adder was designed by the combination of CMOS and transmission gate logic. Transmission gate logic is one of the good techniques for designing logic gates like XOR and XNOR gates. It uses equal number of nMOS and pMOS transistors. Less power consumption is achieved by a smaller number of transistors. Ease fabrication and reduced area are some notable advantages.

Transmission full adder (TFA) circuit is also composed of three modules as displayed in Fig. 4. The first module is composed of 8 transistors to generate XOR and XNOR functions. This is major difference between 14T circuit and TFA circuit. The first module, which consists of one inverter and transmission gates will generate XOR function using transmission theory.

Since the module has more transistors than 14T therefore TFA circuit a bit slower. But once the XOR function (T) is generated it is fed into an inverter to generate XNOR function (T'). These two signals T and T' are further fed into two different modules to generate sum and cout simultaneously. Like 14T the total circuit delay depends on the first module.

6 Results and Discussion

The simulation of one-bit adder cell is carried out and compared with the other adder designs. The aim of this work is to check the performance analysis of adder. Four adder configurations are designed and compared with each-other. In first configuration, Static CMOS logic uses only 28 transistors so it consumes more power compare to other configurations. Table 1 shows the performance analysis of 1-bit adder using various design configurations for 45 nm technology. Table 1 depicts that 14 transistor based adder provides the good performance in terms of power consumption, speed and transistor count. These design configurations are used to check the functionality of adder and analysed with various CMOS technologies such as 45, 32, 22, and 16 nm.

Figure 5 displays the static power consumption of 1-bit adder for various CMOS technologies such as 45, 32, 22, and 16 nm. Static current is the key source of static power dissipation in small scale CMOS technology. It is calculated when the circuit is in the stand-by mode. The static power dissipation increases with scaling of the

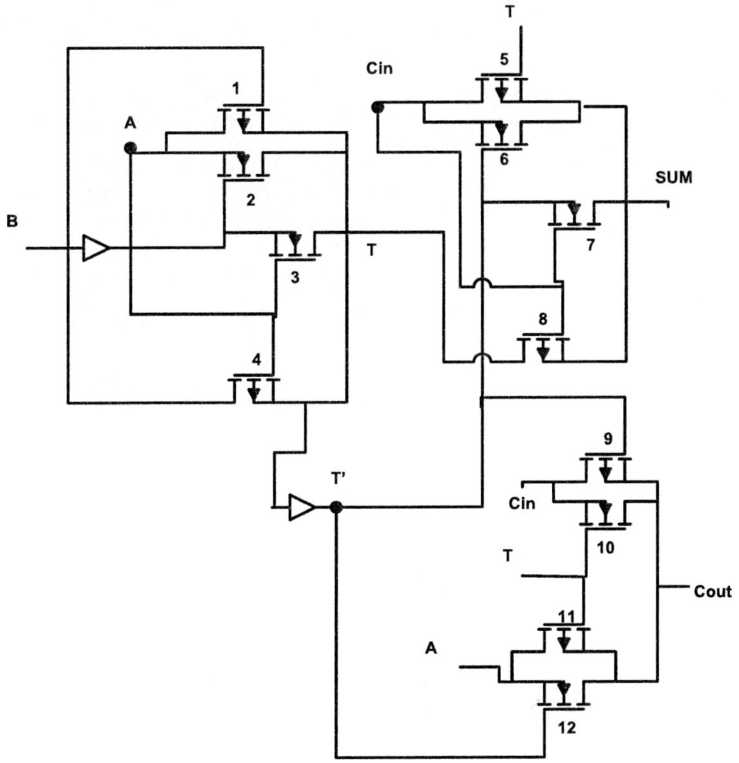

Fig. 4 TFA

Table 1 Performance analysis of 1-bit adder using 45 nm CMOS technology

Design styles	No. of transistors	Total power consumption (μw)	Static power consumption (pw)	Dynamic power consumption (μw)	Delay (ps)	Power delay product (aJ)
Static CMOS	28	2.07	65.4	2.07	42.5	870975
Modified CPL	32	3.69	81.2	3.69	48.5	160.515
TFA	16	1.56	42.8	1.56	17.3	26.988
14T	14	1.44	31.8	1.44	12.8	18.432

CMOS technology. The short channel effects in deep submicron CMOS technology are responsible for increasing the leakage current. It is found that static CMOS logic and modified complimentary pass transistor logic consume large leakage current.

Dynamic power consumption has been calculated for various CMOS technologies as displayed in Fig. 6. With the scaling of CMOS technology, the supply voltage needs

Fig. 5 Static power consumption analysis of 1-bit adder for various technologies

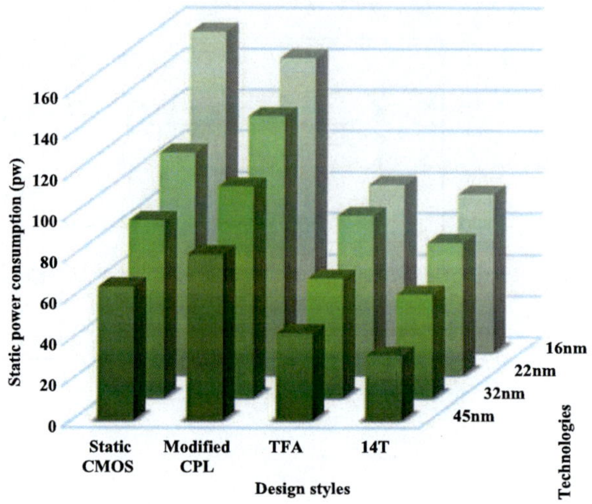

Fig. 6 Dynamic power consumption analysis of 1-bit adder for various technologies

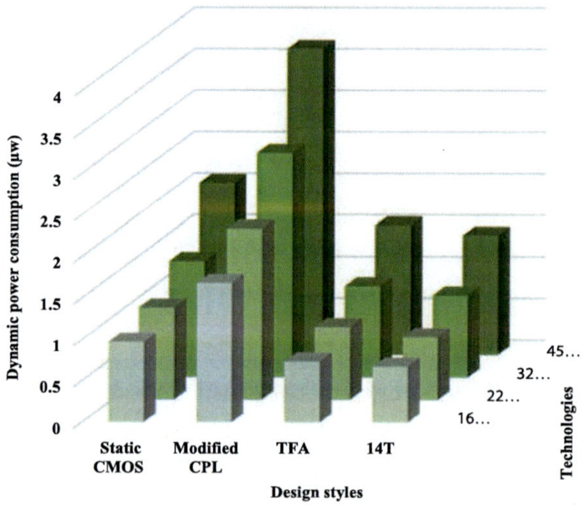

to be reduced, as supply voltage is a function of the dynamic power. So, it reduces as the technology scales down. As shown in Fig. 6, 14 Transistor based adder using 16 nm CMOS technology is reduced the dynamic power consumption up to 65% as compared to the 45 nm CMOS technology.

Delay is one of the important parameters, which is considered for measuring the performance of the adder. Propagation delay is average of τ_{pHL} and τ_{pHL}. Fig. 7 demonstrates the propagation delay for 1-bit adder using various configurations.

Circuit performance of the designs have been calculated for the power consumption and delay calculation. Power delay product is calculated by the product of the total power loss and delay of the circuit. Figure 8 demonstrates the power delay

Fig. 7 Delay analysis of 1-bit adder for various technologies

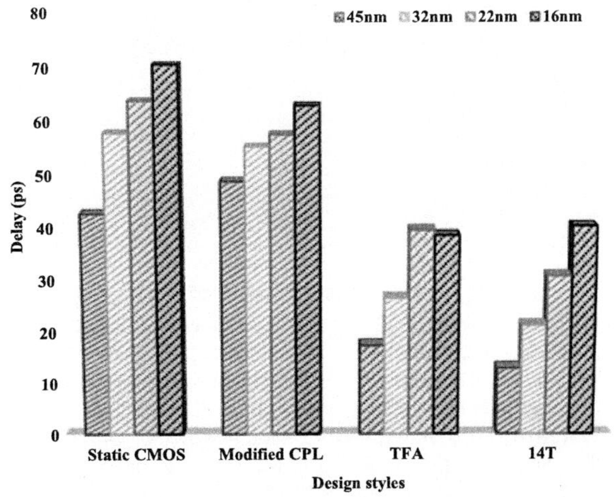

Fig. 8 Power delay product analysis of 1-bit adder for various technologies

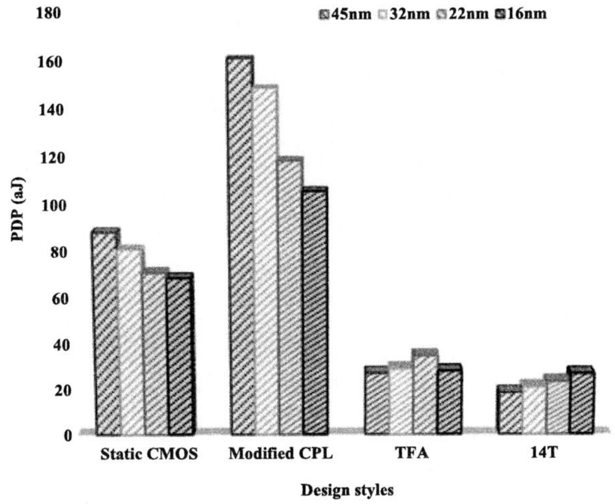

product for adder. Therefore, adder using 14 transistor configuration is efficient than other designs for low power applications.

7 Conclusions

Adder has been designed and compared using different configurations. It is found that 14 transistor based configuration provides better response compare to other configurations. Performance analysis has been calculated for adder using various CMOS technologies such as 45 nm, 32 nm, 22 nm, and 16 nm. Less transistor consumes low static and dynamic power therefore 14 transistor based design shows 70% improvement compared to other designs.

References

1. Shah, A.P., Neema, V., Daulatabad, S.: DOIND: a technique for leakage reduction in nanoscale domino logic circuits. J. Semicond. **37**(5), 55001 (2016)
2. Anis, M.H., Allam, M.W., Elmasry, M.I.: Energy-efficient noise-tolerant dynamic styles for scaled-down CMOS and MTCMOS technologies. IEEE Trans. Very Large Scale Integr. Syst. **10**(2), 71–78 (2002). https://doi.org/10.1109/92.994977
3. Sharma, T., Sharma, K.G., Singh, B.P.: High performance full adder cell: a comparative analysis. In: TechSym 2010–Proceedings 2010 IEEE Students' Technol. Symposium, pp. 156–160 (2010). https://doi.org/10.1109/techsym.2010.5469170
4. Chuang, P., Li, D., Sachdev, M.: Constant delay logic style. IEEE Trans. Very Large Scale Integr. Syst. **21**(3), 554–565 (2013). https://doi.org/10.1109/TVLSI.2012.2189423
5. Bansal, D., Engineering,C., Singh, B.P., Engineering, C., Kumar, A.: A novel low power keeper technique for pseudo domino logic. pp. 1–6 (2015)
6. Zhuang, N., Wu, H.: A new design of the CMOS full adder. IEEE J. Solid-State Circ **27**(5), 840–844 (1992)
7. Mehrabani, Y.S., Eshghi, M.: Noise and process variation tolerant, low-power, high-speed, and low-energy full adders in CNFET technology. IEEE Trans. Very Large Scale Integr. Syst. **24**(11), 3268–3281 (2016)
8. Navarro-botello, V., Montiel-nelson, J.A., Nooshabadi, S., Dyer, M.: Low power arithmetic circuits in dynamic CMOS logic feedthrough, pp. 709–712 (2006)
9. Moradi, F., Vu Cao, T., Vatajelu, E.I., Peiravi, A., Mahmoodi, H., Wisland, D.T.: Domino logic designs for high-performance and leakage-tolerant applications. Integr. VLSI J. **46**(3), 247–254 (2013). https://doi.org/10.1016/j.vlsi.2012.04.005
10. Kaizerman, A., Fisher, S., Fish, A.: Subthreshold dual mode logic. IEEE Trans. Very Large Scale Integr. Syst. **21**(5), 979–983 (2013). https://doi.org/10.1109/TVLSI.2012.2198678
11. Garg, S., Gupta, T.K.: Low power domino logic circuits in deep-submicron technology using CMOS. Eng. Sci. Technol. Int. J **21**(4), (2018). https://doi.org/10.1016/j.jestch.2018.06.013
12. Mehrotra, S., Patnaik, S., Pattanaik, M.: design technique for simultaneous reduction of leakage power and contention current for wide fan- in domino logic based 32:1 multiplexer circuit. no. ICT, pp. 905–910 (2013)

Survey on Captcha Recognition Using Deep Learning

Mohit Srivastava, Shreya Sakshi, Sanghamitra Dutta, and Chitrapriya Ningthoujam

Abstract With the increasing dependency of people and evolving technology, security has become an issue on online platforms. Earlier days CAPTCHA was introduced to provide protection and security in web but lately due to the new bulk of web bots, now security becomes a major concern as it could be easily breached. This paper discusses about the technique involved that can help in optimizing the performance of a captcha breaking model and also provide a moot that can help to withstand the guard of captchas.

Keywords CAPTCHA · Turing tests · Pre-processing · Segmentation · Deep learning · CNN · Convolutional neural network · Text-based CAPTCHA

1 Introduction

The increase in web-based security challenges has emerged the need for inevitable security mechanism, CAPTCHA (Completely Automated Public Turing test to tell Computers and Humans Apart Reference [1]) also called Human Interactive Proofs (HIP Reference [2]), can protect and secure software. It has been successfully applied to major websites liable for protecting the data of millions of people. The research on CAPTCHA breaking is a major concern for improving its quality, so it becomes

M. Srivastava (✉) · S. Sakshi · S. Dutta · C. Ningthoujam
Department of Computer Science and Engineering, Sikkim Manipal Institute of Technology, Majitar, Gangtok, India
e-mail: mohitsri27@gmail.com

S. Sakshi
e-mail: shreyasingh7478@gmail.com

S. Dutta
e-mail: duttasanghamitra1327@gmail.com

C. Ningthoujam
e-mail: chitrapriya.n@smit.smu.edu.in

S. Borah et al. (eds.), *Soft Computing Techniques and Applications*, Advances in Intelligent Systems and Computing 1248,
https://doi.org/10.1007/978-981-15-7394-1_26

important to verify how secure CAPTCHAs are and that is why the breaking technology was introduced. So for checking the reliability of CAPTCHAs we used methods like image segmentation, neural networks, and many other disciplines. Firstly, CAPTCHA breaching can validate the reliability of existing CAPTCHAs and it can stimulate ways to boost the security of CAPTCHA designing. Further, the research on breaking CAPTCHA can also be applied in other fields such as speech recognition and image labelling. This paper will review the future aspects for enhancement of the CAPTCHAs.

The rear section of this research paper is structured as: It briefly introduces the text-based CAPTCHAs and gives an introduction about the paper giving an overview of the plot also talking about the background of the paper which includes literature survey, applications, classification types and type of captcha we've discussed. In later sections, methodology and strategy to be followed are described leading us to the results and discussion. Finally, there is converse about the conclusion obtained from the survey and ending the paper with references.

2 Background

2.1 Literature Survey

In 2004, in a research paper proposed by Bansal et al. was based on the technique of segmentation on connected regions and recognition using distorted evaluation with a 98% accuracy was proposed for EZ-Gimpy CAPTCHAs in which they first preprocesses the given CAPTCHA and then segmented its characters, which was then matched with the character's features in a characteristic table and then used HMM to recognize the characters of the CAPTCHA. Results justify the effectiveness of the approach.

The Captchacker project by Jean-Bapsite Fiot and Remi Paucher in 2009 exploits the potential of Support Vector Machine to break captcha. To remove the background, they used thresholding and for segmentation, they used connected components using 4-connectivity and 8-connectivity. The segmentation failed once out of 3000 tests. They used repeated SVM (Support Vector Machine), models, for CAPTCHA recognition. The method is robust and gives good results with a specific training set.

In the year 2013, Ian good fellow and Julian ibarz published a paper [3] in which their approach was simplified into three steps by using deep CNNs for training on high-quality images which produced wonderful results. It was seen that the performance of the strategy increases when the depth of CNN increases. This approach obtained an accuracy of over 96% in recognizing the distorted strings. Improvement in the accuracy was seen when per-digit recognition task was performed till, 97.8%. They evaluated this approach by performing on various data sets and could observe a good rate of accuracy obtained.

2.2 *Applications*

Breaking of these randomly generated characters can act as an asset in the field of research and its applied applications.

It can help in verifying the threshold and guard (security) of existing platforms using CAPTCHA's and then can help in the advancements of new CAPTCHA design techniques.

Research of breaking CAPTCHA can also be used in image labelling, vehicle number plate detection, avoiding web crawling.

2.3 *Classification Types*

Captchas are used as a Turing test for the users or humans, they are classified based on distorted images, texts, and numbers. Some of the captcha types are given below:

i. CAPTCHA's based on text
ii. CAPTCHA's based on images
iii. CAPTCHA's based on audio
iv. CAPTCHA's based on video
v. CAPTCHA's based on puzzle.

In this paper, we have discussed the most widely used, Text-based captcha only, and techniques to break it.

2.3.1 Text-Based CAPTCHA's

Text-based CAPTCHA's mostly consist of alphabets and numeric in the form of distorted string, adhesion, and overlap. It is very effective but requires a large database to store the information (Table 1).

3 Methodology

3.1 *Techniques to Follow for Breaking Text-Based CAPTCHA's*

Back in the times, non-adherent characters were constituents of the captcha's based on texts. The traditional practice applied in break them included locating a number or character area in an image and identify a single character after the process of segmentation [5].

Table 1 Different types of text-based CAPTCHA [4]

Type	Features
Solid CAPTCHA	Character independent, texture background, some interference
	Many interference lines and noise points
	Multiple strings, overlap, distortion
	Unfixed length, distortion, adhesion
	Double-string, unfixed length, uneven thickness, tilting, adhesion
Hollow CAPTCHA	Hollow, shadows, interference shapes
	Hollow, adhesion, interference lines
	Hollow, virtual contours, distortion, adhesion, interference lines
	Hollow, shadows, interference lines, noise points
Three-dimensional CAPTCHA	Grids, protrusion, distortion, background and character blending
	Colourful, shadow, rotation, zoom
	Special characters
Animation CAPTCHA	Multiple characters jumping
	Multilayer character images blinking transformation

New technology proposes new advancements, nowadays most text-based Captcha's use Crowded Character Together (CCT).

Thus, new breaking frameworks came into existence:

Pre-processing + recognition,

Pre-processing + recognition + post-processing,

Pre-processing + segmentation + recognition.

This paper discussed the framework involving pre-processing, segmentation, and recognition for surpassing the Turing test. However, if we want to use other frameworks, the basic flow involves is discussed in Fig. 1.

3.1.1 Preprocessing Method

First step involves pre-processing images before feeding it to the next stage. The aim is to filter or remove unwanted interfering data in the images, Ref. [6]. For avoiding image breaking, noise and other interferences should be removed.

Some noises are generated during binarization and grayscale. Some methods are shown in Table 2 for denoising an image. Best one is selected according to actual requirement and situation.

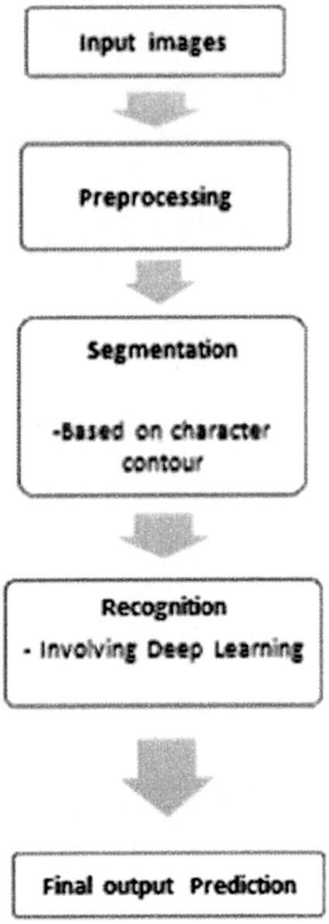

Fig. 1 Flow of the strategy

Table 2 Denoising methods based on filter in the spatial domain

Algorithm used	Implementation	Advantage	Disadvantage
Average filter	Monochrome value of pixel is replaced by taking mean of its neighbour grey pixels.	unrelated details and gaps are removed	Image becomes blur.
Median filter	Monochrome value of target pixel is substituted by taking the median of its neighbour pixel's grayscale	Removes salt and pepper noise	Not applicable to images with many lines and dots.

3.1.2 Segmentation Method

The main target is to get individual components or character elements. Segmentation method using contours against each character helps in analyzing geometric features of character contour so that it might be able to find appropriate segmentation lines.

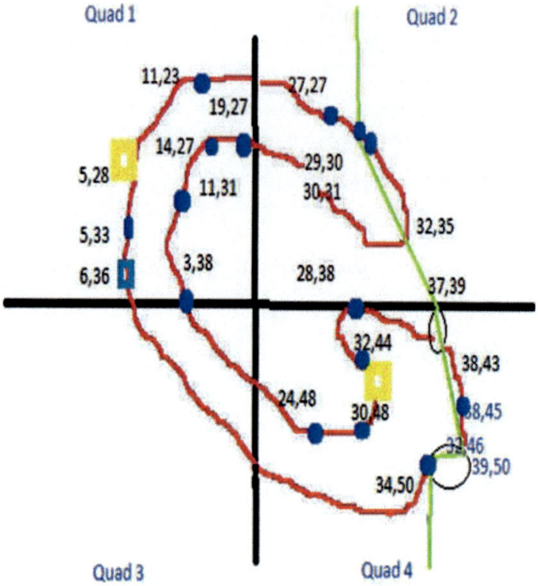

Fig. 2 optimal segmentation line on alphabet 'G'

Connected edge points connected across two integrated points helps in determining the optical segmentation line by confidence, Fig. 2 shows the discussed part by plotting a letter 'G' on the graph.

According to the study of Rabih Al. Nachar (Ref. [7] Fig. 2), he stated that when contours are made all over the image then plotted by dividing the image into four quadrants and because of integrating of consecutive characters, chances of few edge fragments are likely to disappear. In Fig. 2 the parts circled between coordinates (39, 50) and (34, 50) have been vanished and can conduct to the incorrect count of crossed edges. Thus we do not consider crossed fragments or edges for the lines that are crossing the left or right borders of the plot.

3.1.3 Recognition Method

Recognition approach based on Machine Learning makes extensive use of different kinds of ML methods and algorithms to classify characters with maximum accuracy.

(a) **Methods involving traditional techniques**

Some established classifiers include Support Vector Machine (SVM) and *K* Nearest Neighbour (KNN). SVM separated different classes using a hyperplane. Kernel function plays a crucial role in allocating original features into a high magnitude space in a non-linear fashion which helps in improvising the separability of data. comparison of various kernel functions is given in Reference [8]. KNN uses the foundation of categorizing the closest *K* number of clusters to tell about the category of a subset or the population. In the source [9, 10]. We can check for the comparison of different classifiers with their success rates.

(b) Methods involving Neural Network

The system which have ability to imitate human behavior to solve complex problems prove themselves as the most suitable ones. When applying neural networks, character extraction must be done first because the standard of extracted features restricts the final recognition cost to definite boundary. Yan et al. [11] successfully segmented the Microsoft CAPTCHA and recognized it by multiple classifiers, but the recognition rate was only 60% which was not suitable enough to determine the correct outcomes.

(c) Methods involving Deep Learning

In the past decade, Deep learning acquired many accomplishments in recognition areas of image, text, audio, and some others. The commonly used models were RNN, LSTM-RNN, Convolutional neural network, and others.

CNN doesn't perform feature extraction while recognizing images and have a certain degree of deracination and derangement. The advantages of CNN depend on the size of the training data. When training data in not enough, verifying codes using CNN becomes difficult. But according to Research [12], the method involving the use of CNN can considerably improve the accuracy of recognition of a neural network when training data is small at the initial stage.

In Fig. 3, the captcha is segmented into individual characters and then passed into the network as individual characters/numbers so that computations in a neural network could be performed which incorporates a input layer which takes in the inputs, hidden layer(s) which carry out computations and output layer each carrying out different functionality to compute and classify the correct result in a faster way. In the existing research outcomes, CNN is widely used in [14–18] having a good

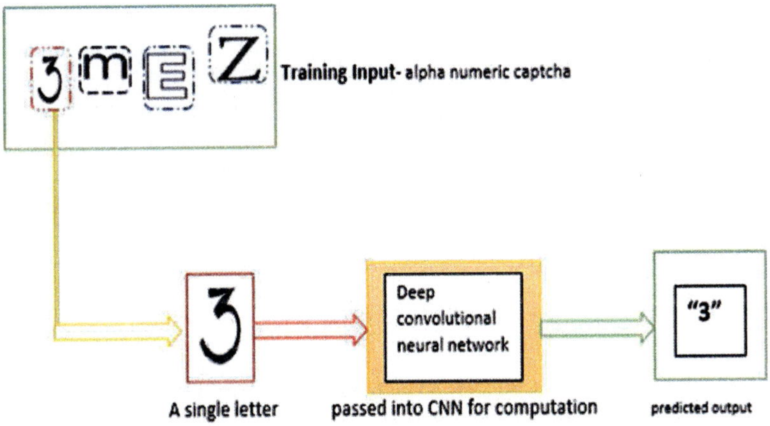

Fig. 3 Recognition in Neural network, Ref. [13]

Table 3 Comparison of recognition methods based on machine learning

Methods	Advantages	Disadvantages
SVM	Best for two kinds of classification Strong approximation with high accuracy	Not applicable when having infinite sample space
KNN	Helps in solving problems related to imbalance samples and suitable for overlapping samples	Computation is complex Cannot determine domain when small training set
CNN	Can accept input image directly, automatically extracts features with high recognition accuracy	Lacks in time dimensions
LSTM-RNN	Effective in preventing gradient disappears.	Unable to extract features automatically

accuracy score. Good fellow reported in Research [19], it uses deep convolutional networks and acquires 98% accuracy in identifying captchas which displays a good result (Table 3).

4 Results

Although there are various types of CAPTCHA's in existence and most commonly used one is text-based captcha. On one hand, Captchas are user friendly and convenient for website users, and on the other, we know that captchas are vulnerable and not so secure for online platforms. This research has tried to check the reliability of existing captchas by seeking different approaches involving the use of deep learning to produce accurate results. For improvising the Captcha's we've proposed some of the solutions or advancements that could be made to improve the security.

The framework discussed in the paper shows that Captcha's can easily be broken to surpass online security which is a major concern for commercial websites that carry out their work online which is secured by text-based captcha. These captchas are common in all platforms since they all are the same and stored in a common database, so it becomes quite easier to hack them.

It is seen that through the use of CNN it becomes quite easy to break captchas without feeding the model with large datasets to train and gaining accurate results.

5 Discussion

It is evident that types of CAPTCHAs we have referred, none of them are foolproof and some approaches that could help in making a guard wall before their security are:

Website Reference Approach: It can be an improvement of text-based CAPTCHAs. In traditional text CAPTCHAs, the type of logical questions asked to humans were the same to all websites and stored in a database. If instead of asking logical questions which are same on all platforms, CAPTCHAs can ask questions related to data available on the visiting website for example a website related to C programming can ask a question like which header file should be included to run the sample basic C program given and user types his answer. A website related to content about sports can ask questions for verification such as, which of the sports are played with a ball, name them.

Combination of Existing CAPTCHAs: The new technique combining different Captchas could act as a new approach and which could not be understood by online bots but can easily understand by humans. For example, system producing Hindi audio captcha (word) and the user must hear and write the corresponding English translation in the text box as a Turing test and pass the verification.

6 Conclusion

In today's fast and developing world, everything is dependent on the internet. But with the advancement in technology, a great risk also exists. CAPTCHAs were introduced to avoid the risk of intervention by hackers with their bots to our services. In this paper, we have discussed how the need for CAPTCHA originated and how they are used now. We have also discussed the breaking techniques of the CAPTCHA. In this paper, we have provided an accurate method with higher accuracy to break CAPTCHAs and proposed methods to improve their security. We also discussed the possibility of more secure and hard to break CAPTCHAs. Hence, we conclude with a solution of reCaptchas and a combination of existing captchas.

7 Future Scope

According to the different studies and methodologies, the best possible solution and approach that could help in captcha breaking are discussed in the paper and the same to be implemented for the next presentation which could help in contributing to the world of captcha security advancements.

References

1. Von Ahn, L., Blum, M., Langford, J.: Telling humans and computers apart automatically. Communications of the ACM, pp. 56–60 (2004)

2. Chellapilla, K., Simard, P.Y.: Using machine learning to break visual human interaction Proofs (HIPs) of advances in neural information processing, pp. 265–272 (2004)
3. Goodfellow, I., Yaroslav, Ibarz, J., Sacha, A., Shet, V.: Multiple digits number recognition from street view images using deep convolutional neural networks (2013)
4. Luo, X., Guo, Y., Zhang, Y., Gong, D.., Jun, C.: Survey on breaking techniques of text based captchas (2017)
5. Chen, J., Luo, X., Yanqing.: Security on communicatins and network (2017)
6. Fiot, J.-B., Paucher, R.: The Captchacker project. Simulation and captcha based methods to build model (2009)
7. Nachar, R.A., Inaty, E., Bonnin, P.J., Alayli, Y.: Breaking of Captcha using fuzzy logic segmentation and edge corners technique, pp. 3995–4012 (2015)
8. Jean-Baptiste, F., Paucher, R.: The captchacker project (2009)
9. Lecun, Y.: The mnist database consisting of handwritten digits algorithm results, text based captcha strength and weakness (2012)
10. Gao, H., Yan, J., Cao, F., et al.: A simple generic attack on text Captchas, pp. 1–14, San Diego, Calif, USA (2016)
11. Yan, J. Ahmad, ASE.: A low-cost attack on a microsoft CAPTCHA. In: Proceedings of the ACM Conference on Computer and Communications Security (2008)
12. Xu, D., Wang, B. XiaoJiang, D.: Verification code for recognition system based on actice deep learning (2019)
13. https://www.academia.edu/12145865/Technique_to_CAPTCHA_Recognition
14. Chellapilla, K., Simard, P.Y.: Using machine learning to break visual human interaction proofs (HIPs). In: Advances in Neural Information Processing Systems, pp. 265–272 (2004)
15. Yan, J., Ahmad, A.S.E.: A low-cost attack on a microsoft CAPTCHA. In: ACM Conference, pp. 543–554 (2008)
16. Marshall, L.: The robustness of a new CAPTCHA, pp. 36–41 (2010)
17. Yan,J. Tayara, M.: The robustness of Google CAPTCHAs (2011)
18. Gao, H., Wang, W., Qi, J., Wang, X., Liu, X.: The robustness of hollow CAPTCHAs. In: ACM Conference, pp. 1075–1085 (2013)
19. Goodfellow, I.J, Bulatov, Y.J.I., et al.; Multi-digit number recognition from street view imagery using deep convolutional neural networks (2014)

The Sattriya Dance Ground Exercise Video Dataset for Dynamic Dance Gesture Recognition

Sumpi Saikia and Sarat Saharia

Abstract Dynamic gesture recognition is a very active research area in computer vision for the last few decades. Feature selection and extraction is one of the most important phases in gesture recognition since it greatly affects the recognition. In this work it is aimed to focus on some discriminate video features that lead to good recognition of dance gestures considering full-body movement of the dancer. Since human body is a highly articulated structure, it is an important issue to extract features that best describe the articulation. In this paper, a novel full body gesture video dataset is introduced, containing 560 video sequences of 28 ground exercises of Sattriya dance as well as annotations of those sequences and class label of every ground exercise. The purpose of creation of this dataset is to develop a computer vision system to classify each ground exercise. As well as this dataset can be useful for benchmarking a variety of computer vision and machine learning methods designed for dynamic dance gesture recognition.This paper also presents a method for dynamic gesture recognition on the Sattriya dance dataset that we have developed.

Keywords Dance gesture recognition · Video dataset · Full-body gesture

1 Introduction

In the arena of computer vision, full-body gesture recognition plays a significant role including dance gesture recognition. In recent years, the rapid growth of technology brought a lot of contribution in this domain. However, it is still a challenging and interesting task. This paper introduces a video dataset of ground exercises of Sattriya dance. This dataset contains 560 video sequences of 28 ground exercises from 20

S. Saikia (✉) · S. Saharia
Department of Computer Science and Engineering, Tezpur University, Tezpur, India
e-mail: sumpiss@tezu.ernet.in

S. Saharia
e-mail: sarat@tezu.ernet.in

S. Borah et al. (eds.), *Soft Computing Techniques and Applications*,
Advances in Intelligent Systems and Computing 1248,
https://doi.org/10.1007/978-981-15-7394-1_27

individuals. To impart training in Sattriya dance, some basic exercises called ground exercises are essential and are also used as a basic dance unit. These are about 64 in number. Most of these are practised to make the body flexible. In this dataset, 28 commonly used ground exercises are included which have been recorded from some Sattriya dance training center and various Sattras of Assam.

1.1 Contribution

In this paper, we have introduced a novel Sattriya dance ground exercise video dataset. This dataset is used for dynamic dance gesture recognition. This dataset can also be used for any computer vision-based gesture recognition system. Secondly, we have developed some important features from the image sequences for classification. Also, we have shown the classification accuracy of the ground exercises by using some well-known machine learning algorithms.

1.2 Organization of the Paper

The organization of the rest of the paper is described as follows. The state of the art is briefly explained in Sect. 2. Section 3 presents the description of the dataset. The different phases of gesture recognition are shown in Sect. 4. The experimental results are shown in Sect. 5. Finally, we have discussed conclusion and future direction in Sect. 6.

2 State of the Art

There are very limited video dataset for full-body dynamic dance gesture recognition in the literature. Heryadi et al. [1] proposed a syntactical modeling and classification for performance evaluation of Bali traditional dance considering the full-body gestures. Another full-body gesture recognition system demonstrated by Nussipbekov et al. [2] for recognizing Kazakh traditional dance gesture where Microsoft Kinect camera is used to obtain human skeleton and depth information. A view-invariant video-based full-body gesture recognition system proposed by Peng et al. [3] which can be applied in gesture-driven HCI is the first one addressing full-body human gesture recognition from video without the recovery of body kinematics or the 3D volumetric reconstruction.

In the literature, it has been observed that there is no available video dataset of Satriya dance for dance gesture recognition.

2.1 Motivation

As dance gesture recognition is an application of general gesture recognition task, many of the traditional machines learning methods are addressed here. A number of approaches have been proposed for dance gesture recognition of different dance forms. However, full-body gesture recognition approaches are very limited in the literature. Any video dataset of Sattriya dance has not been found in the literature till date. Moreover, dance gesture recognition considering full-body movement in this dance form is also not addressed yet. Dance gesture recognition has applications in performance evaluation of dances, e-learning of this 500 year old living dance heritage of Assam. Since ground exercises are the basic requirement to learn Sattriya dance, these are necessary to store in video form. This has motivated us to introduce a video dataset of ground exercises of Sattriya dance considering full-body movement for validation of dance gesture recognition methods.

3 Dataset Description

Sattriya dance is a classical dance of India introduced in the fifteenth century A.D. by the great saint and reformer of the state Assam, Mahapurusha Sankaradeva. Sattriya dance is originated in the eastern state of Assam and has remained a living tradition since its creation.

To impart training in Sattriya dance, some basic exercises called ground exercises are essential and are also used as a basic dance unit. Ground exercises are the structural grammar to learn Sattriya dance. These include all the features of Sattriya dance—basic body positions, body bending, body movements, foot movements, various jumps, turns, gaits, hastas (hand movements), head movements, neck movements, eye movements, etc. [4]. These are about 64 in number. Most of these are practised to make the body flexible. In this dataset, 28 commonly used ground exercises are included which have been recorded from some Sattriya dance training center and various Sattras of Assam.

3.1 Scope of the Dataset

We have created the Sattriya dance ground exercise video dataset to develop a computer vision system to classify each ground exercise, as well as this dataset can be useful for benchmarking a variety of computer vision and machine learning methods designed for dynamic dance gesture recognition.

Fig. 1 Basic poses of the ground exercises

3.2 Class Labels

There are 28 ground exercises, i.e., classes in the dataset. Each ground exercise has a standard name, which will be the required classes for classification. Each ground exercise has a basic pose, which is shown in Fig. 1.

3.3 Capturing System

We have captured the video sequences using a 13 Megapixel camera keeping at a fixed position. The front view of the dancers is captured under uniform background. After capturing the video sequences, these are saved in uncompressed 'AVI' file format.

3.4 Captured Gestures

The ground exercises that have been recorded are: 1. Ora, 2. Orat boha utha, 3. Ora sata, 4. Bohi sata, 5. Kati sata, 6. Udha sata, 7. Sanmukholoi sata, 8. Etiya sata, 9. Jalak, 10. Singha jalak, 11. Tintiya jalak, 12. Charitiya jalak, 13. Haar bhonga, 14. Haat pokua, 15. Pasala tula, 16. Paani sisa, 17. Gerua sua, 18. Haat bhori salana, 19. Pada salana, 20. Sitika, 21. Thiyo muruka, 22. Bohi muruka, 23. Jatoni 24. Ketela, 25. Salana, 26. Athua, 27. Satrawali and 28. Murupa. Some of the video frames of some ground exercise have been shown in Fig. 2.

Name of Ground Exercise	Frames			
Ora				
Orat boha utha				
Ora sata				
Tintiya jalak				

Fig. 2 Some frames of the ground exercises

4 Phases of Gesture Recognition

The various phases of gesture recognition problem are described below:

4.1 Data Acquisition

The first phase is the data acquisition which refers to the collection of ground exercises of Sattriya dance from different dancers. To create this dataset, 560 video sequences of 28 ground exercises have been recorded from 20 individuals. In this dataset, each ground exercise has been performed by 20 dancers.

4.2 File Naming Convention

We define a simple file naming convention for the original data and 2D silhouette data for user's convenience. From the file name, the information such as gesture index,

Table 1 File naming convention

*G_Di*_F_T.Ext	
Symbol	Meaning
G	Name of gesture/class
Di	Dancer index {1:20}
F	Frame number {0:30—image data, dd—video data}
T	Data type {N-Normal, S-Silhouette}
Ext	File extension {PNG, AVI}
Example: Image data: *Ora*_D1_1_N.png Video data: *Ora*_D1_dd_N.avi	

dancer index, type of image, i.e., normal or silhouette image, and frame number can be obtained by the user. Here, all the frames of a video sequence are not considered, only key frames are stored. Moreover, the video sequences in uncompressed 'AVI' file format are also stored in the dataset separately. The basic poses of each gesture are also stored in the dataset.

The file naming convention has been given in Table 1.

4.3 Pre-processing

After data acquisition, data are pre-processed. In this phase, frames are extracted from the video sequences. The number of frames in the video sequences ranges from 150 to 400. Then, 30 key frames are extracted from those frame sequences. Background subtraction has been carried out using vibe [5] to get the 2D silhouettes. The minimum bounding rectangle (MBR) of each frame has been found out.

4.4 Feature Extraction

In order to recognize gestures, feature vectors play a major part. This makes the extraction of feature vectors extremely important in any gesture recognition related work. In this dataset, these features take into account with each gesture: height: width ratio of minimum bounding rectangle (MBR) of each frame, inter frame energy difference, and inter frame entropy difference.

(a) *Height: width ratio of MBR:*

From the 2D silhouettes, the MBR of each frame has been extracted. Height: width ratio of each frame is calculated which is considered as a feature. After finding out the MBR for a video, i.e., image sequences, the height-to-width ratio is calculated. It is observed that this ratio is smaller for the class "Facing front" for almost all the image frames and for the ground exercises "facing front back side," this ratio is found to be larger for the frames where the dancer is facing side since width becomes very small. From this observation, we can classify these two broad classes of ground exercise.

(b) *Energy difference:*

To determine the object movement, inter frame difference is normally used. But, it is not the best choice for extremely low luminance images. An alternative way can be the use of energy information of the images—the inter frame difference of energy. Energy is defined based on a normalized histogram of the image. Energy shows how the gray levels are distributed. When the number of gray levels is low, then energy is high. This feature is extracted from the video frames.

(c) *Entropy difference:*

Entropy is a statistical measure of randomness that can be used to characterize the texture of the input image. It gives the randomness or uncertainty of information to represent an image. Inter frame entropy difference feature is used in this work. To calculate entropy,

$$\mathrm{H} = -\sum_{k=1}^{K} p(e_k) \log p(e_k)$$

Here, $e_1, e_1, \ldots e_k$ is the set of possible events of a random event E with probabilities $\{p(e_1), p(e_1), \ldots p(e_k)\}$

These three features are extracted from each frame of the image sequences and get the feature vector for that image sequence. Each ground exercise video is represented by the computed feature vector,

$$\text{Feature vector} = [EN, EP, h : w]$$

The features for all images were extracted by using the above step and obtain the feature vectors for the entire training dataset and obtain the feature matrix.

The feature matrix, mathematically, can be defined as 2D matrix, where the row indicates features vector of each image sequence and column indicates the number of features. Using this feature matrix, mean feature vector for each person was calculated, and this was used as a template for each ground exercises in the testing phase to verify the ground exercise in the recognition phase.

$$\text{feature\_matrix} = \begin{bmatrix} EN_1\ EP_1\ h : w_1 \\ EN_2\ EP_2\ h : w_2 \\ \cdots\cdots\cdots\cdots\cdots\cdots \\ EN_n \quad EP_n \quad h : w_n \end{bmatrix}_N$$

where EN_i is the energy difference of $(i+1)$th and ith frame, EP_i is the entropy difference of $(i+1)$th and ith frame, $h : w_i$ is the height-to-width ratio of minimum bounding rectangle of the ith frame, and N is the number of image sequences.

We have created four dataset

1. Dataset consisting of features EP and EN
2. Dataset consisting of features EP and HW
3. Dataset consisting of features EN and HW
4. Dataset consisting of features EN, EP and HW

The feature matrix for all the four dataset are given below:

$$\text{feature\_matrix}_1 = \begin{bmatrix} EN_1 \; EP_1 \\ EN_2 \; EP_2 \\ \cdots\cdots\cdots \\ EN_n \quad EP_n \end{bmatrix}_N$$

$$\text{feature\_matrix}_2 = \begin{bmatrix} EN_1 \; HW_1 \\ EN_2 \; HW_2 \\ \cdots\cdots\cdots \\ EN_n \quad HW_n \end{bmatrix}_N$$

$$\text{feature\_matrix}_3 = \begin{bmatrix} EP_1 \; HW_1 \\ EP_2 \; HW_2 \\ \cdots\cdots\cdots \\ EP_n \quad HW_n \end{bmatrix}_N$$

$$\text{feature\_matrix}_4 = \begin{bmatrix} EN_1 \; EP_1 \; h : w_1 \\ EN_2 \; EP_2 \; h : w_2 \\ \cdots\cdots\cdots\cdots\cdots \\ EN_i \quad EP_i \quad h : w_i \end{bmatrix}_N$$

4.5 *Classification*

This is the final stage in any pattern recognition problem. The total 28 ground exercises are classified into two subclasses

- Facing front (FF)
- Facing front back side (FFBS)

We have classified these two broad classes of ground exercises based on height:width ratio of the MBRs. For the class 1 (FF), this ratio is smaller for almost all the image frames since width is larger. For the class 2 (FFBS), this ratio is found to be larger for the frames where the dancer is facing side since width becomes very smaller. Height is almost same for all GEs. From this observation, we have classified these two broad classes of ground exercises.

19 GEs are classified to class 1, and remaining 9 GEs belong to class 2.

To classify class 1 and class 2 gestures, 70% of the data are used for training, and the remaining 30% is used for testing accordingly.

For performance evaluation, we have used SVM, Bayesian network, decision tree, and HMM.

5 Experimental Result

In this work, the features that we have adopted for classification of ground exercises from the Sattriya dance dataset are height-to-width ratio of MBR (HW), inter-frame energy difference (EN) and inter-frame entropy difference (EP). The average classification accuracy of the different ground exercises that we have achieved by taking into consideration of all these features are shown in four tables consecutively according to the four datasets that we have developed. We have selected the machine learning classifiers SVM, Bayesian network, decision tree, and HMM for classification of the ground exercises of Sattriya dance. Tables 2, 3, 4, and 5 show the classification accuracy of the (EN, EP) features, (EP, HW) features, (EN, HW) features, and (EN, EP, HW) features, consecutively.

From the above experimental result, it is observed that the classification accuracy is good for class FF, and HMM results better in comparison with the other classifiers. The analysis of classification accuracy for the different feature sets for class FF and class FFBS are shown in Figs. 3, and 4, respectively.

Table 2 Accuracy of energy and entropy feature in dataset 1

Classifier	Total no of gestures	Correctly classified	Average recognition rate (%)
SVM	19 (FF)	15	79
	9 (FFBS)	7	77.7
Bayesian network	19 (FF)	14	73.68
	9 (FFBS)	7	77.7
Decision tree	19 (FF)	15	79
	9 (FFBS)	6	66.66
HMM	19 (FF)	17	89.47
	9 (FFBS)	8	88.88
Average recognition rate (EN, EP) features for FF class			80.28
Average recognition rate (EN, EP) features for FFBS class			77.735

Table 3 Accuracy of energy and height: width feature in dataset 2

Classifier	Total no of gestures	Correctly classified	Average recognition rate (%)
SVM	19 (FF)	16	84.21
	9 (FFBS)	6	66.66
Bayesian network	19 (FF)	15	79
	9 (FFBS)	6	66.66
Decision tree	19 (FF)	17	89.47
	9 (FFBS)	7	77.77
HMM	19 (FF)	18	94.73
	9 (FFBS)	8	88.88
Average recognition rate (EP, HW) features for FF class			86.85
Average recognition rate (EP, HW) features for FFBS class			75

Table 4 Accuracy of entropy and height: width feature in dataset 3

Classifier	Total no of gestures	Correctly classified	Average recognition rate (%)
SVM	19 (FF)	14	73.68
	9 (FFBS)	5	55.55
Bayesian network	19 (FF)	13	68.42
	9 (FFBS)	6	66.66
Decision tree	19 (FF)	16	84.21
	9 (FFBS)	7	77.77
HMM	19 (FF)	16	84.21
	9 (FFBS)	7	77.77
Average recognition rate (EN, HW) features for FF class			77.63
Average recognition rate (EN, HW) features for FFBS class			69.4

From Fig. 3, it has been observed that the classification accuracy is better for the feature set (EN, EP, HW) for class FF. HMM shows good performane of about 100% in dataset 4.

Since the FFBS class contains the mostly articulated human motion ground exercises, the classification accuracy is not up to the mark. From Fig. 4, it has been observed that the classification accuracy is better for the feature set (EN, EP, HW). HMM shows good performane of about 88.88%.

Table 5 Accuracy of energy, entropy, and height: width feature in dataset 4

Classifier	Total no of gestures	Correctly classified	Average recognition rate (%)
SVM	19 (FF)	17	89.47
	9 (FFBS)	7	77.77
Bayesian network	19 (FF)	16	84.21
	9 (FFBS)	6	66.66
Decision tree	19 (FF)	17	89.47
	9 (FFBS)	7	77.77
HMM	19 (FF)	19	100
	9 (FFBS)	8	88.88
Average recognition rate (EN, EP, HW) features for FF class			90.8
Average recognition rate (EN, EP, HW) features for FFBS class			78

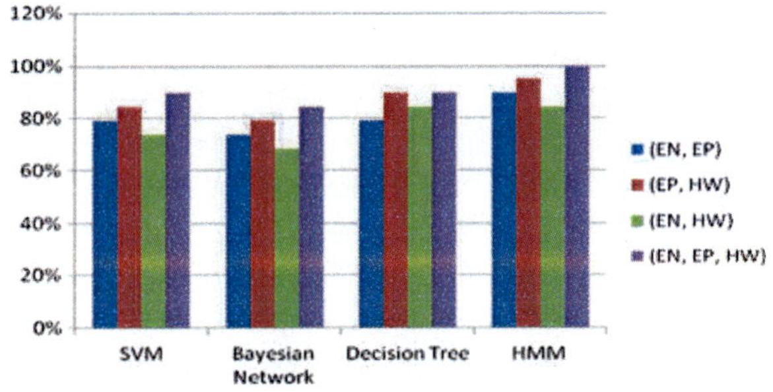

Fig. 3 Comparative analysis of classification accuracy for FF class

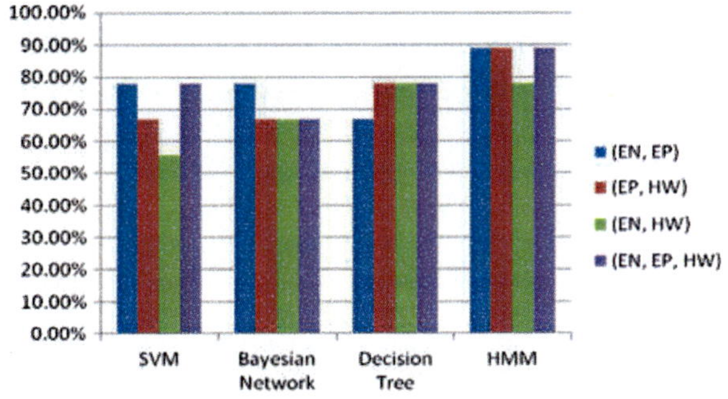

Fig. 4 Comparative analysis of classification accuracy for FFBS class

6 Conclusion

In this paper, a novel video dataset is introduced for dynamic dance gesture recognition. The purpose of creation of this dataset is to develop a computer vision system to classify each ground exercise, as well as this dataset can be useful for benchmarking a variety of computer vision and machine learning methods designed for dynamic dance gesture recognition. Also by extracting the features, we have classified the ground exercises using some well-known machine learning techniques. Since the FFBS class is not classified well, we may extract more discriminating features for better classification.

7 Availability

If anyone is interested in downloading the dataset, they can contact the author by e-mail at sumpiss@tezu.ernet.in.

Acknowledgements We would like to acknowledge Mr. Shiva Hazarika, Mr. Atul Kumar Bhuyan, and Mr. Gopal Chandra Bordoloi the experts in this field for their kind help and cooperation in collecting the data.

References

1. Heryadi, Y., Fanany, M., Arymurthy, A.: A syntactical modeling and classification for performance evaluation of bali traditional dance. In: Advanced Computer Science and Information Systems (ICACSIS'12), International Conference on IEEE, pp. 261–265
2. Nussipbekov, A., et al.: Kazakh traditional dance gesture recognition. J. Phys: Conf Ser **495**(1), 012036 (2014)
3. Peng, B., Qian, G., Rajko, S.: View-invariant full-body gesture recognition from video. In: Pattern Recognition 2008 (ICPR'08) 19th International Conference on IEEE, pp. 1–5 (2008)
4. Internet source. http//www.Itsallfolk.com
5. Barnich, O., Van Droogenbroeck, M.: ViBe: a universal background subtraction algorithm for video sequences. Trans. Img. Proc. **20**(6), 1709–1724 (2011). https://doi.org/10.1109/TIP.2010.2101613

Performance Analysis of Nearest Neighbor, K-Nearest Neighbor and Weighted K-Nearest Neighbor for the Classification of Alzheimer Disease

Olimpia Borgohain, Meghna Dasgupta, Piyush Kumar, and Gitimoni Talukdar

Abstract Alzheimer's disease (AD) has become a major health problem over the past few decades. AD can be defined as a neurodegenerative disorder that causes the brain cells to degenerate and die. AD is the most popular cause of dementia. AD is likely to be more observant in elderly people that are people above 65 years of age. Memory loss is one of the prominent symptoms of AD. The people suffering from AD tend to repeat questions or statements, or eventually may forget the names of their family members. Detection of AD has already been done using several machine learning techniques like supervised learning, unsupervised learning, and reinforcement learning as well as deep learning techniques. In this paper, we have tried to detect AD using Nearest Neighbor, K-Nearest Neighbor, and Weighted K-Nearest Neighbor algorithm. In the area of AD detection using machine learning techniques, KNN has already been applied but a detailed computational overview of different variants of the nearest neighbor algorithms has been performed in this paper which will add a new dimension in the medical domain. We have considered four classes for our work namely Normal, Very mild, Mild, and Moderate. We have evaluated the performance of our system using three performance metrics namely Precision, Recall, and $F1$ score.

Keywords Alzheimer disease (AD) · Minkowski · Euclidean · Distance metrics · Weighted KNN · KNN

O. Borgohain (✉) · M. Dasgupta · P. Kumar · G. Talukdar
Department of Computer Science and Engineering, Royal Global University, Guwahati, India
e-mail: olympiaborgohain@gmail.com

M. Dasgupta
e-mail: meghnadasgupta2108@gmail.com

P. Kumar
e-mail: piyushkumar7399@gmail.com

G. Talukdar
e-mail: gitimoni.talukdar@rgi.edu.in

S. Borah et al. (eds.), *Soft Computing Techniques and Applications*, Advances in Intelligent Systems and Computing 1248,
https://doi.org/10.1007/978-981-15-7394-1_28

1 Introduction

The increased manifestation of Alzheimer's disease (AD) in aged people of the community has led to numerous advancements and researches in the early diagnosis of the disease. AD is a progressive and irreversible disease that is it cannot be cured and it aggravates over time. So far, no treatment of AD has been developed. AD is the most general cause of dementia [1]. As a common form of dementia, AD counts for 60–80% of cases. So, the early detection of this disease is critical in order to delay the progress. The main stages of AD are Normal, Very mild, Mild, and Moderate.

Researchers have used varied techniques for early diagnosis of AD. From the 1970s to 1990s rule-based techniques were used [2]. Robustness, portability, and cost of maintenance were the main struggle in the case of rule-based techniques. From 1990 onwards, various supervised machine learning models like SVM, Random Forest, Decision Tree, etc. were used [2]. Machine learning techniques are very well known due to its low rate of maintenance and easy portability. The main demerit with the machine learning technique was the phase of feature extraction which is the most time-consuming part. Recently, deep learning techniques are high in demand due to its high accuracy rate. It overcomes the demerit of other machine learning by extracting features on its own and providing high-quality results accordingly. The main disadvantage of deep learning is the huge amount of training data required especially in the case of the medical field, as in this area data can be costly and confidential making the data acquisition process critical.

This paper delivers a detailed analysis of how various nearest neighbor techniques can be used for the early detection of AD. The main advantage of Nearest Neighbor algorithms are its simplicity, robustness and validity towards noisy data. Due to various pitfalls of KNN like high computational cost, costly memory requirements and uncertainty deciding the value of K, modified versions of KNN techniques are used. The accuracy of KNN depends on the input parameter K and distance metrics used. The different types of distance metrics used are Euclidean distance, Manhattan distance, Cosine distance, Minkowski distance and Chi-square distance [3, 4]. Smaller value of K leads to high variance and larger value of K leads to lower bias [3]. The variants of KNN give better results in comparison to standard KNN. When the value of K is 1 it becomes NN that is Nearest Neighbor. In this case, the class of the unknown data is the class of the nearest training data [3]. Weighted-adjusted KNN algorithm first learns weights for different attributes and based on the weights assigned, each attribute affects the process of classification. In this paper we have used performance metrics of accuracy, precision, recall and $F1$ score. The formulas for the calculation of the respective performance metrics are described below:

$$P = T_{\mathrm{p}}/\left(T_{\mathrm{p}} + F_{\mathrm{p}}\right)$$
$$R = T_{\mathrm{p}}/\left(T_{\mathrm{p}} + F_{\mathrm{n}}\right)$$
$$F1 = 2 \times \{(P \times R)/(P + R)\}$$

where

P Precision
R Recall
T_p Number of true positives
F_p Number of false positives
F_n Number of true positives (T_p) + Number of false positives (F_p)

This paper is further divided into five sections namely Related Work wherein we have given the detailed survey in this area followed by a section of Features where we have discussed the various features that have been used in our system. The next section includes Methodology where we have explained the detailed working of the three algorithms followed by the Results and Discussion section which highlights the computational results of the three variants of Nearest Neighbor techniques and finally, we have concluded our paper in the last section.

2 Related Work

Machine learning and deep learning techniques are very prominent for the prediction of Alzheimer's disease. In [5] a description of Deep Convolution Neural Network is presented to identify Alzheimer's Disease and Dementia from 3D MRI images. The following work tries to device a solution for the problems associated with supervised classification techniques. The dataset used in [5] is the OASIS dataset that holds information of 416 subjects. A stochastic optimization function Adam Optimizer is used to optimize the neural network with a learning rate of 0.001. Floydhub's GPU is used to train the model. The model gives an accuracy of 80.25%. Again from [6] which has also used a deep learning technique for detection of Alzheimer's Disease based on resting-state brain network it is observed that there is a 31.21% improvement in the prediction rate if deep learning techniques are used. The models namely LDR, LR, SVM, and autoencoder networks are used to bring out the discriminative brain features and classify the subjects. The accuracy of autoencoder, LDR, LR, SVM are 65%, 59%, 59%, and 62%, respectively.

Another deep learning architecture is used in [7] containing autoencoders and SoftMax output layer to help detection of Alzheimer's Disease (AD) and the early symptom of the disease that is Mild Cognitive Impairment (MCI). The deep learning method proposed an accuracy of 87.76% in the binary classification of AD. Machine Learning algorithms such as SVM, Gradient Boosting, Neural Network, K-Nearest Neighbor, Random Forest are also used in the detection of the said disease as mentioned in [8]. Processed data that are procured from neuroimaging technologies helps in the detection of Alzheimer's disease using the above-mentioned machine learning algorithms. The accuracy of each of the algorithms are 97.56%, 97.25%, 98.36%, 95.00%, 97.86%, respectively.

The supervised machine learning algorithms like Nearest Neighbor has several variants. The different variants of Nearest Neighbor are used for classification in many different fields but in the classification of Alzheimer's disease like in [3] a survey is done on KNN and its modified versions that are weighted KNN, Adaptive KNN, etc. The accuracy of normal KNN can be enhanced by using its variants and changing some parameters. In [9] Random Forest, Gradient Boosting, SVM, Lasso is used for recognition. Random Forest and Gradient Boosting show best accuracy of 97.94% for the prediction of AD using the main eight attributes that is MMSE, ASF, eTIV, Age, nWBV, educ, gender and SES, 15 combinations were derived keeping Mini-Mental State Examination(MMSE), Estimated total intracranial volume (eTIV), ASF and nWBV attributes constant. 120 records were taken as training data and the remaining 116 records as test data. In [4] the effect of the distance metrics on medical datasets like blood, breast cancer, *Escherichia coli,* etc. was evaluated. The different distance metrics used are Cosine, Euclidean, Minkowski, and Chi-square. Out of the four, the Chi-square metric performed the best. The Euclidean distance function performed exceptionally well over other numerical and categorical datasets but did not work fine for mixed types of datasets. In [10] a 3D convolutional neural network (CNN) model was used to distinguish between normal patients and patients having AD. Data augmentation was used to add robustness and diversity on the comparatively small training dataset and so the difficulty of using deep learning was removed.

3 Methodology

Our project starts with the data acquisition stage where we have collected data of 373 patients from [11]. The dataset is divided into two subsets out of which the first subset comprising the data of 298 patients is considered for training and the remaining data of 75 patients in the second subset is used for testing.

The next step in the training phase involves extracting the features which in turn will help the model to learn for future classification purposes in the testing phase. In the feature extraction phase to deal with missing values, there are two options, either the entire row can be dropped or the mean of the entire column where the missing value is present can be found out and then that value of the mean can be inserted in the place of missing values to make the computation easier. In the testing phase, we have used three supervised machine learning algorithms such as Nearest Neighbor, K-Nearest Neighbor, and Weighted K-Nearest Neighbor. For the K Nearest Neighbor, we have considered different values of K ranging from 2 to 13. $K = 1$ value is not considered because it automatically corresponds to Nearest Neighbor. We have also analyzed our K-Nearest Neighbor algorithm with different distance metrics such as Euclidean, Manhattan, and Minkowski distance metrics. The K-weighted Nearest Neighbor algorithm is analyzed using the kernel function of inverse distance and also using the above three distance metrics. Finally, our output is test data that is classified as one of the four categories as has been mentioned earlier. We have evaluated our system with three evaluation measures namely precision, recall, and $F1$ score. The entire work of our supervised machine learning system is depicted in Fig. 1.

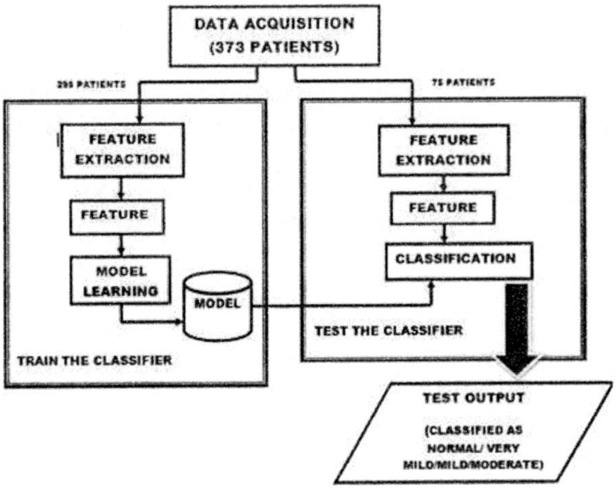

Fig. 1 Workflow diagram of our system

4 Dataset and Features

The OASIS longitudinal dataset consists of 373 rows and 15 columns. The dataset is a collection of 150 persons aged between 60 and 96. The subjects are both male and female and all are right-handed. The patients were examined on two or more visits, separated by a minimum of one year. 72 patients were designated as non-demented, 64 of the included victims were characterized as demented at the time of their first visits and remained the same for their successive visits, including 51 people with mild to moderate AD. Another 14 patients were diagnosed as non-demented at the time of their first visit and were later recognized as demented at their subsequent checkups.

The features which we have selected for the classification part are MMSE, eTIV, nWBV, SES, and EDUC. The Mini-Mental State Examination (MMSE) is a 30-point value. It is generally used to search for dementia and to measure cognitive impairment. A value greater than or equal to 24 points out normal cognition. Value in between 19 and 23 accounts for mild, 10–18 for moderate, and less than or equal to 9 indicates severe. However, MMSE alone is not a strong feature.

eTIV(Estimated total intracranial volume)or ICV value is used to assess AD, Mild cognitive impairment. This feature estimates the age-related change in the structure of the brain. It points to the approximate volume of the cranial cavity. EDUC stands for years of education. SES (socio-economic status) values range from 1 to 5—1 being the highest status and 5 being the lowest status. nWBV stands for Normalized whole-brain volume area. The classification is done based on CDR (clinical dementia rating) value. CDR is a 5-point value which depicts functionalities of AD and other types of dementia. A patient's level of impairment/dementia is designated by this score. 0 stands for Normal, 0.5 for Very Mild, 1 for Mild,2 for Moderate, and 3 for Severe.

5 Results and Discussions

The result of the performance analysis of Nearest Neighbor and its variants are shown in Tables 1, 2, and 3. Table 1 shows the performance of Nearest Neighbor. It can be observed that for Nearest Neighbor the highest accuracy of 82.67%, the precision of 80.00%, and $F1$ score of 79.26% is obtained using Euclidean distance and highest recall of 81.00% is obtained using Minkowski distance.

In K-Nearest Neighbor the values of K are taken from $K = 2$ to the maximum value of $K = 13$. The performance of the minimum value of K that is taken to be 2 and the maximum value of K that is taken to be 13 is shown in Table 2. For the minimum value of K which equals to 2, it can be concluded that the highest accuracy of 76.0% is obtained using both Euclidean and Manhattan distance and highest recall of 73.00%, highest precision of 75.76%, and highest $F1$ score of 72.34% is obtained

Table 1 Summary of nearest neighbor algorithm

Algorithm	Distance metrics	Accuracy (%)	Precision (%)	Recall (%)	$F1$ score (%)
Nearest neighbor	Euclidean	82.67	80.00	79.67	79.26
	Manhattan	77.33	72.51	73.33	72.77
	Minkowski	81.33	78.00	81.00	78.00

Table 2 Summary of K-nearest neighbor algorithm

Algorithm	Value of K	Distance metrics	Accuracy (%)	Precision (%)	Recall (%)	$F1$ score (%)
K-nearest neighbor	$K = 2$	Euclidean	76.00	75.76	73.00	72.34
		Manhattan	76.00	71.02	69.00	69.27
		Minkowski	74.67	73.40	67.1	68.7
	$K = 13$	Euclidean	69.3	61.7	57.8	58.07
		Manhattan	74.7	73.4	67.00	69.00
		Minkowski	74.67	73.4	67.1	68.7

Table 3 Summary of weighted K-nearest neighbor algorithm

Algorithm	Value of k	Distance metrics	Accuracy (%)	Precision (%)	Recall (%)	$F1$ score (%)
Weighted K-nearest neighbor	$K = 2$	Euclidean	82.7	80.00	79.7	79.3
		Manhattan	76.00	72.8	68.00	68.9
		Minkowski	82.7	79.7	79.7	79.3
	$K = 13$	Euclidean	76.00	72.8	68.00	68.7
		Manhattan	76.00	72.8	68.00	68.9
		Minkowski	76.00	72.8	68.00	68.7

using Euclidean distance. For the maximum value of K that equals to 13, the highest accuracy of 74.7% is obtained using Euclidean distance, highest precision of 73.4% is obtained using both Manhattan and Minkowski distance, highest recall of 67.0% and highest F1 score of 69.0% is obtained using Manhattan distance.

In Fig. 2, the accuracy of the three distance metrics is shown with respect to different values of K for KNN. It can be seen that the accuracy for Minkowski distance remained constant with increasing values of K. Similarly, in Figs. 3 and 4 the precision and $F1$ score for Minkowski distance remained the same.

The accuracy, $F1$ score, and precision of Weighted KNN are plotted in Figs. 5, 6, and 7 for all three distance metrices used with reference to raising values of K. The accuracy curve for Euclidean and Minkowski distance overlaps with each other for varying values of K. For Manhattan distance the curve remained the same. The same pattern has been plotted for the other two evaluation metrics.

In Fig. 8 the distribution of Male and Female patients is graphed with respect to

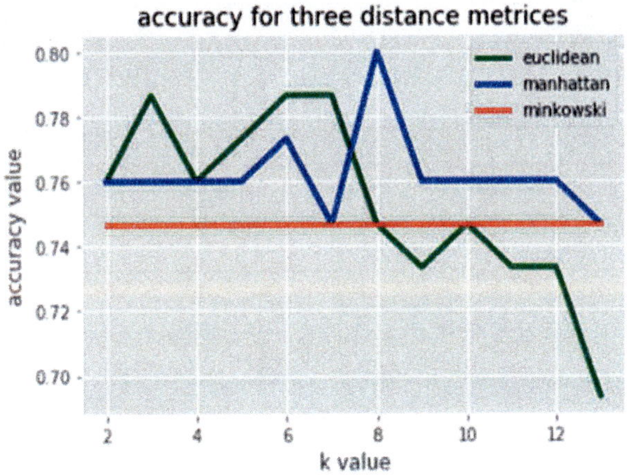

Fig. 2 Comparison of accuracy for different distance metrics for KNN

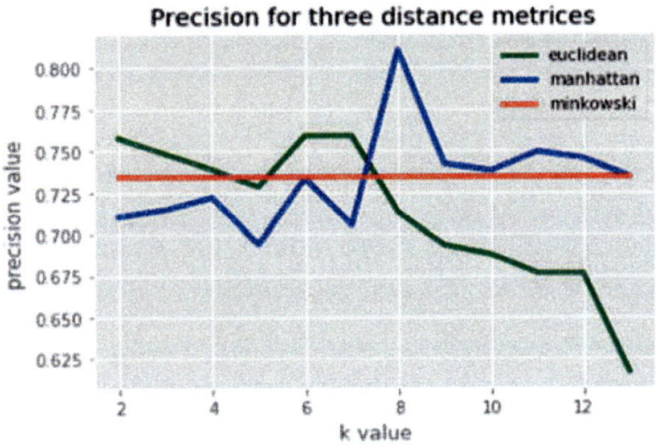

Fig. 3 Comparison of $F1$ score for different distance metrics for KNN

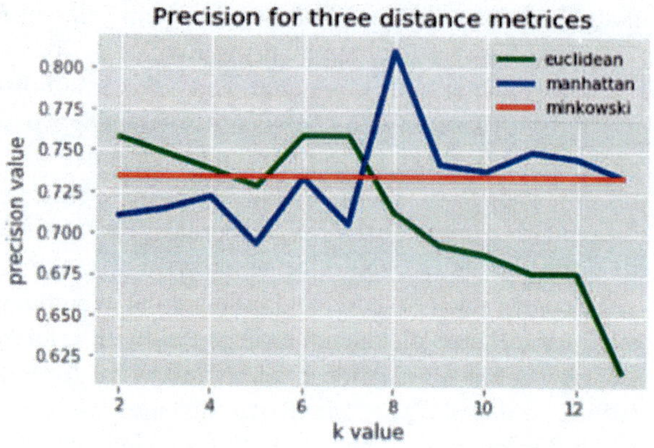

Fig. 4 Comparison of precision for different distance metrics for KNN

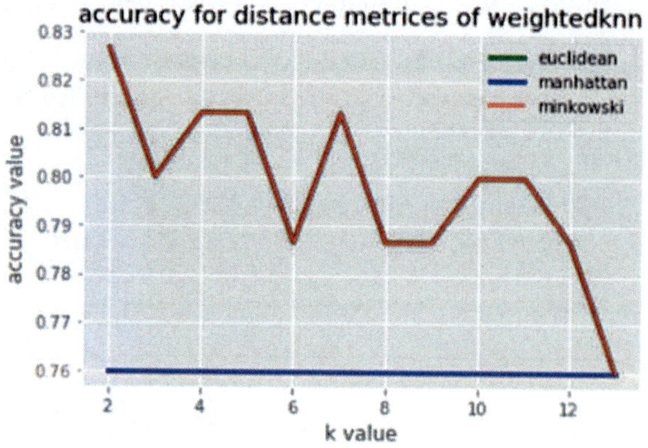

Fig. 5 Comparison of accuracy for different distance metrics for W-KNN

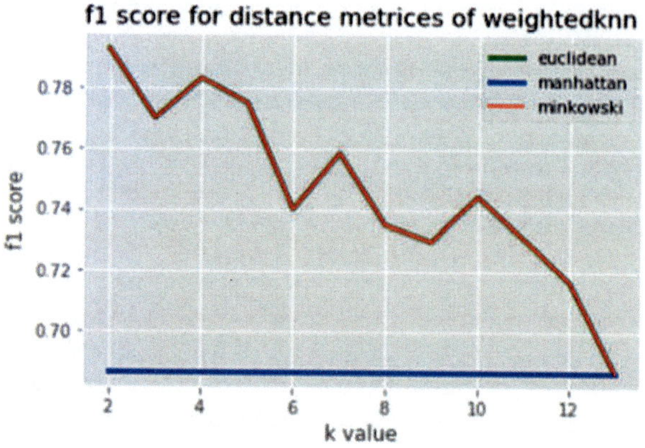

Fig. 6 Comparison of $F1$ score for different distance metrics for W-KNN

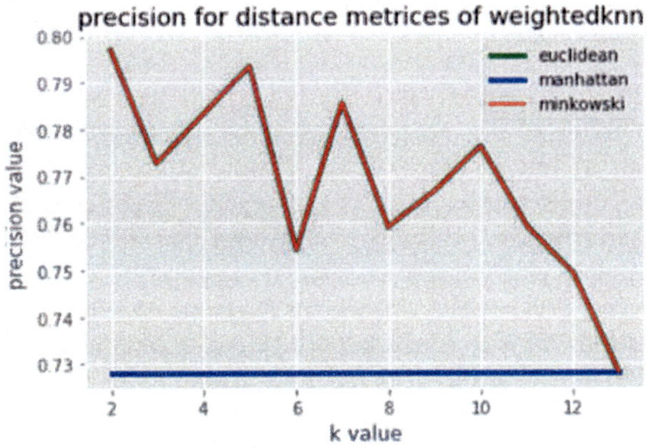

Fig. 7 Comparison of precision score for different distance metrics for W-KNN

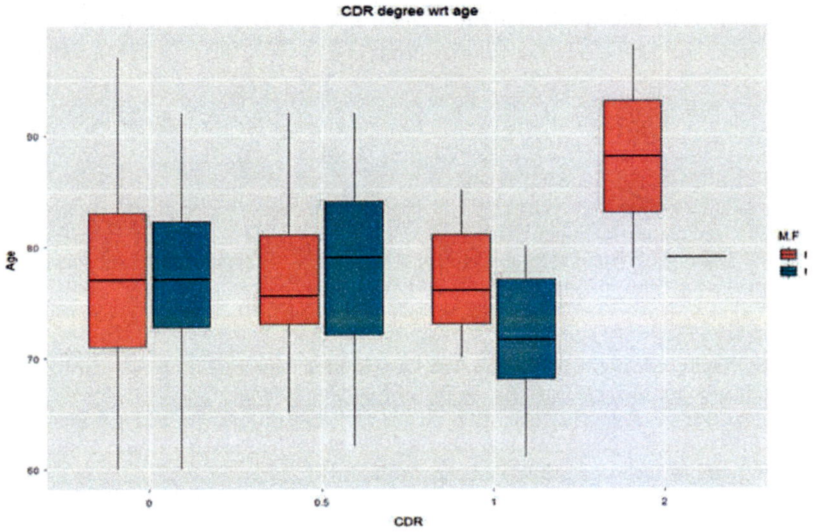

Fig. 8 Gender distribution of patients

their ages and corresponding CDR values. The two different colors correspond to Male and Female patients, respectively.

In Weighted KNN using inverse distance as the kernel function, the performance analysis has been noted. Also, the value of K here is taken ranging from a minimum value of $K = 2$ to a maximum value of $K = 13$. It can be easily noted from the table below that for $K = 2$ the highest accuracy of 82.7%, highest recall of 79.7%, and $F1$ score of 79.3% is acquired using both Euclidean and Minkowski distance. Highest precision of 80.0% is obtained using Euclidean distance. For the maximum value of K that is taken to be 13, it can be observed from the table below that, for the three distance metrics, the same value of accuracy, precision, recall, and $F1$ score is obtained.

6 Conclusion

In this paper, we have tried to give a detailed comparative study of performance analysis of the three variants of the supervised Nearest Neighbor machine learning algorithms for the detection of different stages of Alzheimer disease. In the above algorithms, three distance metrics namely Euclidean distance, Manhattan distance, and Minkowski distance have been considered. We have achieved the maximum accuracy of 82.67% in Nearest Neighbor, 76.00% in K-Nearest Neighbor for the value of $K = 2$ and 74.7% for the value of $K = 13$ and 82.7% for the value of $K = 2$ and 76.0% for the value of $K = 13$ in Weighted K-Nearest Neighbor. In the recent literature, various deep learning techniques have been found to give a more effective performance in the area of Alzheimer disease detection so this work can further be expanded by coming up with a hybrid approach of deep learning and other machine learning techniques such as other variants of Nearest Neighbor such as cluster-based KNN, density-based KNN, fuzzy logic-based KNN, SVM based KNN, etc. and also other unsupervised machine learning approaches.

References

1. Razavi, F., Tarokh, M., Alborzi, M.: An intelligent Alzheimer's disease diagnosis method using unsupervised feature learning (2019)
2. Islam, J., Zhang, Y.: Brain MRI analysis for Alzheimer's disease diagnosis using an ensemble system of deep convolutional neural networks (2018)
3. Lambaand, A., Kumar, D.: Survey on KNN and its variants. Int. J. Adv. Res. Comput. Commun. Eng. **5**(5) (2016)
4. Hu, L., Huang, M., Ke, S., Tsai, C.: The Distance Function Effect on K-Nearest Neighbor Classification for Medical Datasets. SpringerPlus, Article Number 1304 (2016)
5. Ullah, H., Onik, Z., Islam, R., Nandi, D.: Alzheimer's disease and dementia detection from 3D brain MRI data using deep convolutional neural networks. In: 3rd International Conference for Convergence in Technology (I2CT), Pune, India (2018)
6. Ju, R., Hu, C., Zhou, P.: Early diagnosis of Alzheimer's disease based on resting-state brain networks and deep learning. IEEE/ACM Trans. Comput. Biol. Bioinform. (2017)
7. Liu, S., Liu, S., Cai, W., Pujol, S., Kikinis, R., Feng, D.: Early diagnosis of Alzheimer's disease with deep learning
8. Lodha, P., Talele, A., Degaonkar, K.: Diagnosis of Alzheimer's disease using machine learning. In: 2018 Fourth International Conference on Computer Communication Control and Automation (ICCUBEA). IEEE (2014)
9. Naidu, C., Kumar, D., Maheswari, N., Sivagami, M., Gang Li.: Prediction of Alzheimer's disease using oasis dataset. IJRTE **7**(6S3) (2019)
10. McCrackin, L.: Early detection of Alzheimer's disease using deep learning (2018)
11. Kaggle Homepage: https://www.kaggle.com/. Last accessed 10 Jan 2020
12. Sheshadri, H., Bhagya Shree, S., Muralikrishna.: Diagnosis of Alzheimer's disease employing Neuropsychological and classification techniques. Int. J. Innov. Technol. Explor. Eng. (IJITEE) (2015)

Divide-and-Conquer-Based Recursive Decomposition of Directed Acyclic Graph

Anushree Dutta, Nutan Thapa, Santanu Kumar Misra, and Tenzing Mingyur Bhutia

Abstract Many real-time tasks can be divided into several similar subtasks based on their dependencies and these subtasks can be represented with the help of a Task Dependency Graph. In this paper, we have identified the recursive decomposition of a Directed Acyclic Graph to achieve faster execution time and maximize performance. A Recursive Decomposition technique is used for the problems which can be solved using a divide-and-conquer algorithm. A given task is divided into subtasks and topological sorting is applied which gives the sequence in which the subtasks are to be performed. These subtasks are then decomposed recursively using similar division attribute until we reach the last dependent task. After performing the computation, the subtasks are combined recursively to attain the result of the main task. The tasks and their dependencies are provided during compilation. Recursive decomposition helps in achieving the concurrency of tasks as the subtasks are solved in parallel.

Keywords Concurrency · Divide-and-Conquer · Granularity · Recursive decomposition task dependency graph

A. Dutta · N. Thapa · S. K. Misra (✉) · T. M. Bhutia
Department of Computer Science and Engineering, Sikkim Manipal Institute of Technology, Sikkim Manipal University, Gangtok, Sikkim, India
e-mail: santanu.m@smit.smu.edu.in

A. Dutta
e-mail: anushree_201600143@smit.smu.edu.in

N. Thapa
e-mail: nutan_201600208@smit.smu.edu.in

T. M. Bhutia
e-mail: tensing_201600266@smit.smu.edu.in

S. Borah et al. (eds.), *Soft Computing Techniques and Applications*, Advances in Intelligent Systems and Computing 1248,
https://doi.org/10.1007/978-981-15-7394-1_29

1 Introduction

In today's fast-growing world, the need for faster and accurate computation of tasks is considered most crucial. The more the computation of task takes time, the less efficient the method is said to be. To increase the efficiency of the execution of a real-time task, parallel computing plays a major rule. The execution of tasks or computation of tasks concurrently is known as Parallel Computing. To achieve parallel computing a task needs to be divided so that its subtasks can be scheduled to multiple processing units where each processing unit executes different subtasks of the main task acknowledging the task timing parameters. Except for the input of the task, several other parameters need to be considered like execution time, data dependencies between the subtasks and priority of the subtasks. Depiction of subtasks is done by a data structure known as a task dependency graph. It is a Directed Acyclic Graph (DAG) which consists of vertices representing tasks and directed edges representing the dependencies.

The directed acyclic graph has a distinct property that is it already has a topological ordering which means that it is already ordered in the sequence of its execution and the nonexistence of a cycle makes it easier to schedule tasks. And every directed acyclic graph can be decomposed with different methods. The decomposition of a directed acyclic graph is important so that effective analysis of the relations between the tasks can be done. A directed acyclic graph makes it easier to identify the tasks to be scheduled reducing the overall execution time of the task. It is used for many real-world problems like PERT graph for project management, route navigation, and spreadsheets. A directed acyclic graph can be decomposed using four different techniques such as Recursive decomposition, Data decomposition, Exploratory decomposition, and Speculative decomposition.

The key benefit of using these methods is that they provide a framework for dealing with larger action items at the lowest possible level through creating hierarchical structures and breakdowns. Such a framework provides simplicity in planning and performing various task jobs. Another benefit is that we can use sub-tasks as the smallest and simplest activities to perform smaller goals. Özkaya et al. [1] claim that the DAG task partitioner is even more powerful in putting together the highly complex tasks and reduces overall inter-processor communication. Several task partitioning techniques are well established in the field of DAG partitioning. Some of them also focus on multi-level partitioning techniques but the challenge they face is to maintain the additional acyclicity constraints on each level explained by Sanders and Schulz [2].

In this work, we have focused on Recursive decomposition of a directed acyclic graph. The inputs to the task can be taken dynamically as well as during compile time. Recursive decomposition is a widely used technique for a wide variety of problems. The method consists of recursively decomposing the tasks until we reach the last dependent task followed by the computation of subtasks and combining them recursively to attain the result of the task. Recursive decomposition is appropriate for problems that can be solved using divide-and-conquer paradigm.

We initially take a task as input and identify its subtasks. Further, the task and its subtasks are represented using a directed acyclic graph which involves edges between the dependent tasks and no edges between the independent tasks. Topological ordering is done on the graph. Recursive decomposition is implemented on the graph which further depicts the relations between tasks according to the dependencies between them. The independent subtasks are computed concurrently and the dependent tasks are computed according to the dependency of the task on another task.

The following section of the paper discusses the related works and Sect. 3 explains the fundamentals of directed acyclic graphs and various decomposition techniques. Section 4 describes the decomposition technique used in this work along with a simple example to describe the method. Section 5 depicts the implementation of the mentioned approach. Section 6 discusses the results, analysis of the proposed method. Section 7 illustrates the conclusion of the proposed work.

2 Related Works

The decomposition of tasks and their scheduling has been a part of the research for a long time. The emphasis has always been given to the real-time scheduling of a task to maximize the concurrency and performance of a system. The problem of real-time scheduling of parallel tasks having pre-determined execution requirements has been discussed by Saifullah et al. [3]. The authors represent the task in the form of a directed acyclic graph and then decomposed the DAG into sequential tasks providing them the ease to execute the tasks sequentially or non-sequentially. The paper mainly focused on executing the tasks in the specified execution time rather than emphasizing on the dependencies between them. The objective of the paper was to reduce the overall execution time of the task graph which contradicts the dependency constraint of a directed acyclic graph.

There is a high possibility of a chance that the graph produced for a task may not be acyclic, the graph may turn out to be a cyclic graph. Even for the graphs which contain a cycle, Drexler and Arató [4] propose a graph decomposition method that promises a loop-free graph as the result and the proposed algorithm can be used to find the optimal graph decomposition solution. The loop-free graph can be further divided into sequential tasks as described by Saifullah et al. [3].

Decomposition of DAG has been a matter of research for a long time. In another study, Drexler and Arató [5] proved mathematically an inertial method for a loop-free decomposition of DAG. It assigns a one dimensional coordinate to each node which is considered as the depth of the node following which the partition of the nodes is done on the basis of the value of depth. The drawback of this modified method is the difficulty in assigning a coordinate to a non-numerical problem.

Zeckzer [6] proposes a mathematical approach to decompose a DAG in such a manner that it covers all the nodes with minimum time complexity of $O(n^2)$, where n

is the number of nodes in a DAG. Chen and Chen [7] specifies a recursive algorithm to decompose a DAG into its families.

Further progress has been made as decomposing a DAG in few cases eliminates the properties of a DAG. Qamhieh et al. [8] discussed a DAG stretching algorithm where the tasks in the DAG are converted into sequential independent tasks with the intermediate deadlines. The method implemented in that literature stresses on the completion of the subtasks within the prescribed minimum execution time and provides a value for the resource augmentation bound of stretched subtasks. It introduces a new algorithm where a particular graph is decomposed into a directed acyclic subgraph. The paper mainly focuses on the topological decomposition of the graph.

Naghibzadeh [9] further advanced the research by proposing a new method to effectively execute a task. It has proposed a task model that takes both the interaction and precedence of tasks into account. It can communicate data not only at the end of the execution but also during the execution. It also checks whether the proposed task model produces a cyclic or acyclic graph and transforms the graph into an acyclic graph if it is not.

The basic concepts of decomposition and task graphs have been explained by Grama et al. [10]. The various decomposition techniques have been discussed with the simplest examples and explain granularity and task interaction graphs. It discusses the various decomposition methods along with real-life problems. It also discusses the mapping of tasks into multiple processors and scheduling them.

Scheduling a task is as important as decomposing it into its subsequent subtasks. Many researches have been done on scheduling the tasks efficiently considering the task size, processor capacity, and dependencies. Kaur and Nagpal [11] optimized the traditional algorithm to balance load at cloud for scheduling of directed acyclic graphs. The authors focused on job placement as well as reducing the delay. An algorithm has been designed to select jobs to reduce energy and cost consumption. Tariq et al. [12] also propose a new heuristic for task scheduling based on the order of tasks, cost of communication, and availability of resources. The main objective of the heuristic is to schedule the tasks effectively on multiple processors rather than decomposing the graph keeping the dependencies between subtasks in consideration. Several literature of the same category have also been focused since past when Kwok and Ahmad [13] discussed 27 static scheduling algorithms and the various problems faced during scheduling a DAG. It also explained the trends and optimizations made in this area. Ilavarasan et al. [14] proposed another task scheduling algorithm for heterogeneous computing systems and measures the performance by comparing the speedup, schedule length, and other time parameters with the existing algorithms.

Recently Taufer and Rosenberg [15] proposed the scheduler for DAG structured workflows on a cloud instance. It presents a static scheduler which incurs a lower cost and has more effective performance than a dynamic scheduler. But there are very few literature which focus exclusively on the decomposition of DAG satisfying proper dependencies among subtasks.

Quite earlier, one such literature had been introduced by Xie and Geng [16] on recursive decomposition method of a directed acyclic graph mathematically. They have discussed decomposition algorithms with proper theory, displayed experimental results, and finally compared the algorithm with other conventional algorithms. The drawback of the algorithm was that as the number of vertices increases in the graph, finding the separator variables becomes more complex.

After going through the existing research works, it is clear that as problem complexity increases its cost of solving the problem in terms of execution time and energy consumption also increases.

Therefore the main objective is to minimize the cost of solving. This is where decomposition plays an important role. The number of subproblems into which a problem be decomposed is also crucial. For this purpose recursive decomposition plays a suitable role.

3 Directed Acyclic Graph

A graph is a data structure which is composed of vertices and edges where the edges connect the different vertices. Graphs can be directed and undirected as well as cyclic and acyclic. Graphs are used for solving many real-life problems. They are mainly used to represent networks, relations and finding paths, etc. [17]. Directed Graphs are graphs whose edges are directed. The directed edges describe which vertex is related to which vertex and in what direction. For example, if a vertex x has a directed edge towards vertex y, it shows that y has some kind of dependency on x. A Directed Acyclic Graph is a graph that has a definite number of vertices and directed edges and contains no cycles which means the graph cannot be traversed in a loop.

DAG is used in real-life problems. One of the most ancient implementations is the family tree. Family trees can be made with the help of DAG, where the child node does not loop back to the grandparent or its predecessor node. Another important and common field it is used is task decomposition and scheduling. In multiprocessor systems, the tasks are represented in the form of DAG and then the DAG is scheduled to different processors. Other than these DAG can also be used for Data processing networks, Data compression, and GPS (Fig. 1).

Every Directed Acyclic Graph has a topological sorting. A topological ordering is an ordering of the vertices such that the starting vertex is positioned at the earliest in

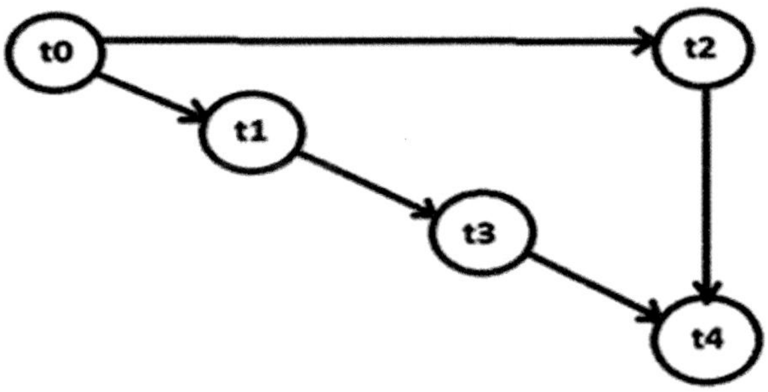

Fig. 1 A task t represented into subtasks in a DAG

the graph and the ending vertex at the end of the graph. This gives a perfect sequence which increases the readability of the graph. The problem of finding the topological ordering of a graph is known as the topological sorting. A topologically ordered graph does not have any cycles. Therefore, an equivalent definition of a DAG can be given as, "A graph which consists of a topological ordering". Each Directed Acyclic Graph consists of at least one topological order.

For a high-level synthesis problem, decomposition of tasks into various similar subtasks has been an important step. It helps to achieve parallelism which reduces the overall execution time of the task. If a graph is decomposed in a way that the nodes representing tasks and the edges indicate that the output of one node will be considered as the input of the successive node, it is known as the task dependency graph.

There are commonly four techniques to decompose a graph: Recursive Decomposition, Exploratory Decomposition, Data Decomposition, and Speculative Decomposition. Recursive decomposition is commonly implemented on problems that can be solved using divide-and-conquer strategy. The task is divided into subtasks and all the subtasks are computed recursively until the desired granularity is achieved. In data decomposition, the data on which computations are done is first identified and the data is partitioned across various tasks. A partition of the output decomposes the graph naturally. Exploratory decomposition involves the problem which consists of a search of state space of solutions. Speculative decomposition is used when a graph may take any one of the possible branches depending upon the outcome of the computation that preceded it.

4 Recursive Decomposition

It is implemented on the problems which can be computed using divide-and-conquer strategy. Quicksort is a problem of divide-and-conquer strategy. Initially, a pivot y is selected and the number sequence P is partitioned into two subsequences $P0$ and $P1$. $P0$ contains the numbers which are smaller than the pivot element y and $P1$ contains the elements which are greater than the pivot element y. This is the divide step. Now, each of the subsequence is divided into further subsequences by calling quicksort. The algorithm is run until each of the subsequence contains only one element (Fig. 2).

At the start, we have a single sequence and by using the divide step we partition the single subsequence into two subsequences. These two subsequences are further partitioned and this step is recursively and concurrently executed. As we move down the graph, the concurrency increases.

In the above diagram, we have chosen 4 as the pivot element and have divided the sequence P into two subsequences $P0$ and $P1$ where $P0$ contains the elements less than 4 and $P1$ contains the elements greater than 4 in the order, they are present in P. The subsequence $P0$ is further decomposed into two subsequences $P00$ and $P01$ where $P00$ contains two elements and P01 contains one. Since $P01$ has only one element, it is not further decomposed but for $P00$ it is decomposed further until the

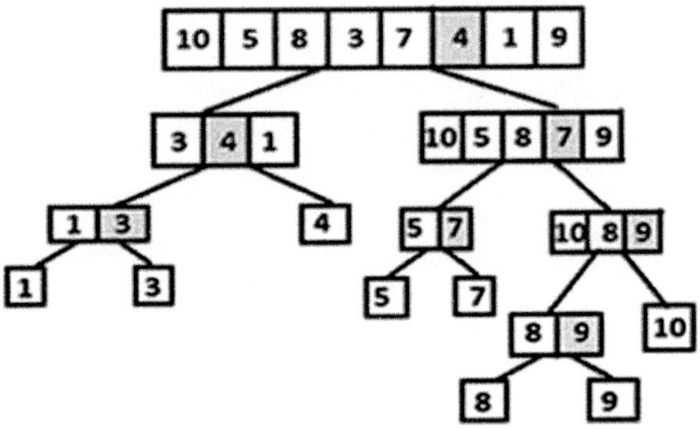

Fig. 2 Recursive decomposition on quicksort algorithm

resulting sequences only have one element. The same procedure is repeated for $P1$ subsequence until all the subsequences have only one element and cannot be further decomposed.

5 Methodology

Decomposition of a DAG is followed by the topological sorting of the DAG. It is a linear ordering of the tasks in such a way that for every directed edge AB from vertex *A* to vertex *B*, vertex *A* should come before vertex *B* in the graph where the vertex represents the tasks and edges represent the dependencies. For each directed edge from vertex *A* to vertex *B*, vertex *A* is dependent on vertex *B*. Using topological sorting, one can find the sequence of execution of the tasks.

We have first selected the root node of the graph as visited and stored the result in a stack. As the graph is traversed along the adjoining vertices of the graph, the visited nodes are marked as visited and stored in the stack. If a node is marked unvisited then the function is recursively called and it is executed until all the nodes are marked as visited. The stack provides us the linear execution of the nodes according to their dependencies represented in DAG. Using the stack, the DAG is decomposed. If there is no incoming directed edge towards a node then it is represented as a separate DAG. The direction of the edge determines the decomposition of the graph along with the node.

Algorithm 1: Depth-First-Search Based Topological Sorting

Input: *Directed Acyclic Graph*
Stack: A empty list that contains the sorted nodes
Output: *Generated subgraphs*

Begin:

```
Step1: function TopoUtil(G.a,visited,stack)
mark node a with a permanent mark
        for each  node b connected to node a do
                if temporary mark on b then
                Recursively call TopoUtil(G.a,visited,stack)
        add node to the stack

Step2: function TopoSort(Graph G)
        mark all nodes with temporary mark
        create empty stack
                for each node i in Graph G do
                        if i is temporary marked then
        TopoUtil(G.i,visited,stack)
```

End

We consider a graph *G* consisting of different nodes connected with directed edges. This algorithm ensures that whenever a node *A* is added to the result list, all the dependents on node *A* have already been added to the output list and are already visited. This Depth First Search algorithm has been devised by Cormen et al. Since all the nodes are visited once, this algorithm has linear time complexity. It is possible that there is multiple topological ordering of a given directed acyclic graph.

Graph is constructed based on the output stack produced by topological sorting of the given acyclic graph. Recursive decomposition is implemented on the output graph and is decomposed based on their dependencies. The graphs are constructed based on the number of incoming and outgoing edges on the nodes and the nodes which have no incoming edges or simply stated the independent tasks are represented in a separated acyclic graph. The application of recursive decomposition produces one or more than one separate graphs. In this technique, a graph is first decomposed in a set of nodes on some criteria and the same criteria are used to break down the decomposed graph into another subgraph. The subgraphs are decomposed repeatedly until all the nodes cannot be broken down anymore and the dependency is integrated between the nodes. The principal focus in this work is that the dependency constraints between the tasks are not disturbed and is maintained throughout the decomposition of the graph. These different graphs can be distributed to different processors which can then be parallelly executed thereby increasing the throughput and decreasing the overall execution time.

6 Result and Discussion

We take an arbitrary graph during the runtime along with the dependency of the tasks between each other. The system generates a random given the number of nodes and dependency relationship (Fig. 3).

In the above graph, a directed edge from vertex 0 to vertex 1 shows that vertex 0 is dependent on vertex 1. The given directed acyclic graph is chosen for verifying the working of the decomposition technique. The output produced after implementing recursive decomposition is shown in Fig. 4.

Since vertex 2 has no incoming edge, it can be represented in a separate graph and scheduled to a different processor. The recursive algorithm was applied to 17 different DAGs with a varying number of nodes and dependencies. Table 1 describes the number of subgraphs produced after implementing recursive decomposition on different graphs.

Below are the different subgraphs which have been shown above in the tabulated form. The graphs on the left have been developed randomly and the graphs on the right are the decomposed subgraphs of the original graph after undergoing recursive decomposition. It is comprehensible from the representations that a directed acyclic graph may have more than one subgraph on the implementation of this algorithm. The figures also make it clear that if a task is independent, it can be represented as a detached node without any connected edges.

As described in Table 1, we have taken 10 nodes in Graph 1 which has divided the graph into two subgraphs of nine and one number of nodes, respectively. Similarly, in

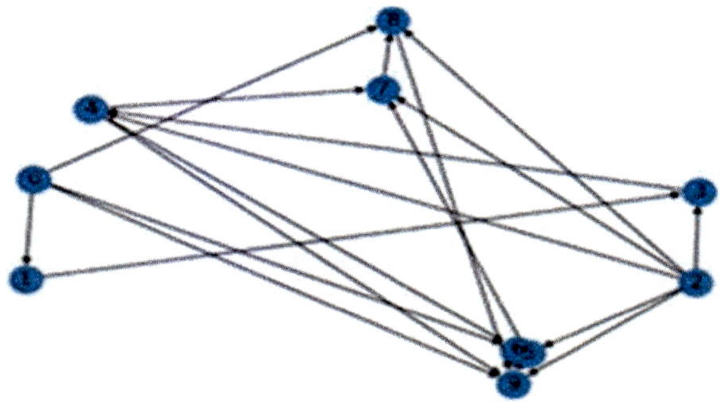

Fig. 3 The graph taken as input and dependencies are shown by the edges

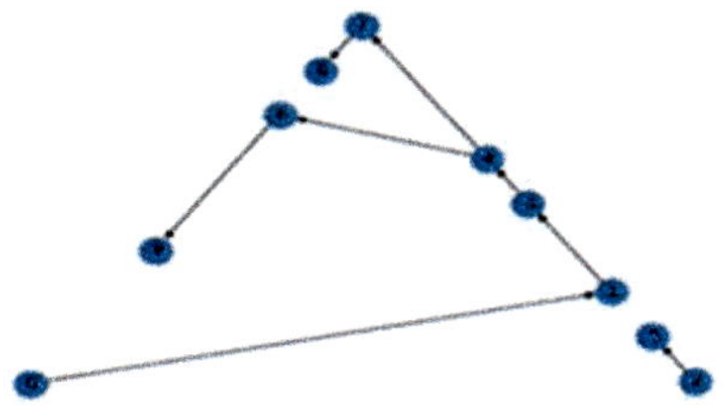

Fig. 4 Directed acyclic graph after recursive decomposition

Table 1 Number of subgraphs produced with various numbers of nodes

Graph No.	No. of nodes	No. of sub-graphs with no. of nodes
Graph 1	10	Subgraph 1: 9 Subgraph 2: 1
Graph 2	10	Subgraph 1: 10
Graph 3	10	Subgraph 1: 6 Subgraph 2: 3 Subgraph 3: 1
Graph 4	10	Subgraph 1: 9 Subgraph 2: 1
Graph 5	10	Subgraph 1: 8 Subgraph 2: 1 Subgraph 3: 1
Graph 6	10	Subgraph 1: 9 Subgraph 2: 1
Graph 7	15	Subgraph 1: 15
Graph 8	15	Subgraph1: 12 Subgraph 2: 1 Subgraph 3: 1 Subgraph 4: 1
Graph 9	15	Subgraph 1: 14 Subgraph 2: 1
Graph 10	15	Subgraph 1: 9 Subgraph 2: 1
Graph 11	15	Subgraph 1: 11 Subgraph 2: 4
Graph 12	15	Subgraph 1: 10 Subgraph 2: 2 Subgraph 3: 1 Subgraph 4: 1 Subgraph 5: 1
Graph 13	15	Subgraph 1: 7 Subgraph 2: 8
Graph 14	15	Subgraph 1: 9 Subgraph 2: 3 Subgraph 3: 1 Subgraph 4: 2
Graph 15	15	Subgraph 1: 8 Subgraph 2: 6 Subgraph 3: 1
Graph 16	15	Subgraph 1: 11 Subgraph 2: 4
Graph 17	20	Subgraph 1: 14 Subgraph 2: 4 Subgraph 3: 2

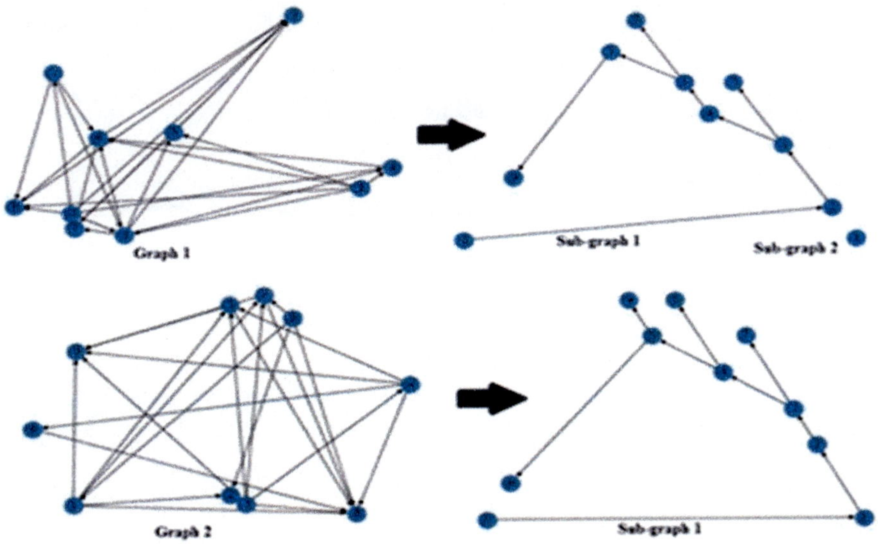

Fig. 5 Decomposition of graph

Graph 2, a single graph has been taken and it has transformed into a single subgraph of ten number of nodes after the execution of the recursive decomposition algorithm with a sequential order of tasks taking place. The same algorithm has been applied on Graph 3, 4, 5, and 6 with the graph consisting of ten nodes. Graph 3 has been decomposed into two subgraphs of six, three, and one node, respectively. Graph 4 into two subgraphs of nine and one node, Graph 5 into one subgraph of eight and two subgraphs of a single node without any connections with other nodes, Graph 6 into two subgraphs of nine and one node, respectively (Figs. 5, 6, 7, 8, 9, 10 and 11).

The work has been further extended to a graph containing 15 nodes as shown in Graph 7 which has been transformed into a sequential order of tasks. Similarly, in Graph 8 after the execution of the algorithm, it has decomposed into four subgraphs one of twelve nodes and three other subgraphs of one node. Continuing the execution of algorithm of Graph 9, 10, 11, 12,13, 14, 15, 16 we get two, two, two, five, two, four, three, two subgraphs respectively. Carrying out the same algorithm on Graph 17 with 20 nodes has broken down the graph into three subgraphs each of fourteen, four, and two nodes, respectively.

7 Conclusion

In this paper, we have implemented recursive decomposition on a given acyclic graph. The graph is first topologically ordered and a sequence is found for the execution of

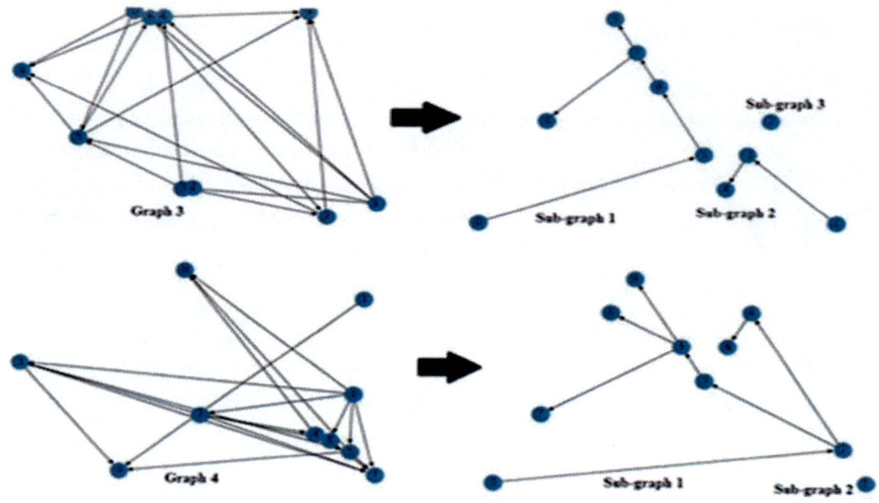

Fig. 6 Decomposition of graph

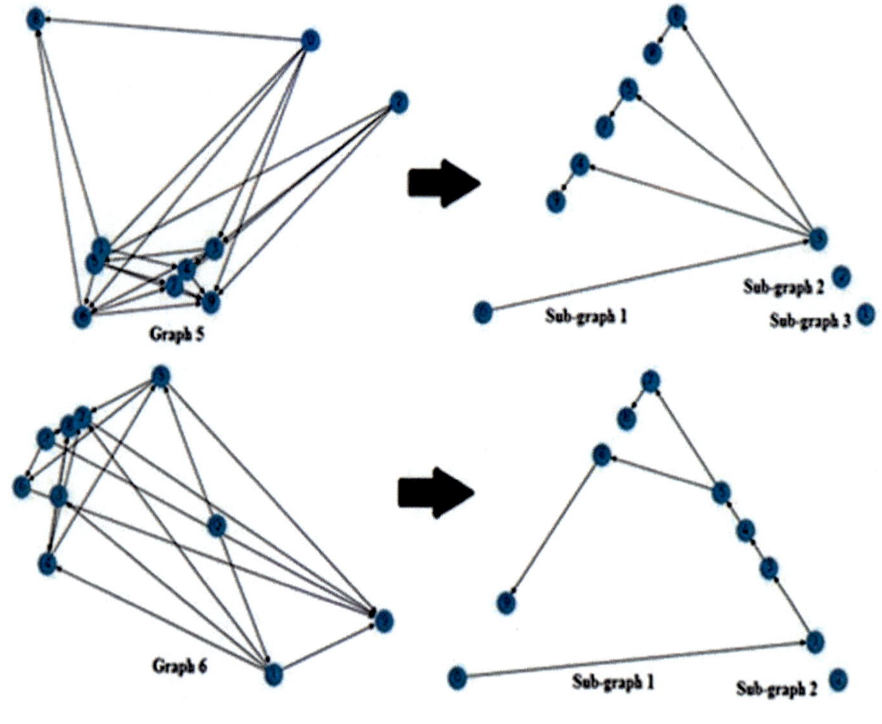

Fig. 7 Decomposition of graph

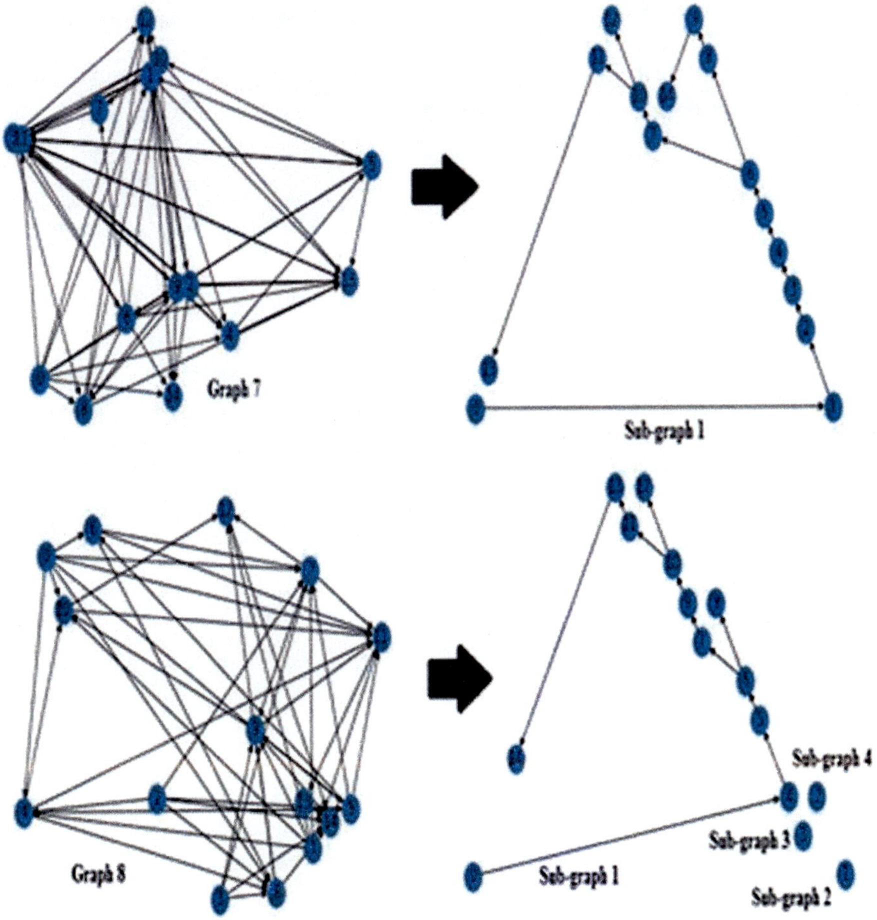

Fig. 8 Decomposition of Graph

tasks based on the dependencies between them. We have implemented the topological sorting based on the depth-first search which gives a linear runtime complexity. Further, the graph is decomposed and the result which is obtained may have one or more than one decomposed graph. The graphs can be supplied to different processors for execution. The work can be further extended by implementing different decomposition techniques on the same graph and finding out the optimal decomposition technique which can be used for parallel processing.

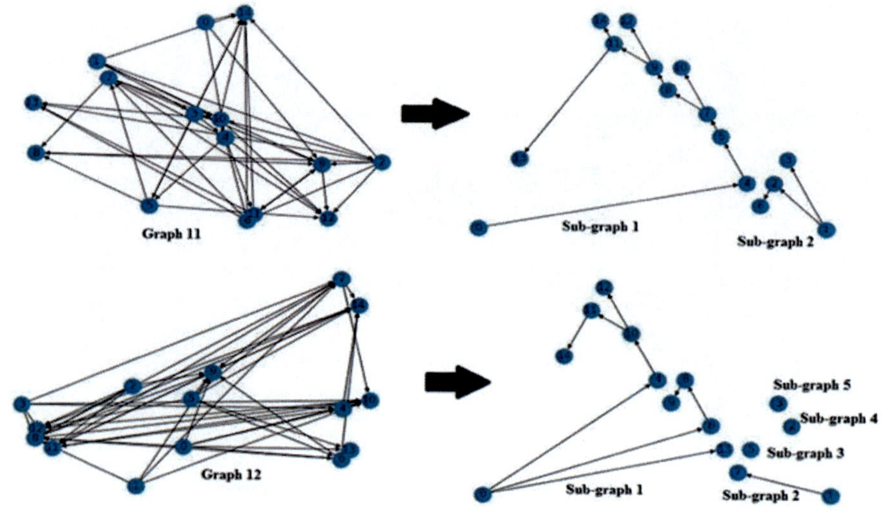

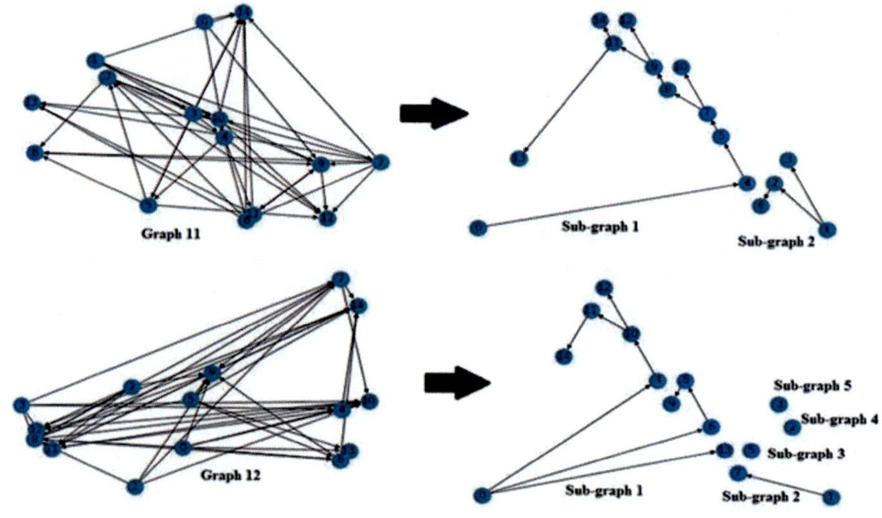

Fig. 9 Decomposition of graph

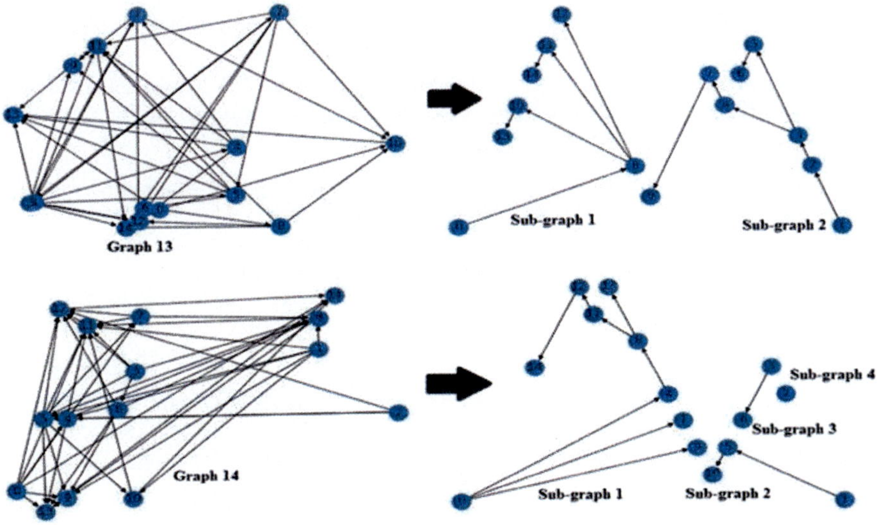

Fig. 10 Decomposition of graph

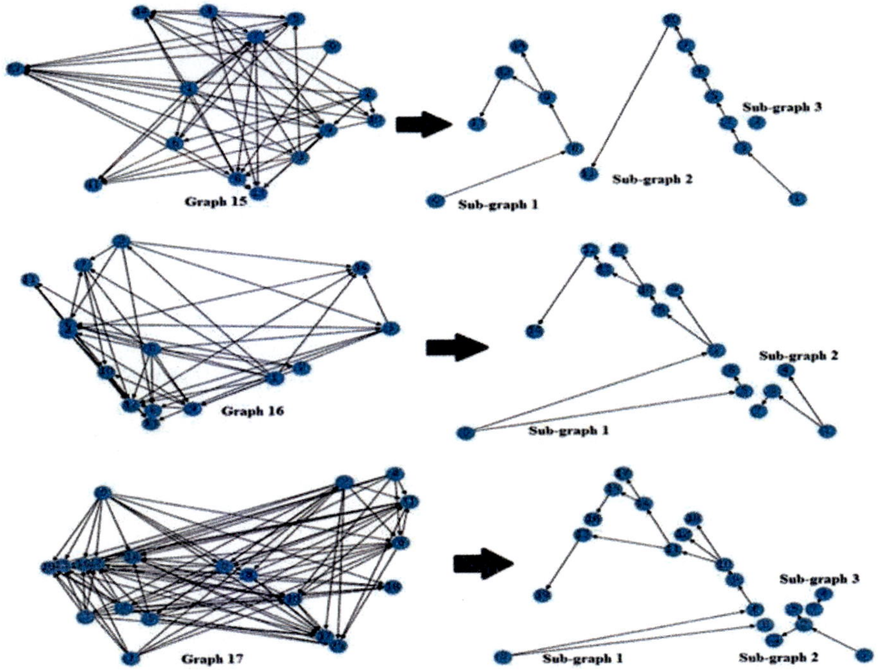

Fig. 11 Decomposition of graph

References

1. Özkaya, M.Y., Benoit, A., Uçar, B., Herrmann, J., Çatalyürek, Ü.V.: A scalable clustering-based task scheduler for homogeneous processors using DAG partitioning. In: 2019 IEEE International Parallel and Distributed Processing Symposium (IPDPS), pp. 155–165. IEEE (2019)
2. Sanders, P., Schulz, C.: Engineering multilevel graph partitioning algorithms. In: European Symposium on Algorithms, pp. 469–480. Springer, Berlin (2011)
3. Saifullah, A., Ferry, D., Li, J., Agrawal, K., Lu, C., Gill, C.D.: Parallel real-time scheduling of DAGs. IEEE Trans Parallel Distrib Syst **25**(12), 3242–3252 (2014)
4. Drexler, D.A., Arató, P.: Loop-free decomposition of directed acyclic graphs. In: WAIT: Workshop on the Advances in Information Technology, Budapest (2015)
5. Drexler, D.A., Arató, P.: A modified inertial method for loop-free decomposition of acyclic directed graphs. MACRo 2015, **1**(1), 61–72 (2015)
6. Zeckzer, A.A.D.: Topological decomposition of directed graphs. J. Graph Algorithms Appl. **21**(4), 589–630 (2017)
7. Chen, Y., Chen, Y.: On the graph decomposition. In: 2014 IEEE Fourth International Conference on Big Data and Cloud Computing, pp. 777–784. IEEE (2014)
8. Qamhieh, M., George, L., Midonnet, S.: A stretching algorithm for parallel real-time dag tasks on multiprocessor systems. In: Proceedings of the 22Nd International Conference on Real-Time Networks and Systems, p. 13. ACM (2014)
9. Naghibzadeh, M.: Modeling workflow of tasks and task interaction graphs to schedule on the cloud. Cloud Comput. **2016**, 81 (2016)
10. Grama, A., Gupta, A., Karypis, G., Kumar, V.: Principles of parallel algorithm design. In: Introduction to Parallel Computing, 2nd ed. Addison Wesley, Harlow (2003)
11. Kaur, S., Nagpal, P.: Efficient directed acyclic graph scheduling in order to balance load at cloud. Int. J. Emerg. Trends Technol. Comput. Sci. **6**(5) (2017)
12. Tariq, R., Aadil, F., Malik, M.F., Ejaz, S., Khan, M.U., Khan, M.F.: Directed acyclic graph based task scheduling algorithm for heterogeneous systems. In: Proceedings of SAI Intelligent Systems Conference, pp. 936–947. Springer, Cham (2018)
13. Kwok, Y.K., Ahmad, I.: Static scheduling algorithms for allocating directed task graphs to multiprocessors. ACM Comput. Surv. (CSUR) **31**(4), 406–471 (1999)
14. Ilavarasan, E., Thambidurai, P., Mahilmannan, R.: Performance effective task scheduling algorithm for heterogeneous computing system. In: The 4th international Symposium on Parallel and Distributed Computing (ISPDC'05), pp. 28–38. IEEE (2005)
15. Taufer, M., Rosenberg, A.L.: Scheduling DAG-based workflows on single cloud instances: High-performance and cost effectiveness with a static scheduler. Int. J. High Perform. Comput. Appl. **31**(1), 19–31 (2017)
16. Xie, X., Geng, Z.: A recursive method for structural learning of directed acyclic graphs. J. Mach. Learn. Res. **9**, 459–483 (2008)
17. Gajbhiye, A., Singh, S.: Resource provisioning and scheduling algorithm for meeting cost and deadline-constraints of scientific workflows in IaaS clouds (2018). arXiv preprint arXiv:1806.02397

An Improved On-Policy Reinforcement Learning Algorithm

Moirangthem Tiken Singh, Aninda Chakrabarty, Bhargab Sarma, and Sourav Dutta

Abstract The paper aims to find the paths for a mobile agent over a stochastic environment. The stochastic environment is chosen to mimic the real world as the agent's actions do not uniquely determine the outcome. We placed the agent in the initial state and allowed to traverse each state, which has different rewards function with the goal to maximize its reward. Various algorithms, viz. SARSA and Q-Learning, etc., are used by many scholars to evaluate the path for the agent. Here, a reinforced learning algorithm is proposed in collaborating with the idea of information theory. The motive was to make the algorithm explore and learn. After that, it decides the best possible action to take based on policy that it learned. The learning rate of the algorithm is kept varying to check the performance of the proposed algorithm.

Keywords Reinforcement learning · Markov decision process · Stochastic environment · Mobile agent

1 Introduction

Reinforcement learning (RL) [1, 2] is concerned with a software agent ought to take actions in an environment to maximize the notion of cumulative reward. The agent is the conceptual entity that interacts with an environment. The agent acts with its actuators, which makes an impact on the environment. It takes percepts

---

M. T. Singh · A. Chakrabarty · B. Sarma (✉) · S. Dutta
Department of Computer Science and Engineering, DUIET,
Dibrugarh University, Dibrugarh, Assam 786004, India
e-mail: sarmabhargab202@gmail.com

M. T. Singh
e-mail: tiken.m@dibru.ac.in

A. Chakrabarty
e-mail: anindachakrabarty4@gmail.com

S. Dutta
e-mail: d.sourav714@gmail.com

S. Borah et al. (eds.), *Soft Computing Techniques and Applications*,
Advances in Intelligent Systems and Computing 1248,
https://doi.org/10.1007/978-981-15-7394-1_30

from the environment using its sensors. In this study, we used the agent as a utility-based learning agent. An environment is a place where the agent acts to complete a particular task. Particularly in this study, the environment is a stochastic one, which is a collection of many states like a Markov decision process (MDP), where reward (utility) is a local heuristic value that an agent gets after transitioning from an old state to a new state. This utility indicates how good an agent's action is taken from the particular state .

The performance of an agent using reinforcement learning depends on balancing exploration and exploitation of the environment. A greedy algorithm [3] like an optimal ϵ adaptive greedy algorithm [4] selects the actions which have the highest estimated rewards with a probability. However, the ϵ greedy action selection does not give any preference for the nearly ravenous or uncertain state. So, specific situations like the environment proposed here, it is sometimes efficient to select a non-greedy action-state that has the potential of providing maximal rewards.

This work defines an algorithm for a mobile agent in a stochastic environment. The proposed algorithm uses the advantages of information theory to select a suitable action–state pair other than the greedy approach that helps the agent to reach its final destination with maximum utility. To simulate the behavior of the algorithm, we build a test environment and use this environment to improve the decision-making abilities of the agent to reach its objective. It is argued that for solving such problems, we could have used uninformed searching strategies [5] from the domain of artificial intelligence; however, when the environment is not deterministic, those strategies failed [6].

The rest of the paper proceeds as follows. Section 2 describes background and related work. In Sect. 3.1, we formulate the environment which is used for our algorithm. Section 3.2 describes the method used by the agent to explore the environment. In Sect. 4, the numerical simulation of the model is presented, and finally Sect. 5 concludes.

2 Literature Survey

The reinforcement learning algorithm is divided into model-free and model-based. The model-free includes the popular off-policy like Q-learning algorithm [7]. The model-based algorithm includes on-policy based like state–action–reward–state–action (SARSA) [8], which is based on temporal difference learning [8]. All the algorithms of RL have to maintain a balance between exploration and exploitation during the learning process. There are many approaches to balance the two concepts. Many researchers have used policy based on the greedy approach [9], and some used strategy based on non-greedy methods like softmax [10], Upper-Confident Bound (UCB) [11], etc.

In greedy exploration [12], the agent exploits the current information that maximized the immediate rewards. It never evaluates inferior actions as it will provide less compensation. So, the agent is locked in a sub-optimal action every time it explores the environment. An alternative approach is to include some actions that are picked up randomly independent of the action–value estimate in addition to the greedy approach. This is how epsilon-greedy algorithm [9] work. However, such methods have disadvantages of overlooking some actions that may be better. Such greedy strategy explores randomly and is inefficient because there is a high probability that the actions with lesser reward may be selected. So the exploration should be done with the steps for which the reward values are uncertain or have a potential of giving higher reward values.

The softmax algorithm [13] balances the exploration and exploitation of the epsilon-greedy strategy. But when it explores, it chooses equally among all actions. So, it is similarly likely to select the worst as well as the next-to-best action.

UCB [14] explores in such a way that it reduces the uncertainty over time. The average value of the rewards for UCB is better than other policy selection algorithm. UCB action selection is one approach to measure this potential by calculating an upper bound value. It is not dependent only on the state value but also dependent on the calculated upper bound value. This strategy prioritizes or gives more importance to the actions which are less explored and more optimistic in producing reward values. However, UCB leads to over-exploration [15], and therefore, applies only to multi-arm bandit problems.

The on-policy-based SARSA algorithm is an improvement from the off-policy-based Q-learning algorithm. The original SARSA algorithm is a slow learning algorithm due to its over-exploration. If the environment has less number of states, then it takes more time to converge. If the environment contains more states, the algorithm converges but with an average reward which is not maxima. In our work, we design our own environment and propose a model-free reinforcement learning algorithm which is comparatively better than the SARSA learning algorithms in that environment.

3 Methodology

3.1 The Environment and Agent

We consider a mobile agent like a car that is moving around in an environment. The mobile agent needs to reach the final destination without crashing and as safely and as fast as possible from its initial position. We assume that the environment is fully observable, stochastic, sequential, static, discrete and has only one single agent. We assume that the environment has only one final state without an object to hurdle the movement of the mobile agent. We also assume that the mobile agent knows its location within the environment. In the sequential environment, when the agent takes action, it will affect future actions. We assume that the agent will go through the safest path(without crashing). Any deviation from that path may make the mobile agent obtain fewer rewards. We represent the environment as Markov decision process.

Definition 1 A Markov decision process is a tuple $\langle S, A, P, R\rangle$ consisting of set of states S, with as set of finite action A. Where $P_a(s, \bar{s})$ is the probability of changing from the state $s \in S$ to another state $\bar{s} \in S$ when an agent took an action A giving a reward $r \in R$.

3.2 *Proposed Model-Free Algorithm*

Given the MDP, we allowed the agent to explore the environment based on the entropy of the states. From [16], we derive the information content of the state. Accordingly, we have the average information of all the states

$$H(S) = E[I(s_i)] = \sum_{s=0}^{n-1} P(s_i) * I(s_i) = \sum_{s=0}^{n-1} P(s_i) * \log_2\left(\frac{1}{p(s_i)}\right) \quad (1)$$

From the Eq. (1), the average information content is calculated for a state. In our case, we only need the information content of a single state–action pair. Therefore, we define the information for a single state x using Eq. (1)

$$I(x) = \log_2\left(\frac{1}{p(x)}\right) \quad (2)$$

According to Eq. (3), we conclude that the more are uncertain about the state $S = x$, more is the information content of the state. It is a trivial fact that since the outcome of the state is known with less certainty, we gain more information from finding the consequences of that state. So, the more the information content of a state–action pair, the more is the chances of exploiting. For our model, instead of finding the information of the independent states (i.e., $I(s)$), we check the information for state–action pair (i.e., $I(Q(S, A))$). Accordingly, we specify the algorithm of the reinforcement learning as shown below:
Algorithm (2) (Modified ϵ—Greedy Algorithm) uses the concepts of information theory for improving performance. It works by exploring at the beginning but increases the exploitation drastically as the time proceeds. In the algorithm, instead of selecting the next state based on the Q value of the state, we choose the future state based on the heuristic value (Optimistic) given by equation (3).

$$H(S') = Q(S', A') + \beta I(X = (S', A')) \quad (3)$$

We select the state transition from S' where H is maximum and $A' \in S(A)$ where the $S(A)$ is the set of actions accessible from S'. The effect of the algorithm is not only to explore using ϵ strategy but also to include exploitation, which results in the

Algorithm 1 Our RL Algorithm (First Part)

Initialize the Hyper-Parameters: α, ϵ_{min}, ϵ_{max}, ϵ_{decay}, γ
Initialize the Q-Table such that $Q(s, a) = 0\ \forall s \in S$ and $a \in A(s)$
while All the Episodes are not executed **do**
  Initialize the Initial State S
  Initialize $\epsilon = \epsilon_{\text{min}} + (\epsilon_{\text{max}} - \epsilon_{\text{min}} * e^{-\epsilon_{\text{decay}} * Current\_Episode})$
  Choose Action A from S using our modified ϵ—Greedy Strategy
  while The Episode does not end **do**
    Take the Action A from S
    Observe the new State S' and Reward R
    Choose Action A' from S' using our modified ϵ—Greedy Strategy
    $Q(S, A) \leftarrow (1 - \alpha)Q(S, A) + \alpha[R + \gamma Q(S', A')]$
    $S \leftarrow S'$
    $A \leftarrow A'$
  end while
end while

Algorithm 2 Policy Updation Algorithm

Let v be some value sampled from the Random variable X from Unform Distribution Curve
if $v < \epsilon$ **then**
  Randomly Select Action A from State S
else
  Find the Heuristic value H(S,A) = Q(S,A) + $\beta I(X = (S, A))$ for the all the State—Action Pair
  Select the Action A' for which the Heuristic Value is Maximum
end if

exploration of each step of learning. Exploration by the proposed algorithm seems like an exploration done by an educated guess. The educated guess is because of the presence of information content.

4 Numerical Simulations

For the numerical simulation, we used an environment of the form as shown in Fig. 1.

Figure 1 is a simple demonstration of the environment with three states. The mobile agent starts from the start state to reach the goal state. We also include the crash state, where the agent entered the state if it crashed. Figure 1 also is a demonstration of Markov decision process (MDP) where there are actions and states, and each state is linked to one another through stochastic transitions and based on the type of transition with each reward for the transition. The transition table for Fig. 1 is represented in Table 1.

By using the concepts shown in Fig. 1, we represent the environment in a grid form as shown in Fig. 2.

Here, the environment is a collection (Fig. 2) of states in the form of a grid where each grid cell is a position and can be represented in the way of coordinate values

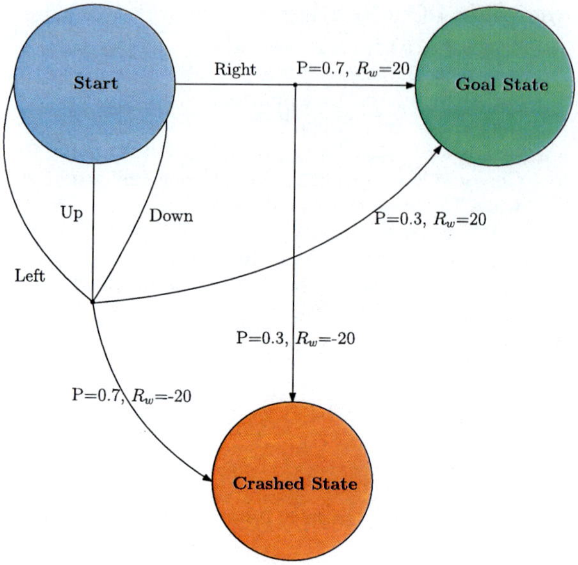

Fig. 1 Stochastic environment

Table 1 State–action transitions with rewards and probability

S_t	A_t	S_{t+1}	R_{t+1}	P
Start	Left	Crashed state	−20	0.7
Start	Left	Goal state	+20	0.3
Start	Right	Crashed state	−20	0.3
Start	Right	Goal state	+20	0.7
Start	Up	Crashed state	−20	0.7
Start	Up	Goal state	+20	0.3
Start	Down	Crashed state	−20	0.7
Start	Down	Goal state	+20	0.3
Goal state	Φ	Φ	Φ	Φ
Crashed state	Φ	Φ	Φ	Φ

(2D Representation) or can be numbered serially from 0 to $N - 1$ where N is the total number of states in the grid. From Fig. 2, the environment has dimensions 4×4 (actually 6×6 for the boundary states), so there are a total of 36 states, including the crash states. The environment implementation has a feature where we can change the dimensions of the state depending on the current needs. Most importantly, from a particular state, we have the option to transition only to the neighboring states, and the agent cannot jump to the non-neighboring states. Apart from the states discussed in the previous section (start, goal and crashed state), the environment also contains intermediate states which lie between the start and the goal state.

Fig. 2 Our grid environment

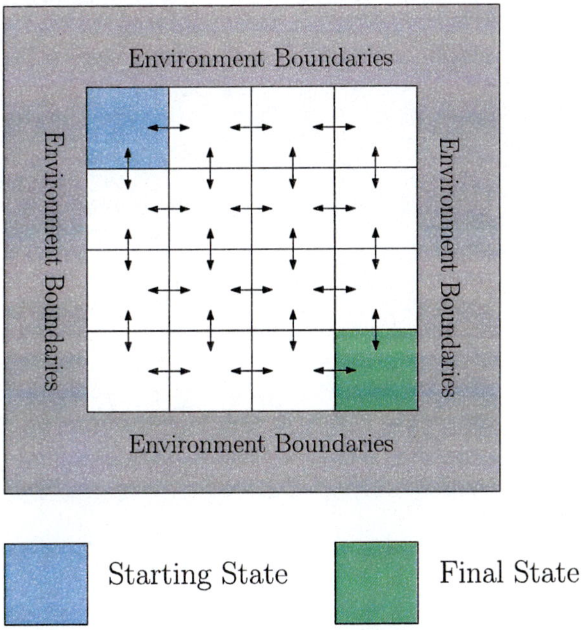

Considering the MDP mentioned above, we compare the performance of our algorithm with the SARSA and Q-learning algorithm. Each figure plotted here represents the average rewards over time and the average rewards over 1000 stages (episodes). We tested our algorithm starting from 4×4 to 11×11. Each algorithm is run for 40,000 episodes.

From each Figs. 3, 4, 5, 6 and 7, we obtained that when we compare the SARSA and the Q-learning algorithm, SARSA has good average rewards. However, at the end of the episode, both algorithms have the same maxima. It is because the ϵ greedy algorithm does not have any new information to explore in the stochastic environment.

If we compare our algorithm with the other algorithms, we find our proposed reinforcement learning algorithm explores more compared to the different algorithms mentioned above. This is evident in figures plotted here. Initially, our algorithm has a depression in the curve line, which indicates that the exploration is taking place. The depression in the graph depends on the exploration required for the environment. As the dimensions of the environment are increased, the depression also increases. This is because the exploration is happening more in our algorithm compared to the other algorithms. A steep rise follows the depression in the learning of the graph. This behavior is since the occurrence of pure exploitation.

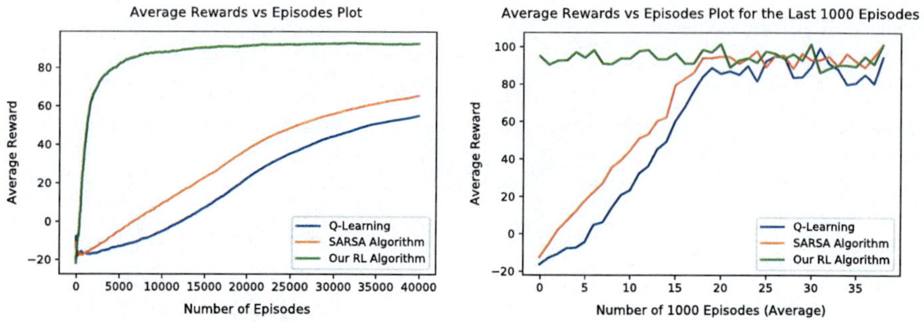

Fig. 3 Results for 4 × 4 environment

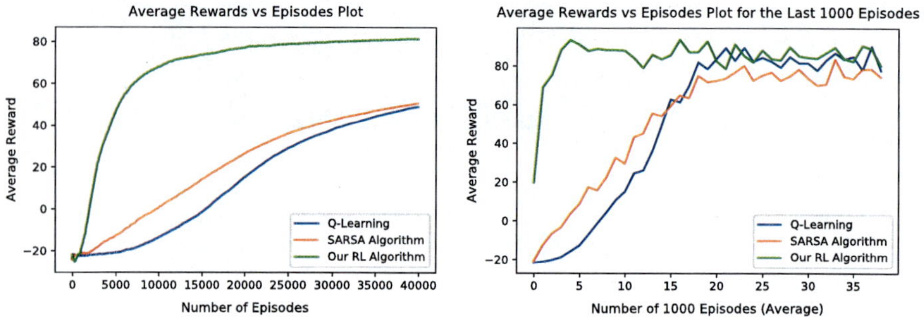

Fig. 4 Results for 6 × 6 environment

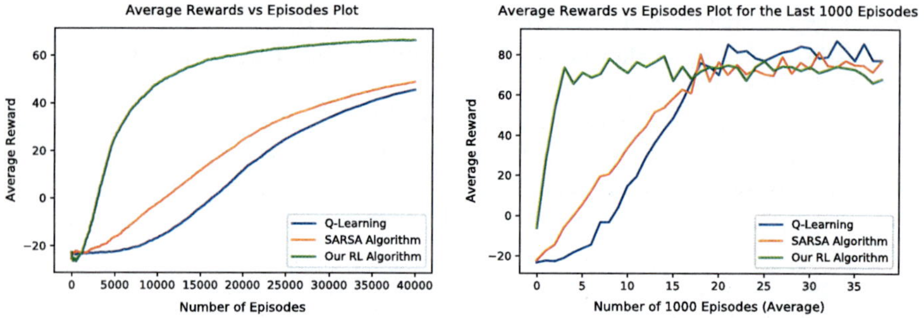

Fig. 5 Results for 7 × 7 environment

5 Conclusion and Future Work

Based on the simulations, our algorithm, though model-free, has shown promising results and is evident from the simulation. But, as the number of states increases, as our algorithm uses Q-table for maintaining records, the learning process becomes a very time-consuming process. It is quite evident that as the number of states increases, the size of the table also increases, thereby making the learning process slower since,

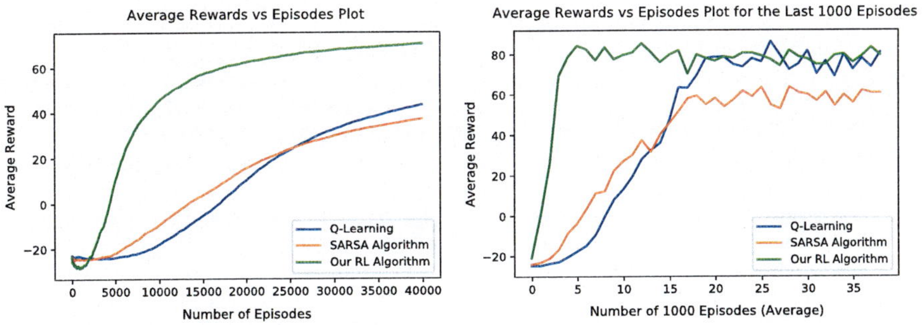

Fig. 6 Results for 8 × 8 environment

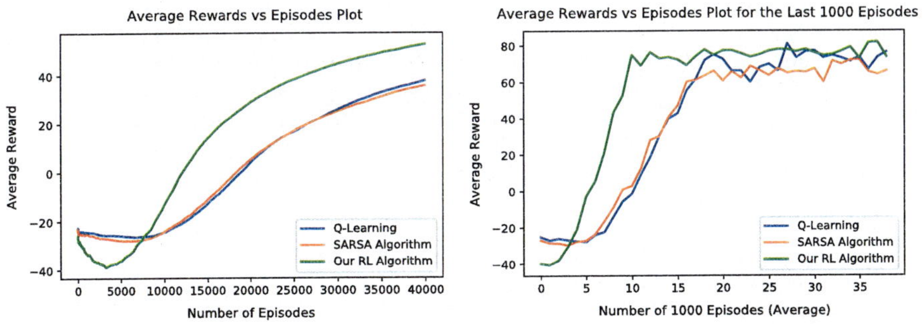

Fig. 7 Results for 11 × 11 environment

for this case, we will need more episodes. In our future work, we will be trying to fix this drawback by combining our model-free algorithm with neural networks so that the learning process can be further enhanced, and it consumes less time. If possible, we will also try to use continuous states rather than discrete states if we see significant improvements. We will also try to improve our heuristic value for our algorithms to balance the exploration and exploitation rates further.

References

1. Arulkumaran, K., Deisenroth, M.P., Brundage, M., Bharath, A.A.: Deep reinforcement learning: a brief survey. IEEE Signal Process. Mag. **34**(6), 26–38 (2017). https://doi.org/10.1109/MSP.2017.2743240. Nov
2. Kaelbling, L.P., Littman, M.L., Moore, A.W.: Reinforcement learning: a survey. CoRR **cs.AI/9605103** (1996). https://arxiv.org/abs/cs/9605103
3. Hinojosa, G., Torres, D.: Introduction to algorithms. Auris Reference (2017)
4. Tokic, M.: Adaptive ϵ-greedy exploration in reinforcement learning based on value differences. In: Dillmann, R., Beyerer, J., Hanebeck, U.D., Schultz, T. (eds.) KI 2010: Advances in Artificial Intelligence, pp. 203–210. Springer, Berlin (2010)

5. Pearl, J.: Heuristics: Intelligent Search Strategies for Computer Problem Solving. Addison-Wesley Longman Publishing Co., Inc, New York (1984)
6. Koenig, S.: Exploring unknown environments with real-time search or reinforcement learning. In: In Proceedings of the Neural Information Processing Systems, pp. 1003–1009 (1999)
7. Watkins, C.J., Dayan, P.: Q-learning. Mach. Learn. **8**(3–4), 279–292 (1992)
8. Sutton, R.S., Barto, A.G.: Reinforcement Learning: An Introduction. MIT Press, Cambrdge (2018)
9. dos Santos Mignon, A., de Azevedo da Rocha, R.L.: An adaptive implementation of epsilon-greedy in reinforcement learning. Procedia Comput. Sci. **109**, 1146–1151 (2017). https://doi.org/10.1016/j.procs.2017.05.431, http://www.sciencedirect.com/science/article/pii/S1877050917311134. 8th International Conference on Ambient Systems, Networks and Technologies, ANT-2017 and the 7th International Conference on Sustainable Energy Information Technology, SEIT: 16–19 May 2017. Madeira, Portugal (2017)
10. Syafiie, S., Tadeo, F., Martinez, E.: Softmax and epsilon-greedy policies applied to process control. IFAC Proc. Vol. **37**(12), 729 – 734 (2004). https://doi.org/10.1016/S1474-6670(17)31556-2. http://www.sciencedirect.com/science/article/pii/S1474667017315562. iFAC Workshop on Adaptation and Learning in Control and Signal Processing (ALCOSP 04) and IFAC Workshop on Periodic Control Systems (PSYCO 04), Yokohama, Japan, 30 Aug–1 Sept 2004
11. Chapelle, O., Li, L.: An empirical evaluation of thompson sampling. In: Shawe-Taylor, J., Zemel, R.S., Bartlett, P.L., Pereira, F., Weinberger, K.Q. (eds.) Advances in Neural Information Processing Systems 24, pp. 2249–2257. Curran Associates, Inc. (2011). http://papers.nips.cc/paper/4321-an-empirical-evaluation-of-thompson-sampling.pdf
12. Efroni, Y., Dalal, G., Scherrer, B., Mannor, S.: Beyond the one step greedy approach in reinforcement learning (2018)
13. Syafiie, S., Tadeo, F., Martinez, E.: Learning control application to nonlinear process control. In: Proceedings World Automation Congress, vol. 16, pp. 260–265 (2004)
14. Abbasi-yadkori, Y., Pál, D., Szepesvári, C.: Improved algorithms for linear stochastic bandits. In: Shawe-Taylor, J., Zemel, R.S., Bartlett, P.L., Pereira, F., Weinberger, K.Q. (eds.) Advances in Neural Information Processing Systems 24, pp. 2312–2320. Curran Associates, Inc. (2011). http://papers.nips.cc/paper/4417-improved-algorithms-for-linear-stochastic-bandits.pdf
15. Hao, B., Abbasi-Yadkori, Y., Wen, Z., Cheng, G.: Bootstrapping upper confidence bound (2019)
16. Shannon, C.E.: A mathematical theory of communication. Bell Syst. Tech. J. **27**(3), 379–423 (1948)

Data Augmentation and CNN-Based Approach Towards the Classification of Melanoma

Prem Dhar Dwivedi, Simran Sandhu, Md. Zeeshan, and Pratima Sarkar

Abstract Melanoma is one of the most risky skin cancers that may cause death. It can of two types Benign and malignant. Early detection of melanoma may help in the recovery of disease and reduces the pain of a patient. It is very important to detect melanoma by taking less number of biopsy samples and early detection of it. Dermoscopy lesion images are used to detect malignant melanoma and analysis of the work. In the proposed approach data augmentation techniques are used to increase number of dermoscopy lesion images so that high accuracy can be achieved. The increased dataset is used for classifying the melanoma using Convolutional Neural Network (CNN). Lesions are classified into two types "malignant melanoma" and "benign". In the result accuracy level with respect to a different number of epochs are measured. The performance of the work is evaluated using Receiver Operating Curve (ROC).

Keywords Skin cancer · Convolutional neural networks · Melanoma classification

1 Introduction

Melanoma is a skin cancer that occurs due to sunlight ultraviolet rays. This ray has more effect on the fair complex people because their skin burns faster and rarely tans. Now a day's excessive pollution is a major cause of ozone layer depletion. Melanoma started growing from the pigment of a cell and letter it spreads on whole

P. D. Dwivedi · S. Sandhu · Md. Zeeshan · P. Sarkar (✉)
Sikkim Manipal Institute of Technology, Sikkim Manipal University, Gangtok, Sikkim, India
e-mail: psmoon2@gmail.com

P. D. Dwivedi
e-mail: dwivedipremdhar@gmail.com

S. Sandhu
e-mail: simranksandhu0@gmail.com

Md. Zeeshan
e-mail: zaanshaz7@gmail.com

S. Borah et al. (eds.), *Soft Computing Techniques and Applications*, Advances in Intelligent Systems and Computing 1248,
https://doi.org/10.1007/978-981-15-7394-1_31

body parts [1]. In the last two decades effect of melanoma got increased by two times. In all types of cancer, it is very important to detect disease in an earlier stage so that recovery is possible. Even surgical treatment survival depends on which stage melanoma detected. One of the most dangerous melanoma is malignant melanoma. Without any technical support detection of melanoma achieved 60–65% accuracy by dermatologists [2].

Dermoscopy images are generated by a high-resolution camera with a high amount of magnifying capability. While capturing this type of images controlled environment required like amount lights required and so many filters are used so that reflection of lights can be reduced [3]. The combination of visual observation by experts and dermoscopy image increased detection accuracy by 75–84%.

Research for detection of melanoma before 2016 was used machine learning techniques to learn the features and consists of preprocessing, image segmentation, feature extraction, and classification. Melanoma detection accuracy is totally dependent on the quality of the selected features. Sometimes due to incorrect feature selection, we miss some of the information. Also, incorrect segmentation may cause losing of information and incorrect area detection [4].

After 2016, deep neural network is used to the classification of melanoma. One of the most popular deep neural networks is Convolutional Neural Network (CNN). It has many application areas like face recognition, object detection, self-driving cars, robot navigation, etc. [5].

CNN needs a large amount of label dataset to train the network. This network is good for locating and recognizing a particular type of cancer so segmentation is also done by CNN. CNN makes a relationship between localization and classification of melanoma detection.

CNN hidden layers are used to extract feature and fully connected layers are used to classify the different types of melanoma [5]. The main difference between a neural network and CNN is that in a neural network each of the previous layer neurons is connected with the next layer neuron but in CNN selective number of previous layer, neurons are connected with current layer neuron [6].

2 Related Work

This section describes different state-of-the-arts related to work, i.e., melanoma classification techniques.

The author pre-processed the images before training; they have resized the images into 256 × 256 and the value of each pixel subtracted from the mean of all pixel value. The main reason is to make all intensity value zero centric [1]. The processed image passed into the LightNet i.e., one of the variation for CNN. The dataset used by the work is ISBI 2016 dataset. They have analyzed accuracy of the work for both segmented and not segmented images. The accuracy level is higher in the case of segmented images, i.e. 85%.

The author used half training and half trial method for the detection of melanoma. Here small dataset is used to train a system. Data augmentation technique is used to increase data size [5]. VGG-16 is a model used to detect melanoma and it generates almost 138 million features. To train this type of network required a vast amount of data. To overcome this problem author used fined, tuned technique one of them is regularization technique. The accuracy level achieved by the work is 80%.

In the work author used U-Net-based segmentation technique to segment image and for classification VGG16 CNN model is used [7]. The author first segmented the image, extracted features, and classifies it. Here two different experiments are made by author classification with segmented image and classification without a segmented image. As a result of the experiment it was found with segmentation it gives higher accuracy, i.e., 83.18%.

In the paper, author divided whole task into two stages training and testing of the model [8]. The experiment was done on 154 training and 44 test images only. With 100 epochs author got 100% accuracy. LeNet-5 architecture is used to train the model. With different number of epochs, different accuracy is identified. It is important to identify optimal number of epoch so that overfitting does not occur. Accuracy level also depends on the following parameters such as the size of input pixels used during, number of layers, and learning rate.

3 Solution Strategy

This work aims at training a model such that it can detect melanoma and classify whether the melanoma is Benign in nature or malignant. To increase the dataset size and accuracy we have used the following data augmentation techniques:

i. Horizontal and vertical shift
ii. Random rotation of images
iii. Random zooming of images
iv. Increasing blurriness in images.

The original size of the dataset was 3297 and after data augmentation, it got increased by 4297. The dataset we have taken from the Kaggle web site. We have used the above dataset to train our model in CNN and the following are the steps used in classification and diagram shown in Fig. 1:

- Provide input image into convolution layer and used 6 convolutional layer.
- Choose parameters before execution of convolution layer. Here parameters are Asymmetry, border, colour, dark, and evolving.
- We have passed all images through convolution layer and use ReLU activation to the matrix.
- Five Maxpooling layers are used to reduce dimensionality size.
- Flatten matrix is used as input to the fully connected layer (FC Layer) and used five fully connected layer to classify images.

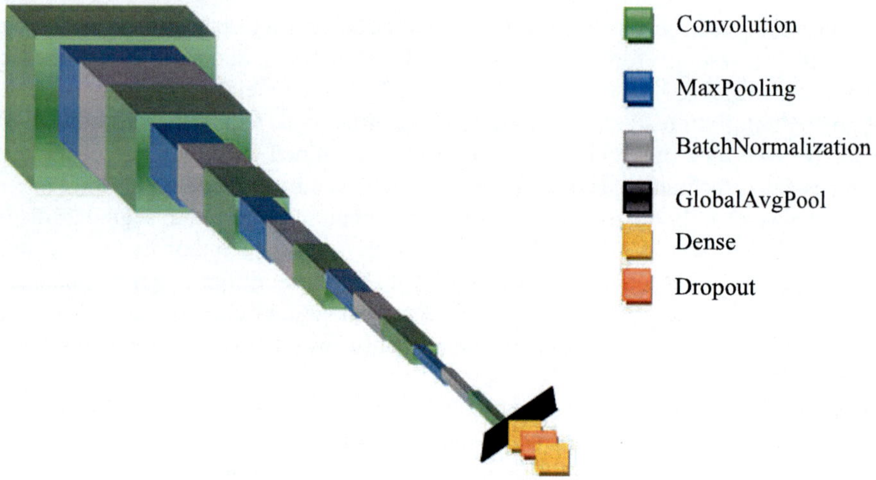

Fig. 1 Convolutional neural network of proposed model

- Softmax function is used to do classification of the images.

3.1 Convolution Layer

Following attributes we have to consider while implementing convolutional layer:

- For tensor input shape is (number of images) × (image width) × (image height) × (image depth).
- It uses filters to convolve also termed as convolve features. Total six kernels are used out of the first kernel of 16 × 3 sizes, second kernel 32 × 3, third kernel size is 64 × 1, forth one is 128 × 3, fifth one is 256 × 3 and sixth one is 256 × 3. ReLU activation function is used.
- This layer is used to extract prominent features from image.

Rectified Linear Unit (ReLU) is an activation function is used convolution layer to get positive value in each pixel. By using the following function it removes negative values:

$$R(z) = \max(0, z)$$

The ReLU function replaced all negative value by 0.

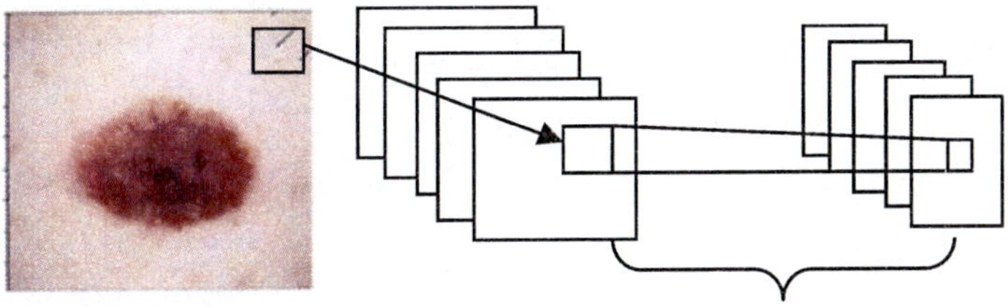

Fig. 2 Convolution and Maxpooling layer

3.2 *Max Pooling*

Pooling layers are used to reduce size of feature map generated by convolution layer. It basically select the maximum valued element from the window of 3 × 3. That means 9 pixel value is mapped into one pixel. Figure 2 shows convolution and maxpooling layer of CNN.

3.3 *Fully Connected Layer*

Fully connected layer of CNN is in which all the neurons of the previous layer must be connected with the next layer. This process is the same as Multi-Layer Perceptron (MLP) [9]. Classification of image done by fully connected layer by using flattened matrix. Softmax activation function is used after fully connected layers to convert numbers into probabilities so that sum is 1. Softmax function id as follows:

$$s(y_i) = \frac{e^{y_i}}{\sum_j e^{y_j}}$$

y_i and y_j one value from the set of j values. $S(y_i)$ is a probability of a value y_i taken from set of j values. After fully connected layer batch normalization is performed that applies a transformation so that activation mean is close to 0 and activation standard deviation is close to 1. After that dropout is required to reduce the unit in a neural network so it reduces computation time. This function used to classify between benign or malignant. Figure 3 is the diagram of fully connected layers:

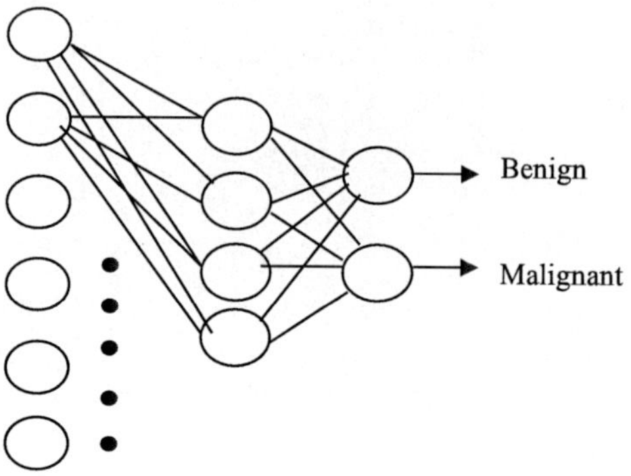

Fig. 3 Fully connected layer of CNN

4 Implementation Details

We use CNN to classify any image given by the user into malignant or benign melanoma [10]. CNN is a type of artificial neural network, in which network can generate feature map by itself no need to extract feature explicitly.

As it is very difficult to select features for from dermoscopic images so we have chosen deep neural network, i.e., CNN to classify melanoma.

Following are the steps for implementing the classification technique:

- We have collected dataset from Kaggle website that is of size 3279.
- Each image size is 224 × 224 pixel.
- The applied data augmentation techniques to increase dataset size and to increase accuracy level.
- We take a dataset of 4297 dermoscopic images of patients that either had melanoma or suspected to have it.
- Out of it 1074 are test data and rest 3223 is used as training data.
- Then, we train our model using Convolution Neural Network with learning rate 0.1435.
- The trained model is then saved to use it for testing.
- Then we pass an image or a set of images as a test case.
- The presence of melanoma is detected with approximately 86% accuracy.

5 Performance Metrics

Work of a classifier is to locate region of interest and classify that location is benign or malignant. While classifying objects it may happen that object classified incorrectly and in absence of object it classifying etc.

Before performing the classification of a particular object we must know the possible different classes. After classification, we should aware of the original class of the object to judge the quality of the work.

The classification is of four types, which are given as follows:

i. True positive (TP): classified correctly and predicted right class.
ii. True negative (TN): classified correctly predicts the wrong class.
iii. False positive (FP): the classified incorrectly predicts the right class.
iv. False negative (FN): the classified incorrectly predicts the wrong class.

Based on the above value, we have calculated accuracy of the work.

$$\text{Accuracy} = \frac{\text{TP} + \text{TN}}{\text{TP} + \text{TN} + \text{FP} + \text{FN}} \times 100\%$$

The above value is specifying percentage of time melanoma is classified correctly.

6 Result and Discussions

To train the CNN model it is required to run number of epochs, i.e., the training dataset mainly generates feature map and make classification possible. We had run the epochs at various values like 10, 70, and 150. In first experiment, we have choosen number of epochs as 10 and as a result, we found detection accuracy is very less so on the next experiment we have an increased number of epochs. From Fig. 4 ROC we can observe that the accuracy level is very low that is 45.5% where x-axis contains accuracy percentage and y-axis number of epochs.

In Fig. 5 when we increased the epoch to 70 we saw that the model could actually classify the images partially, and even the accuracy did increase to 68.6% and from ROC graph it has been observed that the increase in epochs with respect to previous also increases accuracy level of the work.

Figure 6 depicts that for set of 8 test images 4 images are correctly classified for 70 number of epochs and we observe this number of epochs not giving much accuracy so it is important to increase number of epochs.

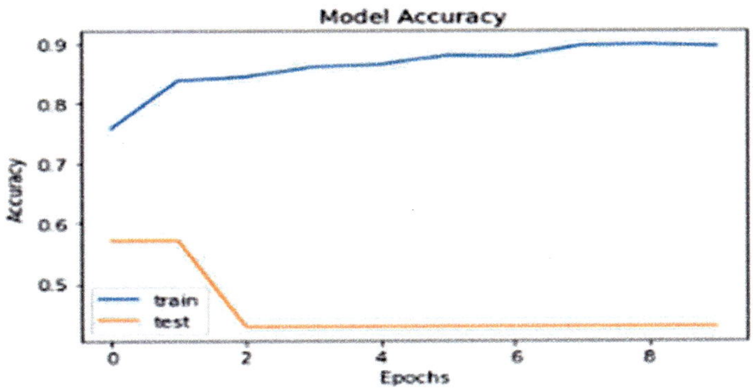

Fig. 4 Epoch versus accuracy graph for 10 epochs

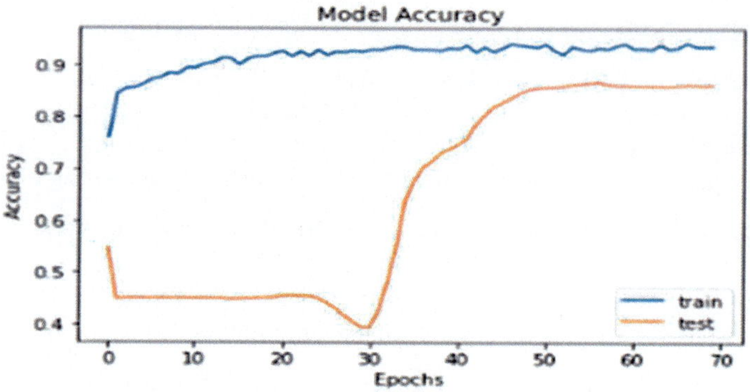

Fig. 5 Epoch versus accuracy graph for 70 epochs

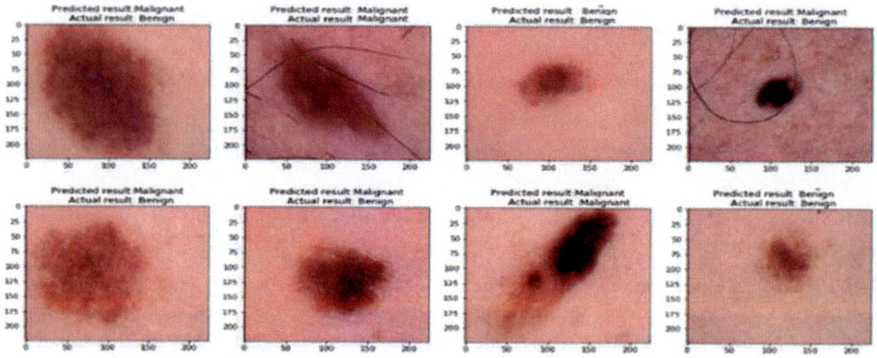

Fig. 6 A set of test cases with actual and predicted results after 70 epochs

During the testing phase, we saw that as we increased the epochs the accuracy rate got better. As we can see from the ROC that the more is the epoch the closer it gets to the train data set and after a while the test- line is almost parallel to the train-line and hence we don't perform an epoch for more than 150 in Fig. 7 as the result proves that the epochs set are enough to validate the model. If we increase more number of epochs then overfitting arises and accuracy level decreased. Here we got accuracy is 86%.

From Fig. 8, we can visualize that out 8 test images all are classified correctly and accuracy is also good. After increasing more number of epochs it was observed that accuracy level decreased due to overfitting occurs with increasing number of epochs.

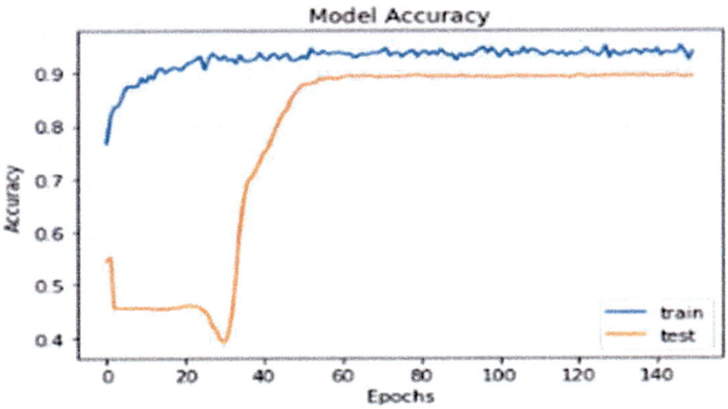

Fig. 7 Epoch versus accuracy graph for 150 epochs

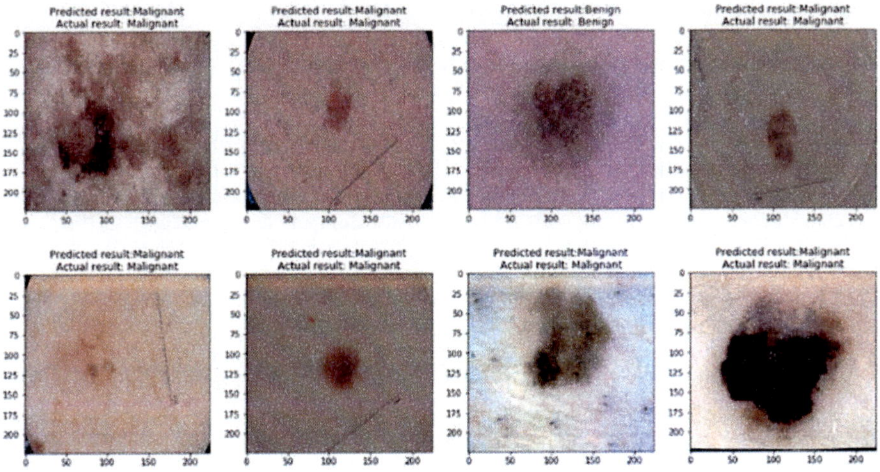

Fig. 8 A set of test cases with actual and predicted results after 150 epochs

7 Conclusions

This work aimed at detection of melanoma on the epidermal level, based on lesions that may have occurred due to various reasons and resulted in melanoma. In this work, we took the dataset from Kaggle and performed data augmentation to increased dataset. CNN is used for classification of Melanoma. Then trained a CNN model with a training data image set using various numbers of epochs and as we increased the number epochs we noticed that the efficiency did increase but after a certain number of epochs the graph of accuracy rate was almost constant hence we did not increase the epoch. We even had to keep an eye on the fact that over-fitting of the model had to be avoided so that our model was trained on how to classify the images rather than memories the data. The model identifies the test case under various labels and returns the result.

References

1. Ali, A. A., Al-Marzouqi, H.: Melanoma detection using regular convolutional neural networks. In: 2017 International Conference on Electrical and Computing Technologies and Applications (ICECTA), Ras Al Khaimah, pp. 1–5 (2017)
2. Ali, A.R.A., Deserno, T.M.: A systematic review of automated melanoma detection in dermatoscopic images and its ground truth data. Proc. SPIE Int. Soc. Opt. Eng. **28**(8318), 1–6 (2012)
3. Kittler, H., Pehamberger, H., Wolff, K., Binder, M.: Diagnostic accuracy of dermoscopy. Lancet Oncol. **3**(3), 159–165 (2002)
4. Brinker, T.J., Hekler, A., Utikal, J.S., Grabe, N., Schadendorf, D., Klode, J., Berking, C., Steeb, T., Enk, A.H., von Kalle, C.: Skin Cancer Classification Using Convolutional Neural networks: systematic review. J. Med. Internet Res. **20**(10), e11936
5. Yu, C., Yang, S., Kim, W., Jung, J., Chung, K.-Y., Lee, S.W., Oh, B.: Acral melanoma detection using convolutional neural network for dermoscopy images
6. Salido, J.A.A., Ruiz Jr, C.: Using deep learning to detect melanoma in dermoscopy images. Int. J. Mach. Learn. Comput. **8**(1) (2018)
7. Seeja, R.D., Suresh, A.: Segmentation and classification using deep learning. Int J Technol Explor Eng. **8**(12) (2019). ISSN:2278-3075 (2019)
8. Refianti, R., Mutiara, A.B., Priyandini, R.R.P.: Classification of melanoma skin cancer using convolutional neural network (IJACSA). Int. J. Adv. Comput. Sci. Appl. **10**(3) (2019)
9. Brinker, T.J., Heckler, A., Enk, A.H., Berking, C., Haferkamp, S., Hauschild, A., Weichenthal, M.: Deep neural networks are superior to dermatologists in melanoma image classification. National Center for Tumor disease(NCT), German Cancer Research Center(DKFZ), Heidelberg, Germany, Published on 28 May 2019
10. Bisla, D., Choromanska, A., Berman, R.S., Stein, J.A., Polsky, D.: Towards automated melanoma detection with deep learning: data purification and augmentation. In: CPRV Workshop IEEE Open Access

Design and Characterization of a Multilayer 3D Reversible "Full Adder-Subtractor" by Using Quantum Cellular Spin Technology

Rupsa Roy, Swarup Sarkar, and Sourav Dhar

Abstract Quantum Cellular Automata (QCA) is an emerging technology in the current nano-technical world where electrons are used in the quantum cells for information storing and transmitting. The analytical and numerical design of different types of combinational and sequential circuits with the help of QCA spin technology is now under process. In this paper, the design of a multilayer 3D Reversible "Full Adder-Subtractor" by using Quantum Cellular Spin Technology ('QCST') is explored under the consideration of different trade-offs. The trust area of this paper is to reduce the cell-complexity, occupied unit-area, and clock-zone (latency of the design). The dissipated power also calculated under the presence of multilayer wire crossing with reversible logic. In this proposed structure reversible 3 × 3 "Feynman Gate" is used and the garbage outputs are utilized to design carry out and borrow out in the proposed design. The 3 × 3 "Modified QR Gate" (Reversible gate also) is also used to provide a new extra logical expression that is used for super-computing. In this also cited the logical design and validate by Verilog Code using Xilinx.

1 Introduction

In the modern technological era, the "Complementary Metal Oxide Semiconductor" (CMOS) technology is a renowned technology where Moore's Law is followed to select the number of transistors in VLSI design [1]. But this technology faces various types of limitations like high power leakage, power consumption, and design complexity when the numbers of transistors are increased in a single die based on

R. Roy (✉) · S. Sarkar · S. Dhar
Department of ECE, Sikkim Manipal Institute of Technology, Sikkim Manipal University, Gangtok, Sikkim, India
e-mail: onerupsa@gmail.com

S. Sarkar
e-mail: swarup.s@smit.smu.edu.in

S. Dhar
e-mail: sourav.d@smit.smu.edu.in

S. Borah et al. (eds.), *Soft Computing Techniques and Applications*,
Advances in Intelligent Systems and Computing 1248,
https://doi.org/10.1007/978-981-15-7394-1_32

full-custom structure. The design based on CMOS also effected by processing delay. To overcome the above-said limitations a new technique needs to be implemented which can reduce power dissipation, latency, and power consumption. In this regards the nano-structural designs can only the solution to above-mentioned problem. The quantum cellular automata spin technology is the key resource of nano-structure design. This novel nano-scaled technique, i.e., 'QCST' is used to design multilayer 3D reversible "Full Adder-Subtractor". In this technique, quantum-dots are used to store and transmit information by electron-trapping [3]. In this proposed technology four quantum dots are used in a quantum cell (18 nm cell). Quantum cells are placed one after another in this chosen technology to get a quantum wire and the interaction between the quantum cells help to flow the information from source to destination with very low power-consumption [3] and leakage-current flow in the internal part of the circuit.

Figure 1 represents a basic 90° quantum cell-design where the polarization of the cells can be calculated based on electro-spin concept. Conventional digital logic bits are two types '1' and '0'. But, qubit or quantum bit is utilized in this technology where two bits are required for a single qubit: '+1' and '−1' polarization (depends on the electrons' rotation in a clockwise direction and in anti-clockwise direction, respectively). So, this technology can access double bit-rate than other nano-technologies like CMOS. It can also work in THz frequency range and a high density (50 Gb per cm^2) and low power-leakage (100 W per cm^2) are the advantages of this suggested technique [5]. This technology can use multiple layers to convert a conventional 2D structure to multilayer 3D structure [9] very easily. This multilayer structure not only reduces the unit-area and used clock-zone [10] of the nano-structure but also reduces the overlapping complexities of quantum-wires in single layer structural formation. A different gate type (reversible gate) is used in this technology widely in this present digital world. This technology used a different type of clock-zone-phase, which is called "Bennett clock-scheme" [6]. In this clock-scheme, the information can be copied before erase. A reversible gate can be understood by its truth-table where the inputs also achieved from the output unlike the conventional logic gates' operation (outputs depend on the inputs only). This N: M ($N = M$, N = no. of inputs and M = no. of outputs) logical gate type can also be used to reduce the area, complexity, energy-dissipation (by reducing the chance of information-loss) and also the latency of the design.

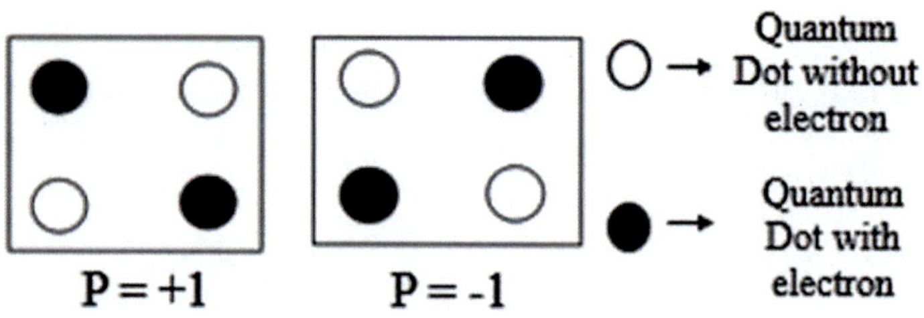

Fig. 1 Basic quantum cell-design with different polarization

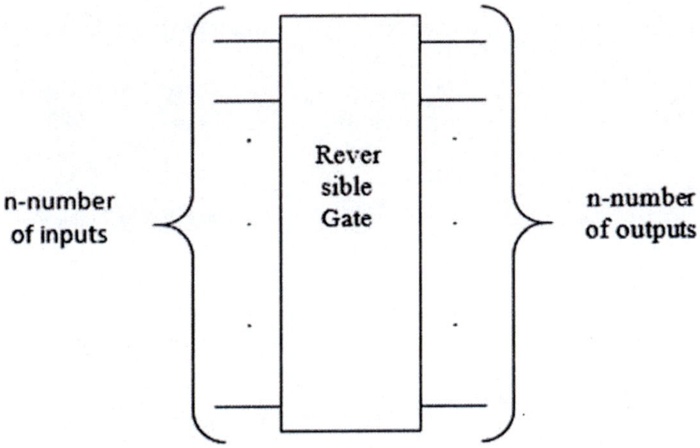

Fig. 2 Block diagram of reversible gate

The block-diagram of the reversible gate is clearly shown in Fig. 2, where the number of inputs and outputs are the same and different outputs present different logical expressions. Basically, reversible gate can explore the QCA-Advantages in terms of area, delay, design complexity, and dissipated-energy optimization. This paper also focuses on the logic level design of the proposed device and also the time scaling waveform is presented of this proposed reversible multilayer design by using Verilog Code in Xilinx. The number of used logic gates in logic-level design also helps to understand the occupied area of the proposed nano-sized device-circuit by using transistor-level technologies. The delay and also the power calculation can be achieved from this software.

2 Suggested Device and Suggested Reversible Gates

In this section, a novel design of a "Full Adder-Subtractor" is proposed which can also represent an extra logical expression and this expression (which is an AND expression of three inputs after inverting the third input) can be used for supercomputing. A logic-level design is given in this paper from which the number of transistors, delay, and power of the proposed design can be calculated. This logic level design, where used logic gates and the connections are clearly shown is developed by using Verilog Code in Xilinx.

Figure 3 explores the logic level design of this proposed structure. In this proposed structure 10 logic gates (7 AND gates and 3 OR gates), 3 inverters, and 1 three-input XOR block are required where total 83 transistors are needed. But the inter lapping complexity of 83 CMOS transistors and the various types of limitations in CMOS-scaling, which is already discussed in the introduction part of this paper, limits this

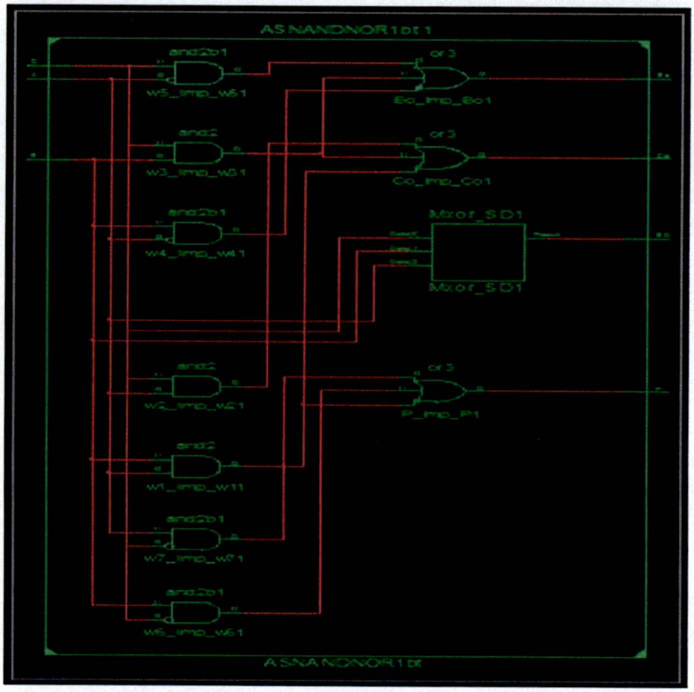

Fig. 3 Logic level design of proposed "Full Adder-Subtractor" design

proposed design to 2D structure. But, the area, power, and delay of the proposed design can be optimized by using a more advanced QCA technology with spintronic electrons. The multilayer structure with reversible gates can be designed in this above technology very easy to reduce the area, power, and delay of the design compare to the CMOS design. The suggested "Full Adder-Subtractor" device of this paper-work was already designed by using different reversible gates in different years. But, in this paper-work, the single-layer design is converted into multilayer 3D design and the cell-complexity, occupied unit-area, latency, and calculated power-leakage are compared to the other proposed designs of other papers (in a different year) in this portion. In 2017, two 'QR' reversible gates are used with a reversible 'CNOT' gate to form this proposed design in a single layer platform. But three hundred and ninety-nine (399) cells are used and two garbage outputs are presented here [4]. In 2018, two "Modified Fredkin Gate" with a "majority voter Gate" are used to design this "Full Adder-Subtractor" formation where also two garbage outputs are required. This single structure presents a quantum wire crossing in this paper [7] with 121 cell-complexity and 4 clock Zones. In a 2019s paper, 83 cells and 4 clock zones are required to design a single-layer proposed device [8]. In this paper a 3 × 3 "Feynman Gate" with a 3 × 3 "Modified QR Gate" is used to form the circuit level designs of our proposed multilayer structure where five layers are used.

Figure 4 presents the block diagram of our proposed reversible structure where one

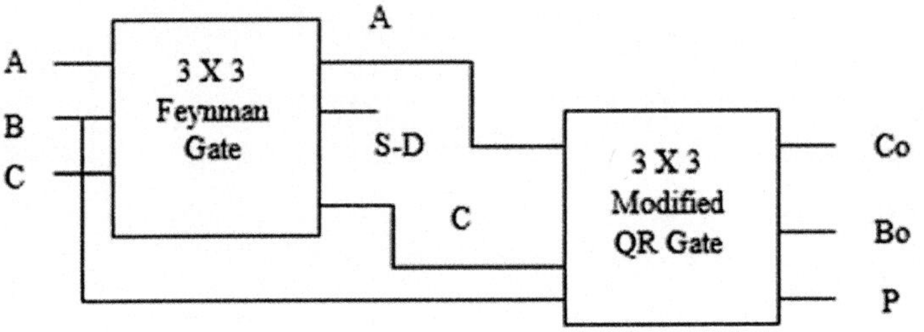

Fig. 4 Block diagram of proposed reversible "Full Adder-Subtractor" structure

extra output is present with a different logic expression ("AND operation" of three inputs) if the three inputs are *A*, *B*, and *C*. In this figure, the Sum-Diff output presents by 'S-D' which is a 3 input 'XOR' operation of *A*, *B*, and *C*. The 'Co' and 'Bo' are the carryout and borrow out, respectively. Here output 'P' presents a different logic expression which is $AB + BC' + AC'$. The truth table of these proposed reversible logic gates (in Table 1A and B) help to understand the proper reversible logic where

Table 1 **A** Truth table for reversible 3 × 3 "Feynman Gate". **B.** Truth table for reversible 3 × 3 "Modified QR Gate"

A	*B*	*C*	*A*	*A* XOR *B* XOR *C*	*C*
A					
0	0	0	0	0	0
0	0	1	0	1	1
0	1	0	0	1	0
0	1	1	0	0	1
1	0	0	1	1	0
1	0	1	1	0	1
1	1	0	1	0	0
1	1	1	1	1	1

A	*B*	*C*	$AB + BC + CA$	$A'B + BC + A'C$	$AB + BC' + C'A$
B					
0	0	0	0	0	0
0	0	1	0	1	0
0	1	0	0	1	1
0	1	1	1	1	0
1	0	0	0	0	1
1	0	1	1	0	0
1	1	0	1	0	1
1	1	1	1	1	1

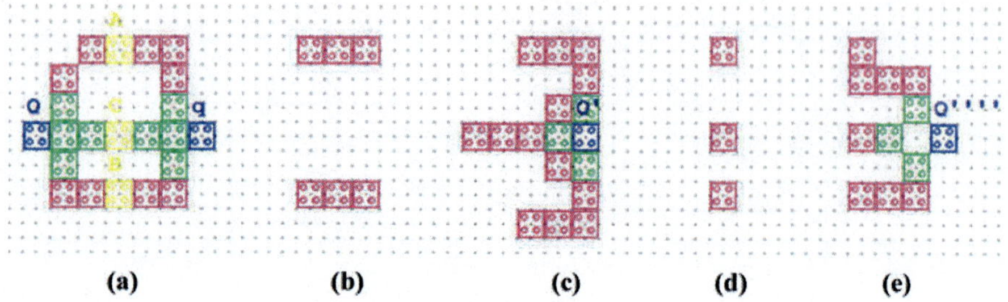

Fig. 5 Five layers of proposed multilayer 3D reversible "Full Adder-Subtractor" design by using 'QCST'

not only the outputs depend on the inputs but also inputs depend on the output [4, 10].

The single-layered structures have four used clock-zones but the clock-zones can be reduced with less complexity, low unit-area, and less power-leakage. The five layers are presented here in Fig. 5a–e (Different layers are shown separately). Here output

$$Q' = S - D = A\,XOR\,B\,XOR\,C$$
$$q = Co = AB + BC + AC$$
$$Q = Bo = A'B + BC + A'C$$
$$Q'''' = \text{Extra output } (P) = AB + BC' + AC'$$

Figure 5a presents the carry output (Q) and borrow output (q) in first layer of this proposed design. Figure 5b shows the second layer of this presented structure which is used as a transmission line between first layer and third layer. The third layer is given here in Fig. 5c. The third layer gives the sum/diff output (Q) of proposed "Full Adder-Subtractor". Figure 5d presents the fourth layer which is used as a transmission line between third layer and fifth layer. Figure 5e presents the extra logical expression (Q'''') of the 3 × 3 "Modified QR Gate". Ultimately, the 3 × 3 "Feynman Gate" is presented by using first layer, second layer, and third layer. The 'A' and 'C' outputs of this reversible gate are used to design another 3 × 3 "Modified QR Gate" by using first layer, fourth layer and fifth layer in this presented structure.

3 Simulated Outcomes and Characterization

The time-scaled waveform of this proposed design by using Xilinx is given in this portion and also the simulated outputs of the 5 layered reversible proposed design by using quantum cellular spin technology.

The time-scaled waveform of the proposed "Full Adder-Subtractor" design by using Verilog Code in Xilinx platform is shown in Fig. 6. In this figure A, B, and

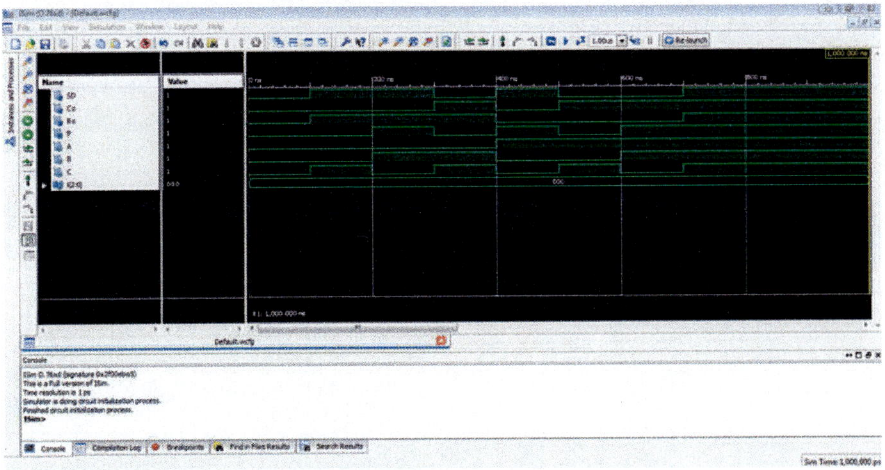

Fig. 6 Time-scaled waveform of proposed "Full Adder-Subtractor" design

C these three inputs are given the outputs are SD, Co, Bo, and P, which are related to the QCA outputs (discussed in the previous section of this paper). The outputs follow the given truth tables in Table 1 A and B, which are also given in the previous section of this paper.

Figure 7 presents the simulated outcomes of 5 layered proposed Multilayer 3D reversible "Full Adder-Subtractor" design by using 'QCST', where the three inputs are A, B, and C. The 1st output (Q) presents the AND operation of input A′ (inverting output of input A), B, and C. This is a borrow output of full subtractor. Then the 2nd output (q) presents the carry output of full adder, which is an AND operation of input A, B, C. Third output (Q') gives the sum/diff operation of full adder and subtractor.

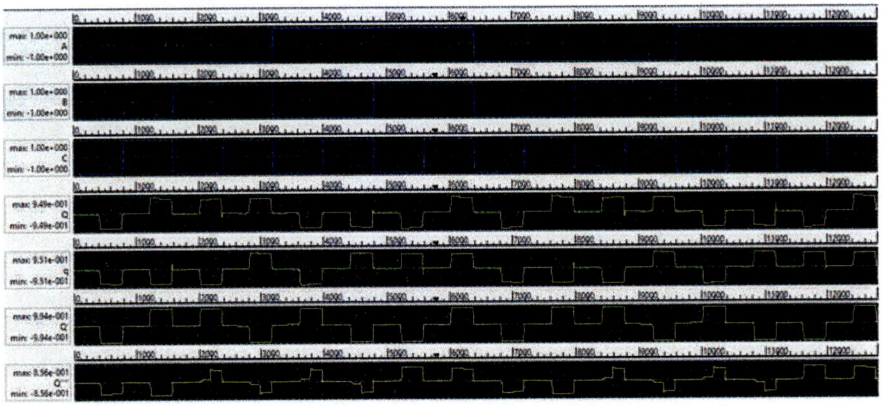

Fig. 7 Simulated outcomes of 5 layered proposed multilayer 3D reversible "Full Adder-Subtractor" design by using 'QCST'

Table 3 Table of compared outcomes of different proposed designs ("Full Adder-Subtractor") of different years

Publication years	Cell-complexity	Occupied unit area (μm^2)	Latency (clock-delay)	Calculated power-leakage
2017 [4]	399	0.50	8 clock-phases (2 clock-delay)	0.5 μW
2018 [7]	121	0.14	5 clock-phases (1.25 clock-delay)	0.14 μW
2019 [8]	83	0.09	5 clock-phases (1.25 clock-delay)	90 nW
2020 (proposed design)	60	0.05	4 clock-phases (1 clock-delay)	50 nW

Table 4 Table of calculated parameters after simulating the logic level design of the proposed structure

Full ADDER-subtractor	Number of CMOS transistors	Delay (ns)	Power-lLeakage
	83	5.895	81 mW

This is a three-input XOR operation of these given three inputs in this presented design. The extra output of used 3 × 3 "Modified QR Gate" is represented here by the 4th output (Q''''), which is the AND operation of input A, B and C' (inverting output of input C). In this figure, maximum of four clock-phases of a clock cycle is required to get the next output after the previous output.

The compared outcomes of different designs of this proposed device by using different gates are enlisted in this paper in Table 3 and the calculated values of different parameters of this design also present here in Table 4 where Xilinx is used for these calculations.

4 Conclusion

After the successful completion of the computational design and physical design of the proposed "Full Adder-Subtractor" it is observed that the area, power, and delay are reduced compared to other CMOS designs. The 27% area reduction, 83% delay reduction, and more than 90% power reduction of this proposed design can possible by using 'QCST' with reversible logic and multiple layers compare to CMOS technique. In this proposed multilayer reversible structure, the unit-area, propagation delay, cell-complexity, and power leakage (which depends on the area of the design) optimization also can be possible up to 44%, 20%, 27%, and 44%, respectively compared to the previous structure in the given 2019's paper. The reduction of these given parameters reduces the cost (Area X Power X Delay) [7] of the design and efficient complexity (depends on the area, number of layers, and cell-complexity) [2] of the design also. Multi-bit proposed structure can be developed also by using these proposed reversible gates in multi-layer platform in the future and also physical level design with proper validity checking and design rule checking can be performed in the future before fabrication of this proposed structure. These proposed arithmetic and logical expressions can be used in the arithmetic unit and logical unit of ALU (arithmetic and logic unit) in the future.

References

1. Beiki, Z., Shahidinejad, A.: An introduction to quantum cellular automata technology and its defects. Rev. Theor. Sci **2**(4):1–9 (2014)
2. Fengbin, D., Xie, G., Yongqiang, Z., Fei, P., Hongjun, L.: A novel design and analysis of comparator with XNOR gate for QCA. Microprocess. Microsyst. **55**, 131–135 (2017)
3. Roy, S.S.: Generalised quantum tunneling effect and ultimate equations for switching time and cell to cell power dissipation approximation in QCA devices. In: The SSR Digital Library, pp. 1–11 (2017)
4. Kianpour, M., Sabbaghi-Nadooshan, R.: Novel 8-bit reversible full adder/subtractor using a QCA reversible gate. J. Comput. Electron **16**:459–472 (2017)
5. Laajimi, R.: Nanoaechitechture of Quantum-Dot Cellular Automata (QCA) Using Small Area for Digital Circuits. Advanced Electronics Circuits–Principles, Architectures and Applications on Emerging Technologies, pp. 67–84 (2018)
6. Sarvaghad-Moghaddam, M., Orouji, A.A.: New symmetric and planar design of reversible full-adder/subtractor in quantum-dot cellular automata. Eur. Phys. J. 1–13 (2018)
7. Ahmad, F., Ahmed, S., Kakkar, V., Mohiuddin Bhatt, G., Bahar, A.N., Wani, S.: Modular design of ultra-efficient reversible full adder-subtractor in QCA with power dissipation analysis. Int. J. Theor. Phys. **57**:2863–2880 (2018)
8. Zoka, S., Gholami, M.: A novel efficient full adder-subtractor in QCA nanotechnology. Int. Nano Lett. **9**, 51–54 (2019)
9. Babaie, S., Sadoghifar, A., Bahar, A.N.: Design of an efficient multilayer arithmetic logic unit in quantum-dot cellular automata (QCA). IEEE Trans. Circ. Syst II. 66(6), 963–967 (2019)
10. Oskouei, S.M., Ghaffari, A.: Design a new reversible ALU by QCA for reducing occupation area. J. Supercomput. **75**, 5118–5144 (2019)

A Study of Recent Security Attacks on Cognitive Radio Ad Hoc Networks (CRAHNs)

Debabrata Dansana and Prafulla Kumar Behera

Abstract Cognitive radio (CR) is a change in technology due to which it is possible for users who are unlicensed to use the unused and underutilized spectral areas effectively without posing any interference from licensed users. This concept essentially engages the development of dynamic spectrum access (DSA), and this spectral sensing acts as the main interest region for CRNs. Cognitive radios are sensitive to small malicious attacks and need to be secured at various levels to stop the disruption of service. However, not much work has been done keeping in mind the security aspect of CR networks (CNR) and still is an area of research. This paper effectively discusses various threats both layer and feature specific. Emphasis has also been given on the security attributes suggested in standard security protocols. This paper has finally discussed the effects of CR threats on an ad hoc network.

Keywords Cognitive radio (CR) · Dynamic spectrum access (DSA) · Unlicensed users · Licensed users · Malicious attacks

1 Introduction

This instruction a cognitive radio is defined as a flexible radio that has a capacity to change the transmitter parameters on the basis of the type of interactions it has with the surrounding environment. The name cognitive suggests that the radio has an inherent ability to sense and gather various information like bandwidth and frequency from the surrounding environment. This radio also has the feature of fast adaptation to change in surrounding conditions for maintaining optimum system performance. CRs basically use the spectrum that has been temporarily out of function also called as the spectrum hole or white space. In case of occurrence of any primary user using

D. Dansana (✉) · P. K. Behera
Department of Computer Science, Utkal University, Bhubaneswar, Odisha, India
e-mail: debabratadansana07@gmail.com

P. K. Behera
e-mail: p_behere@hotmail.com

S. Borah et al. (eds.), *Soft Computing Techniques and Applications*, Advances in Intelligent Systems and Computing 1248,
https://doi.org/10.1007/978-981-15-7394-1_33

the spectrum hole, the secondary users effortlessly shift their occupancy to another white space or sometimes stay in the same band with different transmission powers. This avoids interference with the primary users. Cognitive radio networks basically have the following four functions: [1] spectrum detection that senses all the available spectrum holes to avoid any sort of interference [2]; sharing of spectrum information within neighboring nodes [3]; facilitates smooth communication between various nodes [4]; decisive quality of choosing appropriate spectrum holes for seamless communication. Design of a cognitive radio ad hoc network (CRAHN) comes with a large limitation which makes the system vulnerable to threats. Security, delay tolerance and poor packet delivery ratio are some major concerns of design protocol of CRAHNs [5]. The main reasons behind these issues are mainly improper zone selection that resulted in the unlicensed users to quit the spectral band. This also might have resulted in interference with the licensed users and hence termination of their services. Cognitive protocols vary entirely different than the traditional network protocols [6]. The main objectives of this paper are to study the recent approaches regarding the CRAHN security and investigation of the attacks that are targeted just for a CRAHN system.

2 Literature Review

The paper [7] elaborately discusses a routing protocol for cognitive ad hoc networks, thereby not allowing the interference of the transmission of the licensed users. This routing protocol was improved, and a new cross-routing along with dynamic spectrum allocation was proposed in [8]. The paper [9] discusses a new routing protocol, namely the reactive routing protocol for the mobile ad hoc networks. The paper [10] discusses a new connectivity-based routing skill called the Gymkhana for the communication of mobile ad hoc networks. The paper [11] effectively discusses the steps taken to improve the performance of wireless ad hoc networks through proper designing of MAC layer. In paper [12], the importance of software designed radio in communication system was effectively portrayed. Banerjee et al have elaborately penned down all the possible threats that can be applicable to any ad hoc network in [13]. The paper [14] has proposed an analytical model for primary user emulation attacks in cognitive radio networks. The paper [15] has proposed methods for securing cognitive radio networks against primary user emulation attacks. Randomized RANSAC with sequential probability ratio test is proposed effectively in [16]. The paper [17] suggests the first wireless standard base for cognitive radio the IEEE 802.22. The paper [18] suggests the future work that has to be done in the cognitive wireless network keeping in mind the security aspects (Table 1).

Table 1 Different layers Dos attacks and their feasible defense mechanism

Layer	Attacks	Defense mechanism
MAC	Collision Exhaustion	Error-correction code Rate Limitation
NETWORK	Selective forwarding Sinkhole Sybil Wormhole Hello Flood	Authentication monitoring Redundancy checking Authentication monitoring Authentication Packet leashes by using geographic and temporal information
TRANSPORT	SYN flooding De-synchronization	Client puzzle
APPLICATION	Logical error buffer overflow	Trusted computing
PHYSICAL	Jamming	Spread spectrum priority message
PRIVACY	Traffic analysis, attack on data privacy and location privacy	Homomorphic encryptions

3 Architecture of Cognitive

The structure of a cognitive radio can be classified into three sub-categories [19]:

a. *Hardware*: The radio consists of a normal radio embedded with cognitive software mainly defined as the hardware component.
b. *Software*: The software section of a cognitive radio can be sorted into three sub-categories, namely the front end that consists of the RF portion, the modem and the receiver–transmitter module (Fig. 1).

Entire radio architecture is basically a layered structure which comprises the MAC layer, physical layer, transport layer and network layer. In physical layer, the main task involves searching of spectral holes and to investigate the opportunity of proper functioning. This layer also involves estimation of any chances of interference with the primary users which might result in termination of services (Fig. 2).

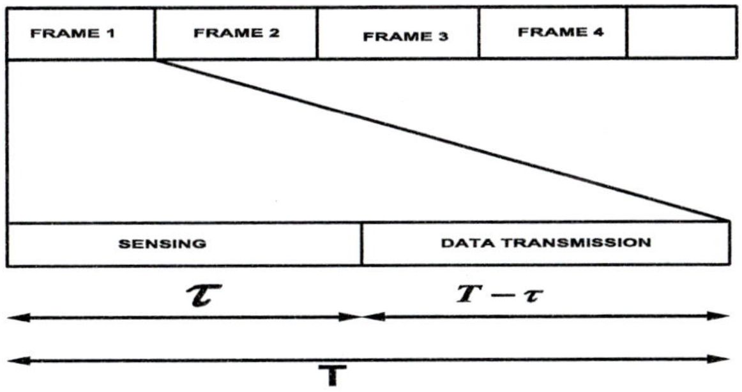

Fig. 1 Cognitive radio network

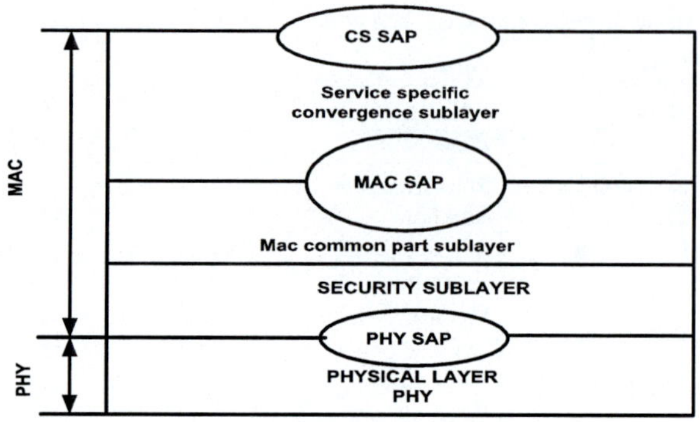

Fig. 2 Structure of MAC layer

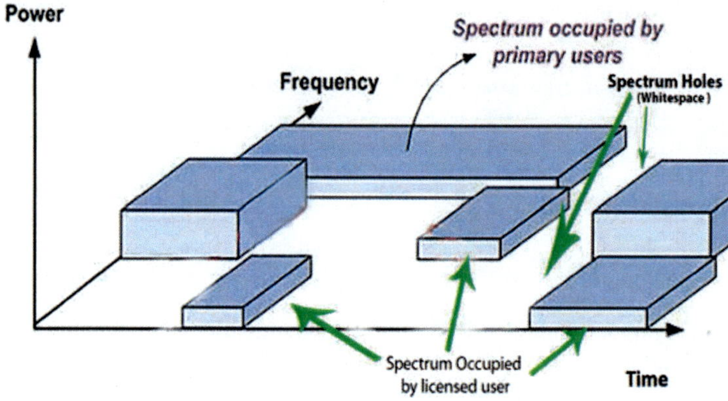

Fig. 3 White hole representation

The MAC is layer basically a decision-making layer and decides the course of data transmission by sensing the availability of spectral holes and to ensure proper utilization of the spectral holes (Fig. 3) [11].

This layer also deals with the effective investigation of various techniques of spectrum sharing with other CRs. This layer also contains information regarding channel occupancy and transmission parameter synchronization. The MAC layers make it possible for the secondary users to sense channels and allow access to unused spectrums.

Network layer—the main responsibilities of this layer are to manage routing and address management. This topology comprises detection of spectrum and sensing the presence of neighbors in the vicinity (Fig. 4).

Transport layer—this layer mainly deals with the congestion problems which affect the performance of MAC protocol and the spectral nature. Proper techniques of spectrum management ensure the total productive outcome of the transmission layers protocol. The top layers that deal with the waveform manipulation are the house of the cognitive radio applications for execution.

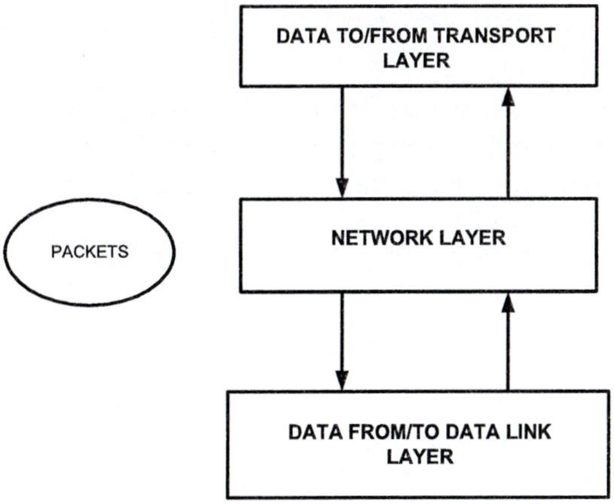

Fig. 4 Structure of network layer

The term software-defined ratio (SDR) [12] mainly defines the trend to implement a combination of software and hardware in cases where the age-old hardware has been replaced by modifiable hardware and software that manages the entire operations. SDR is incorporated with sensors that are input devices and exercise certain multi-dimensional functions. These inputs and outputs demand the CR engine to hold huge databases. Training and adaptation are mainly done in the anticipation stage. The adaptation of any CR network is basically governed by the environment in which it operates. These adaptations of any CRs can be easily infiltrated by the attackers making it an urgent need to modify the operating parameters which degrade the CR performance.

4 Physical Layer Threat

Proper coordination between various physical layers and effective coordination between multiple functioning layers are of utmost necessity of the proper deployment of a CRN [13]. The physical layer holds the complete responsibility of maintaining proper contact among different devices and manages transmission strength and bandwidth and coding-related aspects. The DL layer mainly comprises two child layers, the logical link control (LLC) and medium access control (MAC). This layer mainly supervises the location of proper spectral holes to unlicensed users and ensuring accessibility based on the presence of resources. The network layer deals with proper management for data transfer from source to destination and others higher spectral layers.

4.1 Primary User Emulation (PUE) Attack

The PUE attack can be inflicted to the system by the attacker in several ways. One such way includes exact replication of the energy of signal and faking the destination of the base user. Due to this, the users believe that the malicious attackers are originally primary users. The PUE attacks are:

4.1.1 Selfish Attack

This attack involves maximum utilization of the spectral space by the attacker. This thereby prevents the other users to access the spectral band.

4.1.2 Malicious Attack

On the occurrence of a malicious attack, legitimate secondary users' DSA processes are suspended and they are prevented from using empty **licensed** spectral bands also called denial of service (DoS). The spectral usage in minimum unlike the selfish attack in this occurrence [14].

4.2 Objective Function Attack

Any cognitive radio network executes a particular task based on some parametric values like power of transmission, bandwidth and central frequency. These parametric values need to be adjusted for the optimal performance of the system based on the learning process of some objective functions. These objective functions are manipulated by the attackers and result in sub-optimal system performances. This type of attack is also known as belief manipulation attack (BOA) [20].

4.3 Denial-of-Service (DoS) Attacks

This is mainly governed by total spectrum use caused by the users who are unlicensed during DSA function [13]. This attack can be inflicted from the node inside CRN or from an external source. The main consequences are misleading other nodes to occupy some other spectral areas that may not be empty. This creates disturbance in the CR communication [21].

5 Threat of Data Link Layer

This layer generally comprises two layers: the LLC sub-layers that mainly monitor the traffic density and the fault occurrence probability and the MAC layer deals with resource management that allows multiple accesses efficiently. The DSA takes help of the spectral sensing results and then allocates channels among all the CRNs. As per multi-hop CRN, none can be trusted for the major distribution. This makes it possible for the attackers to modify MAC frames. The data link layer attacks can be classified into:

5.1 Spectrum Sensing Data Falsification (SSDF) Attack

The secondary nodes in CR system mainly pass the data from the nodes to the collector center for processing. The main aim of an attack is to create a false data transmission set and send to the collector center that results in false decisions. The information sensing therefore should be fast in nature so that it can diagnose false spectrum and mitigation is possible. The sequential probability ratio test (SPRT) [22] technique is applied that returns multiple results for local spectrum sensing. These multiple outcomes can then be superimposed at the fusion center to improve data transmission and mitigate inaccuracies.

5.2 Control Channel Saturation Attack

When the availability of CR nodes for a given spectrum increases, it might be difficult for the channel to entertain all the CRN nodes within a specific time scale. This might also head to loss of data in transferring process [16]. This technique makes it easy for an attacker to send many of packets and neutralize the channels. This might head toward hard attack which deprives loyal users from gaining channels access. A channelization technique is introduced for mitigating this attack by shifting the free channels to ignore saturation by the packets [23].

6 IEEE 802.22 WRAN Standards

The security standard that effectively defines the CR technology is the IEEE 802.22 standard. The level is thereby explained in brief.

This standard ensures that the primary and secondary users exist in coherence to one another with minimum interference among each other, thus allowing coexistence and self-coexistence [24]. The above discussed security threats degrade the performance of the CRNs by increasing the interface level. The security layer that is incorporated in WRAN ensures the authenticity, confidentiality and integrity of data of MAC. The process of encapsulation in the security layer maintains a privacy key management protocol. For the detection of an incumbent signal, the base station schedules a quiet period at some standard time period during the span of which each and every network traffic is put on hold and there are a total of 802.22 entities that scan for the incumbent signal. This technique is successful in protecting the intra-cell messages but fails for inter-cell messages against integrity attacks. The attackers successfully insert false message within the cells that leads to the wrong decision taken by the secondary users to sense white space or spectrum holes. There is third-party software inserted by the attackers which mess with the normal CR behavior and is termed as beacon falsification attack. For the prevention of DoS attacks, message authentication codes are written by the WRAN security sub-layer for maintaining the communication of the messages [17]. There is a main infrastructure of the management that is implied here for maintaining message integrity and authenticity. The inter-cell security keys are effectively handled by an access control router (ACR). This in turn has direct communication with the Internet which is then made public. This implies to a distributed key management scheme, where the central node holds responsibility for the generation of each keys.

7 Conclusion

This paper has effectively illustrated some of the security threats that can be inflicted to a cognitive radio network (CRN). The security threats at physical level as well as data link level are discussed, and some discussions have been made regarding the mitigation of these threats. The security feature of the IEEE 802.22 is discussed, and all the investigation has been done in terms of the threats to privacy in a CR network. Future work of this research work can be the solutions to avoid different attacks to make the communications more secure.

References

1. Akyildiz, I.F., Lee, W.Y., Chowdhury, K.R.: CRAHNs: cognitive radio ad hoc networks. Ad Hoc Networks J. **7**(5), 810–836
2. Kolodzy, P.: Interference avoidance. In: Spectrum Policy Task Force. Federal Communications Commission, Washington, DC, pp. 147–158 (2002)
3. Haykin, Simon: Cognitive radio: brain-empowered wireless communications. IEEE J. Sel. Areas Commun. **23**(2), 201–220 (2005)
4. Akyildiz, I.F., et al.: NeXt generation/dynamic spectrum access/cognitive radio wireless networks: a survey. Comput. Networks **50**(13), 2127–2159 (2006)
5. Feng, J., et al.: Supporting secure spectrum sensing data transmission against SSDH attack in cognitive radio ad hoc networks. J. Network Comput. Appl. **72**, 140–149 (2016)

6. Sarkar, D., Narayan, H.: Transport layer protocols for cognitive networks. In: 2010 INFOCOM IEEE Conference on Computer Communications Workshops. IEEE (2010)
7. Deva Priya, S., Kannan, N.: Enhanced spectrum aggregation based frequency-band selection routing protocol for cognitive radio ad-hoc networks. Concurr. Comput. Practice Experience **31**(14), e4911 (2019)
8. Mansoor, N., et al.: RARE: a spectrum aware cross-layer MAC protocol for cognitive radio ad-hoc networks. IEEE Access **6**, 22210–22227 (2018)
9. Khurana, S., Upadhyaya, S.: An assessment of reactive routing protocols in cognitive radio ad hoc networks (CRAHNs). In: Next-Generation Networks. Springer, Singapore, pp. 351–359 (2018) (Appendix: Springer-Author Discount)
10. Ramesh, D., Venkatram, N.: Current state of benchmarking spectrum sensing and routing strategies in cognitive radio ad-hoc networks. J. Theor. Appl. Inform. Technol. **96**(11) (2018)
11. Kaynia, M., Jindal, N., Oien, G.E.: Improving the performance of wireless ad hoc networks through MAC layer design. IEEE Trans. Wireless Commun. **10**(1), 240–252 (2010)
12. Raut, R.D., Kulat, K.D.: SDR design for cognitive radio. In: 2011 Fourth International Conference on Modeling, Simulation and Applied Optimization. IEEE (2011)
13. Banerjee, A., Das, S.: A review on security threats in cognitive radio. In: 2014 4th International Conference on Wireless Communications, Vehicular Technology, Information Theory and Aerospace &Electronics Systems (VITAE). IEEE (2014)
14. Anand, S., Jin, Z., Subbalakshmi, K.P.: An analytical model for primary user emulation attacks in cognitive radio networks. In: 2008 3rd IEEE Symposium on New Frontiers in Dynamic Spectrum Access Networks. IEEE (2008)
15. Yu, R., et al.: Securing cognitive radio networks against primary user emulation attacks. IEEE Network **29**(4), 68–74 (2015)
16. Matas, J., Chum, O.: Randomized RANSAC with sequential probability ratio test. In: Tenth IEEE International Conference on Computer Vision (ICCV'05), vol. 1 and 2. IEEE (2005)
17. Cordeiro, C., et al.: IEEE 802.22: the first worldwide wireless standard based on cognitive radios. In: First IEEE International Symposium on New Frontiers in Dynamic Spectrum Access Networks (DySPAN 2005). IEEE (2005)
18. Burbank, J.L.: Security in cognitive radio networks: the required evolution in approach to wireless network security. In: 2008 3rd International Conference on Cognitive Radio Oriented Wireless Networks and Communications (CrownCom 2008). IEEE (2008)
19. Parvin, S., et al.: Cognitive radio network security: a survey. J. Network Comput. Appl. **35**(6), 1691–1708 (2012)
20. Bhattacharjee, S., Rajkumari, R., Marchang, N.: Cognitive radio networks security threats and attacks: a review. Int. J. Comput. Appl. **975**, 8887 (2014)
21. Schuba, C.L., et al.: Analysis of a denial of service attack on TCP. In: IEEE Symposium on Security and Privacy (Cat. No. 97CB36097). IEEE (1997)
22. Mirkovic, J., et al.: Internet Denial of Service: Attack and Defense Mechanisms (Radia Perlman Computer Networking and Security). Prentice Hall PTR (2004)
23. Lo, B.F.: A survey of common control channel design in cognitive radio networks. Phys. Commun. **4**(1), 26–39 (2011)
24. Ma, L., Shen, C.-C., Ryu, B.: Single-radio adaptive channel algorithm for spectrum agile wireless ad hoc networks. In: 2007 2nd IEEE International Symposium on New Frontiers in Dynamic Spectrum Access Networks. IEEE (2007)

A Proficient Deep Learning Approach to Classify the Usual Military Signs by CNN with Own Dataset

Md. Ekram Hossain, Md. Musa, Nahid Kawsar Nisat, Ashraful Hossen Thusar, Zaman Hossain, and Md. Sanzidul Islam

Abstract Everyday, around the world crimes, like kidnapping or forced to do something to enemy's command, are happening. General people are being the main victim in most cases. Hostage people are usually rescued by military or special force sometimes. The best way to build communication between hostage and military is by using the basic sign language of that military or special force. In this research, we analyzed 2400 images for 24 different basic signs what they use in their real mission. For this analysis, we classified their basic signs by convolutional neural network (CNN) algorithm. This research will help general people to take decision on hostage circumstances so that they can easily communicate with the military who have gone there to rescue them. We used multiple convolutional layers and get 92.50% accuracy. Using our model in any real system, a novice member of the military or special force can learn and validate his sign.

Keywords Convolutional Neural Network (CNN) · Deep learning · Military sign · Image classification

Md. Ekram Hossain (✉) · Md. Musa · N. K. Nisat · A. H. Thusar · Z. Hossain · Md. Sanzidul Islam
Department of Software Engineering, Daffodil International University, Dhanmondi, Dhaka 1207, Bangladesh
e-mail: ekram35-1936@diu.edu.bd

Md. Musa
e-mail: musa35-1870@diu.edu.bd

N. K. Nisat
e-mail: nahid35-1889@diu.edu.bd

A. H. Thusar
e-mail: ashraful35-1908@diu.edu.bd

Z. Hossain
e-mail: zaman35-1903@diu.edu.bd

Md. Sanzidul Islam
e-mail: sanzid.swe@diu.edu.bd

S. Borah et al. (eds.), *Soft Computing Techniques and Applications*, Advances in Intelligent Systems and Computing 1248,
https://doi.org/10.1007/978-981-15-7394-1_34

1 Introduction

In our everyday life machines are considered to be an important part. They put a great impact on our everyday life efficiently and effectively. Image classification and pattern recognition theory are an important topic in computer vision. For example, vehicle detection, signature verification, face recognition, etc. In this project, we select military sign database which has 24 different classes of image, such as sniper, dog, cover, rifle, enemy, pistol, etc. Due to low resolution and image of normal people, this problem is treated as a complex issue in this area. We wish to deeply find out how neural networks will work for image classification. Moreover, we made a great experiment about how convolutional neural network works.

Image classification is a process where classifies an image base on visual content. Identifying objects still now a great challenge in computer vision. We know a military person use their hand sign in different operation. But it's only between them. Our think that if a normal person used that sign to military, it's very helpful for military person to rescue hostage. Hostage people can be fast rescue from enemies.

Hostage rescue is the most sensitive operation in military cases. Hostage people can be suffering for long time. Here some case of hostage report (Table 1).

By this military sign, people can be rescued in less time and it's a help for military people to know about the enemy inside.

Military sign language is the silent communication model in any operation. If normal people know about those sign it will help military people to rescue hostage people. Hostage people will confident and believe on the military to rescue them in less time.

Data Collection Process

Data collection is the process of assembling and aligning information on targeted variable in an efficient way. We create some poster and invite normal people to click images in 24 different poses (military sign language). We camped in our university to click image of those signs. We made our goal to take picture of 100 people that's mean 2400 images are clicked (Fig. 1).

Table 1 Hostage crisis record

Incident	Location(s)	Start	Duration
Escape of Latifa	UAE, India	2018 March	Unknown
Yerevan hostage	Yerevan, Armenia	17 July 2016	15 days
Dhaka attack	Gulshan, Dhaka, Bangladesh	1 July 2016	11 h
Nightclub shooting	Orlando, Florida, United States	12 June 2016	3 h
Kunduz-Takhar highway	Kunduz province, Afghanistan	31 May 2016	Unclear end date
Bamako hotel attack	Bamako, Mali	20 November 2015	9 h

Fig. 1 Data collection process

2 Literature Review

Deep learning techniques are now apperceiving a huge development in their field of application mainly because of the great computational power that arrived by the latest GPUs. Moreover, the presence of big datasets [1–4] has built it possible to train neural networks and nearly perform like human levels, when tested against functions like image classification [5], object detection [6] and face recognition [7].

The range of huge applications, image classification is the hottest in computer vision. Neural network and a series of derivative algorithms [8] are used for a long time to solve image classification problems. The whole network has multiple layers of neurons and those can be a single unit processing.

The idea of neural network is encouraged by the biological neural [9]. The back-propagation algorithm is famous for multiple layers neural networks. Bryson and Yu-Chi [10] were introduced it and Werbos (1974), Rumelhart [11] improved it. Convolutional neuron network has more properties like convolution, down-pooling in place of amplification between input neurons and their weights (Simard et al. 2003).

Rapid growth of computing platforms like GPUs. Now, researchers and scientists are applying this algorithm to solve complicated image classification.

3 Methodology

3.1 Data Pre-processing

Data pre-processing is the transformation of raw data. It is important for training model because raw image will probably bad classification performance. Here firstly we transform our image into greyscale than those again convert to threshold.

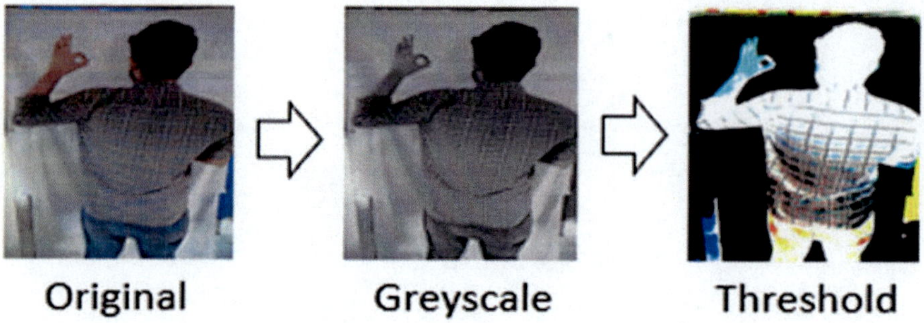

Fig. 2 Thresholding

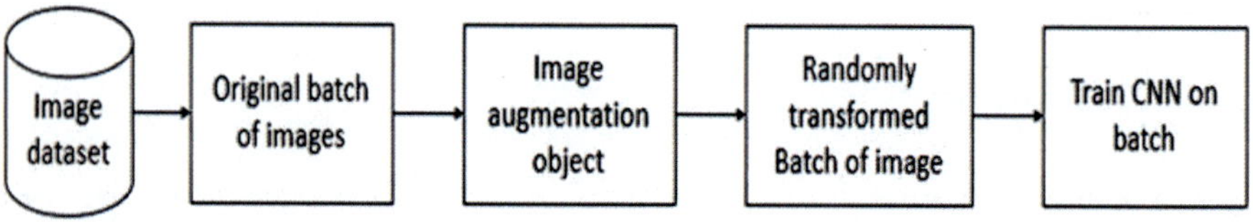

Fig. 3 Data augmentation process

Thresholding is image segmentation, where pixels of an image change for making easy analysis. In thresholding, an image form greyscale is converted into binary image (Figs. 2 and 3).

After that, we apply data augmentation to increase generalizability of the model. It is a process to spread the length of training dataset by modifying images in the dataset. It is increase accuracy of model. Here, we use *on-the-fly* data augmentation to build a great model.

The artificial neural network is a machine learning method that evolved from the idea of simulating the human brain. The data explosion in modern drug discovery research requires sophisticated analysis methods to uncover the hidden casual relationships between single or multiple responses and a large set of properties. Data argumentation plays a strategic and significant role to increase data diversity of training models, without collecting new data. Cropping, padding, and horizontal flipping techniques are commonly used to train large neural networks. CNN commonly used to several transfer functions to discover strong data and arguments data policies.

4 Model Architecture

In neural network, convolutional neural network is mainly used for image classification, object detection, recognition faces, etc. in those area CNNs are widely used. CNN image classifications pick up an input image, process it, and classify it given categories. Base on image resolution, computer sees an image like array of pixels.

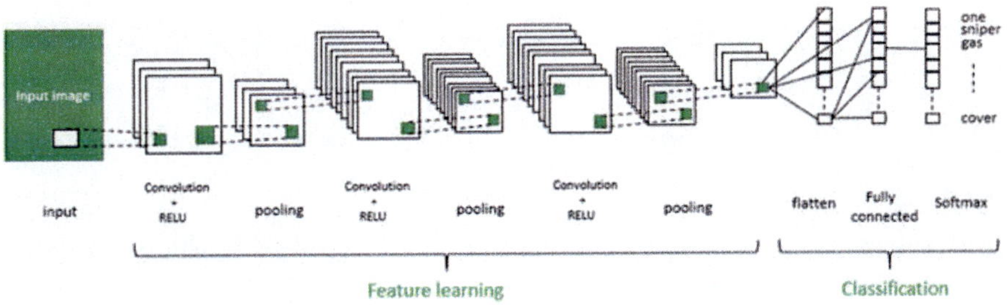

Fig. 4 CNN model architecture

Image resolution depends on $h * w * d$ (Hight = h, Weight = w, Dimension = d). Every input image will pass in a series of convolutional layers with filters, polling, fully connected layers then apply softmax function to classify images (Fig. 4).

4.1 Convolutional Layer

Convolutional layer is a process of extract features form an input image. Mainly, it's reduced the size of image for easy training. This is a mathematical operation that takes image matrix and a filter or kernel. Convolutional function is:

$$(f * g)(t) \stackrel{\text{Def}}{=} \int_{-\infty}^{\infty} f(\tau)g(f - \tau)\mathrm{d}\tau$$

An image matrix of element $(h * w * d)$, a filter has $(fh * fw * fd)$, output of volume dimension $(h - fh + 1) * (w - fw + 1) * 1$. A feature map is a function which maps a data vector to feature space. CNN is the convolution layer which produce a feature map. It is a popular deep learning architecture. Convolution defines the mathematical operation to merge two sets of information. Convolution applied on input data and filtered to produce feature map and visualize them one by one (Fig. 5).

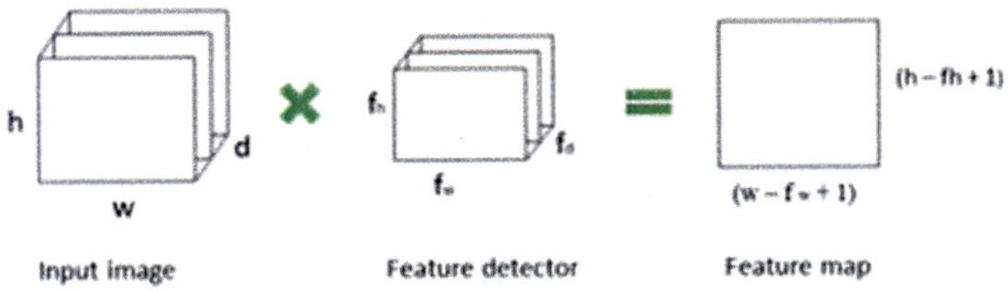

Fig. 5 Feature map

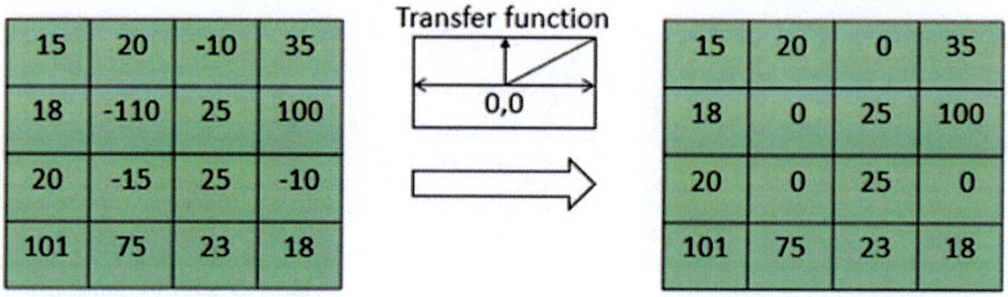

Fig. 6 ReLU operation

Convolution of an image with many kinds of filter such as edge detection, blur, sharpen, emboss, can perform operation.

4.2 ReLU

ReLu- Rectified Linear Units is a function model of ANN. Nowadays ReLu is used to all kinds of deep learning models. The rectified linear activation function is a piecewise linear function that will output the input directly if is positive, else, it will output zero. It is easier to train and often achieves better performance. It is a limitation of hidden layers of a Neural Network Model. To get output layers, here use a softmax function for classified to regression problem simply use a linear function. This function use for non-linear operation. The output is $f(x) = \max(0, x)$ by this function, we will increase non-linearity of an image because images themselves are highly non-linear (Fig. 6).

There are many non-linear functions such as tanh or sigmoid functions can be used to replace of ReLU but most of the data scientist use relu function for best performance [12].

4.3 Pooling

Max pooling has been used for this research. Max pooling helps to get brighter pixels from the images and that is why it is widely used. Max pooling is appropriate for this research because the background of our dataset's images are black and we are only hoping for brighter pixels. Max pooling also helps to reduce computation cost. In this section, reduce the number of parameters when images are large. Spatial pooling is called subsampling which reduce the dimensionality of map. There are different type of spatial pooling (Fig. 7) [13].

Here we use max pooling for reducing the parameters. Max pooling take hold of largest amount of element from the rectified feature map.

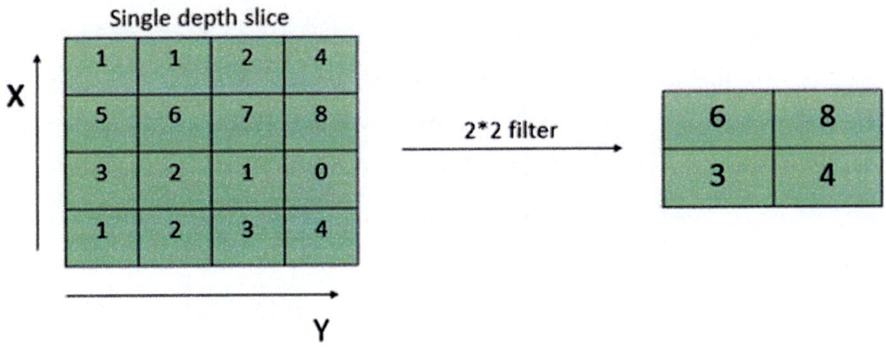

Fig. 7 Max pooling

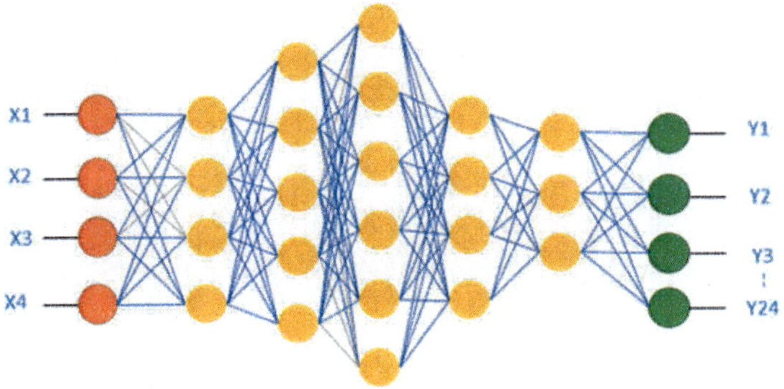

Fig. 8 Fully connected layer

4.4 Fully Connected Layer

This section is called as FC layers. After max pooling, we flattened our matrix into vector feed it into fully connected layer like neural network. The feature map matrix will convert as vector ($x1$, $x2$, $x3$, ...). We combined these features to create an efficient model. Finally, we applied the activation function to classify the outputs as one, sniper, cover, gas etc. (Fig. 8)

5 Model Evaluation

5.1 Result

In our dataset, we have 2400 images of 24 different classes and divided the dataset in 80% for training set and 20% for test set. For image dimension, used 128 * 128

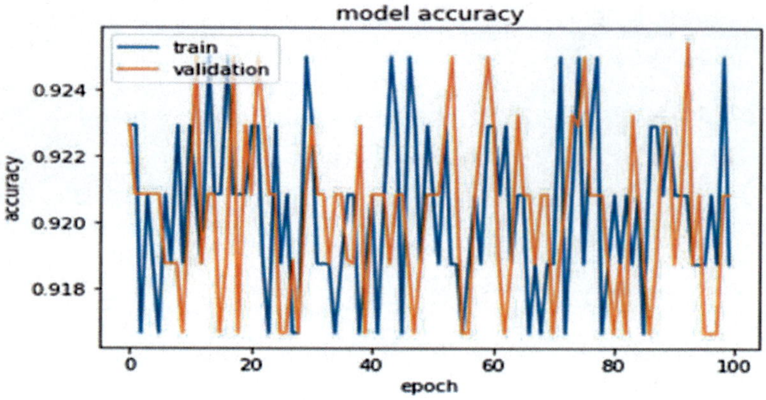

Fig. 9 Model accuracy

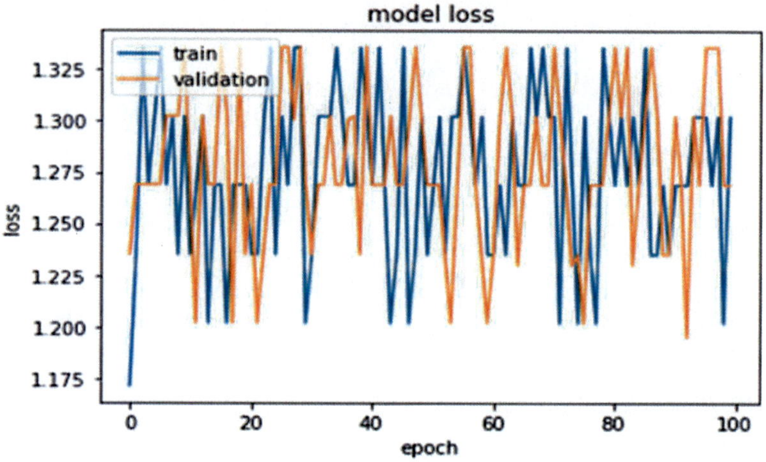

Fig. 10 Model loss

dimensions of image and filter 3 * 3. In first layer, total filters are 32 then 64 and in last layer 128. For this training epochs size was 100. We used four convolutional layers for training and accuracy was 92.50% and the validation result was 92.29% (Figs. 9 and 10).

6 Conclusion and Future Work

In this research paper, we mainly focused on creating an image dataset for ordinary military hand signs for particular meaningful gestures they use. Generally, it takes a huge time and it's difficult to collect accurate hand gestures from general people. We also applied some art of state techniques for our image data processing and training the classification model.

In the research work, we tried image masking and segmenting the background from the upper layer (main object) of 2D image and then gave as the CNN model input. Then we get better accuracy than the previous attempt which wasn't tried as masked image as input.

From the experimental point of view, initially, we collected some ordinary signs (24 classes) of military hand gestures and classified them with the method of Convolutional Neural Network model architecture. But there are some other signs which they use rarely and also there are some gestures which act as continuous movement of hands and body. Initially, we didn't include the rest of the part in our research work.

References

1. https://en.wikipedia.org/wiki/List_of_hostage_crisesİ
2. Babenko, A., Slesarev, A., Chigorin, A., Lempitsky, V.: Neural codes for image retrieval. In: European conference on computer vision, pp. 584–599. Springer, Berlin (2014)
3. Bansal, A., Nanduri, A., Castillo, C.D., Ranjan, R., Chellappa, R.: Umdfaces: an annotated face dataset for training deep networks. In: 2017 IEEE International Joint Conference on Biometrics (IJCB), pp. 464–473. IEEE (2017)
4. Cao, Q., Shen, L., Xie, W., Parkhi, O.M., Zisserman, A.: Vggface2: a dataset for recognizing faces across pose and age. In: 2018 13th IEEE International Conference on Automatic Face & Gesture Recognition (FG 2018), pp. 67–74. IEEE (2018)
5. Girshick, R., Donahue, J., Darrell, T., Malik, J.: Region-based convolutional networks for accurate object detection and segmentation. IEEE Trans. Pattern Anal. Mach. Intell. **38**(1), 142–158 (2016)
6. Guo, Y., Zhang, L., Hu, Y., He, X., Gao, J.: Ms-celeb-1m: a dataset and benchmark for large scale face recognition. In: European Conference on Computer Vision, pp. 87–102. Springer, Berlin (2016)
7. Grm, K., Pernus, M., Cluzel, L., Scheirer, W., Dobrišek, S., Štruc, V.: Face hallucination revisited: An exploratory study on dataset bias. arXiv preprint arXiv:1812.09010 (2018)
8. Mitchell, T.M.: Machine Learning, vol. 45, No. 37, pp. 870–877. McGraw Hill, Burr Ridge, IL (1997)
9. Aleksander, I., Morton, H.: An Introduction to Neural Computing, vol. 38, pp. 133–155. Chapman and Hall, London (1990)
10. Bryson, A.E.: Applied Optimal Control: Optimization, Estimation and Control, pp. 503– 512. CRC Press (1969)
11. Rumelhart, D.E., Hinton, G.E. Williams, R.J.: Learning representations by (1988)
12. https://towardsdatascience.com/activation-functions-and-its-types-which-is-better-a9a5310cc8f
13. https://medium.com/@bdhuma/which-pooling-method-is-better-maxpooling-vs-minpooling-vs-average-pooling-95fb03f45a9
14. He, K., Zhang, X., Ren, S., Sun, J.: Deep residual learning for image recognition. In: Proceedings of the IEEE conference on computer vision and pattern recognition, pp. 770–778 (2016)
15. https://medium.com/@danqing/a-practical-guide-to-relu-b83ca804f1f7

MediNET: A Deep Learning Approach to Recognize Bangladeshi Ordinary Medicinal Plants Using CNN

Md. Rafiuzzaman Bhuiyan, Md. Abdullahil-Oaphy, Rifa Shanzida Khanam, and Md. Sanzidul Islam

Abstract Medicine is the only thing by which we use where we feel bad condition of our body's physical and mental illness. Most of medicines are made of specific plants from our nature. These plants are also known as a medicinal plant. All the traditional Bangladeshi medical systems, namely Ayurveda, Unani, Homeopathy, prominently use medicinal plants. So, it is important to classify the right plant for medical preparation. The ability to identify these plants automatically is needed in recent days. For this, we proposed a renowned algorithm called convolutional neural network for recognizing the plants from leaf image. Our algorithm got 84.58% accuracy. We developed this. We believe that in the future the individuals who do not distinguish medicinal plants will recognize using this methodology.

Keywords Medicinal plant · Plant recognition · Convolutional neural network

1 Introduction

Plants are recognized through plant parts like roots, stem, leaves, flower and fruits. But in many cases, these all attributes do not show much differences mathematically and computationally. However, leaves are reached resources for species identification

Md. R. Bhuiyan (✉) · Md. Abdullahil-Oaphy · R. S. Khanam
Department of Computer Science and Engineering, Daffodil International University, Dhanmondi, Dhaka 1207, Bangladesh
e-mail: rafiuzzaman15-9655@diu.edu.bd

Md. Abdullahil-Oaphy
e-mail: abdullahil15-9868@diu.edu.bd

R. S. Khanam
e-mail: rifa15-9780@diu.edu.bd

Md. S. Islam
Department of Software Engineering, Daffodil International University, Dhanmondi, Dhaka 1207, Bangladesh
e-mail: sanzid.swe@diu.edu.bd

S. Borah et al. (eds.), *Soft Computing Techniques and Applications*, Advances in Intelligent Systems and Computing 1248,
https://doi.org/10.1007/978-981-15-7394-1_35

Table 1 Dataset information

No.	Scientific name	Local name	No. of samples	Actual size	Preprocess size
1.	*Aloe indica*	Aloe vera	150	1920 × 1040	128 × 128
2.	*Adhatoda vasica*	Bashok	150	1920 × 1040	128 × 128
3.	*Bryophyllum pinnatum*	Pathorkuchi	150	1920 × 1040	128 × 128
4.	*Centella asiatica*	Thankuni	150	1920 × 1040	128 × 128
5.	*Ocimum sanctum*	Tulsi	150	1920 × 1040	128 × 128
		Total	750		

and some leaves are also used for medical treatment which is well known as medicinal plants. These plants are also known as an asset of elements which are used for our medical treatment. Its leaves play a vital role in our life. World Health Organization—WHO—showed that nearly 80% of world population rely on herbal medicines. They also added that around 21,000 of different plant species are used for treatment [1].

We have collected 750 images from different nurseries in Dhaka city. The images have five different classes. We use local name to collect it. They are aloe vera, bashok, pathorkuchi, thankuni and tulsi. At first, we crop the images with a certain amount. Then, we apply data augmentation to make the dataset more large to reduce overfitting problem. Our dataset has different shapes. So, we need to resize them to make all the images into fixed pixels. After that, we use our own developed CNN architecture for training purpose. Later, we evaluate our training model for unseen data. Table 1 shows the whole information of our collected data.

In Bangladesh, there are almost 75% people who use medicinal plant leaves as their first priority of their primary health care [2]. Around 500 medicinal plants are found here. Three hundred of them are currently used in the arrangement of conventional drugs [2]. If we can increase the use of these plants, this will eventually decrease the cost of medicine. We know that there is some side effects of medicine, but leaves haven't. So, it will also decrease the side effects of medicine. So, our present work is to build a computer vision-based automated system that differentiates various types of medicinal plant leaves.

The goal of medicinal plant characterization is to help non-botanist clients to recognize the plants consequently. Manual recognizable process is excessively long, tedious, untrustworthy and progressively costly. Accelerating this procedure has a few advantages, for example, minimal effort, less time, less exertion for carrying out their responsibilities. This recognition framework can be utilized in making field guides intelligence, nursery the board, instructive reason and agribusiness rehearses. This innovation additionally can be applied into medicinal services, retail and some more.

2 Literature Review

Many works have been done in medicinal plant images. We have summarized some of them below:

In 2012, Sandeep Kumar proposed a strategy for identifying of medicinal plants dependent on some significant characteristics that extricated from a leaf image. This is where the plant can be recognized dependent on their characteristics, for example region, color histogram and edge histogram [3].

In 2012, Kumar et al. implemented the first portable mobile application using automated visual recognition tools for plant species identification. The framework, called Leafsnap, distinguishes plant species from photographs of leaves. The main point of this framework is to extract features that represent the curvature of leaf border on multiple scales. The system achieves remarkable performance on the actual image [4].

In 2015, Aakif proposed an algorithm that recognizes a plant in three independent stages: (i) preprocessing, (ii) feature extraction and (iii) classification. Various characteristics of leaf, for example morphological characteristics, Fourier descriptions and a new characteristic, are proposed in terms of characterizing shape features that are extricated. These extracted features become fed into an artificial neural network (ANN) [5].

In 2007, Wu proposed a neural network for automatic leaf recognition task. They used probabilistic neural network (PNN) for that. Their main upgrades are extraction and the classifier. They used 12 leaf for extracting features and orthogonality into 5 factors which comprise the info vector that are fed to the PNN. Over 1800 leaves of 32 classes their algorithm trained on [6].

In 2017, Ghazi proposed deep convolutional neural network for recognizing variety of plants. They used LifeCLEF 2015 dataset for evaluating their system. Here, the group utilized three popular transfer learning models, i.e., AlexNet, VGGNet and GoogLeNet [7].

In 2018, Too proposed a method which achieves an accuracy of 99.57% which is really impressive. Their work mainly focuses on fine-tuning technique. They analyzed deep convolutional architecture and also transfer learning models like VGG16, ResNet50, InceptionV4 and DenseNet for plant identification [8].

In 2019, Amuthalingeswaran proposed a renowned CNN-based study with over 8000 images from four different classes [9]

In [10–12] different approach leaf classification and recognition works has been shown.

In our paper, we propose a convolutional neural network for medicinal plant classification through leaf images.

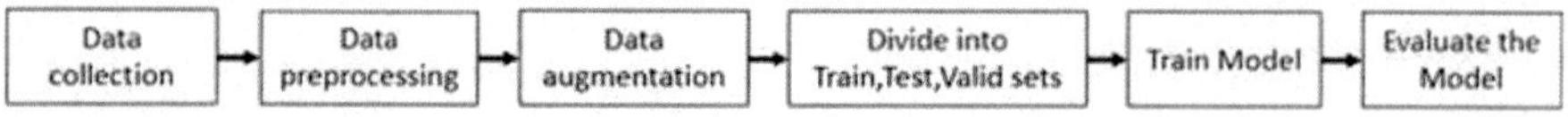

Fig. 1 Work flow

3 Proposed Methodology

The data collection process, preprocessing technique, our proposed model, etc., are all described in Fig. 1.

3.1 Dataset Collection

The variety of medicinal leaves are collected from different nurseries in Dhaka city. This dataset contains 2500 images from 5 different classes. We use 500 images for testing and rest 2000 images for training and validation. Now each of them contains 500 images. We have used local names for our work. The classes of our dataset are as follows (Fig. 2):

i. Aloe vera.
ii. Bashok.
iii. Pathorkuchi.
iv. Thankuni.
v. Tulsi.

3.2 Data Preprocessing

3.2.1 Data Augmentation

Overfitting is a common problem when the dataset is limited. As far our dataset is limited, we may get trouble in overfitting. For eliminating overfitting, we implement data augmentation. It actually artificially expanded the dataset. This technique creates modified versions of images that are not present in our dataset. We augmented our dataset in four different ways.

i. Rotate left 90°.
ii. Height shift range.
iii. Shear by certain amount.
iv. Flip horizontally.

3.2.2 Data Preparation

After augmentation process, these data have different shapes. Since our proposed model need a fixed shape for training & test purpose. All the data were resized into 128 × 128 pixels. We use RGB so that it was used to get a decent accuracy. At first, the dataset has 150 images, and after augmentation it increases to 500 images. Now we separate 100 images for testing and 400 images for training and validation per class, respectively.

3.3 Convolutional Neural Network

Convolutional neural network is an artificial neural network which is commonly used for working with two-dimensional data. Yann LeCun, a well-known scientist young as developed it in 1998 [13]. It consists many types of layers. In here, we take an image from our dataset. The image dimension is 128 × 128 × 3. We are passing into a ConvNet.

Fig. 2 Sample of our dataset

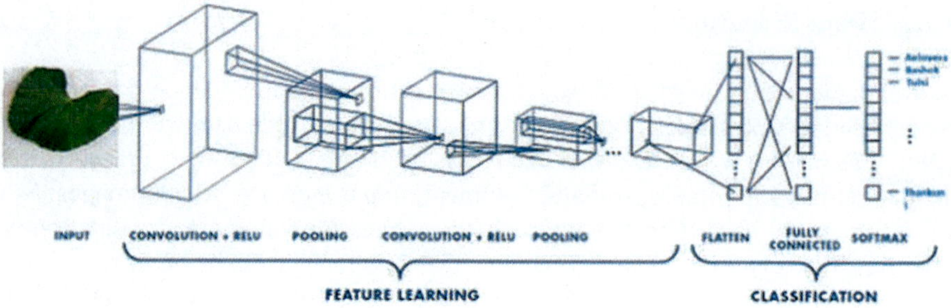

Fig. 3 Structure of CNN

Fig. 4 Max pool

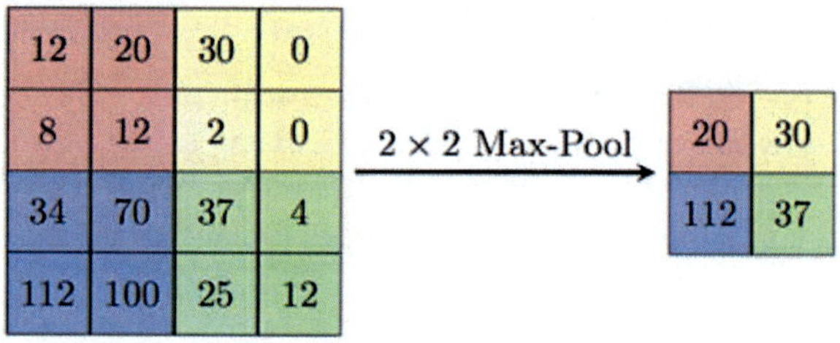

Input Layer: Raw image that we are passing is in input layer, dimension of 128 × 128 × 3.
Convolution Layer: It is the heart of the structure of CNN. It contains filter and also calculates the dot product between width, height of the input layer and the filters. If filter number in here is 24 and input image dimension is 128 × 128 × 3, then output dimension is 128 × 128 × 24.

Figure 3 shows an architecture of CNN.
Activation Function Layer: Activation is some sort of function that gives a corresponding output that is fed from an input. There are linear, nonlinear activation functions out there. In our task, we used a nonlinear activation function called ReLU. It simply converts negative numbers by setting them to zero.

$$f(x) = \max(0, x) \quad (1)$$

Pool Layer: After activation function layer, there comes pool layer. Many types of pool layer out there. We are using max pooling which is used mostly. It actually reduced the dimension of the images. If pool size 2 × 2 and stride 2 are using the max pooling layer, then the output dimension is 64 × 64 × 24. Figure 4 shows a max pooling layer operation.
Fully Connected Layer: Last layer of CNN. It consists of the full connection of all the previous layers. From the previous layer, it takes input and then makes into 1D vector.

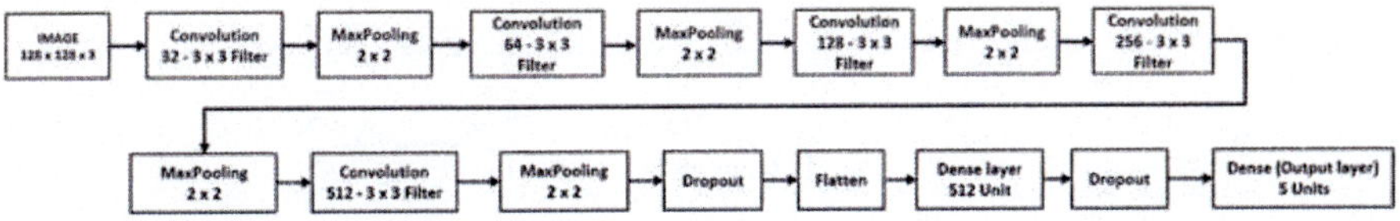

Fig. 5 Proposed model architecture

3.4 *Proposed Model*

We developed an architecture for recognizing medicinal plants. It is totally 15-layered network. The layers are
Convolution layer: We use five convolution layers in our model given below.

i. Convolution 32—3 × 3 filter.
ii. Convolution 64—3 × 3 filter.
iii. Convolution 128—3 × 3 filter.
iv. Convolution 256—3 × 3 filter.
v. Convolution 512—3 × 3 filter.

Pooling Layer: After convolutional layer their sits pooling layer. With a pool size of 2 × 2, we use five max pooling layers.

Dropout Layer: Dropout is dropping out some units in a neural network. For avoid overfitting dropout use. With layer rate of 0.25 and 0.5, our proposed model has two dropout layers.

Flatten Layer: After pooling, we will get a pooled feature map of 2D vector. This vector will then create a continuous 1D vector where the process is called flattening. Our model has one flatten layer.

Dense Layer: A dense layer is used for a classifier. We have used 512 units and 5 units of dense layer. This measures the output of our 5 classes also called output layer.

Figure 5 shows architecture of our proposed model.

3.5 *Training the Model*

We use renowned optimizer called Adam [14] optimizer to train our model with a smaller learning rate. It is one of the most used optimizers which is faster and more reliable to use. Categorical cross-entropy is used for loss function as the problem is multiclass classification. First we need to compile our model & then using fit( ) method for start training our model for 15 epochs with a batch size of 35. From 2000 images, 80% of them are for training and 20% used for validation.

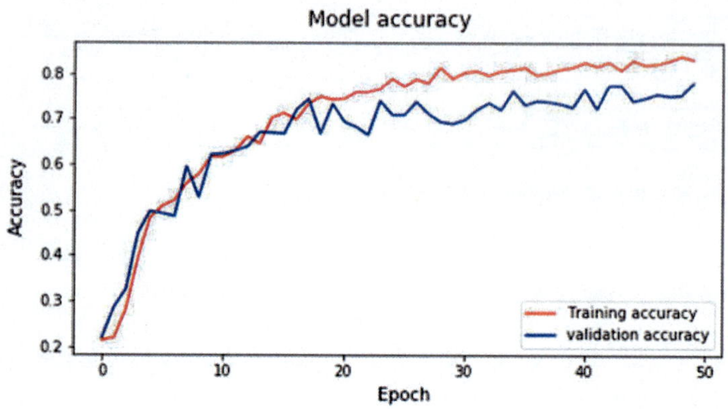

Fig. 6 Training versus validation accuracy

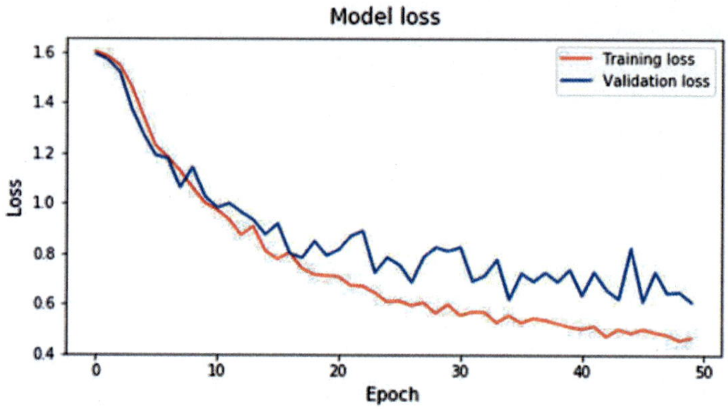

Fig. 7 Training versus validation loss

4 Performance Evaluation

Evaluating the model is really important in machine learning. It shows how effective the scoring of a dataset has been by a prepared model. Finishing training and validation, we acquire our results. When we apply the model into training data is called training accuracy while we apply the model with different classes of test data is called validation accuracy.

Figure 6 shows the training accuracy of our model is 84.58%, and we maintain our validation accuracy from 78.84 to 74.85%.

Figure 7 shows training loss and validation loss of our model. Our validation loss is 0.621.

Table 2 Classification report of our model

Class	Precision	Recall	$F1$-score
Aloe vera	0.78	0.78	0.78
Bashok	0.80	0.81	0.83
Pathorkuchi	0.71	0.88	0.78
Thankuni	0.74	0.84	0.83
Tulsi	0.86	0.84	0.78
Avg.	0.78	0.78	0.78

Table 3 Confusion matrix

	Aloe vera	Bashok	Pathorkuchi	Thankuni	Tulsi
Aloe vera	**88**	15	1	2	4
Bashok	2	**84**	4	6	4
Pathorkuchi	10	2	**83**	1	4
Thankuni	4	2	7	**82**	5
Tulsi	11	3	4	3	**79**

5 Result and Discussion

Completing all of our training, we test our model with 500 data. In Table 2, we see the classification report that is calculated from our model. We evaluated precision, recall and $F1$-score.

In Table 3 shows the confusion matrix of our model.

6 Future Work

There are many kinds of medicinal plants found all over the world. Each of them has its own unique qualities which could be a cure of many types of diseases. So, we need to in the future make the dataset bigger with various medicinal plant leaf images and proposed a CNN with more improvements. Using new technologies and advancement in computer vision, we make better solution for our treatment. In the future, people used more medicinal plant leaves for their treatment rather than medicine.

7 Conclusion

In this paper, our proposed model shows decent accuracy for medicinal plant leaf classification. We proposed a model by our own. This methodology performed well on field images that we are tested. This work assists non-botanist to recognize their desire medicinal plants. Further, it could be used to classify any sort of medicinal leaf and applicable for areas where leaf classification is required.

References

1. Introduction and Importance of Medicinal Plants and Herbs: Introduction and Importance of Medicinal Plants and Herbs. National Health Portal of India. https://www.nhp.gov.in/introduction-and-importance-of-medicinal-plants-and-herbs_mtl. Last Accessed 21 Nov 2019
2. pharmabiz.com/ArticleDetails.aspx?aid=77992&sid=21. Last accessed 23 Nov 2019
3. Kumar, S.: Leaf color, area and edge features based approach for identification of Indian medicinal plants (2012)
4. Kumar, N.: Leafsnap: a computer vision system for automatic plant species identification. In: Fitzgibbon, A., Lazebnik, S., Perona, P., Sato, Y., Schmid, C. (eds.) Computer Vision—ECCV (ECCV 2012). Lecture Notes in Computer Science, vol. 7573. Springer, Berlin (2012)
5. Aakif, A., Khan, Muhammad: Automatic classification of plants based on their leaves. Biosyst. Eng. **139**, 66–75 (2015). https://doi.org/10.1016/j.biosystemseng.2015.08.003
6. Wu, S.G., Bao, F.S., Xu, E.Y., Wang, Y., Chang, Y., Xiang, Q.: A Leaf recognition algorithm for plant classification using probabilistic neural network. In: IEEE International Symposium on Signal Processing and Information Technology, Giza, pp. 11–16 (2007). https://doi.org/10.1109/ISSPIT.2007.4458016
7. Ghazi, M.M., Yanikoglu, B., Aptoula, E.: Plant identification using deep neural networks via optimization of transfer learning parameters. Neurocomputing **235**, 228–235 (2017). https://doi.org/10.1016/j.neucom.2017.01.018
8. Too, E.C., Yujian, L., Njuki, S., Ying-chun, L.: A comparative study of fine-tuning deep learning models for plant disease identification. Comput. Electron. Agric. **161**, 272–279 (2019)
9. Amuthalingeswaran, C., Sivakumar, M., Renuga, P., Alexpandi, S., Elamathi, J., Hari, S.S.: Identification of medicinal plant's and their usage by using deep learning. In: 2019 3rd International Conference on Trends in Electronics and Informatics (ICOEI), Tirunelveli, India, pp. 886–890 (2019). https://doi.org/10.1109/ICOEI.2019.8862765
10. Sainin, M.S., Alfred, R.: Feature selection for Malaysian medicinal plant leaf shape identification and classification. In: 2014 International Conference on Computational Science and Technology (ICCST). IEEE (2014)
11. Jain, G., Mittal, D.: Prototype designing of computer-aided classification system for leaf images of medicinal plants. J. Biomed. Eng. Med. Imaging **4**(2), 115–115 (2017)
12. Vijayashree, T., Gopal, A.: Leaf identification for the extraction of medicinal qualities using image processing algorithm. In: 2017 International Conference on Intelligent Computing and Control (I2C2). IEEE (2017)
13. Bengio, Y., Lecun, Y.: Convolutional Networks for Images, Speech, and Time-Series (1997)
14. Kingma, D.P., Ba, J.: Adam: A Method for Stochastic Optimization. CoRR, abs/1412.6980 (2014)

A Review on Automatic Speech Emotion Recognition with an Experiment Using Multilayer Perceptron Classifier

Abdullah Al Mamun Sardar, Md. Sanzidul Islam, and Touhid Bhuiyan

Abstract Human–machine interaction is becoming popular day by day; to interact with machine, speech emotion recognition is as important as human to human interaction. In this research, we demonstrate a speech emotion recognition system which takes speech as input and classify emotions that the speech contains. We choose multilayer perceptron (MLP) classifier to do this task. Features that we have extracted from speech are mel-frequency cepstral coefficients (MFCC), chroma and mel-spectrogram frequency. RADVES dataset has been used and we have got 73% accuracy.

Keywords Speech emotion recognition · MLP classifier · MFCC · Chroma · Mel-spectrogram frequency

1 Introduction

Emotion has a huge impact in our day to day conversation. When we human interact with another human being, we understand them properly with proper understanding of our emotional state of conversation. Problem arises when a human is interacting with machine, because without proper training, machine will not be able to extract emotion from human speech. In current era, when artificial intelligence is having impact in our day to day life, it is necessary to train machine in a way that they can understand human emotion while dealing with speech. Speech emotion recognition system conducted with the knowledge of speech signal, extracting features from

A. A. M. Sardar (✉) · Md. Sanzidul Islam · T. Bhuiyan
Department of Software Engineering, Daffodil International University, Dhanmondi, Dhaka 1207, Bangladesh
e-mail: mamun35-1930@diu.edu.bd

Md. Sanzidul Islam
e-mail: sanzid.swe@diu.edu.bd

T. Bhuiyan
e-mail: touhidbhuiyan@gmail.com

S. Borah et al. (eds.), *Soft Computing Techniques and Applications*, Advances in Intelligent Systems and Computing 1248,
https://doi.org/10.1007/978-981-15-7394-1_36

speech and then finally recognizing emotions [1]. Several models are there to do this task, for example, CNN model-based SER system [2], deep belief and SVM-based SER system [3], etc. In this work, we have presented a system for recognizing six-acted emotional states (disgust, anger, fear, calm, surprised and happy).

2 Literature Review

Speech emotion recognition can be described as classification problem from machine learning point of view [4]. In prior research, many acoustic features have been investigated. Among all those feature some are common and very effective in speech emotion recognition and they are mel-frequency cepstrum coefficients (MFCC), linear predictor coefficient (LPC), chroma, etc.

Mel-frequency cepstrum coefficients (MFCC): For voice signal processing, MFCC is the most used spectral property and it is the best for speech recognition (SR) because it considers the perception sensitivity of human with respect to frequencies [4]. Figure 1 shows the extraction and calculation process of MFCC.

Some common machine learning and deep learning algorithms have been experimented in this field. In some research, deep neural network (DNN) and extreme learning machine (ELM) have been used to extract high level features [5] shown in Fig. 2.

For utterance level emotion classification, instead of DNN, extreme learning machine (ELM) has been used which is a single hidden layer neural network. SER is also known as pattern recognition system, because emotions are classified as a pattern. Both pattern recognition and SER have some common process [6].

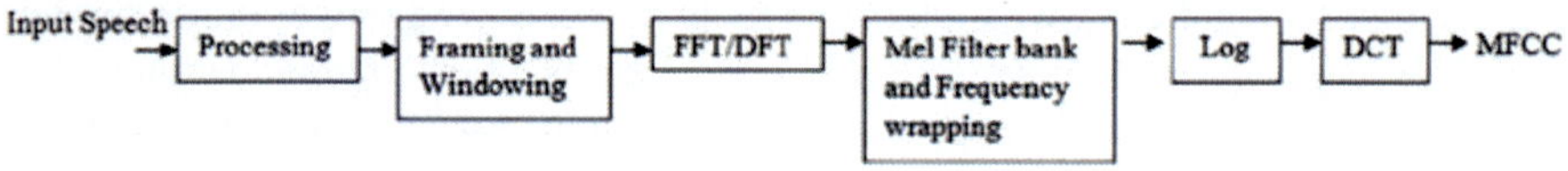

Fig. 1 Extraction of MFCC [4]

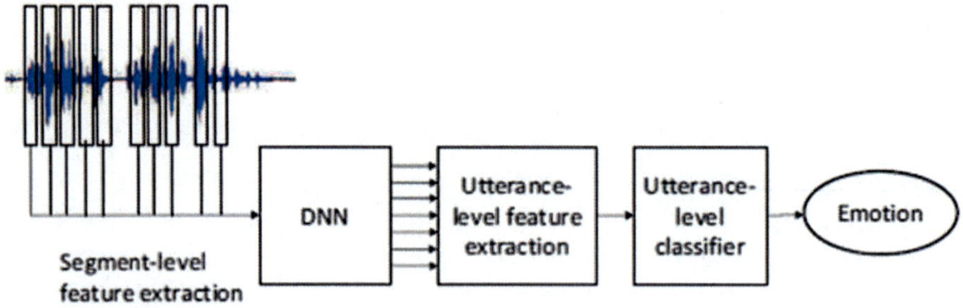

Fig. 2 Model of SER using DNN and extreme learning machine [5]

Convolutional neural network (CNN) is another recognized model for SER. This model is being used in SER to extract emotion from raw speech. It is also being used for sentiment classification. Interactive dialogue systems have also been using CNN to recognize emotion from real-time speech [7].

Some other machine learning algorithms like support vector machine (SVM) and k-nearest neighbor (KNN) have been used in the field of SER. SVM has developed based on the theory of statistics and structural risk reduction principle [3]. SVM is mapped to low-dimensional feature vector to high-dimensional feature vector space to solve nonlinear separable problem [3]. Support vector machine is also being used in the field of emotion recognition system. SVM use two methods to solve classification problem, i.e., (1) one to all and (2) one to one. According to a research which has been done on SER shows that one to one has higher accuracy [3]. SVM is good for limited training data and RNN is good for learning time series data [8]. In some other research, SVM has been used as a binary classifier which is an approach to solve SER problem by building seven different SVMs for seven different emotions. The process was to set each SVM to a class (emotion) and then recognizing the class which gives the highest output score. If the output score was negative, then a sample cannot be classified [1].

K-nearest neighbor (KNN) is a machine learning algorithm which is very popular in pattern recognition field. As SER can also be described as a pattern recognition system [6], KNN has also been experimented in speech emotion recognition field. KNN is the nearest neighbor technique that classify unknown sample. Figure 3 is representing a general process of speech emotion recognition.

There are some other algorithms that are good for feature extraction and classifying emotion from speech. Algorithms like recurrent neural network (RNN) and multivariate linear regression (MLR) have been used quite a lot in speech emotion recognition. [4]. RNN has been used to learn both short-term low-level descriptions like voicing probability, energy, MFCC, harmonics to noise ratio and long-term aggregations [8].

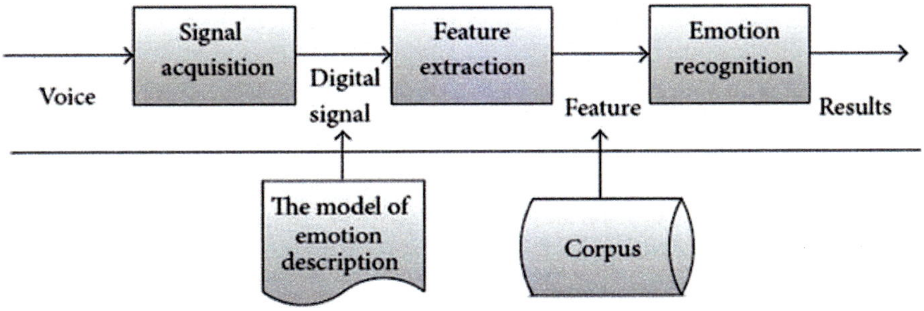

Fig. 3 Block diagram of SER [3]

3 Librosa for SER

Librosa is a Python library which has been built to deal with music signal processing and it being used in emotion recognition system to extract features that are important for classifying emotion. In librosa, a signal from audio is represented as a NumPy array which is one dimensional. Several methods are there that are important to extract essential features from speech. To extract spectral features, librosa has `librosa.feature`[9]. For representing audio signal, a scale is used called mel frequency. Librosa has a method to deal with it called `librosa.feature.melspectrogram` [9]. For speech smotin recognition, MFCC is the most used feature and librosa is used `librosa.feature.mfcc`to deals with it. Another feature called pitch class or chroma that has been used quite also in the field of SER. Librosa use `chroma_stft`(STFT stand for short-time Fourier transform) and `chroma_cqt`(CQT stand for constant Q transform) [9] to deal with chroma. However, in this research, we have only used`chroma_stft`[9]. The code segment given below has some practical example of these features [9].

```
>>> filename = librosa.util.example_audio_file()
>>> y, sr = librosa.load(filename,
...                      offset=25.0,
...                      duration=20.0)
>>> spectrogram = np.abs(librosa.stft(y))
>>> melspec = librosa.feature.melspectrogram(y=y,
...                                          sr=sr)
>>> chroma = librosa.feature.chroma_cqt(y=y,
...                                     sr=sr)
>>> tonnetz = librosa.feature.tonnetz(y=y, sr=sr)
```

4 Proposed System of Speech Emotion Recognition

4.1 Proposed Model

We took speech from RADVES database and then we extracted MFCC, chroma and mel-spectrogram frequency. Then, we classified six emotions using MLP classifier. Model and details are given in Fig. 4.

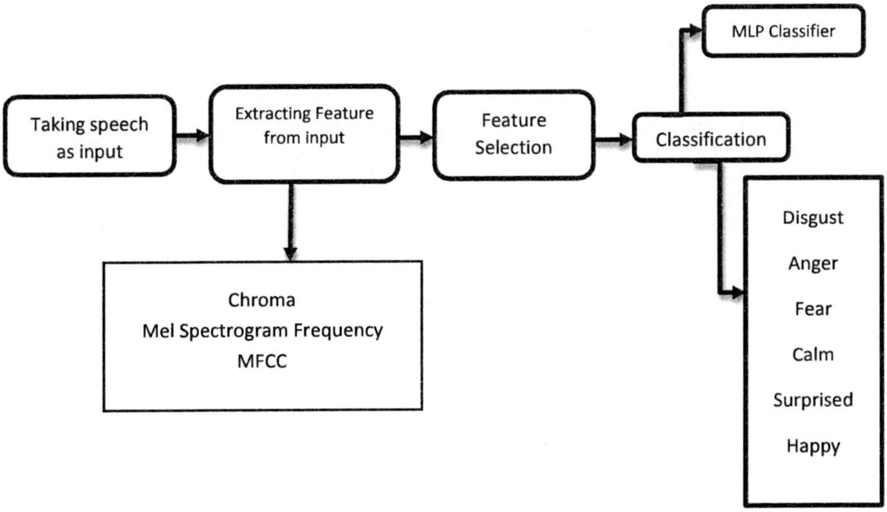

Fig. 4 Proposed model for this research

4.2 Experiment Setup and Feature Extraction

In our system, we have set up the task as classification. It means that each segment has been classified as a particular class of emotion or not. We have split the dataset into two parts—80% data has been taken as test data and 20% as test data. Feature extraction and selection are the process that comes after taking speech as input. In some recent research of speech emotion recognition, researchers extracted some common features and they are very effective. Those features include linear prediction coefficient (LPC), mel-frequency cepstrum coefficients (MFCC). However, we extracted three features and they are mel-frequency cepstrum coefficients (MFCC), mel-spectrogram frequency and chroma.

4.3 Classification Method

The goal of a classification algorithm is to learn from training data and apply them to test data. In literature review section, we described some classification algorithm but now we are describing the algorithm that we used in this research.

4.3.1 Perceptron

A single layer of a neural network is called perceptron. It is also known as a linear classifier. It is used to classify a given input that whether a given input belongs

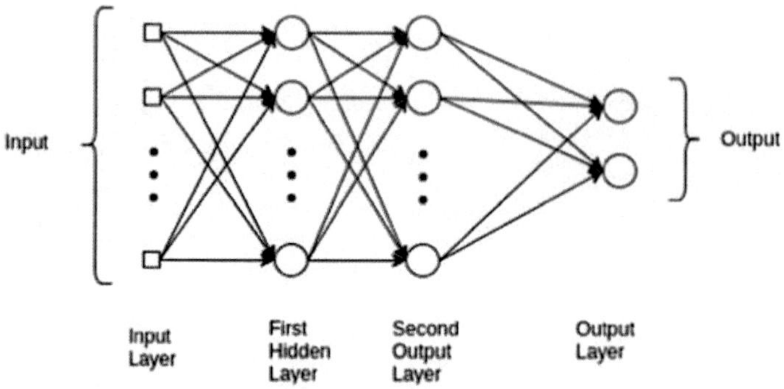

Fig. 5 Multilayer perceptron

to a particular class or not. A perceptron first takes some input and then multiply it with weights and then weighted sum add up these multiplied values. Then that summation value is applied in an activation function which activate based on the range of summation value.

$$Y = \sum (\text{weight} * \text{input}) + \text{bias}$$

4.3.2 Multiplayer Perceptron (MLP)

Multilayer perceptron is built upon more than one perceptron. It is finite acyclic graph. A layer is made by one or more perceptron. Receiving the signal is done at input layer, in this research, output layer classifies the emotion, and there can be one or more hidden layer in MLP. A typical structure of MLP has given in Fig. 5.

4.3.3 Process of MLP

In input layer, MLP takes the data as input and then calculates the weight and then passes it into an activation function. Neurons in ith layer become the input of $i + 1$th layer. Using activation functions, MLP do the calculation in each layer. Many activation functions are there like sigmoid function, rectified linear units (ReLU), etc. In this research, we used ReLU as activation function (Fig. 6).

Here, b is the bias, n is the number of inputs and w is the weight. This function becomes activated when value is greater than 0, otherwise, it does not activate.

MLP push the output at the current layer using any of these activation functions. After pushing, the calculated output through activation function into next layer repeat the process until the output layer is arrived. At the output layer, the calculations are

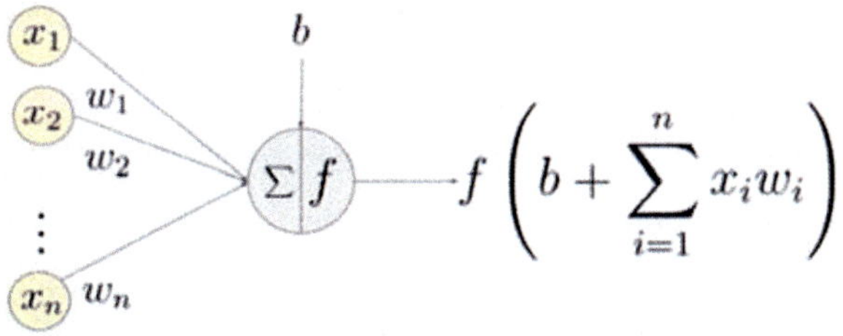

Fig. 6 Activation function, f

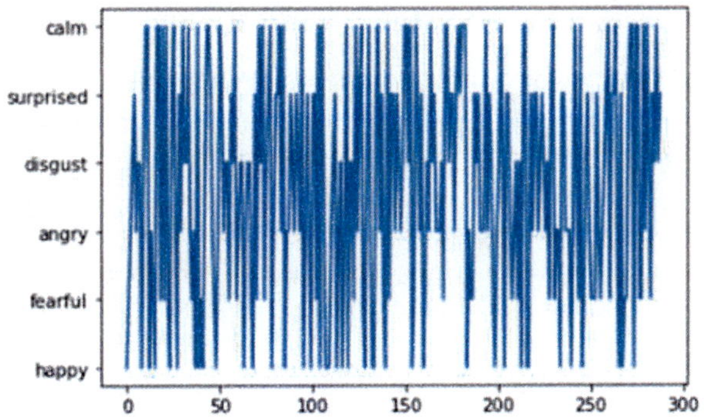

Fig. 7 Classification of emotion

either used for backpropagation algorithm or a final decision will be made based on the output.

5 The Experiment and Analysis

We have used RADVES dataset. The dataset contains 7356 audio files which were performed by 24 actors. MFCC scale value was 60. In this experiment, instead of epoch we used batch_file and the range was 256. Hidden layer of MLP classifier was 300. Using the RADVES dataset and applying them on MLP classifier, we got 73% accuracy. Emotions that we have classified have shown in Fig. 7.

6 Conclusion

In this research, we have presented a speech emotion recognition system which is based on MLP classifier. We have presented some other algorithms that are also very impactful in the field of speech emotion recognition. Speech emotion recognition is a vital part of human–machine interaction. As the uses of artificial intelligence are increasing day by day, the necessity of SER is also increasing. By combining those algorithms and extracting more effective features, maybe, we can have a better experiment and could get more accuracy.

References

1. Khan, M., et al.: Comparison between k-nn and svm method for speech emotion recognition. Int. J. Comput. Sci. Eng. **3**(2), 607–611 (2011)
2. Tripathi, S., et al.: Deep learning based emotion recognition system using speech features and transcriptions. arXiv preprint arXiv:1906.05681 (2019)
3. Huang, C., et al.: A research of speech emotion recognition based on deep belief network and SVM. In: Mathematical Problems in Engineering (2014)
4. Kerkeni, L., et al.: Automatic speech emotion recognition using machine learning. In: Social Media and Machine Learning. IntechOpen (2019)
5. Han, K., Yu, D., Tashev, I.: Speech emotion recognition using deep neural network and extreme learning machine. In: Fifteenth Annual Conference of the International Speech Communication Association (2014)
6. Ingale, A.B., Chaudhari, D.S.: Speech emotion recognition. Int. J. Soft Comput.Eng. (IJSCE) **2**(1), 235–238 (2012)
7. Bertero, D., et al.: Real-time speech emotion and sentiment recognition for interactive dialogue systems. In: Proceedings of the 2016 Conference on Empirical Methods in Natural Language Processing (2016)
8. Mirsamadi, S., Barsoum, E., Zhang, C.: Automatic speech emotion recognition using recurrent neural networks with local attention. In: 2017 IEEE International Conference on Acoustics, Speech and Signal Processing (ICASSP), pp. 2227–223. IEEE, 15 Mar 2017
9. McFee, B., et al.: Librosa: audio and music signal analysis in python. In: Proceedings of the 14th Python in Science Conference, vol. 8 (2015)

Classifying the Usual Leaf Diseases of Paddy Plants in Bangladesh Using Multilayered CNN Architecture

Md. Abdullahil-Oaphy, Md. Rafiuzzaman Bhuiyan, and Md. Sanzidul Islam

Abstract More than 130 million people in Bangladesh depend on rice as their main food. Half of the employment of the rural area and the agricultural GDP of Bangladesh depend on rice production. Nearly more than 10 million farmer families cultivate rice in Bangladesh. Almost 10% of rice cultivation is depreciated by different types of rice plant diseases caused by pests. This is the reason why we worked on detecting rice plant ***(Oryza sativa)*** disease by visual observation(images). 3265 images of rice plant disease have been collected for this study which belongs to four classes: hispa, brown spot, leaf blast and the healthy ones. The images of a diseased leaf are collected from rice fields. We have used a pre-trained model for classification and extracting features. By using this model, we got incentive results with the accuracy rate of 93.21%.

Keywords Rice plant diseases · Deep learning · Convolutional neural network

1 Introduction

The main food in Bangladesh is rice. Not only in Bangladesh but also half of the world population depend on rice as their main food. As diseases of rice have a subversive payoff on manufacturing of rice, that is why it is a key impendence for world food safety. That is why identification of rice disease is necessary for

Md. Abdullahil-Oaphy · Md. R. Bhuiyan
Department of Computer Science and Engineering, Daffodil International University, Dhanmondi, Dhaka 1207, Bangladesh
e-mail: abdullahil15-9868@diu.edu.bd

Md. R. Bhuiyan
e-mail: rafiuzzaman15-9655@diu.edu.bd

Md. S. Islam (✉)
Department of Software Engineering, Daffodil International University, Dhanmondi, Dhaka 1207, Bangladesh
e-mail: sanzid.swe@diu.edu.bd

S. Borah et al. (eds.), *Soft Computing Techniques and Applications*, Advances in Intelligent Systems and Computing 1248,
https://doi.org/10.1007/978-981-15-7394-1_37

ensuring the food safety because every year crops in a huge quantity are damaged by diseases. Globally, microorganism and epidemic reasons rice production damage upto 25–41%. It causes a great loss to the world economy and the farmers each year. Detecting rice diseases in a traditional way requires a specialized laboratory, expert's knowledge and experience that is why it is pretty much expensive and not efficient for rural farmers.

In our country, farmers have to use a lot of pesticides while cultivating rice to protect rice plants (paddy) from various kinds of disease. Use of pesticides in rice cultivation depends on the condition of stiffness of the disease. If farmers can classify the disease in the initial stage, then the problem can be handled by using less amount of pesticides, but if they cannot detect the disease in the initial stage and it gets into serious stage, then farmers have to use a lot of pesticides and it costs a lot to our poor farmers. It not only hampers our rice production but also hampers the farmers financially. But if the diseases can be classified in initial stage, then the extra cost of using a lot of pesticides can be optimized, and it will help our rice production and farmers financially.

On the other hand, in the rural areas where farmers cannot get help from government agricultural officers or agricultural experts easily, they face more problems. Sometimes, they cannot understand the symptoms of the disease, and they use wrong pesticides which is more harmful to our crops. It damages our crops even more. It hampers both the rice production and poor farmers miserably. The main vision of this thought is to improve a mechanized system to classify different types of rice disease just by the images of rice leaf using deep learning (CNN). By this, our poor farmers can classify rice diseases in initial stage without any help of experts, and it will help them in their rice production. As this technology does not require a laboratory with the experience of experts to detect and classify rice diseases, that is why it is not expensive. To improve the technology and its accuracy, researchers have been working for a long time on machine-driven rice disease detection such as by pattern identification methods [1], advanced image processing [2] and computer vision [3]. This approach will help farmers to classify rice disease within a few moments, and it will be very much helpful for the farmers from remote areas.

2 Literature Review

Naik and Sivappagari in the year 2016 used SVM and HSV features to detect plant leaf diseases. In this paper, they used genetic algorithm to optimize loss function [4]. Olmschenk et al. proposed a model which can classify images.They used GAN to classify images automatically [5]. Hadayat et al. used convolutional neural network (CNN) method in their study to detect corn plant diseases. They classified three types of corn plant diseases [6]. Qiang et al. Inception-V3 neural network to identify crop diseases [7]. Ramesh and Vydeki proposed a model to identify blast disease of rice crops in its early stage, thus the loss could be minimized as much as possible. They used KNN and ANN in their methodology [8]. A methodology was proposed by Zhang et al. to classify apple leaf disease using image processing and pattern recognition method [9]. A CNN model was introduced by Mohanty et al. to classify 14 crop species and 26 crop diseases [10].

3 Methods Research Framework

Figure 1 designates the study structure. The study starts by identifying a problem and proposing a solution for that problem (rice plant diseases). Then next step is to analyze the system requirements and study previous works on rice plant disease classification to get knowledge about previous works. The initial work is to collect dataset of images of rice leaf disease and preprocessing the dataset. The final move is to design a CNN model and run that one on the dataset. Then the eventual outcomes obtained are the enumeration of rice leaf disease classification.

3.1 Dataset

For this study, we have prepared a dataset with 3265 images of rice leaf. In this dataset, we have images of three rice leaf diseases, namely hispa, brown spot and leaf blast along with the healthy ones. Training data of the images of healthy and diseased leaf are used so that the model could learn about the characteristics of diseases and could detect them. On the other hand, we used testing data to examine what is the accurateness of the model of classifying and detecting diseases from rice leaf image based on the characteristics of the leaf. Each image of the dataset was RGB image and resized into 128 × 128 pixels to use fixed shaped images in the training and testing of our model (Fig. 2).

3.2 Convolutional Neural Network

Convolutional neural network (CNN) is designed for data processing which is an artificial neural network. Now convolutional neural network is also used in agricultural field to enrich this field scientifically [11].

Figure 3 shows a model of convolutional neural network. It has levels to examine and categorization. Initially, data are inputted to make the model understand and

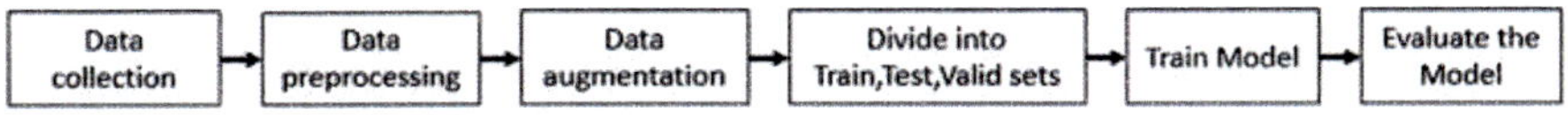

Fig. 1 Study structure

Fig. 2 Sample of dataset

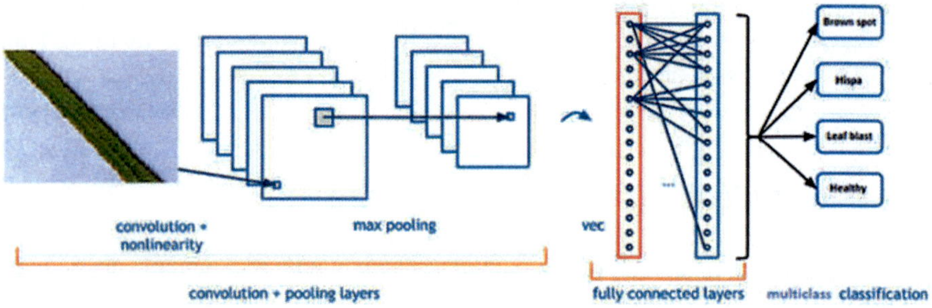

Fig. 3 Structure of CNN

recognize the clues are there in the data. The model goes ahead with activation of ReLu, and then pooling is committed. Then the neurons of a stage get connected to neurons of another stage.

Convolutional Layer In a CNN architecture, convolution layers are the major parts. Convolution operations are performed on the output of other functions in this stage.

The output function from image is applied as a featured map. Convolution layer extracts features from the input images. Linear transformations are produced by convolution from the input data followed by the information available in the data. Which convolution layer is used is differentiated by the weights on that layer.

Activation Function The activation function ReLu is used in our model to increase the nonlinearity. Three by three sized kernel is used for each convolution layer. It is used to increase the accuracy of classification and speeding up the training process.

Pooling Pooling is used to preserve the features present in an image. When we will use max pooling in an image, upto 75% of the information present in an image but not important for our image will be reduced. It will also prevent overfitting. Max pooling method is the one which is commonly used in CNN. One of max pooling layer example is shown in Fig. 4.

Flattening After pooling, we will get a pooled feature map. The results of each row of the feature map will be entered into a single column for further processing, and it is called flattening (Fig. 5).

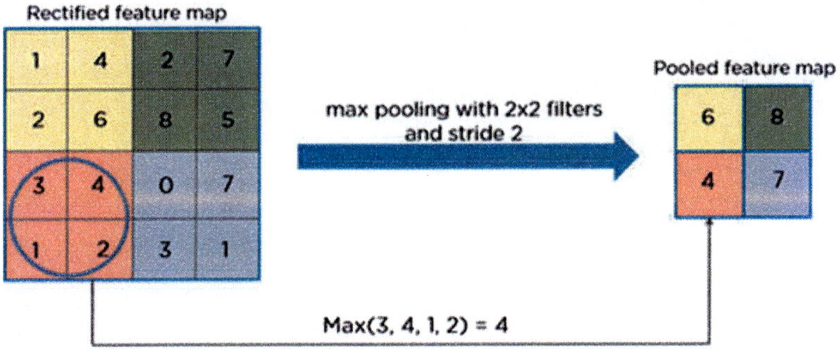

Fig. 4 Max pooling

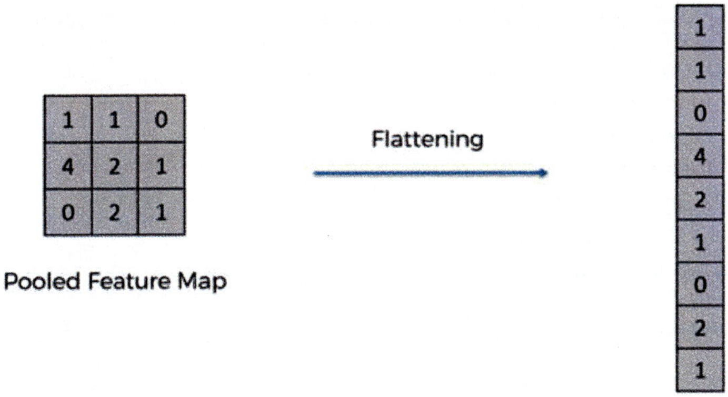

Fig. 5 Flattening

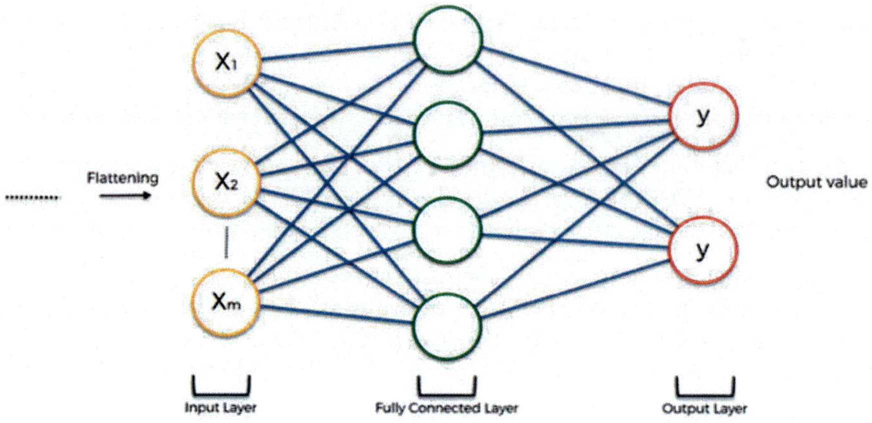

Fig. 6 Full Connection

Full Connection After the flattening layer, the final column of parameters we got will be used as the input layer in fully connected layer. Features of the input layer will be combined with more features so that the model could predict the class even better (Fig. 6).

Output When the training and categorization method is completed, then the output layer is the final step. Three types of rice leaf diseases are there, and they are hispa, brown spot and leaf blast.

3.3 Prooposed Model

For this thought, an architecture has been developed by us to classify rice plant diseases which is a network of 15 layers. They are as follows.
Convolution Layer Five convolution layers have been used in our model which are

i. Convolution 16—3 × 3 filter.
ii. Convolution 32—3 × 3 filter.
iii. Convolution 64—3 × 3 filter.
iv. Convolution 96—3 × 3 filter.
v. Convolution 128—3 × 3 filter.

Pooling Layer After convolution layer 5, max pooling layer has been used of size 2 × 2.

Dropout Layer To avoid overfitting, we have used dropout layer of layer rate 0.50.

Flatten Layer When pooling is finished, we get a two-dimensional array as a result, and then in flatten layer it is converted into a single linear vector. Our model has a flatten layer which is a 1D vector.

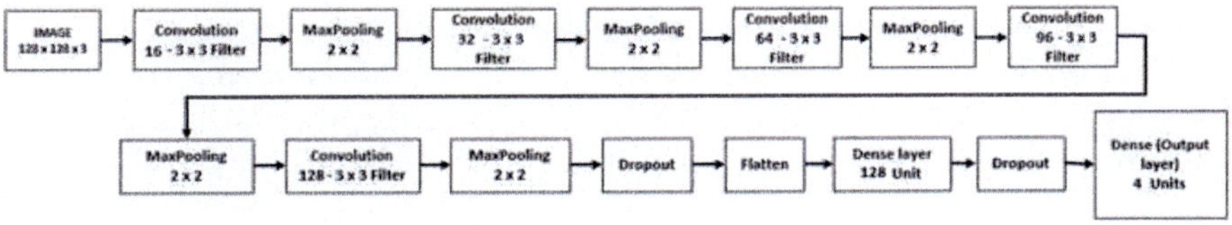

Fig. 7 Proposed model architecture

Dense Layer It is the last layer which classifies the diseases. Here dense layer classifies the four classes of our study and gives us the output.
Figure 7 shows architecture of our proposed model.

4 Results and Discussion

4.1 Hardware Specifications

Hardware machines that are utilized for the work are displayed in Table 1.

4.2 Training

The process which makes a system learn existing features of images and classify them is training process [12]. Here we used 3265 images of rice leaf of three types of rice leaf diseases, namely hispa, brown spot and leaf blast along with the healthy ones.

In Table 2, accuracy gained by our model is shown when the dataset is partitioned in different amount of training and testing data.

The accuracy results of the used CNN model are visualized in Fig. 8. Accuracy on train is 0.9321, and validation is 0.9239.

This graph in Fig. 9 visualizes loss outcomes of the CNN model. Loss on the train of the model is 0.2108, and on test is 0.2170.

Table 1 Hardware specification

Name	Description
CPU	Intel Core i5-8265U
GPU	Nvidia MX250 2GB
RAM	8GB DDR4
HDD	2 TB

Table 2 Classification accuracy with different training testing partition

Training data (%)	Testing data (%)	Accuracy (%)
75	25	90.78
80	20	93.21

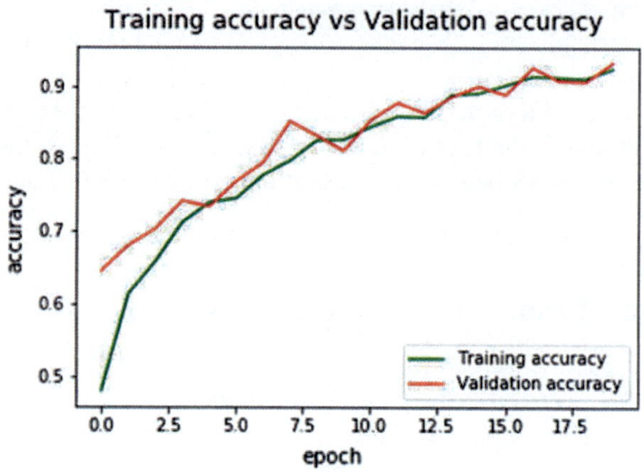

Fig. 8 Model accuracy

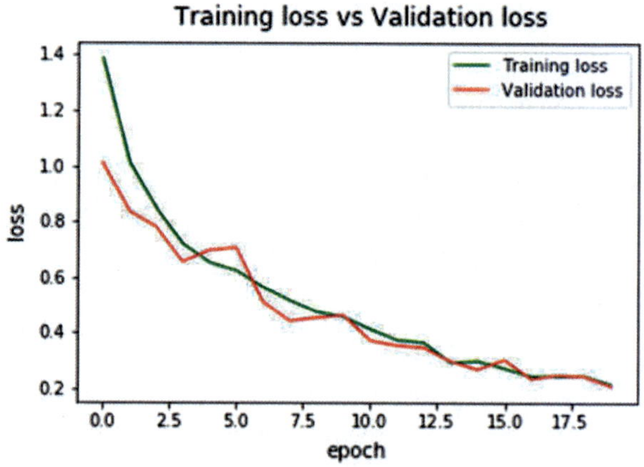

Fig. 9 Model loss

5 Conclusion

Not much work has been done in Bangladesh on automated rice disease classification. Convolutional neural network has been explored in this paper to detect rice plant diseases by their images. The proposed model was tested on three types of diseases for each plant. In the experiment, the dataset was partitioned into 80–20% training–testing set. By larger dataset of images of rice plant diseases, the accuracy of the approach can be improved. In the applications of smart phone, this proposed approach can be used and it will be more convenient for all kind of users.

References

1. Phadikar, S., Sil, J.: Rice disease identification using pattern recognition techniques. In: Proceedings of the IEEE International Conference on Computer and Information Technology (ICCIT), Khulna, Bangladesh, pp. 420–423 (2008)
2. Barbedo, J.G.A.: Digital image processing techniques for detecting, quantifying and classifying plant diseases. SpringerPlus **2**(1), 660–672 (2013)
3. Asfarian, A., Herdiyeni, Y., Rauf, A., Mutaqin, K.H.: A computer vision for rice disease identification to support integrated pest management. Crop Prot. **61**, 103–104 (2014)
4. Naik, M.R., Sivappagari, C.: Plant Leaf and Disease Detection by Using HSV Features and SVM. IJESC **6**(12) (2016)
5. Olmschenk, G., Tang, H., Zhu, Z.: Crowd counting with minimal data using generative adversarial networks for multiple target regression. In: Applications of Computer Vision (WACV), 2018 IEEE Winter Conference on Computer Vision (WACV) (2018)
6. Hadayat, A., Darussalam, U., Irmawati.: Detection of disease on corn plants using convolutional neural network methods (2019)
7. Qiang, Z., He, L., Dai, F.: Identification of plant leaf diseases based on inception V3 transfer learning and fine-tuning. In: Wang, G., El Saddik, A., Lai, X., Martinez Perez, G., Choo, K.K. (eds.) Smart City and Informatization. iSCI,: Communications in Computer and Information Science, vol. 1122. Springer, Singapore (2019)
8. Ramesh, S., Vydeki, D.: Application of machine learning in detection of blast disease in south Indian rice crops. J. Phytol. **11**(1), pp. 31–37. https://doi.org/10.25081/jp.2019.v11.5476
9. Chuanlei, Z., Shanwen, Z., Jucheng, Y., Yancui, S., Jia, C.: Apple leaf disease identification using genetic algorithm and correlation based feature selection method. Int. J. Agricult. Biol. Eng. **10**(2), 74–83 (2017)
10. Mohanty, S.P., Hughes, D.P., Salathé, M.: Using deep learning for image-based plant disease detection. Front. Plant Sci. **7**, 1419 (2016)
11. A. Kamilaris, Prenafeta-Boldú, F. X. : Deep learning in agriculture: a survey. In: Computers and Electronics in Agriculture (2018)
12. Shafira, T.: Implementasi Convolutional Neural Network untuk Klasifikasi Citra Tomat Menggunakan Keras. Skripsi. Fakultas Matematika Dan Ilmu Pengetahuan Alam. Universitas Islam Indonesia: Yogyakarta (2018)

Tropical Cyclones Classification from Satellite Images Using Blocked Local Binary Pattern and Histogram Analysis

Chinmoy Kar and Sreeparna Banerjee

Abstract Classification of tropical cyclone (TC) using cloud pattern is an evolving area of research. The classification result may use for intensity detection of a TC to mitigate the damage. TC classification using image processing technique is a stimulating task due to complexity of the problem. This paper attempted to classify TC images using a modified local binary pattern (LBP). The LBP of an image describes its local structure in an easy and efficient manner. The proposed blocked LBP (B-LBP) is an improvised approach to generate central pixels form an input image. The input image is distributed into 3 × 3 identical blocks and the central pixels of each block are used to construct a new image pattern. This new image further analyzed by histogram and box plot to classify images. This paper shows a significant improvement in result toward classification. The B-LBP approach can improve the performance of the intensity prediction through classification with 91% accuracy.

Keywords Tropical cyclone · Intensity detection · Histogram analysis · Image processing · Local binary pattern · Box plot

C. Kar (✉)
Department of Computer Science and Engineering, Sikkim Manipal Institute of Technology, Sikkim Manipal University, Sikkim, India
e-mail: Chinmoy.k@smit.smu.edu.in

S. Banerjee
Maulana Abul Kalam Azad University of Technology, Kolkata, West Bengal, India
e-mail: sreeparnab@hotmail.com

S. Borah et al. (eds.), *Soft Computing Techniques and Applications*, Advances in Intelligent Systems and Computing 1248,
https://doi.org/10.1007/978-981-15-7394-1_38

1 Introduction

Tropical cyclone (TC) intensity estimation through cloud images pattern proposed by V. Dvorak in year 1975 [1]. Since then the Dvorak technique is successfully applied on TC images to find its intensity level in all over the world. Later, this technique improvised and automated by Velden et al. [2]. Subsequently, many scales are developed to estimate the intensity levels of a cyclone. Indian Meteorological Department (IMD) TC intensity scale World Meteorological Organization's TC intensity scale are few of them. Mainly these scales are used to classify TCs to predict the intensity of a cyclone to reduce the loss of lives and property. The image-based analysis is a challenging task as it is directly based on real-time images. TC classifications from wind speed [19] or cloud pattern [1] are familiar processes. Image-based TC intensity prediction or classification becomes emerging field of research since last few years [14–18]. This work proposes a TC type classification method using LBP and histogram analysis of a TC image. LBP of an image is an efficient feature descriptor ever since it was proposed. Initially, LBP was used for texture analysis [3], after that many researches carried out their research work on LBP and its variants [4]. LBP and its variants applied mostly on texture [5] and face recognition [6–8]. The use of LBP further extended to the field medical image analysis and recognition [9], forensics [10] and other pattern recognition and image processing related work.

Histogram matching is a process for object detection [11]. Histograms are also used for classification and clustering. B-spline histogram [12] and spatial histogram features [13] are some popular techniques, among others. Formation of a spatial histogram is a very important step toward object identification. Here, the LBP of a TC image represented by a histogram of images and further used to finding box plots of 255 gray values (1–255, excluding 0). This paper focuses on an image processing-based approach for TC classification by finding its histogram of LBP images.

2 Data Source

In this research, 130 images are taken from the archive Cooperative Institute for Meteorological Satellite Studies, Space Science and Engineering Center, University of Wisconsin–Madison and National Satellite Meteorological Center.

3 Methodology

3.1 Flowchart of the Proposed Method

Flowchart of the proposed system shown in Fig. 1.

Preprocess images

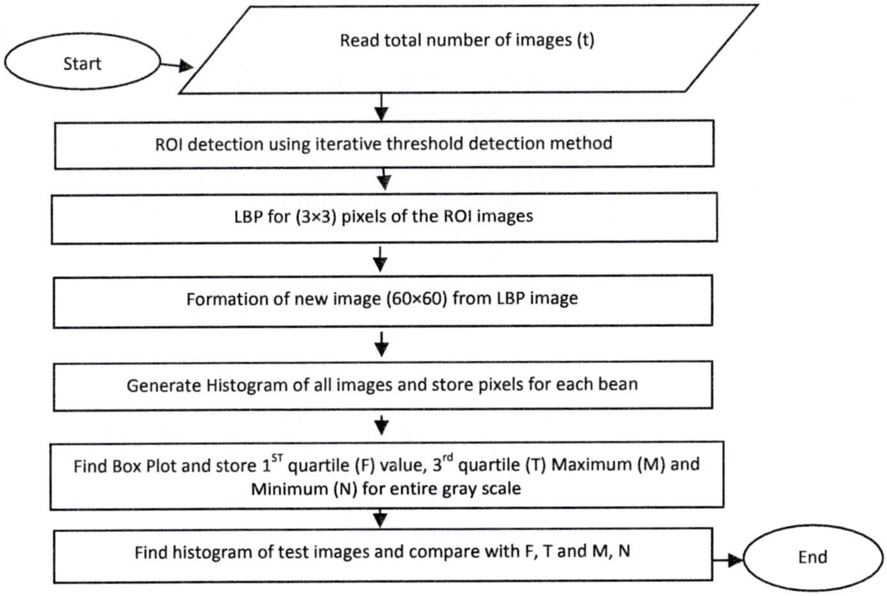

Fig. 1 Flowchart of the proposed system

- The entire image database is divided into training set and test set. A simple method is applied to divide the images into training and test sets.
  - Read each image from this master file and generate a random number between 0 and 1 for each image.
  - If the randomly generated number is less than 0.75, then write the name of the image in training file, else write in test file.
  - The above-mentioned method divides 130 images into 98 for training and 32 for test sets (75% for training set and 25% for test set).
- Each image from training set is resized to 180 × 180 pixels from the above-mentioned sources.
- Region of interest (ROI) of resized image (M, size = $m \times n$) selected automatically using the following methods which is similar to the method explained in [20]:
  - Calculate average intensity of M:

$$x_k = \frac{\sum_{i=1}^{m} \sum_{j=1}^{n} I(i, j)}{m \times n} \tag{1}$$

Here, x is the calculated threshold value of M. $I(i, j)$ is the intensity value of the image at I, jth pixel.

- Segmentation of an image based on above-mentioned threshold.

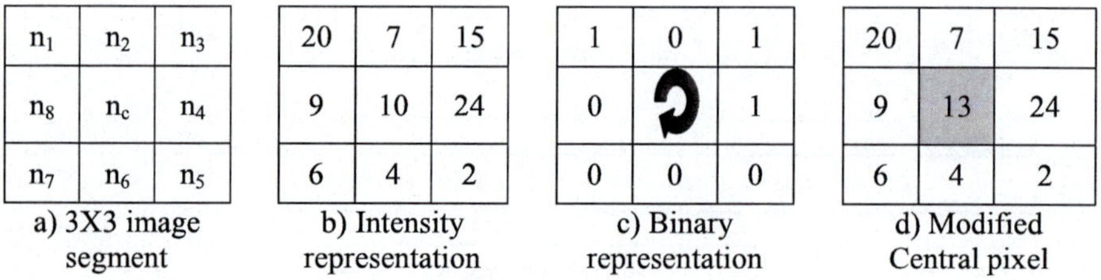

Fig. 2 LBP calculated in a local 3 × 3 neighborhood pixels

$$I_{(i,j)} = \begin{matrix} 0 & \text{if } I_{(i,j)} < x \\ I_{(i,j)} & \text{Otherwise} \end{matrix} \quad (2)$$

If the value of $I(_{i,j})$ is less than x, then assign 0 at ith, jth pixel of $M(_{i,j})$, otherwise, the intensity value remains same.

The above-mentioned steps will continue until $|x_k - x_{k+1}| < T$, where T predefine threshold value.

3.2 *Construction of Blocked Local Binary Pattern*

LBP of a set of pixels [3] is computed by comparing center with its neighbor pixels. A LBP description of a center pixel in 3 × 3 image is generated by using Eq. 3.

$$I_c = \sum_{i=1}^{8} 2^{i-1} I(n_i - n_c)$$
$$I(x) = \begin{matrix} 1 \text{ if } x > 0 \\ 0 \text{ Otherwise} \end{matrix} \quad (3)$$

where I_c is an output value computed using LBP. n_c is the center pixel and $n_i\{n_1, n_2, \ldots n_8\}$ neighbor pixels. A more detail description is given in Fig. 2.

The above-mentioned method is applied on the entire image (M) of $m \times n$ matrix to find blocked LBP. The I_c values of each 3 × 3 blocks are stored in another matrix L of $(\frac{m}{3} \times \frac{n}{3})$. Here, L is B-LBP of the original image M. The relations between these two matrices are shown in Fig. 3.

3.3 *Construction of Histogram*

Histogram of blocked local binary pattern 60 × 60 (L of size $\frac{m}{3} \times \frac{n}{3}$) representation of original image.

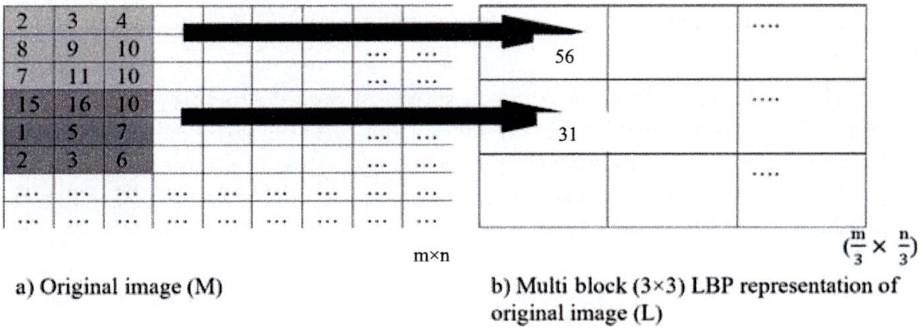

Fig. 3 Original image and corresponding B-LBP

Histogram of a LBP image constructed based on its values at each bean. An example is given below:

$$[PX, B] = H(A) \tag{4}$$

PX Set of pixel counts over set *B*
B Gray values (bean).

3.4 Box Plot

A box plot or whisker plot is a useful tool for analyzing purpose [21]. Here, the box plot attributes are shown in Fig. 4. The box length is the interquartile range which is calculated from a given data set. Parallel box plots are very useful tool to observe multiple samples. Here, in our experiment, we have used parallel box plots for 255 gray values each from 1 to 255 (excluding 0). Each box plot counts the number of pixels in between first quartile (*F*) and third quartile (*T*). Based on this count value, the proposed algorithms try to conclude the relationship between 130 TC images based on this value.

$$\text{Box length (BL)} = \text{3rd quartile-1st quartile} \tag{5}$$

$$\text{Length}(L) = \text{Maximum}-\text{Minimum} \tag{6}$$

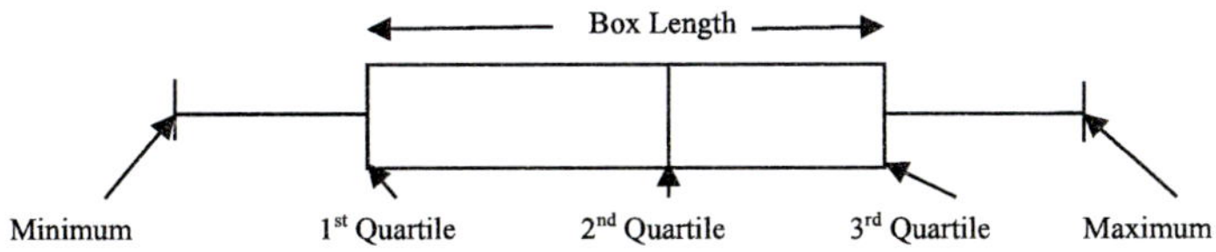

Fig. 4 Box plot attributes

4 Result Analysis

TC image and corresponding ROI image shown in Fig. 5. It is shown that the total number of training and test images are taken from METEOSAT satellites Table 1.

First quartile (*F*), third quartile (*T*), maximum (*M*) and minimum (*N*) values are constructed from box plot for each beans of grayscale (1–255 total 255) and stored in first quartile, third quartile, max and min files, respectively. Some test cases are given in Table 2 to explain how the accuracy is calculated by the proposed method.

Above-mentioned policy shown in Table 2 is used for accuracy detection. Table 3 shows that the number of observation between *F* and *T*, *M* and *N*.

As the classical LBP and proposed B-LBP both are belongs to the same technical domain, Table 4 is used to describe the performance measures between LBP results with blocked LBP.

The result shows that the blocked LBP is much more effective than classical LBP.

Fig. 5 TC image and corresponding ROI image and its histogram

Table 1 TC images used for training and testing

Number of Images used		
Name of the satellite	Training	Testing
METEOSAT-7	45	13
METEOSAT-8	53	29
Total (130)	98	32

Table 2 List of some test cases

List of some test cases							
B	*PX*	*F*	*T*	*M*	*N*	True if $PX \geq F$ and $PX \leq T$	True if $PX \geq N$ and $PX \leq M$
1	10	6	12	16	2	True	True
5	25	20	24	18	28	False	True
8	24	12	18	10	20	False	False

Table 3 Total number of successful observation

Comparison between two methods with respect to accuracy		
Observation method	Box length (T-F)	Maximum to minimum
LBP	69	87
Proposed LBP	91	94

Table 4 Comparison between LBP and proposed blocked LBP method

Total number of successful observation			
Parameters	Total number of successful observations (true)	Total number of observations	Percentage of accuracy (approximate)
Between first quartile (F) and third quartile (T) (if $PX \geq$ F and $PX \leq T$)	7635	$32 \times 255 = 8160$	91
Between maximum (M) and minimum (N) if $PX \geq N$ and $PX \leq M$	7950		94

5 Conclusion

This paper focuses on the classification and detection of TC from its satellite images. Here, satellite images (130 nos.) of cyclonic storm (CS) are collected from IMD archive and processed. The proposed B-LBP method is merging histogram analysis and LBP method. The classification accuracy of TC images (CS type) better than other similar types of approaches. It has also observed that the B-LBP method works better than classical LBP (shown in Fig. 6).

The proposed method is simple, easy to implement in addition it is doing well in comparison with other methods. The analysis on accuracy of the proposed B-LBP is shown in Table 5.

The proposed work has implemented only on CS images only and the same schema can be applied on VSCS, ESCS and SCS images extension of this work may open

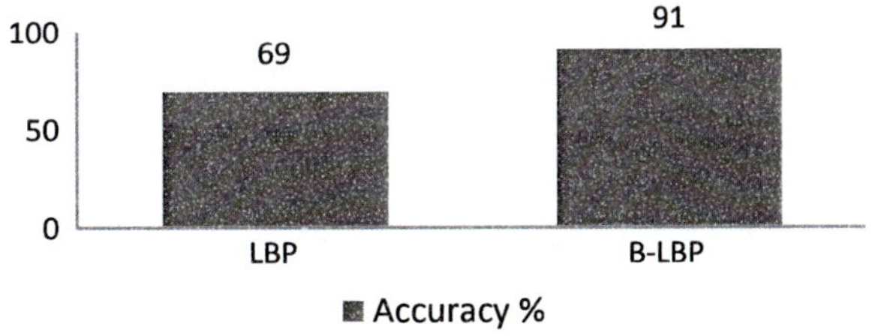

Fig. 6 Comparison between accuracy between LBP and B-LBP

Table 5 Comparison between various methods

Method	Overall accuracy (%)
Active contour model (ACM) [22]	82
B-spline with histogram with multi-scale transforms [12]	88.57–95.87
Elliptic Fourier descriptors [17]	83
Feature vector analysis [14]	83
Multilayer perceptron [15]	84
Proposed LBP	91

a new direction toward the advancement of image processing-based approaches for TC intensity detection and classification.

Acknowledgements National Satellite Meteorological Center, Indian Meteorological Department and Cooperative Institute for Meteorological Satellite Studies, Space Science and Engineering Center for TC images archive.

References

1. Dvorak, V.F.: Tropical cyclone intensity analysis and forecasting from satellite imagery. Mon. Weather Rev. **103**, 420–430 (1975)
2. Velden, C.J., Daniels, D., Stettner, D., Santek, J., Key, J., Dunion, K., Holmlund, G., Dengel, W.B., Menzel, P.: Recent innovations in deriving winds from meteorological satellites. Bull. Amer. Meteor. Soc. **86**, 205–223 (2005)
3. Ojala, T., Pietikäinen, M., Mäenpää, T.T.: Multiresolution gray-scale and rotation invariant texture classification with local binary pattern. IEEE Trans. Patt. Anal. **24**(7), 971–987 (2002)
4. Meena, K., Suruliandi, A.: Local binary patterns and its variants for face recognition. In: 2011 international conference on recent trends in information technology (ICRTIT), Chennai, Tamil Nadu, 2011, pp. 782–786
5. Guo, Z., Zhang, L., Zhang, D.: A completed modeling of local binary pattern operator for texture classification. IEEE Trans. Image Process. **19**(6), 1657–1663 (2010)
6. Datta Rakshit, R., Chandra, S.N., Kisku, D.R.: An improved local pattern descriptor for biometrics face encoding: a LC–LBP approach toward face identification. J. Chin. Inst. Eng. **40**(1), 82–92 (2016)
7. Huang, Di, Shan, C., Ardabilian, M., Wang, Y., Chen, L.: Local Binary Patterns and Its Application to Facial Image Analysis: A Survey. IEEE Trans. Syst Man Cy C. **41**(6), 765–781 (2011)
8. Lu, J., Liong, V.E., Zhou, J.: Simultaneous Local Binary Feature Learning and Encoding for Homogeneous and Heterogeneous Face Recognition. IEEE Trans. Patt. Anal. **40**(8), 1979–1992 (2018)
9. Wan, S., Lee, H., et al.: Integrated local binary pattern texture features for classification of breast tissue imaged by optical coherence microscopy. Med. Image Anal. **38**, 104–116 (2017)
10. Agarwal, S., Chand, S.: Blind forensics of images using higher order local binary pattern. J. Appl. Secur. Res. **13**(2), 209–222 (2018)
11. Zhang, H., Gao, W., Chen, X., Zhao, D.: Learning informative features for spatial histogram-based object detection. IEEE IJCNN Montreal Que. **3**, 1806–1811 (2005)

12. Zhang, C., Wang, X., Duanmu, C.: Tropical cyclone cloud image segmentation by the B-spline histogram with multi-scale transforms. Acta. Meteorol. Sin. **24**(1), 78–94 (2010)
13. Zhang, H., Gao, W., Chen, X., Zhao, D.: Object detection using spatial histogram features. Image Vision Comput. **24**, 327–341 (2006)
14. Kar, C., Banerjee, S.: An image processing approach for intensity detection of tropical cyclone using feature vector analysis. Int. J. Image Data Fus. **9**(4), 338–348 (2018)
15. Kar, C., Kumar, A., Banerjee, S.: Tropical cyclone intensity detection by geometric features of cyclone images and multilayer perceptron. SN Appl. Sci. **1**, 1099 (2019). https://doi.org/10.1007/s42452-019-1134-8
16. Kar, C., Banerjee, S.: An approach towards automatic intensity detection of tropical cyclone by weight based unique feature vector. In: IEEE International Conference on Computational Intelligence and Computing Research (ICCIC), Chennai, pp. 1–4 (2016). https://doi.org/10.1109/iccic.2016.7919616
17. Dutta, I., Banerjee, S.: Elliptic fourier descriptors in the study of cyclone cloud intensity patterns. Int. J. Image Process. **7**(4), 402–417 (2013)
18. Pradhan, R., Aygun, R., Maskey, S., et al.: Tropical cyclone intensity estimation using a deep convolutional neural network. IEEE Trans. Image Process. **27**(2), 692–702 (2018). https://doi.org/10.1109/TIP.2017.2766358
19. Yu, X., Chen, Z., Chen, G., Zhang, H., Zhou, J.: A tensor network for tropical cyclone wind speed estimation. In: 2019 IEEE International Geoscience and Remote Sensing Symposium, Yokohama, Japan, pp. 10007–10010 (2019)
20. Kar, C., Kumar, A., Konar, D., Banerjee, S.: Automatic region of interest detection of tropical cyclone image by center of gravity and distance metrics. In: 2019 Fifth International Conference on Image Information Processing (ICIIP), Shimla, India, pp. 141–145 (2019)
21. Praveen, V., Narendran, T.D., Pavithran, R., Thirumalai, C.: Data analysis using box plot and control chart for air quality. In: 2017 International Conference on Trends in Electronics and Informatics, Tirunelveli, India (2017). https://doi.org/10.1109/icoei.2017.8300877
22. Lee, R.S.T., Lin, J.N.K.: An elastic contour matching model for tropical cyclone pattern recognition. IEEE Trans. Syst Man Cy B. **31**(3), 413–417 (2001). https://doi.org/10.1109/3477.931532

Predicting the Appropriate Category of Bangla and English Books for Online Book Store Using Deep Learning

Md. Majedul Islam, Sharun Akter Khushbu, and Md. Sanzidul Islam

Abstract At the era of this technology, we are seeking every stuff online first. Book was the best friend to us, still, it is. But this changed the way and medium by which we are being engaged with the book. Nowadays bookselling is more popular online than physically from the store. So books categorizing correctly is a very important problem. But there are many category books available like—Novel, Fiction, non-Fiction, etc. So manually categorizing books was a big deal for everyone. For that having an automatic book category system that uses a book title to categorizing books will help many people. Here, a method is proposed where Long short-term memory (LSTM) technique is used for categorizing books using books title. This model was trained on 1500 English and Bangla books title of four categories. The model reported promising results with training accuracy was 95.08% for English and 83.81% for Bangla. Different preprocessing techniques such as removing numeric data, null value removal, repeat data remove are used. In the Long short-term memory (LSTM) networks activation function ReLU is used in the hidden layer and softmax for the output layer.

Keywords Books categorizing · Long short-term memory (LSTM) · Deep learning

1 Introduction

Book categorization is a motive of imposing unknown book name document into predefined known classes depending on its content [1, 2]. Using sequential model

Md. Majedul Islam (✉) · S. A. Khushbu · Md. S. Islam
Department of CSE, SWE, Daffodil International University, Dhaka, Bangladesh
e-mail: majedul15-6784@diu.edu.bd

S. A. Khushbu
e-mail: khusbhu15-6083@diu.edu.bd

Md. S. Islam
e-mail: sanzid.swe@diu.edu.bd

S. Borah et al. (eds.), *Soft Computing Techniques and Applications*,
Advances in Intelligent Systems and Computing 1248,
https://doi.org/10.1007/978-981-15-7394-1_39

categorization of books has been placed more attention in the research field, owing to its ability to lead and forming always increasing book documents over the internet [3, 4]. Main applications of Book Categorization: It assistance to discover necessary information in query engine websites, Document Indexing automatically, Recovery Information, Repeated name filtering, etc. In literature, expansive diversity approaches have been displayed for effective different fields of categorization. Over the last few years, researcher are developed a sequence model but those are their specific different area. This computational power becoming place in different forms of cloud computing and grid computing. Neural networks have a deep history for categorized data, generally in summation with hidden Markov models [5, 6]. In some research paper, they had claimed RNN is an alternative model for HMM [7–9]. Many workers have been designed architecture based on LSTM for sequential model and they pre-trained real-world image data, one researcher they have learned magazine of image data collected by several websites and they obtained good score [10]. They have built their approach to their work how fast and convenient with RNN which through we have used this RNN sequential model. Approaches of the Categorization can be extensively divided into batch and sequentially based on the type of learning approach accepted [11]. Various iterations are completed for training data is trained with batch learning process. Neural Network depends on text classifiers that gives the greatest existing with batch learning process. Our model is performing better with existing batch learning Neural Network Classifiers. Rather classifiers abide by limitations: (i) full dataset is required before execution of learning process. For that reason, Bangla and English book real-time documentation fully trained is difficult. (ii) The network structure has to be declared in before to learning process. Fixed network structure which is maintaining in before may reason the distortion in the performance of the classifier. (iii) This long book dataset takes very high computation time. Even if a new training dataset comes, re-training will be required [12–14]. To address the aforementioned boundaries, the community of research adventure sequential model in ANN approached for categorization [15]. Training Datasets are displayed one-by-one and only once at a time to the network for sequential model. After completing process training documents are discarded in the network. With the signification, sequential model automatically discover minimal network framework [16]. On the other side, many researchers did many work on their datasets for better efficiency of RNN in LSTM. "Recurrent Neural Network Language Model Training using Natural Gradient" on this topic they success to prove RNN is widely used speech recognition using RNNLMs with developing gradient method [17]. Expectation-Maximization (EM) technique with RNN depending on the classification of integration clustering. It reduces the dimensionality of any feature in dataset. Our Dataset an online based dataset which is holding a different type of book with combination of Bangla and English. We have processed with "Rokomari Book Dataset" for sequential model in keras with obtaining significant performance. In the research area, the categorization on Bangla and English book dataset using the sequential model is tinny apart from other work. RNN reaches to give optimization of neural network approaches in the future. Thereby using this dataset we can develop a sequence model for categorizing the book data and sequentially optimize the result. RNN (Recurrent Neural Network) is a mighty model for performing to achieve difficult learning tasks. DNN also well performer for larger dataset but DNN cannot identify the sequences to sequences. We present a sequential model who is pre-trained the data categorized the data with high accuracy. RNN using potential correct array as training focus. RNN can be faster and define superior technique in next period for cloud database thereby we have used an online database or dataset. Using RNN concept N-gram model for better and non-adversarial result already have been done greatly on Regularization technique "Knowledge distillation for Recurrent Neural Network Language Modeling with Trust Regularization" on their work reduced model size they have to use small parameter after checking performance then figure a speech recognition dataset and measure final efficiency [18]. RNN with LSTM substantially over fitted for reduction to make variable to sequence [19]. Speech Recognition, language to language translation end to end process these are the completed in past years. One did work for sequence learning to convert English to French language using LSTM with RNN [20]. Using TMIT dataset for speech recognition another researcher has been done a work using bidirectional LSTM with RNN in end to end process [21]. Sequence Categorization is a fateful modeling issue where input is time or over space will be predicted for the sequence. Sequences are depended on length. Input symbols can be anything text, name, word which may require the model to learn the Long-term context or dependencies in the input sequence. Lots of libraries are more popular rather we make sequence models with keras libraries. Keras work with categorization who is the type of supervised machine learning. Classify and grouped the data after that elaborate sequence to sequence if exist or not. It measures three crucial components which are input layer, Hidden layer and the output layer. Keras high-Level API for python give significant output. Make a model is too easy and fast. Inside the keras is capable to drive with CPU and GPU. Our categorization method and technique are strictly different from other work because our dataset (Table 1) contains contain exclusive data which through using this dataset haven't any work before in research area.

Table 1 Dataset

Data	Type	Quantity	Label
English	English text	1500	'Computer Science and Engineering', 'Fiction', 'Non-Fiction' and 'Novel'
Bangla	Bangla Text	1500	"সমকালীন উপন ◌াস", "বাংলা কিবতা", "ছড়া", "িচরায়ত উপন ◌াস"

2 Literature Review

In the area of Book Categorization using deep learning technique are not much available which give better efficiency with "Rokomari Book Dataset". There are small number of similar works are available for categorization using Sequential models in keras ahead of different object with different data set. Predicting correct category from online book store for both Bangla and English book full procedure starts with data collection highly contrast with other past research work based on data and proposed methodology. Our approaches can be regularized to avoid overfitting also models can easily modify or update new forms rather than other previous approaches. Most researcher works on a simple one dataset but in here online data set given a large amount of dataset for that reason machine can be slow execution possibility. Some literature has been reported on their own work efficiencies using their own dataset.

In the past years for instance 'Building powerful image classification models using very little data' Archives et al. [20] have used two thousand datasets per class hold with thousand datasets based on cats and dogs image finally after pre-trained the data they had found effective score around 80% accuracy. They give a detailed outlook on how to work with keras and evaluate highest accuracy with large datasets.

In another research work Accurate Emotion Classification on Deep Dataset also a deep learning work which gives also much better performance on detect 'Deep Dataset' on behalf of classifying emotion. Samarth Tripathi et al. [21] had used two different neural models one is deep neural network another one convolutional neural network but not use Sequential models in keras because our efficiency level is very much high.

With different type of data in another work Arman Cohan et al. [22] have used LSTM (Long Short Term Memory Networks) And RNN (Recurrent Neural Network) for classifying safety events with two large scale datasets developed neural attention architecture and great two accuracy results are 78 and 88%. "A Neural Attention Model for Categorization Patient Safety" developed their significant results on behalf of medical imaging error with text processing and categorization.

"Predicting Domain Generation Algorithms with Long Short-Term Memory Network"; they classify the DGAs using LSTM with sequential models in keras with four significant classifiers with their features and achieved a good precision result around 90%. Apart from Jonathan Woodbridge et al. [23] proposed HMM (Hidden Markov Model).

In another research work, Chuanhai Dong et al. [24] presented "Character-Based LSTM-CRF with Radical-Level Features for Chinese Named Entity. They proposed BLSTM on behalf of LSTM-CRF based on Neural Network (NN) model which can identify both character level and radical-level. With their work, they found a better performance of around 90%.

In another paper, Dixon et al. [25] "Sequence Classification of the Limit Order Book Using Recurrent Neural Networks" which is a sequential based classification or categorization for Short sequence of observations of limited book order has used

RNN (Recurrent Neural Network) who is a part of artificial neural network predicting book price. This work they used RNN technique which is similar to our work but only being a part for RNN we develop a sequential model using RNN and LSTM for 'Rokomari Book Dataset' after finding a good accuracy level in book categorization.

Similarly, Takeru Miyato et al. [26] "Adversarial Training Methods for Semi-Supervised Text Classification predict text categorization on five types of datasets with defining RNN (Recurrent Neural Network) and LSTM for label and unlabel dataset. They had complete 512 hidden units for LSTM. In their experiment, they found adversarial and virtual adversarial training with better results in sequence models on text categorization.

Apart from previous research in another work Ashraf Elnagar et al. [27] construct supervised and unsupervised state-of-the-art polarity sentiment classifiers models. "An Annotated Huge Dataset for Standard and Colloquial Arabic Reviews for Subjective Sentiment Analysis" worked on sentiment analysis a large dataset for classifying the data had made a temporary data model with 91% accuracy. They had defining on their paper full review of RNN, CNN, SVM, and LSTM. With the section maintaining procedure elaborately explain how they achieved accuracy for supervised and unsupervised sequence model.

3 Proposed Methodology

The proposed "Book Categorization" sequential model has many steps as described below.

3.1 Data Collection and Dataset Properties

Our research focus to make book categorization automatically on web or any machine. There are many E-commerce who sell-out books online. Rokomari is one of them and the largest one in Bangla therefore we scrape data are using beautiful soup using a python script. Hassle dataset has 154347 book information with 16 class labels.

3.2 Data Preprocessing

As hassle data from the web and we have to work on the English book and as well as separately for Bangla book have to Categorize. Consequently, we remove all 14 classes without book title and category class before that we separate Bangla book and English book data in different CSV files.

Dataset for English books: English book data has many categories we take only four-category books they are “Computer Science and Engineering”, ‘Fiction’, ‘Non-Fiction’, and ‘Novel’ each category has 1500 data. We shuffle our data for the balance test and train set. After that, we divide dataset 70% for training and 30% for test.

Dataset for Bangla books: In the sake of Bangla preprocessing is challenging work as many characters do not recognize by machine. On those purpose, we remove extra space, symbolic character form in between text. Then we take four classes: “সমকালীনউপন ◌াস”, “বাংলা কিবতা”, “ছড়া”, “িচরায়ত উপন ◌াস”each category has 1500 data. Still there we shuffle our data for balance test and train set. With repeated we divide dataset 70% for training and 30% for test.

Tokenizer: As input, we have to prepare data for Neural Network, Therefore mostly needed to convert into a numeric representation which is understandable by machine. Keras gives some built-in class to preprocess our data, for example, tokenizer utility class, Label-Encoder, etc. Using tokenizer makes our text token based on word frequency. In dataset, all data are text thereby after tokenizing data to convert text to the numeric value for model readable we use Label-Encoder which is Scikit learn library in python. Label-Encoder behaves as normalize the label, convert column data, and replace text data to encoded data. We use Label-Encoder for both Bangla and English.

3.3 Proposed Model

Sequential model is used for machine translation converted sequential data like Language Modeling, Speech recognition [21], Text categorization [4]. In a recent year for semantic sentences researchers have used a sequential encoder-decoder model for translate semantic sentences [29] with the knowledge we proposing this model for Bangla and English both text sequences.

Bangla:

There are two types of layers in proposed behalf of the sequential model they are: three dense and three activation layers. Dense derive proper connection of neural network. In the proposed model the size of first two dense layers are 512 and 256, last one is output so its size is 4. Using ADAM proposed model should optimize [28]. Propose model use Relu activation function in between first two dense input layers, softmax activation in output layer proposed model flow chart shown in Fig. 1.

$$\mathrm{ReLU}(X) = \mathrm{MAX}(0, X) \tag{1}$$

English:

There are two types of layers in the proposed model behalf of sequential model they are four dense and four activation layers. Dense derives proper connection of neural networks. In the proposed model the size of first three dense layer are 512 and 256 and 128, last one is output so its size is 4. Proposed model use Relu activation

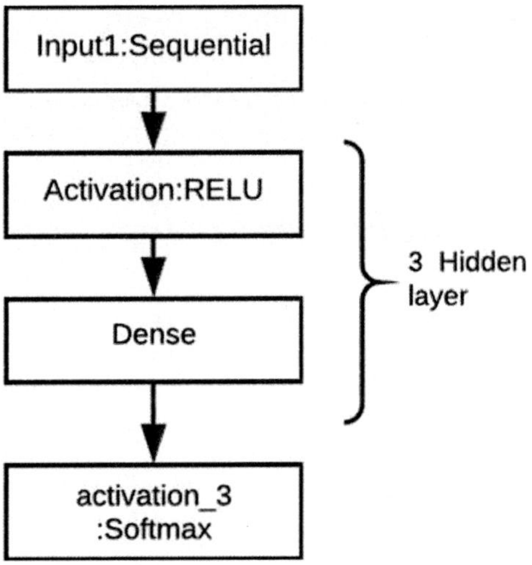

Fig. 1 Proposed model for Bangla

function in between first two dense input layers, softmax activation in output layer proposed model flow chart shown in Fig. 2.

The error metric Categorical Cross-Entropy (3) was used.

$$\sigma(Z)_j = \frac{e^{zi}}{\sum_{K=1}^{K} e^{zk}} \quad \text{for } j = 1, \ldots k \tag{2}$$

$$L_i = -\sum_j t_{i,j} \log(p_{i,j}). \tag{3}$$

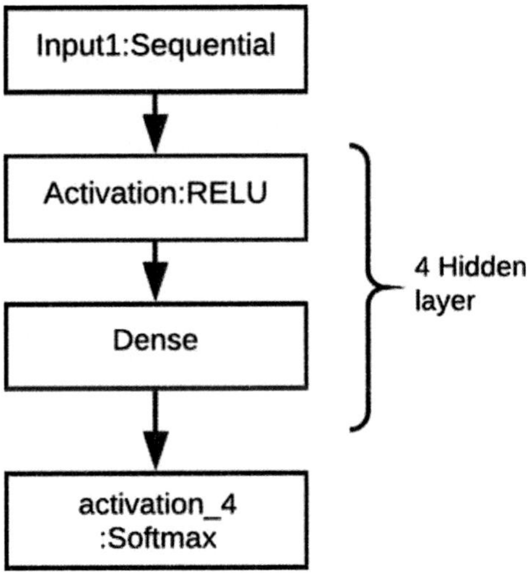

Fig. 2 Proposed model for English

4 Experiment and Results

The experiments and results are discussed below.

4.1 Model Training

Training set and dataset generalize validation defining on trained by the proposed model. Batch size was set to 24. After 50 epochs, manually the accuracy was monitored. For 5–10 epoch repeatedly model was trained to achieve good accuracy.

4.2 Model Evaluation

Here, the model proposed was trained and tested on the dataset. We have been found promising accuracy compared to other related. Below we have fully described datasets of the results.

4.3 Train, Test Sets

The proposed model has been trained on four categories of book names. Dataset were divided into two sets. Before dividing the dataset, we shuffle our dataset to make sure all categories are available in both sets. For training, the model of Set 1 had 70% and Set 2 had 30% for testing the model. Each category of book 1500 data contains each book. Therefore, we have 3521 data for training set and 1510 for testing.

4.4 Model Performance

Many objective functions are sum of sub-functions with composing evaluate the data in different sub-functions to subsamples. Taking radiant steps for each sub-functions result should more efficient [29]. This Adam optimization (4) algorithm are straight forward, efficient with computation, and requires less memory that is the reason have chosen for processing neural network application. In the training process, parameters are not actual. The hyperparameters are belonging learning rate of the model, iteration/epoch, batch size, etc. In the other models are also connected, the hyper-parameter learning rate is set manually other hyperparameters were set intuitive manner. The proposed model after training it gives excellent performance for 10 epoch model gets accuracy 94% for English and 82% for Bangla. The model gives resulted accuracy of 95.08% for English and 83.81% for Bangla ran on 50 epoch. With approaching work procedure the confusion matrix and accuracy graph are shown in with four figures. Figure 3 confusion matrix given for Bangla. Figure 4 Model Accuracy Evaluation for Bangla and also good prediction and bad prediction table are shown in Table 2. In the same way, we predict also English books for categorization. There Fig. 5 are shown in the confusion matrix with Fig. 6 Model Accuracy Evaluation for English. Also prediction with good and bad English with Table 3.

$$v_t = (1 - \beta_2) \sum_{i=1}^{t} \beta_2^{t-i} \cdot g_i^2 \tag{4}$$

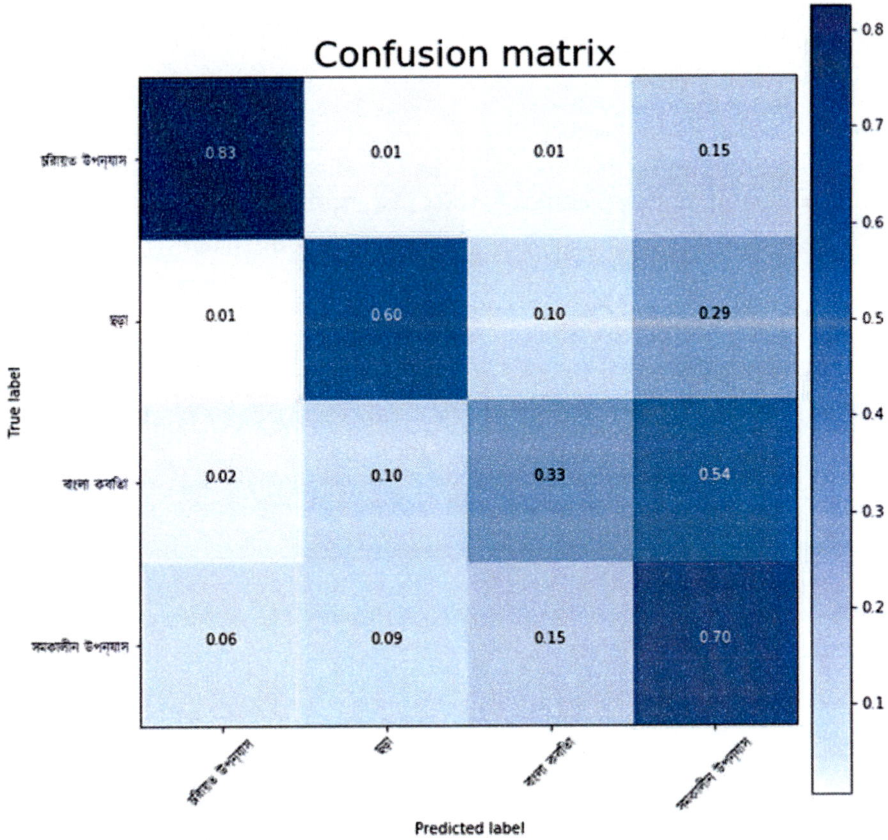

Fig. 3 Confusion matrix for Bangla

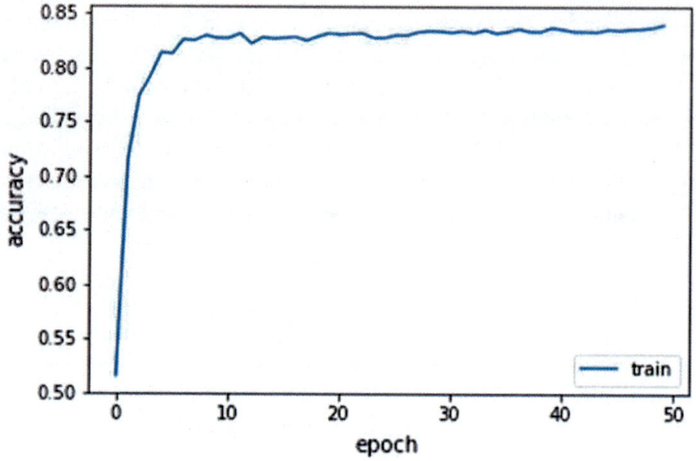

Fig. 4 Model accuracy evaluation for Bangla

Table 2 Prediction table for Bangla

Good prediction	Bad Prediction
সবপ্ন িবলাে◌স ে◌বাধন... Actual label: বাংলা কিবতা Predicted label: বাংলা কিবতা	তু িম এমনই ... Actual label: বাংলা কিবতা Predicted label: সমকালীন উপন ◌াস
যুগান্তর ... Actual label: সমকালীন উপন ◌াস Predicted label: সমকালীন উপন ◌াস	ধপ্লাস ... Actual label: ছড়া Predicted label: সমকালীন উপন ◌াস
বাংলা ভাষার পর্থম ছড়ার বই ... Actual label: ছড়া Predicted label:ছড়া	

5 Conclusion

This research work presented an alternative sequential model which performs best categorization accuracy in the several datasets for both train and validation set for low epochs and low computation time compared with the other sequential model.

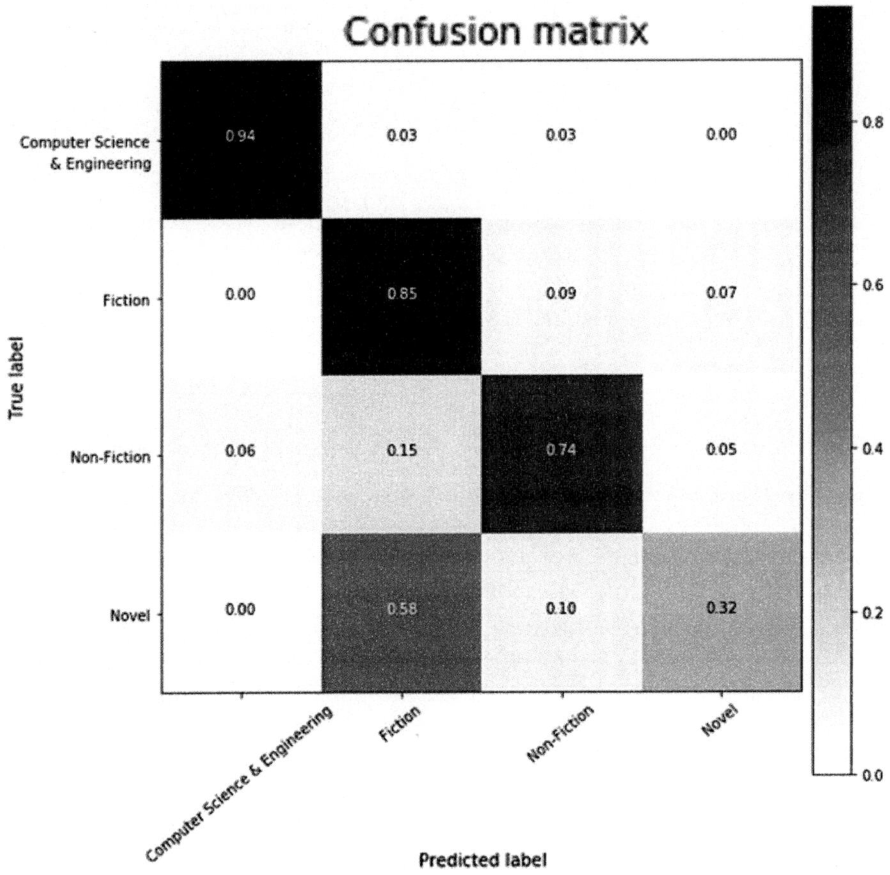

Fig. 5 confusion matrix for English

In general sequential model little much costly but more effective deep learning models. Therefore, fast convergence should be displayed as a significant front of research. Moreover, Improving and achieved good results.

Future work can predict book value in step with page number and author demand inside this dataset. Testing the model in various in public accessible datasets and measuring accuracy. Additionally rising the model's performance on alternative datasets is associated with nursing extension of this work.

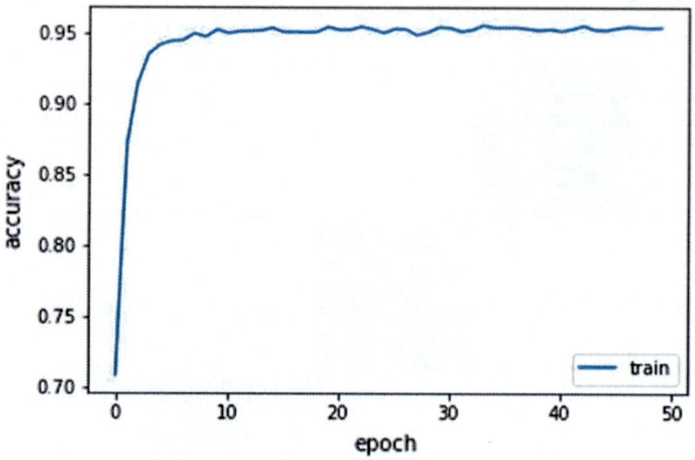

Fig. 6 Model accuracy evaluation for English

Table 3 Prediction table for English:

Good prediction	Bad prediction
The rescue… Actual label: fiction Predicted label: fiction	Mole concept for it—JEE… Actual label: computer science and engineering Predicted label: fiction
Synthesis and optimization of digital circuits… Actual label: computer science and engineering Predicted label: computer science and engineering	
Antifragile (paperback)… Actual label: non-fiction Predicted label: non-fiction	

References

1. Zaremba, W., Sutskever, I., Vinyals, O.: Recurrent Neural Network Regularization. ICASSP 2019 (2019). arXiv: 1409.2329v5 [cs.NE]
2. Sutskever, I., Vinyals, O., Le, Q.V.: Sequence to Sequence Learning with Neural Networks (2014)
3. Labani, M., Moradi, P., Ahmadizar, F., Jalili, M.: A novel multivariate filter method for feature selection in text classification problems. Eng. Appl. Artif. Intell. **70**, 25–37 (2018)
4. Kang, M., Ahn, J., Lee, K.: Opinion mining using ensemble text hidden Markov models for text classification. Expert Syst. Appl. **94**, 218–227 (2018)
5. Bourlard, H.A., Morgan, N.: Connectionist Speech Recognition: A Hybrid Approach. Kluwer Academic Publishers (1994)

6. Zhu, Q., Chen, B., Morgan, N., Stolcke, A.: Tandem connectionist feature extraction for conversational speech recognition. In International Conference on Machine Learning for Multimodal Interaction, MLMI'04, pp. 223–231. Springer, Berlin, Heidelberg
7. Robinson, A.J.: an application of recurrent nets to phone probability estimation. IEEE Trans. Neur. Netw. **5**(2), 298–305 (1994)
8. Vinyals, O., Ravuri, S., Povey, D.: Revisiting recurrent neural networks for robust ASR, in ICASSP (2012)
9. Mass, A., Le, Q., O'Neil, T., Vinyals, O., Nguyen, P., Ng, A.: Revisiting recurrent neural networks for robust ASR. In: Interspeech (2012)
10. Miikkulainen, R., Liang, J., Meyerson, E., Rawal, A., Fink, D., Francon, O., Raju, B., Shahrzad, H., Navruzyan, A., Duffy, N. and Hodjat, B.: Evolving Deep Neural Networks. University of Texas at Austin (2017). arXiv: 1703.00548v2 [cs.NE]
11. Runxuan, Z.: Efficient sequential and batch learning artificial neural network methods for classification problems. Singapore **2**, 825–845 (2005)
12. Kaufmann, P.: Supervised learning with complex valued neural networks. Neur. Netw. (2016)
13. Yu, L. Wang, S., Lai, K.K.: Foreign-Exchange-Rate Forecasting with Artificial Neural Networks. Springer Science & Business Media, p. 107 (2010)
14. Skovajsova, L.: Text document retrieval by feed-forward neural networks. Inf. Sci. Technol. Bull. ACM Slovakia **2**, 70–78 (2010)
15. Babu, G.S., Suresh, S.: Meta-cognitive neural network for classification problems in a sequential learning framework. Neurocomputing **81**, 86–96 (2012)
16. Zhang, Y., Er, M.J.: Sequential active learning using meta-cognitive extreme learning machine. Neurocomputing **173**, 835–844 (2016)
17. Yu, J., Lam, M.W., Chen, X., Hu, S., Liu, S., Wu, X., Liu, X., Meng, H.: Recurrent Neural Network Language Model Training Using Natural Gradient. The Chinese University of Hong Kong SAR, China, Elsevier (2019)
18. Shi, Y., Hwang, M.Y., Lei, X., Sheng, H.: Knowledge Distillation for Recurrent Neural Network Language Modeling With Trust Regularization. ICASSP, IEEE (2019). arXiv: 1904.04163v1 [cs.CL]
19. Graves, A., Mohammed, A., Hinton, G.: Speech Recognition with Deep Recurrent Neural Networks. Department of Computer Science, University of Tornoto (2013). arXiv: 1303.5778v1
20. Archives, G.: Building Powerful Image Classification Models using Very Little Data (2017)
21. Arman Cohan, R.R.: A neural attention model for categorization patient safety. In: European Conference on Information Retrieval p. arXiv: 1702.07092v1 [cs.CL]. Springer (23 Feb 2017)
22. Dong, C., Zuang, J.: Character-Based LSTM-CRF with Radical-Level Features for Chinese Named Entity. Springer (2016)
23. Dixon, M.: Sequence Classification of the Limit Order Book Using Recurrent Neural Networks (2017), pp. arXiv:1707.05642v1 [q-fin.TR]
24. Jonathan Woodbridge, H.S.: Predicting Domain Generation Algorithms with Long Short-Term Memory Network (2 Nov 2016), p. arXiv:1611.00791v1 [cs.CR]
25. Samarth Tripathi, S.A.: using deep and convolutional neural networks for accurate emotion classification on deap dataset. In: AAAI Conference on Innovative Applications. IEEE (2017)
26. Takeru Miyato, A.M.: Adversarial Training Methods for Semi-Supervised Text Classification. ICLR2017 (2017) (p. arXiv:1605.07725v3 [stat.ML])
27. Elnagar, A., Lulu, L., Einea, O.: An annotated huge dataset for standard and colloquial Arabic reviews for subjective sentiment analysis. In: 4th International Conference on Arabic Computational Linguistics, pp 182–189 (2018)
28. Kingma, D.P., Ba, J.: Adam: A Method for Stochastic Optimization. CoRR, abs/1412.6980 213 (2014)
29. Patro, B.N., Kurmi, V.K., Kumar, S., Namboodiri, V.P.: Learning Semantic Sentence Embeddings using Pair-wise Discriminator, India, arxiv:1806.00807v5 [cs.Cl] (2019)

Brazilian Forest Fire Analysis: An Unsupervised Approach

Sadia Jamal, Tanvir Hossen Bappy, and A. K. M. Shahariar Azad Rabby

Abstract Forest fire is one of the most dangerous natural hazards of planet earth now. Here presenting an approach where it has been trying to analyze the danger of fire for the forests of Brazil. The dataset contains data from 1998 to 2017 for all the states of Brazil where a forest fire has been caught throughout the year. An unsupervised approach like—*K*-Means clustering, Fuzzy *C*-Means, and Apriori algorithm was used to do so. This is a large dataset containing 6454 unlabeled data, to build a model with them *K*-Means clustering seems helpful. It tries to build subgroups (clusters) of similar data points from a large group. Fuzzy *C*-Means is also an unsupervised algorithm and it's working process is similar to *K*-Means clustering. By using *K*-Means clustering, Fuzzy *C*-Means, and Apriori method the regions which are in risk of fire danger were detected.

Keywords *K*-means clustering · Fuzzy logic · Fuzzy *C*-Means · Confusion matrix · Classification report · Apriori algorithm

1 Introduction

Since January 2019, Brazil has experienced more than 1 million fires. Among the places, Amazon is the most suffered victim of the fires, according to the country's National Institute for Space Research [3]. Amazon rain forest fire was one of the most talking issues of 2019. It has been said that this fire is bigger than any point

S. Jamal (✉) · T. H. Bappy · A. K. M. Shahariar Azad Rabby
Department of Computer Science and Engineering, Daffodil International University, Dhaka, Bangladesh
e-mail: sadia15-8084@diu.edu.bd

T. H. Bappy
e-mail: tanvir15-7706@diu.edu.bd

A. K. M. Shahariar Azad Rabby
e-mail: shahariar.cse@diu.edu.bd

S. Borah et al. (eds.), *Soft Computing Techniques and Applications*, Advances in Intelligent Systems and Computing 1248,
https://doi.org/10.1007/978-981-15-7394-1_40

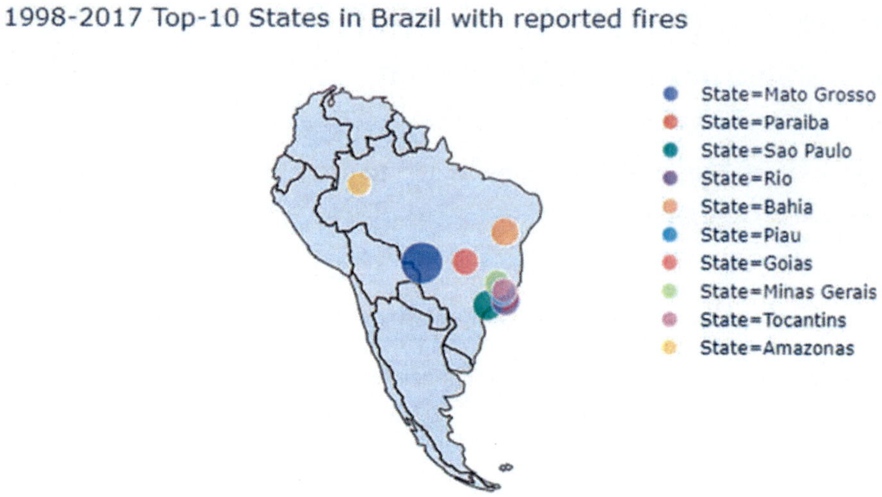

Fig. 1 Top 10 fire-affected states of Brazil in map

since 2010. By the report of national geography, 7000 square miles were burning that time and the ratio is growing bigger from time to time [2]. Thousands of trees, animals were burning along with inhabitant's residences. The ecosystem was badly damaged. Due to deforestation, the climate of the world is going through a hard time. For the years the farmers and ranchers are cutting down trees and leave them in the ground to dry. In the dry season, they set fire on them. The researchers say that this is one of the main reasons for a forest fire.

In concern of those issues, this study will try to give a visualization of the forest fire and cluster the most affected areas. *K*-means cluster, Fuzzy *C*-Means, Apriori method are used to do so. As the dataset contains unlabeled data points so supervised techniques can not apply here as well as no output will return. A confusion matrix and classification report will provide using *K*-Means clustering.

In Fig. 1, there is a map which shows the states of Brazil that reports a fire incident.

2 Background Study

Few researchers have done research on this field regarding different countries like around the world. In this section, the work of those authors will be discussed.

Jafarzadeh A. A., Mahdavi A., Jafarzadeh H. "Evaluation of forest fire risk using the Apriori algorithm and fuzzy *C*-means clustering." [4]. These authors have done interesting research based on the Apriori algorithm and the Fuzzy *C*-Means algorithm. Considering 12 input methods they calculated the fire risk in a province of Iran. They said that their result produces minimum support and minimum confidence and that shows a strong relationship between some variables. They also calculated the wildfire risk from very low to very high in addition to danger. Finally, they generate some maps regarding areas where necessary steps should be taken to prevent the wildfire.

Esra Erten, Vedat Kurgun, Nebiye Musao_lu," Forest Fire Risk Zone Mapping From Satellite Imagery and GIS a Case Study" [5]. In their work, they used satellite image data to predict which areas are prone to a forest fire. Using a Geographic information system (GIS), they combined the factors which are causing forest fires. For this, LANDSAT satellite images were used. And by processing these data with ArcGIS software they found some areas which are previously burnt and which are at risk of fire. By calculating some factors like-(rainfall, slope, forest type map, temperature, and others) they made a list from very high-risk areas to a very low-risk area.

David Radke, Anna Hessler, and Dan Ellsworth," FireCast: Leveraging Deep Learning to Predict Wildfire Spread" [6]. The authors represent FireCast techniques along with GIS. FireCast is a technique that combines deep learning algorithms with satellite images (GIS). A 2D Convolutional Neural Network (CNN) was used here. The CNN contains two convolutional layers of 32 and 64 hidden nodes. The input images are collected based on the atmosphere and geographic situation of a particular time. The FireCast model will produce output that will give a prediction of a highly risking area. They found testing accuracy of 87.7% for FireCast, 50.4% for the Random predictions, 67.8, and 63.6%.

Mr. Ashwin Kumar Thumma, Dr. K. Mruthyunjaya Reddy, Dr. S. Vidyavathi," A proficient forest fire identification using Optimized ANFIS classification" [7]. The authors have used several techniques to classify the forest fire and identify fire territory using satellite images. To do so, they used a modified Fuzzy *C*-Means Clustering (MFCM) to cluster the images. Then from that clustered images they extracted some useful features and used them to improve utilizing the cuckoo search optimization algorithm and after that, the enhanced features are used as input to the ANFIS classifier. This classifier categorized the images into fire or no-fire images. They also showed a comparison between several algorithms like- SVM, Artificial Neural Network, ANFIS, proposed Optimized ANFIS analyze the best performance. From that analysis, they concluded that the proposed ANFIS classifier works best than all.

K. S. N Prasad, Prof. S. Ramakrishna, "An Autonomous Forest Fire Detection System Based on Spatial Data Mining and Fuzzy Logic" [8]. In their research, they proposed a model that will automatically detect the fire. The digital images collected from the satellite are converted into CIELab. *K*-Means clustering is used to cluster the spots of fire. A Fuzzy logic set was built with color space values. From this logic set, the Autonomous Forest Fire Detection System was built.

Saleh Arekhi and Ali Akbar Jafarzadeh, "Forecasting areas vulnerable to forest conversion using ANN and GIS" [9]. The paper aims to predict the rate of deforestation in western Iran. For this, they used an MLPNN with a Markov chain model and form a map. They took LANDSAT satellite images as input data. They claim that their validated map was 94% accurate. In the future, the authors wish to work on predicting forest cover losses for 2020.

Rui Chen, Yuanyuan Luo, Mohammad Reza Alsharif, “Forest Fire Detection Algorithm Based on Digital Image” [10]. This work is based on digital image processing with the help of *K*-Means clustering. They used color space ($L * a * b$ *, YCbCr) along with *K*-Means clustering. They can detect the fire by using flame pixel characteristics. When the fire caught in an area then the fire alarm will report. But sometimes the system couldn’t give the expected output. They concluded by saying that the performance can be improved by using deep learning techniques like Artificial Neural Network.

Paulo Cortez and Anibal Morais, “A Data Mining Approach to Predict Forest Fires using Meteorological Data” [11]. The authors have used several data mining techniques to the burnt areas of the forest fire. SVM, Random Forests, multiple regression, DT, RF, Neural Network, and some distinct feature selection setups (using spatial, temporal, FWI components, and weather attributes), were tested on current real-world data. Data was collected from the local workstations. The authors have said that SVM works better than other algorithms. It can predict the fire burnt off a small area.

Zuoning Wang, Pengfei Liu, Tiejun Cui, “Research on forest flame recognition algorithm based on image feature” [12]. In this work, digital image processing with *K*-Means clustering is used. The authors propose a feature extraction algorithm based on YCrCb and *K*-Means. In the beginning, they analyzed the forest fire image sample. Using the *K*-Means model, with the YCrCb color space and HSI color model, the forest flame is prepared. If any suspected areas found then studied and extracted the area. They concluded by saying that *K*-Means clustering performs satisfactorily by making the clusters instantly and correctly from the flame image and give a satisfactory denotation result.

Begoña C. Arrue, Aníbal Ollero, and J. Ramiro Martinez de Dios, “An Intelligent System for False Alarm Reduction in Infrared Forest Fire Detection” [13]. The authors have done their research for reducing the false fire alarm. They proposed a conventional real-time infrared-visual system by means of image processing and Artificial Neural Network and data collected from meteorological sensors. They also used a geographical information database and design a Fuzzy logic system to develop an intelligent function.

Lei Wang, Qingjian Zhao, Zuomin Wen, and Jiaming Qu, “Raffia: Short-term Forest Fire Danger Rating Prediction via Multiclass Logistic Regression” [14]. In their work, the authors have proposed a method for short-term ratings of forest fire prediction with MLR. They said that the model acquires 98.71% precision and an RMSE value of 0.081. The Multiclass Logistic Regression method is more effective than the Least Square Fitting Regression (LSFR) and Random Forest prediction model.

3 Methodology

This work is for analyzing the danger of forest fire for all states of Brazil. So, the authority can take necessary steps where needed. To do all these we needed to take some steps aside with applying the algorithms. In this section, the description of the processes will be provided.

3.1 Data Processing

Data is the main driving force of research work. For this work, the dataset was taken from the official website of the Brazilian Government [1]. The dataset contains approx. 6454 data points which is pretty much large. From the year 1998 to 2017, the number of forest fire for all states are provided here. To get better output from model cleaning of data is very much important. To remove the missing values of data, the mean and median are used here.

The dataset was written in Brazilian language so it was translated into English for better understanding.

3.2 Statistical Learning

In Fig. 2, there is the working process of this paper.

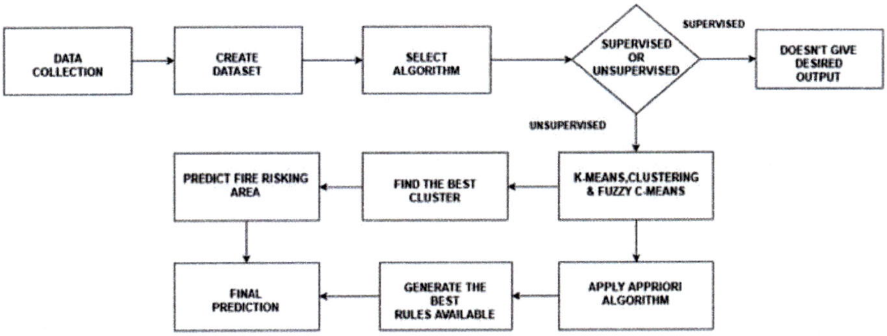

Fig. 2 Flowchart for analysis

3.3 Experimental Result and Discussion

After preparing the dataset, the main concern was to implement it. To implement algorithms in the dataset proper knowledge about that algorithm is required. As the data is unlabeled and there are no testing points so supervised techniques won't work well here. Linear regression, Logistic Regression, and Random forest method were tested with this dataset but they didn't work well here. So, K-Means Clustering method, Fuzzy C-Means, and Apriori method were implemented here. From K-Means Clustering and Apriori method, the confusion matrix and classification report were generated. They gave a satisfactory result; the detailed description will be added in the next section.

4 Experimental Output from Various Algorithms

4.1 K-*Means Clustering*

Clustering refers to group some similar data points from a scattered dataset. K-Means is one of the unsupervised algorithms that's try to group some unlabeled data based on their similarity. The approach of K-Means is to keep outer clusters as far as possible. And also, it tries to keep the inner-cluster data points as similar as possible. With less variation between inner data points, an ideal claustration can be done. To cluster the areas who are at the risk of a forest fire, K-Means is used here.

The equation of K-Means is described as

$$A = \sum_{i=1}^{n} \sum_{k=1}^{k} w_{ik} \left\| x^i - \mu_k \right\|^2 \tag{1}$$

If (cluster) belongs to K then the value of x^i will be 1, otherwise 0, and is the center of the cluster.

After differentiating A with respect to w_{ik} the equation will become,

This step is computing the SSE of the data points.

Then the n (center of cluster) will become.

$$\frac{\partial A}{\partial \mu_k} = 2 \sum_{i=1}^{n} w_{ik} \left(x^i - \mu_k \right) = 0 \tag{2}$$

And

$$\mu_k = \frac{\sum_{i=1}^{n} w_{ik} x^i}{\sum_{i=1}^{n} w_{ik}} \tag{3}$$

Elbow Method:

To build a K-Means model the most crucial thing is to determine the K value. The performance of the model is dependent on the value of K. To find the value of K there are some methods like-Elbow method and Silhouette analysis. For this work, the Elbow method was used. This method selects K based on Sum squared distance between data points and cluster centroids. Elbow methods select the value where SSE starts getting low and form an elbow. Here the value is 2, shown in Fig. 3.

After using $K = 2$, how the method makes the cluster is shown in Fig. 4.

From the visualization, there are some outlier values in the clusters. As all the values are important for this work, so outlier handler cannot be used here.

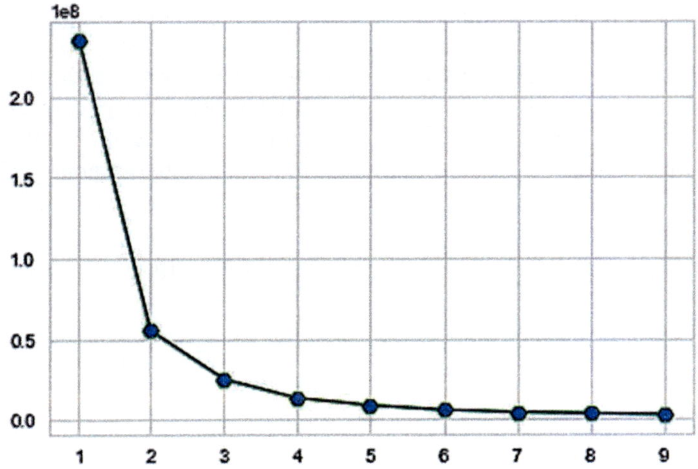

Fig. 3 Elbow method

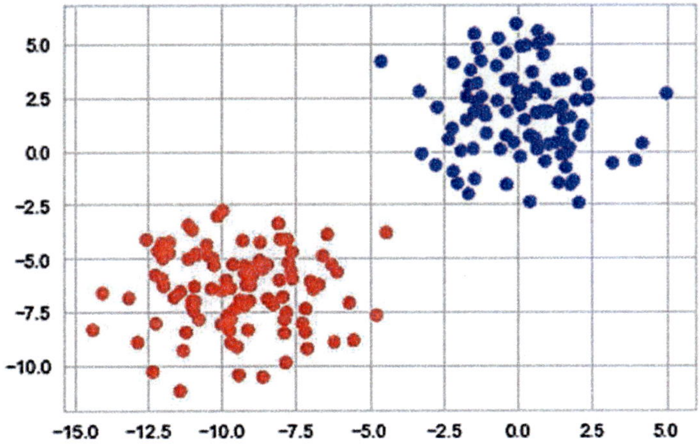

Fig. 4 Clustered values

Table 1 Classification report of *K*-Means clustering

Precision = 90%
Recall = 89%
*f*1-score = 94%
Accuracy = 89%

Table 1 shows the classification report after applying *K*-Means clustering to the model.

4.2 *Fuzzy C-Means*

Fuzzy logic is a type of logic that contains many values but the truth values can be any real number between 0 and 1. Fuzzy *C*-Means is also known as soft clustering or soft *K*-Means, where each data point can belong to more than one cluster. And the graph shows that Fuzzy *C*-Means can handle the data pretty accurately. There are thousands of data points are in this dataset, some of them are at the same risk of a forest fire. To make a cluster of those areas Fuzzy *C*-Means are used here. In Fig. 5 there's a visualization about taking test points. For this purpose, 200 test data were taken along with three clusters.

Clustering with Fuzzy *C*-Means:

Above was the test data for 200 data points. As there is no exact method for clustering in Fuzzy *C*-Means, so for clustering the data points need to select very carefully. After experimenting with a lot of points, it seems that this dataset can

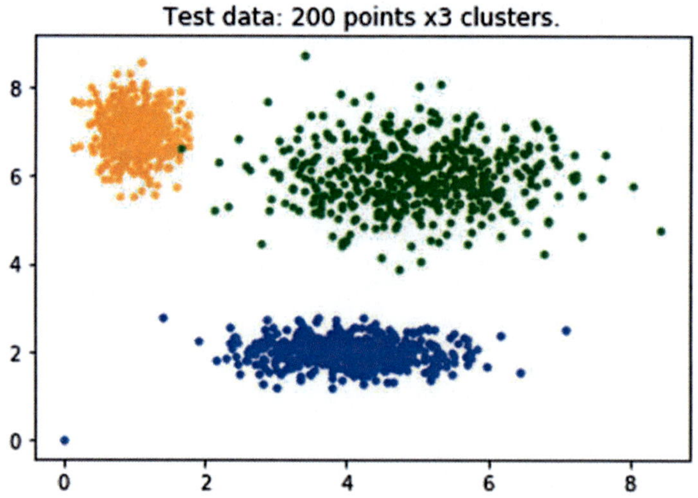

Fig. 5 Testing the data with three clusters

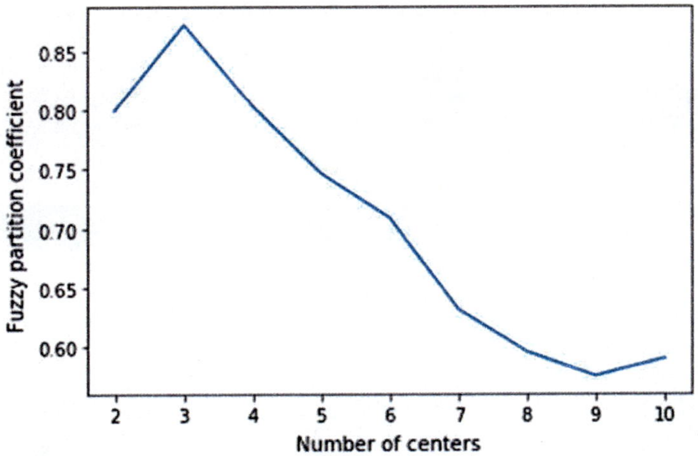

Fig. 6 Clustering between the data points

make a good cluster between data points 2 and 9. Figure 6 shows how the clusters are performing.

FPC Method:

It is called the Fuzzy Partition coefficient which has a range of 0 and 1. It is a scale that how accurately the data are described by the model, where 1 is called the best fit. From Fig. 7 the FPC of this dataset is nearly 0.9 which is a good fit indeed.

Prediction using Fuzzy *C*-Means Clustering:

As the data is already clustered in the model, so now prediction will be made by creating new data points. If the newly created points fit well in the cluster then it can be called a good prediction. The research aims to predict the points where the risk of fire is high. Fuzzy *C*-Means also shows a good cluster between points.

This graph in Fig. 8 visualizes the prediction by Fuzzy *C*-Means according to cluster the areas which are at the same risk.

4.3 Apriori Method

Apriori method is one of the main approaches of Association rules mining. Association rules describe the concept of putting the same data into similar groups. It gives the benefit of using it into larger datasets. Association rules make the groups based on these three factors-Support, Confidence, and Lift.

The implementation of the Apriori algorithm refers to creating itemset from a list of items first. To make a set of frequent items (not all) there needed a threshold—minsup. If the item list contains support > minsup then the itemset can be created. After that, the algorithm uses the re-combination technique to generate all possible rules from frequent itemset.

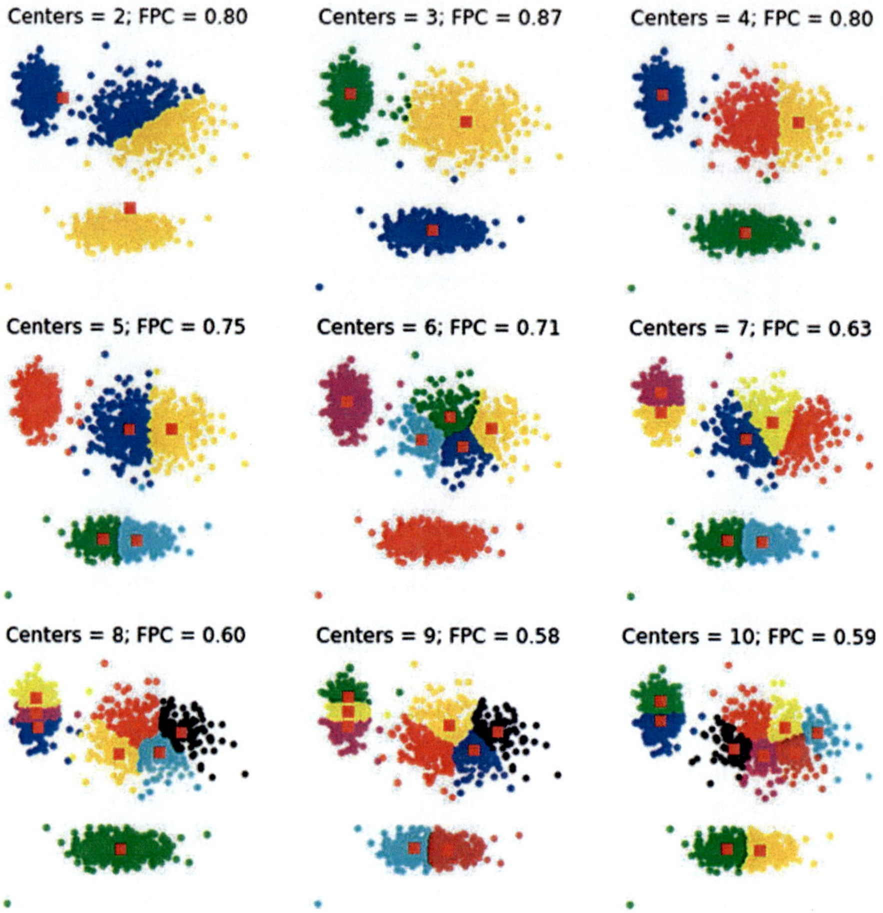

Fig. 7 FPC method

The rules which provide minimum support with maximum confidence is said to have a strong relationship among them. After implementing the dataset into the Apriori method, it gives minimum support of 15% with maximum confidence of 90%. This shows a strong relationship between several forest fires and the states of Brazil which are at risk of a forest fire. After performing 17 cycles on the dataset, the method generates four large item sets followed by the size as ($L1 = 22$, $L2 = 216$, $L3 = 391$, $L4 = 11$).

Apriori generates the best rules available with 12 attributes and 2 classes–$c0$ and $c1$.

Table 2 will show some rules generated by the Apriori algorithm.

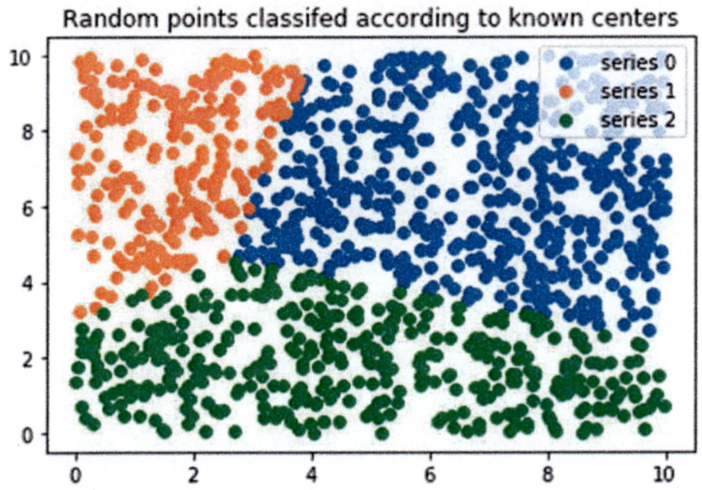

Fig. 8 Accuracy of train and test clusters

Table 2 Rules generated Apriori

Class	Confidence	Lift	Lev.	Conv.
*A*0	1	1.52	0.06	6.46
*A*1	1	1.79	0.08	8.16
*A*3	1	1.52	0.07	6.8
*A*5	1	1.52	0.07	7.14
*A*6	1	1.79	0.09	8.8

5 Result Evaluation

To get our desired prediction—three methods used here followed by (*K*-Means clustering, Fuzzy *C*-Means, and Apriori algorithm). Among them, *K*-Means clustering and Apriori shows a strong relationship between attributes and the risks of fire in Brazil.

The visualization in Fig. 9 shows the number of fires in all states of Brazil from 1998 to 2017.

The graph shows that the states of Mato Grosso and Rio experienced most of the forest fires.

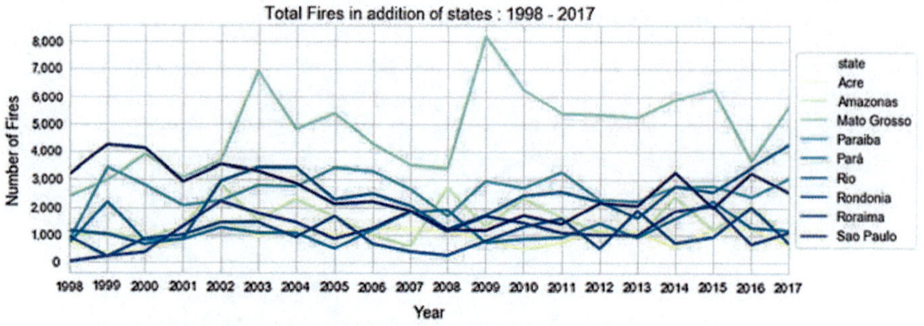

Fig. 9 Total fires in all states of Brazil

6 Future Work and Conclusion

This work is based on analysis of the forest fire and makes some predictions about the fire by means of some unsupervised techniques. So future work will be included by processing the signs and images of fire disasters and try to predict the fire before it happens. Deep learning techniques like Convolutional Neural Network (CNN) can be used to study the images with clustering the signs with the K-Means cluster.

References

1. Dados.gov.br.: Sistema Nacional de Informações Florestais—SNIF—Portal Brasileiro de Dados Abertos. [online] Available at http://dados.gov.br/dataset/sistema-nacional-de-informacoes-florestais-snif. Accessed 17 Dec 2019
2. Nationalgeographic.com.: See how much of the Amazon is burning, how it compares to other years. [online] Available at https://www.nationalgeographic.com/environment/2019/08/amazon-fires-cause-deforestation-graphic-map/. Accessed 17 Dec 2019
3. Phys.org.: Amazon rainforest fires: Everything you need to know. [online] Available at https://phys.org/news/2019-10-amazon-rainforest.html. Accessed 17 Dec 2019
4. Jafarzadeh, A.A., Mahdavi, A., Jafarzadeh, H.: Evaluation of forest fire risk using the Apriori algorithm and Fuzzy C-Means clustering. J. Forest Sci. **63**, 370–380 (2017)
5. Erten, E., Kurgun, V., Musao_lu, N.: Forest Fire Risk Zone Mapping from Satellite Imagery and GIS a Case Study. Turkey
6. Radke, D., Hessler, A., Ellsworth, D.: FireCast: leveraging deep learning to predict wildfire spread. In: Proceedings of the Twenty-Eighth International Joint Conference on Artificial Intelligence (IJCAI-19). Colorado
7. Thumma, A.K., Mruthyunjaya Reddy, K., Vidyavathi, S.: An proficient forest fire identification using optimized ANFIS classification. Int. J. Pure Appl. Mathem. 239–258 (2018)
8. Prasad, K.S.N., Ramakrishna, S.: An autonomous forest fire detection system based on spatial data mining and fuzzy logic. IJCSNS Int. J. Comput. Sci. Netw. Secur. (2008)
9. Arekhi, S., Jafarzadeh, A.A.: Forecasting areas vulnerable to forest conversion using artificial neural network and GIS. Iran (2012)
10. Chen, R., Luo, Y., Alsharif, M.R.: Forest fire detection algorithm based on digital image. J. Softw. **8**, 8 (2013)

11. Cortez, P, Morais, A.: A Data Mining Approach to Predict Forest Fires using Meteorological Data. Portugal
12. Wang, Z., Liu, P., Cui, T.: Research on forest flame recognition algorithm based on image feature. Int. Arch. Photogramm. Rem. Sens. Spat. Inf. Sci. (2017)
13. Arrue, B.C., Ollero, A., Martinez de Dios, J.R.: An intelligent system for false alarm reduction in infrared forest-fire detection. IEEE Intell. Syst. (2000)
14. Wang, L., Zhao, Q., Wen, Z., Qu, J.: Raffia: Short-term Forest Fire Danger Rating Prediction via Multiclass Logistic Regression. www.mdpi.com/journal/sustainability (2018)

Water Quality Based Zonal Classification of Rajasthan Using Clustering

Umesh Gupta, Sonal Jain, and Dharmendra Kumawat

Abstract Clean Water and Sanitation being a sustainable development goal is the prime focus of Government, but the question remains whether the process being adopted is really leading to success. This question is the motivation of this study. This paper is an effort to lead in the systematic direction of creating GIS-based WaterMap integrated with geographical boundaries built on Ground Water Quality Assessment parameters. The objective of this study is to analyze water quality in each district and then classifying these districts. Several data mining techniques are being used for water analysis, including clustering and classification, ANN, SVM, and Regression. In this study, total nine water quality parameters—Total Dissolved Solids (TDS), Chloride (Cl), Sulphate (S), Fluoride (F), Nitrate (N), Total Hardness (TH), Calcium (Ca), Magnesium (Mg) and Iron (Fe) are considered. Three years' data from 2014–15 to 2016–17 has been referred from Ground Water Year Book (GWYB) of Central Ground Water Board, Government of India. The data represents the percentage of samples falling beyond the permissible limit in each district of Rajasthan. *K*-means clustering algorithm is performed for the clustering of the district based on similar chemical and water quality parameters. ANOVA has been performed before applying the clustering method to check the equality of average percentage of samples beyond the permissible limit. It would be helpful in formation of area focused water safety plans, which WHO and UNICEF are supporting for better health indicators under SDG Goal 6.

Keywords Groundwater quality · Sustainable development goals · *K*-means clustering · ANOVA

U. Gupta (✉) · S. Jain
JK Lakshmipat University, Jaipur, Rajasthan, India
e-mail: ugupta.maths@gmail.com

S. Jain
e-mail: sonaljain@jklu.edu.in; drsonalamitjain@gmail.com

D. Kumawat
Birla Institute of Technology, Mesra Extension Center, Jaipur, Rajasthan, India
e-mail: kumawatdhamendra@gmail.com

S. Borah et al. (eds.), *Soft Computing Techniques and Applications*,
Advances in Intelligent Systems and Computing 1248,
https://doi.org/10.1007/978-981-15-7394-1_41

1 Introduction

Rajasthan, a state of India, comprises of 33 districts covering a geographical area of 342,239 km^2. The ground water is being monitored through 1170 National Hydrograph Network Stations (NHS) four times a year simultaneously.

Several researchers analyzed the water quality in different parts of Rajasthan. Mitharwal et al. [1] analyzed it in Jhunjhunu and observed higher nitrate value as a cause of concern. Mathur and Gupta [2] determined water quality index of Jaipur and mentioned the need for water treatment before consumption so as mentioned by Meena et al. [3] for Pali. Saurabh et al. [4] investigated drinking water quality on bus stands and railway stations in Rajasthan and concluded that the state is suffering from very poor quality of drinking water (Fig. 1).

Water quality testing is conducted in laboratories established by Union and State Governments throughout the country. These labs are producing huge dataset that has been attracting researchers to further analyze beyond the labs. Researchers have

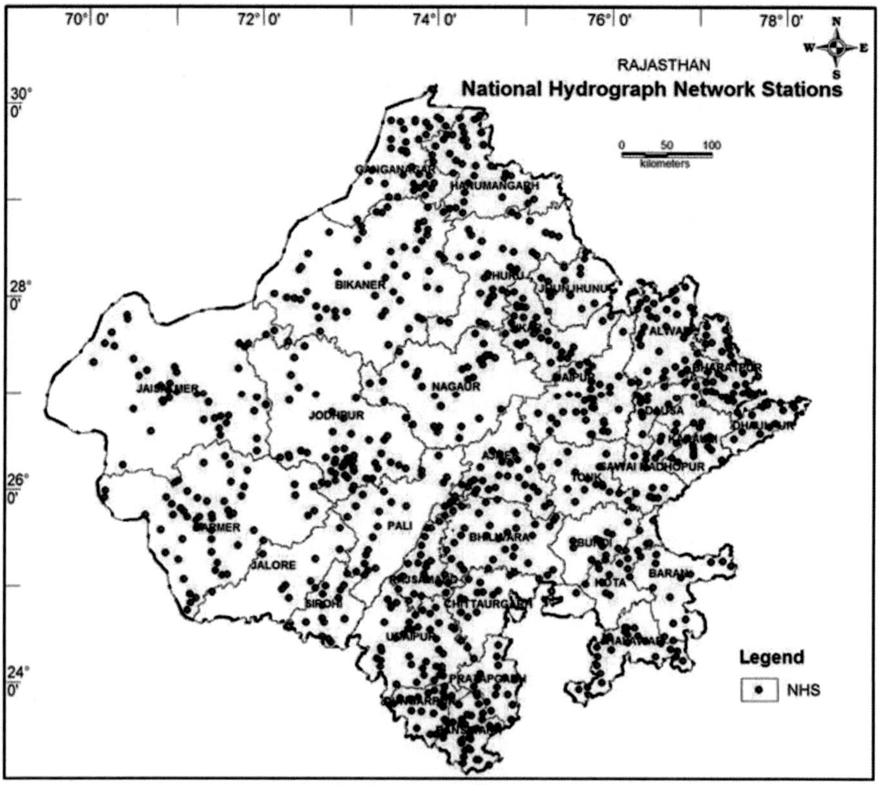

Fig. 1 NHS locations in Rajasthan. *Source* Ground Water Year Book 2016–17, Rajasthan State, CGWB, Western Region, Jaipur

applied various statistical methods for analyzing the water-related data. Parmar and Bhardwaj [5] applied a time series analysis to predict the water quality parameters of Yamuna river. Lešcešen et al. [6] used statistical methods to analyze the water quality parameters in the rivers of Serbia. Hada et al. [7] used goal programming with multiple non-linear regression for seasonal evaluation of water in Tonk district of Rajasthan. Kumar and Singh [8] used multivariate statistical analysis to interpret the water quality parameters of Sanganer Tehsil of Jaipur in Rajasthan. Recently, in a study by Pujar et al. [9] on Krishna river, it has been concluded that a one-way analysis of variance is effective in water quality analysis.

With the advancement of statistical methods to data science, computational approaches brought more accuracy to the analysis. The methods such as Clustering analysis, classification, artificial neural networks are used by data scientists in their research supporting policymakers. Shrestha and Kazama [10] used clustering and classification for surface water quality assessment of the Fuji river basin in Japan. Hierarchical clustering methods were applied by Emad et al. [11] on the dataset of the Euphrates River.

k-means clustering is a popular clustering technique in data mining which was used first by MacQueen [12]. Researchers have used clustering techniques on water quality datasets of various parts of Rajasthan. Mehta and Duggal [13] applied this technique on groundwater contamination in Sanganer Tehsil of Jaipur in Rajasthan. Gupta et al. [14] applied the clustering method to divide 152 water reservoirs of Jaipur into three parts based on water quality. There is a lot of work done on various parts of Rajasthan. This study is not limited to any specific part but it is covering Rajasthan as a whole where the percentage of samples in which water quality parameters were observed beyond permissible limit has been considered.

2 Methodology

2.1 Dataset

Groundwater quality data of Rajasthan is referred from the groundwater yearbook that is released every year by Central Ground Water Board of Government of India. 1170 NHS are setup in 33 districts of Rajasthan which comprises 724 dug wells and 446 piezometers.

Total 561 samples in 2014–15 and 2015–16, 646 in 2016–17 were chemically analyzed by CGWB for evaluation of hydro-chemical status and distribution of various chemical constituents in groundwater in 33 districts of Rajasthan. Percentage of samples showing 9 water quality parameters measured in parts per million (ppm), beyond its permissible limit in each district, has been considered for analysis in this study (Tables 1, 2 and 3).

Table 1 Descriptive statistics of water quality dataset (2014–15)

	TDS	Cl	S	F	N	TH	Ca	Mg	Fe
Mean	23.36	9.12	16.21	26.66	41.93	19.14	5	15.21	37.52
Std. error	3.7	1.99	2.76	3.15	3.5	2.39	1.09	2.45	4.24
Median	18.18	0	12.5	27.78	39.29	18.18	3.7	11.76	38.89
Mode	0	0	0	0	33.33	0	0	0	43.75
Std. deviation	21.23	11.44	15.87	18.09	20.12	13.71	6.26	14.06	24.35
Sample var.	450.7	130.94	251.94	327.09	404.82	187.98	39.15	197.81	593.11
Minimum	0	0	0	0	12.5	0	0	0	0
Maximum	75	33.33	53.85	70.59	83.33	57.14	23.08	53.57	91.67
Sum	770.93	300.9	535.03	879.69	1383.76	631.46	164.98	501.93	1238.2

Table 2 Descriptive statistics of water quality dataset (2015–16)

	TDS	Cl	S	F	N	TH	Ca	Mg	Fe
Mean	25.74	12.53	12.55	26.2	42.54	19.6	5.21	14.84	29.5
Std. error	3.6	2.25	2.01	3.73	3.59	2	1	1.91	3.42
Median	21.05	9.09	10	22.22	40	19.05	4	14.29	27.59
Mode	0	0	0	0	55.56	0	0	0	0
Std. deviation	20.71	12.91	11.56	21.43	20.64	11.48	5.74	10.97	19.65
Sample var.	428.84	166.56	133.53	459.39	425.9	131.7	32.91	120.38	386.14
Minimum	0	0	0	0	12.5	0	0	0	0
Maximum	80	44.83	42.31	100	81.82	44.83	20	48	68.42
Sum	849.3	413.34	414.25	864.59	1403.88	646.78	171.86	489.85	973.62

Table 3 Descriptive statistics of water quality dataset (2016–17)

	TDS	Cl	S	F	N	TH	Ca	Mg	Fe
Mean	22.54	11.22	12.14	24.64	34.47	17.41	4.82	13.13	35.12
Std. error	4.13	3.43	2.08	3.24	3.17	2.73	1.08	1.9	3
Median	14.29	5.56	9.09	18.18	31.25	15.79	4.17	11.76	32
Mode	0	0	0	0	25	0	0	0	50
Std. deviation	23.71	19.71	11.97	18.64	18.23	15.68	6.19	10.92	17.26
Sample var.	562.25	388.64	143.34	347.44	332.33	245.8	38.32	119.28	297.76
Minimum	0	0	0	0	5.88	0	0	0	0
Maximum	100	100	48.65	62.5	81.25	80	22.22	47.5	75
Sum	743.72	370.13	400.78	813.26	1137.43	574.63	159.07	433.28	1158.99

2.2 Analysis of Variance (ANOVA)

Analysis of variance is a statistical method to test the equality of means of more than two samples. This technique was developed by Ronald Fisher. This is the form of hypothesis testing in the analysis of experimental data. It has various types of viz. one-way, two-way with replication, and without replication.

Objective to use ANOVA in this study is to test whether there is a significant difference across the districts in 2014–15, 2015–16 and 2016–17. One-way analysis of variance is used for this test. The level of significance used in this study is 0.05 for which the critical value of F statistic is 3.091.

The hypothesis for this analysis is:

$$H0_{14-15}: \mu_{\mathrm{TDS}} = \mu_{\mathrm{Cl}} = \mu_{\mathrm{S}} = \mu_{\mathrm{F}} = \mu_{\mathrm{N}} = \mu_{\mathrm{TH}} = \mu_{\mathrm{Ca}} = \mu_{\mathrm{Mg}} = \mu_{\mathrm{Fe}}$$

$$H0_{15-16}: \mu_{\mathrm{TDS}} = \mu_{\mathrm{Cl}} = \mu_{\mathrm{S}} = \mu_{\mathrm{F}} = \mu_{\mathrm{N}} = \mu_{\mathrm{TH}} = \mu_{\mathrm{Ca}} = \mu_{\mathrm{Mg}} = \mu_{\mathrm{Fe}}$$

$$H0_{16-17}: \mu_{\mathrm{TDS}} = \mu_{\mathrm{Cl}} = \mu_{\mathrm{S}} = \mu_{\mathrm{F}} = \mu_{\mathrm{N}} = \mu_{\mathrm{TH}} = \mu_{\mathrm{Ca}} = \mu_{\mathrm{Mg}} = \mu_{\mathrm{Fe}}$$

2.3 Clustering

k-means clustering has been used for making clusters of districts based on Euclidean distance. Three clusters have been formed using the data of 9 water quality parameters taking all three years into consideration. Three years are considered to bring more accuracy in cluster formation.

$$\text{Minimize } J = \sum_{j=1}^{k}\sum_{i=1}^{n}\sqrt{(x_i - c_j)^2} \tag{1}$$

where the Euclidian distance $\sqrt{(x_i - c_j)^2}$ is the distance of n data points from their respective cluster centers and objective is to find out cluster centers c_j for minimum distance.

2.4 Graphical Analysis

2-D column charts showing water quality levels per cluster in each district have been prepared to critically analyze the water quality. The objective of this analysis is to

Table 4 Percentage of samples beyond permissible limit

Quality parameter	Limit	2014–15	2015–16	2016–17
Total dissolved solids (TDS)	2000	25.49	25.67	24.46
Chloride (Cl)	1000	10.52	12.83	11.46
Sulphate (S)	400	17.11	13.73	14.70
Fluoride (F)	1.5	27.45	25.31	26.32
Nitrate (N)	45	42.78	40.64	34.67
Total hardness (TH)	600	20.14	20.30	17.70
Calcium (Ca)	200	5.35	5.17	6.04
Magnesium (Mg)	100	16.76	15.90	14.70
Iron (Fe)	0.30	38.15	27.63	35.30

identify the districts with critical stage of water quality parameter so that despite making cluster-based policy, the individuality of districts is not ignored.

3 Results and Discussions

3.1 Quality Assessment

Permissible limit considered for each water quality parameter is as per Table 4.

3.2 Analysis of Variance (ANOVA)

The results using one-way analysis of variance on each of the 9 water quality parameters are shown in Fig. 2.

By comparing the calculated F values with critical value, it is clear that the null hypothesis on equality of means cannot be rejected for all the years, i.e., the average number of samples showing higher values of specific water quality parameter is not significantly different across the years. This test suggests that the clustering can be applied together for all three years so that small variations are also taken care of while clustering.

3.3 K- Means Cluster Analysis

The result of *k*-means cluster analysis divides all 33 districts of Rajasthan into 3 different clusters named Cluster 0, 1, and 2. This is based on the assumption that

TDS						
Source of Variation	*SS*	*df*	*MS*	*F*	*P-value*	*F crit*
Between Groups	182.1149	2	91.05746	0.189469	0.827707	3.091191
Within Groups	46137.02	96	480.594			
Total	46319.14	98				

TH						
Source of Variation	*SS*	*df*	*MS*	*F*	*P-value*	*F crit*
Between Groups	87.57549	2	43.78775	0.232306	0.793148	3.091191
Within Groups	18095.23	96	188.4919			
Total	18182.8	98				

Cl						
Source of Variation	*SS*	*df*	*MS*	*F*	*P-value*	*F crit*
Between Groups	194.9763	2	97.48813	0.426249	0.654184	3.091191
Within Groups	21956.35	96	228.7119			
Total	22151.32	98				

Ca						
Source of Variation	*SS*	*df*	*MS*	*F*	*P-value*	*F crit*
Between Groups	2.483299	2	1.241649	0.033745	0.96683	3.091191
Within Groups	3532.378	96	36.79561			
Total	3534.862	98				

S						
Source of Variation	*SS*	*df*	*MS*	*F*	*P-value*	*F crit*
Between Groups	331.2355	2	165.6177	0.939572	0.394359	3.091191
Within Groups	16921.86	96	176.2694			
Total	17253.1	98				

Mg						
Source of Variation	*SS*	*df*	*MS*	*F*	*P-value*	*F crit*
Between Groups	81.40317	2	40.70158	0.279117	0.757063	3.091191
Within Groups	13998.98	96	145.8227			
Total	14080.38	98				

F						
Source of Variation	*SS*	*df*	*MS*	*F*	*P-value*	*F crit*
Between Groups	73.49216	2	36.74608	0.097219	0.907447	3.091191
Within Groups	36285.35	96	377.9724			
Total	36358.84	98				

Fe						
Source of Variation	*SS*	*df*	*MS*	*F*	*P-value*	*F crit*
Between Groups	1117.564	2	558.782	1.31272	0.273873	3.091191
Within Groups	40864.07	96	425.6674			
Total	41981.63	98				

N						
Source of Variation	*SS*	*df*	*MS*	*F*	*P-value*	*F crit*
Between Groups	1334.13	2	667.0651	1.720637	0.184424	3.091191
Within Groups	37217.75	96	387.6849			
Total	38551.88	98				

Fig. 2 ANOVA results on water quality parameters

the clusters should be homogeneous within the clusters and heterogeneous from the other clusters.

Percentage of samples with water quality beyond permissible limits in each district of the clusters formed using *k*-means clustering are given in Tables 5, 6, and 7.

The grand average of water quality parameters is self-explanatory to explain the logic for dividing these clusters. Descriptive statistics in Tables 1, 2 and 3 and Grand average represented in the Tables 5, 6 and 7 corresponding to each cluster clearly indicates that Nitrate, Iron, TDS, and Fluoride is the major cause of concern in the State.

Table 5 clearly indicates a focus needed on improving on Nitrate, TDS, and Iron in the districts lying in first cluster, i.e., cluster 0. This is also evident that the water quality in 18 districts of cluster 0 is very poor. This cluster covers more than 50% population of Rajasthan and includes 4 division headquarters out of 6 viz. Jaipur, Bikaner, Jodhpur and Ajmer.

Table 6 showcases the need to focus on total hardness, TDS, Magnisium, and Iron in the districts of cluster 1.

Table 7 showcases the need to focus on nitrate and Iron in the districts of cluster 2.

2-D column chart for each cluster describes the district water quality more clearly. It is evident from Fig. 3 that TDS, Fluoride, Nitrate, and Iron are affecting water quality in cluster 0 which is covering 18 districts of Rajasthan. Nitrate is the major cause of concern whereas TDS, Fluoride, and Iron are almost equally affecting

Table 5 Percentage of samples beyond permissible limits (Cluster 0)

Districts	TDS	Cl	S	F	N	TH	Ca	Mg	Fe
Ajmer	25.07	9.65	12.23	60.52	36.38	23.02	6.68	14.22	67.26
Barmer	54.57	34.72	32.11	24.75	60.65	41.39	14.15	31.9	29.86
Bhilwara	17.89	8.85	6.47	51.19	31.78	20.37	3.71	16.57	42.91
Bikaner	20.64	10.83	10.83	36.06	19.44	18.39	2.67	13.5	36.5
Chittorgarh	18.67	6.67	2.56	16.92	40.26	21.67	9.44	9.44	36.45
Churu	51.13	19.28	28.08	39.15	71.8	30.98	1.45	34.68	25.7
Dausa	27.43	7.9	15.57	43.96	26.54	19.77	1.39	16.74	57.68
Dholpur	15.97	5.56	9.72	22.92	40.97	17.36	4.86	19.45	68.75
Hanumangarh	14.18	6.95	14.24	26.35	26.57	18.76	4.63	16.09	29.52
Jaipur	19.56	14.11	6.78	46.67	35.11	16.89	1.33	15.56	15.56
Jaisalmer	46.08	20.03	44.42	19.89	49.55	19.27	11.82	14	18.61
Jalore	74.81	46.67	17.78	28.89	65.19	40.74	0	17.04	10.37
Jodhpur	39.49	20.04	25.79	37.9	63.49	21.43	6.94	18.85	22.82
Nagaur	75.89	39.98	33.83	36.21	81.05	25.89	6.94	21.43	15.48
Pali	49.81	21.25	19.69	35.67	33.82	26.8	12.67	21.44	44.25
Rajsamand	19.39	8.89	16.05	21.95	59.16	20.87	7.22	9.42	41.18
Sirohi	18.67	7.37	5.41	57.73	73.77	14.32	1.96	6.95	33.6
Tonk	18.12	12.98	8.31	24	45.28	16.15	5.51	19.7	15.69
Grand avg.	33.74	16.76	17.22	35.04	47.82	23	5.74	17.61	34.01

Table 6 Percentage of samples beyond permissible limits (Cluster 1)

Districts	TDS	Cl	S	F	N	TH	Ca	Mg	Fe
Baran	14.63	0	20.05	4.04	17.97	22.13	3.33	14.63	23.75
Bharatpur	51.62	29.88	24.21	32.33	40.26	46.55	12.76	49.69	38.86
Bundi	17.68	3.03	17.42	20.2	23.48	20.45	6.06	17.68	24.24
Ganganagar	23.01	10.26	27.32	26.54	12.64	26	9.65	21.58	28.93
Grand avg.	26.74	10.79	22.25	20.78	23.59	28.78	7.95	25.9	28.95

parameters. In this cluster, Nagaur district is the area in which immediate attention is needed on Nitrate affect, may districts have shown improvement in the past three years. Dausa, Rajsamand, and Sirohi have improved significantly in these. The condition in Hanumangarh and Jaisalmer got poorer as far as nitrate is concerned. Though Nitrate is visualized as the most affecting parameter but if the TDS condition is observed, the condition is getting poorer in many districts that raise alarm for policymakers and so are fluoride and iron. Jalore is a district showing a very high percentage of samples with chloride, TDS, and Total hardness beyond the permissible limit. Jaisalmer district needs to be focused on Sulphate in addition to Nitrate that has already been discussed.

Table 7 Percentage of samples beyond permissible limits (Cluster 2)

Districts	TDS	Cl	S	F	N	TH	Ca	Mg	Fe
Alwar	11.84	5.26	6.88	23.64	17.86	7.34	3.67	7.34	50.93
Banswara	0	0	0	7.78	33.16	2.22	0	0	57.73
Dungarpur	0	0	0	9.49	29.86	8.33	0	8.33	48.84
Jhalawar	7.43	0	12.73	4.51	36.32	16.7	5.56	6.92	16.22
Jhunjhunu	7.02	1.75	1.75	15.64	32.75	7.02	0	3.51	18.13
Karauli	18.34	5.36	10.82	7.53	46.29	12.68	7.42	12.38	25.79
Kota	2.08	0	6.25	0	16.67	2.08	0	2.08	33.33
Pratapgarh	0	0	0	14.85	19.02	2.78	0	0	56.41
S. Madhopur	12.3	1.85	7.04	20.92	36.55	14.15	5.37	7.03	39.6
Sikar	8.85	0	3.33	24.62	58.33	3.33	3.33	3.33	13.97
Udaipur	5.83	2.34	2.34	9.7	26.39	11.79	4.76	3.53	34.68
Grand avg.	6.7	1.51	4.65	12.61	32.11	8.04	2.74	4.95	35.97

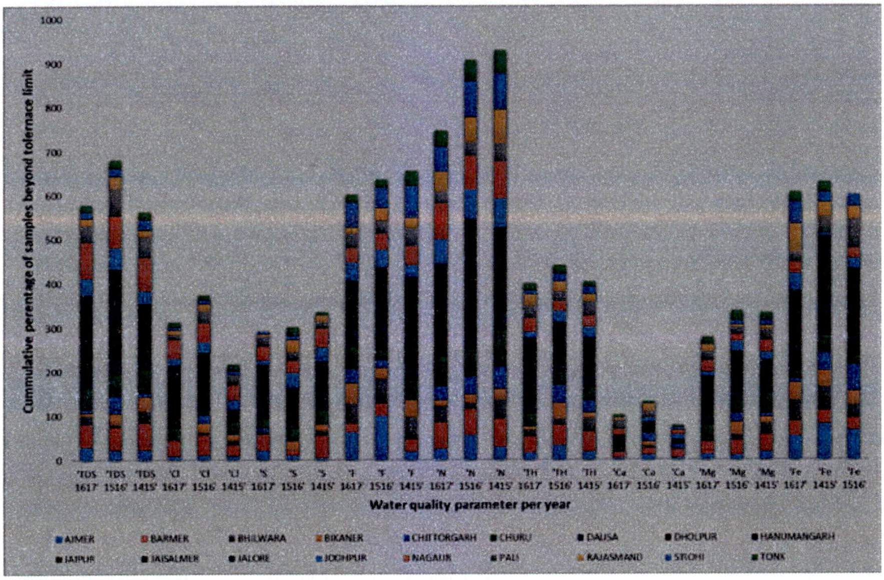

Fig. 3 District wise water quality status in 3 years (cluster 0)

Figure 4 is the graph visualization for the districts lying in the second cluster, i.e., cluster 1. The districts in cluster 1 showed significant improvement in all the parameters. However, magnesium, total hardness, and iron are the water quality parameters to be taken care of while making any policy for this cluster. Bharatpur in this cluster seems to be critical in terms of all the water quality parameters as a higher number of samples for these districts are beyond the permissible limit. Baran

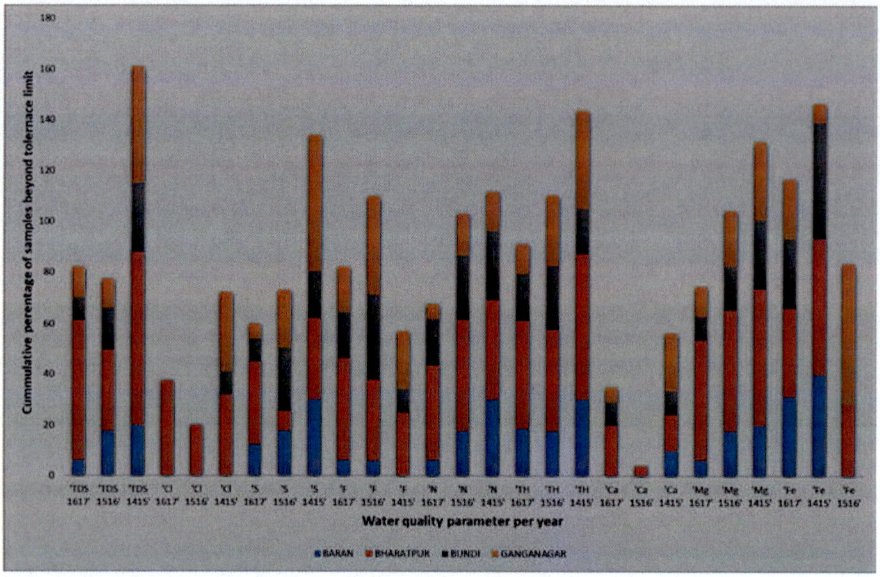

Fig. 4 District wise water quality status in 3 years (cluster 1)

has started showing fluoride beyond permissible limit, It will the right time to take care of this fact there.

Figure 5 is the graph visualization for the districts lying in third cluster, i.e., cluster 2. The districts in cluster 1 showed higher number of samples showing beyond permissible limit in the case of Nitrate and Iron. More than 50% of samples are showing higher Nitrate values in Sikar and Karauli whereas higher iron values are visible in water quality in Dungarpur, Banswara, Pratapgarh, and Alwar. Therefore, while making a common policy on water quality for this cluster, individual quality characteristics in these districts need to be taken care of.

4 Conclusion

Clustering is proven to be a very effective tool of data analysis on water quality. This paper concludes that groundwater quality is improving in many areas in Rajasthan due to the ongoing efforts from Government and awareness among society. However, there are many areas that are needed to be given more attention.

This was also evident from the analysis that the average number of samples showing higher values in three years has not changed. But, while doing analysis on individual districts in each cluster, it was found that the average will not be the only method to analyze the data. *k*-means clustering divides the data into three

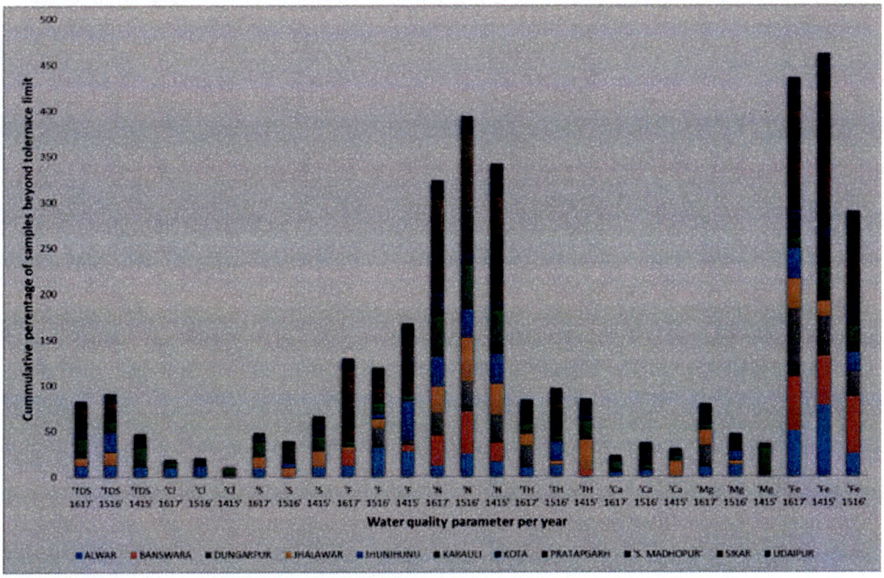

Fig. 5 District wise water quality status in 3 years (cluster 2)

zones (clusters). These clusters would help in the identification of potential areas demanding focus in terms of water quality importance in Rajasthan.

It is concluded that 18 districts out of 33 in Rajasthan need immediate attention in the design and adaptation of water treatment technology to ensure the supply of clean drinking water as per standards. It may be due to over-exploitation of groundwater or due to population density or rainfall. But it requires investigation and proper policy framework for these 18 districts.

Through this study, it is clear that geographical boundaries based on clustering with support of other statistical techniques may be helpful in creating GIS-based WaterMaps and it will further be helpful in designing a more focused policy on groundwater quality of the state.

Acknowledgements The authors would like to acknowledge Central Ground Water Board for publishing Ground Water Year Books which was referred for this data analysis.

References

1. Mitharwal, S., Yadav, R.D., Angasaria, R.C.: Water quality analysis in Pilani of Jhunjhunu district (Rajasthan)—the place of Birla's origin. Rasayan. J. Chem. **2**(4), 920–923 (2009)
2. Mathur, A., Gupta, U.: Assessment of ground water quality of Jaipur Rajasthan India using WQI (water quality index). Int. Bull. Mathem. Res. **2**(1 Special), 83–86 (2015)

3. Meena, A.K., Rajagopal, C., Bansal, P., Nagar, P.N.: Analysis of water quality characteristics in selected areas of Pali district in Rajasthan. Indian J. Environm. Protect. **29**(11), 1011–1016 (2009)
4. Saurabh, S., Singh, D., Tiwari, S.: Drinking water quality of Rajasthan districts. J. Basic Appl. Eng. Res. **1**(10), 105–109 (2014)
5. Parmar, Kuldeep Singh, Bhardwaj, Rashmi: Water quality management using statistical analysis and time-series prediction model. Appl. Water Sci. **4**, 425–434 (2014)
6. Leščešen, I., Dolinaj, D., Pantelić, M., et al.: Statistical analysis of water quality parameters in seven Major Serbian Rivers during 2004–2013 Period. Water Resour. **45**, 418–426 (2018)
7. Hada, D.S., Gupta, U., Sharma, S.C.: Seasonal evaluation of hydro-geochemical parameters using goal programming with multiple nonlinear regression. Gen. Math. Notes **25**(2), 137–147 (2014)
8. Kumar, Manish, Singh, Yashbir: Interpretation of water quality parameters for villages of Sanganer Tehsil, by using multivariate statistical analysis. J. Water Resour. Protect. **2**, 860–863 (2010)
9. Pujar, P.M., Kenchannavar, H.H., Kulkarni, R.M., et al.: Real-time water quality monitoring through internet of things and ANOVA-based analysis: a case study on river Krishna. Appl. Water Sci. **10**, 22 (2020)
10. Shrestha, S., Kazama, F.: Assessment of surface water quality using multivariate statistical techniques: a case study of the Fuji river basin, Japan. Environ. Model Softw. **22**, 464–475 (2007)
11. Emad, A.M.S., Ahmed, M.T., Eethar, M.A.O.: Assessment of water quality of euphrates river using cluster analysis. J. Environm. Protect. **3**(12), 1629–1633 (2012)
12. MacQueen, J.B.: Some methods for classification and analysis of multivariate observations. In: Proceedings of 5-th Berkeley Symposium on Mathematical Statistics and Probability, Berkeley, University of California Press, vol. 1, pp. 281–297 (1967)
13. Mehta, Anurika, Duggal, Rakesh: Data mining of Sanganer Tehsil, Jaipur (Rajasthan) with clustering techniques of groundwater contamination. Int. J. Eng. Sci. Adv. Technol. **5**(5), 474–485 (2015)
14. Gupta, U., Jain, S., Kumawat, D., Sharma, R.: Clustering of water reservoirs based on water chemical analysis. Int. J. Res. Advent Technol. (IJRAT) **7**(6S), 274–277 (2019)

A Meta-heuristic Optimization Algorithm for Solving Renewable Energy System

C. Shilaja

Abstract Energy conversion and distribution is one of the drastically growing and leading research nowadays. Renewable energy systems are incorporated into the Grid system to fulfill the demand dispatch effectively. Various earlier approaches are used for scheduling the energy system incorporated into the grid, but the efficiency of the energy scheduling needs to be improved. The main aim of this paper is to reduce the operation cost using Grey Wolf Optimization (GWO) algorithm for providing an efficient solution for energy scheduling problems in the hybrid energy microgrid system. The hybrid energy system comprises of solar, wind, thermal and electrical vehicles, and analyzed by conducting a simulation. It is found that the proposed GWO model is efficient and represent that this algorithm has a better global search ability than the exiting PSO and MPSO approach.

Keywords Energy conversion · Solar energy · Wind energy · Optimal scheduling

1 Introduction

There are various life-threatening problems to human beings in the twenty-first century includes that the environmental pollution, lack of precise resources, the annoyance of human being society's reliance on energy, energy disaster, which all started threatening on the 1970s. So there is a need for prevention and in quest of new resources for energy is the vital task for enhancing the efficacy of energy consumption [1, 2]. An electric power network that has the micropower grid-connected with distributed power generation units and the end-users. This micropower grid can able to optimize and enhance energy consumption and also it can provide elasticity, controllability, and financial competence of power system function. It is very essential in the fast growth of distributed energy incorporation of power system network

C. Shilaja (✉)
Department of EEE, Kalasalingam Academy of Research and Education, Krishnankoil, Srivilliputhur, Virudhunagar, India
e-mail: shilaja.research@gmail.com

S. Borah et al. (eds.), *Soft Computing Techniques and Applications*,
Advances in Intelligent Systems and Computing 1248,
https://doi.org/10.1007/978-981-15-7394-1_42

for improving efficiency. In the northern region of China, they are using wind and solar energy because they are available huge in that area and it demands better.

The main aim of this work is to minimize the operation costs and considered each outcome of the distributed power source is put as an optimization variable. To enhance the universal, seek potential, the multi-objective Grey Wolf Optimization (MGWO) algorithm is proposed here and simulated this in a case system to prove the achievability of the method suggested. The entire contribution of the paper is carried out into two different stages such as:

- Design and construct a micro-grid system to schedule the optimization problem.
- MGWO and PSO algorithms are implemented and experimented in MATLAB software and the results are compared for performance evaluation.

2 Background Study

To get an optimal solution from the group of solutions obtained with certain criteria, there is a need for computational optimization comprised of several mathematical techniques. From the domain with a set of constraints given with few objective functions, searching for the best solution from the available values is called optimization. For optimization, several techniques of applied mathematics are used. Banos et al. suggested the mathematical techniques for computerized optimization, operational research for designing the system, computerized algorithm for design, analysis, and software engineering process for the implementation [3]. Diwerak [4] illustrated that the process of optimization is an iterative procedure that has the optimizer in each level of the modelling the system. For the given problem of optimization, modelling is the process of finding the objective function, variables and constraints [5]. To find the values of the decision variables which are used to calculate the objective function with criteria is the function of optimizer. This will be again used to a new set of decision variables with the criteria. Thus, this function iteratively doing the same process of optimization until criteria are satisfied [4]. Iterative and heuristics methods in computational optimization techniques are suggested here. There are different types of optimization algorithms are available and depending upon the problem type, the optimization algorithm will be chosen for the best solution. Hence most of the researchers proposed the approximation techniques to provide an optimized solution to the problem.

To search the best solution among the huge set of feasible solutions in less computational time with less complexity, Heuristic methods are best than the other optimization techniques [5]. This is very helpful to provide the optimal solution when the other classical optimization techniques can't do. Meta-heuristics are also available to provide the solution from the distinct search-space. The bio-inspired methods [6] also used for the classification of the meta-heuristic algorithms.

- For altering and enhancing the single candidate solution while searching, Trajectory meta-heuristics use a single-solution method is used and it offers the single

optimized solution. SA, TS, GRASP, VNS and ILS are mainly used meta-heuristic methods.
- There are a fixed number of iterations in the population-based meta-heuristics method which returns the population-based solution while the stopping condition is achieved. GA and PSO are an example of this type of algorithms.
- The techniques which imitate nature for solving the optimization problems in the bioinspired meta-heuristics algorithm. Binitha and Shatia [6] categorized these algorithms into three types that are Evolutionary algorithms, Swarm intelligence and Ecology-based algorithms.

Other kinds of meta-heuristics can be viewed as, for example, hybrid and parallel meta-heuristics. The hybrid meta-heuristic consolidates other enhancement approaches with the meta-heuristic one. In the meantime, the parallel meta-heuristic is a calculation that runs numerous meta-heuristics look in parallel by utilizing parallel processing methods. At times, the multifaceted nature of the issues to illuminate is high to the point that no heuristic neither meta-heuristic technique can acquire exact arrangements in sensible runtimes. Subsequently, parallel processing turns into an intriguing method to get great arrangements with diminished runtimes. Parallel registering is a type of calculation in which vast issues can be partitioned into littler ones, completing numerous counts at the same time. Regular sorts of issues found in parallel figuring for microgrid applications are Monte Carlo Simulation [7–9] and Dynamic Programming [10, 11]. As to brief prologue to computational advancement, it could be stated that an all-encompassing genuine miniaturized scale network arranging issue can be viewed as obliged, stochastic, and multi-objective. Be that as it may, a few researchers have connected distinctive ways to deal with microgrid designing issues. Those issues will be audited in the subsequent sections, together with advancement methods connected to compute them.

2.1 Limitations and Motivation

From the above discussion, it is noticed that still there is a need for scheduling in the grid. One of the existing optimization algorithms, such as PSO is described in [10], used for optimal scheduling in microgrid. In [11], PSO is simulated in Monte Carlo Simulation (MCS-PSO) and it has been used for employing uncertainty in the network and optimal scheduling. It has already proved that PSO is a suitable method for finding local optima in optimization problems. Not only PSO, but several optimization methods can also provide only local optimal value and not able to get global optimal value speedily or with high accuracy.

But this paper proposed Grey Wolf Optimization algorithm, which can be used for local as well as global optimal searching speedily and produce high accuracy in scheduling in large scale grid networks. The entire contribution of the paper is constructing a hybrid energy system incorporated with gird. Create a theoretical model for schedule ng on the grid. Then implement GWO algorithm for scheduling optimization and performance is compared with the existing PSO and MTPSO algorithm.

3 Hybrid Renewable Energy Incorporated with Microgrid

In this paper, the proposed hybrid energy system with grid network comprises three different renewable energy sources such as Solar, Wind, and Thermal. Solar PV module and thermal module are connected with DC–DC converter, whereas the wind turbine module is connected with AC–DC converter which is shown in Fig. 1. Then the converter is connected into IEEE bus system and linked to the converters of grid, load, microgrid or battery system.

3.1 Optimal Scheduling Modelling

The main objective of this paper is to minimize the total operation cost of the hybrid energy system. One of the methods is optimizing the scheduling of power sources where it can fulfill the power demands. Hence this paper proposed an optimization method for scheduling and minimising the total operation cost, which is described below in detail.

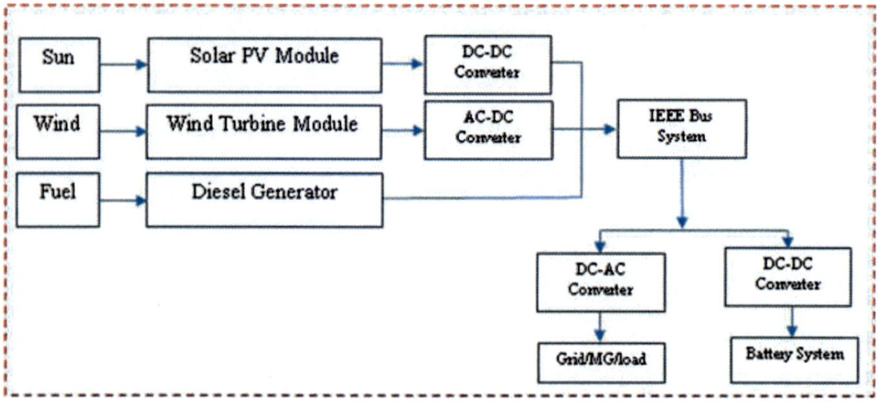

Fig. 1 Hybrid energy conversion model (solar, wind, fuel)

3.2 Objective Functions

In the proposed Heat-Power Hybrid Energy (HPHE) System, the total operation cost includes fuel, power purchase, charging–discharging of (electrical-vehicle) battery costs. After the installation of solar and wind plants, the maintenance and operation costs depend on the output of the HPHE system. In this paper, the objective function includes operation and maintenance without power loss. Hence the Objective Function (OF) is written as:

$$C = \min\left(C_{\text{gs}} + C_{\text{ep}} + C_{\text{om}} + C_{\text{so}}\right) \tag{1}$$

where, C represents the total operation cost, C_{ep} and C_{gs} represents the power and gas purchasing costs, C_{om} represents the maintenance cost, and C_{so} represents the power storage cost. The following formula is used for calculating the purchase cost as:

$$C_{\text{ep}} = \sum_{1}^{H}\left(C_{\text{ec}}^{t} P_{\text{grid}}^{t} \Delta t\right) \tag{2}$$

where t is the time duration, H, P_{grid}^{t}, C_{ec}^{t} are calculated. H taken for charging-discharging is represents by H, P_{grid}^{t} represents the amount of power purchased from the grid during time t, (where $t = 1$ to H) and C_{ec}^{t} is the electricity cost at time t. The cost consumed for only gas is calculated using the following equation as:

$$C_{\text{gs}} = c_{\text{gs}} \sum_{1}^{H}\left(W_{\text{MT}}^{t} + W_{\text{boiler}}^{t}\right) \tag{3}$$

$$C_{\text{om}} = C_{m}\frac{\gamma(1+\gamma)^{m}}{(1+\gamma)^{m}-1} - C_{s}\frac{\gamma}{(1+\gamma)^{m}-1} + C_{\text{op}} \tag{4}$$

$$C_{\text{so}} = C_{\text{storage}} + C_{\text{ev}} \tag{5}$$

where C_m is the amount consumed as initial investment of planting the hybrid system, C_{op} is the maintenance and operation cost, and C_{s} is the cost of sunk. Also, C_{storage} is the storage cost and C_{ev} is the vehicle charging (battery) cost. The above given objective function is solved by satisfying the following constraints are based on the following factors such as:

- Electric Power Balancing (EPB)

This EPB constraint is expressed as

$$P_{\text{G}}(t) + P_{\text{st}}(t) + P_{\text{EV}}(t) + P_{\text{Grid}}(t) = P_{\text{D}}(t) \tag{6}$$

- Distributed Generation (DG)

The constraint of DG is written as

$$P_{\text{Gi, min}} \leq P_{\text{Gi}}(t) \leq P_{\text{Gi, max}} \tag{7}$$

- Ramping

Ramping constraint is written as:

$$-R_i^{\text{down}} \leq P_{\text{Gi}}(t) - P_{\text{Gi}}(t-1) \leq R_i^{\text{up}} \tag{8}$$

- Start-off-time

The Start-off-time constraint is written as:

$$\begin{cases} P_{\text{ch}}(t) \leq P_{\text{ch,max}} \\ P_{\text{dis}}(t) \leq P_{\text{dis,max}} \end{cases} \tag{9}$$

- Storage-Battery

There are two constraints involved with the SB constraints as

Capacity constraints and Charge-Discharge power constraints are written as:

$$E_{\text{st, min}} \leq E_{\text{st}}(t) \leq E_{\text{st, max}} \tag{10}$$

And

$$\begin{cases} P_{\text{ch}}(t) \leq P_{\text{ch,max}} \\ P_{\text{dis}}(t) \leq P_{\text{dis,max}} \end{cases} \tag{11}$$

- Thermal Energy

Finally, the thermal energy constraint is written as:

$$Q_{\text{bl}}(t) + Q_{\text{MT}}(t) = Q_{\text{D}}(t) \tag{12}$$

In Eq. (6), $P_{\text{D}}(t)$ represents the load demand, $P_{\text{G}}(t)$ represents the power generated from wind, PV and gas, $P_{\text{st}}(t)$ represents the charging power and $P_{\text{EV}}(t)$ represents the discharging power in battery, and $P_{\text{Grid}}(t)$ is the output power generated during time t in micro grid. In Eq. (7), $P_{\text{Gi}}(t)$ represents the output power generated from ith generator during t. The lower and upper bound values of the power generator are represented as $P_{\text{Gi, min}}$ and $P_{\text{Gi, max}}$ respectively. In Eq. (8), R_i^{down} and R_i^{up} are the two boundary values of the generator speed. Similarly from Eq. (10), $E_{\text{st, min}}$ and $E_{\text{st, max}}$ represents the lower and upper boundary values of the battery capacities, from Eqs. (9 and 11), $P_{\text{ch}}(t)$ and $P_{\text{dis}}(t)$ represents the charging and discharging power, and from Eqs. (9 and 11), $P_{\text{ch,max}}$ and $P_{\text{dis,max}}$ represents the maximum value of the charging and discharging power. In Eq. (12), $Q_D(t)$ represents the load demand, $Q_{bl}(t)$ and $Q_{MT}(t)$ represents the power output obtained from thermal and gas, respectively, during time t.

4 Grey Wolf Optimization Algorithm for Optimal Scheduling on Hybrid Energy Sources

One of the latest and very effective optimization methods provides high accuracy in optimization than other swarm intelligent algorithms and genetic algorithms is Grey Wolf Optimization (GWO) algorithm. GWO algorithm has been developed, biologically inspired algorithm imitating social behaviour using hierarchical operations. It follows hierarchical operations for hunting grey wolf family. Wolves are gregarious animals create packs to build a hierarchy pyramid. Each pack has twelve number of wolves. In the pyramid hierarchy five labelled wolves such as α, β, δ and ω wolves. The α-alpha wolves are placed at the top of the pyramid having highest priority, considered as a leader for other wolves. Also α-alpha is responsible in decision making and predation where other wolves follows the α-alpha wolves commands. The wolves placed in the second level to the alpha wolves in the pyramid is beta-β wolves. The main responsibility of the beta-β wolves is to assists the α-alpha wolves to take accurate decision. This beta-β wolves have lower priority to the α-alpha wolves. Beta-β wolves control other individual wolves in the pack and feed information again to the α-alpha wolves. The wolves present in the third level of pyramid is delta-δ wolves. The main responsibility of Delta-δ wolves is implementing the decision of alpha-α and beta-β wolves. The priority level of the Delta-δ wolves is higher than the omega-ω wolves which are located at the bottom of the pyramid responsible for help prey. Omega-ω wolves are responsible for tracking and encircling for capturing the prey, but, the other three α, β and δ wolves are responsible for attacking the prey. The pyramid hierarchy of wolves is illustrated in Fig. 2.

GWO algorithm can also be represented mathematically to solve the hierarchy-based optimization problems. For example, the direction of the hunting process is started from the highest priority wolves to lowest priority wolves, like α, β, δ, and ω wolves. The process of hunting needs to follow comprises of various stages before its bouts the victim. All the wolves surround and stop the victim (see Fig. 3) and it is given in the following equation as:

$$D^{\rightarrow} = \left|C^{\rightarrow} \cdot X_p^{\rightarrow}(t) - X^{\rightarrow}(t)\right| \tag{13}$$

$$X^{\rightarrow}(t+1) = X_p^{\rightarrow}(t) - A^{\rightarrow} \cdot D^{\rightarrow} \tag{14}$$

In the above equation, the vector coefficients are represented as $A^{\rightarrow}$ and $D^{\rightarrow}$, $X_p^{\rightarrow}$ represents the position of the victim, the number of iterations is represented as t and $X^{\rightarrow}$ is the position of the grey wolf. Also, the coefficients $A^{\rightarrow}$ and $D^{\rightarrow}$ are obtained

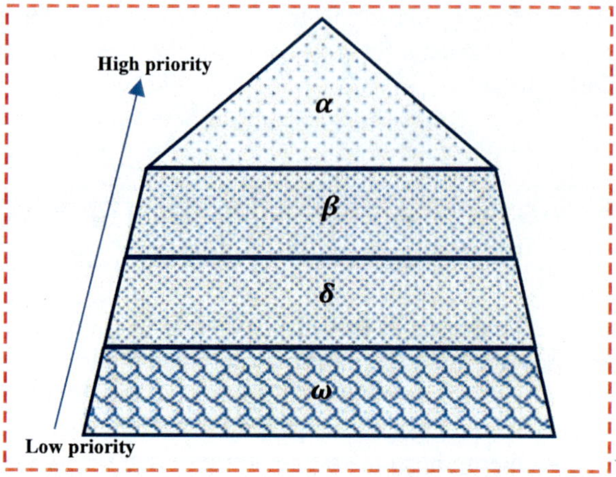

Fig. 2 Hierarchy pyramid of grey wolf optimization algorithm

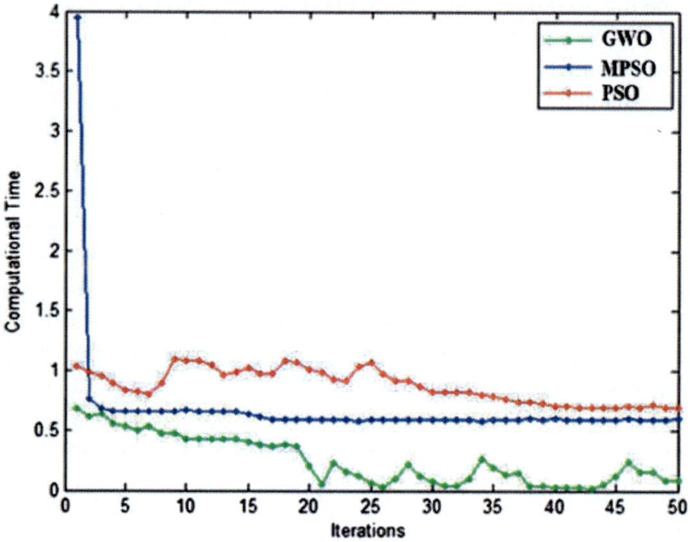

Fig. 3 Convergence rate comparison

by,

$$A^{\rightarrow} = 2a^{\rightarrow} \cdot r_1^{\rightarrow} - a^{\rightarrow} \tag{15}$$

$$C^{\rightarrow} = 2 \cdot r_2^{\rightarrow} \tag{16}$$

$$a = 2 - t * \frac{2}{\text{max\_ iteration}} \tag{17}$$

In Eq. (15), the value of $a^{\rightarrow}$ varies from 0 to 2 in each iteration, the value of r_1 and r_2 varies from 0 to 1. At each iteration, the best value (answer) is calculated from the above three-level wolves and stored. The lowest level wolves increase the position in accordance with the best answers of the other wolves. It can be obtained using the following Eqs. (18)–(20) as:

$$D^{\rightarrow}_{\text{Alpha}} = \left|C^{\rightarrow}_1 \cdot X^{\rightarrow}_{\text{Alpha}} - X^{\rightarrow}\right| \tag{18}$$

$$D^{\rightarrow}_{\text{Beta}} = \left|C^{\rightarrow}_2 \cdot X^{\rightarrow}_{\text{Beta}} - X^{\rightarrow}\right| \tag{19}$$

$$D^{\rightarrow}_{\text{Delta}} = \left|C^{\rightarrow}_3 \cdot X^{\rightarrow}_{\text{Delta}} - X^{\rightarrow}\right| \tag{20}$$

In accordance with the position of the first three-level wolves, the position of the victim is calculated as:

$$X^{\rightarrow}_1 = \left|X^{\rightarrow}_{\text{Alpha}} - A^{\rightarrow}_1 \cdot D^{\rightarrow}_{\text{Alpha}}\right| \tag{21}$$

$$X^{\rightarrow}_2 = \left|X^{\rightarrow}_{\text{Beta}} - A^{\rightarrow}_2 \cdot D^{\rightarrow}_{\text{Beta}}\right| \tag{22}$$

$$X^{\rightarrow}_3 = \left|X^{\rightarrow}_{\text{Delta}} - A^{\rightarrow}_3 \cdot D^{\rightarrow}_{\text{Delta}}\right| \tag{23}$$

$$X^{\rightarrow}(t+1) = \frac{X^{\rightarrow}_1 + X^{\rightarrow}_2 + X^{\rightarrow}_3}{3} \tag{24}$$

The complete searching process and implementation process of grey wolf agents is depending on the parameter A. By adjusting the value of A, searching constraint is given as $|A| \geq 1$ and implementation constraint is given as $|A| < 1$. The entire functionality of the GWO algorithm is given in the form of pseudo-code and in Fig. 2, respectively.

5 Experimental Results and Discussion

The above-described GWO algorithm is implemented in MATLAB software and the results are verified. A sample microgrid is used for testing and verifying the effects, validating the scheduling and cost optimization using GWO. Normal and fault operations are experimented in the microgrid to verify the robustness of the proposed GWO algorithm. Then IEEE-75 bus system is applied to identify the effectiveness and feasibility of the optimal scheduling model. GWO is implemented in MATLAB 2013b in Intel Pentium Core i7 computer with 2.9 GHz and 8 GB RAM CPU. The set of all main parameters such as population size is 100 and the maximum number of iterations is 500 initialised in the program. Obtained results are given in the following tables.

Table 1 depicts the combined Utilization Factors (UFs) of certain generators which support the real power transportation and supplying the reactive power with respect to each load.

The proposed GWO algorithm is evaluated by calculating the total cost and compared with the various existing algorithms, is given in Table 2. From the comparison, it is found that the proposed GWO is better than the other approaches.

Table 1 Utilization factors of 75-bus system generators

Load bus	Gen-1	Gen-2	Load bus	Gen-1	Gen-2
LB-16	0.0026	0.0158	LB-55	0.01	0.018
LB-20	0.0229	0.01	LB-56	0.0279	0.0260
LB-24	0.036	0.0360	LB-57	0.0163	0.0227
LB-25	0.044	0.0032	LB-58	0.0403	0.053
LB-27	0.049	0.0444	LB-59	0.0178	0.0122
LB-28	0.018	0.0201	LB-60	0.0209	0.0128
LB-30	0.4100	0.016	LB-61	0.0094	0.0164
LB-32	0.0132	0.001	LB-62	0.0089	0.0153
LB-34	0.0172	0.0152	LB-63	0.0033	0.0008
LB-37	0.0209	0.0237	LB-64	0.0038	0.0054
LB-39	0.0103	0.0142	LB-65	0.0310	0.0120
LB-42	0.0560	0.02	LB-66	0.002	0.0039
LB-46	0.0234	0.0634	LB-67	0.0109	0.033
LB-47	0.012	0.0127	LB-68	0.0044	0.0072
LB-48	0.0088	0.0024	LB-69	0.0133	0.00139
LB-49	0.0117	0.0103	LB-70	0.0107	0.0036
LB-50	0.0286	0.0428	LB-71	0.0760	0.0118
LB-51	0.0107	0.099	LB-72	0.0292	0.019
LB-52	0.0469	0.0108	LB-73	0.0279	0.044
LB-53	0.0207	0.0127	LB-74	0.0381	0.0299
LB-54	0.0509	0.0106			

Table 2 Cost for energy sources (Yuan/kWh)

Energy source	Power generation cost	Operating-maintenance cost	Downtime maintenance cost
Solar	1.5	0.08	0.002
Wind	0.6	0.15	0.011
Thermal	1.7	0.02	0.000

From the above tables and figures, it is identified that the proposed GWO is better than the existing PSO and MPSO algorithms. The optimal values obtained using GWO are fast than the other algorithms, which is shown in Fig. 3. Hence the proposed GWO is decided as a better algorithm.

6 Conclusion

In this paper, GWO optimization algorithm is used for solving the OPF problem. The efficiency is verified by testing on IEEE-75 bus system. The proposed GWO algorithm result shows that it is more effective in cost reduction and power loss minimization. The proposed GWO is compared with other optimization algorithms such as PSO and MPSO. From the comparison, it is found that the proposed GWO algorithm is efficient and suitable for solving complex OPF problems in hybrid renewable energy system.

References

1. Ding, J.X., Somani, A.: A long-term investment planning model for mixed energy infrastructure integrated with renewable energy. Proc. Green Technol. Conf. **15**(16), 1–10 (2010)
2. Jin, C.R., Sheng, X., Ghosh, P.: Energy efficient algorithms for electric vehicle charging with intermittent renewable energy sources. Proc. Power Energy Soc. Gen. Meet. (PES) **21**(25), 1–5 (2013)
3. Banos, R., Manzano-Agugliaro, F., Montoya, F.G., Gil, C., Alcayde, A., G6mez, J.: Optimization methods applied to renewable and sustainable energy: a review. Renew. Sustain. Energy Rev. **15**(4), 1753–1766 (2011)
4. Diwekar, U.: Introduction to applied optimization. Springer Optimization and Its Applications, 2nd edn (2008)
5. ErdincO, Uzunoglu M.: Optimum design of hybrid renewable energy systems: overview of different approaches. Renew. Sustain. Energy Rev. **6**, 1412–1425 (2012)
6. Binitha, S., Sathya, S.: A survey of bio inspired optimization algorithms. Int. J. Soft Comput. Eng. **2**, 137–151 (2012)
7. Rueda-Medina, A.C., Padilha-Feltrin, A.: Pricing of reactive power support provided by distributed generators in transmission systems. IEEE Trondheim Power Tech, pp. 1–7 (2011)
8. Paschalidis, I.C., Li, B., Caramanis, M.C.: Demand-side management for regulation service provisioning through internal pricing. IEEE Trans. Power Syst. **27**, 1531–1539 (2012)
9. Paschalidis, I.C., Li, B., Caramanis, M.C.: A market-based mechanism for providing demand-side regulation service reserves. In: Proceedings of IEEE Conference on Decision and Control and European Control Conference, pp. 21–26 (2011)
10. Kanchev, H., Francois, B., Lazarov, V.: Unit commitment by dynamic programming for microgrid operational planning optimization and emission reduction. ACEMP, pp. 502–507 (2013)
11. Berka, J.S.M.: Microgrid energy management based on approximate dynamic programming. In: 4th IEEE PES Innovative Smart Grid Technologies, pp. 1–5 (2013)

Biomarkers for Detection of Parkinson's Disease Using Machine Learning—A Short Review

Moumita Pramanik, Ratika Pradhan, and Parvati Nandy

Abstract Detection of Parkinson's disease (PD) from the symptom of motor oriented and non-motor oriented anomalies is a very crucial task. One of the reasons behind this disease is the deficiency of dopaminergic neurons in the brain that leads to various neurodegenerative disorders in the human being mostly in an aged person. Vocal impairments to tremors, difficulty in walking are the prominent symptoms found in Parkinson's disease. Medical scientists and practitioners introduced many biomarkers for ease of diagnosis of PD. This article provides a detailed analysis of various biomarkers such as acoustic, handwriting, Electroencephalography (EEG), and gait signals along with the associated machine learning approaches of PD subjects. This paper also enlightens the symptoms of PD in its various stages and delivers the information about the popular rating scales mostly referred by the medical practitioners during the diagnosis process.

Keywords Parkinson's detection (PD) · Dopaminergic · EEG · Gait · Acoustic · Support vector machine (SVM) · K-nearest neighborhood (KNN) · Naïve Bayes (NB) · Random forest (RF) · Principal component analysis (PCA)

M. Pramanik (✉) · R. Pradhan
Department of Computer Applications, Sikkim Manipal Institute of Technology, Sikkim Manipal University, Majitar, Sikkim, India
e-mail: moumita.pramanik@gmail.com

R. Pradhan
e-mail: ratika.p@smit.smu.edu.in

P. Nandy
Department of Medicine, Sikkim Manipal Institute of Medical Sciences, Sikkim Manipal University, Gangtok, Sikkim, India
e-mail: drparvaty@gmail.com

S. Borah et al. (eds.), *Soft Computing Techniques and Applications*,
Advances in Intelligent Systems and Computing 1248,
https://doi.org/10.1007/978-981-15-7394-1_43

1 Introduction

Parkinson's disease is one of the neuro-related diseases often found in aged people. PD is mainly associated with motor signals of individuals that result from the damage of dopaminergic neurons in the brain [1]. Dopamine act like a messenger between the neurons in the human brain. It helps the brain to send signals to various parts for its proper function, especially in body movements and speech delivery. The symptoms of Parkison's disease arise when significant numbers of neurons are gets damaged, or the level of dopamine is improper. This nervous system problem is progressive in aged people.

Common symptoms of this disease tremor(vibration), rigidity in the form of stiffness, bradykinesia in form of slow movement, postural instability leads to balance problems and problems with walking/gait and most importantly variation in speech [1–4]. Initially, it begins with distressing either part of the body and spread over time to another part of the body.

Medical practitioner often suspects the subject having the Parkinson's disease on the basis of motor test, if the subject response to positive indication in any two symptoms among the three namely, akinesia that leads to difficult to initiate a movement, rigidity that shows an inability to turn or twist and rest tremor are observed. Unluckily, these motor indications are witnessed once 70% of dopaminergic neurons are damaged [5]. This is the main reason to detect Parkinson's disease at its earlier stage so that the patient can take the necessary preventive measures to deal with the disease. As Parkinson's disease considered being not curable, hence early detection of the disease helps the patient a lot to prevent its severe affect.

1.1 Rating Scale of Parkinson's Disease

The effect of Parkinson's disease is experienced by individuals in different ways. All the symptoms do not appear in the same order to all PD individuals. There are some scales used for rating the disease to measure the phases of Parkinson's disease. Various symptoms that primarily suggested by Hoehn and Yahr rating scale and another rating scale namely Unified Parkinson's Disease Rating Scale for PD are discussed in the following section.

1.1.1 The Hoehn and Yahr Rating Scale

The Hoehn and Yahr rating scale describes the Parkinson disease in five stages, as mention below [6, 7]:

Initial Stage: In this initial stage, the people have very few symptoms that do not show adverse effects in daily life. Tremors and slow movement can be observed, particularly only in either part of the body. Mere alterations are also witnessed in the time of walking, posture, and face.

Second Stage: In this, the effect of symptoms visible more than the initial stage. The tremor, stiffness, and movement problem affect both parts of the patient body. The problem of walking and posture instability is visible. Although during this stage, people can able to do their daily activity completion of works needs added difficulty and time.

Third Stage: This is the middle stage of PD. In this, balance instability and slowness of movements are significantly observed. The most common symptom experienced by the patient during this stage is falling; though the patient still can works independently in their daily activities like eating and dressing.

Fourth Stage: In this stage, the symptoms are more prominent and the effect is very severe. In this phase, the patient could stand alone but walking needs support. The patient needs assistance to do their daily activities and difficult to live alone.

Fifth Stage: This is the most hostile stage of the disease. Rigidity in the limbs may lead to non-movement during standing and walking. The person needs a wheelchair or confinement to bed. The person often tends to fall, once trying to upright or turning and also become freeze or stumble when walking. The patient may experience illusions and misbeliefs during this period.

1.1.2 The Unified Parkinson's Disease Rating Scale (UPDRS)

The Unified Parkinson's Disease Rating Scale (UPDRS) [6] is a wide-ranging tool used to measures the severity level of Parkinson's disease. It describes the disease in five categories such as Zero (0)—indicates normal or no problem, One (1)—indicates minimum problems, Two (2)—indicates mild problems, Three (3)—indicates moderate problems and Four (4)—indicates severe problems.

The Movement Disorder Society (MDS) informed the UPDRS rating scale as UPDRS-MDS that includes the intellectual function, mood, behavior, daily living activities, motor examination, and motor complications.

1.2 Various Signals Used for Parkinson's Disease

There are many potential methods proposed by different researchers using various signal to detect Parkinson's disease, such as voice signal [1–3, 8, 9] gait signals [10, 11] handwriting signals [12, 13], EEG [14, 15], and MRI [16]. Among all these mechanisms, the acoustic analysis of voice is one of the most useful methods for early detection of PD, as the vocal impairment is found to be in an individual from the beginning of the PD and even years before a medical diagnosis can be made [5, 17–19].

1.2.1 Acoustic Signal

The acoustic voice samples of PD patients have vocal impairments, referred to as hypokinetic dysarthria [5], which distresses compound levels of speech such as phonation, an articulation that helps to create a pure speech and distinguish the sounds of speech, prosody, and resonance. It affects first of all the laryngeal function that works with vocal folds, pitch, and volume, which is necessary for phonation; second, the articulatory part that is anxious with the creation of speech sounds and also the respiratory muscles.

PD patient's voice is more monotonous, the flow of speech transformed and the patient makes more disfluencies. According to [5, 20], in PD patient consonant articulation is inaccurate, mostly for the consonants like *p, t, k, b, d, g,* and vowel delivery is also compromised, there are significant alterations are found between vowels and sensitively tend to decrease. In the case of the sustained vowel, the phonation, pitch, and intensity are unstable and the tone becomes rough. The authors in [2, 3, 5, 8, 20, 21] use the acoustic signals to detect Parkinson's disease using various features selection and classification mechanisms.

1.2.2 Gait Signal

Another symptom due to motor disorder is walking/gait problems. The existence of Parkinson's disease can be perceived by the fluctuation in a person's gait in slow speed, freezing of gait (FOG), the tendency to falls, short steps, etc. [22]. The gait data can be used as a dynamic activity for the analysis of numerous abnormal activities of human beings. In [22–26], the authors used the gait data to analyze Parkinson's disease with machine learning approaches.

1.2.3 Handwriting Signal

The handwriting of the PD patient is observed to be abnormal as compared to normal individuals. Approximately 5% micrographia arises in the PD patient before the motor symptoms visible significantly [27]. There are significant benefits of using the handwriting signal as these are simple, natural, and less invasive and need not required special infrastructure to capture the sample. The various features that are extracted from the handwriting include Kinematic, non-linear dynamics, and neuro-motor features, pixel density ration, etc. Most of the tasks involved by several strokes and the strokes are fragmented on observation of the pen-down and pen-up actions. Other important parameters such as size reduction, ink consumption, variant of pixel density, etc. are also recorded from the handwriting sample. Authors in [27–30] use various handwriting and graphics sample to the identification of Parkinson disease.

1.2.4 Electroencephalography (EEG) Signal

Electroencephalography (EEG) is a technique used to capture the electrical movements of the human brain. It is often used to detect the abnormalities of cognitive activities in the brain due to the neurodegenerative sickness like Parkinson's disease. For the study of impulsive neural behavior, the EEG plays a brilliant role. EEG features are time-variant in nature and used to investigate the brain dynamics change using the central, frontal, occipital, and parietal EEG sites of Parkinson's patients [15]. The authors in [15, 31–33] used EEG signal to detect Parkinson's disease.

1.3 Machine Learning in Parkinson's Disease Detection

Machine learning algorithms play a prominent role in the classification of Parkinson subjects from control subjects. Various machine learning based approaches such as Logistic Regression (LR), Support Vector Machine (SVM), Neural Networks (NN), Decision Trees, etc. and an ensemble of these approaches are the mastermind behind the modern Parkinson detection engines. In [1, 3] authors used Minimum average maximum (MAMa) tree and Singular Value Decomposition and K-nearest neighborhood (KNN), in [8] authors used singular value decomposition, neighborhood component analysis(NCA) and 1NN with logistic regression for features selection and detection mechanism with acoustic data signals. In [24], authors used ensemble classification techniques that include SVMs with RBF kernels, Multiplayer Perceptron and Naïve Bayes (NB), Random Forest(RF), KNN and logistic regression algorithm to detect the Parkinson's disease.

There are numerous machine learning algorithms used with several kinds of signals such as acoustic, gait, handwriting sample that shows the prominent result in the detection of Parkinson's disease. A wide range of machine learning methods is discussed in the subsequent section for identification of Parkinson's disease.

2 Literature Reviews

Lately, the machine learning process for the detection of Parkinson's disease has been explored extensively. In this section, various literature related to Parkinson's disease proposed by eminent researchers over time are discussed briefly.

Gómez-Vilda et al. [2], explain the articulation dynamics of the kinematic activities of the jaw-tongue of human. Classifying the speech from the individual and used for the detection of Parkinson's disease. In this investigation, a Random Least Squares Feed-Forward Network (RLSFN) along with the Monte Carlo simulations has been carried out for detection of Parkinson's disease, which shows 99.4% accuracy in detection with 0.6% error rates. In [3], the authors used various forms of voice recording and handwriting samples for feature extraction, and optimal features were selected using the decision tree and KNN method. They used the optimized version of the traditional cuttlefish algorithm for the classification model to detect the disease. Successful implementation of the detection scheme shows impressive detection accuracy with 94%.

Sakar et al. in [8] used Tunable Q-Factor Wavelet Transform (TQWT) on the voice signal data for higher frequency feature extraction. Next, optimum features were selected using the Minimum Redundancy Maximum Relevance algorithm on extracted higher frequency features. Total of 50 optimum features was selected for further standardization and classification. The important selected features then standardized and fed to the ensemble classifiers comprises of Linear SVMs, RBK kernels, Multilayer Preceptor, Logistic Regression, Naïve Bayes, Random forest and KNN. The authors also use the Mel-Frequency Cepstral Coefficients (MFCCs) for feature extraction and comparison of the result has been carried out for PD detection. It has been noticed that SVM (RBF) performs better with 86% accuracy with all features subsets. In [1], the author uses a combination of MAMs and SVD techniques for feature extraction from the vowel signals. The popular relief feature selection method had been used to select 50 most unique features from it and KNN classification applied for detection mechanism, which shows 92.46% accuracy in detection.

Solana et al. [34] proposed an approach of PD detection with reduced features sets of vocal signals. The authors used 8–20 numbers of effective features out of 754 vocal features. After the feature selection, four classifiers such as KNN, Multilayer Perceptron, Random Forest, and SVM were used for detection of the Parkinson's disease. Among the four classifiers, the SVM shows the highest accuracy with 94.7% and specificity with 92.68% and sensitivity with 98.4%.

Researchers also work on gait signal for detection of Parkinson's disease where emphasized are given on vision-based and sensor-based data of the human gait. In the vision-based, cameras are used to capture the movement of human beings from different angles, and in sensor-based methods, number of sensors are put on both leg of the individual and human kinematics and kinetics data are captured.

Pun et al. [25] used the gait signals and achieved 90% detection accuracy. In this, optimized features have been selected, then visualization and formula integration carried out considering the proportion relationship among various parameters such as swing ration difference, double support interval stride length, etc. In [23], the u-shaped sensing platform was used to capture the sensor-based data. Three different types of gait factors were extracted, such as demographic data, spatio-temporal data, and turning gait data. The step and stride length, speed, cadence, stance time, time of swing and pre-swing, etc. features are extracted from the sensor data of the gait cycle associated with the two feet of the individual. The authors used nine classifiers, including Naive Bayes, KNN, SVM, random forest, and others to detect the disease. Random Forest shows better results with 92.49% accuracy among the other classifiers. In [24], the authors used the dataset provided by Phisonet and analyzed the vertical ground reaction forces. The data compression has been carried out to capture the various features and different classification techniques such as Decision Trees, Random Forrest, SVM, and Logistic Regression and KNN. Logistic Regression displayed 90.06% accuracy in the PD detection.

In [35], the authors used a deep learning approach to find out the severity level of Parkinson's disease using gait data signals. The researcher used Long short-term memory (LSTM) and CNN (Convolutional Neural Network) to study the spatial behavior of gait data. The accuracy that achieved using the LSTM and CNN is 98.61%.

Similarly, a deep learning-based PD detection engine explored by Maachi et al. in [36] with gait information. 1D convolutional neural network has been used to form a deep neural network classifier. Firstly, 1D convolution neural network was built where each network put data from a vertical ground reaction force signal and passed it to the succeeding four layers. Subsequently, relationships among the spatial features were evaluated and finally the output layer builds with 5 neurons to predict Parkinson's disease. The experiment shows the 98.7% accuracy to detect the disease.

Juliana et al. [37], measured different gait information for detection of Parkinson's disease. A detailed analysis of time series of gait data such as stride intervals, stance time, stance interval, swing intervals, stride to stride parameters were considered as parameters for the detection of the disease. Related features on the basis of fluctuation magnitude and fluctuation dynamics were obtained and SVM, KNN, Naïve Bayes, LDA (Linear Discriminant Analysis), and DT methods were used in the detection model. The experiment results in 96.8% of average accuracy in detection.

Ly et al. [15] represent a classification technique using EEG data for sensing turning freezing events. EEG features were divided using independent component analysis and entropy bound minimization. S-Transformation used for features selection and Bayesian Neural Networks as a classifier that provides 86.2% accuracy for the detection of turning freezing (TF).

Yuvaraj et al. [38] used EEG signal analysis with higher-order spectra feature extractor. Efficient bispectrum features were extracted and ranked accordingly. The higher ranked feature has been considered for the classification analysis. Multiple classifiers such as decision tree, KNN, Fuzzy-KNN, NB, Probabilistic neural network, and SVM has been used. Among the various classifiers, the SMV shows the average accuracy of 99.62%.

In [31], Lyapunov exponents (LE) one of the constructive features of EEG for identification of abnormalities of the neural system, had played a crucial role in the PD detection process. In the human brain, two important parts, such as the temporal brain and frontal brain, behave differently for PD patients and normal individuals. In a normal person, it is observed that the frontal brain is less than the temporal brain, whereas, in the case of PD patients the frontal brain is higher than the temporal brain. EEG signals often used for early diagnosis of PD [39] and also used for predicting the stage of the disease [40], where the authors achieved phenomenal results.

Ly et al. [41] used a large volume of EEG data for PD detection, where wavelet transformation used for feature extraction and SVM used for classification. Data dimension further reduced using Principal Component Analysis and independent analysis using entropy bound minimization was applied for refining the detection of gait initiation failure in PD. This experiment shows 86.3% PD detection accuracy.

Authors of paper [30] investigated handwriting samples and used micrographia as an indicator of Parkinson's disease (PD). A numerical investigation of handwriting samples is carried out that helps in observing micrographia in the detection of PD. Handwriting morphology improving the purpose of the detection procedure. Numerous metrics with minimal sensitivity are capture from the handwriting pieces, such as character size-reduction, ink utilization, and pixel density within a writing sample. Pixel density analysis shows the significant differences among the features set for the identification of PD.

In [12], the authors use the dynamics of handwriting data and suggested a recurrence plot to map the handwriting signal to the image area. Convolution neural network has been used for learning and detection mechanism. This experiment showed 87% accuracy.

Loconsole et al. [42], offered a handwriting based PD detection mechanism. Two dynamic features and two static features were extracted using Electro-MyoGraphy(EMG) signal processing techniques. Detection classification is carried out by ANN using Genetic Algorithm (mono) that provides 95.81% detection accuracy.

Extending the work recently Donato et al. [43] proposed a Parkinson detection approach using Artificial Neural Networks and a Multi-Objective Genetic Algorithm (MOGA). The combination of ANN and MOGA proved to be a good detector with an accuracy rate of 96.85%. Similarly, Castrillon et al. [27] used handwriting tasks as a detection signal, where, different features set were extracted from the signal such as neuromotor, kinematic, and non-linear dynamic and use machine learning approaches like SVM, KNN and multilayer perceptron. The accuracy achieved by the author is ranged in-between 81 and 97%.

3 Bio-signal Analysis in Parkinson's Detection

At this point in time, we have seen variations in bio-signals that play an important role in Parkinson's disease detection. Now in the following section, an effort has been made to analyze detection results due to bio-signals. This analysis of results obtained using bio-signals; is very necessary because it will provide the overall idea of bio-signals and help to choose a suitable one among these for different stages of Parkinson's disease detection. A summary of literature reviewed has been summarized in Tables 1, 2, 3, and 4 for conducting the analysis.

Table 1 represents the Parkinson's detection technique using acoustic signals. From the table it can be seen that the Gómez-Vilda et al. [2] method shows the highest level of accuracy; whereas, the Gupta et al. [3] approach for voice data shows moderately low 83.48% of accuracy. On the other hand, the Tuncer et al. [1] approach which reveals 92.46% accuracy seems to be the most appealing method of detection, since the detection scheme has been designed on a good number of samples. This makes the concerned system the most realistic approach among others.

Table 1 Summary of literature where acoustic signals are used as a biomarker for detection

Authors	Signal	Techniques	Size of dataset/sample	Detection accuracy
Gómez-Vilda et al. [2]	Vowel pronunciation signal	Random least squares, feed-forward network, Monte Carlo simulations	Total: 142 subjects 51 normal (25 male + 25 female), 91 PD (53 Male + 38 female)	99.4%
Gupta et al. [3]	Speech data, voice data	KNN, decision tree algorithm	Speech dataset (31 normal + 23 PD subject), voice data (28 PD Subjects)	92.194% (speech dataset), 83.48% (voice dataset)
Sakar et al. [8]	Vowel pronunciation signal	Tunable Q-factor wavelet transform (TQWT), Mel-frequency Cepstral coefficients (MFCCs), minimum redundancy maximum relevance, ensemble classifiers (linear SVMs, RBK kernels, multilayer preceptor, logistic regression, Naïve Bayes, random forest and KNN.)	Total: 188 subjects, 107 PD (107 men + 81 women), 64 normal (23 men + 41 women)	86% (SVM-RBF)
Tuncer et al. [1]	Vowel pronunciation signal	Minimum average maximum (MAMa) tree, Singular value decomposition, Relief feature selection, k-nearest neighbor	Total: 756 signals, 252 people, 64 subjects (23 male +41 women), 188 PD (107 men + 81 women)	92.46%
Solana et al. [34]	Vowel pronunciation signal	KNN, multi-layer perceptron, SVM, random forest	PD: 188 (107 male and 81 female) with ages ranging from 33 to 87, control subjects: 64 (23 female and 41 female) with ages from 41 to 82 years old	94.7%

Table 2 Summary of literature where gait signals are used as a biomarker for detection

Authors	Signal	Techniques	Size of dataset/sample	Detection Accuracy
Pun et al. [25]	Gait signal	Proportion relationship	Total: 25 (9 PD + 16 normal subjects)	90%
Wu et al. [23]	Gait signal	Naïve bayes (NB), KNN, SVM, C4.5—decision tree, Linear discriminant analysis (LDA), quadratic discriminant analysis (QDA), Adaboost	Total: 386, 168 PD (88 men + 80 female), 218 normal (103 men + 115 female)	92.49%
Mittra et al. [24]	Gait signal	Decision trees, random forrest, SVM, logistic regression, KNN	310 files per 12,118 records of data in each file	90.06%
Zhao et al. [35]	Gait Signal	Combination of Long short-term memory (LSTM) and convolutional neural network (CNN)	93 PD and 73 control subjects, measurement of vertical ground reaction force (VGRF)	98.61%
Maachi et al. [36]	Gait signal	1D convolutional neural network	93 PD and 73 control subjects 18 1D-signals coming from foot sensors measuring the vertical ground reaction force	98.7%

Considering gait signals of PD and control subjects, it is observed that, the Zhao et al. [35] methods of detection engine shows a better accuracy of detection. Not only the method shows the best results among its peers, but also, the approach is based on the cutting edge Convolutional Neural Network (CNN) having the ability to work smoothly on large datasets. However, in this case, Zhao et al. [35] methods work on a very little dataset of 166 subjects. In the same guideline, it can be seen that the Wu et al. [23] approach works on 386 subjects making it the most practical method among other detection approaches.

For EEG signal, all the methods reviewed are tested on a very little number of subjects (Table 3). Among all the approaches provided by Yuvaraj et al. [38] shows highest amount of accuracy that too with maximum number of samples of 40 subjects. The improvement was mainly due to the ability of Higher-Order Spectra (HPS) to

Table 3 Summary of literature where EEG signals are used as a biomarker for detection

Authors	Signal	Techniques	Size of dataset/sample	Detection Accuracy
Ly et al. [15]	EEG signal	Independent component analysis, S-transform, entropy bound minimization, bayesian neural networks	6 PD subjects and normal subjects	86.2% accuracy (TF)
Yuvaraj et al. [38]	EEG Signal	Higher order spectra, SVM classifier, PD diagnosis index	Total: 40 subjects (20 PD subjects, 20 control subjects)	99.62%
Oh et al. [39]	EEG signal	Convolutional neural network	Total: 40 subjects (20 PD subjects, 20 control subjects)	88.25%
Naghsh et al. [40]	EEG signal	Independent component analysis (ICA), source localization	Total: 20 subjects (7 females + 13 males)	95%
Ly et al. [41]	EEG Signal	Wavelet transformation, SVM, principal component analysis, entropy bound minimization	5 PD subjects	86.3%

extract non-Gaussian and nonlinear characteristics in the complex patterns of EEG signal.

Finally, for handwriting signals, it is observed that the Castrillon et al. [27] approach shows promising accuracy up to 97%. But the method proposed by Gupta et al. [3] has been tested on the highest 632 instances; thus, making the system the most stable and a practical approach.

At the end, a minimum, maximum analogy has been created to understand the most prominent biomarkers for designing PD detection engines. Though it is understood that each biomarker has its own importance and weightage in the detection process, but this analogy will provide a glimpse of research direction in the field of diagnostic of Parkinson. The analogy of the detection results of the approaches discussed so far has been presented in Fig. 1.

It is observed from Fig. 1 that, the Gait signal followed by Handwriting signals have a low variation in detection accuracy; thus, making these two biomarkers to be more prominent as compared to acoustic and EEG signals. Nevertheless, EEG and

Table 4 Summary of literature where handwritings are used as a biomarker for detection

Authors	Signal	Techniques	Size of dataset/sample	Detection Accuracy
Afonso et al. [12]	Handwriting	Recurrence plot techniques for signal to images, convolution neural network	Total: 35 Control-21 PD-14	87%
Loconsole et al. [42]	Surface electro Myo graphy (sEMG) signals of handwriting	Artificial neural network, a multi-objective genetic algorithm	Total: 11 Subjects (all males), PD: 4, control: 7	95.81%
Gupta et al. [3]	Handwriting	KNN, decision tree algorithm	Meander test (632 instances), spiral test (632 instances)	87.12% (Meander dataset), 88.46% (Spiral dataset)
Castrillon et al. [27]	Handwriting	SVM, KNN and multilayer perceptron	Total: 149 (55 PD, 94 normal)	81–97%
Donato Cascarano et al. [43]	Myo Armband muscle activation signals while writing	Artificial neural network, a multi-objective genetic algorithm	Total: 32 subjects (PD: 21 subjects, control: 11 subjects)	Maximum: 96.85%

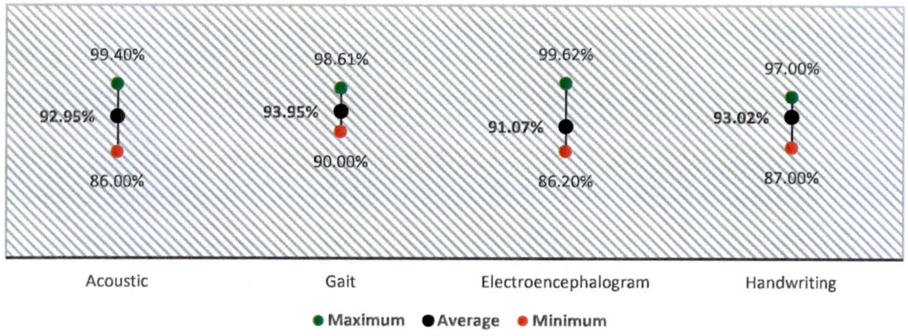

Fig. 1 The detection results obtained due to various biomarkers

acoustic signals have the highest amount of accuracy and therefore mostly cited by the researchers and clinical practitioners.

4 Conclusion

It is well-known fact that there is no reversible cure for Parkinson. Therefore, early diagnosis of the disease can delay the neurodegenerative process. In this article, an effort has been made to understand the severity of Parkinson among human being. Further, two widely used rating scales to identify the stages of Parkinson have been discussed in brief. Moreover, various biomarkers responsible for the effective detection of Parkinson along with the associated state-of-the-art PD detection systems have been highlighted. Finally, an analogy graph is designed to understand the weightage and importance of biomarkers in the PD detection process. However, it is advised that the actual usage, effectiveness of the biomarkers must be identified in consultation with the medical practitioners.

References

1. Tuncer, T., Dogan, S., Acharya, U.R.: Automated detection of Parkinson's disease using minimum average maximum tree and singular value decomposition method with vowels. Biocybern. Biomed. Eng. **40**(1), 211–220 (2020)
2. Gómez-Vilda, P., et al.: Parkinson disease detection from speech articulation neuromechanics. Front. Neuroinform. **11**, 56 (2017)
3. Gupta, D., et al.: Optimized cuttlefish algorithm for diagnosis of Parkinson's disease. Cogn. Syst. Res. **52**, 36–48 (2018)
4. Bourouhou, A., Jilbab, A., Nacir, C., Hammouch, A.: Comparison of classification methods to detect the Parkinson disease. In: 2016 international conference on electrical and information technologies (ICEIT), pp. 421–424 (2016)
5. Jeancolas, et al.: Automatic detection of early stages of Parkinson's disease through acoustic voice analysis with mel-frequency Cepstral coefficients. In: 2017 International Conference on Advanced Technologies for Signal and Image Processing (ATSIP), pp. 1–6 (2017)
6. ParkinsonsDisease.net, "Parkinson's Rating Scale." [Online]. Available https://parkinsonsdisease.net/diagnosis/rating-scales-staging/. Accessed 25 Feb 2020
7. Parkinson's Foundation. Stages of Parkinson's. [Online]. Available https://www.parkinson.org/Understanding-Parkinsons/What-is-Parkinsons/Stages-of-Parkinsons. Accessed 25 Feb 2020
8. Sakar, C.O., et al.: A comparative analysis of speech signal processing algorithms for Parkinson's disease classification and the use of the tunable Q-factor wavelet transform. Appl. Soft Comput. **74**, 255–263 (2019)
9. Mostafa, S.A., et al.: Examining multiple feature evaluation and classification methods for improving the diagnosis of Parkinson's disease. Cogn. Syst. Res. **54**, 90–99 (2019)
10. Joshi, D., Khajuria, A., Joshi, P.: An automatic non-invasive method for Parkinson's disease classification. Comput. Methods Programs Biomed. **145**, 135–145 (2017)
11. Zeng, W., Liu, F., Wang, Q., Wang, Y., Ma, L., Zhang, Y.: Parkinson's disease classification using gait analysis via deterministic learning. Neurosci. Lett. **633**, 268–278 (2016)
12. Afonso, L.C.S., et al.: A recurrence plot-based approach for Parkinson's disease identification. Futur. Gener. Comput. Syst. **94**, 282–292 (2019)
13. Rios-Urrego, C.D., Vásquez-Correa, J.C., Vargas-Bonilla, J.F., Nöth, E., Lopera, F., Orozco-Arroyave, J.R.: Analysis and evaluation of handwriting in patients with Parkinson's disease using kinematic, geometrical, and non-linear features. Comput. Methods Programs Biomed. **173**, 43–52 (2019)

14. Yuvaraj, R., Murugappan, M., Acharya, U.R., Adeli, H., Ibrahim, N.M., Mesquita, E.: Brain functional connectivity patterns for emotional state classification in Parkinson's disease patients without dementia. Behav. Brain Res. **298**, 248–260 (2016)
15. Ly, Q.T. et al.: Detection of turning freeze in Parkinson's disease based on S-transform decomposition of EEG signals. In: 2017 39th Annual International Conference of the IEEE Engineering in Medicine and Biology Society (EMBC), pp. 3044–3047 (2017)
16. Cigdem, O., Beheshti, I., Demirel, H.: Effects of different covariates and contrasts on classification of Parkinson's disease using structural MRI. Comput. Biol. Med. **99**, 173–181 (2018)
17. Harel, B., Cannizzaro, M., Snyder, P.J.: Variability in fundamental frequency during speech in prodromal and incipient Parkinson's disease: a longitudinal case study. Brain Cogn. **56**(1), 24–29 (2004)
18. Postuma, R.B., Lang, A.E., Gagnon, J.F., Pelletier, A., Montplaisir, J.Y.: How does parkinsonism start? Prodromal parkinsonism motor changes in idiopathic REM sleep behaviour disorder. Brain **135**(6), 1860–1870 (2012)
19. Rusz, J., et al.: Quantitative assessment of motor speech abnormalities in idiopathic rapid eye movement sleep behaviour disorder. Sleep Med. **19**, 141–147 (2016)
20. Viswanathan, R. et al.: Efficiency of voice features based on consonant for detection of Parkinson's Disease. In: 2018 IEEE Life Sciences Conference (LSC), pp. 49–52 (2018)
21. Aich, S., Younga, K., Hui, K.L., Al-Absi, A.A., Sain, M.: A nonlinear decision tree based classification approach to predict the Parkinson's disease using different feature sets of voice data. In: 2018 20th International Conference on Advanced Communication Technology (ICACT), pp. 638–642 (2018)
22. Kour, N., Arora, S., et al.: Computer-vision based diagnosis of Parkinson's disease via gait: a survey. IEEE Access **7**, 156620–156645 (2019)
23. Wu, X., Chen, X., Duan, Y., Xu, S., Cheng, N., An, N.: A study on gait-based Parkinson's disease detection using a force sensitive platform. In IEEE International Conference on Bioinformatics and Biomedicine (BIBM), pp. 2330–2332 (2017)
24. Mittra, Y., Rustagi, V.: Classification of subjects with Parkinson's disease using gait data analysis. In: International Conference on Automation and Computational Engineering (ICACE), pp. 84–89 (2018)
25. Pun, U.K., Gu, H., Dong, Z., Artan, N.S.: Classification and visualization tool for gait analysis of Parkinson's disease. In: 2016 38th Annual International Conference of the IEEE Engineering in Medicine and Biology Society (EMBC), pp. 2407–2410 (2016)
26. Ortells, J., Herrero-Ezquerro, M.T., Mollineda, R.A.: Vision-based gait impairment analysis for aided diagnosis. Med. Biol. Eng. Comput. **56**(9), 1553–1564 (2018)
27. Castrillon, R. et al.: Characterization of the handwriting skills as a biomarker for Parkinson's disease. In: 2019 14th IEEE International Conference on Automatic Face & Gesture Recognition (FG 2019), pp. 1–5 (2019)
28. Thomas, M., Lenka, A., Kumar Pal, P.: Handwriting analysis in Parkinson's disease: current status and future directions. Mov. Disord. Clin. Pract. **4**(6), 806–818 (2017)
29. Smits, E.J., et al.: Graphical tasks to measure upper limb function in patients with Parkinson's disease: validity and response to dopaminergic medication. IEEE J. Biomed. Heal. Inf. **21**(1), 283–289 (2015)
30. Zhi, N., Jaeger, B.K., Gouldstone, A., Sipahi, R., Frank, S.: Toward monitoring Parkinson's through analysis of static handwriting samples: a quantitative analytical framework. IEEE J. Biomed. Heal. Inf. **21**(2), 488–495 (2016)
31. Saikia, A., Hussain, M., Barua, A.R., Paul, S.: Significance of Lyapunov exponent in Parkinson's disease using electroencephalography. In: 2019 6th International Conference on Signal Processing and Integrated Networks (SPIN), pp. 791–795 (2019)
32. Ly, Q.T. et al.: Detection of gait initiation failure in Parkinson's disease patients using EEG signals. In: *2016 38th Annual International Conference of the IEEE Engineering in Medicine and Biology Society*(EMBC), pp. 1599–1602 (2016)

33. Handojoseno, A.M.A. et al.: An EEG study of turning freeze in Parkinson's disease patients: the alteration of brain dynamic on the motor and visual cortex. In: 2015 37th Annual International Conference of the IEEE Engineering in Medicine and Biology Society (EMBC), pp. 6618–6621 (2015)
34. Solana-Lavalle, G., Galán-Hernández, J.-C., Rosas-Romero, R.: Automatic Parkinson disease detection at early stages as a pre-diagnosis tool by using classifiers and a small set of vocal features. Biocybern. Biomed. Eng. (2020)
35. Zhao, A., Qi, L., Li, J., Dong, J., Yu, H.: A hybrid spatio-temporal model for detection and severity rating of Parkinson's disease from gait data. Neurocomputing **315**, 1–8 (2018)
36. El Maachi, I., Bilodeau, G.A., Bouachir, W.: Deep 1D-Convnet for accurate Parkinson disease detection and severity prediction from gait. Expert Syst. Appl. **143**, 113075 (2020)
37. Félix, J.P. et al.: A Parkinson's disease classification method: an approach using gait dynamics and detrended fluctuation analysis. In: 2019 IEEE Canadian Conference of Electrical and Computer Engineering, CCECE (2019)
38. Yuvaraj, R., Rajendra Acharya, U., Hagiwara, Y.: A novel Parkinson's disease diagnosis index using higher-order spectra features in EEG signals. Neural Comput. Appl. **30**(4), 1225–1235 (2018)
39. Oh, S.L. et al.: A deep learning approach for Parkinson's disease diagnosis from EEG signals. Neural Comput. Appl. 1–7 (2018)
40. Naghsh, E., Sabahi, M.F., Beheshti, S.: Spatial analysis of EEG signals for Parkinson's disease stage detection. Signal, Image Video Process. **14**(2), 397–405 (2019)
41. Ly, Q.T. et al.: Detection of gait initiation failure in Parkinson's disease based on wavelet transform and support vector machine. In: Proceedings of the Annual International Conference of the IEEE Engineering in Medicine and Biology Society, EMBS, pp. 3048–3051 (2017)
42. Loconsole, C. et al.: Computer vision and EMG-based handwriting analysis for classification in Parkinson's disease. In Lecture Notes in Computer Science (including subseries Lecture Notes in Artificial Intelligence and Lecture Notes in Bioinformatics), vol. 10362 LNCS, pp. 493–503 (2017)
43. Donato Cascarano, G. et al.: Biometric handwriting analysis to support Parkinson's disease assessment and grading. pp. 15–18 (2019)

In Perspective of Combining Chaotic Particle Swarm Optimizer and Gravitational Search Algorithm Based on Optimal Power Flow in Wind Renewable Energy

C. Shilaja

Abstract Optimizing the power flow is one of the recent and emerging problems need to solve immediately to enhance the power flow in any power system applications. To do this, various earlier researchers have used many artificial intelligence methods or optimization methods. The obtained results are not cost and time effective. Still the real-time power system applications need a solution with cost and time effective. This paper aimed to integrate the chaotic particle swarm optimizer (CPSO) and gravitational search algorithm to solve the OPF problems. The proposed methods are simulated in MATLAB software, tested on IEEE buses of 30 and 57, and the results are verified. From the obtained results, it is identified that the GSA method is concluded as better method for solving OPF problems in power system applications.

Keywords Chaotic particle swarm optimizer (CPSO) · Gravitational search algorithm (GSA) · Optimal power flow (OPF) and wind energy

Nomenclature

NPV	Number of voltage controlled buses
NPQ	Number of PQ buses
NTL	Number of transmission lines
P_{G}	Active power output of generators at PV bus
$V_{\mathrm{Li}}^{\min}$ and $V_{\mathrm{Li}}^{\max}$	Minimum and maximum load voltage of *i*th unit
S_{li}	Apparent power flow of *i*th branch
cr_1 and cr_2	Chaotic variables

C. Shilaja (✉)
Department of Electrical and Electronics Engineering, Kalasalingam Academy of Research and Education, Krishnankoil, Virudhunagar, India
e-mail: shilaja.research@gmail.com

S. Borah et al. (eds.), *Soft Computing Techniques and Applications*, Advances in Intelligent Systems and Computing 1248,
https://doi.org/10.1007/978-981-15-7394-1_44

V_G	Terminal voltages at generation bus bars
Q_C	Output of shunt VAR compensators
T	Tap setting of the tap regulating transformers
Q_{Ci}^{min} and Q_{Ci}^{max}	Minimum and maximum V_{ar} injection limits of ith shunt capacitor
T_i^{min} and T_i^{max}	Minimum and maximum tap settings limits of ith transformer
S_{li}^{max}	Maximum apparent power flow limit of ith branch
x^{lim}	Limit value of the dependent variable x
G	Initial value (G_0) and time (t)

1 Introduction

Optimal power flow is one of the optimization methodologies where it solves power system problems based on different constraints. It is well defined with hierarchical constrained optimization and experimentally substantial. OPF is represented using equality constraints, and its problem has been introduced in 1979 [1]. It is an extended problem of best economical dispatch creation in any power systems. So, the equations of economic dispatch are also integrated. In general, OPF problems are solved using metaheuristic approaches. Later according to the extension of power systems with renewable energy sources, equations of uncertainties are also included in the OPF calculation. Authors in [2, 3] have introduced minimizing the optimal solution using first-order gradient methods associated with constraints. Author in [3] proposed OPF solutions for solving irregulations of power system applications. Different kinds of approaches, methods and algorithms have been proposed for solving optimal power flow problems in normal as well as hybrid power systems. For example, in real-time power industries, mathematical methods such as linear, nonlinear, gradient [4], integer and newtons [5] models. Gradient methods have used for solving penalty based OPF problems. It solves the problem associated with the stability constraints by incorporating adjoint equations [4]. Newtons methods have used for OPF with control variables including with non-separable objective functions. It also solves issues and problems related to load and current-load flow by using numerical methods solving the constraints. Nonlinear programming methods are used to solve various reactive and active power dispatch in large-scale power systems [6].

Dommel and Tinney right off the bat displayed the detailing ideal power stream [5]. At that point, this theme has been dealt with by numerous specialists. The OPF issue has been fathomed by utilizing conventional and transformative-based calculations. Regular streamlining strategies, for example, inside point strategy, direct, nonlinear, linear, and quadratic-programming methods have been represented for caring OPF issues [7–12]. In any case, the dis-preferred standpoint of these procedures is that it is unimaginable to expect to utilize these methods in useful frameworks as a result of nonlinear characteristics, for example, valve point impacts, denied working zones and piecewise quadratic cost capacity. In this way, it ends up important to improve the enhancement strategies that are equipped for overcoming these hindrances and dealing with such challenges [13]. A portion of the populace-based strategies have been proposed for taking care of the OPF issue achievement completely, with the end goal that hereditary calculation [14], improved hereditary calculation [15], TABU-search method [9], Swarm intelligence [16], higher-order differential calculation [17], reproduced strengthening [15], transformative programming.

Metaheuristic methods are flexible and powerful in search the optimal solution with respect to various constraints. They have found lot of success stories in obtaining optimal solution is various real-time applications under various kinds of fields. Both heuristic and metaheuristic can provide better solutions with least computational time and cost. In power system problems also, various earlier researchers have used heuristic and metaheuristic approaches for solving different kinds of electrical problems. One of the popular and highly required problems need to solve in power systems is optimal power flow problems. For example, author in [16] has used improved genetic algorithm for solving OPF problem where it is a nonlinear and largescale. Author in [17] implements evolutionary algorithms to resolve power planning problems in reactive power problems. The optimization is decomposed into two modules as P and Q, where the algorithms optimize both the modules repeatedly a greater number of times to get the optimal solution. One of the popular algorithms is PSO used for OPF problems in [18]. Author in [19] has used the PSO and proposed improved PSO and hybrid PSO for finding optimal solutions in reactive power systems. In addition to this, most of the researchers have focused on solving renewable energy power systems. For example, author in [20] applied PSO algorithm for solving OPF problems in DG systems. Author in [21] has discussed about microgrid problems in power systems and provide solution for energy management. A smart energy management system has been proposed to obtain the optimal solution for power production and distribution in DG systems [22]. Similar to that author in [23] proposed adaptive modified PSO algorithm for solving multi-operational power management problems using multi-objective solutions. Author in [24] has focused on microgrid incorporated with renewable energy saving system for obtaining the optimal solutions for storage capacity.

One of the novel and emerging heuristic algorithms is gravitational search algorithm where it follows the newtons law for solving optimal problems. GSA is reported as a brilliant algorithm since it has been implemented and verified by the authors in [25–29]. GSA follows the calculative methodologies of gravitational force and makes use of it for searching the optimal solution. Based on this, GSA will start from a lower limit of gravitational value into upper limit of the same and search the optimal solution in the search space [31, 32]. Also, this algorithm consumes lesser memory and computational time to do any calculations [33, 34]. Hence, this paper aimed to solve OPF problems using GSA in various IEEE bus systems and the results verified. The execution of GSA is tested on standard IEEE 30/57-bus test system. From the simulation outputs, it comes to know that the GSA gives entirely amazing outcomes to taking care of the OPF issue.

The CPSO proposed here receives the turmoil seek strategy [35]. The tumult instatement and the annoyance of the disarray look technique is utilized rather than the arbitrary introduction and the irregular irritating. In the introduction stage, CPSO upgrades the underlying particles as indicated by the qualities of mix advancement issues. By means of thing order, comparative things are gathered into a similar classification, along these lines decreasing the quantity of blends. Along these lines, it is conceivable to list all blend conspires and improve the hunt proficiency. In the tumult bothering stage, another arrangement of annoying tenets is intended to irritate speeds and places of particles adequately, so that CPSO has great worldwide inquiry capacity and flexibility, and the untimely union issue of particles is likewise viably understood. In the over two stages, CPSO controls the quantity of chose things in every class to guarantee the assorted variety of the last mix conspire. The wellness capacity of CPSO uses the idea of the customized imperatives and general obliges to get a customized interface, which can be utilized to illuminate the relating customized combinatorial advancement issue.

2 Problem Formulation of OPF

In this paper, OPF is considered as a nonlinear problem, need to be solved. The basic objective is to assign the values of the control factors as far as a specific target work exposed to different uniformity and imbalance imperatives. As a rule, the OPF issue can be scientifically detailed as pursues:

$$\text{Min } F(x; u) \tag{1}$$

$$\text{subject to } g(x; u) = 0 \tag{2}$$

$$h(x; u) < 0 \tag{3}$$

F denotes the output capacity to be controlled, x denotes the ward vector values and u represents the control variables. The value of x also includes active power at P_{G1}, load voltage V_{L}, and reactive power Q_{G}. Transmission line loading S_{l}. x can be represented as:

$$\mathrm{x}^{\mathrm{T}} = [P_{\mathrm{G1}}; V_{\mathrm{L1}} \ldots V_{\mathrm{LNPQ}}; Q_{\mathrm{G1}} \ldots Q_{\mathrm{GNPV}}; S_{\mathrm{l1}} \ldots S_{\mathrm{lNTL}}] \tag{4}$$

Similar to the above Eq. (4), u can be written as:

$$u^{\mathrm{T}} = [P_{\mathrm{G2}} \ldots P_{\mathrm{GNG}}; V_{\mathrm{G1}} \ldots V_{\mathrm{GNG}}; Q_{\mathrm{C1}} \ldots Q_{\mathrm{CNC}}; T_1 \ldots T_{\mathrm{NT}}] \tag{5}$$

The set of all notations given in the Eq. (1) to (5) is described in nomenclature.

2.1 Various Conditions Considered

2.1.1 Equality Conditions

Following Eq. (6) and (7) shows the equality constraints of typical load flow:

$$P_{Gi} - P_{Di} - V_i \sum_{j=1}^{NB} V_j\left[G_{ij}\cos(\delta_i - \delta_j) + B_{ij}\sin(\delta_i - \delta_j)\right] = 0 \quad (6)$$

$$Q_{Gi} - Q_{Di} - V_i \sum_{j=1}^{NB} V_j\left[G_{ij}\sin(\delta_i - \delta_j) + B_{ij}\cos(\delta_i - \delta_j)\right] = 0 \quad (7)$$

In the above equations, *NB* represents the number of buses in the system, the symbols V_i and V_j represents the voltages obtained from the corresponding buses as i and j. P_{Gi}/Q_{Gi}, P_{Di}/Q_{Di} represents the active-reactive power and demand of active-reactive power obtained at the corresponding buses as i and j. Similarly, G_{ij}, B_{ij}, and d_{ij} represents the conductance, susceptance and phase-difference obtained at the corresponding buses i and j.

2.1.2 Inequality Constraints

The notation "*h*" denotes the set of all inequality conditions and are:

Generator constraints:

the voltage obtained from the generator is controlled within a min and max boundary, is:

$$V_{Gi}^{\min} \le V_{Gi} \le V_{Gi}^{\max}, \quad i = 1, 2, \ldots\ldots, NPV \quad (8)$$

$$P_{Gi}^{\min} \le P_{Gi} \le P_{Gi}^{\max}, \quad i = 1, 2, \ldots\ldots, NPV \quad (9)$$

$$Q_{Gi}^{\min} \le Q_{Gi} \le Q_{Gi}^{\max}, \quad i = 1, 2, \ldots\ldots, NPV \quad (10)$$

In the above equation, $V_{Gi}^{\min}$ and $V_{Gi}^{\max}$ represent the voltage generated minimum and maximum level at the generator—i, $P_{Gi}^{\min}$ and $P_{Gi}^{\max}$ represents the minimum and maximum output power generated at the unit—i, and $Q_{Gi}^{\min}$ and $Q_{Gi}^{\max}$ represents the minimum and maximum level of reactive power generated at the unit-i

Transformer constraints:

Constraints of transformer are given below.

$$T_i^{\min} \le T_i \le T_i^{\max}, \quad i = 1, 2, \ldots\ldots, NT \quad (11)$$

where $T_i^{\min}$ and $T_i^{\max}$ represents the min and max bounds of the transformer unit-i.

Shunt VAR compensator constraints:

Constraints applied on the shunt VAR compensator are given here as:

$$Q_{ci}^{\min} \leq Q_{ci} \leq Q_{ci}^{\max}, \quad i = 1, 2, \ldots\ldots, NC \tag{12}$$

In the above equation, $Q_{ci}^{\min}$ and $Q_{ci}^{\max}$ are the lower and upper values of VAR-injection obtained at shunt capacitor-i.

Security constraints:

The constraints applied for improving the security over voltage and power flow are given here. Each time the voltage of load and power is restricted within a lower and upper bound value as:

$$V_{Li}^{\min} \leq V_{Li} \leq V_{Li}^{\max}, \quad i = 1, 2, \ldots\ldots, NPQ \tag{13}$$

$$S_{li} \leq S_{li}^{\max}, \quad i = 1, 2, \ldots\ldots, NTL \tag{14}$$

where $V_{Li}^{\min}$ and $V_{Li}^{\max}$ represents the lower and upper limits of the voltage-load at unit-i, and S_{li} and $S_{li}^{\max}$ represents the normal and maximum value of the apparent power flow obtained at unit-i. Based on the inequality constraints, the power system generates optimal power flow to fulfill the power industry applications. Also, a scientific representation of the punishment capacity is given as:

$$\begin{aligned} \text{Jmod} = & \sum_{i=1}^{NPV} F_i(P_{Gi}) + \lambda_P \left(P_{G1} - P_{G1}^{\lim}\right)^2 + \lambda_V \sum_{i=1}^{NPQ} \left(V_{Li} - V_{Li}^{\lim}\right)^2 \\ & + \lambda_Q \sum_{i=1}^{NPV} \left(Q_{Gi} - Q_{Gi}^{\lim}\right)^2 + \lambda_S \sum_{i=1}^{NTL} \left(S_{li} - S_{li}^{\max}\right)^2 \end{aligned} \tag{15}$$

where λ_P, λ_V, λ_Q and λ_S represents the set of all penalty factors. And $x^{\lim}$, $x^{\min}$, and $x^{\max}$ represents normal, minimum, and maximum values of the s and can be written as:

$$x^{\lim} \begin{cases} x^{\max}; & x > x^{\max} \\ x^{\min}; & x > x^{\min} \end{cases} \tag{16}$$

3 CPSO and GSA

Here, it is elaborated the both algorithm of CPSO and GSA.

3.1 Chaotic Particle Swarm Optimizer (CPSO)

The PSO calculation faces up to untimely assembly since data can be traded between particles rapidly and the particles are getting close to one another quickly. Consequently, the scattering of particles diminishes in the inquiry space and it is hard to escape from neighborhood optima. The exhibited strategies for this issue have endeavored to control the scattering of particles in the hunt space [17–19]. In this paper, disordered frameworks are connected to improve the assorted variety of the molecule swarm in the pursuit space to abstain from getting caught in neighborhood optima.

So as to build a populace's decent variety, confused frameworks were utilized to introduce the particles' populace and speed. Along these lines, D-diverse clamorous factors are produced by chosen tumultuous frameworks with a given introductory esteem and after that the disordered factors () are changed over to the relating scopes of streamlining factors, that is, the comparing *j*th part of enhancement factors can be characterized by

$$x_{ij} = x_{\min_j} + \left(x\max_j - x_{\min_j}\right) cx_{ij} \quad i = 1, \ldots.., N \quad j = 1, \ldots.., D \tag{17}$$

Here, x minimum and x maximum are the search boundaries of xj. Thus, the particle's position is $\overrightarrow{x}i = (xi1, \ldots, xii)$. A similar approach can be used to initialize the velocity. In CPSO, a sequence generated by selected chaotic systems substitutes the random parameters $r1$ and $r2$ in PSO. The velocity update equation for CPSO can be formulated as:

$$v_i(t+1) = wv_i(t) + c_1 cr_1(p\text{best}_i(t) - x_i(t)) + c_2 cr_2(g\text{best}_i(t) - x_i(t)) \tag{18}$$

In (18), cr_1 and cr_2 are the chaotic variables.

3.2 Gravitational Search Algorithm (GSA)

This GSA algorithm is a popular stochastic algorithm used for hunt computation proposed by Rashedi et al. [29]. The computational method applied on Newtons law of gravity and mass. Most of the scientists are using this mass estimation to exhibit the estimation over power, load, voltage and capacity. GSA is reported as a brilliant algorithm since it has been implemented and verified by the authors in [25–29]. GSA follows the calculative methodologies of gravitational force and makes use of it for searching the optimal solution. Based on this, GSA will start from a lower limit of gravitational value into upper limit of the same and search the optimal solution in the search space [31, 32]. The mathematical representation of the mass in terms of directions is:

$$X_i = \left(x_i^1 \ldots x_i^d \ldots x_i^n\right) \forall i = 1, 2, \ldots, N \tag{19}$$

The search space is defined with the dimension of n. Each position in the search space is defined as x_i^d. The notation i says the agent and d says the dimension of i. At the beginning, based on the newtons law and gravitational value, the searching process has started randomly. The mass j is applied on mass i during time t can be expressed as:

$$F_{ij}^d(t) = G(t) \frac{M_i(t) \times M_j(t)}{R_{ij}(t) + \in} \left(x_j^d(t) - x_i^d(t)\right) \tag{20}$$

where M_i represents the ith object's mass, M_j represents the jth object's mass, $G(t)$ is the gravitational value during time t, e is a small constant, and $R_{ij}(t)$ represents the Euclidian distance among the objects i and j, is calculated as:

$$R_{ij}(t) = \left\| X_i(t), X_j(t) \right\|_2 \tag{21}$$

The total force of ith object is calculated as:

$$F_i^d(t) = \sum_{j \in kbest, j \neq i}^{N} \text{rand}_j F_{ij}^d(t) \tag{22}$$

where rand_j is generated as a random integer between 0 and 1 and the best value obtained using K operators as k best based on the greatest wellness and mass. In request to discover the speeding up of the ith specialist, at t time in the dth measurement law of movement is utilized specifically to ascertain. In accor-hit the dance floor with this law, it is relative to the power following up on that operator, and conversely corresponding to the mass of the specialist.

$$a_i^d(t) = \frac{F_i^d(t)}{M_{ii}(t)} \tag{23}$$

Also, the best value can be searched according to the velocity and position values of the agent. It can be continuously assessed and finding the best values using a function by comparing with the current velocity and position values with the previous values after calculating the acceleration. Velocity and position values are calculated using the following formula as:

$$v_i^d(t+1) = \text{rand}_i \times v_i^d(t) + a_i^d(t) \tag{23}$$

$$X_i^d(t+1) = X_i^d(t) + v_i^d(t+1) \tag{24}$$

where $v_i^d(t)$ and $X_i^d(t)$ represents the current velocity and position values of the agent during time t of dimension d. The rand$_i$ obtains the random number from 0 to 1 using randomization function. The gravitation value G is always constant. Here, it is initialized as $G0$ and change the value based on time t for controlling the searching accuracy.

3.3 *Hybrid of CPSO and GSA*

```
Step 1. Recognize the solution space.
Step 2. Set population, lower and upper range of the
control variables.
Step 3. Compute the fitness function for each agent
Step 4. Calculate and update the values of G(t),
best(t), worst(t) and Mi(t) for i = 1, 2,…, N.
Step 5. Compute total force for various directions.
Step 6. Compute the velocity and movement speed.
Step 7. Update the values of velocity and movement
speed.
Step 8. Do from step-3 to step-7 until end.
Step 9. Stop.
```

4 Result and Discussion

Standard IEEE 30 Bus System

GSA algorithm is programmed in MATLAB software and the results are verified. To do that, the experiment is tested in IEEE 30-bus system and the performance parameters are examined. The obtained results are compared with the performance parameters obtained in earlier systems [15]. The highest but least voltages of all buses are viewed as 1.05–0.95 in p.u. The proposed methodology has been connected to tackle the OPF issue for various cases with different target capacities. For each situation, 50 trials were performed for taking care of the OPF issue utilizing GSA. The value of G is calculated by initializing $G0$ is 100, T as 10, used in Eqs. (22) and (23). The experiment is carried out 200 number of times iteratively to obtain best values. The experiment is done in MATLAB software, complied in 2.63 GHz Pentium system with the RAM size 512 MB-RAM. The experimental results are given and explain in detail below.

4.1 Quadratic Cost Function (Case-1)

The cost is calculated using the following equation, based on the objective function as:

$$J = \sum_{i=1}^{N} F_i(P_{Gi}) = \sum_{I=1}^{NG} \left(a_i + b_i P_{Gi} + C_i P_{Gi}^2\right) \tag{25}$$

In Eq. (25) F_i and P_{Gi} represents the ith generator's fuel cost and output, respectively, the cost-coefficients a_i, b_i and c_i of ith generator, and the total number generators is represented as *NG*. Table 1 shows the cost coefficients' values and the optimized control variable values. From Table 1, it has been proved that GSA reduced the cost as much as possible, which is better than other algorithms. The convergence rate of the GSA algorithm is depicted in Figure 1, at the time the optimal value is obtained.

Table 1 Simulation results of case-1

33 Method	Min	Average	Max	Time
CPSO-GSA	797.5246	797.2465	798.4635	10.1253
BBO [31]	799.111	799.198	799.204	11.02
PSO [15]	800.41	NA	NA	NA
Improved GA [14]	800.80	NA	NA	NA
MDE [2]	802.37	802.38	802.40	23.25
Gradient method [30]	804.85	NA	NA	4.324
Enhanced GA [18]	802.06	NA	802.14	76

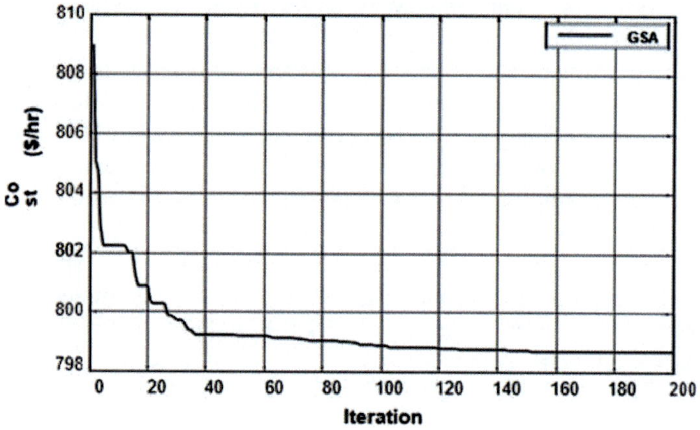

Fig. 1 Cost calculation for IEEE-30-bus system

Table 2 Performance evaluation of case-2

Methods	Min	Average	Max	Times
CPSO-GSA	0.0813	0.08354	0.0856	10.42
BBO [29]	0.1020	0.1105	0.1207	13.23
DE [1]	0.1357	NA	NA	NA
PSO [15]	0.0891	NA	NA	NA

From the comparison results, it is noticed that the proposed method is better than existing methods.

4.2 Voltage Profile: Case-2

The security of the power system is considered as voltage of the bus systems. Feasible solution is the cost and time consumption, not the voltage profile. Hence, this paper considered two objectives as the main objectives need to be solved by reducing the cost, voltage [1] and enhancing the voltage profile. So, the objective function can be written as:

$$J = \sum_{i=1}^{NG} \left(a_i + b_i P_{Gi} + C_i P_{Gi}^2\right) + n \sum_{I=1}^{NG} |V_i - 1.0| \tag{26}$$

In equation (26), the notation g represents the appropriate weight factor selected by the customer. Here, in the experiment, the g value is selected as 100, referred from [1, 31]. All the results obtained using GSA is given in Tables 1, 2, 3 regarding control variables, voltage profile for GSA, and voltage profile for case-1 and case-2, respectively. From the comparison, it is identified that the proposed GSA is concluded as better method for power systems.

4.3 Voltage Stability: Case-3

The output value of line 2–6 is simulated based on the contingency stage and the voltage stability is verified according to the objective function given in equation-(36). The obtained results and the comparison results are given in Table 1 and Table 3, respectively. From the results, it is noticed that the proposed GSA is better method which can improve the voltage stability, is depicted as $L_{\max} = 0.093073$, which is lesser than the existing DE algorithm.

To evaluate the performance of GSA, it is tested on large-scale problem solving like IEEE-75 bus test system. It has 80 transmission lines, 7 generators at various buses such as 1, 2, 3, 6, 8, 9 and 12, and 15 branches under various load constraints.

Table 3 Comparison of the simulation results for case 3

Control variables	EGWO-DA
PG1 (MW)	1.05921
PG2 (MW)	0.99915
PG5 (MW)	0.92547
PG8 (MW)	0.74582
PG11 (MW)	4.25687
PG13 (MW)	0.11923
V1 (p.u.)	1.05974
V2 (p.u.)	1.06000
V5 (p.u.)	1.05999
V8 (p.u.)	1.04590
V11 (p.u.)	0.90000
V13 (p.u.)	0.90000
T11 (p.u.)	0.90208
T12 (p.u.)	0.32548
T15 (p.u.)	0.02563
T36 (p.u.)	0.02548
QC10 (MVar)	0.01227
QC12 (MVar)	0.05999
QC15 (MVar)	0.84999
QC17 (MVar)	0.99921
QC20 (MVar)	0.05921
QC21 (MVar)	0.93243
QC23 (MVar)	1.09587
QC24 (MVar)	0.14430
QC29 (MVar)	0.15241
FUEL COST($/h)	790.56

The power sources for shunt reactive power at various buses are 18, 25 and 53. The total load demand of the proposed system is 1250.8 MW and 336.4 MVAR.

5 Conclusion

The main objective of this paper is to implement a novel optimization algorithm for solving nonlinear optimal power flow problems in power systems. In order to do that this paper used two different heuristic algorithms such as GSA algorithm and CPSO algorithm. Both algorithms are explained, problem formulated and implemented in MATLAB software. The simulation is verified on IEEE test bus systems such as 30 and 57 bus. From the results it is identified that GSA algorithm is better algorithm for solving OPF under various conditions. It is identified that GSA can also be applied for large size power systems application in real time industries. GSA reduces the cost and computational time.

References

1. Carpentier, J.: Optimal power flows. Int. J. Electr. Power Energy Syst. **1**(1), 3–15 (1979)
2. Dommel, H.W., Tinney, W.F.: Optimal power flow solutions. IEEE Trans. Power Appar. Syst. **87**(10), 1866–1876 (1968)
3. Momoh, J.A., Koessler, R.J., Bond, M.S., et al.: Challenges to optimal power flow. IEEE Trans. Power Syst. **12**(1), 444–447 (1997)
4. Sun, Y., Xinlin, Y., Wang, H.F.: Approach for optimal power flow with transient stability constraints. IEE Proc. Gener. Trans. Distrib. **151**(1), 8–18 (2004)
5. Sun, D.I., Ashley, B., Brewer, B., Hughes, A., Tinney, W.F.: Optimal power flow by Newton approach. IEEE Trans. Power Appar. Syst. **103**(10), 2864–2880 (1984)
6. Wei, H.H., Kubokawa, S.J., Yokoyama, R.: An interior point nonlinear programming for optimal power flow problems with a novel data structure. IEEE Trans. Power Syst. **13**(3), 870–877 (1998)
7. Abido, M.A.: Optimal power flow using tabu search algorithm. Electr. Power Compo. Syst. **30**(5), 469–483 (2002)
8. Yan, X., Quantana, V.H.: Improving an interior point based OPF by dynamic adjustments of step sizes and tolerances. IEEE Trans. Power Syst. **14**(2), 709–717 (1999)
9. Habiabollahzadeh, H., Luo, G.X., Semlyen, A.: Hydrothermal optimal power flow based on combined linear and nonlinear programming methodology. IEEE Trans. Power Appar. Syst. PWRS **4**(2), 530–537 (1989)
10. Burchet, R.C., Happ, H.H., Vierath, D.R.: Quadratically convergent optimal power flow. IEEE Trans. Power Appar. Syst. PAS **103**, 3267–3276 (1984)
11. Momoh, J.A., El-Hawary, M.E., Adapa, R.: A review of selected optimal power flow literature to 1993 II Newton, linear programming and interior point methods. IEEE Trans. Power Syst. **14**(1), 105–111 (1999)
12. Huneault, M., Galina, F.D.: A survey of the optimal power flow literature. IEEE Trans. Power Syst. **6**(2), 762–770 (1991)
13. Roy, P.K., Ghoshal, S.P., Thakur, S.S.: Biogeography based optimization for multi-constraint optimal power flow with emission and non-smooth cost function. Expert Syst. Appl. **37**, 8221–8228 (2010)
14. Deveraj, D., Yegnanarayana, B.: Genetic algorithm based optimal power flow for security enhancement. IEE Proc. Gener. Transm. Distrib. **152**(6), 899–905 (2005)
15. Lai, L.L., Ma, J.T., Yokoyama, R., Zhao, M.: Improved genetic algorithms for optimal power flow under normal and contingent operation states. Int. J. Electr. Power Energy Syst. **19**(5), 287–292 (1997)
16. Abido, M.A.: Optimal power flow using particle swarm optimization. Electr. Power Energy Syst. **24**, 563–571 (2002)
17. Varadarajan, M., Swarup, K.S.: Solving multi-objective optimal power flow using differential evolution. IET Gener. Transm. Distrib. **2**(5), 720–730 (2008)
18. Bakirtzis, A.G., Biskas, P.N., Zoumas, C.E., Petridis, V.: Optimal power flow by enhanced genetic algorithm. IEEE Trans. Power Syst. **17**(2), 229–236 (2002)
19. Lee, K.Y.: Optimal reactive power planning using evolutionary algorithms: a comparative study for evolutionary programming, evolutionary strategy, genetic algorithm, and linear programming. IEEE Trans. Power. Syst. **13**(1), 101–108 (1998)

20. Abido, M.A.: Optimal power flow using particle swarm optimization. Int. J. Electr. Power Energ. Syst. **24**(7), 563–571 (2002)
21. Esmin, A.A.A., Lambert-Torres, G., Zambroni de Souza, A.C.: A hybrid particle swarm optimization applied to loss power minimization. IEEE Trans. Power Syst. **20**(2), 859–866 (2005)
22. Amanifar, O., Hamedani Golshan, M.E.: Optimal distributed generation placement and sizing for loss and THD reduction and voltage profile improvement. Tech. Phys. Prob. Eng. (IJTPE) **3**(2) (2011)
23. Sanseverino, E.R., Di Silvestre, M.L., Ippolito, M.G., De Paola, A., LoRe, G.: An execution, monitoring and replanning approach for optimal energy management in microgrids. Energy **36**(5), 3429–3436 (2011)
24. Chen, C., Duan, S., Cai, T., Liu, B., Hu, G.: Smart energy management system for optimal microgrid economic operation. IET Renew. Power Gener. **5**(3), 258–267 (2011)
25. Moghaddam, A.A., Seifi, A., Niknam, T., Alizadeh Pahlavani, M.R.: Multi-objective operation managemen to farenewable MG (micro-grid) with back-up micro-turbine/fuel cell/ battery hybrid power source. Energy **36**(11), 6490–6507 (2011)
26. Sharma, S., Bhattacharjee, S., Bhattacharya, A.: Grey wolf optimisation for optimal sizing of battery energy storage device to minimise operation cost of microgrid. IET Gener. Transm. Distrib. **10**(3), 625–637 (2016)
27. Madhad, B., Srairi, K., Bouktir, T.: Optimal power flow for large-scale power system with shunt FACTS using efficient parallel GA. Electr. Power Energ. Syst. **32**, 507–517 (2010)
28. Li, C., Zhou, J.: Parameters identification of hydraulic turbine governing system using improved gravitational search algorithm. Energy Convers. Manage. **52**, 374–381 (2011)
29. Zibanezhad, B., Zamanifar, K., Nematbakhsh, N., Mardukhi, F.: An approach for web services composition based on QoS and gravitational search algorithm. In: Proceedings of the Innovations in Information Technology Conference, pp. 340–344 (2010)
30. Hassanzadeh, HR., Rouhani, M.: A multi-objective gravitational search algorithm. In: Proceedings of the Communication Systems and Network Conference, pp. 7–12 (2010)
31. Balachandar, S.R., Kannan, K.: A meta-heuristic algorithm for set covering problem based on gravity. Int J Comput. Math. Sci. **4**, 223–228 (2010)
32. Duman, S., Güvenç, U., Yörükeren, N.: Gravitational search algorithm for economic dispatch with valve-point effects. Int. Rev. Electr. Eng. **5**(6), 2890–2895 (2010)
33. Ceylan, O., Ozdemir, A., Dag, H.: Gravitational search algorithm for post-outage bus voltage magnitude calculations. In: International Universities Power Engineering Conference, Wales (UK), 31 August–3 September (2010)
34. Rashedi, E., Nezamabadi-pour, H., Saryazdi, S.: BGSA: Binary gravitational search algorithm. Nat. Comput. **9**, 727–745 (2010)
35. Rashedi, E., Nezamabadi-pour, H., Saryazdi, S.: GSA: A gravitational search algorithm. Inform. Sci. **179**, 2232–2248 (2009)

Bengali News Headline Generation on the Basis of Sequence to Sequence Learning Using Bi-Directional RNN

Abu Kaisar Mohammad Masum, Md. Majedul Islam, Sheikh Abujar, Amit Kumer Sorker, and Syed Akhter Hossain

Abstract The newspaper provides significant information every day. Each news describes every event largely, but the headline contained a summary of the news. A meaningful headline is very good for a reader who can understand the gist of corresponding news. Text generation is a language modelling where the machine can generate text automatically. Predict the next correct sequence of a text in the main concept of text generation. Using the text generation approach, automatic headline generation is the possible solution for any language. Many experiments have already been completed for English text generation but few in the Bengali language. Here, it has a big scope to generating an automatic news headlines generator for Bengali text using deep learning concept with bi-directional recurrent neural network (RNN). Bengali text summarization is our recent research work in Bengali NLP. During this research, Bengali text generation for text summarizer is challenging work. Thus, to solve these types of problem, automatic text generation is a good solution.

Keywords Headline generation · Deep learning · Recurrent neural network · Automatic text generation · Text summarization

A. K. M. Masum · Md. Majedul Islam · S. Abujar (✉) · A. K. Sorker · S. A. Hossain
Department of CSE, Daffodil International University, Dhaka, Bangladesh
e-mail: sheikh.cse@diu.edu.bd

A. K. M. Masum
e-mail: mohammad15-6759@diu.edu.bd

Md. Majedul Islam
e-mail: majedul15-6784@diu.edu.bd

A. K. Sorker
e-mail: amit15-1618@diu.edu.bd

S. A. Hossain
e-mail: aktarhossain@daffodilvarsity.edu.bd

S. Borah et al. (eds.), *Soft Computing Techniques and Applications*,
Advances in Intelligent Systems and Computing 1248,
https://doi.org/10.1007/978-981-15-7394-1_45

1 Introduction

Text analysis and processing help to extract main or core information long text and text document. Core information provides a better solution for any kinds of a language processing problem. Nowadays, natural language processing uses to solve any language modelling problems not only for the English language but also other languages. Since Bengali is the fifth most usable language in the world. So, it is imperative to concentrate on the development of tools and procedure utilizing natural language processing to process Bangla language [1, 2]. This research work talked about the Bengali newspaper headlines text generation automatically based on the next Bengali word prediction. A newspaper headline corpus was making for predicting its next correct sequence.

Text generation is such type of problem where accurate next sequence of word prediction [3] is its output. In previous research work, several methods introduce to solve the sequence prediction problem. But, deep learning algorithms provide a great solution for predicting the next sequence of the word. All of the deep learning approach work on the basis of neural network where each network creates artificially. A large number of the dataset is required for the trained algorithm to find accurate output. In sequential text processing, long short-term memory (LSTM) is working very efficiently and gives absolute output. Word embedding of the whole dataset is the input of this model.

2 Literature Review

Many research works were done previously in these fields. Different research work introduces a different method and algorithm, and all of those work are focused on increasing accuracy and current result of text generation. Understanding the human writing pattern, [4] is important for text analysis. In this section, some of those research work will be discussed shortly.

Producing content language demonstrating is the most valuable procedure in regular language preparing. That model is prepared to foresee the following expression of a book succession. Marc' Aurelio Ranzato [5] tackled giving an estimation which direct improves at test time. Using the figuring gives the best execution to an avaricious age. Two significant issues need to lessen inclinations and misfortune work which assists with improving the content age general methodologies.

Martin Sundermeyer [6] introduced two different methods for machine translation one is word-based and another is phase-based. Additionally, introduce bi-directional RNN to solve the machine translation problem. All approach provides an improvement of over the baseline approaches.

Alex Sherstinsky [7] worked on LSTM and RNN, which utilized to create a perplexing sequence. Methods are shown for text and handwriting where the text contains a discrete and handwriting contains a real-valued. All of the results predict the output as the next text sequence. In another work, he indicated Vanilla LSTM [8] which was a logical contention. It changes the RNN to understand difficulties to unravel in training time.

Sanzidul Islam et al. [9] introduce sequence to sequence method using LSTM for Bengali sentence generation. Applying deep learning approaches, he predicts the next sequence of Bengali text word or sequence. After training, using model can generate full Bengali sentences. The model can generate Bengali full sentences [2] after training.

LSTM gives a better output for sequential data. All news headline words follow a logical order. In this research work, we propose a method about automatic newspaper headline sequence generation using the bi-directional RNN.

3 Methodology

After the invention of machine translation, several language modelling problems can solve easily. In language handling, a significant issue, for example, content examination, discourse to content transformation, content outline, picture to content change, language to language interpretation, everything is tackled utilizing machine interpretation procedure. Text generation is also an important part of machine translation. The automatic paragraph generates automatic document writing and sentence generation which are the main use of an automatic text generator. Since Bengali is one of the most used languages in the world, need is to make an automated text generator to reach research and development area in this language, given below a workflow for this research works (Fig. 1).

3.1 Dataset Properties and Data Collection

Newspaper headline generation is the main focusing section of this research work. A profound learning model needs countless information to give a precise yield. We collect data from Bengali newspaper "prothom alo" using Web scraping technique with Python scripting. Dataset properties look like

i. Total 9757 news headline.
ii. Total words 19,510.
iii. 39,315 total character.
iv. All types of news headlines.

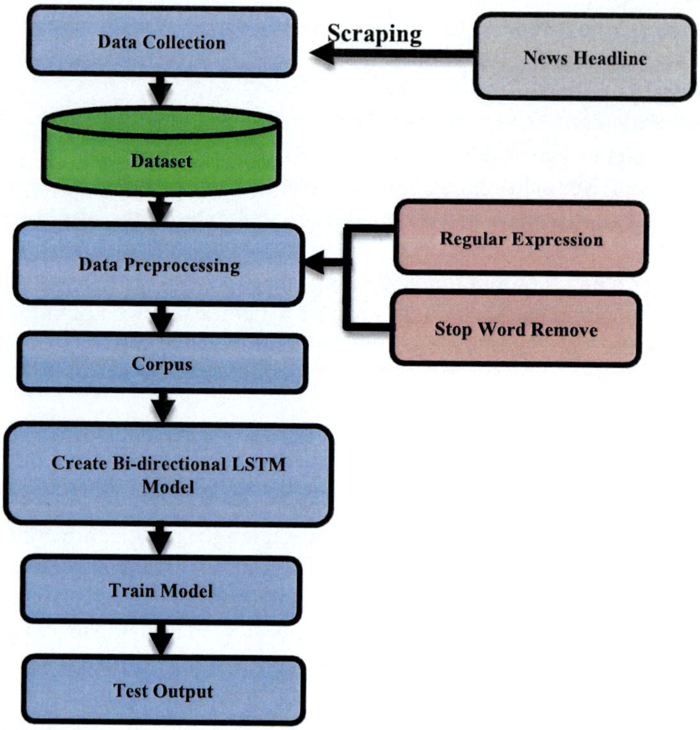

Fig. 1 News headline generation working process

3.2 *Data Preprocessing*

The process of Bengali text data is difficult from the process of other languages data. The machine could not identify Bengali language characters or symbols automatically. To remove an unwanted character, space letter, or digit, Bengali punctuation needs to define Bengali Unicode of the characters. The range of Bengali Character Unicode is 0980-09FF. Then, need to remove space from the line and remove the stop words.

3.3 *Add Pretrained Word Embedding*

In this experiment, we used Bengali pretrained word embedding. Generally, word embedding contains vector presentation of large document words. All words are presented by a numeric value. Word implanting can change the sentences or word over tokenize word to a vector where every vector tends to the jargon of substance accounts.

Table 1 Pad sequence generation

Input words	Next words
২০০ ডাকছে	২০০ ডাকছে তামিম
২০০ ডাকছে তামিম	২০০ ডাকছে তামিম রাঙাতে
২০০ ডাকছে তামিম রাঙাতে	২০০ ডাকছে তামিম রাঙাতে পারবেন
২০০ ডাকছে তামিম রাঙাতে পারবেন	২০০ ডাকছে তামিম রাঙাতে পারবেন তো

3.4 Sequence Generation

This section for sequence generation needs to tokenize the dataset headline text. Keras provides built-in library to tokenize the word easily. Then, need to generate n-gram sequence where each n-gram contains a corresponding numeric value of dataset each word. After this, pad sequence generation is required. Pad sequence helps to predict the next sequence, given below an example of pad sequence generation (Table 1).

3.5 Model

RNN is part of a neural network. Solving the sequential data problem, RNN shows better performance. Each RNN works using the loop. The sequence of data is an input of an RNN, then out of this data is an input of the next cell. This process continues to end of the neuron layer. So, RNN can easily detect the next of a given data [10, 11]. Bi-directional RNN has two directions forward and backward direction [12]. Forward direction contains past to future word prediction and backward contains future to past word prediction, but the output will be the same. Generally, two hidden layers have used to store both information.

In this experiment, we used bi-directional LSTM with 512 dense layers, where it predicts the sequence of the word which is the input of the model. A total of 39,315 predicting sequences are the input of the model according to the dataset. ReLU activation function is used in each input layer. For reducing, the overfitting needs to define dropout value 0.5. In the yield layer, all-out word is the yield with a softmax enactment work. For misfortune work, we utilized scanty straight out cross-entropy with Adam enhancement work. Fig. 2 shows the perspective on bi-directional RNN model utilizing LSTM.

In mathematical term, this model will be,

$$h_{\mathrm{f}} = \sigma(W_{\mathrm{f}} * X + h_{\mathrm{f}} + b_{\mathrm{f}}) \tag{1}$$

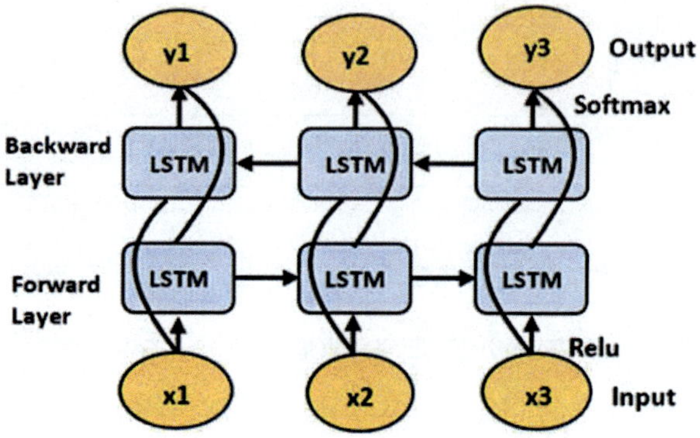

Fig. 2 View of bi-directional model with LSTM

$$h_b = \sigma(W_b * X + h_b + b_b) \tag{2}$$

$$y = (h_f W_f + h_b W_b + b) \tag{3}$$

Here, σ contain activation function
h_f is forward hidden layer
h_b is backward hidden layer
W is Weight and b is bias
X is input, and y is output

First equation input value x is multiplied with the value of weight in the forward layer. Then, added with the forward hidden layer value and biased. Whole value is multiplied by activating function "ReLU." The second equation calculates the value similarly to the backward layer. In the output layer, forward layer value and backward layer value both are addition with biased using activation function "softmax." This section shows the deep learning model algorithm for headline generation using bi-directional RNN.

***Algorithm1** for Bengali headline generation using Bi-directional LSTM*

```
1:Set function model create(max sequence length, total words):
2:    input length= max sequence length-1
3:    declare Sequential()
4:       add(Bidirectional(LSTM(units, activation),input shape)
5:    add(Dropout(0.5))
6:    add(Dense(total words, activation))
7:    compile(loss function, optimizer)
8:    return model
9: create model(max sequence length, total words)
```

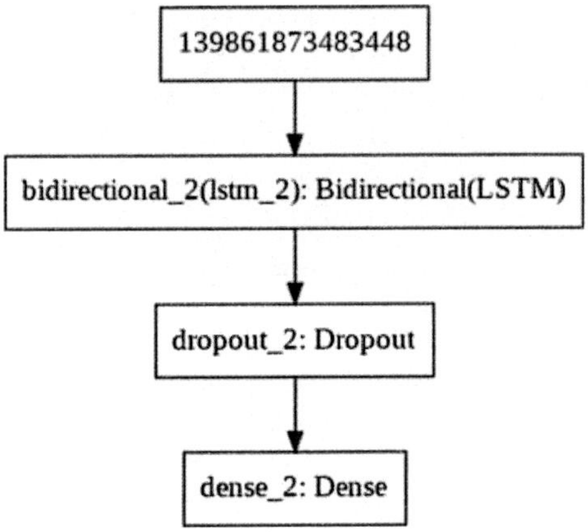

Fig. 3 Graphical structure of model algorithm

Step one defines the model to create a function with parameters maximum sequence length and total word. From maximum sequence length, decrease value minus one and put them into the input sequence length variable. In step three, declare the sequential function. After this in step four, add bi-directional LSTM with 512 unit layer and input shape variable. 512 unit contains a bi-directional LSTM that has a single input and single output 512 parameters. In input shape, it used a total of 39,315 sequences with pretrained word2vec model which embedded size is 300. Reducing overfitting set the dropout value 0.5 and add with the model in step five. Step six defines output layer with total word and activation function. In step seven, define a loss function in compile time with optimization function. In this portion, we show our model calculation graphical view. Here, exceptional id is the commitment of the methodology which will continue to the dense or yield layer (Fig. 3).

i. **Long short-term memory:**

Two primary confinements are evaporating and detonating of the slope in the neural system which is associated. To take care of this issue, LSTM was created. Numerous actuation capacities and three doors with cell-expressed works like a memory. The condition of an LSTM cell will be

$$i_t = \sigma\left(w_i\left[h_{t-1}, x_t\right] + \mathrm{b}_i\right) \tag{4}$$

$$f_t = \sigma\left(w_\mathrm{f}\left[h_{t-1}, x_t\right] + b_\mathrm{f}\right) \tag{5}$$

$$o_t = \sigma\left(w_\mathrm{o}\left[h_{t-1}, x_t\right] + b_\mathrm{o}\right) \tag{6}$$

$$c_t = f_t * c_{t-1} + i_t * \sigma\left(w_c\left[h_{t-1}, x_c\right] + b_c\right) \tag{7}$$

$$h_t = o_t * \sigma(c_t) \tag{8}$$

Here, i_t is input gate′s
f_t is forget gate's,
o_t is output gate,
c_t is cell state,
h_t is hidden state,
σ is activation function

ii. **Activation function:**

This examination needs to two actuate work. One in the info layer, and another in the yield layer. For the info layer, "ReLU" initiation work utilizes and "softmax" for the yield layer. ReLU is working with always zero to maximum value where softmax is working with all real number.

4 Result Analysıs

Finding the aftereffect of the model needs to set up the model before the training. The model is fitting with the present and upcoming next word. In whole dataset, 80% data is for the training, and 20% data is for the testing. Set epoch size 200 for the dataset and set verbose 1 by using the fit limit with regards to preparing model. Train model fitting around 4 h, and it shows a better precision of 81% with setback 0.10. The model train exactness diagram of the model is shown below (Fig. 4).

The major target of this experiment is to build the accompanying arrangement of Bengali words. Therefore, in the yield, we have made a breaking point where we describe Bengali preportray word embedding for a set each word with a vector which describes the related word in a report with a numeric worth and seed content for appearing. Test the model response that needs to define a function. The functions structure will be

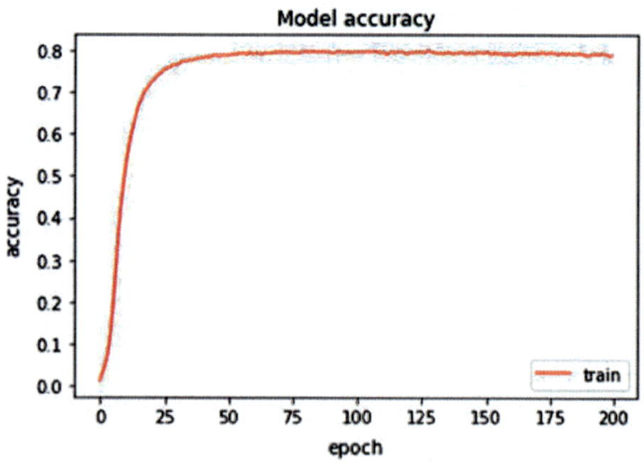

Fig. 4 Accuracy curve for model

generate_headline(*input, next_words, model, maximum sequence length)*

Test sample1:

```
generate_headline("অপয়া", 7, model, max_sequence_len)
```

Output1:

অপয়া সাদা জার্সি যে কারণে ফিরিয়ে আনল ব্রাজিল

Test sample2:

```
generate_headline("অবিশ্বাস্য মাহমুদউল্লাহকে", 5, model, max_sequence_len)
```

Output2:

অবিশ্বাস্য মাহমুদউল্লাহকে ছাড়িয়ে রথী মহারথীদের পাশে সাকিব

Test sample3:

```
generate_headline("২০০ ডাকছে", 4, model, max_sequence_len)
```

Output3:

২০০ ডাকছে তামিম রাঙাতে পারবেন তো

Previously several experiments were done for making automatic text generation in a different language. For this experiment, we train our dataset using existing approaches such as text generation using LSTM and bi-directional LSTM. Finally, build the model using bi-directional approach for automatic Bengali news headline generation. Only the LSTM model provides a good out for text generation. But for increasing the efficiency of work and finding a better result, we used bi-directional LSTM in this research work. Accuracy comparison of both approaches is given in Table 2.

Table 2 Accuracy comparison with the existing approach

Approach	English text generation accuracy (%)	Bengali news headline generation accuracy (%)
LSTM	90	79
Bi-directional LSTM	93	81

Find a good result for a Bengali text is very challenging in any language modelling problem. This experiment result also shows that Bengali headline generation accuracy is less than the English text generation approach (Table 2). Since no model provides an accurate result, this model for Bengali text provides a meaningful and good result for Bengali news headline generation, which we have seen in the output sample.

5 Conclusion and Future Work

This exploration work proposed a superior strategy for making a changed Bengali news feature age utilizing bi-directional RNN. Since no model gives a precise outcome in any case yet, utilizing this model gives a decent outcome. Utilizing the proposed strategy has effectively made a fixed sequence length and essentialness of full Bengali content. There are a few blemishes in the proposed structure, for example, the need to characterize the number creating word length. Another deformation is expected to depict a cushion token for foreseeing next words. Every research work need a vision to complete the whole work properly. Therefore, in future work, we need to make an altered Bengali book generator which gives an optional length Bengali content without utilizing any movement.

Acknowledgements We would like to thank DIU-NLP and Machine Learning Research Lab for providing all reearch and experiment facilities, also Dept. of CSE, Daffodil International University for all their help.

References

1. Tanaka, H., Kinoshita, A., Kobayakawa, T., Kumano, T., Kato, N.: Syntaxdriven sentence revision for broadcast news summarization. In: Proceedings of the 2009 Workshop on Language Generation and Summarisation, UCNLG + Sum'09, pp. 39–47, Stroudsburg, PA, USA. Association for Computational Linguistics (2009)
2. Islam, Sanzidul, et al.: Sequence-to-sequence Bangla sentence generation with LSTM recurrent neural networks. Procedia Comput. Sci. **152**, 51–58 (2019)
3. Graves, A., Fernández, S., Schmidhuber, J.: Bidirectional lstm networks for improved phoneme classification and recognition. In: Artificial Neural Networks: Formal Models and Their Applications–ICANN 2005, pp. 753–753 (2005)
4. Hochreiter, S., Schmidhuber, J.: Long short-term memory. Neural Comput. **9**(8), 1735–1780 (1997)
5. Sutton, R.S., Barto, A.G.: Reinforcement learning: an introduction, vol. 1(1). MIT Press, Cambridge (1998)
6. Hu, Z., Yang, Z., Liang, X., Salakhutdinov, R., PXing, E.: Controllable text generation. arXiv preprint arXiv:1703.00955 (2017)
7. Abujar, S., Hasan, M.: A comprehensive text analysis for Bengali TTS using unicode. In: 5th IEEE International Conference on Informatics, Electronics and Vision (ICIEV), Dhaka, Bangladesh, 13–14 May (2016)

8. Abujar, S., et al.: A heuristic approach of text summarization for Bengali documentation. In: 8th IEEE ICCCNT 2017, IIT Delhi, Delhi, India, 3–5 July (2017)
9. Schuster, M., Paliwal, K.K.: Bidirectional recurrent neural networks. IEEE Trans. Signal Process. **45**(11), 2673–2681 (1997)
10. Yu, L., Zhang, W., Wang, J., Yu, Y.: Seqgan: sequence generative adversarial nets with policy gradient. arXiv preprint arXiv:1609.05473. (2016)
11. Li, J., Monroe, W., Jurafsky, D: Learning to decode for future success. arXiv preprint arXiv: 1701.06549 (2017)
12. Ho, J., Ermon, S.: Generative adversarial imitation learning. In: Advances in Neural Information Processing Systems, pp. 4565–4573 (2016)

Bengali Accent Classification from Speech Using Different Machine Learning and Deep Learning Techniques

S. M. Saiful Islam Badhon, Habibur Rahaman, Farea Rehnuma Rupon, and Sheikh Abujar

Abstract The work starts with a question "Does human vocal folds produce different wavelength when they speak in different accent of same language?" Generally, when humans hear the language, they can easily classify the accent and region from the language. But the challenge was how we give this capability to the machine. By calculating discrete Fourier transform, Mel-spaced filter-bank and log filter-bank energies, we got Mel-frequency cepstral coefficients (MFCCs) of a voice which is the numeric representation of an analog signal. And then, we used different machine learning and deep learning algorithms to find the best possible accuracy. By detecting the region of speaker from voice, we can help security agencies and e-commerce marketing. Working with human natural language is a part of Natural Language Processing (NLP) which is branch of artificial intelligence. For feature extraction, we used MFCCs, and for classification, we used linear regression, decision tree, gradient boosting, random forest and neural network. And we got max 86% accuracy on 9303 data. The data was collected from eight different regions (Dhaka, Khulna, Barisal, Rajshahi, Sylhet, Chittagong, Mymensingh and Noakhali) of Bangladesh. We follow a simple workflow for getting the ultimate result.

Keywords Bengali accent · MFCCs · Bangla speech

S. M. S. I. Badhon · H. Rahaman · F. R. Rupon · S. Abujar (✉)
Department of Computer Science and Engineering, Daffodil International University, Dhaka, Bangladesh
e-mail: sheikh.cse@diu.edu.bd

S. M. S. I. Badhon
e-mail: saiful15-7878@diu.edu.bd

H. Rahaman
e-mail: habibur15-7761@diu.edu.bd

F. R. Rupon
e-mail: farea15-7707@diu.edu.bd

S. Borah et al. (eds.), *Soft Computing Techniques and Applications*, Advances in Intelligent Systems and Computing 1248,
https://doi.org/10.1007/978-981-15-7394-1_46

1 Introduction

Humans interconnect with particular dialects. People use different languages to contact with each other. Such languages are an excellent way of sharing one person's thoughts with another. Yet their differences are the way they interact with people in different locations [1]. Accent is the way to pronounce a language of a distinct area or group or a country. Nowadays, accent recognition has become one of the mooted topics in NLP. Bengali language is very well known in the world as it is the fifth most spoken local language. And by the overall number of speakers in the world, it is the seventh most spoken language [2]. More than 250 million people talk in the standard Bengali language, except some people who talk the Bengali local language. This research makes an application that can detect a region of a person based on the wave frequency of a voice using Bengali language. The Bengali language has many types of regional accents. Throughout Bangladesh, the main spoken dialects are Dhaka (old), Rajshahi, Chittagong, Mymensingh, Barisal, Sylhet, Rangpur, Khulna and Noakhali. In this research, we worked with all these dialects except Rangpur and added standard Bengali. There are many beneficial applications of voice-sensing accent recognition. Some of the uses are narrated here. Different types of demographical inquiry can be done by accents. This type of application will gather demographic statistical details such as population of that region, educational status, etc. through their voice [3]. Automated accent recognition system can help the forensic speech scientists in speaker profiling tasks. Intelligent investigation system can monitor suspicious criminal activities and facilitate surveillance inquiry of suspects who conceal their identification details deliberately through accent detection. Criminals conduct various forms of offenses via phone calls and voice messages. They can be detected by their accent and brought under trial by this system [3, 4]. In tourists and travel centers or automated international telephone call systems, non-native accents may also be identified. If the listener can recognize the dialect, he will be able to locate a perfect assistant who understands the individual client's first language [5].

2 Related Work

A lot of research has been already done to recognize the accent from different languages. So, identifying an accent from a language is not new research. In various languages, there are a lot of accents and all the researchers tried to classify those accents in a different way. In a research paper, [6] for classifying two regional accents like the northern region versus southern regions, MSA speech rhythm matrices were used. From the speech dataset named ALGerianArabic Speech Database (ALGASD), seven rhythm metric vectors were computed by using both interval measures and control-compensation index algorithms. The machine was trained and tested using different input vectors. The best accuracy of the NN-classifier was accomplished when a mix of all measurements was utilized (88.6%). The mood tests were haphazardly partitioned into preparing (70%), approval (15%) and test (15%) sets. The best acknowledgment rate (84.7%) is accomplished when the info vector incorporated the standardized interim measure—VarcoC—that was registered from terms of consonants. 88.6% accuracy was computed when the entire input vectors were given. Another research paper [1], reflects, language highlight recognition analyses of various sorts of Bangladesh. It shows a methodology to watch Bangladeshi various accents that misuse Mel recurrence cepstral coefficient (MFCC) and repetitive neural system (RNN). The voice of Dhaka (old), Chittagong, Sylhet, Rajshahi, Khulna, Barisal, Mymensingh, Rangpur and Noakhali is assembled. Shrewd voice recorder transformed into and used to report voice tests at 16 kHz test rate and 16-piece quantization. Best accuracy was achieved by 83% for Barisal division. A paper from Nigeria [7], sound information as discourse was gathered from 100 speakers from every one of the three significant Nigerian indigenous dialects, in particular: Hausa, Igbo and Yoruba. Boldness programming was introduced in the chronicle gadget. The proposed 1D CNN LSTM arrange model with the planned calculation had the option to perform an order of speakers into Hausa, Igbo and Yoruba giving a normal exactness of 97.7%, 90.6% and 96.4%, separately. Another methodology [8] for speaker and highlight acknowledgment dependent on wavelets, specifically Discrete Wavelet Packet (DWPT), Dual-Tree Complex Wavelet Packet Transform (DT-CWPT) and Wavelet Packet Transform (WPT)-based non-direct highlights is researched. The outcomes are contrasted and regular MFCC and LPC highlights. K-nearest neighbors (k-NN), support vector machine (SVM) and extreme learning machine (ELM) classifier are utilized to measure the speaker and emphasize acknowledgment rate. The most elevated correct nesses for speaker acknowledgment utilizing English digits were 92.16% and for Malay words was 93.54% accomplished utilizing ELM classifier. A paper [9] introduced a top-to-bottom examination on the order of territorial accents in Mandarin discourse. Examinations are completed on Mandarin discourse information methodically gathered from 15 diverse topographical locales in China for wide inclusion. This paper assessed different kinds of classifiers for the Mandarin highlight distinguishing proof undertaking and proposed the utilization of a bLSTM-emphasized classifier to switch naturally among standard and complemented Mandarin ASR models. The discourse database utilized right now 135 k articulations (84.7 h) from 466 speakers. With I-vector speaker adjustment, this paper got relative CERRs of 13.2%, 15.3% and 14.6% for the gatherings A1, A2 and A3, respectively. Another research [10] contrasts in the complement are because of both prosodic and enunciation qualities, a mix of long haul and momentary preparation is proposed right now. Every discourse test is handled into different discourse sections with equivalent length. The overall accuracy is 51.92%, and the UAR is 52.24%.

3 Methodology

Classification of age or gender from voice is quite common, and machine can easily identify the difference between different age and gender voice. The challenge was detecting different accent of same language by ignoring gender and age. And for that, we made a plan of work which is given in Fig. 1.

3.1 Dataset

The most important material of any research work is dataset, and we work with our dataset which was collected from different people manually, some YouTube video and Google form. The age of speakers is from 20 to 50 years old, and we collected both male and female voice. The duration of voice is mostly 4–7 s. We have exactly 9303 voice of different-region people. The amount of data on comparing different region is given in Fig. 2. And all the audio file is normalized in 16,000 kHz. The number of speakers and number of male female speakers are given in Table 1.

3.2 Feature Extraction

After collecting the raw audio file, we need to represent them in numeric or machine understandable format so that it becomes possible to work with them. For that we

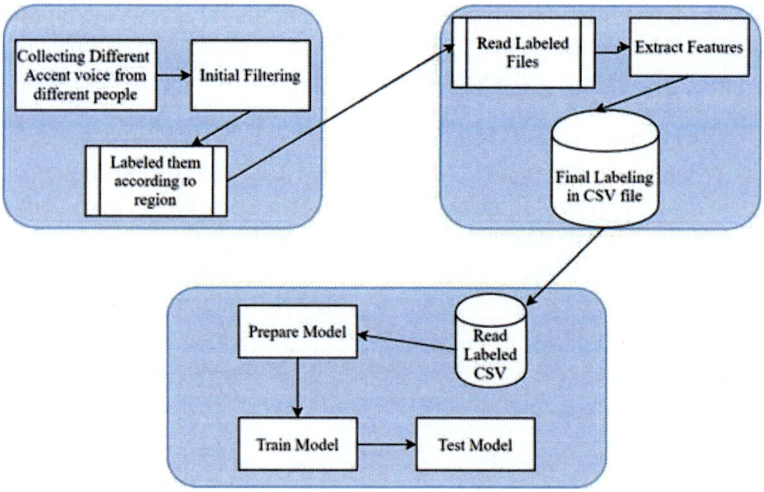

Fig. 1 Workflow

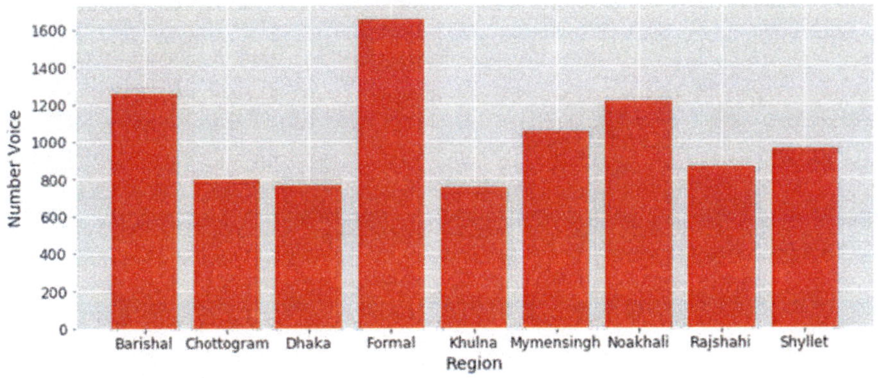

Fig. 2 Region-wise data amount

Table 1 Dataset information

Region	Represented as	Number of speaker	Male speaker	Female speaker	Total data
Barisal	0	67	29	38	1257
Chittagong	1	39	25	14	797
Dhaka	2	26	10	16	763
Formal	3	89	49	40	1652
Khulna	4	52	32	20	750
Mymensingh	5	42	26	16	1053
Noakhali	6	55	20	35	1213
Rajshahi	7	21	13	8	860
Sylhet	8	27	14	13	958

need to extract the features, and the features were chroma feature, root mean square error, spectral centroid, spectral bandwidth, roll off, zero crossing rate and 20 features from MFCCs [11].

3.2.1 Zero Crossing Rate

This represents the changes of sign by time. As different region has its own way of speaking patterns, this feature can be helpful to detect those difference in between them [12]. A variety of zero crossing rate depending on different region is given in Fig. 3.

As the accent is not depending too much on how border or thicker the voices are, that's why difference between them is very minimal.

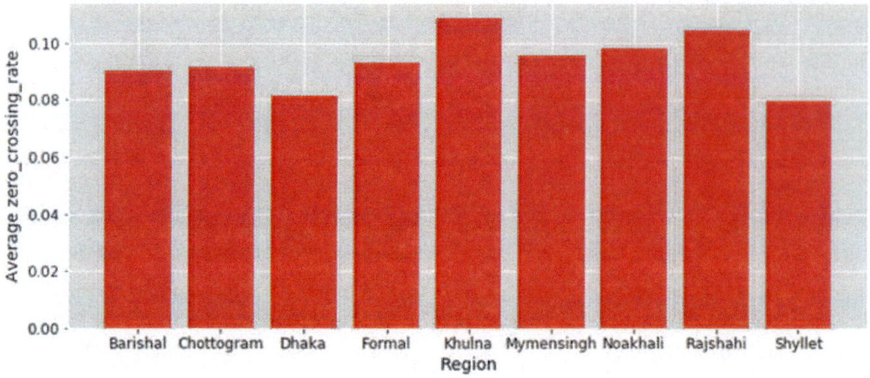

Fig. 3 Variety of zero crossing rate

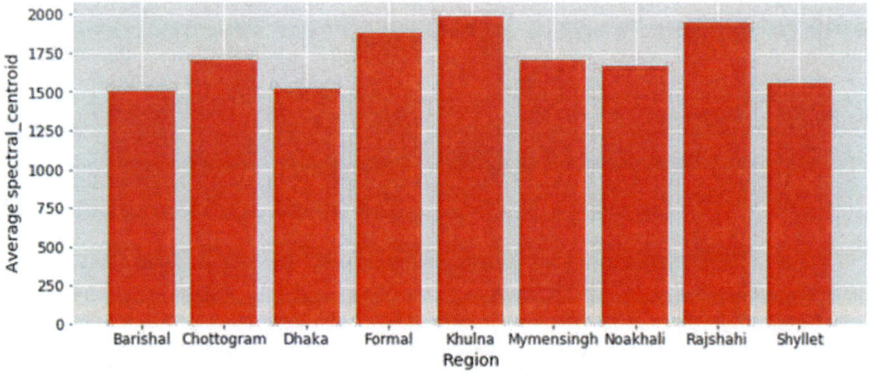

Fig. 4 Variety of spectral centroid

3.2.2 Spectral Centroid

It finds out the center of a voice. Specifically, it calculates the loudness of an audio file by time. With this feature, we can measure loudness of a voice by nature. Khulna-accented people have more loudness than others, and it will be more clearer with the presentation of Fig. 4.

3.2.3 Chroma Feature

This actually focuses on the voice part of the audio signal [13, 14]. Nature vocal folds of all regions are quite similar so it is tough to find out the different between them. But it makes difference when speaker uses their vocal folds differently and exactly its happened when people speak in different accents under same language. The differences are visualized in Fig. 5.

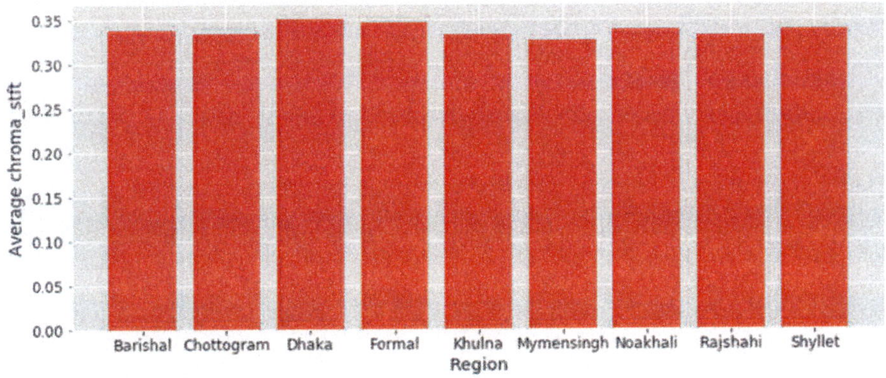

Fig. 5 Variety of chroma feature

Fig. 6 Boxplot representation of spectral bandwidth

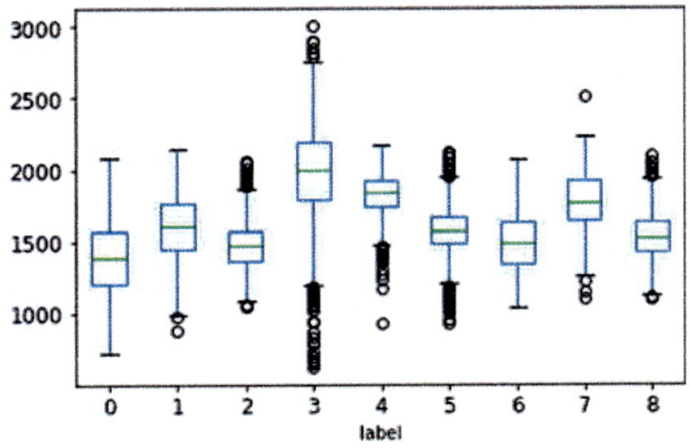

3.2.4 Spectral Bandwidth

This finds out the difference between lower and higher points in a audio signal. In Fig. 6, the boxplot representation of our data will clarify the variety of different region voices.

3.2.5 MFCCs Features

Mel-frequency spectrum coefficient is the most successful and used technique of feature extraction. It is actually a packet of 20 numeric values [15]. For finding out the values, we need to follow some steps:

- Frame the signal.
- Calculate periodogram estimate.
- Apply the Mel filter-bank.
- Keep DCT coefficients 2–13. Discard the rest.
- Take the DCT of all filter-bank energies.
- Take the logarithm of all filter-bank energies.

And we need to convert our audio into Mel scales; below equations will help us for that.

$$M(f) = 1125 \times \ln\left(1 + \frac{f}{700}\right) \tag{1}$$

$$M^{-1}(m) = 700 \times \left(\exp\left(\frac{m}{1125} - 1\right)\right) \tag{2}$$

Equation 1 is the formula of converting frequency to Mel scale, and Eq. 2 is the formula of converting Mel scale to the frequency. And Eq. 3 is for converting into discrete Fourier transformation which was mentioned in above list for extracting MFCCs.

$$Si(k) = \sum_{N}^{n=1} S_i(n)h(n)e^{-i2\Pi kn/N} \Longrightarrow here, 1 \leq k \leq K \tag{3}$$

In Eq. 3, i is the number of frame $Si(k)$ is the power spectrogram of i frames and $Si(n)$ denotes time domain.

4 Experiments and Result

4.1 Experimental Setup

We tested our model with 20% of our total data which is 1861 and trained with 7442 audio signals which are 80% of our total data. And we tried random forest, gradient boosting, K-nearest neighbor classifier, logistic regression and neural network among all of them random forest, gradient boosting and neural network give us better accuracy. In below 7, 8 and 9, no figure will give a better understanding of the algorithms performance by its confusion matrixes (Figs. 7, 8 and 9).

4.2 Result

From all the algorithms, random forest gives us the best accuracy. Table 2 will make it clear with its comparison.

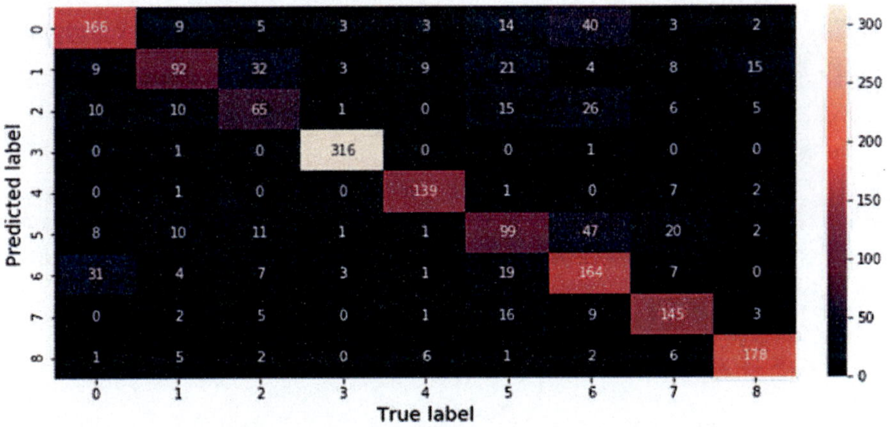

Fig. 7 Confusion matrix of neural network

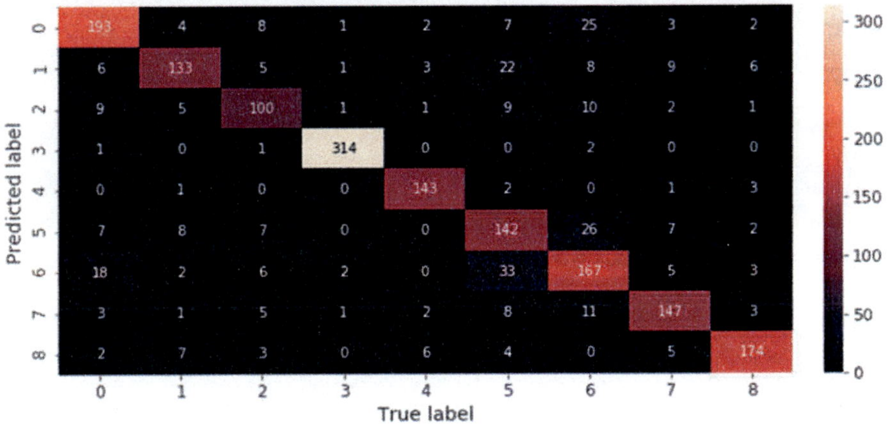

Fig. 8 Confusion matrix of gradient boosting

5 Conclusion

This work tried to detect the accent of Bengali language from Bangladeshi people voice. Bangla is one of the most popular languages, and in Bangladesh and some part of India, Bangla is the native language. When we have native speakers, they have their own way of speaking the native language. Language is always changeable, and Bangla has a huge variety depending on region. The previous works of this sector in Bangla have less data and less accuracy where we made some improvement with the accuracy 86% on 9303 data. This research works with eight different region and the standard Bengali accent. In future, our target is to enrich our dataset with more region and variety of speakers.

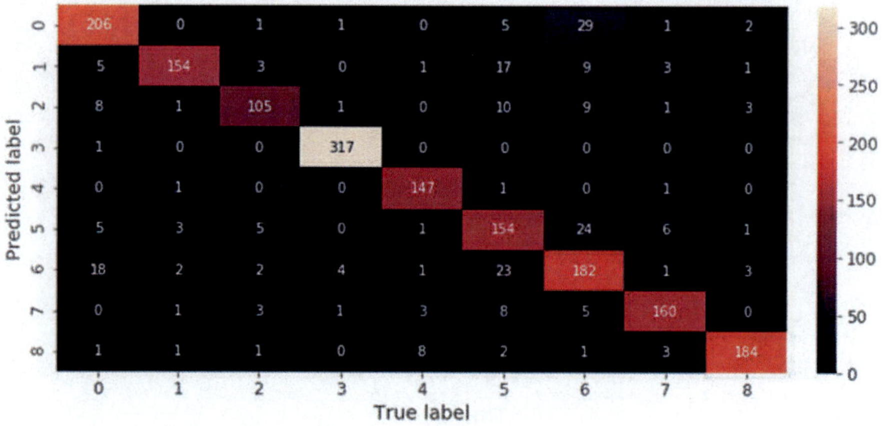

Fig. 9 Confusion matrix of random forest

Table 2 Algorithm performance

Algorithm	Precision	Recall	F1 score	Accuracy
SVM	0.37	0.41	0.38	0.45
Logistic regression	0.47	0.48	0.47	0.52
KNN	0.57	0.57	0.56	0.58
Random forest	0.86	0.85	0.85	0.86
Gradient boosting	0.80	0.80	0.81	0.81
Neural network	0.81	0.81	0.81	0.81

References

1. Mamun, R.K., Abujar, S., Islam, R., Badruzzaman, K.B.M., Hasan, M.: Bangla speaker accent variation detection by MFCC using recurrent neural network algorithm: a distinct approach. In: Saini, H., Sayal, R., Buyya, R., Aliseri, G. (eds.), Innovations in computer science and engineering. Lecture notes in networks and systems, vol. 103 (2020). Springer, Singapore
2. Bengali language. https://en.wikipedia.org/wiki/Bengali_language. Accessed on 4 Apr 2020
3. Lin, F., Wu, Y., Zhuang, Y., Long, X., Xu, W.: Human Gender Classification: A Review (2015)
4. Jiao, Y., Tu, M., Berisha, V., Liss, J.: Accent identification by combining deep neural networks and recurrent neural networks trained on long and short term features. Proc. Interspeech **2016**, 2388–2392 (2016)
5. Patel, I., Kulkarni, R., Yarravarapu, S.R.: Automatic non-native dialect and accent voice detection of south Indian English. Adv. Image Video Process **5**. https://doi.org/10.14738/aivp.51.2749
6. Droua-Hamdani G: Classification of regional accent using speechrhythm metrics. In: Salah, A. A., et al. (eds.), SPECOM 2019, LNAI 11658, pp. 75–81 (2019)
7. Salau, A.O., Olowoyoand, T.D., Akinola, S.O.: Accent Classification of the Three Major Nigerian Indigenous Languages Using 1DCNN LSTM Network Model, (2020). Springer Nature, Singapore Pte Ltd

8. Abdullah, R., Muthusamy, H., Vijean, V., Abdullah, Z., Kassim, F.N.C.: Real and complex wavelet transform approaches for malaysian speaker and accent recognition. Pertanika J. Sci. Technol. **27**(2), 737–752 (2019)
9.] Weninger, F., Sun, Y., Park, Y., Willett, D., Zhan, P.: Deep Learning based Mandarin Accent Identification for Accent Robust ASR (2019) ISCA
10. Jiao, Y., Tu, M., Berisha, V., Liss, J.: Accent identification by combining deep neural networks and recurrent neural networks trained on long and short term features (2019) ISCA
11. Music Feature Extraction in Python (2018). https://towardsdatascience.com/extract-features-of-music-75a3f9bc265d. Accessed on 4 Apr 2020
12. Gouyon, F., Pachet, F., Delerue, O., et al.: On the use of zero-crossing rate for an application of classification of percussive sounds. In: Proceedings of the COST G-6 Conference on Digital Audio Effects (DAFX-00) (2000). Verona, Italy
13. Kattel, M., Nepal, A., Shah, A., Shrestha, D.: Chroma Feature Extraction (2019). https://www.researchgate.net/publication
14. Reith, H.: Why are male and female voices distinctive? (2016) 330796993 Chroma Feature Extraction. https://www.quora.com/Why-are-male-and-femalevoices-distinctive. Accessed on 21 Sept 2019
15. The mel frequency scale and coefficients (2013). http://kom.aau.dk/group/04gr742/pdf/MFCCworksheet.pdf. Accessed on 27 Aug 2019

MathNET: Using CNN Bangla Handwritten Digit, Mathematical Symbols, and Trigonometric Function Recognition

Shifat Nayme Shuvo, Fuad Hasan, Mohi Uddin Ahmed, Syed Akhter Hossain, and Sheikh Abujar

Abstract Scientific methods are mostly based on mathematical solutions. Optical character recognition (OCR) is one of the most required solution for digitalizing and/or processing the handwritten documents into any digital form. Therefore, the movement toward the use of digital documents in the scientific community has significantly increased. Scientific documents and more precise mathematical documents are sometimes initially drafted in handwritten form. In this scope, to let understand the machine about those mathematical equations, so in this paper, a research is done on scripted mathematical symbol recognition model, which was built of 32,400 images dataset. So many research in English handwritten OCR has already achieved very good accuracy, but for Bengali language handwritten OCR, there are a minimum number of research that is done till now. This dataset contains samples of handwritten mathematical symbols and especially Bengali numerical digits. In detail, this dataset captures—Bangla numerical digits, operators of algebra, set-symbols, limit, calculus, symbols of comparison, delimiters, etc. Thus, MathNET is presented as a model that helps to recognize 10 handwritten Bangla digits and 44 other handwritten mathematical symbols. The planned model training accuracy is 96.01% and validation accuracy with test data is 96.50%, which is very good precision for identification of mathematical symbols.

S. N. Shuvo · F. Hasan · S. A. Hossain · S. Abujar (✉)
Department of Computer Science and Engineering, Daffodil International University, Dhaka, Bangladesh
e-mail: sheikh.cse@diu.edu.bd

S. N. Shuvo
e-mail: shifat15-9836@diu.edu.bd

F. Hasan
e-mail: fuad15-9901@diu.edu.bd

S. A. Hossain
e-mail: aktarhossain@daffodilvarsity.edu.bd

M. U. Ahmed
Department of Computer Science and Engineering, Jahangirnagar University, Dhaka, Bangladesh
e-mail: akankha.ahmed@gmail.com

S. Borah et al. (eds.), *Soft Computing Techniques and Applications*, Advances in Intelligent Systems and Computing 1248,
https://doi.org/10.1007/978-981-15-7394-1_47

Keywords Bangla handwritten digits · Data processing · Pattern recognition · CNN · Mathematical symbols · Trigonometrical function · OCR

1 Introduction

The consequentiality of math is in all areas of science like physics, electronics, pharmacy, banking, engineering, medicine, etc. The apperception of handwritten mathematical font or construction and symbols is highly paramount area of scientific experiment. Therefore, MathNET is extremely effective to introduce a recognition scheme for handwritten mathematical archives in Bangla and accelerate the work in this valuable field. Given this Bangla handwritten recognition of mathematical symbols, dataset does not have vigorous model that avails to construct powerful Bangla OCR, since there is still no model that can recognize both character (numeral, symbols). Several studies have been performed, but there is not any single work for mathematical symbols. The past works are only passionate on digits [1] or primary recognition [2]. MathNET is the first approach which can recognize Bangla numerals and mathematical symbols. Convolutional neural network (CNN) is used as numeral's recognizer and symbol identification scheme for gaining the best possible outcome.

2 Literature Review

The identification in Bangla digits and symbols, former study concentrated primarily on Bangla numerals. This territory includes only ten numerals. Few works are accessible for handwritten identification of Bangla numerals which are fixated on only ten digits as well as handwritten characters. In recent years, some literature on Bangla character recognition has been recorded as "A full printed Bangla OCR system" [3], "Bangla Continuous Character and Digit Recognition" [4]. Some of the works investigating the recognition of handwritten digits are "Identification of ungrudging handwritten Bangla digits [5]. A framework for postal automation [6] and handwritten mathematical symbols dataset [7] in this work only focuses on the Arabic, Latin characters and Arabic, Latin numerals. There is no complete work for Bangla numerals and symbols together. Out of this viewpoint, this paper is predominantly proposed to identify only handwritten Bangla numbers and symbols.

3 Proposed Methodology

This latest CNN model that is proposed here called "MathNET" has several phases as illustrated beneath.

Fig. 1 Dataset example

3.1 Datasets

For this mentioned CNN model, 6000 numerals images (0–9) data from "Ekush" [8] and other 44 class mathematical symbols have total 26,400 images were collected. These 44 handwritten symbols were collected from 500 students both male and female. The clarity of the images builds on the size of the character. Every image has less black background padding, and the character is white. The images in this dataset is distortion-free 28 × 28 px, and the edge of the image looks obstructed. Then, concatenate two datasets to get the final dataset of total images 32,400. Example of datasets is shown in Fig. 1.

3.2 Preparation of Dataset

In deep learning, variety of data inside dataset is very important. In this suggested MathNET model, it has used Ekush dataset and the newly collected dataset for mathematical symbols. To reducing the computational cost, it has converted all the images into BINARY_INV. Then, resize the dataset 28 × 28 px, remove the unnecessary black pixel, and convert the whole dataset in csv format for high-speed calculation process. The whole dataset has 785 columns in every row, where 28 × 28 = 784 columns contain the pixel or dot value which represent the image and 785 number columns store the label or class for the digits and symbols. In the time of training also, apply a Minmax (1) normalization to minimize the outcome of light friction. CNN links up faster on the range of value (0–1) rather than (0–255). Transform the 54 labels to one hot encoding.

$$Z_i = \frac{X_i - \text{minmum}(x)}{\text{maximum}(x) - \text{minmum}(x)} \tag{1}$$

3.3 MathNET Architect

MathNET used CNN which is a traditional variant of multilayer perceptron's to recognize Bangla handwritten digits and symbols. This model has Maxpool layer, completely attached dense layer, and used dropout [9] for regularization method (Table 1).

Table 1 MathNET model architect analysis sequentially

Layer (Type)	Output shape	Param#	Layer (Type)	Output shape	Param#
Conv2D-1	(28, 28, 32)	832	MaxPool2d-2	(7, 7, 64)	0
Conv2D-2	(28, 28, 32)	25,632	Dropout-2	(7, 7, 64)	0
MaxPool2d-1	(14, 14, 32)	0	Flatten-1	(3136)	0
Dropout-1	(14, 14, 32)	0	Dense-1	(256)	8,03,072
Conv2D-3	(14, 14, 64)	18,496	Dropout-3	(256)	0
Conv2D-4	(14, 14, 64)	36,928	Dense-2	(54)	13878

The first two convolutional layer has filter size of 32 and kernel_size (5,5) and use activation function ReLU with padding = "same." These two layers are input of next max_pooling2d_1 layer which is connected to 25% dropout_1 layer. The output of dropout_1 goes into layer conv2d_3 and conv2d_4 as an input. These two layers also have same padding and filter size 64 and kernel size (3,3). Max_pooling2d_2 layer takes input from conv2d_3 and conv2d_4 and gives the output as an input to 25% dropout_2 layer. After performing these eight operations, the output goes through flatten_1 layer and attached to a dense_1 layer with 256 backstage units. These units are regularized with 25% dropout_3. Then, the final layer dense_2 is a completely linked layer and also called the output layer for the proposed model. This layer has 54 nodes and use SoftMax [17] activation function in Eq. (2). Figure 2 shows the MathNET architect.

$$\sigma(z)_j = \frac{e^{z_j}}{\sum_{k=1}^{k} e^{z_k}} \quad \text{for } j = 1, \ldots k \tag{2}$$

3.4 Optimizer and Learning Rate

To minimize the glitch of the CNN result, optimizer is used. Optimizers form a central part of the neural network. In this MathNET, model refers to used RMSprop [10, 11] optimizer and sets learning rate value to 0.001. The RMSprop optimizer is equivalent to the momentum gradient descent algorithm. The RMSprop optimizer limits vertical direction of the oscillations. Moreover, this can accelerate the learning rate, and our algorithm will take bigger steps in a more rapidly converging horizontal direction. The distinction between RMSprop (3) (4) and gradient descent is how the gradients are measured. It is commonly used for studying computer vision.

$$W = W - \alpha . \frac{\mathrm{d}w}{\sqrt{v_{\mathrm{d}w}} + \varepsilon} \tag{3}$$

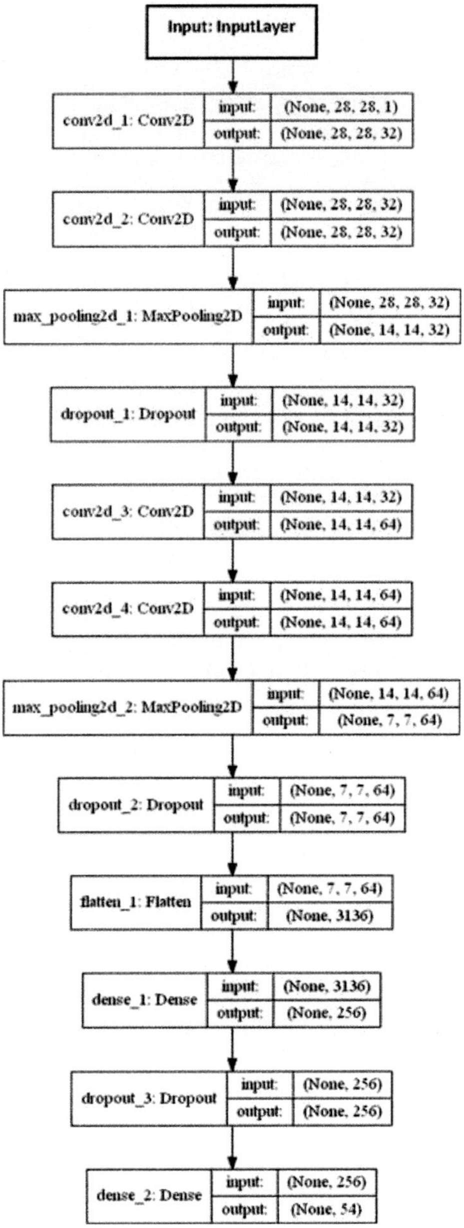

Fig. 2 Architecture of MathNet

$$b = b - \alpha . \frac{\mathrm{db}}{\sqrt{v_{\mathrm{db}}} + \varepsilon} \tag{4}$$

To measure the error, here categorical cross entropy (5) method is used to optimize the algorithm. Recent work illustrates that cross entropy plays better result than any other method, such as mean squared error [12].

$$L_i = -\sum_j t_{i,j} \log(p_{i,j}) \tag{5}$$

Learning rate is most crucial hyper parameter for tuning a convolutional neural network while training. In this proposed model, using self-acting learning rate reduction method [13] we get incredibly good result. Setting the min_lr value to 0.00001 helps to measure validation accuracy and reduces it automatically.

3.5 Data Augmentation

CNN model works better when it finds a lot of data during training time. Here comes the data augmentation method. It helps to generate artificial data, to avoid the overfitting of model. By choosing several augmentation methods, these are: zoom_range set to 0.1, haphazardly shift images horizontally 0.1, haphazardly shift images vertically 0.1.

3.6 Training the Model

The MathNET model is trained with our own collected dataset and some images of Ekush [8] dataset explained in Sect. 3.1. By training the model with batch size 86 and after executing 50 , the model obtains 96.01% on training accuracy as well as 96.50% on validation accuracy. The formula was used for reducing learning rate performed well and the final learning rate obtain to 1e−05.

4 Model Performance

For training and validating the MathNET model, use the collected fresh datasets. Therefore, for testing purpose, separate some of test data from collected data. Finally, it gives incredibly good result in this model.

Fig. 3 Model accuracy

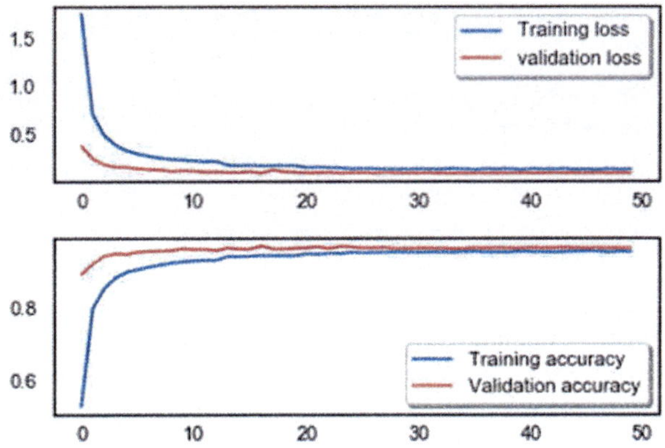

4.1 Train, Test, and Validation Split

To measure the functioning of the model here, split the dataset in three categories one for the training purpose, one for testing, and another one is validation test for the scheme. The training set is work like a trainer for the model with label data. During the training time, validation set check the model functioning good or not. When the model is successfully created, by testing it with the test dataset, get favorable outcome. Testing method runes by the model with collected 10% data from all classes. About 3200 test data performed well in this model and recognize them.

4.2 Model Accuracy

After successfully executing 50 epochs, MathNET has obtained 96.01% on training accuracy and 96.5% on validation accuracy. Analyzing the model with test datasets, from error matrix here found that the model works well and predicts accurately. Figure 3 shows the train-validation loss and train-validation accuracy, respectively.

4.3 Result Compression

There were a few researches which can recognize both Bangla numerals and mathematical symbols. Table 2 shows some analogy between some previous works.

Table 2 Analogy among some past effort

Work	Performance	Recognition object
"EkushNet: Using CNN for Bangla handwritten recognition" [14]	97.73%	Bangla numerals and characters
"Using SVM method handwritten primary character recognition" [15]	93.43%	Only character
"Using convex hull formula Bangla handwritten characters and numerals recognition" [16]	76.86%	Digits and characters
MathNET (proposed)	96.01%(train) 96.50% (validation)	Bangla numerals and mathematical symbols

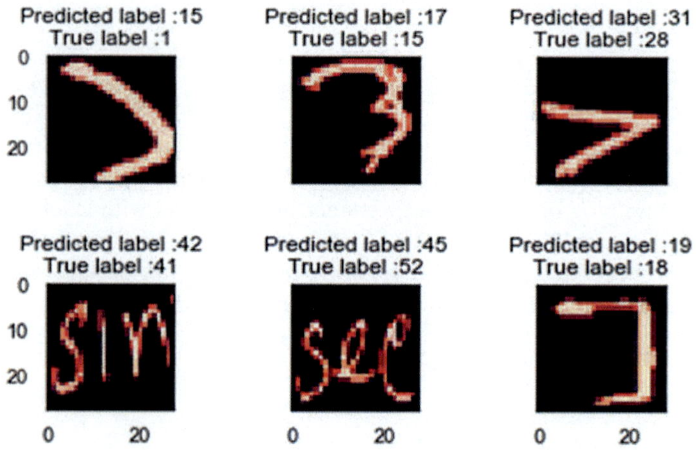

Fig. 4 Some error of our recognition model

5 Error Remark

Finding the delusion from given test set can declare that MathNET has been successfully recognize 97% of the images from test data. On Fig. 4, top six errors have been shown. This is happened because of the wrong labeled data in the test set. And some of the error also confuse us, and this can also be made by human.

6 Conclusion and Future Work

Notifying that, this CNN model has been successfully recognizing all of the test data having good performance. This can build a full OCR system which can recognize full Bangla mathematical equations, and inclusion of more data in this dataset will increase the use in massive level. Though this dataset perfectly worked in some models, it was tested.

References

1. Alom, M.Z., Sidike, P., Taha, M., Tarek, Asari, Vijayan.: Handwritten Bangla digit recognition using deep learning (2017)
2. Rahman, M.M., Akhand, M.A.H., Islam, S., Shill, P.C.: Bangla handwritten character recognition using convolutional neural network. I. J. Image Graph. Signal Proc. **8**, 42–49 (2015)
3. Chaudhuri, B.B. Pal, U.: A complete printed Bangla OCR system. Pattern Recogn. **31**, 531–549 (1998); Sindhushree, G.S., Amarnath, R., Nagabhushan, P.: Entropy-based approach for enabling text line segmentation in handwritten documents (2019)
4. Hasan, F., Shuvo, S.N., et al.: Bangla continuous handwriting character and digit recognition using CNN. In: 7th International Conference on Innovations in Computer Science and Engineering (ICICSE 2019), vol. 103, springer, Singapore, pp. 555–563 (2019)
5. Pal, U. Chaudhuri, B.B.: Automatic recognition of unconstrained offline Bangla hand-written numerals. In: Tan, T., Shi, Y., Gao, W. (eds.), Advances in Multimodal Interfaces, Lecture Notes in Computer Science, vol. 1948, pp. 371–378. Springer, Berlin (2000)
6. Roy, K., Vajda, S., Pal, U., Chaudhuri, B.B.: A system towards Indian postal automation. In: Proceedings of the Ninth International Workshop on Frontiers in Handwritten Recognition (IWFHR-9), pp. 580–585 (October, 2004)
7. Chajri, Y., Bouikhalene, B.: Handwritten mathematical symbols dataset
8. Ekush: A multipurpose and multitype comprehensive database for online off-line Bangla handwritten characters. https://github.com/shahariarrabby/Ekush. Accessed 20 Jun 2018
9. Srivastava, N., Hinton, G., Krizhevsky, A., Sutskever, I., Salakhutdinov, R.: Dropout: a simple way to prevent neural networks from overfitting. J. Mach. Learn. Res. **15**, 1929–1958 (2014)
10. A direct adaptive method for faster backpropagation learning: the rprop algorithm. In: Neural Networks, Riedmiller, Braun (1993)
11. Riedmiller, M., Braun, H.: A direct adaptive method for faster backpropagation learning: the RPROP algorithm. In: IEE International Conference on Neural Networks (1993)
12. Janocha, K., Czarnecki, W.M.: On loss functions for deep neural networks in classification. arxiv abs/1702.05659 (2017)
13. Schaul, T., Zhang, S., LeCun, Y.: No more pesky learning rates. arXiv preprint.\:1206.1106,2012
14. Shahariar Azad Rabby, A., Haque, S., et al.: EkushNet: using convolutional neural network for bangla handwritten recognition. Procedia Comput. Sci. **143**, 603–610 (2016). ISSN 1877-0509
15. Azim, Riasat, et al.: Bangla hand-written character recognition using support vector machine. Int. J. Eng. Works **3**(6), 36–46 (2016)
16. Pal, U.: On the development of an optical character recognition (OCR) system for printed Bangla script. Ph.D. Thesis (1997)
17. Liu, W., Wen, Y., Yu, Z.: Large-margin softmax loss for convolutional neural networks. In: ICML (2016)

A Continuous Word Segmentation of Bengali Noisy Speech

Md. Fahad hossain, Md. Mehedi Hasan, Hasmot Ali, and Sheikh Abujar

Abstract Human voice is an important concern of efficient and modern communication in the era of Alexa, Siri, or Google Assistance. Working with voice or speech is going to be easy by preprocessing the unwanted entities when real speech data contains a lot of noise or continuous delivery of a speech. Working with Bangla language is also a concern of enriching the scope of efficient communication over Bangla language. This paper presented a method to reduce noise from speech data collected from a random noisy place, and segmentation of word from continuous Bangla voice. By filtering the threshold of noise with fast Fourier transform (FFT) of audio frequency signal for reduction of noise and compared each chunk of audio signal with minimum dBFS value to separate silent period and non-silent period and on each silent period, segment the signal for word segmentation.

Keywords Speech preprocessing · Noise cancellation · Word segmentation · Fast Fourier transform

1 Introduction

From the previous few decades, large numbers of research ruled over computer science, artificial intelligence, and linguistic analysis. Human–computer interaction (HCI) introduced the multidisciplinary field of communication where humans can

Md. Fahad hossain · Md. Mehedi Hasan · H. Ali · S. Abujar (✉)
Department of Computer Science and Engineering, Daffodil International University, Dhaka, Bangladesh
e-mail: sheikh.cse@diu.edu.bd

Md. Fahad hossain
e-mail: fahad15-9600@diu.edu.bd

Md. Mehedi Hasan
e-mail: mehedi15-9804@diu.edu.bd

H. Ali
e-mail: hasmot15-9632@diu.edu.bd

S. Borah et al. (eds.), *Soft Computing Techniques and Applications*, Advances in Intelligent Systems and Computing 1248,
https://doi.org/10.1007/978-981-15-7394-1_48

interact with computer devices. Natural Language Processing (NLP) makes the HCI more efficient by providing a lot of applications. The input and output of an NLP system can be speech or written text. As technology is improving day by day, the way of communication is going so easy. Nowadays, technology prefers speech rather than written text which is limited and sometimes, non-existence for performing more efficient communication. However, the problem is working with speech is not as easy as written text. Speech is more sensible than text. Therefore, there are some unwanted entities of voice like noise, silence, and joint of words. Speech preprocessing is an important term of working with speech data. Several types of preprocessing have been done with speech data for more efficient automatic speech recognition (ASR), text to speech (TTS), and any other application related to speech. Kwon et al. [1] performs a preprocessing which increases speech rate, eliminating silent period, and boosting frequency band of old person voice can acquire 12% more accuracy of modern speech recognition.

Almost 200 million people worldwide, 160 million of whom are Bangladeshi speaking Bangla as their first language [2]. This research focuses on noise reduction and word segmentation of Banglavoice which can create a thousand ways for working with Bangla voice. Processing with real data containing a lot of noise and continuous speech is quite difficult for further processing. Therefore, this research has been trying to reduce noise over real data from random people and random places, and word segmentation for further ways of processing this data. When Bangla speech data is collected, it has a lot of noise and the gap between interconnected words (join word) is very low. So it is difficult to split the word from another word for further recognition. Noise is often not that motionless, so noise reduction has been accomplished by using the noise statistic estimation during silent periods. A good number of research on speech signal segmentation has been done so far. Voice processing systems sometimes require the segmentation of speech waveform into the principal word unit. Segmentation is the process of dividing the speech into specific units. This research performs an investigation of segmenting word over speech when a speaker delivers his speech continuously and showed the comparative study of manual segmentation and segmentation using algorithms.

2 Literature Review

A lot of research has been done in the field of noise reduction and word segmentation. Previously performed ASR research field is based on applied linguistic and biomechanics technology. However, there is no related work for Bangla speech preprocessing. The way of working with Bangla speech is limited only for the lack of processed data.

For elderly speech recognition, Igarashi and Hughes [3] describe the non-verbal feature of voice for control of HCI. For robust speech recognition, Hirsch et al.

[4] introduce two new algorithms combined with a nonlinear spectral subtraction scheme, discriminating Pathological voice. Umapathy et al. [5] show a time–frequency associated approach to classify emotional voices of continuous speech signals using acclimatize time–frequency transform algorithm. Maeran et al. [6] investigate the use of a half-cast soft-computing entrance for perceptual distinct unit classification and signal segmentation; Prodeus [7] performed a comparative study of six-noise reduction algorithms containing Wiener filtering, spectral subtraction, logMMSE, and MMSE, Wiener-TSNR and HRNR algorithms. Performing noise reduction, Jeannès and Faucon [8] use a coherence function to regulate a speech/noise classification algorithm based on spectral subtraction and evaluate its consequence; Brueckmann et al. [9] use adaptive noise reduction and synthetic auditory process accomplished for the mobile reciprocal action against robot combining neural voice activity detection. Linhard [10] introduces a method that is used for the elimination of noise and improvement of voice quality [11, 12] and uses a method by which an audio file matches relative speech from a database that contains a huge flock of original recordings and processing of speech from a conference session. Reduction of noise while mixing with speech to become corrupted speech from microphone array insensitive to far-field sounds response for near-field sounds is performed by Kang and Heide [13],and a two-step noise reduction (TSNR) approach was proposed by Plapous et al. [14] for observing the convenience of the decision-directed technique. Dyba et al. [15] perform an efficient noise reduction algorithm based on the Freescale StarCoreTM SC3400 core, which includes subjective evaluation, by expert listeners and Kim et al. [16] propose a way of reducing the noisy signal of intelligent devices for the home environment founded on input-SNR determination process with preprocessing on intelligent home devices. Pramanik and Raha [17] adopted a new process for noise cancellation by calculating the average power of speech signal for each speech composition calculated above certain threshold level while Cheikh-Rouhou and Tilp [18] contributes a new method based on spectral subtraction in combination with pitch-adaptive post-processing for a dual-channel augment of confused speech signals. Agarwal and Kabal [19] use linear prediction coefficient (LPC) estimatio; Xu et al. [20] try to reduce noise from speech and video from oil well environment using enhanced variable-zero attracting (VZA-LMS) algorithm using discrete cosine transform (DCT).

For word segmentation of continuous data Régine [21] design, a process of investigating signal by an AR statistical model to detect a change in the parameter of the model, [22] use an automatic segmentation that allows sharpness of a loudness potential syllabic boundary between the convex hull function and the loudness function. Godard et al. [23] perform an unsupervised word segmentation assuming a pairing between recordings in the UL with translations in a well-resourced language, while an audio word segmentation without phonetic knowledge is done together until the cepstral coefficient variance is low within each new segment by Gold and Scassellati [24]. Yurovsky et al. [25] showed that child-directed speech makes synchronous speech segmentation flexible for human learners. Where this paper proposed a method that provides a noise-free segmented Bengali word from continuous noisy speech.

3 Methodology

Two different experiments have been done over speech in this research for further investigation. Those experiments have been explained below.

A. Noise reduction

There are several algorithms for noise reduction. But in this research, use an algorithm which requires two inputs. The first input is a noisy audio clip containing representative noise of the audio clip and a signal audio clip that encompass the signal and the noise desired to be removed, required for the second input. Algorithms take some steps using these two inputs and the step of the algorithm can be divided into two stages. First of all, this algorithm starts working with a noisy audio clip. It calculates the fast Fourier transform (FFT) value of that audio clip. Based on calculated FFT, some statistics are determined in frequency. Then, based on the statistics, a threshold value is calculated. After that, second stage starts similarly to the first step, by calculating the FFT of the signal audio clip. From the comparison of this second FFT and the threshold value, it determines a mask. This is followed by sleeking the mask with frequency and time. Finally, the algorithm applies this mask on the FFT we got from the audio clip and we get a better noise-free audio signal.

Working process of noise reduction is shown in Fig. 1 by a flowchart.

B. Word segmentation

While working on a continuous voice signal, we observed that the input file of a continuous signal could be quite large to send as an input to a machine. Therefore, this continuous signal should be divided into small segments. Word segmentation can be done either manually or using algorithms. The manual word segmentation process is quite difficult. To split a continuous audio, we need both human and computer software. It is time consuming too. We can make this efficient using algorithms.

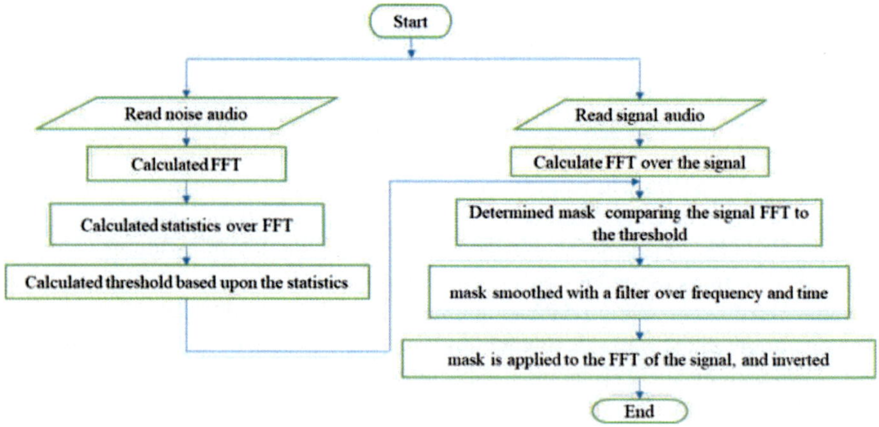

Fig. 1 Process of noise reduction

This way is a bit time consuming but way better than manual splitting. In this paper, we applied an algorithm that works based on silence periods through the whole continuous audio. This algorithm iterates over the audio file and whenever it gets a silent period, it splits the audio. This algorithm works through three parameters: source file, minimum silence period, and silence threshold. The minimum silence period is the minimum length of silence which will be used to split the audio and the silence threshold is the minimum level of dBFS of each splitted chunk. dBFS is the measure of the highest signal level that can be achieved by a digital audio file. For each splitted chunk, if dBFS is smaller than the minimum threshold, then that chunk will be considered as a silent period. Working process of noise reduction is shown in Fig. 2 by a flowchart.

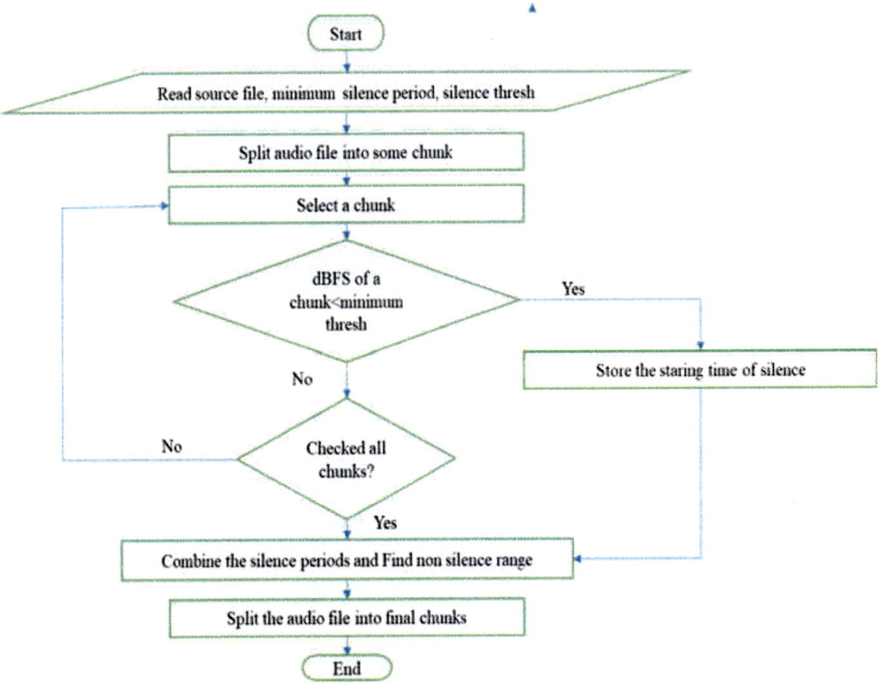

Fig. 2 Process of word segmentation

Algorithm 1: Word Segmentation

Input :	Continous audio file, *A*
Output :	Some audio chunk that contains the word portion of a continuous audio $C_1, C_2, C_3 \ldots C_n$
	Read continous audio,*A*
	Split the whole audio in some chunk of the length of the minimum silence period
	Check every chunk for silence
	FOR every chunk
	IF dBFS of a chunk is less than minimum thresh
	store the starting time of silence
	ENDIF
	IF all chunks have been checked
	BREAK
	ENDIF
	ENDFOR
	combine and store all the silence ranges
	find non-silence ranges using silence ranges
	split the main audio into final chunks using non-silence ranges
	Output some audio chunk that contains the word portion of a continuous audio $C_1, C_2, C_3 \ldots C_n$

In this algorithm, a continuous audio file, A is taken as input. Then the whole signal A is splitted into some temporary chunks. The length of each chunk is equal to the minimum silence period. The dBFS value of each chunk is compared with the minimum silence threshold to find out the silence periods. After storing silence periods, they are combined to find the final silence ranges. Later, non-silence periods are calculated from these silence periods and based on these non-silence periods, the audio is splitted into final chunks $C_1, C_2, C_3, \ldots, C_n$.

4 Result

In this paper, we have shown one sample from our dataset. That sample is a recording of a sentence consisting of seven Bangli words. It was recorded in a noisy place. Those seven words are "পার হয়ে যায় গরু, পার হয় গাড়ি" where word1 represents "পার", word2 represents "হয়ে", word3 represents "যায়", word4 represents "গরু", word5 represents পার", word6 represents "হয়" word7 represents "গাড়ি".The result of algorithms applied to this sentence, is given in the following section.

A. Result of noise reduction

As noise has a remarkable effect on the signal, we can easily point the difference after the reduction of noise from the signal which is clearly visible in Figs. 3 and 4.

The processes of discrimination are based on several statistical indicators, which are root mean square (RMS) and signal-to-noise-ratio (SNR) values of the signal.

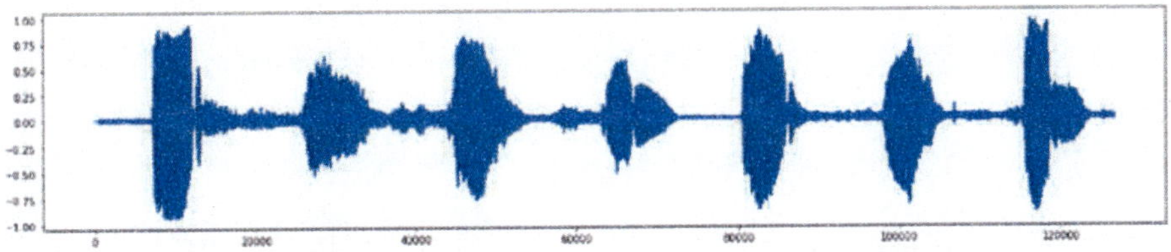

Fig. 3 Wave frequency of noisy speech data

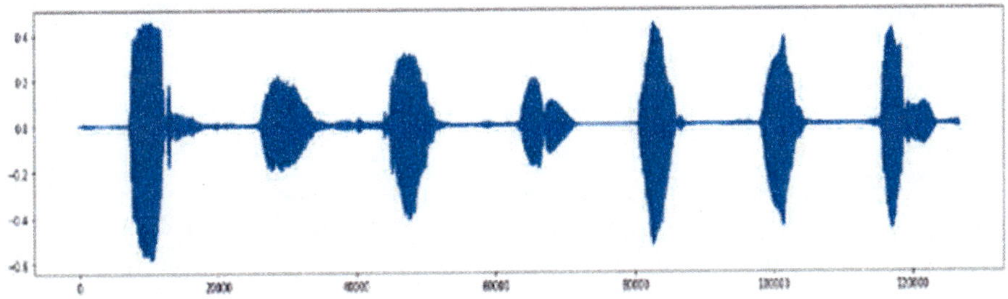

Fig. 4 Wave frequency after reduction

RMS of a signal is the measure of the loudness of the audio. In other words, it represents the strength of a signal. The decrease of RMS means the signal lost some data. SNR is a ratio of the power of the signal to the power of background noise. For digital signals, it can be calculated easily from the numerical representation of signal waves. The increase of SNR means the increase of signal or the decrease of noise. Table 1 shows SNR and RMS value. RMS values show that the signal lost some data; hence, the increase of SNR means the decrease of noise.

B. Word segmentation

In this research, the main focus is to perform word segmentation. Figure 5 shows the wave frequency of the segmented word. More specifically 5. A is the wave frequency of "word1", 5.B is for "word2" and so on.

Pitch is another representation of the frequency of a signal. The human ear understands the frequency as pitch. Table 2 shows the RMS and pitch values of each word before reduction and after reduction almost the same. Pitch value indicates that the necessary data remains the same after the reduction of noise. Noise reduction did not have any effect on the necessary data. The change of RMS values in Table 2 shows that the signal has lost some data which has already been discussed in the previous section.

Table 1 SNR and RMS value of those audio files

File Type	SNR	RMS
Full raw signal	$-2.8708324\text{e-}10^5$	331,974,833
Full reduced signal	$-2.1408927\text{e-}10^5$	162,820,252

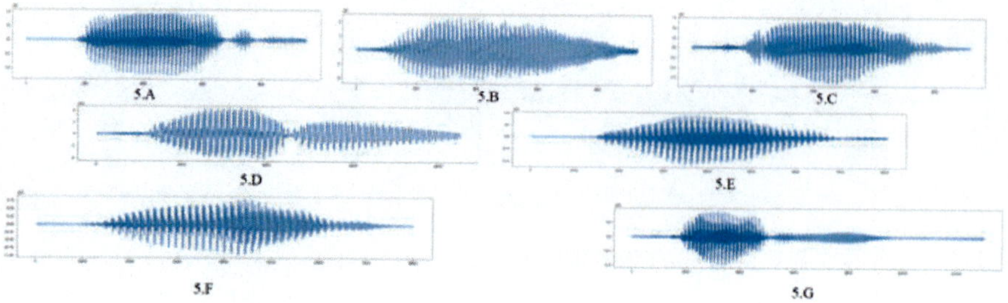

Fig. 5 Wave frequency for every word

Table 2 RMS and pitch values of those words

Word	RMS		Pitch	
	Before noise reduction	After noise reduction	Before noise reduction	After noise reduction
Word 1	630,780,974	325,536,469	159.7826	159.7826
Word 2	373,588,301	153,443,890	150.0	148.99
Word 3	545262953	247,270,660	153.125	153.125
Word 4	362,219,648	135,037,239	154.1958	154.1958
Word 5	537,235,167	280,420,248	158.6330	159.7826
Word 6	455,049,955	221,886,910	147.0	147.0
Word 7	499,832,684	170,830,446	158.6331	158.6331

5 Limitation and Future Scope

There is lots of traffic on the road. For that reason, when a user delivers a voice, the traffic sound is included in the voice, and traffic horn frequency 420–440 Hz where human voices frequency 85–180 Hz, so it would be very difficult to reduce this type of noise. If we try to reduce such noise, then they will lose the original voice. On the other hand, we must be concerned about the user delivery speed. Because if a user delivers voice at too much speed, then this algorithm does not give output exactly by doing word segmentation. According to this current algorithm, users must take a minimum quarter second break between consecutive words. As this algorithm skips a silent chunk which is shorter than 250 ms. Users should maintain a uniform flow of speaking and should speak words clearly. Otherwise, this algorithm will split a large word into segments. A combination of slow and fast speaking may force our algorithm to do an incorrect split. We should try to solve these limitations in the future.

6 Conclusion

In the view of the above investigation, the prime objective of this paper is to reduce noise and segment every word from a continuous speech audio to process Bangla voice over modern technology. This presented algorithm shows a satisfactory result over our data. In this research, experiments have done by our own data collected from a noisy place and continued speech which helps us to perform both of our concerns at single data. This research also discusses manual segmentation and segmentation using algorithms and the problem with the limited way of collecting data. Speech data containing joint words and the ratio of speech and noise are highly influenced by the experiment. Joint word is quite impossible to split into different words and if the ratio of noise is richer than the voice, the above experiment would not work completely.

References

1. Soonil, K., Kim, S.-J., Choeh, J.Y.: Preprocessing for elderly speech recognition of smart devices. Comput. Speech Lang. (2015). 36.10.1016/j.csl.2015.09.002
2. Banglapedia, Bangla Language.: (2016 August 30). Available http://en.banglapedia.org/index.php?title=Bangla_Language
3. Igarashi, T., Hughes, J.: Voice as sound: using non-verbal voice input for interactive control. (2002). https://doi.org/10.1145/502348.502372
4. Hirsch, H.-G., Ehrlicher, C.: Noise estimation techniques for robust speech recognition. In: ICASSP IEEE International Conference on Acoustics, Speech and Signal Processing—Proceedings. vol. 1, pp. 153–156. https://doi.org/10.1109/icassp.1995.479387
5. Umapathy, K., Krishnan, S., Parsa, V., Jamieson, D.: Discrimination of pathological voices using a time-frequency approach. IEEE Trans. Bio-Med. Eng. **52**, 421–430 (2005). https://doi.org/10.1109/TBME.2004.842962
6. Maeran, O., Piuri, V., Gajani, Giancarlo.: Speech recognition through phoneme segmentation and neural classification. **2**, 1215–1220 (1997). https://doi.org/10.1109/imtc.1997.612392
7. Prodeus, A.M.: Performance measures of noise reduction algorithms in voice control channels of UAVs. In: 2015 IEEE International Conference Actual Problems of Unmanned Aerial Vehicles Developments (APUAVD), Kiev, pp. 189–192 (2015)
8. Le Bouquin Jeannès, R., Faucon, G.: Study of a voice activity detector and its influence on a noise reduction system. Speech Commun. **16**, 245–254 (1995). https://doi.org/10.1016/0167-6393(94)00056-G
9. Brueckmann, R., Scheidig, A., Gross, H.-M.: Adaptive noise reduction and voice activity detection for improved verbal human-robot interaction using binaural data. pp. 1782–1787 (2007). https://doi.org/10.1109/robot.2007.363580
10. Linhard, K.: Noise-reduction method for noise-affected voice channels. J. Acoust. Soc. Am. **98** (1995). https://doi.org/10.1121/1.413359
11. Wang, A., Smith, J.: Systems and methods for recognizing sound and music signals in high noise and distortion (2014)
12. Nierhaus, F.P., Vandermersch, P.: Method for background noise reduction and performance improvement in voice conferencing over packetized networks. US 2003/0063572A1, Apr 3, 2003 (2008)
13. Kang, G.S., Heide, D.A.: Acoustic noise reduction for speech communication: (second-order gradient microphone). In: 1999 IEEE International Symposium on Circuits and Systems (ISCAS), Orlando, FL, vol. 4, pp. 556–559 (1999)

14. Plapous, C., Marro, C., Mauuary, L., Scalart, P.: A two-step noise reduction technique. In: 2004 IEEE International Conference on Acoustics, Speech, and Signal Processing, Montreal, Que, pp. I-289 (2004)
15. Dyba, R.A., Su, W.W., Deng, H.: Fixed-point implementation of noise reduction using starcore—SC3400. In: 2009 IEEE 13th Digital Signal Processing Workshop and 5th IEEE Signal Processing Education Workshop, Marco Island, FL, pp. 194–199 (2009)
16. Kim, J., Han, S., Lee, D., Hahn, M.: Adaptive noise reduction algorithm on smart devices in pervasive home environment for voice communication service. In: 2008 5th IEEE Consumer Communications and Networking Conference, Las Vegas, NV, pp. 1210–1211 (2008)
17. Pramanik, A., Raha, R.: De-noising/noise cancellation mechanism for sampled speech/voice signal. In: 2012 Ninth International Conference on Wireless and Optical Communications Networks (WOCN), Indore, pp. 1–4 (2012)
18. Cheikh-Rouhou, F., Tilp, J.: Two-channel noise reduction with pitch-adaptive post-processing. **2** (2000). https://doi.org/10.1109/acssc.2000.911286
19. Agarwal, T., Kabal, P.: Preprocessing of noisy speech for voice coders. pp. 169–171 (2002). https://doi.org/10.1109/scw.2002.1215761
20. Xu, Z., Bai, F., Bang, G.: Improved variable ZA-LMS algorithm based on discrete cosine transform for high noise reduction. pp. 5116–5121 (2016). https://doi.org/10.1109/chicc.2016.7554149
21. André-Obrecht, Régine: A new statistical approach for the automatic segmentation of continous speech signals. Acou. Speech Signal Proc. IEEE Trans. **36**, 29–40 (1988). https://doi.org/10.1109/29.1486
22. Mermelstein, Paul: Automatic segmentation of speech into syllabic units. J. Acou. Soc. Am. **58**, 880–883 (1975). https://doi.org/10.1121/1.380738
23. Godard, P., Zanon Boito, M., Ondel, L., Berard, A., Yvon, F., Villavicencio, A., Besacier, L.: Unsupervised word segmentation from speech with attention. 2678–2682 (2018). https://doi.org/10.21437/interspeech.2018-1308
24. Gold, K., Scassellati, B.: Audio speech segmentation without language-specific knowledge (2006)
25. Yurovsky, Daniel, Yu, Chen, Smith, Linda: Statistical speech segmentation and word learning in parallel: scaffolding from child-directed speech. Front. Psychol. **3**, 374 (2012). https://doi.org/10.3389/fpsyg.2012.0037

Heterogeneous IHRRN Scheduler Based on Closeness Centrality

V. Seethalakshmi, V. Govindasamy, and V. Akila

Abstract Big Data is the current emerging trend in data analytics. MapReduce is the programming model which is widely used for the data intensive applications in the Big Data environment. Scheduling of job attempts to provide faster processing of jobs and to reduce the response time to the minimum by using innovative techniques for allocating system resources depending on the nature of jobs. The default scheduling algorithm takes up Hadoop homogeneous environment. This homogeneity consolidation does not always work in use and restricts the MapReduce efficiency. There is a need for addressing heterogeneous system. Data locality is primarily bringing computation nearer to input information so as to provide faster access. MapReduce does not always test the heterogeneity from the point of view of the data locality. Improvising the data locality for Map Reduce system is an endearing problem to increase the fruition of Hadoop environment. The existing job scheduling method of Hadoop is Self Adaptive Reduce Scheduling (SARS) algorithm. SARS concentrates only on the reducer phase and the algorithm may cause data locality problem. This paper highlights a new concept to further improve the utility of the SARS technique. This paper presents a new technique that efficiently uses closeness centrality—the shortest distance of all the nodes in the cluster from the data node. Node's closeness centrality analyses its average distance to all other nodes. The proposed algorithm is Improved Highest Response Ratio Next scheduling with Closeness Centrality (IHRRN_CC). Considering the Node Processing Capacity

V. Seethalakshmi (✉)
Research Scholar, Department of Computer Science and Engineering, Pondicherry Engineering College, Puducherry, India
e-mail: seethaveeral6@gmail.com

V. Govindasamy
Associate Professor, Department of Information Technology, Pondicherry Engineering College, Puducherry, India
e-mail: vgopu@pec.edu

V. Akila
Assistant Professor, Department of Computer Science and Engineering, Pondicherry Engineering College, Puducherry, India
e-mail: akila@pec.edu

S. Borah et al. (eds.), *Soft Computing Techniques and Applications*, Advances in Intelligent Systems and Computing 1248,
https://doi.org/10.1007/978-981-15-7394-1_49

(NPC), another variant of IHRRN Scheduling—IHRRN_NPC is also proposed. The SARS algorithm is compared with Improved Highest Response Ratio Next (IHRRN) scheduling with Closeness Centrality (IHRRN_CC). The proposed scheduling algorithm is compared to state-of-the-art schedulers utilizing heterogeneous workloads in a heterogeneous Hadoop environment from the Hi-Bench benchmark array. The results of the experiment show that our proposed schedule improves the performance of a wide range environment MapReduce. Further, the proposed systems minimize job execution time and throughput. Comparative analysis of the proposed IHRRN_NPC and IHRRN_CC with the existing Hadoop default scheduler, SARS, and IHRRN algorithm shows the improvement in data locality.

Keywords Scheduling · Hadoop · Big data · MapReduce · Data locality and heterogeneous environment

1 Introduction

Big Data actually implies to a large amount of structured and unstructured information which is very huge [1]. The old technologies that are utilized make the job harder to utilize changeable database and other software systems. In many cases, the quantity of information is extremely enormous. It is exceedingly hasty. It goes beyond the current implementing capacity level. Big Data has the possibility to help the companies to improve the process and to do the intelligent decision much faster than the existing system. Big Data when apprehended, can be manipulated, captured, stored, examined, and arranged, can help an institution to improve its income, actions, acquire, or hold consumers. Big Data applies to information that cannot be considered to utilize habitual process or tools. Nowadays, more organizations are to access and utilize huge information [2]. Big Data reduces issues of these firms. It helps to handle the huge number information within the very short period of time. Big Data saves manpower and amount of working time. Big Data is even more effective since it could be structured and unstructured.

The MapReduce relies on following tasks such as mapping task and reducing task. The map function takes the input pair and creates the mediator key value pair [3]. The reduction function or the deduction function receives an intermediate key of the corresponding values.

In Hadoop, the data locality eventually moves the calculation to the main data area of the node rather than converting the Big Data into computing. In turn it reduces the congestion of network and improves the system throughput.

The goal is to plan an efficacious scheduler. This scheduler is responsible for deciding the machine at any time. The prominent goal of scheduling is to reduce the time required to complete the task to processors. Appropriate task scheduling does not utilize the real dormant of existing system in the clustering [4]. Cross-switch network traffic is the main issue in data intensive computing. It can be minimized by the data locality. Data locality has been delineated as the distance between the task-allotted node and the input data node. The problem of data locality has received a lot of attention from the research community.

The task execution time in each node in the MapReduce system depends on the entire promulgation of workloads is allocated to that node [5]. Hadoop distributes the data blocks to various nodes for the purpose of managing the loads in a cluster by considering disk space availability. For a homogeneous system where nodes are similar, such a technique for the distribution of information could be competent. If load in the heterogeneous nodes has been shared equally, then similar node processing capacity can complete their implementation at various times. Therefore, we have to spread the load among the nodes based on the processing volume of nodes in the clustering to reduce the processing time.

Improvising the data locality and reducing MapReduce processing time, the performance can be improved in the following ways:

- Reducing task processing time as data transfer would consume most of the total task processing time.
- Minimizing of total network traffic in the data center.

In a unified environment, data transmission may not be efficient, as all nodes in a cluster have the same gestalt hardware. However, nodes with high computation can accomplish their local data in a heterogeneous environment faster than low-computing nodes. For this case, if the enormous amount of data transmission is small, the overhead of transferring unprocessed data from slower processing node to faster processing node was high. This quagmire inspires us to establish flexible data locality based on the technique of closeness centrality. In this situation, if the huge amount of transfer of data is high, the overhead of moving the data from the slower processing node to faster processing node is high. These quagmires encourage us to introduce dynamic data locality based on closeness centrality technique. This technique attempts to reduce the amount of data movements between slower and rapid node execution and thus enhances the quality of MapReduce in the heterogeneous environments. A node's closeness centrality tests the total distance to other nodes.

Closeness centrality is a way of finding whether transmission of data through a graph is very effective. A node's closeness centrality calculates the total distance to all other nodes [6]. Nodes with a high score closeness have the shorter ranges to all other nodes. The closeness centrality method evaluates the count of its distances to all other nodes for each node, depending on the computation of the shortest paths among all node pairs. In order to determine the Closeness Centrality ranking for that node, the resulting amount will be reversed. A node's raw closeness centrality is computed using the formula below:

$$\text{Raw Closeness Centrality (node)} = 1/\text{Sum (distance from node to all other nodes)}$$

Normalizing this rating is more familiar to direct the length of the shortest paths relative to their total [7]. This arrangement allows the closeness of nodes of various sizes to be compared. The equation for standardized Closeness Centrality is as follows:

$$\text{Normalized Closeness Centrality (node)} = (\text{number of nodes} - 1) / \text{Sum (distance from node to all other nodes)}$$

1.1 Key Contributions of This Research Paper

- Proposal of information transmission technique that dynamically transmits input information to the nodes of their clustered base processing volume or capacity.
- Development of a data locality-based technique which maps jobs according to the processing capacity of various nodes in the different clusters.
- Comparing our proposed scheduling method with the state of the art schedulers using heterogeneous workflows in the heterogeneous Hadoop system.
- Proposal of scheduling method to minimize information transfer processes in the clustering by increasing the data locality levels as well as reducing the processing time and throughput.

2 Related Works

The main idea behind Apache's Hadoop design is to run a large number of jobs such as Web indexing and log mining. The jobs submitted by the users primarily reach the queue. However, this system makes a lot of changes in reducing time factor as well as a cost factor. It also helps to share a huge data with many systems at the same time in a wider range. Nevertheless, sharing requires the support of the Hadoop job scheduler: the ability to deliver on productive tasks, and a good response time for interactive tasks while allocating stratagems among users.

FIFO scheduling algorithm [8] depends on queue method. When the TaskTracker is available on the nodes, each and every tasks are loaded into the task queue and then sent to the slot generating allotted slots. The issue of resource sharing requires a good scheduling strategy. All jobs must be completed in a timely manner and provide a better response time for each job.

The Fair Scheduling [9] method for controlling access to the Hadoop cluster was built on Facebook. The scheduler's aim is to provide an equal share of cluster resources to each client in a timely manner. The preemptive strategy would make the scheduler demolish tasks as per the scheduling policy. The priority requirements are also allocated to different pools. Here jobs are subject to interleaving scheduling depending on the priority.

Zhuo et al. [10] says about when to begin the reduce tasks. The reducing tasks are the major issues to boost the work of MapReduce. The efficiency of a MapReduce scheduling technique will be affected when the result of map jobs becomes huge in size. This work will explain the purposes of the slot resources waste of the system.

The authors propose an efficient method called Self Adaptive Reduce Scheduling (SARS) on the Hadoop environment to reduce start times of jobs.

Gothai and Balasubramanie [11] recommend a boost partitioning technique using round robin. Round robin architecture is capable to reduce skew through the distribution of records on average when compared with current hash partitioning. Round robin distributes the data uniformly to all the destination data partition. When an amount of partitions divides an amount of items, the skew is likely to be zero. The data splits are applied to the mapper and the outcome is the sorted list.

In Tang et al. [12], the jobs deviation sent to different clusters are increasing. This addresses Longest Approximate Time to End (LATE) MapReduce scheduling technique. The proposed LATE algorithm is depending on three concepts: prioritizing speculative task, selecting rapid nodes running on, and speculative tasks to avert thrashing. To understand these ethics, LATE technique utilizes following parameters: (a) SlowNode Threshold—This is the cap to avert scheduling on gradual nodes. (b) Speculative Cap—It is the cap on the amount of speculative tasks that can be done at once. (c) SlowTask Threshold: This is a progress rate threshold to decide if a task is gradual sufficient to be speculated upon.

R. ThangaSelvi et al. [13] highlights the MapReduce archetype and its uses of handling a huge data for different purposes. The compressing issue of the Big Data is discussed.

He et al. [14] say an enlargeable MapReduce job scheduling method for time limits in HADOOP environment: MTSD. MTSD allows the client to define a task's time limit period and attempt to complete a job before the allocated time. The computing capacity of the algorithm node specified in MTSD can be performed by node classification. This algorithm expands its types into nodes in a wide variety of clusters. With this algorithm, the authors illuminate a new set of data by distributing the model. This model allocates information based on the node's capacity level. MTSD highlights the client's timeout issues.

Pastorelli et al. [15] contributed a method to the allocation of resource problem. The proposed scheduler is known Hadoop Fair Sojourn Protocol (HFSP). HFSP implements a size-based restraint. It satisfies the side by side process, the system about the response, and fairness requirements. HFSP uses a simple and effective model: size is traded for speed and accuracy by introducing virtual time and aging systems.

YiYao et al. [16] tried to develop a novel MapReduce scheduling method to promote the data locality of the map task. Some have tried to mix this method with the default FIFO scheduler and the fair scheduler. This does not involve an elaborate parameter tuning process, as opposed to the delay algorithm. The main reason behind their method is to provide every slave node an equal opportunity to take local tasks before any non-local tasks are given to any slave node. The proposed matchmaking scheduling method attempts to find a match, which is a slave node that includes input data, for each map task that is not normally assigned. Matchmaking algorithm was designed to enhance MapReduce clusters' data locality level and average response time.

Kumar et al. [17] is created to allow groups to transfer Hadoop clusters in a predictable and easy manner. This provides guarantees of capacity for queues while ensuring elasticity for use in the cluster of queues. This will result in a queue's unused capacity being harnessed by loaded queues that requires a lot of time. Here, there is huge count of clients and the aim of this strategy is to check the fair distribution of computing resource among clients. Capacity scheduling tasks allow users to subtract and submit slots to queues with adjustable maps. The Queues that compromise the various tasks assign their adjusted capacity. The remainder of a queue is distributed with the other queues.

Mapreduce is the programming model used in many industries and organization. Completing the process in short period of time is very important. For this process, many scheduling algorithms are used in MapReduce. Even though many scheduling algorithms are used, they are not fulfilling the user needs. All algorithms concentrate only on some parts of the MapReduce. Hence, we need a new algorithm to improve the performance of MapReduce.

3 Limitations of Existing System

The following are the drawbacks in the existing system:

- It concentrates only on reducer part.
- In a heterogeneous environment, the default scheduler does not perform better.
- Time of execution and data locality level decrease for small jobs as compared to the larger jobs.
- If multiple reduce jobs are executed at same time from one of the Task Tracker process, it leads to network I/O rivalry and the rate of communication during the copying stage is low.
- Another limitation is that the existing system does not consider data locality for reduce task.

Various papers dealing with the existing algorithms, problems, advantages, and disadvantages related to MapReduce are discussed. From this, it is inferred that round robin methodology performs well among the existing scheduling algorithms. But, it also has some drawbacks. In order to overcome this drawback, Improved Highest Response Ratio Next (IHRRN) algorithm and Closeness Centrality technique may be used. Because of this study, the performance of MapReduce and scheduling algorithms used in the MapReduce are understood. Each algorithm has some specific functions, advantages, and disadvantages. Further, the tool used to run the MapReduce program is also studied. By this literature review, it is understood that the IHRRN scheduling algorithm can overcome all the drawbacks of the existing scheduling algorithms.

The objective of this paper is to create the IHRRN scheduling algorithm with closeness centrality technique to enhance the efficiency of MapReduce. The IHRRN scheduling technique will improvises the performance of MapReduce by round robin method. Finding the job which has the highest response is one of the objectives of the paper. By finding the response ratio, an algorithm is designed. The Improved Highest Response Ratio Next (IHRRN) algorithm with Closeness Centrality is designed for the performance improvement of the scheduling algorithm.

4 Proposed Data Locality-Based IHRRN Scheduler in Heterogeneous Environment

In the paper, the scheduling method is proposed to solve the issue of information sharing and scheduling of task in the MapReduce system based upon the capacity of processing of the node, in order to improve MapReduce efficiency in the heterogeneous environments.

The Improved Highest Response Ratio Next (IHRRN) algorithm is designed to enhance the efficiency of the scheduling algorithm. In this environment, sometimes handling capability of the nodes vary as a whole. The node with high-processed capacity can process the local data faster than the node with smaller-processed capacity. The way to improve map redundancy efficiency in heterogeneous computing surroundings has to reduce the amount of transmission of data between simple- and complex-processed nodes and minimizing execution time and throughput. To handle the information loads in this environment, we use a distribution method with a dynamic system. This method helps to split the information dynamically based on the cluster's node processing capacity.

The Execution Flow Diagram of the proposed system is shown in the Fig. 1.

The execution flow is explained as follows.

1. The client input data is submitted to the NameNode. Dynamic block is in the partitioner NameNode. According to the processed capacity of node in the clustering, it splits the input information. Dynamic block partitioning receives node Node Processing Capacity (NPC) of nodes from meta data details and this details is updated regularly.
2. The dynamic block partitioner refers the input data blocks to corresponding nodes of data in the clustering based on their processed capacity.
3. Client submits task to the JobTracker. Then, JobTracker chooses a task from the task running queue.
4. In addition, the TaskTracker will obtain the information from the meta data then assigns tasks to the TaskTrackers based on the number of slots (Map or Reduce) existing on the processed node in a heterogeneous Hadoop clustering.

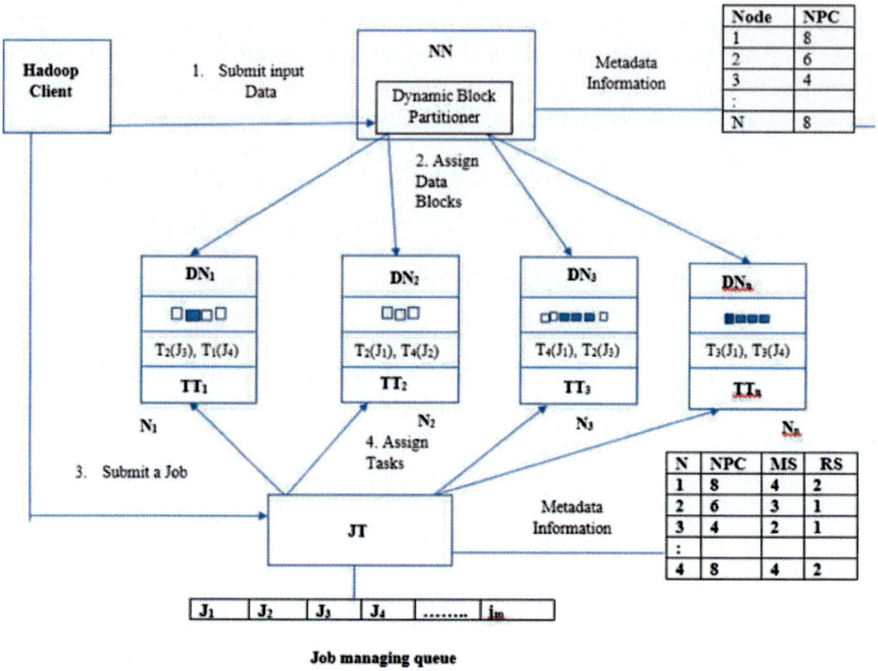

1. NN- NameNode
2. DN- Data
3. Node
4. TT-TaskTracker
5. N-Node
6. J-Job
7. MS- Map Slots
8. RS- Reduce Slots
9. JT- JobTracker

Fig. 1 Execution Flow diagram

4.1 Task Allocation Based on Node Processing Capability

Dynamic block partitioner dynamically divides the different data blocks depending upon the cluster's node processed capacity. Our proposed technique allocates that data blocks to the nodes. The node with the higher processed capacity receives additional data blocks when compared to the cluster processed node.

$$TDB \leftarrow DS/BS \tag{1}$$

The total amount of data blocks used for incoming data is calculated using Eq. (1). Here,

TDB Total amount of Data Blocks
DS Data Size
BS Block Size

The processing capacity of each node '*i*' in heterogeneous Hadoop clustering is estimated using Eq. (2).

$$NPC(i) \leftarrow PDN(i) + ATE(i) \tag{2}$$

Here,

NPC (*i*) - Node processing capacity of *i*th node
PDN (*i*) - Performance of *i*th node
ATE Average process execution time in *i*th node

The NPC of all nodes in the clustering is found out by discussing the following:

1. Average task implementation time of a specific task in that node of the cluster
2. The CPU and usable memory (RAM) of that node are integrated at frequent intervals.

ATE time is found out by assigning fewer jobs to a specific task on that node. ATE has been applied to the NPC buzz, and the work for a variety of nodes in the cluster can be performed at different times. The CPU utilization on each and every node of heterogeneous environment is calculated. Finally, the CPU and free memory utilization in each cluster node are computed using the Eqs. (3) and (4).

$$\text{FreeCPU}(i) = 100\text{--CPU usage} \tag{3}$$

$$\text{FreeRAM}(i) = 100\text{--RAM usage} \tag{4}$$

Here, CPU(i) is processed capacity of *i*th node in the clustering, RAM(i) is the memory capability of *ith* node in the clustering.

Alogorithm1: Alloc_NPC(n,m)	
n: Total number of nodes in cluster	
m: Total number of Map jobs	
B: Burst time of each Map job	
NPC: Node Processing Capacity	
Step1: Begin	
Step2:	Call **NPC** (i, j)
Step3:	Sort descending order of Nodes as per NPC
Step4:	Sort descending order of Map tasks as per B
Step5:	Repeat until j<m
Step6:	**Begin**
Step7:	For i=1 to n
Step8:	**Begin**
Step9:	Assign B[j] to Node[i]
Step10:	i=i+1
Step11:	j=j+1
Step12:	**end** //for
Step13:end //repeat	
Step14:end // Alloc_NPC	

The Algorithm 1 is Alloc_NPC. For a given cluster, the Algorithm 1 accepts as input '*n*' is the number of nodes and '*m*' is number of mapping jobs. This algorithm utilizes the node processing capacity (NPC) function to compute the NPC of all the '*n*' nodes. This Algorithm 1 addresses the allocation of map jobs to the cluster according to the computed node processed capacity of the clustering. The node processing capacity function is executed to compute the NPC of each of the nodes in all the clusters. The next processing step is to sort the nodes according to the NPC. This is followed by sorting the map jobs in descending order according to each of their burst time. Then, all the map jobs are allocated to the corresponding nodes according to the sorted order of burst time of each of map tasks and their NPC of the nodes.

The Algorithm 1 will distribute the data blocks considering the processed capacity of the nodes on the cluster of nodes divided by the input data dynamically, so that the cluster will process all of the local blocks. The data nodes are arranged according to their NPC in a Hadoop heterogeneous environment. First, it is ensured that with the availability of the mapping slots, the processed data node has the highest handling capacity within the Hadoop cluster. Map tasks are assigned to the rapid-processed node in the cluster.

Algorithm2: NPC- Performance Calculation (n,m)

n: Total number of nodes in cluster
m: Total number of Map jobs
B: Burst time of each Map job
CPU(i):Processing capacity of ith node
RAM(i): RAM capacity of ith node
Step1:Begin
Step2: *For* each node in the cluster
Step3: **Begin**
Step4: FreeCPU(i) = 100 – CPU_usage
Step5: FreeRAM(i)=100 – RAM_usage
Step6: Freedisk(i) = Free Disk Space of ith Node
Step7: PDN(i) = FreeCPU(i) + FreeRAM(i)+ Freedisk(i) + MB + IOB
Step8: NPC(i) =PDN(i) + ATE(i)
Step9: **End** //for
Step10:**End** //NPC

The Algorithm 2 deals with the computation of the performance of the cluster by summing up the performance of all the nodes of the cluster. The CPU(i) is the processing ability of *i*th node in the group. Mem(i) is the RAM capacity of *i*th node in the clustering. Used CPU(i) is the amount of CPU used in *i*th node of the clustering. UsedMem(i) is the amount of memory used in *i*th node of the clustering. The efficiency of *i*th data node is found out by considering the free CPU availability and memory from that node.

The CPU usage, the usage of RAM, and the free space in disk for each node are considered. The Performance of the *i*th node PDN(i) is the sum of the CPU usage, the usage of RAM, and the free space in disk for each node. The sum of the performance of the *i*th Node (PDN(i)) together with its average process execution time of the *i*th node (ATE(i)) constitute the node processing capacity (NPC (i)) of the *i*th node.

The average process execution time (ATE) of a node can be calculated by performing few tasks in a specific job. The computation of the NPC of a node considers its ATE. The reason of considering ATE for the computation of NPC is that the job will consume different execution time for each of the heterogeneous nodes in the cluster. The free memory and the CPU utilization are calculated in each node in the cluster following Eqs. (3) and (4).

4.2 *Task Allocation Based on Data Locality and Node Processing Capability*

The data blocks in geographically diverse regions are replicated. The data locality distance (DLD) refers to the capability to move the calculation near to where the real data resides on the node, rather than moving huge information closer to calculation. Data locality distance is then computed.

Algorithm3: Procedure Alloc_NPC_DLD (n, m)

n: Total number of nodes in cluster
m: Total number of Map jobs
B: Burst time of each Map job
Step1: Begin
Step2: Call NPC()
Step3: Sort in descending order of NodeData as per NPC
Step4: Index the NodeData by Data Block Number
Step5: Sort in descending order of Map tasks as per B
Step6: Repeat until j<m
Step7: **Begin**
Step8: For i=1 to n
Step9: **Begin**
Step10: While !endof NodeData
Step11: Begin
Step12: If NodeData.status ==0
Step13: NodeData. Data Block Number== Block
Step14: If NodeData.status=1
Step12: NodeData.Task=Task Number
Step13: Else
Step14: Find a neighbour node with min(DLD)
Step15: If NodeData.status=1
Step16: NodeData.Task=Task Number
Step17: End If
Step18: End If
Step19: **end** //While
Step20: **end** //repeat
22. end // Alloc_NPC_DLD

This Algorithm 3 deals with allocation of node processing capacity based on data locality distance. The node processing capacity is calculated by sorting order of NodeData in downward rule. The NodeData is indexed based on the data block number. Further, the map tasks are segregated in descending order as per their burst time. The status of the NodeData is ascertained that it is 0. In that case, the data is blocked. Further, the status of the NodeData is ascertained that it is 1. Then, the task is set to the task number. Failing which, the neighbor node with the minimum DLD is identified.

We present a IHRRN with closeness centrality algorithm. IHRRN dynamically distributes data blocks and schedules map tasks for a node that has highest computing ability in the system as shown in Algorithm 3.

The above three algorithms define about the allocation of map jobs to the cluster based on the node processing capacity of the group followed by sorted order of burst time. Next the usage of RAM, CPU, and disk is calculated to find the free space in the cluster which helps in computing performance and execution time. Finally, node processing capacity is calculated by sorting order of NodeData in descending order using DLD which sets a status as 0 or 1 which certain blocking and unblocking status of the Node.

4.3 Task Allocation (Shuffle) Based on Closeness Centrality

Closeness centrality (CC) of each node is computed by the following equation.

$$C_C(n_i) = \left[\sum_{j=1}^{g} d(n_i, n_j)\right]^{-1} \tag{5}$$

Here '*d*' is the distance between the given node and all the other nodes. The shuffler and reducer tasks are assigned to nodes with maximum *CC* values.

Proximity is a node of concentration. It is a measure of concentration of a cluster, and is the sum of the interactions measured by the smallest paths between the node and all other nodes. Thus, the more centralized node is closer to all other nodes.

For example, if a node1 has value of 10 and another has value of 5, it signifies that node1 has ability to process information twice as quickly as possible. In homogeneous surroundings where each node has the same storage and disk space, the distribution of data system in Hadoop may not be efficient and effective. If we have a homogeneous Hadoop environment, data block is assigned to nodes.

In Algorithm 1, an example for information allocation is considered. The input data size of 12.8 GB and block size of 128 MB are taken for execution. The maximum amount of data blocks for the given input is 100 blocks (12,800 MB/128 MB = 100). There data blocks are to be shared among the heterogeneous node of the clustering. The node processing capacity of node is assumed as 16 units and total node processed capacity of all nodes in the clustering is taken as 50 units. Then, the number of data blocks to be assigned to node1 is 100 * 6/50 = 12 blocks. As node4 and node8 are fast-processed data nodes, their node processing capacity is 8 units. Thus, the number of data blocks allocated 100 * 8/50= 16 blocks. Compared to other nodes in the clustering, Node3 and node10 are the slowly processed nodes because their node processed capacity is 2 units. These slow processing nodes are assigned 100 * 2/50 = 4 data blocks. Thus, allocating the number of data blocks to each node in the clustering depends upon their processed capacity.

4.4 Performance Evaluation

The proposed data locality-based scheduler performance is evaluated by a series of experiments. The proposed scheduling approach is evaluated using the following metrics:

- **Job execution time**

It is the basic measurable parameter because the purpose of good scheduled algorithm to reduce the job execution time.

Table 1 Hadoop evaluation environment table

S. No.	Node type	Hadoop configuration
1	Master node	8MapSlotsand5ReduceSlots
2	Slave node1	4MapSlotsand2ReduceSlots
3	Slave node2	2MapSlotsand1ReduceSlot
4	Slave node3	2MapSlotsand1ReduceSlot
5	Slave node4	2MapSlotsand1ReduceSlot
6	Slave node5	4MapSlotsand2ReduceSlots

- **Data locality**

  It is actually measured by the total number of tasks executed in the residing data. The data locality rate can be defined as:

$$\text{Data Locality} = \text{Number of the local map task/Total mapping task.}$$

4.5 *Experimental Evaluation*

The experiments are conducted on a heterogeneous Hadoop clustering. The configuration of the Hadoop evaluation environment is illustrated in Table 1. The local test-bed consists Single Master Node (NN and JT) and five slave nodes (DN and TT). The heterogeneity in the Hadoop cluster is measured by considering the disk space, memory size, and CPU types on each node. All nodes in the clustering are co-related with a Gigabit Ethernet switch. The node are executed in Ubuntu15. 10 operating system. In experiments, Hadoop 1.2.1 stable version with a block size of 128 MB and JDK version8 is deployed. Three replicants for each data block in a clustering are maintained to hike the data availability.

The proposed scheduler is evaluated using Hi-Bench benchmark suite. The WordCount should count the appearance of words from the input data. The Random TextWriter generates words from the input data. On average, the proposed scheduler completes tasks 18.5% faster than SRAS. It is 20.7% faster than the Matchmaking scheduler. It is 30.8% faster than the Hadoop default scheduler.

4.6 *Result Analysis*

An essential way of displaying a result is plotting a graph. A graph design-based result is used to show a result comparison with Hadoop default scheduler, SARS, IHRRN, IHRRN_NPC, and IHRRN_CC. It is proved that job execution time by

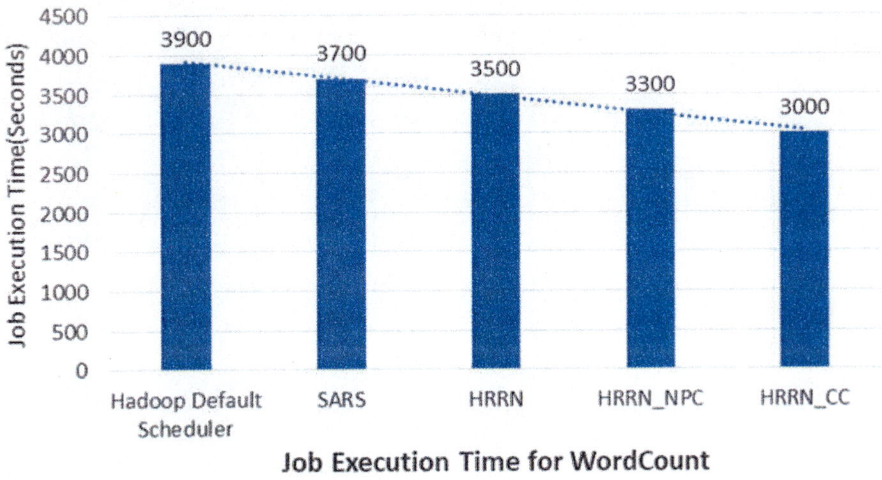

Graph 1 Job execution time for WordCount

proposed scheduler is reduced. The performance comparison of IHRRN with closeness centrality and existing algorithms on Hadoop heterogeneous environment is highlighted in the graph.

The performance of the existing and the proposed technique is depicted in Graph 1. It indicates the comparison of execution time between the existing and the implemented proposed techniques. Graph 2 represents the comparison of Data Locality between the existing and the proposed technique.

Significance of Graph 1:

The Graph 1, depicts the execution of Time in seconds. In the above graph, X-axis depicts the algorithm and the execution time in seconds is denoted in Y-axis. The proposed technique is approximately 300 s better than the existing techniques. This fact is highlighted in the above graph. The better performance of the proposed work with respect to the existing technique is clearly depicted in the Graph 1.

Significance of Graph 2:

The Graph 2 represents the increased percentage of data locality. In the above Graph, the algorithm is represented by X-axis. Y-axis shows the data locality in terms of percentage. We observe that our proposed scheduler has the higher rate of the data locality compared to the Hadoop default scheduler, and SARS.

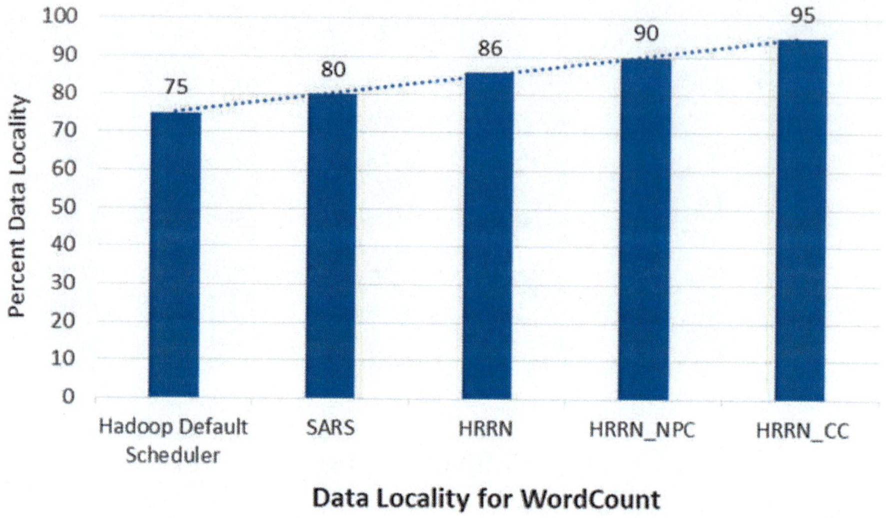

Graph 2 Data locality for WordCount

5 Conclusion

This paper highlights a novel Data Locality-based scheduling algorithm. The proposed algorithm upgrades the MapReduce framework performance in heterogeneous Hadoop clustering. It describes job scheduling in heterogeneous MapReduce Hadoop environments, taking into account both task execution times, data locality and throughput. The reasonable utilization of clusters resources is realized by this heterogeneous MapReduce Hadoop environments. The work has designed an Improved Highest Response Ratio Next (IHRRN) scheduling algorithm to improve the efficiency of MapReduce with the various adaptive methodologies.

References

1. Singh, K., Malik, D., Sharma, N.: An improved K-means algorithm based on MapReduce and grid. IJCEM Int. J. Comput. Eng. Manage. **28–30** (Sep, 2015)
2. Ma, L., Gu, L., Li, B., Ma, Y., Wang, J.: Limitation of K-means algorithm. In: IEEE/ACS 12th International Conference of Computer Systems and Applications (AICCSA) (2015)
3. Chen, Q., Zhang, D., Guo, M., Deng, Q., Guo, S.: SAMR: a self adaptive MapReduce scheduling algorithm in heterogeneous environment. In: IEEE International Conference and Information Technology (2010)
4. Lee, G., Chun, B.-G., Katz, R.H.: Heterogeneity-aware resource allocation and scheduling in the cloud. In: Proceedings of the 3rd USENIX Conference on Hot Topics in Cloud Computing, ACM (2011)
5. Yao, Y., Wang, J., Sheng, B., Tan, C.C., Mi, N.: Self adjusting slot configuration for homogenous and heterogeneous Hadoop clusters. IEEE Trans. Cloud Comput. (2017)

6. Xu, X., Cao, L., Wang, X.: Adaptive task scheduling strategy based on dynamic workload adjustments for heterogeneous Hadoop clusters. IEEE Syst. J. (2016)
7. Caruana, G., Li, M., Qi, M., Khan, M., Rana, O.: GSched: a resource aware Hadoop scheduler for heterogeneous cloud computing environments. Concurrency Comput. Practice Exp. (2016)
8. Seethalakshmi, V., Govindasamy, V., Akila, V.: Job scheduling in big data-a survey. In: International Conference on Computation of Power Energy Information and Communication (ICCPEIC), pp. 023–031. IEEE (2018)
9. Seethalakshmi, V., Govindasamy, V., Akila, V.: To improvise the mapreduce performance using hrrn algorithm. Int. J. Pure Appl. Math. **119**(18), 2057–2063 (2018)
10. Tang, Z., Jiang, L., Zhoua, J., Li, K., Li, K.: A self-adaptive scheduling algorithm for reduce start time. Elsevier **28**(11) (2016)
11. Gothai, E., Balasubramanie, P.: A novel approach for partitioning in Hadoop using Round Robin technique. J. Theor. Appl. Info. Technol. 105–112 (2015)
12. Tang, Z., Zhou, J., Li, K., Li, R.: A MapReduce task scheduling algorithm for deadline constraints. pp. 1–4. Springer (2014)
13. Thangaselvi, R., Aruna, R.: Longest approximate time to end scheduling algorithm in Hadoop environment. Int. J. Adv. Res. Manage. Architect. Technol. Eng. 77–83 (2016)
14. He, C., Lu, Y., Swanson, D.: Matchmaking: a new MapReduce scheduling technique. In: IEEE 3rd International Conference on Cloud Computing Technology and Science (2014)
15. Pastorelli, M., Carra, D., Dell'Amico, M., Michiardi, P.: Bringing size-based scheduling to Hadoop. IEEE Trans. Cloud Comput. **28–30** (2015)
16. Yao, Y., Tai, J., Sheng, B., Mi, N.: A job size-based scheduler for efficient task assignments in Hadoop. IEEE Trans. Cloud Comput. 77–83 (2015)
17. Kumar, A., Sabharwal, Y., Sen, S.: Approximate K-means clustering algorithm. In: IEEE 3rd International Conference on Cloud Computing Technology and Science (October, 2014)

A Study on Various Machine Learning Algorithms Used for Prediction of Diabetes Mellitus

Gaurav Pradhan, Ratika Pradhan, and Bidita Khandelwal

Abstract Diabetes mellitus (DM) has become one of the common disease in middle- and low-income countries. In the recent past, machine learning (ML) and data mining approach has been applied for prediction of diabetes with acceptable accuracy rate. A reliable diagnostic prediction system is thus desired by medical professionals. This paper highlights various machine learning algorithms for prediction of diabetes that include: Decision tree, random forest, Naïve Bayes, artificial neural network, support vector machine, and logistic regression, etc.

Keywords Machine learning (ML) · Diabetes mellitus (DM) · Data mining · Decision tree · Artificial neural network

1 Introduction

Diabetes mellitus is a group of metabolic diseases characterized by hyperglycemia resulting from defects in insulin secretion, insulin action, or both. As per IDF Diabetes Atlas 2019, about 463 million adults were living with diabetes and by 2045, this will rise to 700 million. According to WHO report, India had 73.96 million people living with diabetes in 2019 and nearly 98 million people in India may have type-2 diabetes by 2030. As per National Programme for Prevention and Control of Cancer, Diabetes, Cardiovascular Diseases and Stroke (NPCDCS), screening for diabetes & hypertension, Sikkim reported the highest prevalence of DM (13.67%). The development in

G. Pradhan · R. Pradhan (✉)
Department of Computer Applications, Sikkim Manipal Institute of Technology, SMU, Majitar, Rangpo, East Sikkim, India
e-mail: ratika.p@smit.smu.edu.in

G. Pradhan
e-mail: gaurav_201910010@smit.smu.edu.in

B. Khandelwal
Department of Medicine, Sikkim Manipal Institute of Medical Sciences, SMU, Tadong, Gangtok, East Sikkim, India
e-mail: bidita.k@smims.smu.edu.in

S. Borah et al. (eds.), *Soft Computing Techniques and Applications*,
Advances in Intelligent Systems and Computing 1248,
https://doi.org/10.1007/978-981-15-7394-1_50

the areas of machine learning (ML) and data mining has increased the prevalence of data–driven medical decision support systems. There are various machine learning techniques that are useful for scrutinizing the data into valuable information. This paper highlights various machine learning algorithms for prediction of diabetes that include: Random forest, decision tree, Naïve Bayes, artificial neural network, support vector machine, and logistic regression.

Diabetes mellitus is primarily classified into three types: Type 1, Type 2, and gestational diabetes mellitus. In Type 1 diabetes, the immune system attacks and destroys the beta cells in pancreas and the damage is considered permanent. Type 2 diabetes starts with insulin resistance wherein the body cannot use insulin efficiently. Gestational diabetes mellitus (GDM) occurs due to insulin blocking hormones during pregnancies and may enhance or vanish after delivery.

Recently, numerous ML approaches have been explored in medical analysis to increase diagnostic accuracy, and in addition, it further reduces costs and human resources. Many researchers have chosen ML methods for predicting diabetes and have come out with favorable results.

2 Machine Learning (ML) Algorithms Used in Predicting Diabetes Mellitus

Machine learning algorithm has played an important role in detection of various types of diseases. Most common ML algorithms that has been used frequently by the majority of the researcher for predicting diabetes mellitus include:

(a) **Decision tree learning:** A decision tree learning by far is most popular ML technique used for predicting DM. It is a supervised ML technique makes tree-like graph or model of decision from a training data. In decision tree, leaves represent classifications, non-leaf nodes represent features, and branched represents conjunctions of features that leads to classification. The main challenges lie in building the smallest decision tree. Some of the popular methods for building decision tree include: ID3, C4.5, C5.0, REPtree, classification and regression tree (CART), etc.

(b) **Random forest:** Random forest builds a large number of individual decision trees which act as an ensemble. In random forest, every individual tree spits out a class prediction. The most voted class becomes model's prediction. It works well with very large dataset with huge number of features.

(c) **Naïve Bayes:** Naive Bayes classifier is probabilistic ML model that employs the Bayes theorem. It defines all features as autonomous and unrelated that suites well for the data with issues of imbalance and missing values.

(d) **Artificial neural network:** Artificial neural networks intend to mimic the way humans learn. Neural networks encompass input and output layers, in addition to (in maximum instances) a hidden layer which includes systems that transform the input into something that the output layer can use.

(e) **Support vector machine:** Support vector machine (SVM) is a supervised machine learning algorithm popular for both classification and regression analysis. It builds a hyperplane (or set of hyperplanes) to classify all the inputs in a higher dimensional space by maximizing the margin between the classes' closest points.

(f) **Logistic regression:** Logistic regression is a classification algorithm that models the probability of a discrete outcome given an input variable. The most common logistic regression models a binary outcome (such as True/False). It assumes that the decision boundaries are linear.

3 Related Works

Table 1 summarizes the recent work carried out in predicting DM using ML techniques. The detection type has been broadly classified as supervised approach, ensemble approach, and hybrid approach. Figure 1 shows the results obtained on Pima Indian Diabetes Dataset (PIDD) using various techniques.

4 Conclusion

This paper provides various machine learning algorithms that have been applied for the detection of diabetes mellitus using supervised, hybrid, or ensemble approach. It is a great challenge for the researchers to give the acceptable accuracy rate. It is concluded that ensemble approach provides acceptable accuracy as compared to other two.

Table 1 Summary of literature review

Sl No	Author(s)/year	Detection type	Description and technique used
1.	Maniruzzaman et al. [1]	Ensemble approach	Methodology: random forest, decision tree, naïve bayes and adaboost Dataset used: national health and nutrition examination survey Accuracy achieved: for ML-based system: 90.62% with LR-based feature selection and RF-based classifier: 94.25%
2.	Saru and Subashree [2]	Ensemble approach	Methodology: Naïve bayes, decision trees, and (KNN) Dataset used: Pima Indian Diabetes Dataset (PIDD) Resampling: bootstrapping resampling technique to enhance the accuracy Validation: ten-fold cross validation Accuracy achieved: 90.36% with bootstrapping resampling and 83.72% for decision stump
3.	Li and Fong [13]	Supervised approach	Methodology: coefficient of variation Dataset used: Pima Indian Diabetes Dataset (PIDD) Validation: 10 cross validation Accuracy achieved: 77.0833%
4.	Pei et al. [3]	Ensemble approach	Methodology: J48, AdaboostM1, SMO, bayes net, and naïve bayes Dataset used: Shengjing hospital of China medical University Data splits: Ttraining set 70 and test set 30% Accuracy achieved: J48 outperforms with 95.03%
5.	Deshmukh and Fadewar [22]	Supervised approach	Methodology: convolutional neural network (CNN) with fuzzification Dataset used: Pima Indian Diabetes Dataset (PIDD).: CNN with fuzzification is found better than normal NN
6.	Tejas et al. [23]	Ensemble approach	Methodology: SVM, logistic regression, and ANN

(continued)

Table 1 (continued)

Sl No	Author(s)/year	Detection type	Description and technique used
7.	Manna et al. [4]	Ensemble approach	Methodology: logistic regression and random forest Accuracy achieved: 85.7% using random forest and 86.7% using logistic regression
8.	Samer et al. [14]	Supervised approach	Methodology: artificial neural networks (ANN) Dataset used: association of diabetic's city of Urmia Accuracy achieved: 87.3%
9.	Zou et al. [17]	Hybrid approach	Methodology: decision tree, random forest, and neural network Dataset used: Luzhou dataset and PIDD Dimensional reduction: principal component analysis (PCA) and minimum redundancy maximum relevance (mRMR) Validation: hold-out method and five-fold cross validation Accuracy achieved: 80.84% for Luzhou dataset and 77.21% for PIDD
10.	Kumar et al. [18]	Hybrid approach	Methodology: random forest and gradient boosting classifiers. Dataset used: Pima Indian Diabetes Dataset (PIDD) Data splits: Training set 80 and test set 20% Dimensional reduction: Principal component analysis (PCA) Oversampling: SMOTE- and ADASYN-based oversampling Accuracy achieved: 90%
11.	Hasan [24]	Hybrid approach	Methodology: bayesian regularization (BR), Levenberg–Marquardt algorithm (LM) and scaled conjugate gradient (SCG). Dataset used: PIDD Data splits: 70% as training set, 15%, as validation set and 15% as testing set, respectively Results: The 15 hidden layers performs the best

(continued)

Table 1 (continued)

Sl No	Author(s)/year	Detection type	Description and technique used
12.	Ramana Kumar and Krishnamohan Rao [15]	Supervised approach	Methodology: boosted tree algorithm. Dataset used: PIDD Data splits: training set 70 and test set 30% Accuracy achieved: 90%
13.	Alehegn et al. [5]	Ensemble approach	Methodology: SVM, Naïve Net (NN), decision stump, and proposed ensemble method (PEM) Dataset: PIDD Validation: 10 cross validation Accuracy achieved: 90.36% for PEM
14.	Rathore et al. [6]	Ensemble approach	Methodology: SVM and decision tree Dataset used: PIDD Accuracy achieved: 82% for SVM
15.	Joshi and Alehegn [25]	Ensemble approach	Methodology: KNN, naïve bayes, random forest, and J48 Dataset used: PIDD Results: decision tree provided high accuracy
16.	Chen et al. [19]	Hybrid approach	Methodology: *K*-means and decision tree Dataset used: PIDD Validation: 10 cross validation Accuracy achieved: 90.04%
17.	AIyer et al. [10]	Ensemble approach	Methodology: decision tree (J48) and naïve bayes Dataset used: PIDD Data splits: training set 70 and test set 30% Validation: ten-fold cross validation Accuracy achieved: 74.8% using J48 and Bayes gives an exactness of 79.5% by utilizing 70:30 split

(continued)

Table 1 (continued)

Sl No	Author(s)/year	Detection type	Description and technique used
18.	Prasad and Agarwal [26]	Hybrid approach	Methodology: GBREA with only six indicators of type-2 diabetes. This algorithm trains four different classifiers named DTC, KNN, Naïve Bayes, and LDA Data splits: training set 75 and test set 25% Results: GBREA algorithm performed the best
19.	Sharmila and Manickam [27]	Supervised approach	Methodology: decision tree Dataset used: collected from various laboratories in and around Chennai Results: The size of the decision tree is 10 when complexity parameter (cp) falls to zero
20.	Kaur and Chhabra [28]	Supervised approach	Methodology: Modified J48 Dataset used: PIDD Accuracy achieved: 99.87%
21.	Yasodha and Kannan [29]	Ensemble approach	Methodology: bayes, J48, REP tree, and random tree Validation: 10-overlap cross approval approach Dataset used: PIDD Accuracy achieved: 60.2%
22.	Alexis Marcano-Cedeño et al. [16]	Supervised approach	Methodology: artificial metaplasticity on multilayer perceptron (AMMLP) Dataset used: Pima Indian Diabetes Dataset (PIDD) Accuracy achieved: 89.93%.
23.	Mohammed Khalilia et al. [8]	Ensemble approach	Methodology: random forest, SVM Dataset used: healthcare cost and utilization project (HCUP) dataset Accuracy achieved: 88.79%
24.	Patil et al. [20]	Hybrid approach	Methodology: *K*-means algorithm and decision tree (C4.5) Dataset used: PIDD Validation: 10 cross validation Results: 84.24%

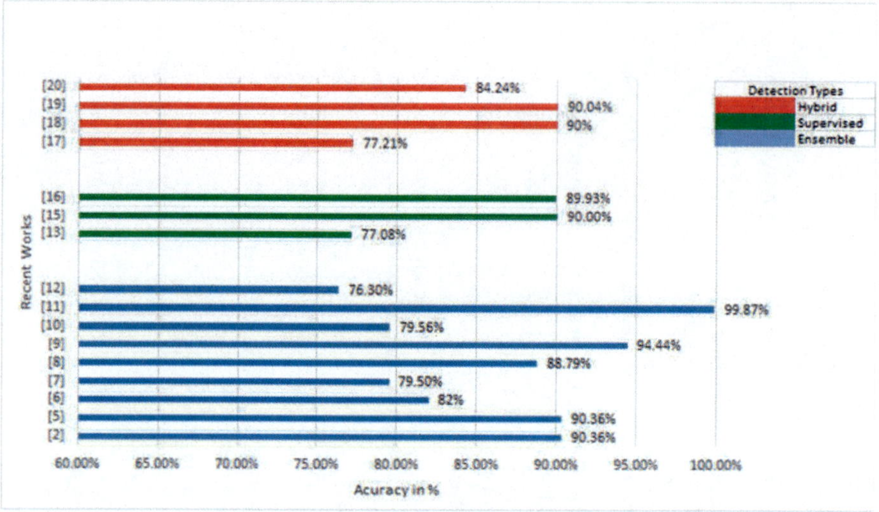

Fig. 1 Comparative study of various ML algorithm used for predicting diabetes

References

1. Maniruzzaman, M., Rahman, M.J., Ahammed, B., Abedin, M.M.: Classification and prediction of diabetes disease using machine learning paradigm. Springer (2020)
2. Saru, S., Subashree, S.: Analysis and prediction of diabetes using machine learning. Int. J. Emerg. Technol. Innov. Eng. **5**(4) (2019) (ISSN: 2394–6598)
3. Pei, D., Gong, Y., Kang, H., Zhang, C., Guo, Q.: Accurate and rapid screening model for potential diabetes mellitus. BMC Medical Info. Decis. Making (2019)
4. Manna, S., Maity, S., Munshi, S., Adhikari, M.: Diabetes prediction model using cloud analytics. In: International Conference on Advances in Computing, Communications and Informatics (ICACCI) (2018)
5. Alehegn, M., Joshi, R., Mulay, P.: Analysis and prediction of diabetes mellitus using machine learning algorithm. Int. J. Pure Appl. Math. **118**(9), 871–878 (2018)
6. Rathore, A., Chauhan, S., Gujral, S.: Detecting and predicting diabetes using supervised learning: an approach towards better healthcare for women. Int. J. Adv. Res. Comput. Sci. **8**(5) (2017)
7. Aljumah, A.A., Ahamad, M.G., Siddiqui, M.K.: Application of data mining: diabetes health care in young and old patients. J. King Saud University-Comput. Info. Sci. **25**, 127–136 (2012)
8. Khalilia, M., Chakraborty, S., Popescu, M.: Predicting disease risks from highly imbalanced data using random forest. Published 29 July (2011)
9. Aada, A., Tiwari, S.: Predicting diabetes in medical datasets using machine learning techniques. Int. J. Sci. Eng. Res. **5**(2), (2019)
10. Iyer, A., Jeyalatha, S., Sumbaly, R.: Diagnosis of diabetes using classification mining techniques. IJDKP **5**(1), 01–14 (2015)
11. Sengamuthu, R., Abirami, R., Karthik, D.: Various data mining techniques analysis to predict diabetes mellitus. Int. Res. J. Eng. Technol. (IRJET) **05**(05) (May, 2018)
12. Sisodia, D., Sisodia, D.S.: Prediction of diabetes using classification algorithms. ICCIDS (2018)
13. Li, T., Fong, S.: A fast feature selection method based on learning. Int. J. Extreme Autom. Connectivity Healthcare **1**(1) (2019)

14. El_Jerjawi, N.S., Abu-Naser, S.S.: Diabetes prediction using artificial neural network. Int. J. Adv. Sci. Technol. **121**, 55–64 (2018)
15. Ramana Kumar, D., Krishnamohan Rao, S.: Health care system: stream machine learning classifier for features prediction in diabetes therapy. Int. J. Appl. Eng. Res. **13**(1), 59–65 (2018). ISSN 0973-4562
16. Marcano-Cedeño, A., Torres, J., Andina, D.: A prediction model to diabetes using artificial metaplasticity. In: Ferrández, J.M., Álvarez Sánchez, J.R., de la Paz, F., Toledo, F.J. (eds.) New Challenges on Bioinspired Applications. IWINAC 2011. Lecture Notes in Computer Science, vol. 6687. Springer, Berlin, Heidelberg (2011)
17. Zou, Q., Qu, K., Luo, Y., Yin, D., Ju, Y., Tang, H.: Predicting diabetes mellitus with machine learning techniques. Published: 06 November (2018)
18. Dasa, S.K., Kumar, A., Mishrab, Roy, P.: Automatic diabetes prediction using tree based ensemble learners. In: Proceedings of International Conference on Computational Intelligence and IoT (ICCIIoT) (2018)
19. Chen, W., Chen, S., Zhang, H., Wu, T.: A hybrid prediction model for type 2 diabetes using k-means and decision tree. IEEE (2017)
20. Patil, B.M., Joshi, R.C., Toshniwal, D.: Hybrid prediction model for Type-2 diabetic patients. Expert Syst. Appl. **37**(12), 8102–8108 (2010)
21. Faruque, M.F., Asaduzzaman, A., Sarker, I.H.: Performance analysis of machine learning techniques to predict diabetes mellitus. arXiv preprint arXiv:1902.10028 (2019)
22. Deshmukh, T., Fadewar, H.S.: Fuzzy deep learning for diabetes detection. In: Iyer, B. et al. (eds.) Computing Communication and Signal Processing, Advances in Intelligent Systems and Computing, Springer Nature Singapore Pte Ltd (2019)
23. Joshi, T.N., Chawan, P.M.: Diabetes prediction using machine learning techniques. J. Eng. Res. Appl. **8**(1) 09–13 (2018). ISSN: 2248-9622 (Part-II)
24. Hasan, S.: Prediction of diabetes based on artificial intelligence technique. Concepts J. Appl. Res. CJAR **03**(09) (2018)
25. Joshi, R., Alehegn, M.: Analysis and prediction of diabetes diseases using machine learning algorithm: Ensemble approach. Int. Res. J. Eng. Technol. (IRJET) **04**(10) (2017)
26. Prasad, B.R., Agarwal, S.: Modeling risk prediction of diabetes—a preventive measure. IEEE, 12 February (2015)
27. Sharmila, K., Manickam, S.: Efficient prediction and classification of diabetic patients from big data using R. Int. J. Adv. Eng. Res. Sci. **2** (2015)
28. Kaur, G., Chhabra, A.: Improved J48 classification algorithm for the prediction of diabetes. Int. J. Comput. Appl. (0975–8887) **98**(22) (2014)
29. Yasodha, P., Kannan, M.: Analysis of a population of diabetic patients databases in R tool. Int. J. Sci. Eng. Res. **2**(5) (2011)

Clustering Algorithms for MANETs: A Review on Design and Development

Sunil Pathak, Sonal Jain, and Samarjeet Borah

Abstract The self-motivated environment of mobile nodes confronts the connectivity and the stability in mobile MANETs. In MANETs, nodes are usually defined as a network which is created using multiple free or autonomous mobile nodes. Composition of mobile node requires self-configuration and lacks the centralized control to form a network. Scalability and stability along with robustness is a further challenge in MANETs, which are easily handled by using clustering. Times to time various clustering algorithms were proposed in the literatures which are responsible for cluster formation and maintenance in MANETs. In this paper, comparative study is carried out for clustering algorithm focusing on design and development issues. The analysis process for numerous clustering algorithms is approved based on IDs, degree, balance battery power, movement and weight of node. Moreover, this paper describes a meticulous depiction of their procedures, and conversation of the topographies and failings of respective algorithm.

Keywords Clustering algorithms · Clusterhead election · Re-clustering · Cluster maintenance · Clustering overhead · Clustermember · Cluster stability

1 Introduction

Mobile MANET (MANET) is an auspicious type of wireless network, in which all moveable devices are connected by ad hoc manner. In MANETs, mobile nodes are

S. Pathak
Amity University, Jaipur, India
e-mail: sunilpath@gmail.com

S. Jain
JK Lakshmipat University, Jaipur, Rajasthan, India
e-mail: sonaljain@jklu.edu.in; drsonalamitjain@gmail.com

S. Borah (✉)
Sikkim Manipal Institute of Technology, Sikkim, India
e-mail: samarjeetborah@gmail.com

S. Borah et al. (eds.), *Soft Computing Techniques and Applications*, Advances in Intelligent Systems and Computing 1248,
https://doi.org/10.1007/978-981-15-7394-1_51

capable for making self-connection without any pre-required infrastructure, self-healing and enabling peer-level communications among numerous movable nodes without prerequisite of consolidated assets or permanent structure [1, 59]. Sending data packets in MANET is completed by multiple routes. Absence of static arrangement and regular connection failure limits the usage of outdated routing algorithms for wired networks to MANETs indirectly, as bounded routing accept the network to be established and overhead will be nearly minor. Wired network either wants the capability to acclimate to dynamic topology of MANETs cause a huge overhead which is responsible to reduce the network performance.

Wireless networks are being progressively used while forwarding data packets between various mobile devices of the most varied types and sizes. Various mobile devices are being used in the field of weather forecasting, research, battle field and other important application in the society and to solved complexes engineering problems [2, 3].

The organization of wireless networks in the absence of infrastructure or required limited arrangement is not dependable. MANETs have been suggested to resolve such complications. A MANET is a gathering of wireless nodes which may deploy dynamically based on demand. In MANETs, each mobile device will interconnected straight to everyone. Mobile nodes are not straightly connected and communicated by sending its traffic flow via its inter-linked nodes, wherein each mobile node behaves as a router [4].

Important key characteristics of MANETs are flexible type of network; it reduce installation charge and provide robustness. MANETs can be set up without difficulty, even in not reachable spaces and can endure to natural catastrophes and war. These characteristics make MANETs well appropriate for providing security, various emergency operations, disaster prevention and retrieval, large-scale community networks and small networks for communication between conference attendees or students in a lecture room.

Like the cell arrange, the MANET can be isolated into a few groups. Every bunch is made out of one group head and numerous typical hubs, and all the bunch heads frame a whole overwhelming set. The group head is accountable for gathering data (flagging, message and so on) and dispensing assets inside its bunch and speaking with other bunch heads. Also, the ordinary hubs speak with each other through their group head, regardless of they are in the same bunch or not. A few unique bunching calculations have been proposed in MANET [5].

2 Related Work on Clustering Algorithms

Various clustering algorithms were used to split the big flat network into reduced logical virtual subsets which fulfill assured property. Such simulated logical subsets are known as clusters. Clustering algorithms normally work into twofold phases. First stage shows that all mobile devices have been divided in various small–small logical sub-division and in second stage, determinations were finished to reservation the structure which was created in previous stage to escape the whole destruction. Additional maintenance suffered implies higher overhead on network. First phase should be highly stable to reduce the number of request call needed for maintenance phase [6]. A node with mobility [60] can easily join the cluster and can leave a cluster which disturbs cluster's stability. In clustering algorithm, primarily a mobile device is selected as a clusterhead and all its neighboring node of the clusterhead acts as clustermember to formed cluster. As the clusterhead has the accountability of handling the cluster, the clusterhead loses energy faster and clusterhead sometimes felt shortage of power. Node with great volume of power starts working as clusterhead which leaks the primary demise of clusterhead. The algorithms usually take an assumption and consider a network with homogeneous nodes that is nodes with same capabilities and resources. But high scalability and capacity is exhibited via heterogeneous models that is mobile nodes with various abilities. The algorithms which will be developed for similar network may not carry matching enactment with the varied network. Clustering algorithms used in MANET can be categorized as given in Fig. 1.

2.1 Introduction to Clustering

Mixture of mobile nodes in such a method that each mobile node will be in the same set named as cluster, which may be much identical to one other than to those which are in other cluster. Cluster investigation himself is not definite algorithm, but this is a universal chore to be resolved. This may have accomplished by numerous procedures that fluctuate significantly of their idea of what constitutes a cluster and the right way to effectively in finding them.

2.2 Clustering Requirement

Number of nodes in cluster may vary. It is separated into two slices: strongly joined and lightly joined clustering. Tightly coupled clustering has a structure in which each wireless device can join cluster as clusterhead, clustermember, or gateway. While in lightly joined clustering that type of structure does not exist. An innovative clustering algorithm is discussed in [7] in order to decrease the number of clusters.

In this paper, we present a detailed review of cluster creation and maintenance procedures in MANETs. Thorough comparative study, various algorithms were compared on various parameters. Introduction of MANETs was discussed in Section 1, Section 2 covers detailed introduction and Section 3 discussed detailed previous work related to various clustering algorithm from MANET and provide the comparative study for separate algorithms. A complete relative review of the algorithms and their strengths and weakness was described in Section 4 and Section 5 accomplishes the paper.

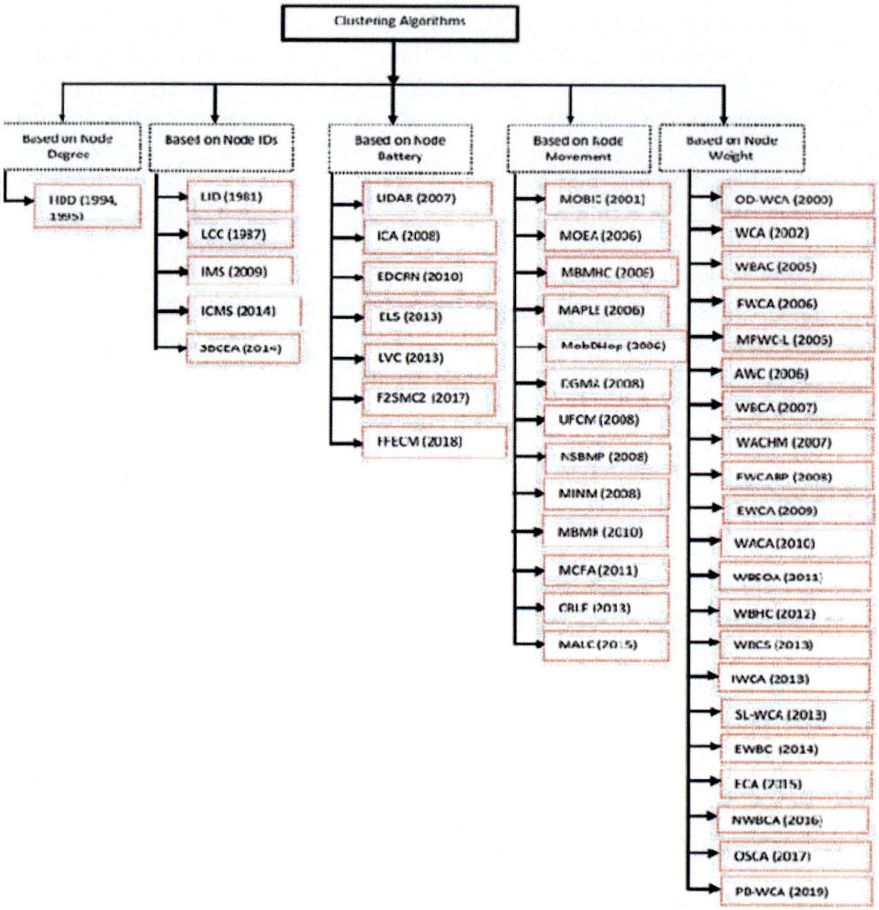

Fig. 1 Classification of clustering algorithms in MANETs

3 Taxonomy of Clustering Algorithms in MANET

This segment contains representative clustering algorithms for MANETs which are categorized into the five groups based on mobile node property.

3.1 Algorithm Based on Mobile Node Degree

In highest-degree algorithm [8, 9], a mobile device x is thought to be a neighbor of another node y if x exists in the transmission scope of y. The node with greatest number of neighbors (i.e., most extreme degree) is picked as a clusterhead. As the quantity of node in a cluster expands, the throughput drops and consequently a slow debasement in the framework execution is brought on. Every one of these disadvantages happens on the clusters.

3.2 Algorithm Based on Mobile Node IDs

In lowest-ID algorithm [10], it allots a one of a kind id to every hub and picks the hub with the base id as a group head. Concerning this calculation, the framework execution is better contrasted and the highest-degree algorithm as far as throughput. In any case, it does not endeavor to adjust the heap consistently over every one of the hubs.

Least clusterhead change: In [11], clustering is breakdown in two stage, in first phase cluster creation is carried out and in second phase is responsible for cluster maintenance. Cluster formation process use LID [10] to designated a node as clusterhead. In maintenance phase is based on event driven and executed in tow steps. In first phase, if two or more than two clusterheads are ling in similar frequency range, then the clusterhead which has lowest ID will remain as clusterhead and additional will act as clustermember within the same cluster. In second phase if there is no clusterhead within the cluster, rebuild a new clusterhead by using LID [10]. In IMS [12], cluster formation mechanism is based on LID [10], where the node with lowest IDs among its neighbors elect as clusterhead, and rest will act as clustermembers. In maintenance phase of IMS, when two clusterheads are within the same radio range, clusterhead changes [61] will be delayed up to delayed time. As delay time expires, it will again check for clusterhead, if they are still in same radio range, they will wait up to Maximum limit and each time IDs of node will be compared and every time delay time will be increase by hello interval. After maximum limit, then node with lowest IDs will be elected as clusterhead and other will act as clustermember. ICMS [13] combine priority factor with LIDs and IMS algorithm. In ICMS, priority of each node is calculated and the node with high priority, elected as clusterhead and other behave as clustermember. In [55], each mobile node within the cluster maintains its status. If a node is moving out from the cluster and want to join other cluster, it can join based on its status. Handling with status for each mobile node, it is need to exchange more messages within the cluster which increase clustering overhead and delay time. Node ID based clustering algorithms are shown chronologically in Fig. 2.

3.3 Algorithm Based on Mobile Node Movement

In [14], authors proposed a distributed clustering algorithm which is based on mobility of mobile node for selection of stable clusterhead as compared with least clusterhead changes. In MOBIC, each node calculates its relative mobility metrics

Fig. 2 Clustering algorithm based on node IDs

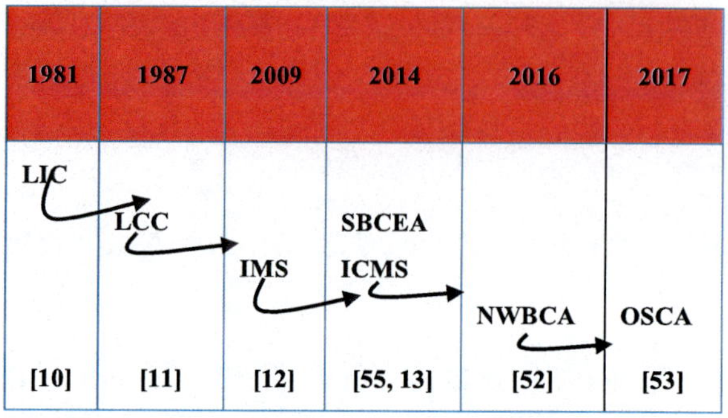

and broad cast hello message with mobility metrics within its 1-Hop neighbor. A node with its lower mobility matrices value elected as clusterhead and other nodes will act as clustermember. In MOEA [15], a stable cluster structure is achieved by taking advantage of the node movement. It provides stable clustered topology, optimal cluster structure in terms of clustering metric for small-scale network topology. Performance of MOEA was compared with WCA and found better performance in terms of stable, optimal clusterhead. Mobility-Based Multihop Clustering [16], analysis of message and time complexity were present for various clustering algorithms such as mobDHop [18], LID [10], DMAC and MCC. It also provides upper bound of time complexities for cluster creation and cluster maintenance. MAPLE [17] focuses the clustering problem at medium access control and facilitate pro-active and energy-efficient clustering based on mobile node mobility information. During cluster formation process, MAPLE uses channel reservation technique. Result shows that stability, load distribution, clusterhead change rate with transmission range, average amount of transmission power consumed per mobile while sending control message, link failure on mobility and control overhead reduced as compared to other clustering approach in the literature. MobDHop [18] is a distributed clustering algorithm. The MobDHop creates dynamic stable multihop cluster. MobDHop requires 1-hop neighbor node information during cluster formation. MobDHop is simple and deserves minimal overhead because each mobile node is required to know its 1-hop neighbors in the cluster. It can change diameter adaptively with respect to node movement pattern which form variable diameter cluster. Performance evaluation of MobDHop is carried out based on simulation Approach and Theoretical Analysis. Primary objective of DMGA [19] is long-drawn-out lifetime of clusterheads, valid in extremely mobile atmosphere and reduced re-clustering in MANET. DGMA works on mobility metric, altitudinal reliance. Unified Framework for Cluster Manager Election [20, 62] discusses the issues on election of cluster manager mechanism and re-clustering. It simplifies clustering processing in respect to feasibility and efficiency. Clustering and clusterhead election process is carried out in multiple stages in terms of extra power, degree and announcement volume. NSBMP [21] guarantees stable cluster structure as other algorithms which are proposed in the literature. In

NSBMP, accurate mobility is forecast of each host mobile node based on its neighbors; therefore, it is responsible for cluster formation. MINM [22] explores the shock of various types of mobility on clustering for mobile nodes. Aim of this paper is to investigate clustering and mobility of mobile node. Energy cost and other metrics related to construction and maintaining a cluster are used as performance parameters in MINM. As per past publication from the literature, multicast routing is not consider with mobility parameter for clustering in ad hoc network. Expected relative mobility of each node and virtual multicast backbone is used in MBMR to elect stable cluster. MBMR [23] performs better as compared with existing protocol in terms of packet delivery ratio, multicast route lifetime and end-to-end delay. In past algorithms, mobility parameter of network (mobile nodes) is fixed but in actual all mobile nodes are able to move randomly within network. In MCFA [24], mobility of nodes is assumed as random variable with unknown distributions. Mobility of clusterhead is estimated in relative with its neighbors by using mobility parameters. It is a distributed algorithm where minimum expected relative mobility node elected as clusterhead. Cluster-based leader election existing algorithms fail to manage the dynamic nature of ad hoc network. In CBLE [25], network is divided into various clusters. Each cluster starts clusterhead selection process. The overlapped nodes in clusters are selected as global leader. The MALC [26] utilizes transmission power of mobile node to select clusterheads. A node with high transmission power can cover a large cluster area; this provides stable clustering while mobile nodes are moving within the cluster. But sending message within such cluster required mover energy for each node. In the absence/movement of clusterhead from the current cluster, maintenance of re-clustering will be higher. Figure 3 contains various influential clustering algorithm for MANET based on node mobility.

3.4 Algorithm Based on Mobile Node Battery

LIDAR [27] is a fast, simple and low-cost clustering algorithm which follows property of LID algorithms for formation and maintenance of a cluster. LIDAR concerns available energy and mobility metrics of a mobile node to extend lifetime and avoid unnecessary re-clustering. This uses simple mobility rate method, reduced control traffic volume, packet forwarding/delivery is fast and cluster formation is pro-active and based on event. In improved clustering algorithm, dynamic change in clusterhead is determined using energy of mobile node and largest energy minimize ID. ICA [28] works on clustering analysis, energy consumption and clusterhead changes. Energy-driven cognitive radio network [29] a dynamic energy-efficient clustering rotation algorithm is introduced which increases the network lifetime. Local load balancing is achieved for each cluster by selecting mobile node with highest energy and such nodes are elected as clusterhead to increase lifetime of network. In Energy-Efficient Location-Based Clustering Scheme [30, 58] thresholds are set for power level, link duration and capacity of mobile node to calculate energy conservation. It is best suitable for dynamic environment and scenarios where direction and moving rate are

Year	Protocol	Performance Metric
2001	MOBIC [14]	Relative Mobility Metrics
	MOEA [15]	Node Movement Proximity.
	MBMHC [16]	Mobility, ID, Distance, cluster distance for cluster.
2006	MAPLE [17]	Node Mobility, Channel Reservation Technique and Pro Active.
	MobDHop [18]	Distributed Clustering, 1-Hop Neighbor node information and Re-Active.
	DGMA [19]	Flexibility Metric, Dimensional Dependency derivative from the lined distance of a node's movement.
2008	UFCM [20]	Extra power, degree and broadcast capacity.
	NSBMP [21]	Used stability of neighbor nodes
	MINM [22]	Energy Cost.
2010	MBMR [23]	Responsible for building and preserving a support for clusterheads
2011	MCFA [24]	Nodes are assumed as random variable with unknown distributions. Distributed algorithm.
2013	CBLE [25]	Tree Based Centrality Method.
2015	MALC [26]	Mobility, High Transmission power of a node, Large Cluster Size.

Fig. 3 Clustering algorithm based on node mobility

not constant. In LVC [31] and ECCRP [32] use wireless network coding to minimize energy consumption while receiving and transmitting data packet to a node. Network coding is applied into CBRP protocol to improve clustering process and to reduce clustering overhead. It takes parameter such as mobility, initial energy of the nodes and packets to be sent. Weighted clustering algorithm is used by ECCRP. Which make sure best clusterhead selection and improved cluster life time by reducing energy consumption of each mobile nodes. In EA-OLSR [33] routing protocols consume less energy and improve transmission delay. Based on mobile nodes density and mobility EA-OLSR provides most important improvements concerning clusterhead election process. The residual energy will be used as a metric to optimize the number of clusters Head and to maximize the network lifetime. In fuzzy approach for clustering [56], a fuzzy approach for clustering is proposed based on various performance metrics. Two fuzzy based systems known as F2SMC1 and F2SMC2 was presented node clustering in MANETs. From the analysis result it was clear that if node is selecting based on small distance and high security values, the nodes are mover closer to the clusterheads. In [57] is a vigorous algorithm which provide safe routing clustering algorithm, energy-aware and development competences for extremely growing QoS in MANET. Some of the premier clustering algorithms developed based on node energy are summarised in Fig. 4.

Year	Protocol	Performance Metric
2001	LIDAR [27]	Remaining Battery Life, Mean Mobility Rate and N1 and N2 are Normalization Factors
2008	ICA [28]	Node Mobile and the Limit of Energy
2010	EDCRN [29]	cluster-head node real-time energy load, energy threshold,
	ELS [30]	Power & Mobility, Re-Active
2013	LVC [31]	Node Type, Euclidean Distance of two node, state, density and Average degree of clusterhead.
2014	ECCRP [32]	Wireless Network Coding, mobility, initial energy packets
2015	EA-OLSR [33]	Nodes Density, Mobility, Transmitted Energy, Receiving Energy, Idle Energy
2017	F2SMC1 [57]	Fuzzy Approach for clustering
2018	FFECM [58]	A robust Algorithm for providing QoS for MANETs.

Fig. 4 Clustering algorithm based on node energy

3.5 Algorithm Based on Mobile Node Weight

On-demand weighted clustering algorithm: Most of the exiting clustering algorithms like HDD, LID and none weighted heuristic not show the way of a finest selection of clusterheads and are invoked periodically. In OD-WCA, [34] is in the category of reactive routing protocol and invoked dynamically based on request. In OD-WCA, *combined weight* metric is calculate founded on the ultimate node degree, broadcast power, mobility and the battery power of the nodes parameters. In OD-WCA, election process for a clusterhead is late as extended as possible to abridged further calculation cost and increase stability of cluster. Repetitive changes in clusterhead election process required high message passing among nodes this leads high communication overhead and consumed battery power of nodes. OD-WCA performs better in terms of less communication overhead, less re-election of clusterhead and less message passing with in a cluster. Weighted clustering algorithm [35] was initially proposed by M. Chatterjee. It contemplates four variables and makes the choice of bunch head and support of group more sensible. As is appeared in condition (1), the four elements are hub degree contrast, remove summation to all its neighboring hubs, speed and remaining battery control separately. Furthermore, their relating weights are w_1 to w_4. Besides, it changes over the bunching issue into an advancement issue and a target capacity is defined.

$$W_i = w_1\Delta_i + w_2 D_i + w_3 V_i + w_4 E_i \quad (1)$$

Notwithstanding, just those hubs whose neighbor number is not exactly a settled limit esteem can be chosen as a group head in WCA. It is not exceptionally alluring in the viable application. For instance, some very much associated hubs whose neighbor number is bigger than the altered edge may be a decent hopeful also. Plus, its vitality model is excessively basic. It treats the bunch head and the typical hubs similarly and its residual power is a direct capacity of time, which is additionally not exceptionally alluring. Thus, we proposed an enhanced grouping calculation as takes after. In [36],

WBACA available node position information is used for clustering approach, which is based on global positioning system. It uses mobility of the node, transmission rate, degree difference, battery power and transmission power of nodes to calculate weight factor. The node with smallest weight is elected as clusterhead others act as clustermember. Weight factor is calculate by $WN = w\sim{}^{*}M + w{:}\ {}^{*}\ \mathrm{B} + w{,}{}^{*}Tx + w\sim{}^{*}$ $D - \mathrm{I} - w\ \sim/TR$ equation where *w*l, *w*2, *w*3, *w*4. And *w*5 are the weighing factors, *M*: Mobility of the node, *B*: Battery power, *Tx*: Transmission power, *D:* Degree difference and *TR*: Transmission rate. Based on number of parameter to calculate weight factor, WBACA found local minima for clusterhead selection and also assures that there is no two clusterhead be one hop neighbors. In [37], FWCA uses actual transmission power, average speed and remaining battery power of the node as parameter for clustering mechanism as inspired WCA [35]. FWCAs center of attention is actual transmission power (P_i) as compared by WCA. Actual transmission power measure better life of a node as compared by cumulative time, which allows additional lifetime to a mobile node. Resulting efficient, stable and low clustering overhead clusterhead is elected. In [38] MPWCA-L local clusterhead election process is happen based on an on-demand distributed clustering algorithm for multihop packet wireless ad hoc network based on distributed mobility predication for weighted clustering algorithm. In MPWCA-L, it added the interval between two messages to limit the overhead. During this interval, mobile nodes try to approximation the movement of their cluster head and then await clusterhead changes to avoid stability of cluster. AWC [39] finds out properties of mobile nodes to decide clusterhead to reduce clustering overhead and provides stable cluster structure. AWC use cluster radius is responsible as a dynamic parameter while cluster formation. Cluster maintenance is based on weight threshold criterion and node movement. In [40], WBCA clusterhead selection is based on election rules. WBCA works for on five steps which are finding the neighbors of each node, calculate the degree difference, calculate the consumed energy, calculate the cumulative time and calculate the weight. In WACHM [41], election of clusterhead and maintenance is depends on degree difference, battery power, average link stability and dependency probability of mobile node. WACHM performs better in terms of scalability and reduced overhead. Flexible weighted clustering algorithm based on battery power [42] is a reactive protocol, works on two phases clustering formation and clustering maintenance. FWCABP uses following preassumptions as a topology of network is static; Each mobile node joins exactly one clusterhead, each packet must received by its neighbors with in specified time, MAC address is used as node ID, ID of node must unique within cluster, cluster ID is combined with node ID to make unique identification of each node, counter is used to count number of node in a cluster to keep track of cluster size. FWCABP selects the cluster heads based on the information of neighbor nodes and maintains clusters locally. It also controls battery power consumption by changing role of node clusterhead to clustermember. Load balancing is improved by introducing the concept of predefined threshold value for nodes, which are created by clusterhead. It reduced average number of cluster formation, stability of cluster and minimum lifespan of mobile node. EWCA [43] is next version of WCA which remove the drawback of WCA, by dividing it into two phase clustering setup and cluster maintenance. Primary focal point of EWCA is to reduce the frequency of re-affiliation, the lifetime of the node and the number of cluster. It also responsible to control battery power consumption for mobile node by changing it status from clusterhead to clustermember. The EWCA [43] cannot be applied for an isolated node who wants to join as a clustermember of an existing cluster or declare itself as a clusterhead separately. WACA [44] removes the disadvantage of EWCA [44]. EWCA calculates combined weight for each node, sum of distances and average of speed. In Weight Based Self-Organizing approach-WBSOA [45], expression points are identified by exchanging signal messages. Expression points are identified at the of cluster formation to get rid of searching process. Working of WBSOA completes as initially any random mobile node start searching for its neighboring nodes, send READY message for all its neighbors, node will reply with OK message, mobile node will exchange message to make connection and decided clusterhead selection process based on reply OK Message. Weight-Based Hierarchical Clustering (WBHC) [46] covers the review or routing protocol and proposed weighted protocol based on hierarchical clustering. A node degree, mobility of mobile nodes and transmission range as parameter to calculate weight factor. This algorithm provides stable and less clustering overhead while creating and maintaining a clusterhead. Weight-Based Clustering Scheme (WBCS) [47] is a weight-based clustering algorithm which takes density, mobility, energy and trust of a mobile node as to select finest clusterhead. It performs better for mobile node resource like bandwidth, less cluster formation and maintenance, increased lifetime. As compared with LID, HD and WCA, performance of WBCS is better for clusterhead election process. In improved weighted clustering algorithm (IWCA) [48], combine weight-based approach is used for cluster formation and clusterhead election based on mobility, degree difference of node, and transmission range and battery poser of node system parameters. It reduces malicious node while clusterhead election process to enhance the clusterhead formation process. Performance results show that performance of IWCA is better as compared by WCA in terms of cluster formation and clusterhead changes. In plans and allowances in weighted clustering algorithms (SL-WCA) [49], a newly stability model was derived mathematically. This paper is extension of WCA where degree difference was used. As compared with WCA its performance in terms of cluster formation and stability is good. It is a non-periodic procedure; thus, it reduces computation and communication costs. SL-WCA removes weakness of WCA in term of computing the cumulative time during a node acts as clusterhead; WCA cannot guarantee of energy consumption. In Enhanced Weight-Based Clustering (EWBC), [50] is extension of WCA which provide clusterhead selection based on battery life time, processing speed, connectivity and mobility of nodes. It also improves hash faction, in-case there is tie when two or more nodes have same weight while selection process of clusterhead. In efficient clustering algorithm (ECA), [51] is an on-demand clustering algorithm which is based on weighted parameters to elect and maintenance cluster stricter on relative movement speed of a node. It improved clusterhead selection and maintenance; every node consumes less energy and increased life time of network. In novel weight-based clustering algorithm [52], NWBCA clustering is divided into two phase: cluster formation and cluster maintenance. In cluster formation, phase node degree and bandwidth requirement is considered. In cluster maintenance, priority is calculated of each node based on node degree and reaming battery. Based on the priority, node with higher priority starts working as clusterhead and other will hand over its charge and act as clustermember. This minimized clustering overhead and less clusterhead change. But if a new clusterhead actually wants to join in a new cluster, he has to wait up to the maximum limit period. In each iteration, priority of each clusterhead (old and new) is calculated and compared. This increased extra overhead and waiting time to join a new cluster will be increased. An optimized stable clustering algorithm (OSCA) for mobile MANETs [53], A new node was introduced which is works as backup node within the cluster. If clusterhead dies or moves out from the cluster, such backup node automatically starts working as clusterhead and responsibility of electing new backup node is now depends upon newly elected clusterhead. In OSCA, the concept of backup node minimized immediate clusterhead election process and reduced clustering overhead within the cluster. Priority-based weighted clustering algorithm, cluster formation is very important and challenging work in MANETs. In [54], a priority-based weighted clustering algorithm is proposed to minimize clusterhead changes and clustering overhead. Whole algorithm is divided into three phases identify clusterhead by calculating initial priority of each neighbors node within the cluster and second and third phase used for cluster maintenance and select new clusterhead without further delay. In PB-WCA, election of clusterhead automatically started when battery power reaches up to the threshold value. Influential node weight based algorithms are summarised in Fig. 5.

4 Conclusion

From literature survey, it is cleared that many open issues are still remain open for further research. Almost all previous clustering algorithms have been designed for electing stable clusterhead using node degree, node IDs, node movement, battery power and weight-based clustering. So there is scope of research in clustering by applying evolutionary technique such as genetic algorithm to find best suitable survival clusterhead, stable cluster and maintenance scheme for a cluster.

As we know that technology is growing very fast and moving to 5G, security is most important challenge for MANETs which always has scope for further research. As clustering algorithms use hierarchical topology for network design and nodes are dynamic in nature, design secure clustering algorithm is also a key concept for further research.

From this paper, it is clear that there are many issues in clustering algorithms such as clusterhead change, providing stable clusterhead election process, mobility issues of mobile node, clustering overhead, battery power, and transmission and cluster maintenance which required further research work.

Year	Protocol	Performance Metric
2000	OD-WCA [34]	Ideal Degree, Broadcast Power, Mobility and Battery Power, On Demand
2002	WCA [35]	Supreme node degree, Transmission Power, Mobility and Battery Power, On Demand
2005	WBACAs [36]	Transmission Power, Transmission Rate, Mobility, Battery Power and Degree of a node, On Demand
2006	FWCA [37]	Node Degree, Remaining Battery Power, Transmission Power, and Node Mobility, On Demand
	MPWC-L [38]	Nodes Mobility, On Demand
	AWC [39]	Combined Weight, On Demand
	WBCA [40]	Mean Connectivity Degree and Battery Power, On Demand
2007	WACHM [41]	Neighbor Type, Degree-Difference, Average Link Stability, Average Dependency Probability, Estimate the Residual Energy and Combined Weight, On Demand
2008	FWCABP[42]	Battery Power, On Demand
2009	EWCA [43]	Ideal node degree, Transmission Power, Mobility and Battery Power, On Demand
2010	WACA [44]	Combined Weight Function, On Demand
2011	WBSOA [45]	Identify and Avoid Formation of Articulation Points, On Demand
2012	WBHCA [46]	Combined Weight Based on Node Degree, Transmission Range and Mobility, On Demand
2013	WBCS [47]	Trust, Density, Mobility and Energy , On Demand
	IWCA [48]	Mobility, Degree difference, Transmission Range and Battery Power, On Demand
	SL-WCA [49]	Degree Difference is Replaced used initially in WCA, On Demand
2014	EWBC [50]	Battery life time, Processing Speed, Connectivity and Mobility, On Demand
2015	ECA [51]	WCA Parameter + Relative Movement Speed, On Demand
2016	NWBCA [52]	WCA+ Priority Factor, On Demand
2017	OSCA [53]	Concept of Backup Node was Introduced to reduce immediate clusterhead changes.
2019	AND OPEN PB-WCA	Clustering is automatically started when clusterhead reaches up to the minimum threshold value.

Fig. 5 Clustering algorithm based on node weight

References

1. Lin, C.R., Gerla, M.: A distributed control scheme in multi-hop packet radio networks for voice/data traffic support. In: Proceedings of IEEE GLOBECOM, pp. 1238–1242 (1995)
2. Pathak, S., Jain, S.: A survey: on unicast routing protocols for mobile ad hoc network. Int. J. Emerg. Technol. Adv. Eng. **3**(1), 204–210 (2013)

3. Yu, J.Y., Chong, P.H.J.: A survey of clustering schemes for mobile ad hoc networks. IEEE Commun. Surv. Tutor. **7**(1), 32–47 (2005)
4. Alinci, M., Spaho, E., Lala, A., Kolici, V.: Clustering algorithms in MANETs: a review. In: Ninth International Conference on Complex, Intelligent, and Software Intensive Systems, pp. 330–335 (2015)
5. Zabian, A., Ibrahim, A., Kalani, F.A.: Dynamic head cluster election algorithm for clustered ad-hoc networks. J. Comput. Sci. **4**(1), 42–50 (2008)
6. Ghaderi, J.: Hierarchical cooperation in ad hoc networks: optimal clustering and achievable throughput. IEEE Trans. Info. Theory 3425–3436 (2009)
7. Shayesteh, M., Karimi, N.: An innovative clustering algorithm for MANETs based on cluster stability. Int. J. Model. Optim. **2**, 239–244 (2012)
8. Parekh, A.K.: Selecting routers in ad-hoc wireless networks. In: Proceedings of the SBT/IEEE International Telecommunications Symposium (1994)
9. Gerla, M., Tsai, J.T.C.: Multicluster, mobile, multimedia radio network. ACM/Baltzer Wireless Netw. **1**(3), 255–265 (1995)
10. Baker, D.J., Ephremides, A.: The architectural organization of a mobile radio network via a distributed algorithm. In: IEEE Transactions on Commicationuns COM-29 11, pp. 1694–1701 (1981)
11. Ephremides, A., Wieselthier, J.E., Baker, D.J.: A design concept for reliable mobile radio networks with frequency hopping signaling. Proc. IEEE **75**, 56–73 (1987)
12. Yadav, N.S., Deosarkar, B.P., Yadav, R.P.: A low control overhead cluster maintenance scheme for mobile ad hoc networks. Int. J. Recent Trends Eng. **1**(1), 1–9 (2009)
13. Pathak, S., Dutta, N., Jain, S.: An improved cluster maintenance scheme for mobile AdHoc networks. In: IEEE 2014 International Conference on Advances in Computing, Communications and Informatics, pp. 2117–2121 (2014). https://doi.org/10.1109/icacci.2014.6968281
14. Basu, P., Khan, N., Little, T.D.C.: A mobility based metric for clustering in mobile ad hoc networks. In: Proceedings IEEE ICDC-SW'01, pp. 413–418 (2001)
15. Cheng, H., Cao, J., Wang, X., Das, S.K.: Stability-based multi-objective clustering in mobile ad hoc networks. In: The Third International Conference on Quality of Service in Heterogeneous Wired/Wireless Networks, pp. 1–8 (2006)
16. Er, I.I., Seah, W.K.G.: Clustering overhead and convergence time analysis of the mobility-based multi-hop clustering algorithm for mobile ad hoc networks. J. Comput. Syst. Sci. **72**, 1144–1155 (2006)
17. Palit, R., Hossain, E., Thulasiraman, P.: MAPLE: a framework for mobility-aware pro-active low energy clustering in ad-hoc mobile wireless networks. Wirel. Commun. Mob. Comput. **6**, 773–789 (2006)
18. Seah I, W.K.G.: Performance analysis of mobility-based dhop (MobDHop) clustering algorithm for mobile ad-hoc networks. Comput. Netw. **50**, 3375–3399 (2006)
19. Zhang, Y., Ng, J.M.: A distributed group mobility adaptive clustering algorithm for mobile ad hoc networks. In: Proceedings of the IEEE International Conference on Communications (ICC), pp. 3161–3165 (2008)
20. Wang, S.C., Pan, H.H., Yan, K.Q., Lo, Y.L.: A unified framework for cluster manager election and clustering mechanism in mobile ad hoc networks. Comput. Stan. Interfaces **30**, 329–338 (2008)
21. Konstantopoulos, C., Gavalas, D., Pantziou, G.: Clustering in mobile ad hoc networks through neighborhood stability-based mobility prediction. Comput. Netw. **52**, 1797–1824 (2008)
22. Ghosh, R., Basagni, S.: Mitigating the impact of node mobility on ad-hoc clustering. J. Wirel. Commun. Mob. Comput. **8**, 295–308 (2008)
23. Torkestani, J.A., Meybodi, M.R.: Mobility-based multicast routing algorithm in wireless mobile ad hoc networks: a learning automata approach. J. Comput. Commun. **33**(6), 721–735 (2010)
24. Meybodi, M.R.: A mobility-based cluster formation algorithm for wireless mobile ad-hoc networks. Clust. Comput. **14**(4), 311–324 (2011)

25. Paul, A., Preetha, K.G.: A cluster based leader election algorithm for MANETs. In: International Conference on Control Communication and Computing, pp. 496–499 (2013)
26. Neethu, V.V., Singh, A.K.: Mobility aware loose clustering for mobile ad hoc network. In: Eleventh International Multi-Conference on Information Processing (IMCIP-2015), Procedia Computer Science, vol. 54, pp. 57–64 (2015)
27. Gavalas, D., Pantziou, G., Konstantopoulos, C. et al.: Telecommun Syst. **36**, 13 (2007). https://doi.org/10.1007/s11235-007-9053-1
28. Huawei, S.J., Gaoping, W.: An improved clustering algorithm based on MANET network. In: IEEE International Symposium on IT in Medicine and Education, pp. 940–944 (2008)
29. Tan, G.X.X., Wei, S., Guo,S., Wang, B.: An energy—driven adaptive clusterhead rotation algorithm for cognitive radio networks. In: IEEE (2010)
30. Singh, K., Awasthi, A.K., Mishra, R.: A stable energy-efficient location based clustering scheme for ad hoc networks. In: Institute for Computer Sciences, Social Informatics and Telecommunications Engineering, LNICST, vol. 115, pp. 75–85 (2013)
31. Yang, P.Z.X., Yu, W., Fu, X.: A loose-virtual-clustering-based routing for power heterogeneous MANETs. IEEE Trans. Veh. Technol. **62**(5) (2013)
32. Kanakala, S., Ananthula, V.R., Vempaty, P.: Energy-efficient cluster based routing protocol in mobile ad hoc networks using network coding. Hindawi Publishing Corporation J. Comput. Netw. Commun. **2014**(351020), 1–12 (2014). https://doi.org/10.1155/2014/351020
33. Loutfi, A., Elkoutbi, M.: A new efficient and energy-aware clustering algorithm for the OLSR protocol. IJCSNS Int. J. Comput. Sci. Netw. Secur. **15**(7), 85–91 (2015)
34. Chatterjee,M., Das, S.K., Turgut D.: An on-demand weighted clustering algorithm (WCA) for ad hoc networks. In: Proceedings of IEEE GLOBECOM, San Francisco, pp. 1697–1701 (November, 2000)
35. Chatterjee, M., Das, S.K., Turgut, D.: WCA: a weighted clustering algorithm for mobile ad hoc networks. Kluwer Academic Publishers, Manufactured in The Netherlands, pp. 193–204 (2002)
36. Dhurandher, S.K., Singh, G.V.: Weight based adaptive clustering in wireless ad hoc networks. In: IEEE International Conference on Personal Wireless Communications (ICPWC), New Delhi, India, 23-25, pp. 95–100 (2005)
37. Bazzal, Z.E., et al.: A flexible weight based clustering algorithm in mobile ad hoc networks. In: IEEE International Conference on Systems and Networks Communications (2006)
38. Bricard-Vieu, V., Nasser, N., Mikou, N.: A mobility prediction-based weighted clustering algorithm using local cluster-heads election for QoS in MANETs. In: WIMOB, pp. 24–30 (2006)
39. Rasheed, T., Javaid, U., Reynaud, L., Agha, K.A.: Adaptive weighted clustering for large scale mobile ad hoc networking systems. In: Wireless Algorithms Systems and Applications, Lecture Notes in Computer Science, vol. 4138, pp. 206–216 (2006)
40. Yang, W.D., Zhang, G.: A weight-based clustering algorithm for mobile ad hoc network. In: IEEE Proceedings ICWMC'07 0-7695-2796-5/07 (2007)
41. Wang, Y., Chen, H., Yang, X.: WACHM: weight based adaptive clustering for large scale heterogeneous MANET. In: Proceedings IEEE 2007 International Symposium on Communications and Information Technologies (ISCIT 2007), pp. 936–941 (2007)
42. Hussein, A.R., Amer, O., Salem, A., Yousef, S.: A flexible weighted clustering algorithm based on battery power for mobile ad hoc networks. IEEE Int. Symp. Indus. Electron. **2008**, 2102–2107 (2008)
43. Li, C., Wang, Y., Huang, F., Yang, D.: A novel enhanced weighted clustering algorithm for mobile networks. In: 5th International Conferences on Wireless Communications Networking and mobile Computing, pp. 24–26 (2009)
44. Nath, I., Chaki, R., Chaki, N.: WACA: a new weighted adaptive clustering algorithm for MANET. In: Recent Trends in Networks and Communications, Communications in Computer and Information Science, vol. 90, pp. 270–283 (2010)
45. Muthuramalingam, S., VasanthaKumar, M., Balaji, C.M.: A new weight based self-organizing approach to clustering in mobile ad hoc networks. In: ACWR'11 Proceedings of the 1st

International Conference on Wireless Technologies for Humanitarian Relief, pp. 489–492 (2011)
46. Sahana, S., Saha, S., Gupta, S.D.: Weight based hierarchical clustering algorithm for mobile ad hoc networks. Int. Conf. Modell. Optim. Comput. Procedia Eng. **38**, 1084–1093 (2012). https://doi.org/10.1016/j.proeng.2012.06.137
47. Bentaleb, A., Harous, S., Boubetra, A.: A weight based clustering scheme for mobile ad hoc networks. In: MoMM'13 Proceedings of International Conference on Advances in Mobile Computing and Multimedia, pp. 161 (2013)
48. Gupta, N., Singh, R.K., Shrivastava, M.: Cluster formation through improved weighted clustering algorithm (IWCA) for mobile ad-hoc networks. In 2013 Tenth International Conference on Wireless and Optical Communications Networks (WOCN), vol. 26–28, pp. 1–5 (2013)
49. Aissa, M., Belghith, A., Drira, K.: New strategies and extensions in weighted clustering algorithms for mobile ad hoc networks. In: The 4th International Conference on Ambient Systems, Networks and Technologies (ANT 2013) Procedia Comput. Sci. **19**, 297–304 (2013). https://doi.org/10.1016/j.procs.2013.06.042
50. Shehzad, D., Ishaq, W., Khan, Z., Iqbal, J.: An enhanced weight based clustering algorithm for mobile adhoc networks. In: International Conference on Computer Science and Information Systems (ICSIS'2014) Oct 17–18, Dubai (UAE), pp. 116–118 (2014)
51. Zhou, H., Zhang, J.: An efficient clustering algorithm for MANET based on weighted parameters. In: 8th International Symposium on Computational Intelligence and Design, pp. 145–148 (2015)
52. Pathak, S., Jain, S.: A novel weight based clustering algorithm for routing in MANET. Wireless Netw. **22**(8), 2695–2704 (2016). https://doi.org/10.1007/s11276-015-1124-8
53. Pathak, S., Jain, S.: An optimized stable clustering algorithm for mobile ad hoc networks. EURASIP J. Wireless Commun. Netw. **1**, 1–11 (2017). https://doi.org/10.1186/s13638-017-0832-4
54. Pathak, S., Jain, S.: A priority-based weighted clustering algorithm for mobile ad hoc network'. Int. J. Commun. Netw. Distrib. Syst. **22**(3), 313–328 (2019). https://doi.org/10.1504/IJCNDS.2019.098872
55. Pathak, S., Jain, S.: A status based clusterhead election algorithms for mobile ad hoc networks. Int. Elixir Netw. Engg. **77**, 29093–29097 (2014). ISSN: 2229-712X
56. Ozera, K., Sakamoto, S., Elmazi, D., Bylykbashi, K., Ikeda, M., Barolli, L.: A fuzzy approach for clustering in MANETs: performance evaluation for various parameters. Int. J. Space-Based and Situated Comput. **7**(3), 166–176 (2017)
57. Marcel, G., Kovalan, A.: A full-featured efficient clustering mechanism for improving QoS in MANETs. Int. J. Netw. Virtual Organ. **19**(4), 105–132 (2018)
58. Kumar, S., Singh, A.K.: A localized algorithm for clustering in cognitive radio networks. J. King Saud Univ. Comput. Inf. Sci. (2018)
59. Alsuhli, G.H., Khattab, A., Fahmy, Y.A.: Double-head clustering for resilient VANETs. Wirel. Commun. Mob. Comput. **2019**, 1–17 (2019)
60. Ge, X., Gao, Q., Quan, X.: A novel clustering algorithm based on mobility for VANET. Int. Conf. Commun. Technol. Proc. ICCT 473–477 (2019)
61. Ahmad, M., Hameed, A., Ikram, A.A., Wahid, I.: State-of-the-art clustering schemes in mobile ad hoc networks: objectives, challenges, and future directions. IEEE Access **7**, 17067–17081 (2019)
62. Ren, M., Labiod, H., Veque, V., Khoukhi, L., Zhang, J.: A unified framework of clustering approach in vehicular ad hoc networks. IEEE Trans. Intell. Transp. Syst. **19**(5), 1401–1414 (2017)

Exploring a Novel Strategy for Detecting Cyber-Attack by Using Soft Computing Technique: A Review

Sona D. Solanki, Jaymin Bhalani, and Nazir Ahmad

Abstract In today's globalized world, the computer-based information system is one of the most significant domains that have played a major role in human life. On the other hand, data has been the superlative valuable stuff of an organization. Organizations are thus more reliant on the information as more data would be collected and distributed on a network-based system. Contravention of network security is one of the severe forms of cyber-attacks, as attackers generally seek complete access to all information and devices. Cybersecurity serves a tremendous amount of security as it is intended to protect the information. Safeguard mechanisms are used to enhance information security. With the development of soft computing, intrusion detection effectively monitors the dubious operation of system behavior and monitoring station that has become an integral part of security execution. Detection of intrusion is a mechanism of intelligently detecting events that occur in the operation of suspicious and malicious device activity and device resources within a network environment. It focuses on identifying possible attacks, capturing data and attempts to investigate. If IDS is implemented with alerts, then a device will be placed under risk. It is an enormous synchronized challenge on the availability of system assets resources that is conducted indirectly through a significant number of unreliable internet computer operators that disrupt services of valid users by consuming the server assets like CPU, hard drives, etc. This academic article addresses several facets of combating DDOS attacks by using soft computing methods for identification, security and prevention, strategies to trace back, relevant queries and research issues using artificial neural network (ANN).

S. D. Solanki (✉) · N. Ahmad
Communication Engineering Department, Babaria Institute of Technology, Vadodara, India
e-mail: solankisona28@gmail.com

N. Ahmad
e-mail: nazirqayoom@gmail.com

J. Bhalani
Computer Science Department, Hewad University, Kabul, Afghanistan
e-mail: jayminbhalani.ec@bitseducampus.ac.in

S. Borah et al. (eds.), *Soft Computing Techniques and Applications*, Advances in Intelligent Systems and Computing 1248,
https://doi.org/10.1007/978-981-15-7394-1_52

Keywords Cyber-attacks · DDOS · Soft computing · Intrusion detection system · Artificial neural network

1 Introduction

Over the past decades, the software-oriented framework has played a significant role in different spheres of human life. Data has been the most valuable commodity of organizations. Since more data is collected and transmitted on network-based systems, organizations have become increasingly reliant on the information. The extensive use of e-commerce has greatly increased the need to safeguard the network. Infringement of network security is one of the worst kinds of data breaches, as it usually requires complete access to all information and devices. Issues of great significance such as storing, distributing and streamlining the transmission of information are assigned to them. The level of protection of the information transmitted can differ from private and industrial to defense and state secrets [1]. The violation of the secrecy, integrity and availability of the data can result in harm to its owners and have substantial undesirable effects. Therefore, several organizations and industries are extremely worried about the issue of cybersecurity for the establishment of security services, which require substantial contributions. Cybersecurity serves a significant part in safeguards. It is intended to protect the information in both static and dynamic conditions. Safety measures may be used to enhance the security of information. Access monitoring, awareness education, inspection and compliance, risk evaluation, vulnerability scanning, risk mitigation as well as safety assessment and authorization are some of the initiatives. With the advancement of soft computing, intrusion detection [2–5] that intelligently tracks suspicious activity device operations and alerts to a monitoring station has become an integral part of the security process. It has a range of "varieties" and it works at identifying unusual traffic in various ways [6–9]. This paper is concerned with identifying the DDOS attack, which is a distributed denial-of-service attack. It is a huge-scale, synchronized threat on the availability of target device or system assets resources that is conducted indirectly via a huge number of vulnerable computer agents on the internet that interrupts services of a rightful user's by consuming the server assets like CPU, memory, hard drive/database and bandwidth [10]. Thus, we can overcome DDOS attacks by using the soft computing method in the intrusion detection framework and by defining a successful mitigation mechanism using artificial neural networks or ANNs.

2 Evolution, Significance and Application of Soft Computing

Computational intelligence is a collection of computing techniques and methods influenced by evolution to solve complex actual-world problems. Intelligent systems that have human-like skills such as domain intelligence, ambiguous reasoning and adjustment to a noisy and time-varying system are essential in addressing realistic computing issues. Soft computing is a concept for optimizing and processing strategies that accommodate the effects of inaccuracy and ambiguity, making them desirable to use when identifying intrusions. Zadeh initially proposed soft computing to develop new era computationally smart hybrid systems comprising of neural networks, fuzzy inference method, indirect reasoning and are used to automate the process of intrusion detection and eliminate human involvement in techniques like evolutionary computing, machine learning and free derivative optimization techniques [11]. The concept behind the implementation of soft computing techniques and especially artificial neural networks in the application of IDSs is to provide an intelligent system capable of revealing latent patterns in unusual and regular connection audit records. Writing Identification, Production and Industrial Management, Image Analysis and Information Compaction, Design, Decision Assistance, Systems Power Technologies, Fuzzy Logic Regulation, Neuro-Fuzzy Structures are a few applications of soft computing [12]. Soft computing is not competitive but a compliment, and it is a key element for the evolving conceptual knowledge sector. Here, Fig. 1 illustrates the classification of soft computing techniques.

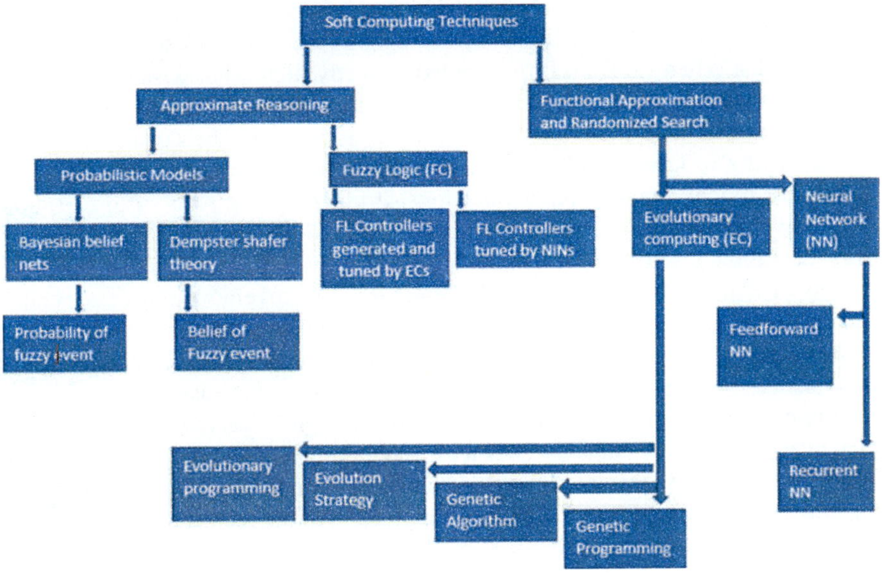

Fig. 1 Soft computing technique (SCT)

3 An Outline of the Intrusion Detection Systems (IDSs)

As users, people rely on the Internet in their everyday lives to manage important tasks such as reviewing e-mails, handling personal and financial details, etc. Sending these data over the Internet often implies that the network has become an increasingly desirable place for attackers. The research community has responded to this threat with a growing interest in intrusion detection, contributing to the construction of new technologies for prompt detection of intruders and mitigation of damage. On a subatomic scale, an intrusion detection system (IDS) [13, 14] is a process of intelligently detecting or tracking events that occur in the operations of suspicious and malicious system activity as well as system resources in a network system and notifying the monitoring station whether such activities are symptomatic of a breach or constitute a legitimate use of the framework. It has a range of "varieties" and it works at identifying unusual traffic in various ways [15]. It concentrates mainly on detecting potential incidents, recording data and attempts to investigate. In organizations, they are used for defining security policy concerns, record current threats and dissuade individuals from breaching security policies. Many types of breaches are feasible, extending from the violation of privileges to the use of threats to manipulate vulnerability in software or protocols. It is primarily directed at ensuring the security, confidentiality and integrity of sensitive networked data systems. They are an essential element of safeguards that defend against the misuse of software systems and networks. When an IDS is implemented correctly, alerts can be issued indicating that a system is under danger. Intrusion detection aims to identify software threats by investigating data logs that are witnessed by the processes on the same system. This detector will also launch probes to activate the evaluation process such as seeking requests of model numbers for applications. It utilizes three sorts of information: long-term information relevant to the method required to detect intrusions, configuration data about the present state of the system and review data explaining the events that are occurring to the system [16]. The detector's job is to remove unnecessary audit trail details. It then represents either a synthetic viewpoint of the security-associated measures taken during standard system consumption or a synthesized approach of the system's existing security state. A judgment is then taken to assess the likelihood that these actions or this state could be perceived as signs of an intrusion or fragility. A countermeasure element then takes remedial measures either to avoid the actions being performed or to restore the device status to a secure condition. Thus, intrusion detection systems have become a crucial supplement of the security infrastructure to identify such attacks before inflicting irrecoverable harm.

3.1 Identification Framework for Intrusion Is Categorized into Two Categories

A. *Host-based attacks*

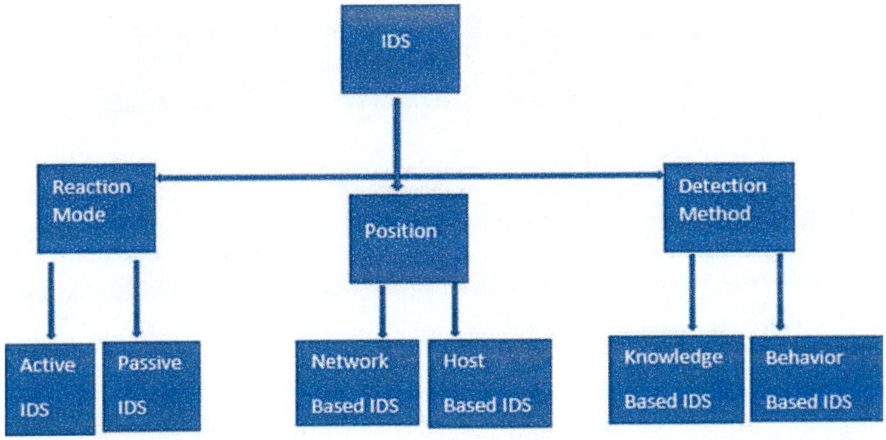

Fig. 2 Categories of IDS

Host-based attack identification routines generally use device call data from an evaluation process that records every network calls made on each user's behalf on a given device. These inspection processes typically execute on every device being tracked.

B. *Network-based attacks*

Network-based threat identification routines typically use network calls made on a specific device on behalf of each client. These inspection operations usually operate on every device being supervised. They usually use network traffic information from a network sniffer (e.g., Wireshark).

4 Soft Computing Methodology in IDS

There are several methods for detecting cyber-attacks in IDS. These essential methods are support vector machine (SVM), neural network (NN), fuzzy logic (FL) and evolutionary computation (EC). Since neural networks are used for further research work, the discussion explicitly addresses the artificial neural network (ANN). Figure 2 gives a glimpse of the categorization of IDS.

5 Association of DDOS Attack and Soft Computing Technique

The Internet is the universal forum that revolutionized the computer and connectivity domain. Cyber-crime is one of the daunting tasks in today's digital era. Many

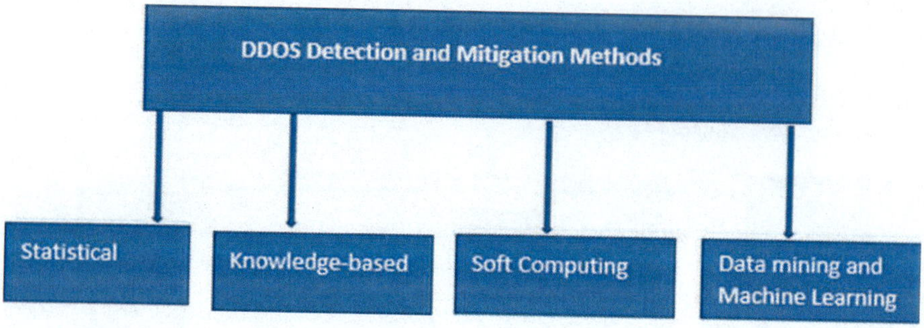

Fig. 3 DDOS detection and mitigation methods

researchers have suggested numerous frameworks for cyber-defense. Still, all these safeguards are menaced by cyber threats thrown by the attackers. For aspiring researchers to tackle the forthcoming cyber-threats, the majority of the existing approaches should be reframed. Intrusion detection systems play a significant role in managing the threats within a framework with many established technologies available. Soft computing strategies are an emerging technology for coping with imprecise and accurate challenges. This article addresses some of the soft computing methods that can be used in the intrusion detection system to tackle network activity with improved efficiency and precision in identifying threats. The study includes two categories of soft computing techniques: artificial neural networks (ANNs) and support vector machines (SVMs) [12]. Since neural networks are used for further research work, the discussion explicitly addresses the artificial neural network (ANN). The strategies of identification and mitigation of a DDOS attack are indicated in Fig. 3.

6 Artificial Neural Network

An artificial neural network (ANN) is a category of statistical data processing model influenced by the way biological neuron structures, such as the brain operates data and is used to precise tasks that can rely on a substantial number of inputs and something unspecified. It is a type of connection-oriented processing where information is acquired and retained by a network of strongly connected computing elements, i.e., neurons, which is essentially a summing component including measured synapses and threshold logical unit working to solve particular issues [1]. The links have numerical weights that can be altered through observation, effectively rendering neural networks and learning. This is especially useful in implementations where data complexity makes the construction of this method unfeasible. The output of every neuron (after imposing the connection-associated weight parameter) is supplied as the input to all the cells in the next layer. The process of this creative learning is a method of optimization where the conditions of the best set of relevance parameters (weights) are identified to fix a problem. ANN methodology tackles the core principle of hybrid devices with accuracy, consistency and specificity to maximize the tolerance for inaccuracy, ambiguity, reduced cost of the solution, reliability and partial truth to gain tractability and better relationship with reality. As ANNs are capable to make multi-class categorizations, the intrusion identification is performed using an ANN.

7 DDOS Cyber-Threat

Denial-of-service (DOS) and distributed denial-of-service (DDOS) threats are generally deliberate efforts to overwhelm the bandwidth of targets or interrupt the connection of authorized users to access its resources. Existing Internet infrastructure is susceptible to DDOS threats, which offers potential for an intruder to control a significant number of infected devices by targeting their flaws in establishing attack systems or Botnets [17]. If an attack network or Botnet is formed, then a hacker provokes a synchronized and massive-scale attack against one or more victims. Due to the ongoing emergence of new threats and an increasing number of susceptible hosts on the network, several methods for DDOS attack identification, mitigation and trace back are developed [10]. In this article, we aim to review various sorts of attacks and DDOS intrusion tactics and their defense mechanism. The importance of this paper is that it addresses several facets of combating DDOS attacks namely identification, security and prevention, strategies to trace back, relevant queries and research issues. The various phases of the DDOS attack can be observed in Fig. 4.

Fig. 4 Phases of performing DDOS attack

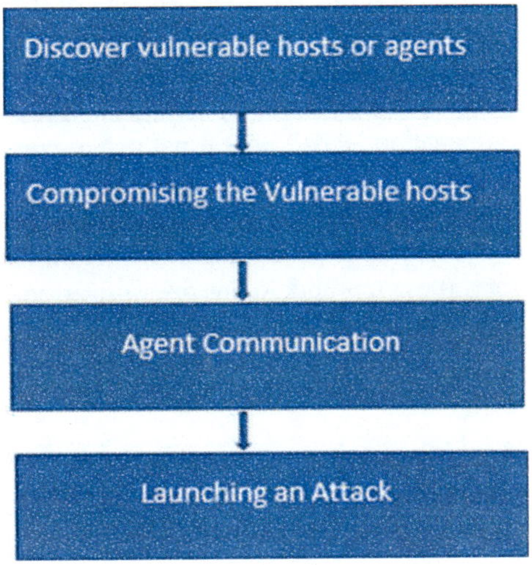

8 Implementation of DDOS Identification

Understanding methodologies such as artificial neural networks (ANNs), radial base structures and genetic algorithms that are commonly used in DDOS threat identification due to their intelligent and automatic classification capability and are useful to recognize and prevent unidentified disruptions in a framework because of its auto-learning feature. SPUNNID is used as a DDOS threat recognition system centered on a preprocessor statistic and an artificial neural network that is not monitored. It uses pre-processing statistics to collect traffic characteristics and utilizes an unmonitored neural network to interpret and identify traffic as intrusion or ordinary traffic. An anomaly driven approach for identifying DDOS using the radial basis function (RBF) neural networks centered on attack packet interpretation functionality [10]. Classification of information as standard or threat categories is implemented. If the received traffic is recognized as threat traffic, the source IP address of the threat traffic is sent to the filtering device and the attack detector unit carries out further measures. If the traffic is standard, then it is driven to the destination. A strategy for monitoring DDOS attacks in open networks based on statistical characteristics computed in brief-time window evaluation of an incoming data packet is executed. To classify the activity of the DDOS attacks, a limited number of statistical parameters are applied. Effective representation is obtained by using neural networks with the radial basis function. DDOS attacks are monitored utilizing decisions tree and the grey functional evaluation method. The mapping of the threat from the neutral state is described as a classification issue. They utilize 15 parameters to track the incoming/outgoing packet/byte rate and to obtain the flag values of TCP, SYN and ACK to describe the flow structure of traffic. The decision tree technique is applied to create a dataset to identify the irregular flow of traffic and to use a unique traffic sequence matching method to classify traffic flow identical to the intrusion flow and trace back the source of an attack. Classifier ensembles deploy the resilient back propagation (RBP) neural net as the core classifier for DDOS detection. They are primarily focused on increasing the efficiency of the base classifier. The RBP boost incorporates the output of the classifier outputs ensemble with the cost reduction technique of Neyman Pearson, for ultimate decision on classification [10]. Therefore, these methodologies are utilized to identify and mitigate DDOS attacks. The SYN flooding DDOS attack is explained in Fig. 5.

9 Conclusion

Over the last few decades, it has been proved that data transmission over the internet has often stated that the network has become an increasingly popular location for hackers. Nowadays, several organizations are concerned about the cybersecurity issue for the establishment of security services. Cybersecurity is an integral part of

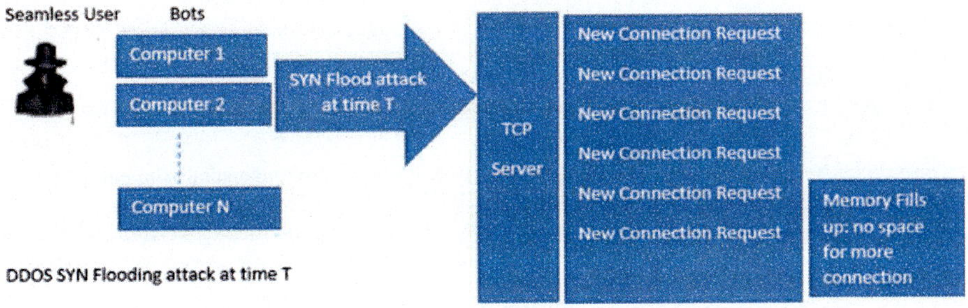

Fig. 5 SYN flooding DDOS attack

safeguarding the data and it is intended to protect it under static and dynamic conditions. With soft computing advancement, intrusion detection that detects dubious performance of system operation has now become a significant part of the security process. The research community is engaged in intrusion detection contributing to the development of new technologies for timely identification of intrusions and risk reduction due to the threat of cyber attackers. This academic article explains a DDOS attack that is a distributed denial-of-service attack. It is a synchronized threat to the availability of target devices that is conducted indirectly through several vulnerable internet computer agents that can easily disrupt the services of the appropriate customer. In this paper, it is addressed that the DDOS attack is overcome by the use of soft computing techniques in the intrusion detection framework by establishing an effective mitigation mechanism utilizing an artificial neural network.

References

1. Chaturvedi, D.K.: In: Soft Computing: Techniques and Its Applications in Electrical Engineering, Springer (2008)
2. Dutt, I., Borah, S., Moitra, I.K.: Survey of bio-inspired techniques based on system resource usage in intrusion detection. Test Eng. Manage. **82**, 5734–5738 (2020). ISSN 0193-4120
3. Prasad, K.K., Borah, S.: Use of genetic algorithms in intrusion detection systems: an analysis. Int. J. Appl. Res. Stud. (IJARS) **2**(8), 01–07 (2013). ISSN: 2278-9480
4. Dutt, I., Borah, S.: Some studies in intrusion detection using data mining techniques. Int. J. Innov. Res. Sci. Eng. Technol **4**(7), 5500–5511 (2015). ISSN: 2347-6710
5. Dutt, I., Borah, S., Maitra, I.: Intrusion detection system using artificial immune system. Int. J. Comput. Appl. **144**(12), 19–22 (2016). ISSN: 0975-8887
6. Panigrahi, R., Borah, S.: A statistical analysis of lazy classifiers using canadian institute of cybersecurity dataset. In: Proceedings of ICDSM-2019, Bhubaneswar, India. Lecture Notes on Data Engineering and Communication Technology, vol. 7, pp. 215–222. Springer (2019). ISSN 2367-4512, ISBN 978-981-15-0978-0
7. Panigrahi, R., Borah, S.: Rank of normalizers through TOPSIS with the help of supervised classifiers. Int. J. Eng. Technol. **7**, 483–490 (2018)
8. Panigrahi, R., Borah, S.: A detailed analysis of CICIDS2017 dataset for designing intrusion detection systems. Int. J. Eng. Technol. **7**(3.24), 479–482 (2018)
9. Panigrahi, R., Borah, S.: Rank allocation to J48 group of decision tree classifiers using binary and multiclass intrusion detection datasets. Procedia Comput. Sci. **132**, 323–332 (2018)

10. Munivara Prasad, K., Rama Mohan Reddy, A., Venugopal Rao, K.: DOS and DDOS attacks: defense, detection and traceback mechanisms-a survey. Global J. Comput. Sci. Technol. E Netw. Web Secur. **14**(7) (2014). ISSN: 0975-4172
11. Zadeh, L.A.: Fuzzy logic, neural networks and soft computing. Commun. ACM **37**(3), 77–84 (1994)
12. Zadeh, L.A.: A definition of soft computing- adapted from eliance spells the end of the road for ICT amateurs, The Australian (May 07 2013)
13. Panigrahi, R., Borah, S.: Dual stage intrusion detection for class imbalance scenario. Comput. Fraud Secur. **2019**(12), 12–19 (2019). ISSN 1361-3723
14. Dutt, I., Borah, S., Moitra, I.K.: Immune system based intrusion detection system (IS-IDS): a proposed approach. IEEE Access **8**(1), 34929–34941 (2020). https://doi.org/10.1109/ACCESS.2020.2973608. ISSN:2169-3536
15. Lekha, J., Mohamad Imran Khan, A., Rohini, R.: Soft computing techniques to detect cyber attack. Int. J. Res. Sci. Eng. Technol. (2015)
16. Scarfone, K., Mell, P.: In: Guide to Intrusion Detection and Prevention Systems (IDPS). Computer Security Resource Center (National Institute of Standards and Technology), pp. 800–894 (2010)
17. Sharma, J., Borah, S.: Botnet detection techniques-an analysis. Int. J. Rec. Technol. Eng. **8**(1), 2130–2137 (2019). ISSN 2277-3878

Comparison Between the Statistical Method Models for Better Time Series Sales Forecasting Model

Theviya Darshini A/P Ponniah, Sharifah Sakinah Binti Syed Ahmad, and Samarjeet Borah

Abstract Time series analysis contains strategies for studying time series data a good way to extract significant facts and different traits of the information. Paperion of the time series is the use of a model to forecast future values based on formerly predicted values. Time series are extensively used for non-stationary statistics, like economic, weather, stock price, and retail income on this put up. This paper is to investigate on comparison between the traditional statistical method models which are ARMA, ARIMA, and ARIMAX model using the dataset provided in Kaggle competition webpage. Three numbers of time series were implemented on the validation, and comparison is done in terms of performance. As a result, the following month's forecast for the total monthly sales for every shop within the next month will be determined and predicted. This paper will provide a step-by-step guide for fitting the traditional statistical method models using Jupyter Notebook. The methodology used in the paper is data analytics lifecycle which is a technique used to understand the data and apply statistics to get insights for an achievable objective. By end of this paper, the main aim to be achieved is to understand and analyze the data, then to do the overall revenue prediction for each store over the next month.

Keywords Time series prediction analysis · ARMA · ARIMA · ARIMAX · Data analytics · Statistical method model

---

T. D. A. Ponniah · S. S. B. S. Ahmad
Universiti Teknikal Malaysia, Melaka, Malaysia
e-mail: theviya1@gmail.com

S. S. B. S. Ahmad
e-mail: sakinah@utem.edu.my

S. Borah (✉)
Sikkim Manipal Institute of Technology, Sikkim Manipal University, Sikkim, India
e-mail: samarjeetborah@gmail.com

S. Borah et al. (eds.), *Soft Computing Techniques and Applications*, Advances in Intelligent Systems and Computing 1248,
https://doi.org/10.1007/978-981-15-7394-1_53

1 Introduction

Analysis of the time series includes methods for the analysis of time series data as a good way to derive relevant details and various knowledge traits [1]. Time series forecasting is the usage of a model to predict future values based on formerly determined values [2]. Time series are extensively used for non-stationary statistics, like economic, weather, stock price, and retail income on this put up [3]. This paper is to investigate on comparison between the traditional statistical method models which are ARMA, ARIMA, and ARIMAX model using the dataset provided in Kaggle competition webpage. The time series dataset, consisting of daily sales data, is provided by one of Russia's largest software companies, 1C Company. 1C Company in Moscow, Russia, was founded in 1991 by Boris Nuraliev. 1C Company is an independent Russian software developer, video game localization and publisher, with its headquarters in Moscow, Russia. 1C Company focuses on improvement, distribution, publishing and help of mass-market software program. 1C Company turned into one of the primary businesses inside the enterprise with over 2000 fulltime personnel. 1C Company has the most important software program income volume in CIS IT industry. By end of this paper, the main aim to be achieved is to understand and analyze the data and then to do the overall revenue prediction for each store over the next month.

2 Methodology

This section primarily discusses the methods used by the data science approach to analyze the data. In data science, methodology is an operation of specific procedures which is used to understand and analyze the problems to be solved in a research. By implementing a methodology, this enables the researcher to evaluate their study. In the methodology parts, there are basic steps which describes on what are the things that needed to be done. The methodology part of the research focuses on the six repetitive steps, and each of the steps is explained on what tasks need to be executed in general. The specifics on what shall be done in each step in details are explained in the next segment.

The methodology used in the paper is data analytics lifecycle. Data analytics lifecycle is a technique used to understand the data and apply statistics to get insights for an achievable objective [4]. It is also intended to address data science and big data issues which function as a strategy to guide its user in solving complication [5]. This repetitive cycle consists of six main phases which are discovery, understand the data, data preparation, model planning, model building and communicate result (Fig. 1).

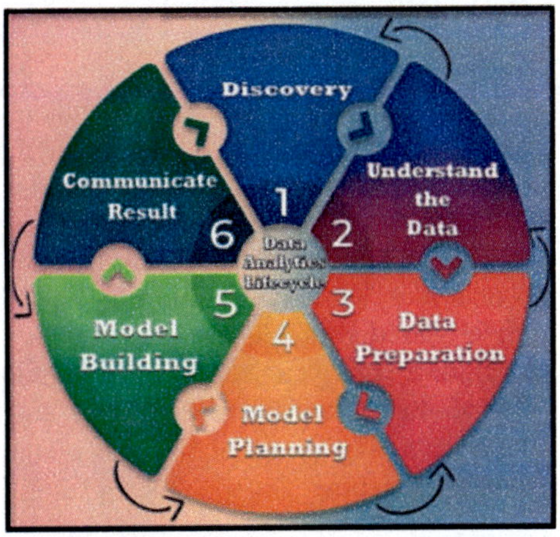

Fig. 1 Data analytics lifecycle

3 Time Series Prediction Using Statistical Method Models

3.1 Introduction

This segment mainly focuses on the proposed methods or steps that are being done in each phase stated in the paper methodology in the previous segment. Discover is the first phase of the data analytics lifecycle. In this phase, the problem to solve will be determined. A major research should be conducted to determine the problems to be solved. The next part of this segment is understanding the data. Major part of the research begins in this phase in order to understand the problem determined better. During the phase of understanding the data, the business trends will be analyzed [6]. Data preparation is the third phase in the data analytics lifecycle. In this phase, data preprocessing will be mainly focused. Data preprocessing involves some processes to be executed on the dataset. The process which should be executed on dataset for data preprocessing is data cleaning which is to eliminate noisy data, detect outliers, and detect change point [7]. Method planning is the fourth phase of the data Analytics lifecycle. Basically, the method to be used is determined on the basis of the hypotheses and data structure that are made during the planning phase of the model. Appropriate techniques and approach need to be decided as it has to be implemented to meet the objectives of the paper to meet the end goal. Next, the fifth step of data analytics lifecycle is model building. This phase consists of four process which is load the data, preprocessing the data which is to check on the validation set, and check on the stationary of data, fit time series models, and predict values on validation set, and check the performance of the model [8]. Communicate results is the final phase of the data analytics lifecycle. In this phase, the success rate of the paper objective will be illustrated through aspects of the results which are identified based on the findings [9]. However, the interpretation and deeper analysis of the results will be done in the following segment [10].

3.2 Discovery

Discovery phase in this paper is to obtain the data. Obtaining data is one of the important parts in this paper. While choosing the dataset from all the resources provided on the open source pages provided, I choose the dataset which provided on the Kaggle competition webpage. Kaggle is the place to do data science papers. Kaggle, owned by Google Inc., is an online community of data scientists and computer learners. Kaggle enables users to find and publish data sets, explore and build models in a Web-based data science environment, collaborate with other data scientists and machine learning engineers, and join competitions to tackle data science challenges [11]. Kaggle started off by offering machine learning competitions and now also provides a public data platform, a cloud-based data science workbench and short form AI coaching. Google revealed their acquisition of Kaggle on 8 March 2017.

3.3 Understand the Data

The dataset has three main files which is the item file dataset, shop file dataset, and item category file. The historical dataset has the information of products sold in every shop from January 2013 to October 2015. The datasets contains 11 attributes. The table below explains the meaning of each attributes.

3.4 Data Preparation

This phase involves some processes to be executed on the dataset. The process which should be executed on dataset for data preprocessing are data cleaning, which is to eliminate noisy data, detect outliers, and detect change point [12]. Besides that, to prepare an appropriate data, a suitable tools or software needs to be used. The data is also filtered from January 2013 to October 2015.

3.5 Understand the Data

The dataset has three main files which is the item file dataset, shop file dataset, and item category file. The historical dataset has the information of products sold in every shop from January 2013 to October 2015. The datasets contains 11 attributes.

4 Analysis of Model Fitting and Comparison Study Using Statistical Method Models

This segment includes all analyses which is conducted to forecast the monthly sales of the following month. As a result, the following month's forecast for the total sales for every store is predicted and determined. The analysis of comparison study is conducted on the statistical method models which is ARMA, ARIMA, and ARIMAX model. There are 60 stores in the prediction frame, which is based on the store ID within 0–59. The comparison between the statistical method models are evaluated based on the performance of the models and prediction values.

4.1 Autoregressive Moving Average Model (ARMA)

ARMA is one of the earliest statistical method model. According to [13], this shows how p values can be approximated on the basis of a regression surface, and dfuller states the p value also [14]. Adding P value up to 5 AR terms, q value is up to 5 MA terms and get the smallest akaike information criterion (AIC). When it is not stationary, Boxcox method is implemented. Boxcox is used for transforming series before using ARMA tools, as ARMA do not have differencing function to make it stationary. To validate the AIC value, output calculation using root mean square error and normalized root mean square error is implemented. No forecast is possible because for the past three months as the shops have no sales which will not be considered in the prediction [15].

4.2 Analysis of ARMA Prediction Value to Actual Value

Prediction results of autoregressive moving average model in comparison with actual sales value is presented in Fig. 2.

From the comparison graph above, we can notice that the ARMA model can predict the sales, especially done well for shop no 5, 45, and 49. The sales predicted was having high offset with real sales from shop no 54–59. Do notice that shop no 9, 20, and 36 was predicted 0, but the sales was higher as the condition for predict the sales data was not full-filled due to shop newly open or do not have past three months' sales data.

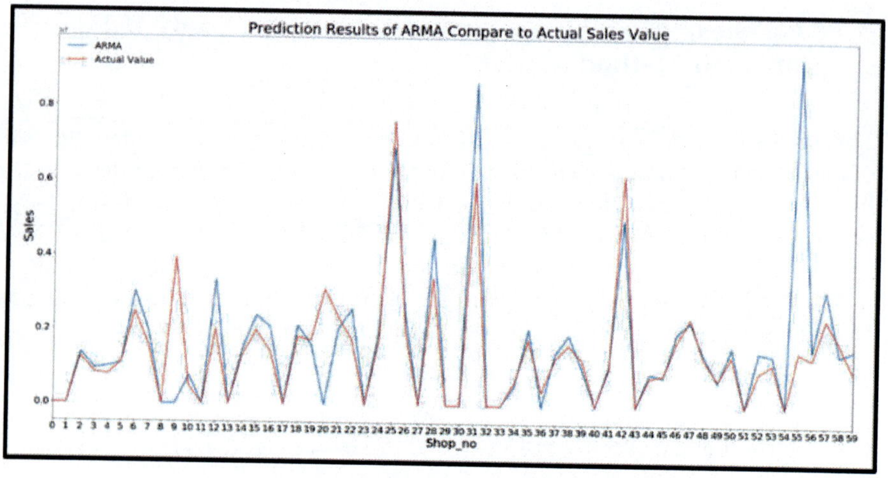

Fig. 2 Prediction results of ARMA in comparison to actual sales value

4.3 Autoregressive Integrated Moving Average (ARIMA)

ARIMA models are the most general class of models for forecasting a time series that can be made stationary by differentiation if required. *D* value is usually set between 0 and 2 to prevent from over differencing. According to MacKinnon (1994), this shows that how *p* values can be approximated on the basis of a regression surface, and dfuller states the *p* value also[14]. *P* value is up to 5 AR terms, and *q* value is up to 5 MA terms. Adding every *p* and *q* value tends to get smallest AIC which will be the best value of (*p*, *d*, *q*). There is no pattern in a stationery sequence, when its variations around its parameters have a constant selling over time. To validate the AIC value, output calculation using root mean square error and normalized root mean square error is implemented. No forecast possible because for the past three months as the shops have no sales which will not be considered in the prediction.

4.4 Analysis of ARIMA Prediction Value to Actual Value

Prediction results of autoregressive integrated moving average in comparison with actual sales value is presented in Fig. 3.

From the comparison graph above, the ARIMA model was predicting well for the last few shops, especially on shop no 42, 47, and 57. The ARIMA model was not predicted well on the first few shops. Do notice that shop no 9, 20, and 36 was predicted 0, but the sales were higher as the condition for predicting the sales data was not full-filled due to shop newly open or do not have past three months sales data. Shop no 5, 48, and 55 were not predicted as the sales data series was not stationary

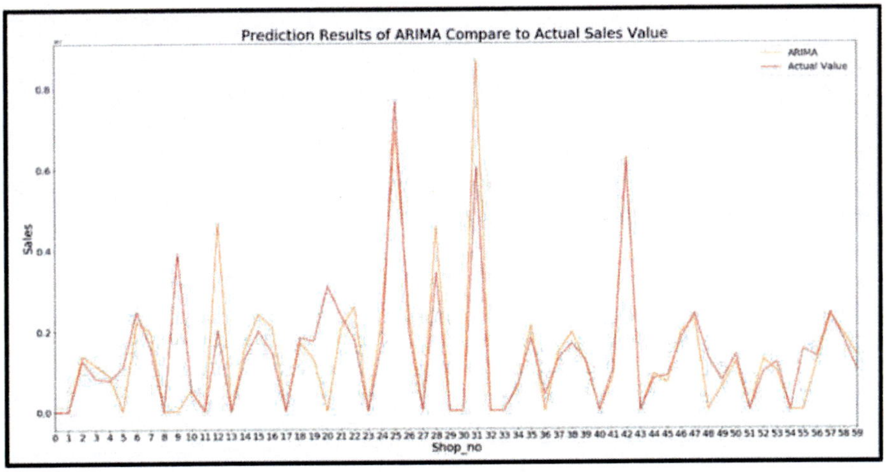

Fig. 3 Prediction results of ARIMA in comparison to actual sales value

after three times of differencing. The ARIMA model has closer line to actual sales value line compared to the ARMA model.

4.5 Autoregressive Integrated Moving Average with Explanatory Variable (ARIMAX)

ARIMAX is simply an ARIMA model with inclusion of exogenous variables. This indicates that whether one or more explanatory regression techniques are added to the forecasting equation. According to MacKinnon (1994), he shows how *p* values can be approximated on the basis of a regression surface, and dfuller states the *p* value also[14]. By adding *P* value up to 7 AR terms, *q* value is up to 7 MA terms and get the smallest AIC. When graph is not stationary, differencing is done. *D* value is usually set between 0 and 2 to prevent from over differencing. After two differencing, the graph is stationary means that there is no prediction done. To validate the AIC value, output calculation using root mean square error and normalized root mean square error is implemented. No forecast is possible because for the past three month the shops have no sales which will not be considered in the prediction.

4.6 Analysis of ARIMAX Prediction Value to Actual Value

Prediction results of autoregressive integrated moving average with explanatory variables in comparison with actual sales value is presented in Fig. 4.

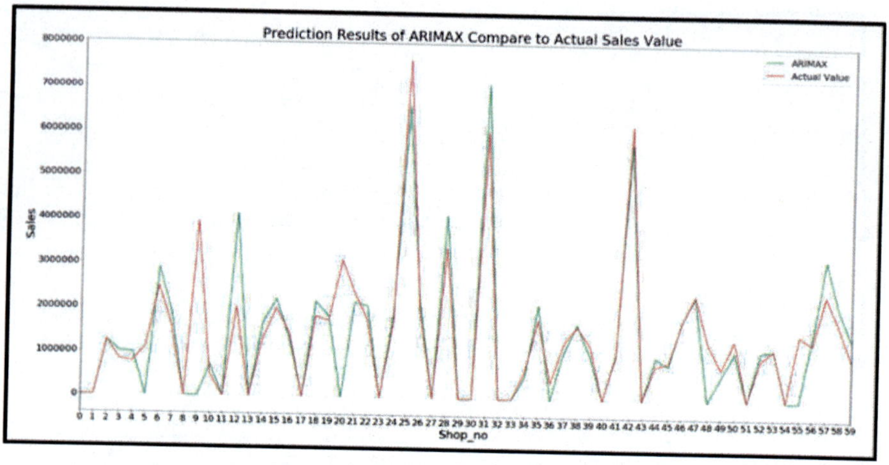

Fig. 4 Prediction results of ARIMAX in comparison to actual sales value

From the comparison graph above, the ARIMAX model was fitted well. The sales was predicted well especially for shop no 2, 16, 24, 27, 41, 45, 46, and 53. Do notice that shop no 9, 20, and 36 were predicted 0, but the sales was higher as the condition for predict the sales data was not full-filled due to shop newly open or do not have past three months sales data. Shop no 5, 48, and 55 was not predicted as the sales data series was not stationary after three times of differencing. The ARIMAX model has more accurate predicted value compared to ARIMA model and ARMA model.

4.7 Analysis Comparison of Statistical Method Models on Validation Set

Prediction results of statistical methods in comparison with actual sales value is presented in Fig. 5.

From the comparison graph above, the ARIMAX model was fitted well. The sales was predicted well especially for shop no 2, 16, 24, 27, 41, 45, 46, and 53. Do notice that shop no 9, 20, and 36 were predicted 0, but the sales was higher as the condition for predict the sales data was not full-filled due to shop newly open or do not have past three months sales data. Shop no 5, 48, and 55 were not predicted as the sales data series was not stationary after three times of differencing. The ARIMAX model has more accurate predicted value compared to ARIMA model and ARMA model.

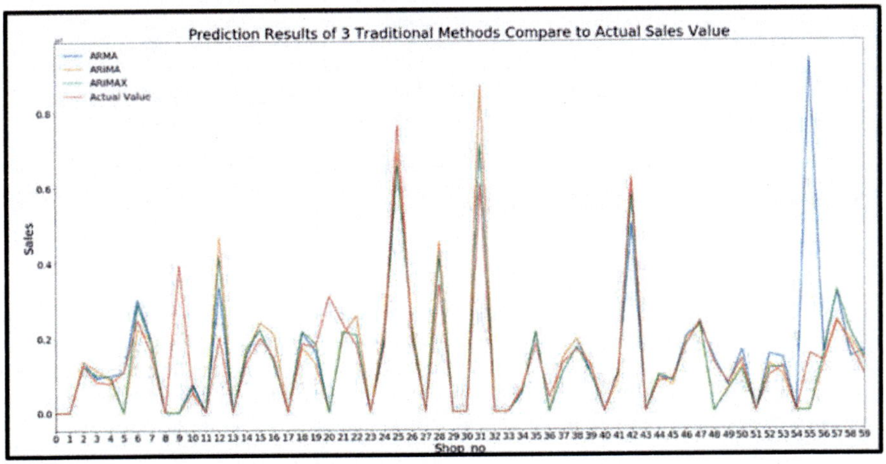

Fig. 5 Comparative performance analysis of traditional statistical methods vs actual sales value

4.8 Analysis Comparison of Normalized Root Mean Square Error (NRMSE) Statistical Method Models on Validation Set

A comparative analysis on all of the above mentioned traditional models namely, ARMA, ARIMA and ARIMAX is given in the Fig. 6 below:

To have a confirmed and high support on the accuracy based on checking the performance of the model, normalized root mean square error is implemented on the three statistical method models outputs. From the bar chart above, in the normalized root mean square error (NRMSE) analysis, the ARIMAX (0.2649) have the best performance across the three traditional method, followed by ARIMA (0.3695) and ARMA (0.7655). The lower the NRMSE, the higher the performance of the model.

Fig. 6 Comparative analysis of ARMA, ARIMA and ARIMAX based on root mean squared error

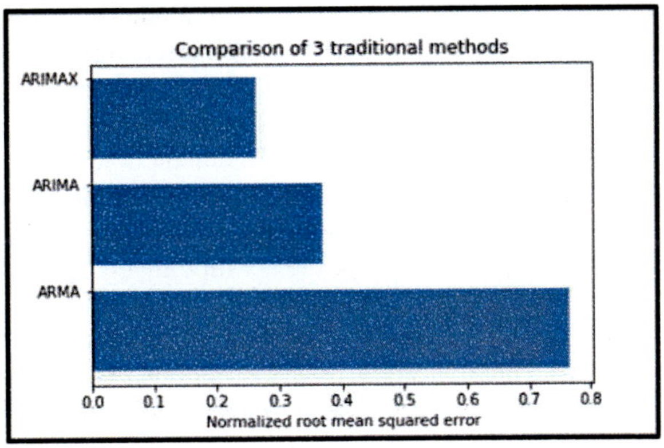

Table 1 Comparison of traditional modelling methods implementation

	Autoregressive moving average (ARMA)	Autoregressive integrated moving average (ARIMA)	Autoregressive integrated moving average with exogenous regressor (ARIMAX)
Methodology	Data analytics lifecycle	Data analytics lifecycle	Data analytics lifecycle
Model building	Data modelling steps	Data modelling steps	Data modelling steps
Stationary method	Boxcox transformation	Differencing ($d \leq 2$)	Differencing ($d \leq 2$)
Performance metric tools	Root mean square error, normalized root mean square error	Root mean square error, normalized root mean square error	Root mean square error, normalized root mean square error
Model performance accuracy	Less accurate	Accurate	Most accurate
Best monthly sales forecasting shops	5, 45, 49	42, 47, 57	42, 31, 28

4.9 Comparison of Traditional Modelling Methods Implementation

A comparative analysis of all the three traditional modelling methods, namely ARMA, ARIMA and ARIMAX is presented in the Table 1. The analysis is carried out based on methodology, model building, stationary method, performance metric tools, model performance accuracy and best monthly sales forecasting shops.

5 Conclusion

In this research work, it can be seen that model selection is a necessary step for any practical machine learning task. The model selection should choose related one to another to find out the advantage and disadvantage of the category of models. In this paper, several techniques were explored and implemented. Three common statistical modelling methods were chosen which is ARMA, ARIMA, and ARIMAX for the forecasting predicting process. The methodology used in the paper is data analytics lifecycle. Data analytics lifecycle is a technique used to understand the data and apply statistics to get insights for an achievable objective. The methodology part of the research focuses on the six repetitive steps. Every model used needed to be verified by the tools by using mathematical modelling to compare the results. From this research, the best results came from ARIMAX model method. The measure of performance strongly indicates the best model from the chosen models. As a recommendation from the best model which is ARIMAX, shop no 42, 31, and 28 have the highest sale predicted value which is more accurate to the actual value.

References

1. Yucesan, G.: Performance, Comparison between ARIMAX, ANN and ARIMAX-ANN Hybridization in Sales Forecasting for Furniture Industry, pp. 1–14 (2018)
2. Skehin, T., Crane, M., Bezbradica, M.: Day Ahead Forecasting of FAANG Stocks Using ARIMA, LSTM Networks and Wavelets, pp. 1–12 (2018)
3. Adebiyi, A.A, Adewumi, A.O., Ayo, C.K.: Comparison of ARIMA and Artificial Neural Networks Model for Stock Price Prediction, pp. 1–8 (2014)
4. Ghofrani, M.M.: Time Series and Renewable Energy Forecasting. Dalam INTECH, Time Series Analysis and Applications, pp. 1–66 (2018)
5. Werngren, S.: Comparison of Different Machine Learning models for Wind Turbine Power Predictions, pp. 1–11 (2018)
6. Peter, D., Silva, P.: ARIMA versus. ARIMAX—which approach is better to analyze and forecast macroeconomic time series? In: 30th International Conference Mathematical method in Economics, pp. 1–5 (2012)
7. Anele, A.O., Hamam, Y., Abu-Mahfouz, A.M., Todini, E.: Overview, Comparative Assessment and Recommendations of Forecasting Models for Short-Term Water Demand Predictions, pp. 1–12 (2017)
8. Anggraeni, W., Vinarti, R.A.: Performance Comparisons Between ARIMA and ARIMAX Method in Moslem Kids Clothe Demand Forecasting: Case Study. In: 3rd Information Systems International Conference (2015)
9. Edward, A.: Forecast Model Using ARIMA for Stock Prices of Automobile Sector, pp. 1–9 (2016)
10. Fong, S.J., Li, G., Dey, N., Crespo, R.G., Herrera-Viedma, E.: Finding an accurate early forecasting model from small dataset: a case of 2019-ncov novel coronavirus outbreak. arXiv preprint arXiv:2003.10776 (2020)
11. Dey, N., Fong, S., Song, W., Cho, K.: Forecasting energy consumption from smart home sensor network by deep learning. In: International Conference on Smart Trends for Information Technology and Computer Communications, pp. 255–265. Springer, Singapore (2017, August)
12. Hu, S., Liu, M., Fong, S., Song, W., Dey, N., Wong, R.: Forecasting China future MNP by deep learning. In: Behavior Engineering and Applications, pp. 169–210. Springer, Cham (2018)
13. MacKinnon, J.G.: Numerical distribution functions for unit root and cointegration tests. J. Appl. Econ. **11**(6), 601–618 (1996)
14. Fong, S.J., Li, G., Dey, N., Crespo, R.G., Herrera-Viedma, E.: Composite Monte Carlo decision making under high uncertainty of novel coronavirus epidemic using hybridized deep learning and fuzzy rule induction. Appl. Soft Comput. 106282 (2020)
15. Lotfi, A., Langensiepen, C., Mahmoud, S.M., Akhlaghinia, M.J.: Smart homes for the elderly dementia sufferers: identification and prediction of abnormal behaviour. J. Ambient Intell. Humaniz. Comput. **3**(3), 205–218 (2012)

Performance of Scalable Data Linked to Internet of Things: A Case Study

Jyoti Prakash Mishra, Samarjeet Borah, and Sambit Kumar Mishra

Abstract Data can be scaled as well as transacted with modularity based on specific applications. The term Internet of things in this domain is conceptualized based on smart virtual applications toward large segment of data with heterogeneity. In such situation, the deployment of different sensors can be applied to different synchronized as well as continuous data stream. The involvement of machine-to-machine applications is of course there while linking real-time analysis with Internet of things. As it is known that the Internet of things permits communicating among devices, the physical devices can communicate to common man about the current status and location. Accordingly, the acquired data from the devices can be properly scaled and filtered to attain relevancy. The better mechanisms with set of rules can be developed to observe the significance of data with varying scales. There is a variation with large scale of data while incorporating the techniques and extracting the relevancy from focused data. In such scenario, it may be desirable to focus toward similar clusters linked with specific domains.

Keywords Data stream · Virtualization · Cluster · Routing protocol · Metaheuristic · Data centers · Query terms

J. P. Mishra · S. K. Mishra
Gandhi Institute for Education and Technology, Baniatangi, Bhubaneswar, Odisha, India
e-mail: jpmishra@gietbbsr.com

S. K. Mishra
e-mail: sambitmishra@gietbbsr.com

Biju Patnaik University of Technology, Rourkela, Odisha, India

S. Borah (✉)
Sikkim Manipal Institute of Technology, Sikkim Manipal University, Sikkim, India
e-mail: samarjeetborah@gmail.com

S. Borah et al. (eds.), *Soft Computing Techniques and Applications*,
Advances in Intelligent Systems and Computing 1248,
https://doi.org/10.1007/978-981-15-7394-1_54

1 Introduction

In general, data are termed as raw facts which are accumulated by the sensor devices toward communicating with centralized servers. Accordingly, the optimality can be expected by enabling the performance and judging the efficiency. The involvement of different heterogeneous networks is a great support toward data transmission as well as enhancement of quality and efficiency linked with Internet of things. While managing the data along with feature extraction, it is very much essential to focus on retrieval of Internet of things data and enabling them to generate continuous streams of data in a scalable way. During this period, the continuity of large stream of data is managed properly with the linked devices. The technological enhancement as well as growth in the digital devices has impacted in positive aspects with the inter-connected objects. The potential growth of the objects connected to Internet of things may also increase the data rate and accumulation of redundant and unnecessary data. In such situation, it is prime concern to manage the large sized non-trivial data considering the analytics as well as security concerns. Accordingly, it is also essential to focus on key requirements to manage and process these category data with parameterization.

2 Review of Literature

Mann et al. [1] in their work have focused on routing protocols. They observed the separation of clusters in case of hierarchical routing protocols along with varying sizes. Also individual cluster may be associated with other categories of clusters. In fact, the clusters are coordinated among themselves, and the linked information may be passed between the clusters.

Vijayalakshmi et al. [2] during their research have concentrated on specific application using particle swarm optimization focusing on consumption of lowest energy and focused toward enhancing the applicability of clusters linked with wireless sensor network. As per their observations, the clusters can be applied through the search and optimization techniques to improve the survival rate of nodes.

Wang et al. [3] in their study have proposed the particle swarm optimization based on wireless sensor network. While incorporating the clusters in virtual machines and focusing on specific optimization criteria, they observed that the implementation mechanism will include the residual energy along with the position of nodes. During inspection of all linked clusters, they observed the selecting of the node whose residual energy is larger than the average residual energy of all nodes in the cluster.

Rao et al. [4] in their work have proposed a specific particle swarm optimization algorithm toward obtaining the weight function as well as methods of formation of clusters.

Tam et al. [5] during their research have conceptualized the mechanism based on fuzzy clustering as well as particle swarm optimization to minimize the linked sensors applied in the network. In their research, they have also focused on the routine associated with the unconnected sensors within the clusters. Based on the objective functions, they have also tried to achieve optimality through simulation mechanisms.

Kalantari et al. [6] in their research tried to obtain efficient 3D layout of DBSs to maximize the sum logarithmic rate of the users. In their research, they have considered the 3D layout of backhaul resource allocation, user associations, user types, and DBSs simultaneously. Accordingly, they observed that the specific wireless back trip bandwidth allocation can be obtained through the decomposition method.

Shakhatreh et al. [7] in their research have focused on provision of wireless coverage implemented toward distribution among users. In their application, they have also focused on particle swarm optimization to obtain the efficient three-dimensional layout. Considering the similar number of users in the application, the actual convergence speed may not be similar among the equally distributed users.

Pan et al. [8] in their work have cited their application on particle swarm optimization toward joint uplink subcarrier distribution as well as power control associated with cellular network in 5G. As per their observation, the 5G network which is massive MIMO being associated with powerful intelligent algorithm is more efficient to solve nonlinear complex problems. Similarly, as it is oriented toward a variety of scenarios, many scenarios may have different requirements toward security as well as network operations.

Wang et al. [9] in their work have focused on optimal routing algorithms to enhance the network ability. The specific algorithm combines the clusters and simulates based on the optimization criteria. It optimizes through heuristic factors to obtain the optimal sink movement and track the data through single hop communication.

Zhou et al. [10] have focused on specific optimization mechanism to enhance the network of clusters. The optimization mechanism can be applied to choose only the target nodes. Accordingly, the protocols may be responsible in enhancing the network structure by choosing the target node implementing the optimization criteria. The particle swarm optimization in this case is used to minimize transmission distance by updating cluster nodes and to optimize network energy consumption.

2.1 Applications of Data Linked with Internet of Things

In many situations, data can be varied in structure as well as in application. In such case, based on the application, decisions can be taken doing extensive data analysis and mostly focusing on values as well as variability. Considering data with heterogeneity, it is of course desirable to classify, process, and take decisions to make viable in commercial application. According to Mhetre et al. [11], the users must be provided with meaningful applications and search techniques. While focusing on generation of smart query plans, it is essential to prioritize entities and segregate the strong entities. Then, the integrity rules can be applied to focus on mechanisms associated with entity integrity as well as referential integrity. In this case, the functional dependencies among the attributes have the major roles to project on fully functional or partial functional dependencies during implementation. Along with the large sized data and the ability to optimize, it is required to implement machine learning as well as pattern recognition techniques. Hassanien et al. [12] during their research realized that to judge the efficiency of the mechanisms in virtual applications, the simulation of these sequenced data can be done by applying heuristic as well as metaheuristic techniques.

2.2 *Issues Associated with Large-Scale Data Linked with Internet of Things*

Generally, massive storage along with huge computing capabilities is required while dealing with large sized heterogeneous data. To minimize this typical problem, the processing capabilities can be properly chained, and scalability and security measures of data should be prioritized. Bhatt et al. [13] in their work have focused on the major issues to handle large sized data along with services linked to data management. While integrating the entities linked to Internet of things, Dey et al. [14] observed some typical Web applications processed to obtain the optimality which are associated with near approachable solutions. Sometimes the implementation mechanisms of Internet of things is a issue as the proper linkages of devices should be unique. In many times, the accumulated data can be more essential and seek privacy. The mechanisms adopting security toward devices should be the prime concern. Also standardization toward devices with interoperability should be monitored. Finally, confidentiality on data along with encryption mechanisms should be ensured.

2.3 *Issues Associated with Data Security*

Considering the challenges, the Internet of things sometimes faces some critical problems linked with secured computations as well as transactions with data centers. Elhayatmy et al. [15] while filtering the redundant data with proper scalability and imposing real-time application, observed that sometimes the multi-layered security system can assist to maintain consistency scatter the unnecessary parts from the network. In such case, the system must obey the network access control policies, and the technologies toward networking must be accumulated with signified protocols to encrypt the multi-point with specific network identification.

3 Application of Metaheuristic Technique

While considering the metaheuristic approach and particularly optimization based on ant colony, it is clear that some partial solution as destination can be achieved through pheromones, and in this situation, each ant may be termed with query terms of each database. As the ants are quite efficient in obtaining their paths through higher range pheromones, in the similar way, the query terms can be processed to obtain near optimal solutions with better efficiency. The main intention of implementation of the metaheuristic technique is to obtain global optimality with feasible solutions.

3.1 Steps to Process Heuristically Clustered Data

```
Step 1: Define the size of databases and link the query
terms to position of ant
Step 2:    Obtain  the  probability  of  transition  of
individual ant linked to specific locations
Step 3:  Specify the pheromone defining the formula and
iterate towards specified position
        Ωij(q + t) = (1 - ρ)* Ωij(q) + δΩij ,
        where Ωij (q) is defined as pheromone on specific
        position  and  δΩij  denotes  the  differential
        parameters
Step 4: Quantify the pheromone associated with the query
terms and determine volatilization value
Step 5: Determine the residual or persistence coefficient
within the range less than 1
Step 6: Determine the cluster and categorize the same as
deterministic as well as random uncertainty
Step 7: Implement the clustering techniques, i.e. static
and  dynamic  clustering  where  the  network  may  be
associated with many clusters
Step 8: Obtain the evaluation time of the linked process
in the network with specified distributed parameters
Step 9: Considering the data transmission rate, obtain
the response time in every stage of transmission
        If  data  transmission  rate  is  ζt ,  then  the
        response time of data during transmission= μt +
        ζt* s +( μt - ζ) , where μt is the average
        parametric  coefficient  of  all  the  linked
        processing elements in the network, s is the no.
        of servers or processing elements in the system
Step 10: Determine the cache coherence and association of
IoT devices
Step 11: Validate each device and obtain the parametric
coefficient of each processing element. If the individual
query terms associated with the cache is validated and is
desirable, then process the upstream requests.
```

4 Experimental Analysis

See Table 1.

While processing multiple tasks in the servers, the allocation of time is automatic, but the priority will be assigned based on status of implementation. Sometimes a few tasks may be alternated during processing. In such situation, it is desirable to focus on turnaround time. It is nothing but the total time taken between the submission of the task toward execution and obtaining the output. In other words, turnaround time can be termed as summation of burst time and waiting time. The waiting time is amount of time for a task to wait in the ready queue toward availing the desired slot. In such situation, the response time is the average time of tasks elapsed from submission until the previous response is generated. Considering Fig. 1, it is obvious that making constancy in the allocation of tasks within the servers, the turnaround time is proportionate toward allocation of servers (Table 2).

Many times it is seen that large sized queries take bit longer times toward execution. In such situation, it is essential to focus on repository as well as efficiency of processing elements. Also it is desirable to concentrate on elapsed time which is nothing but the total time consumed by the queries including processing elements time as well as network waiting time. If the queries run simultaneously, cumulative time may be considered along with the database access time linked to tasks along with fixed sized queries as cited in Fig. 2 (Table 3).

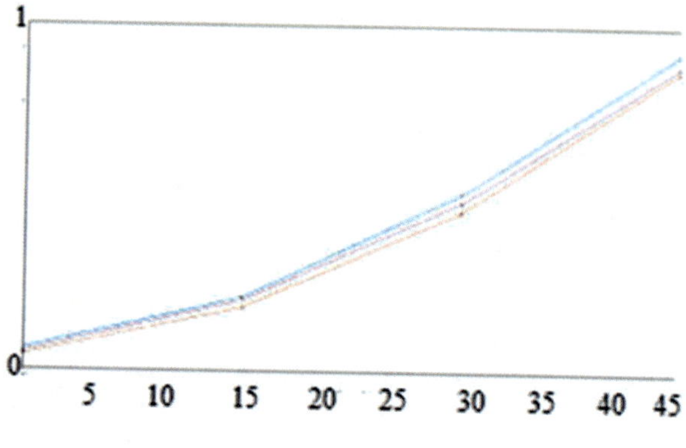

Fig. 1 Number of database servers along with turnaround time of tasks

Table 1 Assigned database servers along with turnaround time of tasks

S. No.	No. of database servers	Tasks assigned to each database servers	Turnaround time (ms)
1	19	280	0.46
2	28	280	0.55
3	37	280	0.66
4	40	280	0.70

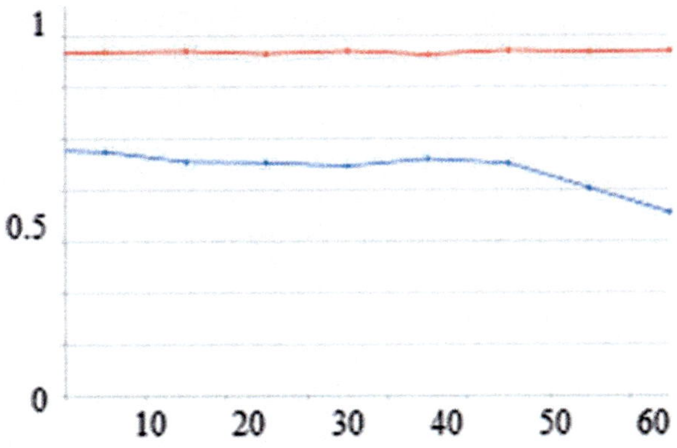

Fig. 2 Query terms (database) with evaluation (ms)

Table 2 Processed queries in databases with elapsed time of query terms

S. No.	No. of databases associated with each server	No. of queries processed	Elapsed time of query terms (ms)
1	29	550	0.74
2	37	550	0.64
3	46	550	0.53
4	55	550	0.51

In general, the data centers are linked with a number of processing elements including the application controllers and network devices which are also very essential toward storage and management of most critical systems. Considering Fig. 3, and as per current scenario, large sized application on data are being processed and managed though cloud data centers implementing the mechanisms of virtual machines which accumulate a large number of processing elements. Therefore, it is

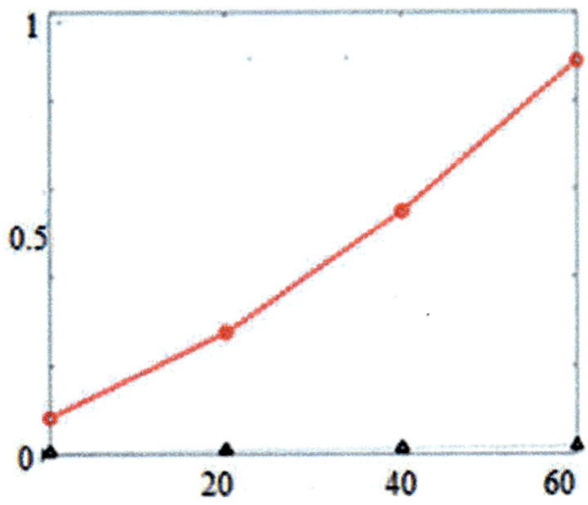

Fig. 3 Data centers versus execution of individual PEs

Table 3 Processed queries in databases with elapsed time of query terms

S. No.	No. of linked data centers	No. of processing elements associated with each data center	Estimated elapsed time of processing elements (ms)
1	28	47	0.46
2	37	47	0.51
3	47	47	0.55
4	55	47	0.64

essential to allocate these data and to enable the control of the resources to enhance efficiency (Table 4).

The clusters are in general implemented toward intensive computing tasks and partly linking database services. The clusters usually have the ability to incorporate the nodes as per their current schedule. Sometimes, it is required to obtain the performance of clusters based on bigger computational assignments as reflected in Fig. 4. While linking new element to the cluster, it must be required to measure the reliability and evaluate the performance of individual processing element.

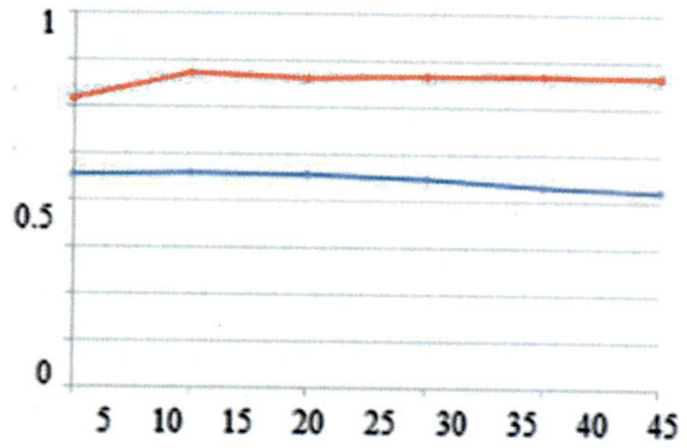

Fig. 4 Number of PEs with response time of clustered elements

Table 4 Clusters associated with processing elements with response time

S. No.	No. of processing elements associated	No. of clusters incorporated with processing elements	Response time of clusters (ms)
1	19	29	0.61
2	29	29	0.53
3	37	29	0.52
4	40	29	0.51

5 Discussion and Future Direction

In many situations while analyzing the data, the databases are estimated along with the storage time. Accordingly, while addressing such databases with heterogeneity, the processing elements are involved toward initiation and retrieval mechanisms toward complex data. As mentioned by Schaller et al. [16], the users linked with the processing elements definitely seek authentication as well as authorization toward analysis and retrieval of data in these platforms. Along with the enhancement in the IoT technologies, the processing elements often face challenges with enhanced capacity as well as minimum latency which makes the system complex. In such scenario, the adaptabilities of the processing elements matter a lot to achieve optimality.

6 Conclusion

Generally alternate approach sometimes is ascertained toward large scaled data linked with databases due to their varying sizes and complexities. In many cases, it points toward the scalability of databases. Dey et al. [17] also observed that during real-time analysis of large scaled data, the applications can be focused toward scalability of data as well as individual processing elements of different domains. While analyzing the performance of IoT systems, it can be seen that the associated routines in many complex applications measure scalability as well as obtain reliable and almost optimal solutions. Accordingly, these will be support to formalize the suitable system for consideration.

References

1. Mann, P.S., Singh, S.: Energy-efficient hierarchical routing for wireless sensor networks: a swarm intelligence approach. Wirel. Pers. Commun. (2017)
2. Vijayalakshmi, K.; Anandan, P.: A multi objective Tabu particle swarm optimization for effective cluster head selection. In: WSN Cluster Computing (2019)
3. Wang, J., Cao, Y., Kim, H., Lee, S.: Particle swarm optimization based clustering algorithm with mobile sink for WSNs. Future Gener. Comput. Syst. (2017)
4. Rao, P.C.S., Jana, P.K., Banka, H.: A particle swarm optimization based energy efficient cluster head selection algorithm for wireless sensor networks. Wirel. Netw. (2017)
5. Tam, N.T., Hai, D.T.: Improving lifetime and network connections of 3D wireless sensor networks based on fuzzy clustering and particle swarm optimization. Wirel. Netw. (2018)
6. Kalantari, E., Bor-Yaliniz, I., Yongacoglu, A., Yanikomeroglu, H.: User association and bandwidth allocation for terrestrial and aerial base stations with backhaul considerations. In: Proceedings of the 28th IEEE Annual International Symposium on Personal, Indoor and Mobile Radio Communications (PIMRC), Montreal, QC, Canada, 8–13 Oct 2017
7. Shakhatreh, H., Khreishah, A., Alsarhan, A., Khalil, I., Sawalmeh, A., Othman, N.S.: Efficient 3D placement of a UAV using particle swarm optimization. In: Proceedings of the 2017 8th

International Conference on Information and Communication Systems (ICICS), Irbid, Jordan, 4–6 Apr 2017
8. Pan, Q., Ni, Q., Du, H., Yao, Y., Lv, Q.: An improved energy-aware cluster heads selection method for wireless sensor networks based on k-means and binary particle swarm optimization. In: International Conference in Swarm Intelligence. Springer, Cham, Switzerland, Oct 2014
9. Wang, J., Cao, J., Sherratt, R.S., Park, J.H.: An improved ant colony optimization-based approach with mobile sink for wireless sensor networks. J. Supercomput. (2018)
10. Zhou, Y., Wang, N., Xiang, W.: Clustering hierarchy protocol in wireless sensor networks using an improved PSO algorithm. IEEE Access (2016)
11. Mhetre, N.A., Deshpande, A.V., Mahalle, P.N.: Trust management model based on fuzzy approach for ubiquitous computing. Int. J. Ambient Comput. Intell. (IJACI) **7**(2), 33–46 (2016)
12. Hassaniien, A.E., Dey, N., Borra, S. (Eds.) Medical big data and internet of medical things: advances, challenges and applications. CRC Press
13. Bhatt, C., Dey, N., Ashour, A.S. (Eds.). Internet of things and big data technologies for next generation healthcare (2017)
14. Dey, N., Hassanien, A.E., Bhatt, C., Ashour, A., Satapathy, S.C. (Eds.). Internet of things and big data analytics toward next generation intelligence, pp. 3–549. Springer, Berlin (2018)
15. Elhayatmy, G., Dey, N., Ashour, A.S.: Internet of things based wireless body area network in health care. In: Internet of Things and Big Data Analytics Toward Next Generation Intelligence, pp. 3–20. Springer, Cham (2018)
16. Schaller, A., Mueller, K.: Motorola's experiences in designing the internet of things. Int. J. Ambient Comput. Intell. (IJACI) **1**(1), 75–85 (2009)
17. Dey, N., Ashour, A.S., Bhatt, C.: Internet of things driven connected healthcare. In: Internet of things and big data technologies for next generation healthcare, pp. 3–12. Springer, Cham (2017)

A Chaotic-Jaya Optimized OSELM Model for Cancer Classification

Prajna Paramita Debata, Puspanjali Mohapatra, Debahuti Mishra, and Samarjeet Borah

Abstract In the rapidly defiled environment, cancer has emerged out as the most threatening disease to human species. So, the diagnosis and prognosis of cancer should be easier and affordable. To help the medical practitioner, machine learning researchers aim to design a robust classification model which can able to classify cancer data with high accuracy and less computational complexity. Here, Chaotic-Jaya (CJaya) algorithm optimized online sequential extreme learning machine (CJaya-OSELM) is proposed for classifying cancer data. To estimate the efficacy of the proposed model, two standard cancer datasets, breast cancer and lung cancer, are used. The result of this model is compared with CJaya optimized multi-layer perceptron (CJaya-MLP), CJaya optimized ELM (CJaya-ELM), and OSELM. Sensitivity, Gmean, Fscore, specificity, precision, and accuracy are taken as performance evaluation measure for these models. The experimental outcome reveals that CJaya-OSELM is significantly better than other models.

Keywords Classification · Cancer data · Online sequential extreme learning machine · Chaotic-Jaya algorithm

---

P. P. Debata · D. Mishra (✉)
Department of Computer Science and Engineering, Siksha 'O' Anusandhan Deemed to be University, Bhubaneswar, Odisha, India
e-mail: debahutimishra@soa.ac.in

P. P. Debata
e-mail: prajnaparamitadebata@soa.ac.in

P. Mohapatra
Department of Computer Science and Engineering, International Institute Information Technology, Bhubaneswar, Odisha, India
e-mail: puspanjali@iiit-bh.ac.in

S. Borah
Sikkim Manipal Institute of Technology, Sikkim Manipal University, Sikkim, India
e-mail: samarjeet.b@smit.smu.edu.in

S. Borah et al. (eds.), *Soft Computing Techniques and Applications*,
Advances in Intelligent Systems and Computing 1248,
https://doi.org/10.1007/978-981-15-7394-1_55

1 Introduction

From dawn to dusk, cancer like deadly diseases have created havoc among the human beings due to the changing lifestyle of human being in polluted environment. In the last two decades, spending a lot on cancer research it has been concluded that if the cancer-causing genes can be detected at the early stage, then some cancers can be effectively cured for lifetime, if not then these can be treatable. Many machine learning researchers have developed different statistical models for cancer data classification, still it is challenging because it is a non-deterministic polynomial-time (NP)-Hard problem. Numerous statistical and soft computing-based classification approaches like Euclidean minimum distance (EMD) [1], Bayesian decision theory, multilayer perceptron (MLP) [2], fuzzy rule-based systems [3], support vector machine (SVM) [4], and k-nearest neighbor (KNN) classifiers [4] have been applied in the classification field [12]. The major drawback of statistical methods [2] is that these methods are not efficient in case of nonlinear problems. Like statistical methods soft computing methods also have some merits and demerits. Two main advantages of soft computing approaches are: (1) they are able to solve nonlinear problems, where mathematical models are not possible and (2) they reflect the knowledge of human like understanding, learning, cognition, recognition, and others into the fields of computing. The drawbacks of soft computing approaches [5] are: training time of high-dimensional datasets is very high, selecting an optimum kernel function is quite difficult, and the final model depends on variable weights which makes it difficult to interpret. In ANN, back propagation algorithm is generally used as learning technique, and the major demerit of it is: very slow rate of convergence. So, various optimization techniques are applied to optimize ANN. In order to achieve high performance, traditional feed forward network tunes the weights and biases at different layers in each iteration. Hence, large number of iteration steps are required. If the learning method uses the gradient descent learning algorithm, then it is easily trapped at local minima due to randomly chosen parameters [2]. Generally, single-hidden layer feed forward neural network (SLFN) needs maximum no. of hidden neurons, randomly chosen input biases, and weights, to be tuned each time. To avoid this problem, Huang et al. [6] introduced a revolutionary model, called as extreme learning machine (ELM) which has good generalization ability and faster learning speed. Till today, many variants of ELM have already been implemented for solving many classification problems [7] still it has some limitations like randomly choosing weights, dealing with more computational complexity, and execution time. Here, one of the variants of ELM, i.e., OSELM [8], is considered as the classifier to classify the cancer data. In OSELM, chunk by chunk, equal, or different size training data is learned. As the input weights are arbitrarily chosen in OSELM, it may produce unstable results. To handle this problem, many optimization algorithms like as fire fly optimization (FFO) [9], cuckoo search optimization (CSO) [2], cat swarm optimization [4], etc., are proposed by the researchers.

The abstract view of the work is depicted in Fig. 1. Here, cancer datasets are first normalized by min–max normalization process [2]. Then, a recently developed Jaya

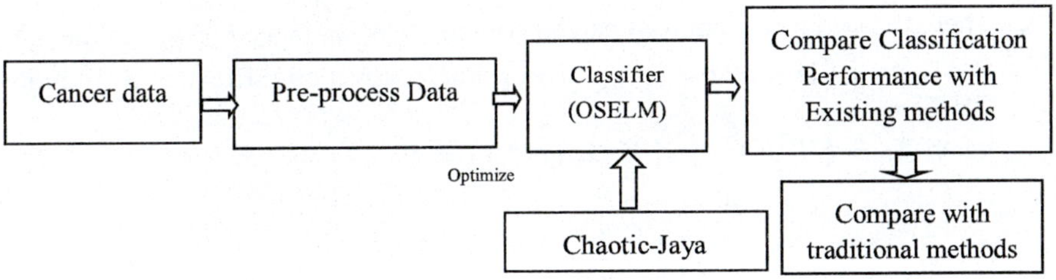

Fig. 1 An abstract view of the proposed work

optimization algorithm [10] is used to optimize randomly taken weights and bias parameters of OSELM. This algorithm is introduced by R. Venkata Rao which can solve both unconstrained and constrained optimization problems. It is designed on the basis of keeping the best one and removing the worst one. In addition to this, chaos concept is used to upgrade the random parameters of Jaya algorithm [11]. In this work, the chaotic Jaya is hybridized with OSELM and called as the chaotic Jaya-OSELM (CJaya-OSELM) algorithm. To assess this model, two standard cancer datasets such as Breast Cancer and Lung cancer are taken. The efficiency of this model is compared with CJaya optimized multilayer perceptron (CJaya-MLP), CJaya optimized ELM (CJaya-ELM), and OSELM models. In Sect. 2, all the supported methodologies are discussed. Section 3 reflects the experimentation and result analysis part, respectively. Section 4 is ended with the conclusion part.

2 Proposed Methodologies

2.1 Online Sequential Extreme Learning Machine (OSELM)

The OSELM algorithm has two parts such as initialization phase and sequential phase.

Steps of initialization phase:

1. The parameters of nodes, the input weights (w), and bias (b) are randomly taken in this algorithm.
2. Here, H_0 is considered as output of hidden layer and calculated as:

$$H_0 = \begin{bmatrix} A(w_1, b_1, y_1) & \cdots & A(w_K, b_K, y_1) \\ \vdots & & \vdots \\ A(w_1, b_1, y_{M_0}) & \cdots & A(w_K, b_K, y_{M_0}) \end{bmatrix} \quad (1)$$

3. Then, the output weight (β_0) can be computed as $\beta_0 = S_0 H_0^{\mathrm{T}} O_0$, and the target is $O_0 = \left[o_1, o_2, \ldots o_{M_0}\right]^{\mathrm{T}}_{M_o \times n}$. The primary aim of OSELM is to minimize $\|H_0\beta_0 - O_0\|$
where $S_0 = \left(H_0^{\mathrm{T}} H\right)^{-1}$ and $K_0 = \left(H_0^{\mathrm{T}} H\right) = S_0^{-1}$.

Steps of sequential phase:

1. Here, $(p+1)$th chunk of data will be found as: $M_{p+1} = \left\{\left(y_j, o_j\right)\right\}_{j=\left(\sum_{i=0}^{p} M_i\right)+1}^{\sum_{i=0}^{p+1} M_i}$
2. H_{p+1} (Partial hidden layer) can be calculated as follows:

$$H_{p+1} = \begin{bmatrix} A\left(w_1, b_1, y_{\left(\sum_{i=0}^{p} M_i\right)+1}\right) & \cdots & A\left(w_K, b_K, y_{\left(\sum_{i=0}^{p} M_i\right)+1}\right) \\ \vdots & & \vdots \\ A\left(w_1, b_1, y_{\sum_{i=0}^{p+1} M_i}\right) & \cdots & A\left(w_K, b_K, y_{\sum_{i=0}^{p+1} M_i}\right) \end{bmatrix}_{M_{p+1} \times K} \tag{2}$$

3. β^{p+1} (Output weight matrix) will be computed as:

$$\beta^{p+1} = \beta^p + L_{p+1}^{-1} H_{p+1}^{\mathrm{T}}\left(O_{p+1} - H_{p+1}\beta^p\right) \tag{3}$$

$$O_{p+1} = \left[o_{\left(\sum_{i=0}^{p} M_i\right)+1}, \ldots o_{\sum_{i=0}^{p+1} M_i}\right]^{\mathrm{T}}_{M_{p+1} \times n} \tag{4}$$

$$L_{p+1} = L_p + H_{p+1}^{\mathrm{T}} H_{p+1} \tag{5}$$

During recursive process, the inverse of K_{p+1}^{-1} should be omitted. Therefore,

$$L_{p+1}^{-1} = L_p + \left(H_{p+1}^{\mathrm{T}} H_{p+1}\right)^{-1} = L^{-1} - L^{-1} H_{p+1}^{\mathrm{T}}\left(I + H_{p+1} L_p^{-1}\right) H_{p+1} L_p^{-1} \tag{6}$$

and $N_{p+1} = L_{p+1}^{-1}$. Therefore, Eqs. (4) and (5) can be written as

$$N_{p+1} = N_p - N_p H_{p+1}^{\mathrm{T}}\left(I + H_{p+1} N_p H_{p+1}^{\mathrm{T}}\right)^{-1} H_{p+1} N_p \tag{7}$$

$$\beta^{p+1} = \beta^p + N_{p+1} H_{p+1}^{\mathrm{T}}\left(O_{p+1} - H_{p+1}\beta^p\right) \tag{8}$$

4. Increase p as $p = p + 1$, then go to step 3.

2.2 Jaya Optimization Algorithm

The major advantage of Jaya optimization [10] algorithm is that it requires less algorithm specific parameters. The steps of Jaya algorithm are summarized as follows:

Step 1 Initialize the size of the population, number of design variables, and the termination criteria.
Step 2 From the population, the best and worst solutions should be found out.
Step 3 The result based on the best and worst solutions would be changed by using Eq. (9).

$$O'_{j,k,i} = O_{j,k,i} + r_{1,j,i}\left[\left(O_{j,\text{best},i}\right) + \left|\left(O_{j,k,i}\right)\right|\right] - r_{2,\text{best},i}\left[\left(O_{j,\text{worst},i}\right) - \left|\left(O_{j,\text{worst},i}\right)\right|\right] \tag{9}$$

During ith iteration, $O_{j,k,i}$ is the value of the ith variable for the kth candidate. Here, k is the population size, i value gives the count of iteration, and j value gives the count of design variables.
Step 4 Then, the existing solution is compared with the modified one, and if it is found that the modified solution is better, then it will be replaced by the previous one, otherwise the previous solution will be kept.
Step 5 The procedure from Step 2 to Step 4 will be repeated till the termination criteria is reached.

2.3 Chaotic Learning Method

Chaotic-Jaya (CJaya) algorithm is mainly based on the chaos theory. This algorithm provides the better convergence speed and better exploration of the search space without trapping in local optima [11]. Mathematically, chaos can be stated as the randomness of a deterministic dynamical system. For applying chaos theory in different optimization algorithms, various chaotic maps with different mathematical equations are used. In this work, logistic map function is applied for generating the chaotic random numbers due to its simplicity. This logistic map function is defined as follows:

$$C_{t+1} = 4 \cdot C_t \cdot (1 - C_t) \tag{10}$$

where C_t is the obtained value of the chaotic map at tth iteration.

The procedures of the CJaya algorithm is same as the Jaya algorithm, but the random numbers in CJaya algorithm are created by adopting a chaotic random number generator. In CJaya algorithm, the two random variables (r_1 and r_2) of Jaya algorithm are interpreted by the logistic chaotic variables. Equation (9) of Jaya algorithm will be modified as:

$$Z'_{j,k,i} = Z_{j,k,i} + C_{t,j,i}\left[\left(Z_{j,\text{best},i}\right) + \left|\left(Z_{j,k,i}\right)\right|\right] - C_{t,j,i}\left[\left(Z_{j,\text{worst},i}\right) - \left|\left(Z_{j,k,i}\right)\right|\right] \tag{11}$$

where t represents the iteration number, C_t is the value of tth chaotic iteration, and the initial value of C_0 is randomly created in between [0, 1].

2.4 The Proposed CJaya-OSELM Algorithm

Here, the random weight of OSELM are optimized by CJaya optimization algorithm. The algorithm of the proposed CJaya-OSELM is illustrated in the following steps:

1. Input the size of the population (PS), no. of iterations, Hidden layer size (HL), and upper bound and lower bound of the variables.
2. Divide the dataset into training data and testing data in the ratio of 7:3.
3. Generate HL no. of random weight population, each having size of 1 × HL for each population (P_i), where $P_i = \left\{W_1^i, W_2^i, W_3^i, \ldots, W_{H_c}^i\right\}$ for $i = 1, 2, 3, \ldots$PS
4. The fitness (error value or miss classification rate) of each population is evaluated by OSELM.
5. The lowest miss classification rate is considered as the global best (g_{best}), and the highest miss classification rate is considered as the global worst (g_{worst}).
6. Find the values of the chaotic map C_m, by applying Eq. (10). Update the two random values of Jaya algorithm, r_1 and r_2 in Eq. (9) using Eq. (10). Then, according to the g_{best} and g_{worst} value, update each population by using Eq. (11).
7. If the fitness of new population is better than the previous one, then replace previous population with new population, else keep the previous one.
8. Repeat Step 4 to Step 7, till the stopping condition reaches.
9. Find the final g_{best} and keep its corresponding population and consider it as the final weight.
10. Determine the classification accuracy using this final weight.

3 Experimentation and Result Analysis

3.1 Dataset Description and Parameter Setting of Algorithms

In this paper, two benchmark binary class cancer datasets such as Breast Cancer (699 samples and 9 features) and Lung Cancer (32 samples and 56 features) are collected from UCI repository for evaluating the proposed model. CJaya algorithm parameters setting: In this study, the value of population size and the number of iterations is taken as 100. MLP parameters setting: In this study, the number of hidden layers is taken as 3 and, in each layer, five nodes are considered.

3.2 Performance Measures

Some performance metrics used in this work to estimate the efficiency of the model are: sensitivity (Sn), accuracy (Acc), Fscore (Fs), Gmean (Gm), specificity (Sp), and precision (Pr).

$$\text{Sn} = \frac{\text{TP}}{(\text{TP} + \text{FN})} \tag{12}$$

$$\text{Sp} = \frac{\text{TN}}{(\text{TN} + \text{FP})} \tag{13}$$

$$\text{Gm} = \sqrt{\text{Sp} * \text{Sn}} \tag{14}$$

$$\text{Fs} = \frac{(2 * \text{Sp} * \text{Sn})}{(Sp\ + \text{Sn})} \tag{15}$$

$$\text{Pr} = \frac{\text{TP}}{(\text{TP} + \text{FP})} \tag{16}$$

$$\text{Acc} = \frac{\text{TP} + \text{TN}}{(\text{TP} + \text{FN} + \text{FP} + \text{TN})} \tag{17}$$

3.3 Experimentation and Result Analysis

In this work, the CJaya-OSELM is proposed for classifying the cancer datasets like Breast cancer and Lung cancer. The comparison of the proposed model with CJaya-MLP, CJaya-ELM, and OSELM is also done in this study. The outcomes are given in Table 1. The efficiency of the proposed CJaya-OSELM model is proved through many performance metrices. Moreover, Figs. 2 and 3 pictorially represent the results in terms of confusion matrix for Breast cancer and Lung cancer dataset, respectively. According to Table 1, in the proposed model, the sensitivity, and specificity of Breast cancer and Lung cancer datasets are 0.9927, 0.9893, 1, and 0.80, respectively, which shows positive and negative samples are well classified. The comparison of testing accuracies among all the models in Table 1 determines that the proposed CJaya-OSELM model outperforms in both Breast cancer and Lung cancer datasets. Figures 4 and 5 also show that the CJaya-OSELM model graph converges faster with less no. of iterations than other models in both Breast cancer and Lung cancer datasets.

Table 1 Various performance measures of the proposed model (CJaya-OSELM) and comparison of testing accuracies % of all the models

Datasets	Performance measures of CJaya-OSELM					Testing accuracies % of all the models			
	Sn	Sp	Fs	Gm	Pr	CJaya-MLP	CJaya-ELM	OSELM	CJaya-OS ELM
Breast cancer	0.9927	0.9893	0.9909	0.9900	0.9927	89.8	98.2	98.7	99.1
Lung cancer	1	0.80	0.8888	0.8949	0.9642	85.7	93.1	93.8	96.9

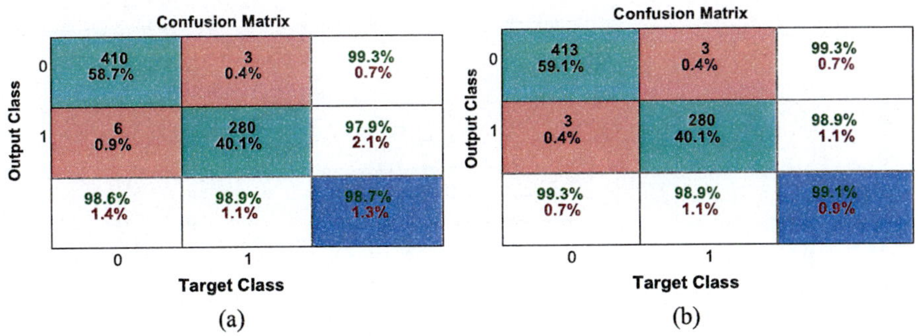

Fig. 2 Confusion matrix of Breast cancer dataset, **a** OSELM, **b** CJaya-OSELM

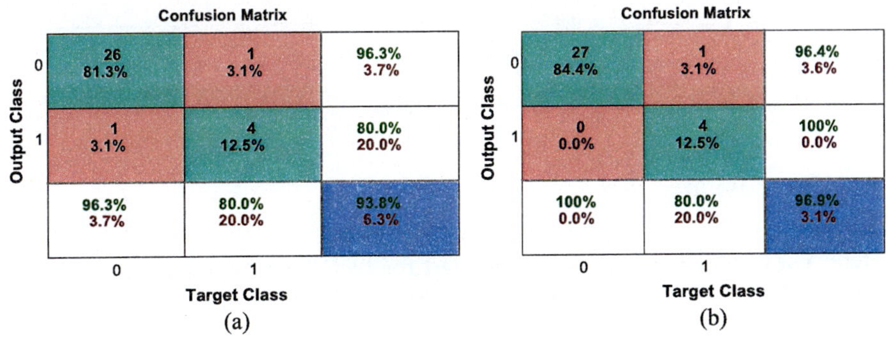

Fig. 3 Confusion matrix of Lung cancer dataset **a** OSELM, **b** CJaya-OSELM

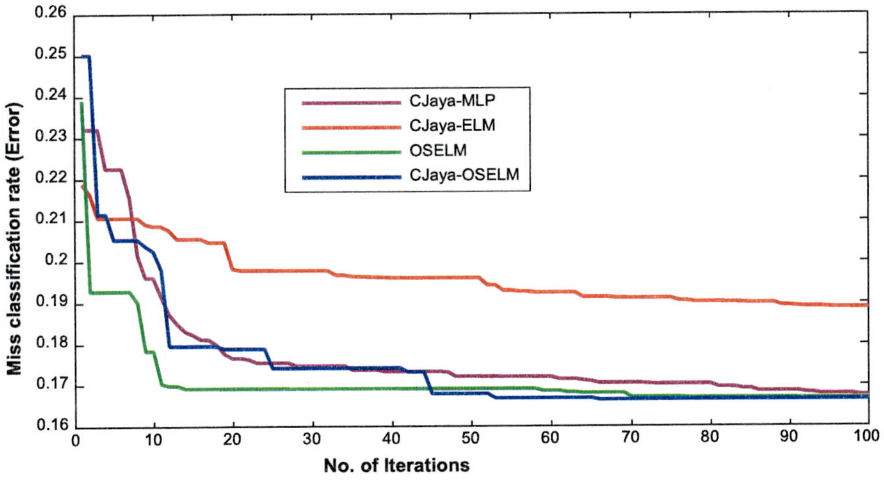

Fig. 4 Comparison of convergence graphs of all the models in breast cancer dataset

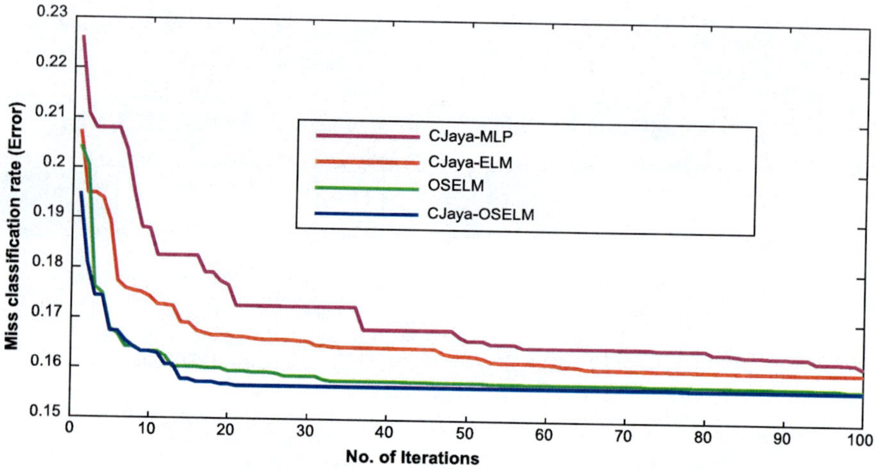

Fig. 5 Comparison of convergence graphs of all the models in lung cancer dataset

4 Conclusion

Here, CJaya-OSELM model is presented for the cancer data classification. The efficiency of the proposed model is compared with CJaya-MLP, CJaya-ELM, and OSELM models. Various performance metrices are used to estimate the performance of the proposed model. The outcomes prove that CJaya-OSELM model can successfully handle the sub-optimal problem and gives better performance in comparison with other models. This study concludes that the proposed CJaya-OSELM model efficiently classifies the cancer data.

References

1. Duda, P.O., Hart, P.E.: Pattern Classification and Scene Analysis. Willey, New York (1973)
2. Mohapatra, P., Chakravarty, S., Dash, P.K.: An improved cuckoo search based extreme learning machine for medical data classification. Swarm Evol. Comput. **24**, 25–49 (2015)
3. Mousavi, S.M., Tavana, M., Alikar, N., Zandieh, M.: A tuned hybrid intelligent fruit fly optimization algorithm for fuzzy rule generation and classification. Neural Comput. Appl. **31**(3), 873–885 (2019)
4. Mohapatra, P., Chakravarty, S., Dash, P.K.: Microarray medical data classification using kernel ridge regression and modified cat swarm optimization based gene selection system. Swarm Evol. Comput. **28**, 144–160 (2016)
5. Zhao, Y., Wang, G., Yin, Y., Li, Y., Wang, Z.: Improving ELM-based microarray data classification by diversified sequence features selection. Neural Comput. Appl. **27**(1), 155–166 (2016)
6. Huang, G.B., Zhu, Q.Y., Siew, C.K.: Extreme learning machine: a new learning scheme of feedforward neural networks. In: 2004 IEEE International Joint Conference on Neural Networks (IEEE Cat. No. 04CH37541), vol. 2, pp. 985–990. IEEE (2004)

7. Huang, G.B., Zhu, Q.Y., Siew, C.K.: Extreme learning machine: theory and applications. Neurocomputing **70**(1–3), 489–501 (2006)
8. Yildirim, Ö.: Ecg beat detection and classification system using wavelet transform and online sequential elm. J. Mech. Med. Biol. **19**(01), 1940008 (2019)
9. Parhi, P., Bisoi, R., Satapathy, P.: A hybridized adaptive fruit fly optimization based online sequential extreme learning machine for bio-medical data classification. In: 2019 International Conference on Information Technology (ICIT), pp. 149–154. IEEE (2019)
10. Rao, R.: Jaya: a simple and new optimization algorithm for solving constrained and unconstrained optimization problems. Int. J. Ind. Eng. Comput. **7**(1), 19–34 (2016)
11. Rao, R.V.: Jaya: an advanced optimization algorithm and its engineering applications. Springer International Publishing, Cham
12. Panigrahi, R., Borah, S.: A statistical analysis of lazy classifiers using canadian institute of cybersecurity dataset. In: Proceedings of ICDSM-2019, Bhubaneswar, India, Springer Lecture Notes on Data Engineering and Communication Technology, ISSN 2367-4512, vol. 37, pp. 215–222 (2019). ISBN 978-981-15-0978-0

A Comprehensive Study of Contemporary IoT Technologies and Varied Machine Learning (ML) Schemes

Mahendra Prasad Nath, Sushree Bibhuprada B. Priyadarshini, Debahuti Mishra, and Samarjeet Borah

Abstract Currently, Internet of things (IoT) is advancing quickly and different applications have turned out from the scholarly community and industry. In this context, quick improvements in equipment and programming-based associated gadgets with corresponding innovations present Internet of things (IoT) in a different look. IoT, the innovation trendy expression of the Web has picked up consideration among specialists, researchers, engineers, programming designers, and the scholarly world. The Internet of things (IoT) incorporates billions of savvy gadgets that can speak with each other with negligible human mediation. It is one of the quickest created fields throughout the entire existence of figuring with an expected 50 billion gadgets before the finish of 2020. With the implementation of machine learning (ML) algorithms, the performance of IoT applications can be enriched. Furthermore, artificial intelligence (AI) systems can perform complex undertakings by gaining from data instead of adhering to pre-modified principles. One of the middle focuses of AI is to prepare PCs to use data or past experience to handle a given issue. Today, AI is employed in various applications and is a focal thought for keen structures. IoT innovation empowers correspondence among genuine items to connect in a wise

M. P. Nath · D. Mishra
Computer Science and Engineering, Siksha 'O' Anusandhan Deemed to Be University, Bhubaneswar, Odisha, India
e-mail: mahendranath272@gmail.com

D. Mishra
e-mail: debahutimishra@soa.ac.in

S. B. B. Priyadarshini
Computer Science & Information Technology, Siksha 'O' Anusandhan Deemed to Be University, Bhubaneswar, Odisha, India
e-mail: bimalabibhuprada@gmail.com

S. Borah (✉)
Sikkim Manipal Institute of Technology, Sikkim Manipal University, Sikkim, India
e-mail: samarjeetborah@gmail.com

S. Borah et al. (eds.), *Soft Computing Techniques and Applications*, Advances in Intelligent Systems and Computing 1248,
https://doi.org/10.1007/978-981-15-7394-1_56

manner. Thus, IoT is about network and robotization that require shrewd calculations to make the brilliant articles cooperate. AI can likewise help machines or a huge number of machines that get together to comprehend what individuals need from the information made by people. Additionally, AI assumes a fundamental job in IoT angle for handling the tremendous measure of data produced by those machines. On this viewpoint, this survey paper intends to talk about the innovation components of IoT and how ML can be implemented on IoT applications.

Keywords Classification · Deep learning · IoT · Machine learning · Pervasive computing · Prediction · WSN

1 Introduction

Artificial intelligence (AI) is a field of concentration that may imply PCs gaining from past comprehension (past data) to improve future execution. The fundamental focal point of this field is the capacity of machines to utilize modified learning strategies. In this connection, learning suggests comprehension or consideration of changes concerning past experiences. Rather than building liberal machines with unequivocal programming, now stand-out calculations are being made accessible which will push the machines to get a handle on the virtual condition and considering their understanding and the machine will take specific choice. This will decline the amount of programming contemplations and the machine will wind up taking choices in isolation. Likewise, machine learning (ML) is utilized as a kind of Web search up for spam channels, recommended structures, credit scoring, stock exchanging, and various applications. A continuous report from the McKinsey Worldwide Institute avows that AI (a.k.a. data mining or perceptive examination) will be the driver of the accompanying gigantic surge of improvements [1]. Regardless, a critical piece of "social data" that is required to adequately make the ML applications not to benefit in totality. A couple of times, various ML calculations take longer than ought not out of the ordinary cases in gaining from a dataset [2, 3]. Broadening the present Internet with every single associated gadget (or alluded to as "Things") and their virtual portrayal has been a developing pattern in decades. This will make loads of possibility in new applications, items and administrations in a wide range of perspectives, for example, savvy homes, brilliant social insurance, car, shrewd vehicle and coordination, and natural checking [IBM12]. The examination right now as of late picked up bunches of consideration, obviously loads of cash, and is bolstered by the joint effort from the scholarly community, industry, and institutionalization bodies in a few networks such as: media transmission, medical coverage organizations, semantic Web, and informatics. This prompts the loads of investments to go with the tide.

For a long time, normal frameworks have been restricted and were intended for explicit purposes with constrained adaptability. This implies when one framework is running, it cannot be changed powerfully and adapt ably. The present activity on presenting the IoT requests application, items, and administration stages which can catch, impart, store, access, and offer information from the physical world, while particularly speaking with the world. This will make new open doors in a major scope of spaces, similar to keen well-being, retail, efficient power vitality, and fabricating, shrewd homes and furthermore, customized end-client applications. Right now, IoT plays increasingly more significant job in everyday life. Basically, we can develop increasingly significant applications, items, and administrations that change our lives consequently and drastically. For instance, readings from meters can be utilized to more likely foresee equalization in power utilization in savvy lattices, investigating mix of traffic, contamination, climate and blockages in tangible information records that can give better traffic. This information change procedure can be better outlined utilizing the notable "information chain of command." We adjust the implications of the layers to the setting of IoT and semantics. The open door that the IoT delivers to people is unending. It is a blend of more than one extraordinary Internet and stuff assortments. These capacities focus the center points in the cloud work in a tantamount way. In this manner, the ensuing advancement of the cloud is known as IoT. These empower clients and the data are put in databases, in addition, information is given by physical world stuff. The idea of IoT empowers physical articles or objects around us to speak with one another in a shrewd manner with the assistance of key segments, for example, radio frequency identification (RFID) marks, processors, sensors, actuators, WAN, WLAN, and so on, with negligible human obstruction. At first, the English technologist Kevin Ashton authored the term "Internet of Things" IoT in 1999. The English technologist Kevin Ashton first coined the word "Internet of Things," IoT in 1999 [4]. Regular articles or things alludes not exclusively to electronic gadgets or mechanical advancements however it additionally incorporate things that we do not envisioned for computing only, for example, nourishment and apparel [4]. IoT empowers items from anywhere to collect, analyze, and distribute among them, whenever, they have availability for anything. CISCO reported that the quantity of regular physical articles and gadgets associated with the web in 2010 amounted to approximately 12.5 billion and is expected to reach 30 billion by 2016 as the quantity of smart gadgets per individual increase and 50 billion by individual increase by 2020 [5]. With a rise in genius products and things, IoT usage is far reaching in various environments, ranging from home robotics in domestics, industrial computerization in the company division to debacle/crisis management in various projects.

2 IoT Technologies

IoT lives on three basic engineering sections to equip the ordinary "stuff" with follow-up and detection [5]. When its "stuff" get appropriately prepared; it permits to accumulate, examine, convey and circulate information among them. The segments incorporate equipment, programming, and introduction instruments.

2.1 *Software*

IoT empowers distinctive specialized devices, gadgets and applications to work together. Middleware is an instrument that empowers knowledge checking tools for on-demand functionality and processing. It serves as an interface and fosters IoT communication between network and "stuff." Middleware tackles issues such as interoperability through heterogeneous gadgets and setting knowledge, gadget disclosure and board, flexibility, data analysis, privacy and security [6]. Middleware enhancement for IoT is itself a major dynamic research area.

2.2 *Hardware*

Equipment segments, for example, in IoT, actuators, WSN, NFC, RFID, Bluetooth, and Wi-Fi are used to differentiate the real world items in the genuine word and are alluded on the network [7]. Coordinating the above-mentioned pieces of the equipments into IoT allows it to be standardized in design.

2.3 *Near Field Communication (NFC)*

It is kind of a convention on radio correspondence that empowers short-go matching between devices. It is formed on benchmarks of RFID, and works like RFID. It incorporates labels and a small transmitter, which send data to other brilliant gadgets close to each other. NFC tag includes a one-size-fits-all ID, which gives enabling correspondence between gadgets. NFC resides within the 13.56 MHz unlicensed RF band [8]. It has a scope of around 10 cm, that is significantly less than bluetooths, where data can be moved up to 10 meters. NFS information propagation speed is only 424 Kb/s. It can speak to various remote technologies and has the option of interacting with a wide range of NFC transmissions. With other technologies, NFC is similarly safe and easy to use.

2.4 *Actuators*

After the information is gathered, it has to be handled, dissected, and reported with the aid of sensors. This is achieved with the gadget help called actuators. It changes to movement over vitality. For IoT, actuators are used wherever the gadget needs to be switched on/off or primed by the application of a control [2].

2.5 Bluetooth

With the introduction of wreckage organize abilities, (Bluetooth 5) is prepared to turn into the general language of IoT because of its four-fold territory, two-fold pace, 800 percent expansion in information broadcasting limit, interoperability and decreased utilization of essentialness [3].

2.6 Wi-Fi

Another reasonable innovation regarding IoT correspondence is Wi-Fi. Wi-Fi for IoT has a common network interface and is used by various IoT practices. Wi-Fi gives a valuable preferential position in terms of limit, inclusion, and usability. Since the keen gadgets are slowly increasing, it will also need an even more expanding Wi-Fi inclusion and cap. We depend on this Wi-Fi halow to be propelled in 2018 for a confirmation procedure. Furthermore, it is anticipated that Wi-Fi's spectrum will double the usability of today.

2.7 Wireless Sensor Network (WSN)

Benefits from a large number of cheap coordinated remote gadgets and circuits fit for calculation and correspondence give approach to become productively easier, less forceful expending gadgets for detecting applications. WSN comprises of an enormous number of insightful sensor, empowering the assortment, handling, examination of information accumulated in an assortment of condition [7]. WSN's sensor central features like: a transducer, small scale, Smartphone, and source of power are essential. The transducer generates electrical signs which are dependent on the physical impacts and wonders that get detected. The processor runs on the microcomputer and stores the sensor yield. The handset gathers directions from a focal PC and sends data to PC. The force is obtained from a battery for each sensor hub. WSN assumes a significant job in broad applications, for example, industry mechanization, robotized and advanced mobile phones, video observation, traffic checking, clinical gadget checking, climate condition checking.

2.8 IoT Data Visualization

Information representation in IoT assists individuals with understanding the noteworthiness of information by bringging it in a visual environment. Increasing the size of information stream in IoT, additionally, includes the necessity for good information processing systems. The ongoing headway in contact picture innovations, for example, advanced mobile phones and surface processing give straightforward representations among the clients who utilizes IOT [7]. Development in 2D and 3D strategies gives information representation to be a big guarantee for the customers.

3 IoT Applications

With an expansion in informative articles and stuff, the utilization of IoT is all-encompassing in numerous practice regions, extending from home mechanization in domestics, industry computerization in the market sector segment to debacle in government ventures. Here, the client territories are divided into three groups like: common government, company, and domestics. Figure 1 depicts the various real-world IoT applications.

The potentiality given by the IoT makes it possible to create an enormous number of uses. The figure above illustrates the IoT advancement applications that are divided into three distinct classifications which are different like: Domestics, Government/Social, and Company. Basically, anyone from any time and wherever will have

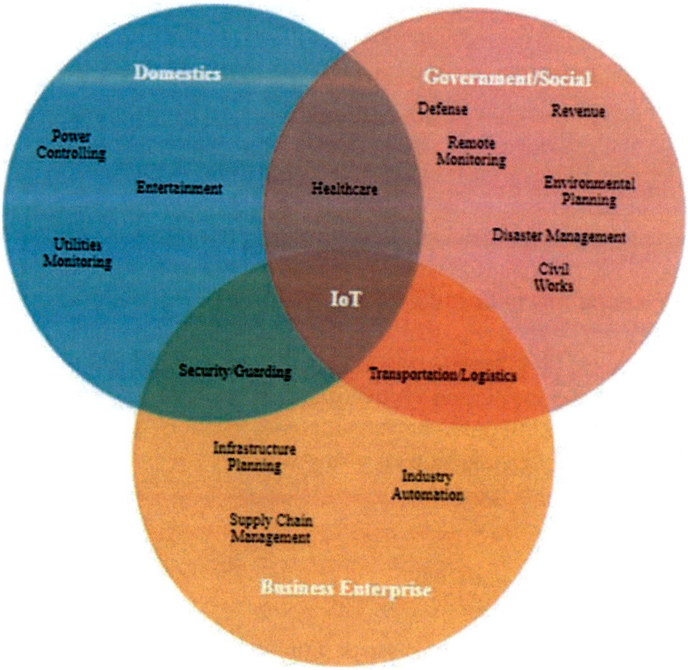

Fig. 1 Applications of IoT

availability for anything and it is common for such connections to evolve and build a fully propelled dynamic network. IoT technologies can also be used to create another idea and a broad space for the production of smart homes to provide insight, comfort, and personal satisfaction. Home computerization can be beneficial to individuals who need to use home computers when away from home and can greatly change disabled people's lives. Home automation will be home, housework, or family movement roboticisation. Home mechanization may include centralized lighting control, HVAC (warming, ventilation, and cooling), appliances, entrance and entrance protection locks, and various systems to provide enhanced accommodation, comfort, vitality efficiency, and health. Home robotics for the elderly and the disabled may offer an increased personal satisfaction to people who may need parental or institutional consideration in some way or another [6]. IoT planning components assume a vital job in the business enterprises range, with lower cost and lower power requirement. Gazing from framework intending to inventory network the executives, the IoT made the working procedure of the framework a lot less difficult and compelling. The job of IoT in transportation gives huge points of interest. IoT made roadways a brighter place by empowering vehicles fitted with labels and sensors that delivers significant data to local traffic control and transport vehicles to the more probable course of traffic, thus, helping to handle stops, providing vacationers with suitable transportation data, and monitoring the status of the goods being transported [8].

The IoT administrations are massive for the Social/Government and the rate of development is rising steadily. It gives numerous advantages in the healthcare area, for example, following patients well-being record, recognizable proof and verification of staffs working in the medical clinic, programmed information assortment and expectation examination, and so forth., Recently, catastrophic events and inadvertent disasters taking place are increasing more regularly. IoT technologies, such as RFID and WSN, will take on a crucial job of alerting tragedy before it starts and restore the calamity after it is locked [7].

4 Machine Learning and IoT

The ultimate fate of a keen IoT is relying on having connected a huge number of savvy gadgets together. In this way, the mere amount of information made by those will be incremented to an unheard of state. Such information may have very important knowledge of what is important and what is not [8]. The dissect of information assumes a critical job in calling attention to the contention that emerges in IoT applications. Surveying and knowing this knowledge and doing it as such with customary methods is imposable to humans. It is preferably robotized to boost velocity and accuracy of information analysis for IoT to fulfill its requirement, and to decide on information investigation and dynamics to be adaptable. One way is with AI's guide to executing. Figure 2 delineates AI measurement appropriateness in IoT. AI is an sub-field that learns and changes by learning naturally. It centers around forecast dependent on realized properties gained from the preparation information. In IoT, AI

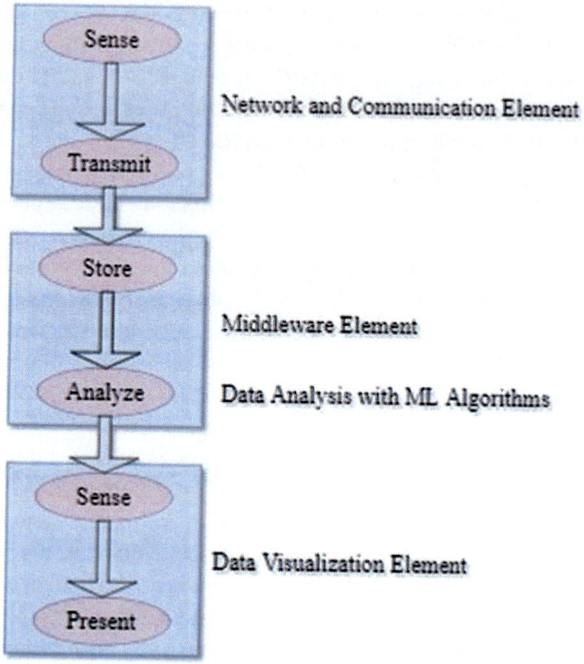

Fig. 2 Machine learning in IoT

can help surveying and dissecting the information gathered from the earth to discover examples or similitudes that can be gained from, the aim is to make better choices. AI gives IoT and IoT-related devices, a cerebrum for thought which is known as "Injected Intelligence" [8].

5 Machine Learning Algorithms

As the volume of information in IoT increases, the main issue is how to classify the information, how to sort it out, and how to use it to decide on mechanized choices. Data order is the establishment for the identifiable informative evidence and the board of enormous details in the system of things.

5.1 Classification Algorithms

It is the establishment within the Web of things for the keen identifiable proof and the board of gigantic data. A classifier is taught from a lot of preparing models of class names in characterization. A classifier's presentation is determined by its precision to order. Grouping approaches are suited for managing a lot of information in information mining. It can predict class marks and organize information that is based on class and set names being prepared, that should be used to order concealed accessible information along those lines [6, 9, 10]. Following are some measurements of classification commonly used in IoT utilizations:

5.1.1 K-Nearest Neighbor (K-NN)

It is one of the supervised machine learning algorithms. It uses the Bayes's theorem which assumes that all features are statistically independent of each other. This theorem is based on the native assumptions that input variables are independent of each other, i.e., there is no way to know anything about the other variables.

5.1.2 Naive Bayes

It is a classification technique based on Bayes theorem. In the event that the autonomy presumption holds, at that point, it works more productively than different algorithms. It requires less preparing information. It is exceptionally versatile. It can make probabilistic expectations [7].

5.1.3 Decision Trees

These are the type of supervised machine learning where data is continuously splitted depending upon certain parameters. Such classifier asks a set of carefully crafted questions regarding to the input attributes. Each time it gets an answer, a subsequent inquiry is posed until it makes an inference about the record's class mark.

5.2 Prediction Algorithms

Prediction resembles classification. First, it constructs a model and uses the model built to anticipate obscure interest. It is special in the sense order in relation to classifier that alludes predicting straight out class mark which is distinct or ostensible. Though, prediction models are called potential nonstop, that is, they predict obscure or lacking qualities [3]. Following are some widely used expectation techniques:

5.2.1 Linear Regression

It is a technique for showing the association between a scalar ward variable y and in any event, one illustrative elements called self-ruling components connoted X [5]. The goal of relapse is to pursue through a dissipate plot to match a solitary thread. It consists of three phases: examining the relation, evaluating the model, and assessing the model's validity and usefulness [6].

5.2.2 Neural Network Model (NNM)

Neural networks model includes programmed gaining of conditions just from estimated information with no compelling reason to include additional data. The neural system is trained from the verifiable information with the intention that it will discover veiled conditions and have the choice of using these for future predictions [7].

6 Use of Machine Learning on IoT Applications

Below the models, any IoT implementation is illustrated in the literature where AI is effectively implemented.

6.1 Example1: Traffic Management

These days, transportation is becoming smarter with a few system administration elements, such as actuators, sensors, and other preparatory forces. Today's city roads are fitted with RFID labels and sensors which are all the more likely to send information to the traffic course to local traffic control and transport vehicles. LarKC [23] engineering abuses systems of AI to give simpler adaptability in the transportation area. Figure 3 represents the entire framework and it recommends a few course to the goal.

6.2 Example 2: In Home Appliances

This is a use to lower an espresso machine's cost of vitality [8]. This revolves around espresso machines in spaces close to those in which individuals for the most part could not care less on sparing vitality, for example, the work environment. The proposed expectation calculation works as proficient as conceivable to lessen the vitality cost expected to work the espresso machine. Figure 4 describes how IoT is used in various home applications.

Fig. 3 IoT and ML in traffic management

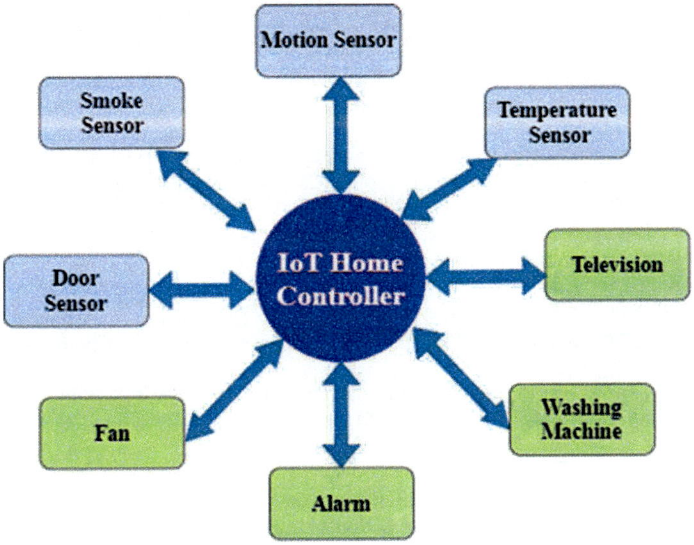

Fig. 4 IoT and ML in home appliances

7 Conclusion

IoT is changing us altogether. AI transforms human machine perception. Moreover, these technologies are becoming central to our lives with the ability to manage big data. IoT consists of countless gadgets with assortments that are connected to each other and transmit huge measurements of knowledge. The aim of this paper is not to hold the conceptual concept of AI in IoT, but to actually create an inspiring domain by simply using programming, libraries, Web administrations, and so on that we need to make and distribute for nothing. The advancement of Internet technology and the emergence of sophisticated web-based applications have transformed the Internet of things into another creation, which will computerize everything from home appliances to secure robot controllers. Once effectively updated, IoT will not control the world with any uncertainty later on. ML researchers' main focus is on making diagramming increasingly gainful (both in reality) learning methodologies that can perform better for spare time and cost in different applications. Right now, as described with IoT, they have tried to give an analysis of the forms of ML, ML undertaking and their applications, etc. Then again, the application might simply make perceptions based on past history, what is to come is still something questionable or other, that's the enhancement.

References

1. Anjomshoa, F., Aloqaily, M., Kantarci, B., Erol-Kantarci, M., Schuckers, S.: Social behaviometrics for personalized devices in the internet of things era. IEEE Access **5**(c), 12199–12213 (2017)
2. Meidan, Y. et al.: ProfilIoT: a machine learning approach for IoT device identification based on network traffic analysis. In: SAC2017 32nd ACM Symposium on Application Comput*ation*, pp. 506–509 (2017)
3. Ganz, F., Puschmann, D., Barnaghi, P., Carrez, F.: A practical evaluation of information processing and abstraction techniques for the internet of things. IEEE Internet Things J. **2**(4), 340–354 (2015)
4. Nath, M.P., Sagnika, S.: Capabilities of Chatbots and Its performance enhancements in machine learning machine learning and information processing. Adv Intell Syst Comput **1101**, 183–192 (2020)
5. Nath, M.P., Goyal, K., Prasad, J., Kallur, B.: Chat Bot—an edge to customer insight. Int. J. Res. Sci. Innov. (IJRSI) **5**(5), 29–32 (2018)
6. Li, S., Da Xu, L., Zhao, S.: The internet of things: a survey. Inf. Syst. Front. **17**(2), 243–259 (2015)
7. Pillonetto, G., Dinuzzo, F., Chen, T., Nicolao, G.D., Ljung, L.: Kernel methods in system identification, machine learning and function estimation: a survey. Automatica **50**, 657–682 (2014)
8. Suthaharan, S.: Big data classification: problems and challenges in network intrusion prediction with machine learning. ACM SIGMETRICS Perform. Eval. Rev. **41**(4), 70–73 (2014)
9. Panigrahi, R., Borah, S.: Rank of normalizers through TOPSIS with the help of supervised classifiers. Int. J. Eng. Technol. **7**(3.24), 483–490 (2018)
10. Panigrahi, R., Borah, S.: A statistical analysis of lazy classifiers using canadian institute of cybersecurity dataset. Proceedings of ICDSM-2019, Bhubaneswar, India, Springer Lecture Notes on Data Engineering and Communication Technology, ISSN 2367-4512, vol. 37, pp. 215–222 (2019). ISBN 978-981-15-0978-0

A Hybrid Strategy Based on Monitored Region Segregation for Redundant Data Minimization (HS-MRS) in Sensor Networks

Sushree Bibhuprada B. Priyadarshini, Debahuti Mishra, and Samarjeet Borah

Abstract This research proffers a hybrid strategy stirred by maximum far-flung scalar premier determination and triangular scalar premier determination for actuation of camera sensors. The whole monitored zone is segregated into a number of logical sub-regions. Afterwards, the scalar premiers get selected intellectually such that, two and three scalar premiers are selected in alternative sub-regions. Whenever, any event occurs, the scalar premiers operate as the representatives of scalars and on behalf of the prevailing event, they publicize the event erudition to their relevant concerned video sensors. Afterwards, the camera sensors get actuated collaboratively through these premiers while avoiding greater amount of probable overlapping among their field of views. Experimental outcomes attained from the experimental inquest assert the clout of our proffered algorithm in comparison to another popular existing method.

Keywords Field of view · Depth of field · Far-flung · Scalar premier

1 Introduction

Currently, video sensors (also called camera sensors) have gained the attention of research community spanning from environmental tracking to battle field monitoring. Basically, the sensors are sprinkled from air planes arbitrarily throughout the targeted monitored zone. Scalar sensors are conjointly deployed with the video sensors since the video sensors consume greater amount of energy, if they are turned on all the

S. B. B. Priyadarshini · D. Mishra (✉)
Siksha 'O' Anusandhan Deemed to Be University, Bhubaneswar, Odisha, India
e-mail: debahutimishra@soa.ac.in

S. B. B. Priyadarshini
e-mail: bimalabibhuprada@gmail.com

S. Borah
Sikkim Manipal Institute of Technology, Sikkim Manipal University, Sikkim, India
e-mail: samarjeetborah@gmail.com

S. Borah et al. (eds.), *Soft Computing Techniques and Applications*, Advances in Intelligent Systems and Computing 1248,
https://doi.org/10.1007/978-981-15-7394-1_57

times as compared to scalars. Generally, the video sensors possess two basic features: Field of view (FOV) and depth of field (DOF). FOV represents "the angle at which a camera can trap the exact photograph of any item" [1]. DOF indicates "the distance within which a camera can capture exact photograph of an item" [1].

At the moment of ensnaring the video through these sensors, covering the event region effectively becomes a crucial issue. At the time of event occurrence, it gets ensnared by the scalar sensors and the scalars disseminate regarding it to the concerned video sensors and the video sensors collectively decide who among them have to undergo actuation. But the problem to be considered is that at the time of considering the video sensor actuation, superimposing triumphs among the FOVs of the video sensors. Hence, the scalars dwelling within the superimposed territories of FOVs communicate the duplicate info in numerous turns giving rise to repeated data transmittal. This gives rise to undesired energy exhaustion, which is undesirable. Hence, our motto is to activate only the aspired and suitable video sensors which can cover the event zone effectively so that the amount of redundant data transmittal will be diminished.

This article is arranged in the following way: in the subsequent section, we present the literature survey. Section 3 collaborates the proffered scheme. Section 4 discusses the implementation and experimental outcomes. At length, in Sect. 5, we wrap up the paper.

2 Literature Review

This section collaborates the thread of previous work that has been carried out relevant to this field. A popular distributed approach known as "Distributed collaborative camera actuation based on scalar count (DCA-SC)" [1] is recently proffered in literature, where the video sensors collectively adjudicate whichever amidst them are to be aroused. In this method, the video sensors are incited hinged on diminishing succession of concerned scalar count values. The work depicted in [2, 3] focuses on directional cover sets problem. Furthermore, Girault [4] proposed a methodology which moves in two passes for repeated data diminishment. The method detailed in [5] suggests a strategy hinged on data alikeness for repeated data removal. Moreover, the approach in [6] introduced an optimal approach for sensor placement. The work in [7] and [8] also discusses similar approaches. Priyadarshini and Panigrahi proposed maximal far-flung approach in [9] that chooses maximal distant premiers in sub-regions. Similarly, the work done in [10] discusses a triangular scheme of scalar premier determination followed by camera sensor actuation.

In this article, we have proffered a Hybrid Strategy based on Monitored Region Segregation for Redundant Data Minimization (HS-MRS) that combines the approaches employed in both [9, 10]. This strategy determines the premiers effectively in every sub-regions of the tracked zone and the premiers communicate the camera sensors about the prevailing event. Consequently, the informed cameras collectively decide who among them have to be activated.

3 Proposed HR-MRS Scheme

The proffered HR-MRS scheme involves actuation of diminished number of video sensors in the sub-regions that are required to give enriched coverage of the prevailing event region than DCA-SC [1] in literature. The main alms of the HR-MRS scheme lies in the following dimensions:

- The HR-MRS scheme employs a hybrid approach involving determination of maximal far-flung (i.e., far-distant) scalar premiers and triangular scalar premiers in alternate sub-regions for video sensor.
- The premier determination is accomplished such that the video sensors actuated owing to those determined scalar premiers can span more distinct portions of the prevailing event region. This gives rise to activation of diminished number of video sensors having minimum amount of superimposing among their respective FOVs, which results in reduced amount of repeated data transmittance.
- The proposed HR-MRS scheme enriches the amount of event area coverage, while curtailing the amount of energy expenditure for camera invigoration and the repeated data transmissions.

3.1 Detailed Procedures Involved in Proposed HR-MRS Scheme

These are the phases involved in our proposed HR-MRS scheme:

3.1.1 Sensor Deployment

At first, the scalars and video sensors are arbitrarily put in the monitored region. Scalars and video sensors send my scalar message (MSM) and my camera message (MCM). MSM depicts a message, broadcasted by individual scalar sensor, that retains the id and position information of concerned scalar. *MCM* indicates a message, sent by each video sensor, that retains the id and position information of corresponding video sensor. The ids of all the cameras are preserved in a catalogue referred to as *waiting list* (*WL*). Likewise, *Event Grade* List (*EGL*) is preserved which reserves the ids of already turned on video sensors. After deployment, the sensors gauge the *Euclidian distance* amidst one another. The Euclidian distances between any two sensors can be interpreted as follows:

$$D(N_i, N_j) = \sqrt{(X_j - X_i)^2 + (Y_j - Y_i)^2}$$

where

$D(N_i, N_j)$ *Euclidian* distance between sensor node N_i and sensor node N_j

X_i The X-*coordinate* of sensor i
Y_i The Y-*coordinate* of sensor i
X_j The X-*coordinate* of sensor j
Y_j The Y-*coordinate* of sensor j.

3.1.2 Monitored Region Segregation and Intellectual Scalar Premier Determination

The whole monitored region to be monitored is considered as rectangular. It gets segregated into squares as illustrated in Fig. 1 so that the length of side of the square is equivalent with (1/10)th of length of rectangular area to be monitored. The figure justifies that the whole field (large square) is divided into total hundred sub-regions (smaller squares). and in each alternate regions, maximal far-flung scheme denoted by SM [6] and triangular scheme denoted by ST are used [7]. In every sub-region, scalar premiers (SPs) are selected as illustrated in Fig. 2, where the selected SPs has been shown in case of both the far-flung and triangular approaches. In far-flung

S_M	S_T	S_M	S_T	S_M	S_T	S_M	S_T	S_M	S_T
S_T	S_M	S_T	S_M	S_T	S_M	S_T	S_M	S_T	S_M
S_M	S_T	S_M	S_T	S_M	S_T	S_M	S_T	S_M	S_T
S_T	S_M	S_T	S_M	S_T	S_M	S_T	S_M	S_T	S_M
S_M	S_T	S_M	S_T	S_M	S_T	S_M	S_T	S_M	S_T
S_T	S_M	S_T	S_M	S_T	S_M	S_T	S_M	S_T	S_M
S_M	S_T	S_M	S_T	S_M	S_T	S_M	S_T	S_M	S_T
S_T	S_M	S_T	S_M	S_T	S_M	S_T	S_M	S_T	S_M
S_M	S_T	S_M	S_T	S_M	S_T	S_M	S_T	S_M	S_T
S_T	S_M	S_T	S_M	S_T	S_M	S_T	S_M	S_T	S_M

Fig. 1 Segregation of monitored dominion into sub-regions

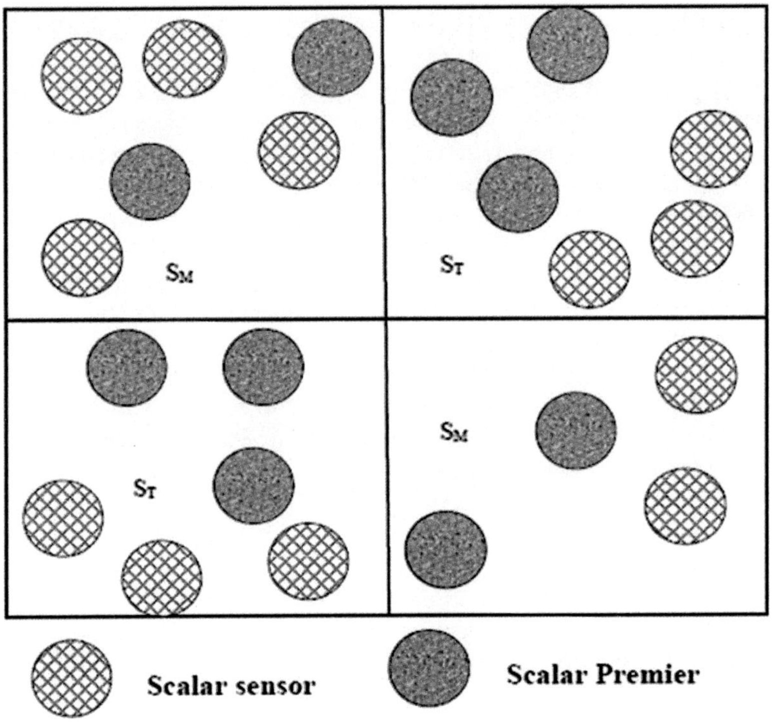

Fig. 2 Determination of scalar premiers (SPs) in alternate segregated compartments

scheme, at first, a SP is determined such that it possesses the least mean distance amidst all the scalars dwelling in concerned sub-region.

Afterwards, another SP gets selected in the same sub-region which is present at maximal distance from the initially selected premier. On the contrary, in the triangular scheme, a scalar premier known as primordial premier gets determined just like that it possesses the least mean distance among the scalars dwelling in the corresponding sub-region. The line connecting the central point of the corresponding sub-region with the coordinate location of primitive premier is called the Stand line.

Then, the subsidiary premier is determined from the primitive one such that it is dwelling across an angle 60° at a maximal distance from the PCP along counterclockwise direction from the coordinate location of PCP around the stand line. Thereafter, a torrential premier is determined that indicates the primordial whose average mean distance from both the primordial and subsidiary premier is the minimal among all the scalars residing within the corresponding sub-region. Thus, three premiers get determined in every sub-region in the triangular scheme. Since our proffered hybrid approach employs the concept of both the far-flung and the triangular approach, it is a hybrid one.

3.1.3 Event Occurrence, Video Sensor Collaboration, and Activation

Whenever, any event prevails, it is first of all ensnared by the scalar premiers pertaining to the concerned sub-regions in lieu of all the scalars [1]. Consequently, the event ensnaring premiers report such information to their corresponding video sensors through broadcasting detection message (DM), that retains the ids and position information of concerned premiers. Subsequently, the corresponding camera updates its event reporting scalar premier table (ERSPT). ERSPT gets maintained by every video sensor that retains the event ensnaring scalar premier ids.

In this step, all the video sensors evaluate the sum of event-communicating scalar premiers dwelling within the purview of their FOVs known as event reporting scalar premier Addition (ER-SPA). Then, the video sensors decide their order of actuation according to the descending sequence of their ER-SPA value while matching their respective premier ids with the ids of SPs pertaining to already actuated video sensors. As and when a video sensor undergoes actuation, its id gets added to EGL.

4 Experimental Evaluation and Performance Assessment

4.1 Simulation Setup

This research article has been implemented using C++. The sensors are randomly deployed. We have changed the number of video sensors (nvs) and marked its repercussion on number of video sensors activated (nvsa) in [1] and HR-MRS by considering the following execution parameters:

- Number of video sensors actuated (nvsa)
- Coverage ratio (cr): It is depicted as "the ratio of area of event camouflaged through already stimulated video sensors to the total area of the prevailing event" [1]. Greater cr indicates enriched coverage of the concerned event zone.
- Event-loss ratio (elr): It is defined as "the ratio of area of event not camouflaged by activated video sensors to the whole area of the concerned event."
- Redundancy ratio (rr): It is depicted as the ratio of whole portions of superimposed zones of FOVs of turned on video sensors spanning the prevailing event zone to the total exclusively distinct portions of the event zone which get spanned through the turned on video sensors.
- Energy consumption for video sensor activation (ecvsa).

4.2 Result Analysis

The number of video sensors (nvs) get changed and the impact was marked on different aforesaid parameters in DCA-SC [1] as well as proposed HR-MRS as

Table 1 Effect of variation of nvs in DCA-SC

nvs	nvsa	cr	rr	elr	eevsa
150	113	0.66	0.63	0.34	144.64
170	118	0.67	0.65	0.33	151.04
190	123	0.71	0.68	0.29	157.44
210	125	0.72	0.70	0.28	160.00
230	129	0.74	0.73	0.26	165.12
250	134	0.76	0.74	0.24	171.52

Table 2 Effect of variation of nvs in HS-MRS

nvs	nvsa	cr	rr	elr	eevsa
150	99	0.77	0.38	0.23	126.72
170	103	0.79	0.39	0.21	131.84
190	106	0.80	0.41	0.20	135.68
210	107	0.82	0.42	0.18	136.96
230	111	0.83	0.43	0.17	142.08
250	116	0.87	0.45	0.13	148.48

illustrated in Tables 1 and 2. The tables state that with a rise in nvs, the nvsa raises in both the strategies since with hike in nvs, the number of event tracking scalars/scalar premiers hikes. The nvsa value obtained in our HR-MRS approach is attained to be lower than that of [1]. This is because SPs are selected intellectually in the sub-regions in proffered scheme. Similar impact is speculated on the ecvsa since it depends on the nvs in both the schemes.

Likewise, the cr is obtained to be increasing with progress in nvs as higher number of scalars/scalar premiers trap the occurring event. In addition, the cr has been again obtained as the highest at 87% in our HR-MRS scheme. This is because SPs are selected intellectually in the proposed HR-MRS in the individual virtual sub-regions. The elr is decreasing with escalation in nvs since more amount of event zone is camouflaged gradually. But, this is obtained as the lowest at 13% in HR-MRS. With rise in nvs, the rr hikes for both the approaches since the amount of superimposing rises. The rr obtained in the proffered HR-MRS is lower than that of [1]. The least rr obtained in case of HASL-CA scheme is 38%.

5 Conclusions and Future Outlook

This article presents a novel strategy, namely HR-MRS which divides the entire geographic zone under speculation into several virtual Sub-regions and selects premiers among the scalar sensors intellectually in each of the logical compartments so that the video sensors turned on by them afford enriched event region coverage along with diminished redundant data transmittal. Experiments have been carried out to assess the efficacy of the proffered strategy over [1]. The investigation outcomes justify the supremacy of HR-MRS over the DCA-SC in terms of lowered camera actuation, enriched coverage ratio, lesser event-loss ratio, reduced redundancy ratio, and lesser energy expenditure for video sensor activation. We have planned to extend our approach to work in three-dimensional scenarios while employing mobile video sensors as a direction toward future investigation.

References

1. Newell, A., Akkaya, K.: Distributed collaborative camera actuation for redundant data elimination in wireless multimedia sensor networks. Ad Hoc Networks **9**(4), 514–527 (2011)
2. Cai, Y., Lou, W., Li, M., Li, X-Y.: Target-oriented scheduling in directional sensor networks. In: Presented at 26th IEEE International Conference on Computer Communications (IEEE INFOCOM2007). [Online]. Barcelona. pp. 1550–1558 (6–12 May 2007)
3. Available http://ieeexplore.ieee.org/xpl/articleDetails.jsp?arnumber=4215764
4. Girault, A.: Elimination of redundant messages with a two-pass static analysis algorithm. Parall. Comput. **28**(3), 433–453 (2002)
5. Available http://www.sciencedirect.com/science/article/pii/S0167819101001508
6. Ghaddar, A., Simplot-Ryl, I., Razafindralambo, T., Tawbi, S., Hijazi, A. (14–17 June 2010) Algorithm for data similarity measurements to reduce data redundancy in wireless sensor networks. In: IEEE International Symposium on a World of Wireless Mobile and Multimedia Networks (WoWMoM), [Online]. Montreal, QC, Canada, pp. 1–6
7. Available http://ieeexplore.ieee.org/xpl/login.jsp?tp=%26arnumber=5534888%26url=http%3A%2F%2Fieeexplore.ieee.org%2Fxpls%2Fabs_all.jsp%3Farnumber%3D5534888
8. Toumpis, S., Tassiulas, T.: Optimal deployment of large wireless sensor networks. IEEE Trans. Inf. Theory [Online] 52(7), 2935–2953 (2006, July). Available http://ieeexplore.ieee.org/xpl/articleDetails.jsp?arnumber=1650348
9. Priyadarshini, S.B.B., Panigrahi, S.: A distributed approach based on maximal far-flung scalar premier selection for camera actuation. In: 12th International Conference on Distributed Computing and Internet Technology (ICDCIT 2016), KIIT University, Bhubaneswar, Odisha, 15th–18th January, pp. 87–91 (2016)
10. Priyadarshini, B.B., Panigrahi, S., Bagjadab, A.B.: A distributed triangular scalar cluster premier selection scheme for enhanced event coverage and redundant data minimization in wireless multimedia sensor networks. Indian J. Sci. Research **14**(2), 96–102 (2017)

Design and Implementation of Ploughing and Seeding of Agriculture Robot Using IOT

S. Poonguzhali and T. Gomathi

Abstract In India, 70% people rely on agriculture for their food that is rice, crop like wheat, barley, sunflower, etc. In this paper, a robot system which helps us to cultivate agricultural lands without the use of man power. In other countries, they do not require man power in agricultural sector instead of that, we can use robot to save man power time and increase in the efficiency of production. IOT is a latest technology to grow the network of everyday object of agriculture. In the proposed system, a robot system is used to cultivate the agriculture land without the help of man power and it is also controlled by using Wi-Fi. Here, the robot is used to automate the process of sowing crops and ploughing in the field and water sprinkling to the crops are also done automatically by the robot and motors are used to move the robot vehicle in the front and back, respectively. The proposed system is used to train the robot by embedded instructing coding to sow, plough and water the crops. Based on a time interval, these works are done automatically without the man power.

Keywords PH · Wi-Fi · Internet of things · Humidity · Automation

1 Introduction

The information regarding humidity, soil fertility, etc., can be collected using Internet of things, IOT [1], technology. The other applications of IOT are in agriculture, scrutinizing of the planted crops [2], finding out the presence of weed, checking the presence of pests and animals in the field. The farm can be joined with the farmer at different places and time with the help of IOT. The various conditions of the farm are controlled by wireless devices [3]. The various techniques are used for controlling [4] the farms. Wireless cameras are used to take the snapshot of the working process.

S. Poonguzhali · T. Gomathi (✉)
School of Electrical and Electronics, Sathyabama Institute of Science and Technology, Chennai, India
e-mail: gomes20@gmail.com

S. Poonguzhali
e-mail: poornidp@gmail.com

S. Borah et al. (eds.), *Soft Computing Techniques and Applications*, Advances in Intelligent Systems and Computing 1248,
https://doi.org/10.1007/978-981-15-7394-1_58

The working conditions of the farm is updated lively with the help of IOT using a smart phone [5]. The production of normal farming has been increased [6] by Internet of things. The robot system [7] is used to maintain the process of cultivating ploughing, seeding and harvesting agricultural land without the use of human beings. The productivity can be increased by reducing the man power and time [8]. The productivity rate for the automated robot works like weeding, harvesting and so on. Here, like ploughing the land, watering the plants, sowing the seed, spraying the fertilizers and to navigate the vehicle motion are done by this robot using its microcontroller [9]. The idea of using a robot in the farm field is to decrease the efforts of human in the agricultural sector and automatic process is applied and robot helps the human to perform the operation like ploughing, seeding and water sprinkling at night as well as hot summer days [10].

2 Existing System

The system which exists normally has disadvantages like man power is required, long distance communication [11] is difficult and in some places, they use animals brutally making them work day and night hard. Also lots of human effort is needed and enormous time and energy required. Number of people required increases labour cost to be too high investment. Efficiency also depends on the sincerity of the labours. So our aim is to automate the process of ploughing the farming lands using agriculture robot.

3 Proposed Methodology

The proposed robot system is used in man-less cultivation in agriculture land. In the proposed system, a robot is used to monitor the system. It consists of a power supply and two motors which are used to move the robot vehicle in the forward direction and backward direction and we are using four types of sensors [12]. They are ultrasonic sensors, temperature sensors, PH sensors and level sensors for sensing required parameters and also microcontroller which is connected to the Wi-Fi [13] to track and get the output of all the sensors and in the robot section, ploughing, seeding and water sprinkling are done automatically [14] (Fig. 1).

3.1 Automatic Motion Control of the Robot

The sensors are used here because robots have no eyes or ears. So sensors [15] are added to the robots with which they sense it and work automatically for an instance like ultrasonic sensors used in this model as they send ultrasonic waves and when

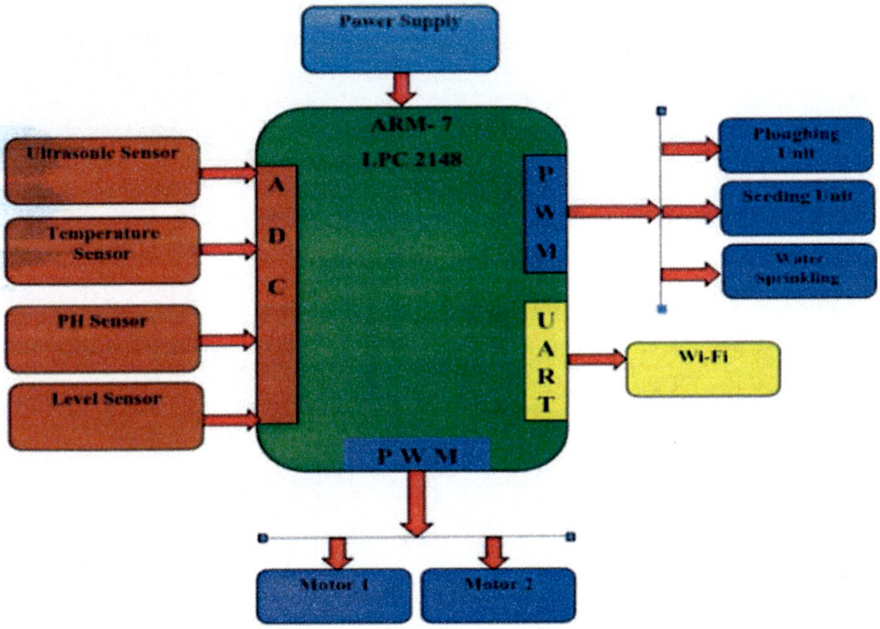

Fig. 1 Robot section

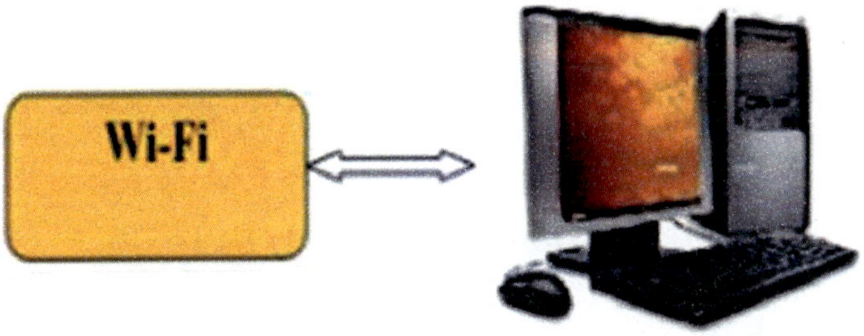

Fig. 2 Wi-Fi connected to PC

the waves reflects back, it calculates the distance between the robot and other objects presence in the field and according to it controls its movement without getting clash or creating any obstacle [16] (Fig. 2).

3.2 *Soil Temperature Measurement*

Temperature sensors are used to collect and send data [17] about soil temperature in different environment. In this system, LM35 temperature sensors whose capability

Fig. 3 LM 35 temperature sensor

is 150c and its readings are measured in centigrade. This sensor initiates process for water spraying and gives information about the field temperature to take necessary actions to make the field suitable for ploughing and seeding process (Fig. 3).

3.3 Field Nutrient and Mineral Analysis

The third one is PH sensors which are used to adjust the soil nutrients and the level of PH as nitrogen and potassium are required for the regulation of PH or else there will be low productivity. Along with these sensors [18], level sensors are used to detect the water level.

3.4 Process Controller

LPC2148 microcontroller is a 16-bit/32-bit ARM7TDMI-S CPU with real-time emulation and embedded trace support, which is a microcontroller with its embedded flash memory of high speed which ranges from 32–512 kB. The name LPC is Low-Power Low-cost microcontroller [19] which is a 32-bit type microcontroller produced and manufactured by Philips semiconductors (NXP). In order to use a less requirement of the key, LPC218 can be used because of the less power consumption and smaller size. Because of its efficient consumption of power and its performance 32-bit ARM 7 microprocessor is used in having a larger efficiency [19]. It includes two instruction sets, namely 32-bit ARM and a 16-bit ARM.

3.5 Coding Storage

In order to store the data and to write coding, flash type of memory is used. The programming can be done in many different various ways and it has been done through serial port. In order to store data and code, an 8 KB SRAM is used. It can also be used for the efficient [20] and effective operation [22–26].

3.6 Working

The sensors are used here because robots have no eyes or ears so sensors; sensors are added to the robots with which they sense it and work automatically for an instance like ultrasonic sensors they used in this model as they send ultrasonic waves and when the waves reflects back, it calculates the distance between the robot and other objects presence in the field and according to it controls its movement without getting clash or creating any obstacle. They help in determining how much more distance is yet to be covered by the robot, and this is done with the help of echo pulse which in simple terms means measuring the time taken for the pulse to leave and return to the sensor. If the waves come across a boundary wall, then it immediately takes 180 degree turn to proceed to the next column of the field and so on till the last boundary all is reached [21]. The distance to be covered is shown on a LCD display that is mounted on the robot. All the above-mentioned sequential functions are given as operational instructions in keil C codes. In order to store the data and for coding purpose, we are using flash memory. Through the serial port, the programming can be done. Here, the motors are used to move the robot vehicle in the forward direction and reverse direction, respectively.

4 Results and Discussions

The robot is designed and it can be used in agriculture. It has been displayed below (Figs. 4, 5 and 6):

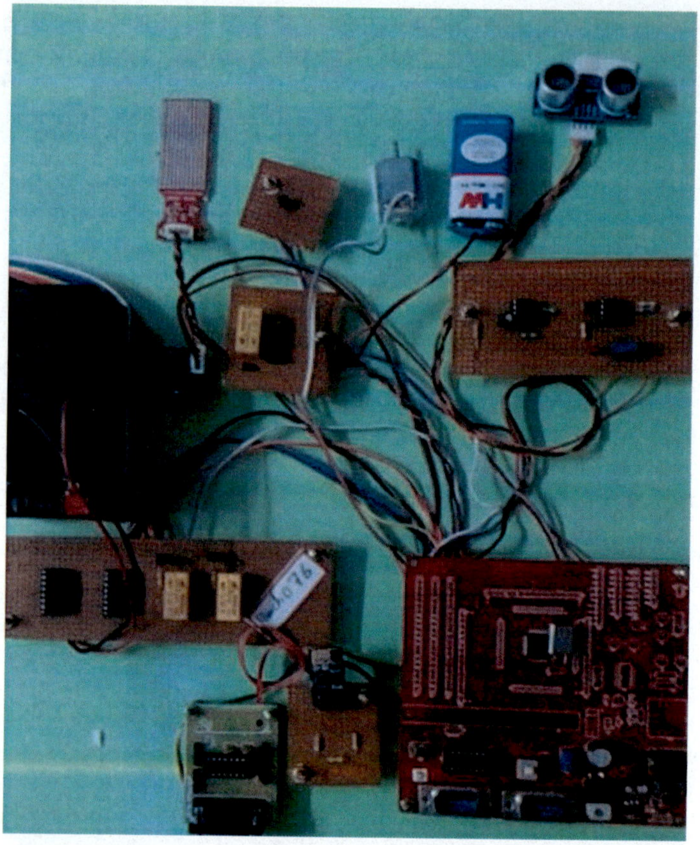

Fig. 4 Hardware setup

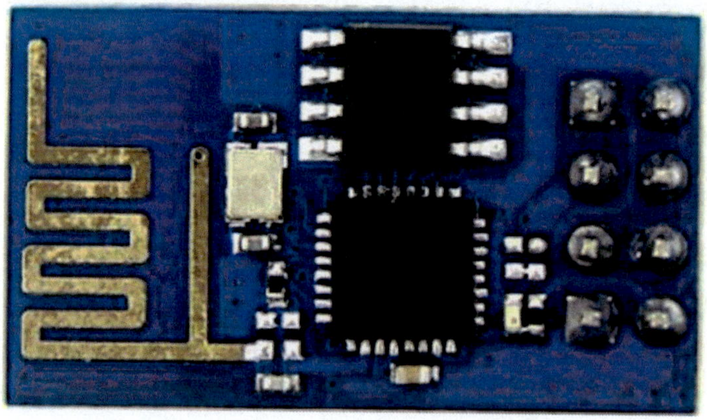

Fig. 5 PCB layout of the system

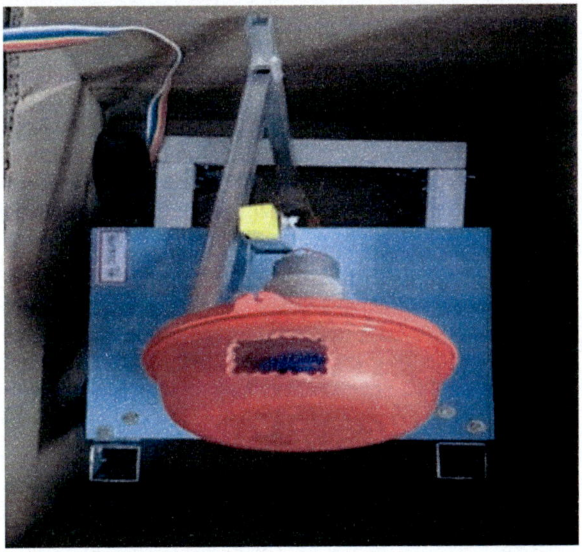

Fig. 6 Robot section

5 Conclusion

The robot is used to automate the process of sowing crops and ploughing in the field and water sprinkling to the crops also done automatically by the robot and reduces time, manpower, capital required for doing agriculture. This model is used to increase the productivity rate of crops and it can be upgraded with IDE like Arduino and Raspberry pi with Python coding for improved performance and additional functionalities of the system.

References

1. Nikkila, R., Seilonen, I., Koskinen, K.: Software architecture for farm management information systems in precision agriculture. Comput. Electron
2. Kaloxylos, A., Wolfert, J., Verwaart, T., Terol, C.M., Brewster, C., Robbemond, R., Sundmaker, H.: The use of future internet technologies in the agriculture and food sectors: integrating the supply chain. In: 6th International Conference on Information and Communication Technologies in Agriculture, Food and Environment, pp. 51–60
3. Ashton, K.: That Internet of Things thing RFID Journal. It can be accessed at http://www.rfidjournal.com/articles/view?496
4. Singh, D., Tripathi, G., Jara, A.J.: A survey of Internet-of Things: Future Vision, Architecture, Challenges and Services in Internet of Things (WFIoT) (2014)
5. Tubaishat, M., Madria, S.K.: Sensor Networks: An Overview. IEEE Potentials 05/2003
6. Arias, F.F.C., Barajas, Y.J.C., Lopez, J.L.R.: Wireless sensor system according to the concept of internet of things. Int. J. Adv. Comput. Sci. Inf. Technol. **3**(3) (2014). ISSN: 2296-1739

7. Tefera, T., Kanampiu, F., De Groote, H., Hellin, J., Mugo, S., Kimenju, S., Beyene, Y., Boddupalli, P.M., Shiferaw, B., Banziger, M..: The metal silo: an effective grain storage technology for reducing post-harvest insect and pathogen losses in maize while improving smallholder farmers' food security in developing countries. The International Maize and Wheat Improvement Center (CIMMYT) **30**(3) (March 2011)
8. Singleton, G.R..: Impacts of rodents on rice production in Asia. IRRI Discussion Paper Series No. 45, 30 pp (International Rice Research Institute: Los Banos, Philippines.)
9. TongKe, F.: Smart agriculture based on cloud computing and IOT. J. Converg. Inf. Technol. (JCIT) **8**(2) (Jan 2013)
10. Sharma, S., Tim, U.S., Gadia, S., Wong, J.: Growing Cloud Density and as-a-Service Modality and OTH-Cloud Classification in IOT Era. Available at www.public.iastate.edu/sugamsha/art (2015)
11. Tsai, C.-W., Lai, C.-F., Vasilakos, A.V.: Future Internet of things: open issues and challenges. Wireless Netw **20**, 2201–2217 (2014)
12. Farooq, M.U., Waseem, M., Khairi, A., Mazhar, S.: A critical analysis on the security concerns of internet of things (IoT). Int. J. Comput. Appl. (0975 8887) **111**(7) (Feb 2015)
13. Khan, R., Khan, S.U., Zaheer, R., Khan, S.: Future internet: the internet of things architecture, possible applications and key challenges. In: 10th International Conference on Frontiers of Information Technology (FIT 2012), pp. 257–260 (2012)
14. Gomathi, T.: Performance analysis of parallel co-operative spectrum sensing using scheduling algorithm in CRNS. Int. J. Appl. Eng. Res. **10**(2), 2733-2747 (2015). ISSN 0973-4562
15. Gomathi, T., Maflin Shaby, S.: Capacitive accelerometer for micro electromechanical applications: a review. In: International IEEE Conference ICCICCT-2016 (2016)
16. Gomathi, T., Maflin Shaby S., Priyadharshini, B.: Instill dental antennas for minimally interfering bio-medical devices. Springer publication ICCVBIC, pp 1049–1060 (2017)
17. Poonguzhali, S.: Wireless sensor network with a novel key distribution for improved four-tier network security. Res. J. Pharmaceut. Biol. Chem. Sci. 6 (4):319–327 (2015) RJPBCS
18. Poonguzhali, S.: An overview of routing algorithms based on energy constraints and network life-span. Int. J. Appl. Eng. Res. **10**(3), 6323–6330 (2015)
19. Poonguzhali, S.: Advanced fetal essential parameters observation using WSN. Int. J. Pharm. Technol. **8**, 21629–21634 (2016)
20. Priya, C., Lakshmi, D., Gowthami, S., Poonguzhali.: Lung pattern classification for interstitial lung diseases using an ANN- back propagation network. In: International Conference on Communication and Signal Processing (ICCSP). IEEE, pp. 1917–1922 (2017)
21. Raja, J., Stanley Karunakaran, W.: Automatic ploughing and seeding robot. In: International Conference on Electrical, Information and Communication Technologies, 2278–8735, pp. 68–73
22. Bhattacharjee, S., Borah, S.: A survey on the application of fuzzy logic controller on DC motor. Int. J. Appl. Innov. Eng. Manag. (IJAIEM), **2**(6), 84–90. June 2013 ISSN 2319-4847
23. Bhuyan, R., Borah, S.: Image segmentation: computational approaches for medical images. Int. J. Comput. Appl. (IJCA) Special Issue on Bioinf. Comput. Biol. **3**, 13–17. May 2014, ISSN: 0975-8887, IF: 0.715
24. Kalita, N., Chitra, R., Samarjeet Borah, S.: eKMP: a proposed enhancement of KMP algorithm. Comput. Intell. Data Min. Smart Innov. Syst. Technol. **33**, 479–487 (2015). Springer, ISSN: 2190-3018
25. Borah, S., Bhattacharjee, D., Vijay, K., Singh, K., Rai, B.: Multithreaded KMP: a proposed DNA . Int. J. Adv. Sci. Eng. Technol. **3**(3), 36–39, (Oct.-2015), ISSN: 2321-9009
26. Borah, S., Jha, R.S., Kumar, S., Kumar, I., Banerjee, D.: A matrix symmetry based method for brain abnormality detection, IJCTA **10**(9), 279–285 (2017). ISSN: 0974-5572

Implementation of Branch and Bound Method for Maximizing Network Capacity of Cognitive Radio Networks

T. Gomathi and S. Poonguzhali

Abstract This paper supplies uninterrupted affiliation between mobile users in real-time radio communication environment. The secondary users are enabled, so that the spectrum cannot be used by others. Generally in many networks, many devices will share a single channel which has been located within a defined network. The different availability of channels is observed by the secondary users in a cognitive network before broadcasting the information about the channels. Hence, the availability of spectrum is not known before the communication begins which has to be considered as a difficult and challenging task. The collision should be efficiently mitigated while broadcasting. The solution to these challenges is to handle traffic-aware issues and capacity-improving mechanism in cognitive radio network. Implementation of branch and bound technique leads to an acceptable improvement in the capacity of overall network as compared to conventional system.

Keywords Radio communication networks · Collision · Branch & bound method · Cognitive radio · Broadcasting · Spectrum

1 Introduction

Cognitive radio ad hoc is a new approach to the radio receiver and radio communications procedure. After the progress of the software-outlined radio, cognitive radio is the basic and important step ahead making a sure mighty radio communications techniques that has been developed for efficient use of the spectrum. There are a number of hand resources; amongst that cognitive radio is the best. The surroundings and the inner mode with capabilities of making and selecting for new and any saved pre-defined element can be done in a cognitive ad hoc network. Commonly,

T. Gomathi · S. Poonguzhali (✉)
School of Electrical and Electronics, Sathyabama Institute of Science and Technology, Chennai, India
e-mail: poornidp@gmail.com

T. Gomathi
e-mail: gomes20@gmail.com

S. Borah et al. (eds.), *Soft Computing Techniques and Applications*, Advances in Intelligent Systems and Computing 1248,
https://doi.org/10.1007/978-981-15-7394-1_59

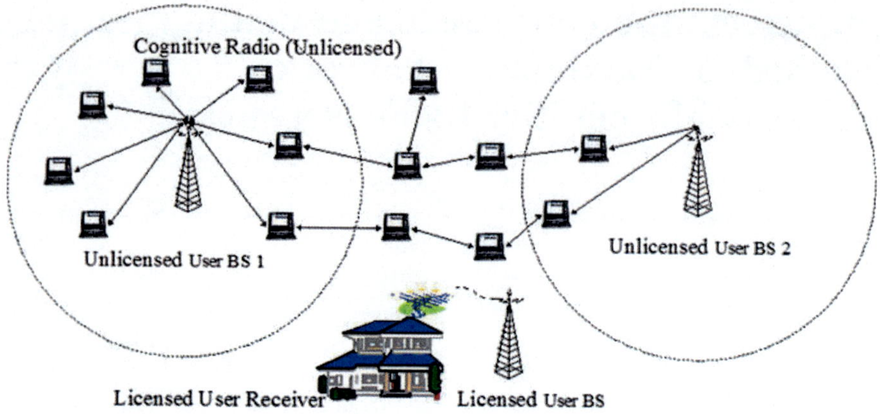

Fig. 1 CR spectrum [1]

the cognitive radio is also anticipated to appear at parameters akin to configure the spectrum, available of free channel, the types of modulation and type of transmitting knowledge which can be used in establishing communication. This additionally is regulated according to the requirements. In order to reconfigure itself, for satisfying, the procedure which helps in transmission the software-defined radio forms the fundamental requirement for any communication. Hence, of these reasons, the tightly linked rule between cognitive radio system and radio with programmed outline is followed which has been shown in Fig. 1.

2 Related Work

In [1] paper, the proposed channel assignment method maximizes spectral efficiency, minimizes transmit power, or maximizes data rate. Based on the proposed method, when a number of available channels are detected, a cognitive radio node can decide a suitable channel to use. These methods are executed distributive in cognitive radio nodes without using central controllers. The drawback for this method is that it was not able to analyse the data rate varying with the node whose transmit power is high.

In [2] paper, implementation of frequency division multiple access scheme for power allocation has been investigated, and arbitrary M base stations (BSs) coordinately allocate their resources to each user equipment (UE). If multi-BS UE (user served by multi-BS) requires the same amount of spectrum from BSs, then the proposed power allocation algorithm minimizes the total power consumption if multi-BS UEs are limited by the resource. Besides this, it has been observed that there occurs a problem in minimizing the total power consumption when the data rate requirements are satisfied.

In [3] paper, the researchers proposed the optimization of green energy-powered wireless networks using energy-efficient cognitive radio techniques. Existing works related to energy-aware spectrum sensing, its management, and sharing have been investigated in detail. The energy-efficient CR-based wireless access network has been discussed in various aspects such as relay and cooperative radio and small cells. It has been concluded that green energy is an important energy resource in the future, and the network performance largely depends on the dynamics of the spectrum available and green energy. Also the interference level incurred by secondary spectrum usage increases the collisions.

In [4] paper, it systematically studies the inter-channel interference and how it affects the link capacity and the performance of multi-channel protocols in LWNs. First, it develops empirical models to characterize the attenuation between channels based on experiments conducted on TelosB motes. It proposed a noble measurement algorithm, which can efficiently reduce the overhead of multi-channel interference measurement by using the spectral power density (SPD) measured at the transmitter. Here, the 802.11 inter-channel interference models are not suitable for low-power wireless networks [5].

In [6] paper, a general framework has been proposed that elaborates the spectrum assignment problem for various criteria including overall system utility. Global optimization is NP-hard that can be shown by reduction of the allocation problem to variant of the graph colouring. It has examined both the centralized strategy and the distributed approach. In centralized strategy, an allocation assignment has been calculated by central server based on environment knowledge, whereas in distributed strategy, devices team up to confer local channel assignments for global optimization. Because of the high intricacy of the BILP algorithm, many researchers have employed genetic evolutionary algorithms to solve the spectrum assignment problem in cognitive radio networks.

In [7] paper, introduces a capacity-aware spectrum allocation model for cognitive radio network. First, it models interference constraints based on the temperature model, and then, the secondary users (SUs) increase their transmission power until the interference temperature of one of their neighbours exceeds its threshold interference temperature. After, knowing the SINR and bandwidth of potential links, the link capacity has been calculated based on the Shannon formula, and co-channel interference is modelled between each channel potential by plotting an interference graph. Next, the spectrum assignment problem is [8] formulated by using binary integer linear programming (BILP) for finding the finest set of concurrently active links. However, in BILP, with every instance, the numbers of operations are increased exponentially and thus resulting in complexity of the system.

3 Existing System

The available of channel is examined by two different techniques. A channel can be selected by the transmitter randomly in order to broadcast the signals [9] likewise the receiver can also select a channel for broadcasting the required information [10]. The likelihood for picking up the identical channel lies in the channel spectrum massiveness.

One more feasible resolution is that the control message can be broadcasted on each channel [11]. If the channel quantity is large, then the broadcast delay will be lengthier. The above method will not give any assurance for receiving messages without any loss in the information [12] as the receiving node can change the channel [13].

4 Proposed System

An excellent headquartered broadcast protocol which has been used for multi-hop network has been put forth in establishing the communication in a cognitive ad hoc network. In this method, ranking-based channel decision algorithm has been used. Two schemes of the ranking protocol are practised. The sensible conditions are considered in which the topology is not known and the secondary users on the channel are not known, some secondary users are not equal,and synchronization of time are not used. For channel assignment, a hopping sequence has been designed for both the secondary users sender and receiver. Thus there is only one channel for the sender and receiver. Therefore, it is very easy for understanding multi-transmitter communication in which the information regarding the blind atmosphere.

4.1 Modules

The proposed system has been subdivided into four different modules. They are listed as follows

(i) Channel design
    (a) Timer setup
    (b) Channel information update
(ii) Design listening state
(iii) Design transmission state.

A. ***Channel designing***

In this section, it is assumed that each consumer consist of multi-transmitter. The system is ready to observe the channel availability through one antenna and information transmission by different antenna. The time is used to set off the channels from one to others. At any time of interval, the node has to inform the changes to the neighbours. By using the timer, the node has to generate the packet which involves in the channel expertise.

B. *Listening state*

In the proposed method with branch and bound algorithm, it is considered that the channel hopping takes place with single antenna and transmission of information with different antenna. So, in this way constantly, one antenna is allocated to channel hopping. This technique uses only one channel, so the node at all times listen the neighbour node availability with the aid of making use of that channel. And the node collects the neighbour channel availability. So, in future, if the node wishes to make communication to neighbour, then it could actually accumulate the channel, and on the other hand, it can get the information from the neighbour desk.

C. *Transmission state*

This section includes the check for the common channel availability. By using the listening mode, the node makes use of the neighbour desk to determine the channels of neighbours and then it has to check its channel availability and then it evaluates the channel matching by using branch and bound algorithm method.

If the record of channel availability is considered as the function of number of channels and number of fields, then channel list in a particular node can be given as

$$L_{o,p} = \bigcup_{i=1}^{n} (l_o \cap l_p)i$$

5 Results and Discussions

The scenario has been considered which consists of eighteen secondary users and three primary users are present evenly in the available area. Next, both traffic-aware routing as well as branch and bound algorithm was applied to rank the primary users according to requirements of secondary users along with avoiding collisions between the channels. Selecting best channel provides less latency in communication network which is given in Fig. 2. It can be observed that proposed system provides a better throughput as the number of packets increases which has been shown in Fig. 3. Also, proposed system decreases the delay time compared to existing system as shown in Fig. 4.

6 Conclusion

Thus, the channel allocation problem can be reduced with the help of the cognitive networks by introducing the concept of secondary users. This cognitive radio ad hoc network enables a way for establishing communication between devices without any

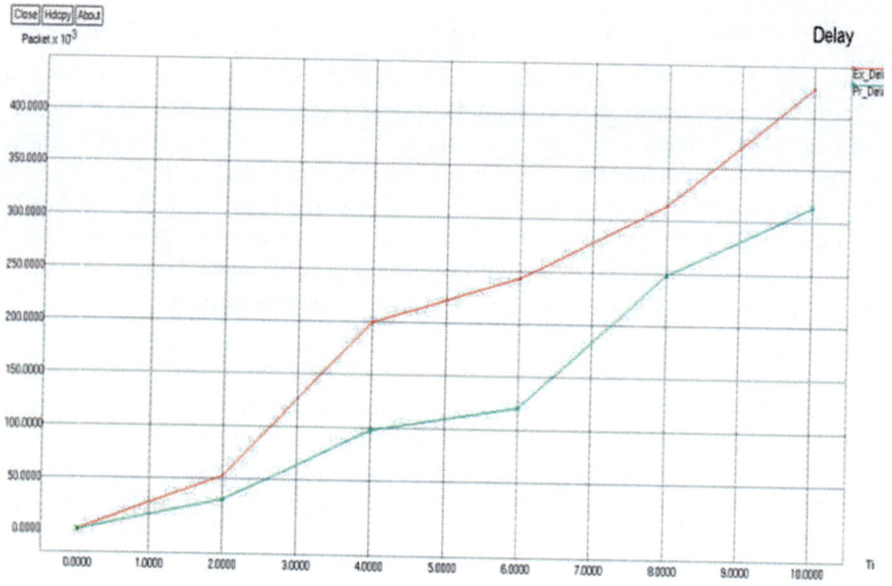

Fig. 2 Delay rate versus time

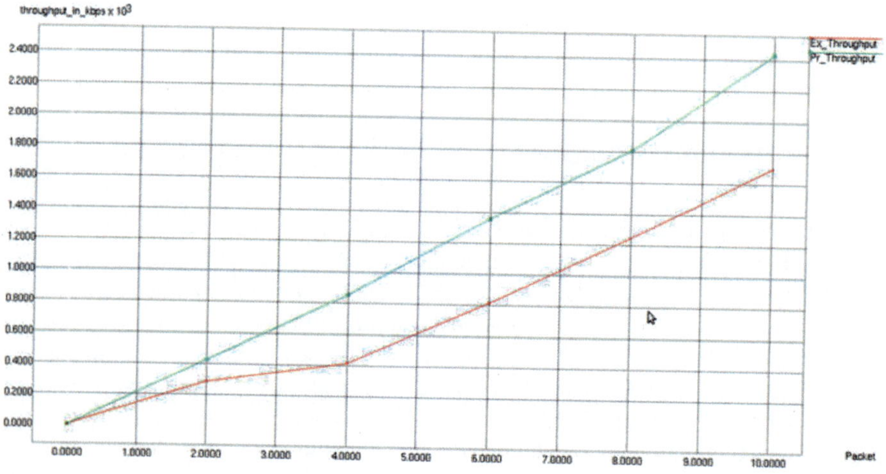

Fig. 3 Throughput versus no. of packets

prior knowledge about the condition of the atmosphere. It also helps to determine whether the requested channel is free for transmission or not and gives information regarding the traffic that exist in that channel. It also provides a very less delay while broadcasting and achieves more success in establishing communication between different users.

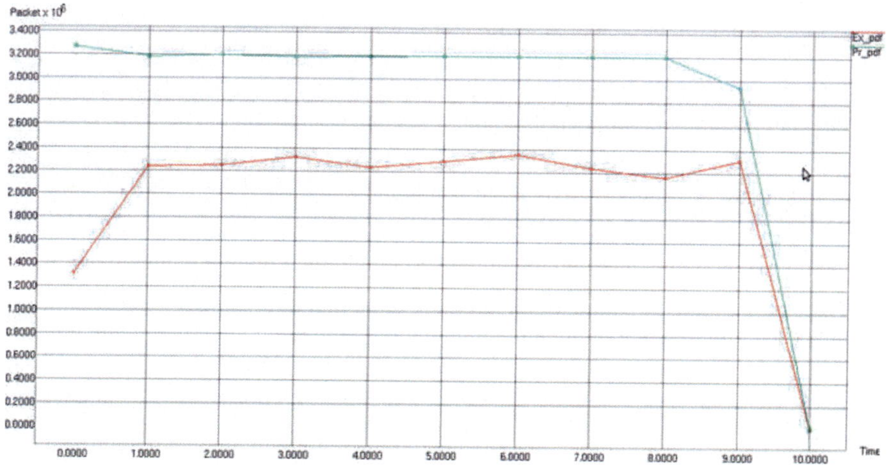

Fig. 4 No. of packets versus time

References

1. Li, X., Zekavat, S.A.R.: Distributed channel assignment in cognitive radio networks. In: Proceedings of International Conference on Wireless Communication, Mobile Computation, Connecting World Wirelessly, pp. 989–993 (2009)
2. Huang, X., Ansari, N.: Joint spectrum and power allocation for multinode cooperative wireless systems. IEEE Trans. Mobile Comput. (2015) https://doi.org/10.1109/tmc.2015.2388478
3. Huang, X., Han, T., Ansari, N.: On green energy powered cognitive radio networks. IEEE Commun, Surveys Tuts. **17**(2), 827–842 (2015)
4. Xing, G., et al.: Multi-channel interference measurement and modelling in low-power wireless networks. In: Proceedings of 30th IEEE RTSS, pp. 248–257 (2009)
5. Sivasangari, A., Ajitha, P., Rajkumar, I., et al.: J Ambient Intell Human Comput (2019)
6. Zhao, Z., Peng, Z., Zheng, S., Shang, J.: Cognitive radio spectrum allocation using evolutionary algorithms. IEEE Trans. Wireless Commun. **8**(9), 4421–4425 (2009)
7. Yousefvand, M., Ansari, N., Khorsandi, S.: Maximizing network capacity of cognitive radio networks by capacity-aware spectrum allocation. Wireless Commun. **14**(9) (2015)
8. Yu, B., Xu, L., Li, Y.: Bluetooth Low Energy (BLE) based mobile electrocardiogram monitoring system. In: Proceeding of IEEE International Conference on Information and Automation, vol. 5(2), pp. 30–37 (2012)
9. Rajasekaran, S., Kumaran, P., Premnath, G., Karthik, M.: Human health monitoring using wireless sensor network (WSN). Int. J. Appl. Innov. Eng. Manage. (IJAIEM) (2013)
10. Poonguzhali, S.: An overview of routing algorithms based on energy constraints and network life-span. Int. J. Appl. Eng. Res. **10**(3), 6323–6330 (2015)
11. Priya, C.L., Gowthami, D., Poonguzhali, S.: Lung pattern classification for interstitial lung diseases using an ANN-back propagation network. In: Proceedings of the 2017. IEEE International Conference on Communication and Signal Processing, ICCSP 2017 (2018)
12. Poonguzhali, S.: Advanced fetal essential security. Res. J. Pharm. Biol. Chem. Sci. **6**(4), 319–327 (2015)
13. Poonguzhali. S.: Wireless sensor network with a novel key distribution for improved four-tier network parameters observation using WSN. Int. J. Pharm. Technol. **8**(4), 21629–21634 (2016)

Energy-Efficient Data Transmission to Detect Pest in Cauliflower Farm

J. Adeline Sneha and Chakravarthi Rekha

Abstract The miscellaneous collection of sensor hubs has an example of continuous implanted plans with fast calculation, correspondence, memory, and vitality assets that are being utilized for a gigantic scope of utilizations where the obsolete model-based system is for the most part infeasible. The sensor hubs are firmly sent in the information of a cauliflower to screen, distinguish, and investigate the physical amount and disburse significant measure of vitality while transmitting the data. It is unfeasible and once in a while agonizing to supplant the battery and to keep up a longer system lifetime. In this way, there is an impediment to the lifetime of the battery force, and vitality preservation is a difficult issue in WSN. From this time forward, so as to build the system lifetime the vitality in the system has been kept up. Subsequently, Energy productive steering calculation is proposed right now powerful pest recognition. The backpropagation neural network is implemented for the pest classification. The soft computing technique adapted in this paper increases the adaptability of the network. The improved form of ADHOC on request separation vector Routing is depicted right now the examination with AODV steering convention is being talked about and dissected utilizing NS2 programming.

Keywords Routing protocol · Energy-efficient routing · Pest identification · EAODV · AODV · Pest

1 Introduction

Farming is one of the chief vital hotspots for human sustenance on Earth. Not exclusively it gives parcel of fundamental nourishment to human presence and utilization anyway also assumes a genuine significant job inside the economy of the nation. A

J. Adeline Sneha (✉) · C. Rekha
Sathyabama Institute of Science and Technology, Chennai, India
e-mail: j.adelinesneha@gmail.com

C. Rekha
e-mail: rekha_2705@yahoo.co.in

S. Borah et al. (eds.), *Soft Computing Techniques and Applications*, Advances in Intelligent Systems and Computing 1248,
https://doi.org/10.1007/978-981-15-7394-1_60

few great money bucks are spent worldwide for the security of harvests, horticultural product, and merchandise, sound yield. It's a matter of worry to protect crops from bio-aggressors, for example, pest and creepy crawlies that in any case cause across the board mischief and loss of yields. In a nation, for example, India (2009), roughly 18% of harvest yield is lost because of vermin assaults each year which is esteemed around 90,000 million rupees [1]. Pest are the major threats for the growth of crops in agriculture [2] and the storage of the consumables. Indeed, the accurate and timely invention of pest in the field is very important for protecting the crops [3]. Manual monitoring of pest and diseases is time-consuming and involve labor charges. Therefore, pesticides are used as an effective tool for the pest elimination from the farms. Environmental exposure to pesticides causes various diseases like thyroid, cancer, etc. in humans [4]. More resilient and sustainable approaches are urgently needed to minimize crop yield losses resulting from pest activity and reduce the impacts of pest management on human health and the environment [5]. Conservatively, manual pest monitoring techniques, sticky traps, blacklight traps are being utilized for Pest monitoring and detection in farms. But these traps cannot eradicate the pest completely from the farms. Therefore, automatic pest identification is preferred everywhere. The wireless sensor networks-based detection of pest is used for automatic detection. In this, the sensor nodes are deployed in the fields without harming the environment. But the problem relay in saving the lifetime of the sensor nodes. The lifetime of the network is increased by lifetime improvement algorithm of wireless sensor networks [6]. Routing of data through the optimal path also improves the network performance [7]. RMER and RMECR are proposed for networks in which either hop-by-hop or end-to-end retransmissions ensure reliability [8]. Dynamic programming is utilized to find the shortest path after which the network is divided into different layers to evaluate energy efficiency [9]. In this network, nodes produce energy from nature to tolerate the power desirable for data transmission and transfer a portion of their gathered energies to neighboring nodes through energy assistance. An iterative algorithm that updates the data flows, energy flows, and power distribution over outgoing data links sequentially [10].

2 Routing

Manual vermin review methods are tedious and difficult to the accessibility of a human to distinguish the equivalent. So as to address these inconveniences, a few present-day bug location and control philosophies exist which incorporate sensor-based nuisance recognizable proof. Right now, novel technique for bug discovery utilizing WSN innovation with NS2 to effectively detection of pest on the leaf bothers, rather than the clingy traps. NS2 is a free access simulation tool that runs on Linux. NS2 is used to simulate network scenarios and analyses its outcomes based on various network factors like packet loss, delay, error, intrusion, etc. It shows the dynamic path of information transmission. Packet loss is common in wireless sensor networks. It could waste a substantial amount of energy, which is one of the most

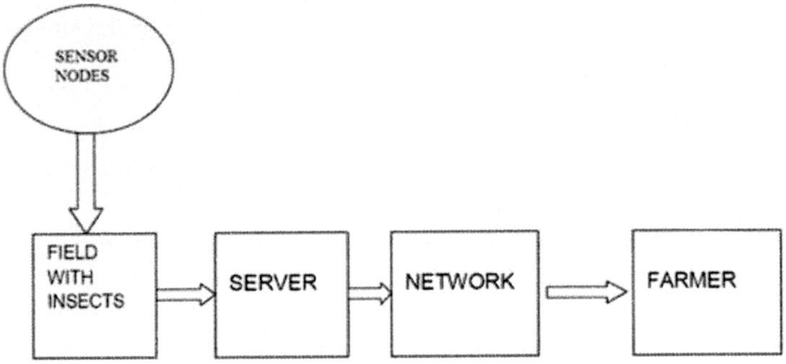

Fig. 1 Sensor node in agriculture

precious resources of sensor nodes. Thus, to extend the operational time of sensor nodes, packet losses should be minimized. Generally, packet losses in wireless sensor networks are caused by three major factors: (1) frame error due to channel interference; (2) frame collision due to channel contention; and (3) packet drop due to buffer.

The routing problem can be tackled by more robust coding and modulation schemes at the physical layer. The second problem can be tackled by optimizing the MAC behavior at the MAC layer, and the last problem can be tackled by congestion control at the transport layer (Fig. 1).

Sensor Networks involves various sensor center points which are obligated for identifying the physical sums from nature. The image sensors accumulate consistent information from the earth. The assembled information is moved to the server. The viable directing convention utilized in moving of continuous data in time which will make the farmer take corrective action immediately if the pest infestation happened on the farm.

In this exertion, AODV routing algorithm and EAODV are discoursed and its evaluation is made. AODV is a kind of reactive protocol which will form the route on the demand. Ad hoc On-Demand Distance Vector (AODV) routing protocol by creation it power-aware and appropriate for sprinkle nets. It enhancements the AODV flooding mechanism by excluding all non-bridge slaves from captivating apart in the AODV route sighting process. In accretion, it expands the AODV route discovery phase by giving consideration to the hop count, the established node's power, and the average traffic strength for each node as metrics for best route assortment [11]. The major challenge of AODV routing protocol is energy. Since the sensor nodes are smaller in size and will be active for a long time the continuous energy is required. In this work, the energy can be sustained by using EAODV protocol. In this algorithm, the sensor nodes cannot be dried completely. The shortest path has been chosen and the data has been sent to the node which is having high energy before it drains the data has been passed to the next node the residue energy of the node can also utilized for sending the data.

The Energy Trust is premeditated based on the residual energy and Geographical Usual Energy of each sensor node

$$\mathrm{GAE} = \sum_{i=1}^{n} \mathrm{RE}_i / n$$

RE_i Residual Energy
n Number of nodes in the region

Routing in specially appointed systems has gotten crucial thought with various fluctuations and delays proposed. In EAODV, the course which burns through less effort and claims a bigger limit is chosen by engineered examination. Use programming reproducing and investigate the execution of it. The recreations demonstrate that this convention is better than AODV in parcel conveyance, throughput, vitality, and less postponement [12].

The EAODV routing protocol increases the lifetime of the network.

3 Pest Detection in Cauliflower Farms

Cauliflower is the most fertilized vegetable since it is affected by many pests at all stages of cauliflower vegetation. The most important pests are Beet armyworm, Cabbage loppers, Aphids. These pests not only infect the cauliflower vegetable but also infects the entire plant of cauliflower. Routinely, manual observing strategies, clingy traps, dark light snares are being used for bothering checking and discovery in ranches. Manual pest checking methods are tedious and abstract to the accessibility of a human master to identify the equivalent. So as to address these impediments, a few present-day bother discovery and control strategies exist which incorporate picture preparing based vermin distinguishing proof and furthermore picture investigation for clingy traps. Automatons watch geometric properties symbolism during the vegetation time frame in cauliflower [13]. cameras are appropriate to make exact 3D models for cauliflower plants [14]. An epic technique for bug identification utilizing WSN innovation with NS2 to effectively personality bugs on the leaves itself, rather than the clingy traps. NS2 is an open-source recreation device that sudden spikes in demand for Linux. It is a tactful test system focused at systems explore [15] and provides substantial support for simulation of routing, multicast protocols, and IP protocols, such as UDP, TCP, RTP networks. NS2 is used to Simulate network scenarios and analyses its outcomes based on various network factors like packet loss, delay, error, intrusion, etc., It shows the dynamic path of the information transmission.

Automatic pest identification is very important for initial identification of pests. The image sensors are fixed at the cauliflower farms. The image sensors continuously capturing the cauliflower images. If the cauliflower head is examined with the pest immediately the information has sent to the farmer's mobile through effective routing. This leads the farmer to take control action in time. The image sensor used in this work is AH5020B23-S1-2Z1 is a USB Video Class (UVC) agreeable camera module with video include, intended for convenient note pad PC picture applications. It is comprised of the accompanying parts, CMOS sensor, focal point, holder, backend, PCB, picture preparing circuit, and connector, to turn out a computerized video gadget. It will be a solid gadget which is implanted in scratchpad PC to move video information through USB interface to PC.

The process flow of this work is indicated in the flow chart Fig. 2. Nodes moving with higher mobility rates will have better recent routes compared to slow-moving nodes which may not be aware of the drastic changes happening in the network [16].

Arrangement of sensors is a basic issue in the sensor organize planning since it has a critical effect on the presentation of the system and decides the expense of sending the data [17]. The sensor nodes are arranged in a grid fashion. The image sensors are used as the sensor nodes for capturing real-time data. The passing of information from one node to another node without loss of energy and packets are the challenging task.

At once the sensor nodes are fixed the routes are defined by choosing the shortest path by counting the number of hops. The minimum energy nodes are chosen for computation, processing, and communication. Therefore, an energy-efficient algorithm has been implemented. The AODV routing protocol uses the shortest path

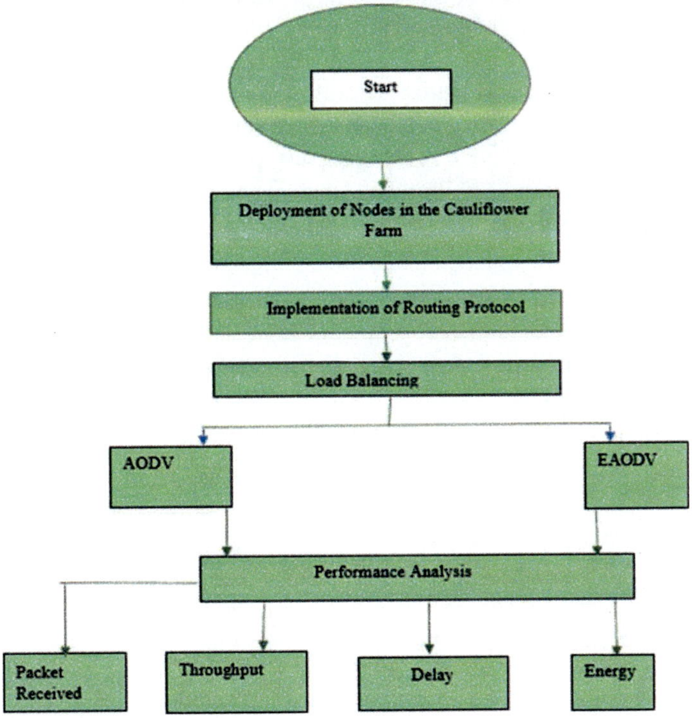

Fig. 2 Flow chart of performance analysis of a network

but maintaining the energy of each node is the challenging task. It does not use any congestion control and traffic avoidance methodology for transferring the data. Hence, EAODV is implemented with highly energy-saving paradigm.

The major benefit of EAODV is the reduction of network overhead. This has to be done by two mechanism route discovery and route selection. The EAODV is implanted in this work and it has been compared with AODV. The performance of both protocol and its role in pest detection has been clearly explained in this work.

The Backpropagation Neural Network technique is implemented for effective pest identification. Smart routing is a crucial phenomenon for enhancing the standard of service of the network. This may increase the performance of the network and reduce the delay in the transmission of knowledge. Thanks to WSN constraints and particularly the sensors' energy scarcity, a sensible routing should be done to balance the energy consumption among nodes. Therefore, prolong the network lifetime and insuring network coverage. While clustering significantly to enhance the network lifetime, we'd like to scale back the consumed energy and increase the scalability of the sensor network. Sensing nodes can be established using ANN and reduce energy ended some series of power measurements and predictions. With the simplicity of NN, strongly agree its compatibility in WSN routing. NN has some challenges that can't deal with the character and therefore the frequent changes in WSN properties.

The Backpropagation neural networks are used for performing training of the sensor nodes and therefore the structure of the communication links present within the network. This system enables the network for locating the shortest path of the nodes to the cluster heads then to the sink node. The clusters within the routing perform routing of the packets that are collected by the sensor nodes to the cluster heads through the cluster member nodes or directly. The cluster head nodes find the shortest path by applying a route discovery process and maintain the route for performing effective routing of the info packets. The network is trained for conditions like traffic level, bandwidth availability, and congestion status using convolutional neural networks.

4 Network Configuration and Simulation Results

The performance analysis of the network is analyzed by NS2 simulator. The execution of AODV and EAODV is done with the Ubuntu operating system and NS 2.34 version.

The nodes are decided in grid fashion. The sensor nodes collect the information from the real-time environment. The simulation parameters used for the analysis are given in Table 1.

The effectiveness of two routing protocols AODV and EAODV have been compared in this paper. The simulation scenario of varying nodes is given in Fig. 3. The crops are represented by green color. There are five image sensors utilized in this simulation which is represented by purple color and the shape given is hexagon. The server is represented in black color. The pests are represented in red color.

Table 1 Simulation configuration utilized in this work

Parameter	Value
Number of nodes	50
Network size	2000 * 2000
Simulation time	50 s
Routing protocol	AODV, EAODV

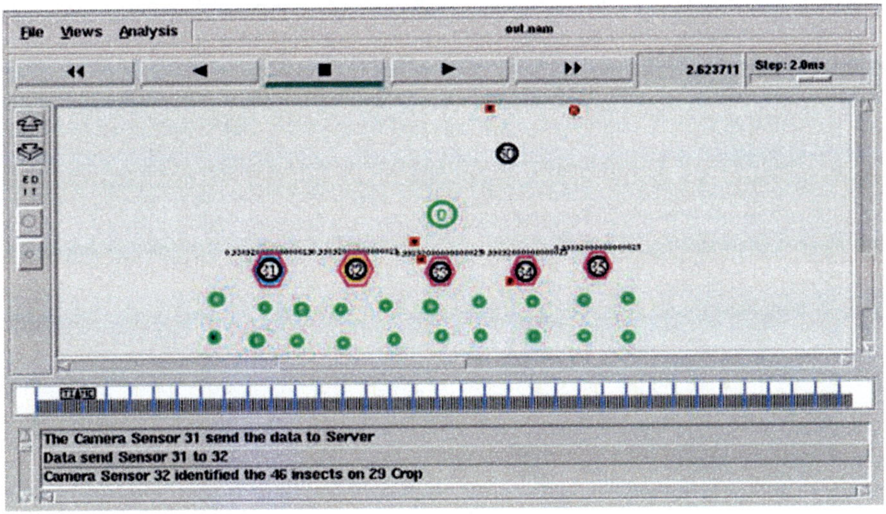

Fig. 3 Simulation scenario indicating the farm

When the pest attacks the crop, the nearest camera which is present near to the crop which is infected by pest detects the pest invasion in the crops, which sends the data to the server continuously. The routing protocol effectively utilizes the energy of each node for transitory the evidence to the base station. The pests started to infect the crops. The nearest image sensor 3 and 4 detects the presence of pest. The pest detected message has been sent to the base station in order to take control action (Fig. 4).

In Fig. 5 the packet received by each node has been represented. Comparing to AODV protocol, EAODV received a greater number of packets with minimum loss of packets. The total quantity of nodes used in the replication environment is 50. The red line represents the response of AODV routing protocol whereas the green line indicates the response of EAODV routing protocol. The packet received gives a brief explanation of how much packets received by the nodes irrespective of the sender (Fig. 6).

Throughput outlines the ratio of the number of packets successfully acknowledged by the receiver within the unit time span. EAODV prove that it receives high amount of data which is represented by green graph.

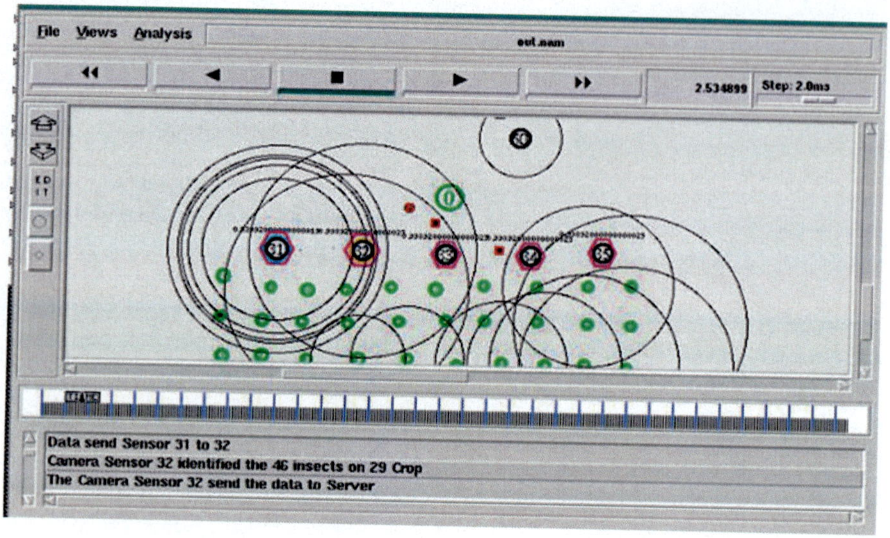

Fig. 4 Simulation of pest attacking the farm

$$\text{Transmission time} = \text{file size/Bandwidth (s)}$$

$$\text{Throughput} = \text{File size/Transmission Time (bps)}$$

Delay is the difference between the time of data send by the sender and received by the receiver. The EAODV receives the data faster than AODV. That response graph of EAODV is given in green color in Fig. 7

$$\text{Delay} = \text{sum of all delay samples/number of samples}$$

The energy of EAODV is higher than the AODV routing protocol which is represented in green color in Fig. 8.

5 Conclusion

The proposed technology detects the pest in its initial stage without harming the crop and environment. This also reduces the usage of pesticides and also saves them time and money from the farmer. In this paper, the pest is detected by the image sensor, the images are processed and the data are effectively transferred by the routing protocols to the base station whereas the control action has been taken. Once the crop is diagnosed with pests, the processed image is passed through the server. The inbuilt routing algorithm EAODV uses the energy of each node effectively in transferring

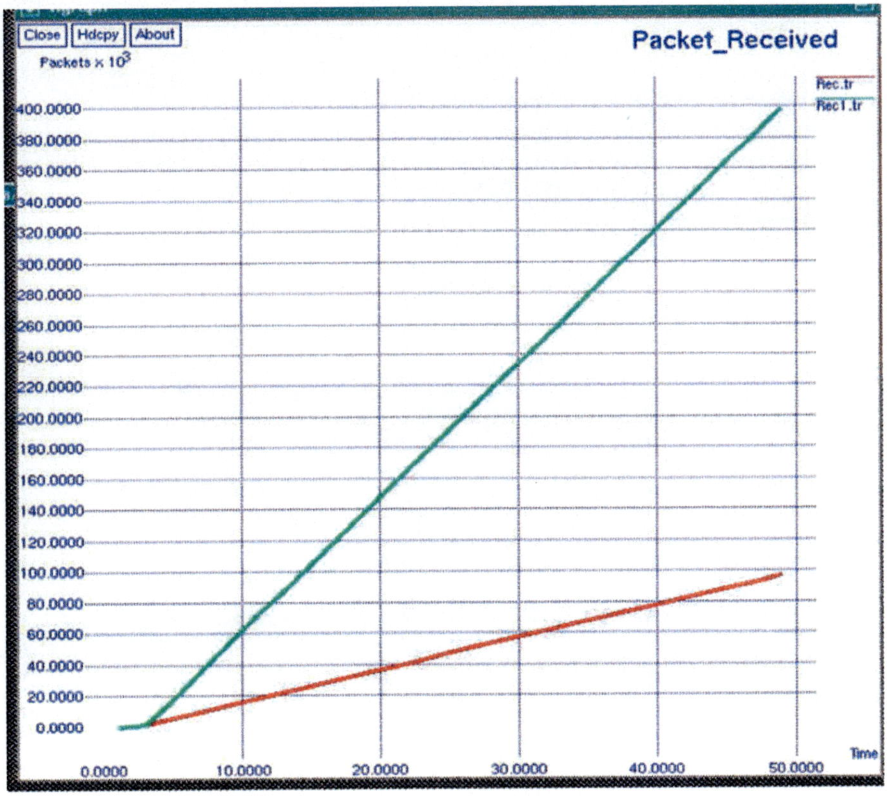

Fig. 5 Performance analysis of packet received using AODV and EAODV

the information to the base station without any loss of packets. According to the message from the base station the effective control action has been taken by farmers and thereby crops are protected. Here the routing protocols AODV and EAODV are utilized for pest detection. The performance of the network has been analyzed by NS2 Simulator. From the results its inferred that packet received ratio, throughput is very high in EAODV routing protocol. The delay, and energy loss are very low in EAODV. In all aspects, EAODV is highly efficient than AODV in pest detection.

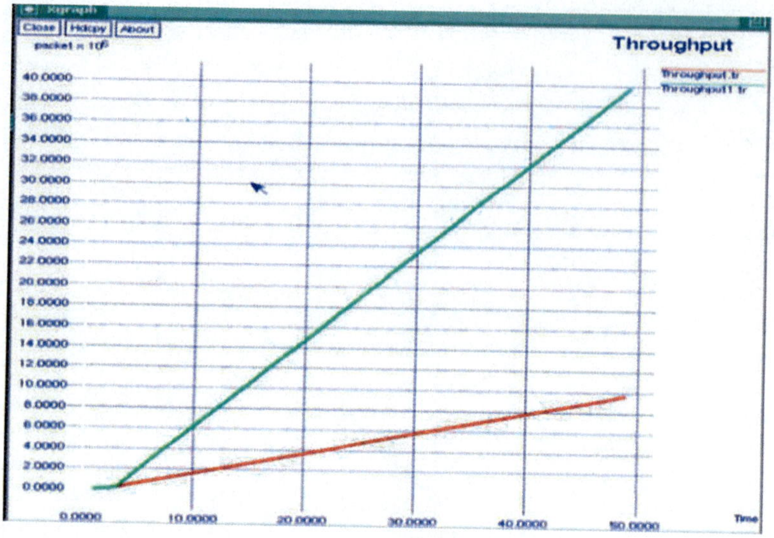

Fig. 6 Performance analysis of throughput using AODV and EAODV

Fig. 7 Performance analysis of delay using AODV and EAODV

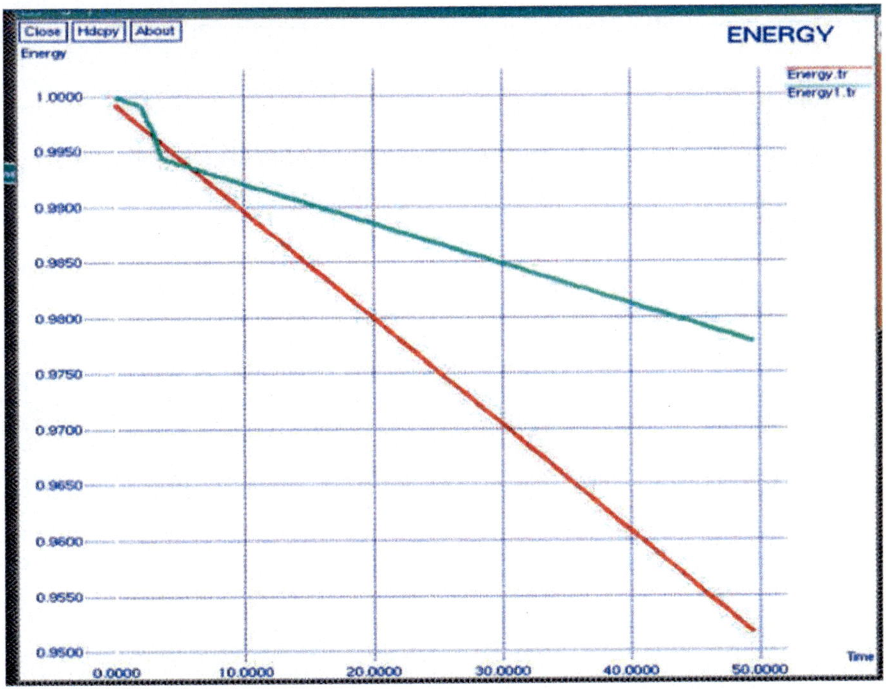

Fig. 8 Performance analysis of energy using AODV and EAODV

References

1. Press Trust of India: India loses Rs 90k-cr crop yield to pest attacks every year [Online] (2009, Feb 26). Available: http://www.financialexpress.comlnews/india-Ioses-rs-90kcr-cropyield-to-pest-attacks-everyyear14280911
2. Cheng, X., et al.: Pest identification via deep residual learning in complex background. Comput. Electron. Agric. **141**, 351–356 (2017)
3. Priya, C., Praveen, K., Srividya, A.: Monitoring of pest insect traps using image sensors & dspic. Int. J. Eng. Trends Technol. **4**(2), 4088–4093 (2013)
4. Requena, M., et al.: Environmental exposure to pesticides and risk of thyroid diseases. Toxicol. Lett. **315**, 55–63 (2019)
5. Baker, B.P., Green, T.A., Loker, A.J.: Biological control and integrated pest management in organic and conventional systems author links open overlay panel. Biol. Control 140 (2020)
6. Guimarães, D.A., Frigieri, E.P., Sakai, L.J.: Influence of node mobility, recharge, and path loss on the optimized lifetime of wireless rechargeable sensor networks. Ad Hoc Networks 102025 (2019)
7. Prithi, S., Sumathi, S.: LD2FA-PSO: a novel learning dynamic deterministic finite AutomatA with PSO algorithm for secured energy efficient routing in wireless sensor network. Ad Hoc Networks 102024 (2019)
8. Vazifehdan, J., Venkatesha Prasad, R., Niemegeers, I.: Energy-efficient reliable routing considering residual energy in wireless ad hoc networks. IEEE Trans. Mob. Comput. **13**(2), 434–447 (2013)

9. Cui, R., Qu, Z., Yin, S.: Energy-efficient routing protocol for energy harvesting wireless sensor network. In: 2013 15th IEEE International Conference on Communication Technology. IEEE (2013)
10. Gurakan, B., Ozel, O., Ulukus, S.: Optimal energy and data routing in networks with energy cooperation. IEEE Trans. Wireless Commun. **15**(2), 857–870 (2015)
11. Al-Jarrah, O., Megdadi, O.: Enhanced AODV routing protocol for bluetooth scatternet. Comput. Electr. Eng. **35**(1), 197–208 (2009)
12. Zhang, Z., Pei, T., Zeng, W.: Modified energy-aware AODV routing for ad hoc networks. In: 2009 WRI Global Congress on Intelligent Systems, vol. 3. IEEE (2009)
13. Grenzdörffer, G.J.: Automatic generation of geometric parameters of individual cauliflower plants for rapid phenotyping using drone images. In: International Archives Of The Photogrammetry, Remote Sensing & Spatial Information Sciences (2019)
14. Andújar, D., et al.: Using depth cameras to extract structural parameters to assess the growth state and yield of cauliflower crops. Comput. Electron. Agric. **122**, 67–73 (2016)
15. Gupta, S. G., et al.: Open-source network simulation tools: an overview. Int. J. Adv. Res. Comput. Eng. Technol. (IJARCET) **2**(4), 1629–1635 (2013)
16. Shobha, K.R., Rajanikanth, K.: Enhanced AODV routing protocol for mobile adhoc networks. In: International Conference on Computer Science and Information Technology. Springer, Berlin (2011)
17. Singh, A.K., Debnath, S., Hossain, A.: Efficient deployment strategies of sensor nodes in Wireless sensor networks." 2016 International Conference on Computational Techniques in Information and Communication Technologies (ICCTICT). IEEE, 2016

Optimizing Energy Efficiency in Device-to-Device Communication Using Intelligent Algorithms

Varadala Sridhar and S. Emalda Roslin

Abstract The need of the communication technology is the device to device for the communication among the mobile phones and one car to another car in vehicular communication and wireless sensor network. This communication is the transfer of data or information or audio and video signals. As the users of devices are increased tremendously, the development of 5G network communication requires the efficient communication, at the same time, the energy requirement is needed to minimize for the longer back up supply. The power consumed for the data transmission is needed to reduce. To minimize the power consumption or maximize the backup power, optimization is required. Intelligent algorithms have proved in the recent year as the best optimization technique for minimization power. In this paper, I am going to proposing different types of intelligent algorithms and hybrid intelligent algorithm in order to reduce energy efficient.

Keywords Intelligent algorithms · 5G communication · D2D communication · Energy efficient optimization · Hybrid intelligent algorithm

1 Introduction

The D2D communication is ever increasing at fast rate due to the recent development of communication technology. Due to Internet of things (IoT), all devices need to communicate with each other in the network. This needs large bandwidth and high speed information exchange between devices [1]. Wired communication has less usage and wireless communication gain more and more application in all fields of D2D communication. Device heterogeneous needs communication among them.

V. Sridhar (✉) · S. Emalda Roslin
ECE Department, Sathyabama Institute of Science and Technology, Chennai, India
e-mail: varadalasri@gmail.com

S. Emalda Roslin
e-mail: roemi_mich@yahoo.co.in

S. Borah et al. (eds.), *Soft Computing Techniques and Applications*, Advances in Intelligent Systems and Computing 1248,
https://doi.org/10.1007/978-981-15-7394-1_61

Most of these devices have data processing capabilities and require more power. But the requirement of the day is to use less or efficient power for the devices [2].

To reduce the energy usage in the devices, it is common to use cellular network and additional energy booster system in the device [3]. This scheme reduces the transmission power and helps to improve transmission efficiency of the communication system. Clustering the devices for the usage of common resources and dived into sub-problem for energy efficient is one of the best methods for finding optimal energy usage for the device communication [4]. Recent development of smart cities and vehicle to vehicle communication uses many device to device communication and process multimedia data such as video and photos. This occupies communication channel and reduces the number of device connectivity. To increase the number of device connections, the problem may dived channel allocation and power control sub-problems. The solution of these sub-problems gives the energy efficient use of communication system [5]. Non-orthogonal multiple access (NOMA) is one of the techniques to reduce interference and improve efficiency and throughput in the device to device communication [6].

Multiple-tier network communication and direct device to device connection without the intervention of base station is one best method of energy efficient operation. There is a possibility for energy harvesting when downlink resources superimpose their messages with the same throughput in two-tier network system [7]. Energy efficient problem may divide into three sub-problems such as routing, channel traffic and energy utilization. Branch and bound algorithm has the capability to find energy efficient solution [8]. SWIPT is one technique used to harvest energy in the communication signal may use for energy efficient. In the initial stage finding, the pairs of devices and in the next stage energy efficient algorithm may use to optimize the use of energy for the devices [9]. Reusing present cellular resources to transfer information from one to another device is one best method for energy efficient. When the paired devices are not able to communicate between them an intermediate relay is essential. For the intermediate relay, energy harvest from the communication channel power splitting protocol and solar energy may utilize [10].

Full duplex device communication along with cellular leads to interference and maximum energy uses. Proper allocation of network resources may improve energy usage. For this application, two-phase approach gives better solution. In the first phase, power allocation among devices may find using monotonic algorithms, and in the second phase, channel assignment among the devices may find using assignment problems solvers [11]. The recent development of 5G technology uses same channel for mobile and device communication spectrum. This challenging complex problem may divide into two parts and may solve iterative to get optimal solution. First part is maximizing energy and second part is control the power in the channel allocation, and Lyapunov method is useful for solving these iterative problems [12].

Device and cellular communication channel energy efficient with resource matching may improve by converting this problem into nonlinear mixed-integer problem. After the problem formulation, many to many device matching is constructed in the first phase, and second phase resource allocation for the paired devices is calculated gives better energy efficient. Cross-layer communication tier and self-organized device communication help for the energy efficient. Resources block for the devices and power allocation based on game theory may give better energy efficient.

In this approach, the problem is divided into sub-problem as power allocation and resource allotment for better energy efficient. Coexistence of device and cellular communication degrade the resource sharing and energy efficiency. Three-mode approach may improve this energy efficiency, the modes are direct, two hop, and cooperative. Three-dimensional resource management and orthogonality approach find device pair and help the three-mode approach for the energy efficient. MIMO system approach tackles the energy efficient communication problem. Machine type communications (MTC) increase the clustering and enable these clusters to use resources efficiently. Channel estimator and liner detectors in the MIMO system works without prior knowledge of channel coefficients and gives better energy efficient solution.

There are lot of existing intelligent algorithms are available to reduce energy efficient. In this paper, introduced intelligent algorithms are to reduce energy efficient. The two main advantages are ability to solve complex optimization and parallelism. For the optimization problem, the solution needs to find in the solution space, when the solution moves beyond the solution space penalty may include in the solution, thereby maintain the constraint solution space. Enhanced hybrid algorithm uses penalty method to discard the non-feasible solutions. Hybrid intelligent algorithm approach needs expertise in the energy efficient problem approach in detail and need to understand the uncertain and unfavorable situation. The optimization of given problem is discovered within the solution space by controlling the control variables of the considered complex problem.

2 Literature Survey Table

See Table 1.

3 System Model

Communication model of D2D and cellular [1] is shown in Fig. 1. The figure shows base station to cellular and D2D communication system. For the cellular communication, it needs to connect through the base station, and for device communication, it may or may not connected through base station.

The spectrum in cellular network has K bands, and W_i is ith band of BW. The letter i in the variables varies from 1 to K. Stochastic geometry gives distribution of spatial random cellular users in the ith band and model in homogeneous Poisson point process $\emptyset_{c,i}$ with density $\lambda_{c,i}$ on the 2D plane. The transmission power of cellular users in the ith band is $P_{c,i}$, and the total transmission power of cellular users is P_c

Table 1 Literature survey table

Author	Problem Identified	Solution	Problems yet to focused
Sridhar et al. (2019)	Energy efficient	Intelligent algorithms and hybrid intelligent algorithm	5G environment for D2D and mobile units
Misra et al. [1]	Reviewed the crossover mix of 4G and mm wave	D2D communications, Web-based gaming, high definition streaming, and distributed	mm wave correspondence innovation in 5G
Chen [2]	Energy efficiency for D2D correspondence underlying heterogeneous networks	Configuring different types of hybrid algorithms and to use the energy components for the heterogeneous wireless networks	Yet the device user equipment gains the best data rate on fading channel
Zhang et al. [3]	Device to device communication on multiple bands	Maximizing the energy efficiency subject to transmission power and non-convexity of this problem	5G wireless networks
Chai et al. (2019)	Obtained the optimal power allocation strategy supported, which we then propose a modified K-means algorithm	Investigated the issue of resource allocation and machine to machine communication systems for efficient resource management	Range allocation
Liu et al. (2019)	Power control algorithm supported the Lambert W function to maximize the EE of a single device to device communication and mobile users are computing the power levels	A channel assignment in view the Gale–Shapley proposed method esteems to preference values to coordinate different sides, which targets at highest S/I ratio of mobile users and EE of device to device communication	5G framework

(continued)

Table 1 (continued)

Author	Problem Identified	Solution	Problems yet to focused
Alemaishat et al. (2019)	Power allocation algorithm for D2D communication dependent on NOMA to maximize the uplink EE	Kuhn–Munkres method to assign a channel for each device to device gathering and figures an OPA issue using Karush–Kuhn–Tucker conditions	Influence of variety number of device to device receivers will be examined
Kuo et al. (2016)	Branch and bound method to effectively take care of this globally optimal ideal issue	Non-convex MINLP problem	Intelligent optimization technique
Zhou et al. (2018)	Increasing the EE execution of D2D sets and the measure of power gathered by CUEs continuously	Gale–Shapley algorithm	Intelligent optimization technique
Salim et al. (2019)	Relay selection-based power allocation EH-aided method	PS factor problematic incentive with ideal connections power designation	Execution of the EH hand-off supported D2D correspondence accepting versatile D2D hubs in heterogeneous systems
Chour et al. (2019)	Worldwide optimal solution can be accomplished by decoupling the original problem into two sub-problems as PA and CA	Polyblock-based algorithm, MARIO, and Khun–Munkers algorithm	Intelligent optimization technique
Hao et al. (2018)	Maximizing the EE and SE with stability constraint	Lyapunov techniques and mixed-integer and non-convex method	Intelligent optimization technique
Meng et al. (2019)	Maximize the sum SE of D2D communications	Mixed-integer nonlinear programming	Intelligent optimization technique
Shahid et al. (2017)	Self-organized RBs and power and energy efficiency enhancement while maintaining the QoS requirements of other heterogeneous parties	Game-theoretic manner and conventional schemes	Intelligent optimization technique

(continued)

Table 1 (continued)

Author	Problem Identified	Solution	Problems yet to focused
Zhang et al. (2019)	Increasing the total EE of D2D communications while ensuring the Qos of both D2D and mobile communications	Independent of different D2D communication and 3D resource allocation algorithm	Intelligent optimization technique
De Figueiredo et al. (2019)	Linear detectors for channel estimator, as for the quantity of number of antennas and uplink transmission power	Linear MMSE channel estimation	Intelligent optimization technique
Orsino et al. (2016)	IoT arrangements under cell availability as in smart city	IoT and modulation and coding scheme (MCS)	Flexibility designs for the IoT gadgets for the radio resource allocation, the transmission mode selection and the power consumption
Ravi et al. [11]	Solution for optimal power flow in electrical engineering	Hybrid intelligent algorithm	D2D and mobile units
Anastasios et al. (2012)	Solution for optimal power flow in electrical engineering	Intelligent algorithm	5G environment for D2D and mobile units
Irfan Mulyawan Malik et al. (2012)	Solution for optimal power flow in electrical engineering	Intelligent algorithm	5G environment for D2D and mobile units
Attia et al. [12]	Solution for optimal power flow in electrical engineering	Intelligent algorithm	5G environment for D2D and mobile units

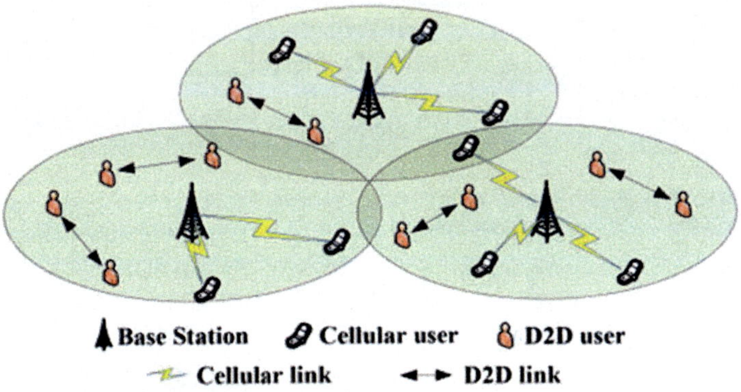

Fig. 1 D2D and cellular communication model

$$\sum_{i=1}^{k} P_{c,i} = P_c \tag{1}$$

ith band D2D spatial random distribution of users can also be modeled as a PPP $\varnothing_{d,i}$ with density $\lambda_{d,i}$ on <. The D2D transmission power of users and the D2D transmission power are

$$\sum_{i=1}^{k} P_{d,i} = P_d \tag{2}$$

4 Proposed Methodology

The research is to identify the D2D communication problems in the latest 5G network as it is the upcoming technology. The identified problem is in terms of real-world practical problem and need to formulate the mathematical model or equation which governs the practical approach. Once mathematical model is ready, conventional or intelligent tools may use to find the solution for the practical problem. The recent research trend is to use intelligent algorithms to solve the practical complex problems. In this research, intelligent algorithms such as genetic algorithm (GA), particle swarm optimization (PSO), spider monkey optimization (SMO), and hybrid of intelligent algorithms are aimed to use for finding solutions D2D communication problems in the latest 5G network (Fig. 2).

Energy optimization in D2D communication is one of the most important problems and may considered in 5G network. This problem might be solving with different

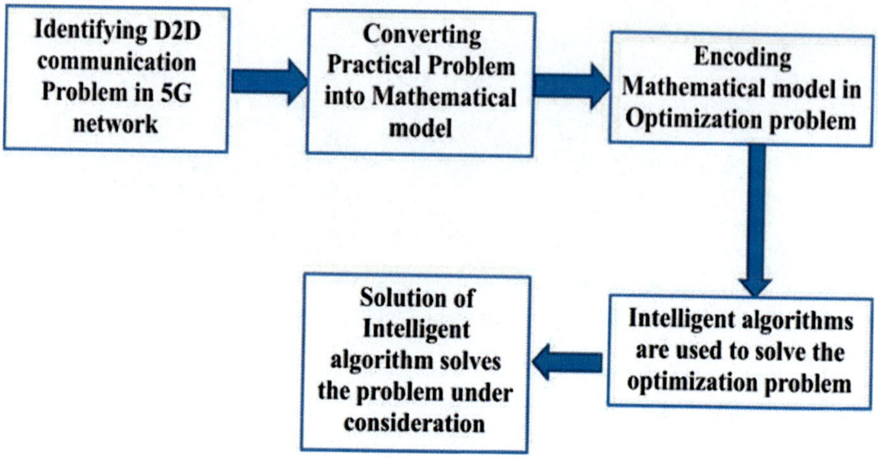

Fig. 2 Proposed methodology in D2D and cellular communication model

intelligent algorithms and from the solutions, suitable intelligent algorithm may identify. Suitable intelligent algorithm provides optimized energy consumption of the devices in the 5G network.

5 Conclusion

Device to device communication has explosive growth and needs energy efficient procedure to enhance the performance. The device identification in the geometry position is essential and communication among the devices should be efficient, reliable, and robust. The energy efficient is the predominant issue in the D2D communication. This is an optimization problem of con convex nature. This problem may solve efficiently by using hybrid intelligent algorithm since it is proved as optimization solver for many complex engineering problems. This hybrid intelligent algorithm finds best solution in solving energy efficiency in 5G networks, and discontinuous, non-convex and complex functions at reasonable time frame.

References

1. Misra, G., Agarwal, A., Misra, S., Agarwal, K.: Device to device millimeter wave communication in 5 g wireless cellular networks (a next generation promising wireless cellular technology). In: 2016 International Conference on Signal Processing, Communication, Power and Embedded System (SCOPES) (pp. 89–93). IEEE (2016)
2. Chen, X.: Efficient device to device communication underlaying heterogeneous networks (2016)
3. Zhang, Y., Yang, Y., Dai, L.: Energy efficiency maximization for device-to-device communication underlaying cellular networks on multiple bands. IEEE Access **4**, 7682–7691 (2016)
4. Chai, R., Liu, C., Chen, Q.: Energy efficiency optimization-based joint resource allocation and clustering algorithm for M2M communication systems. IEEE Access **7**, 168507–168519 (2019)
5. Liu, S., Wu, Y., Li, L., Liu, X., Xu, W.: A two-stage energy-efficient approach for joint power control and channel allocation in D2D communication. IEEE Access **7**, 16940–16951 (2019)
6. Alemaishat, S., Saraereh, O.A., Khan, I., Choi, B.J.: An efficient resource allocation algorithm for D2D communications based on NOMA. IEEE Access **7**, 120238–120247 (2019)
7. Gupta, S., Zhang, R., Hanzo, L.: Energy harvesting aided device-to-device communication in the over-sailing heterogeneous two-tier downlink. IEEE Access **6**, 245–261 (2017)
8. Kuo, W.K., Chu, S.H.: Energy efficiency optimization for mobile ad hoc networks. IEEE Access **4**, 928–940 (2016)
9. Zhou, Z., Gao, C., Xu, C., Chen, T., Zhang, D., Mumtaz, S.: Energy-efficient stable matching for resource allocation in energy harvesting-based device-to-device communications. IEEE Access **5**, 15184–15196 (2017)
10. Salim, M.M., Wang, D., Liu, Y., Elsayed, H.A.E.A., Elaziz, M.A.: Optimal resource and power allocation with relay selection for RF/RE energy harvesting relay-aided D2D communication. IEEE Access **7**, 89670–89686 (2019)
11. Ravi, C.N., Selvakumar, G., ChristoberAsirRajan, C.: Hybrid real coded genetic algorithm—differential evolution for optimal power flow. Int. J. Eng. Technol. (IJET) **5**(4), 3404–3412 (2013)
12. Attia, A.-F., Al-Turki, Y.A., Abusorrah, A.M.: OptimalPower flow using adapted genetic algorithm with adjusting population size. Electr. Power Components Syst. **40**(11), 1285–1299 (2012)

Discovery and Deterrence of Black Hole Attack in Clustering Ad Hoc Networks Based on Software Agents

A. Aranganathan, C. D. Suriyakala, and V. Vedanarayanan

Abstract Wireless technologies allow mobile devices to activate a wireless ad hoc network powerfully interfacing through the remote medium. A mobile ad hoc network is an assortment of versatile hubs with no defined centralized framework which shares a remote channel. Mobile ad hoc network is uncertain by its tendency, the explanation being the nodes reserve the option to join or leave the system with no limitation whenever at any given time. A node in mobile ad hoc network which experienced some assault displays an abnormal or malicious conduct. A node is said to be a malicious node if and just on the off chance that it experiences with at least one of following qualities: transmission capacity utilization, message altering, session catching, battery depletion, counterfeit directing, postponement of data bundles, connection break, parcel drop, copying data and stale data bundles. Packet dropping attack is called as black hole (BH) attacks is one of the routing attacks in network layer which affects the network performance by sending quick reply to the source node with high destination sequence number. This attack can be avoided using proposed Software Agents based Black Hole attacks Detection and Prevention in Clustering Ad hoc On-Demand Distance Vector (SABHDP-CAODV) routing protocol based on the unique identity of legitimate node, location-based distance calculation, threshold time with the simulation metrics of packet delivery ratio throughput, average end-to-end delay and network routing overhead when compared to existing systems.

Keywords Black hole attacks · Mobile ad hoc networks · Agent · Clustering networks · AODV

A. Aranganathan · V. Vedanarayanan
School of Electrical and Electronics, Sathyabama Institute of Science and Technology, Chennai, India
e-mail: arangaece@gmail.com

V. Vedanarayanan
e-mail: veda77etce@gmail.com

C. D. Suriyakala (✉)
School of Ocean Engineering & Underwater Technology, Kerala University of Fisheries & Ocean Studies, Kochi, Kerala, India
e-mail: drcdsk@yahoo.in

S. Borah et al. (eds.), *Soft Computing Techniques and Applications*, Advances in Intelligent Systems and Computing 1248,
https://doi.org/10.1007/978-981-15-7394-1_62

1 Introduction

"A mobile ad hoc network" is one of the emerging trends in wireless networks. The term ad hoc means in Latin word is "for this purpose only" and usually the nodes are mobile (movable) which forms temporary topology. Mobile Ad hoc networks mean the nodes are mobile when compared to a wired network. Wireless ad hoc networks do not have a centralized base station or gateway. Every node acts as a gateway that is interconnected by multi-hop wireless communication paths. The drawbacks of mobile ad hoc network is the quick changes in node movement which results in increased routing overhead and also consumes more energy because all the nodes are involved for data transmission by means of RREQ and RREP. Whole networks are divided into different forms of cluster. Each cluster contains cluster members, head node and a gateway. A group head is elected based on received signal strength. Cluster head acts as an administrator for all the cluster networks which passes all the information from one cluster network to another cluster networks. Cluster head nodes are called as master nodes and cluster members are called as slave nodes as shown in Fig. 1. In this paper, software agents act as an agent for collecting data packets from one network to other networks with reduced network overload; clustering networks with software agents have been implemented for secure routing, delivery of data packets, identifying legitimate nodes, black hole nodes and updating routing tables.

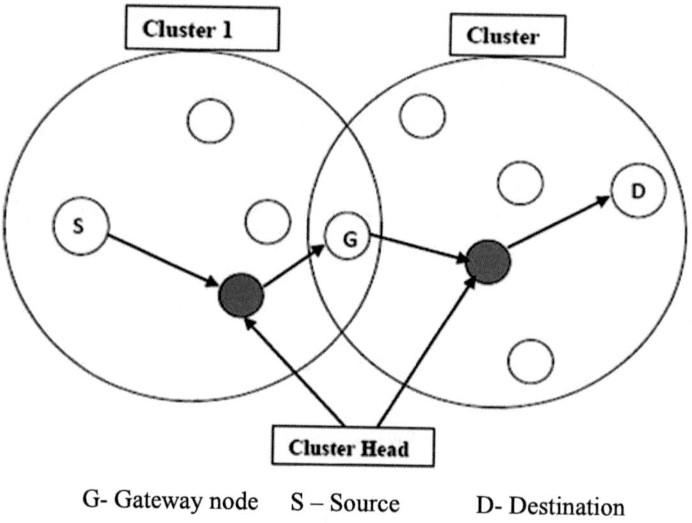

Fig. 1 Clustering ad hoc networks

2 Related Works

Routing attacks in network layer are serious attacks in mobile ad hoc networks. Inside and outside attacks are involved in routing attacks. Many researchers have analysed different routing attacks with various routing protocols for securing AODV in existing systems of Secure Ad hoc On-Demand Distance Vector (SAODV), cluster-based certificate revocation and agent-based detection against black holes for securing the networks. Some of the existing research has been reviewed about black hole using proactive, reactive and hybrid routing protocols as followed.

Kulbhushan and Jagpreet Singh [1] proposed fuzzy logic-based discovery scheme for black hole attacks. Every node can sense the behaviour of their own neighbouring node based on some extracted parameters which were analysed using fuzzy-based algorithm. Once a black hole attack is detected, then transmission of that particular node will be stopped and no packets will be forwarded to that node. Also, alarm packets with IP address of malicious nodes will be spread throughout the network.

Liu et al. [2] proposed a group-based credential disavowal with vindication capacity for improving the precision and unwavering quality of real nodes utilizing limit systems to make secure correspondence against pernicious aggressors. As indicated by node dependability, there are three classes, they are normal node, warned node and revoked node. Normal node with high unwavering quality to send information bundles to the neighbouring node and it might go about as a bunch part or group head. Warned nodes have few limitations for conveying information bundles with its neighbour node. It is a low unwavering quality node with a mix of authentic nodes and scarcely any malignant nodes. Revoked nodes with little unwavering quality can goes about as vindictive nodes. Affirmation authority is for checking the hubs conduct and is utilized for refreshing cautioned rundown and boycott of malevolent nodes. Certification authority is used for monitoring the behaviour and is utilized for refreshing cautioned rundown and boycott of vindictive nodes.

Ali Dorri et al. [3] proposed another system for distinguishing and taking out an agreeable black hole nodes with on-demand routing protocol. This approach involves a source node to verify both the present and the previous node for securing the path. Data routing information tables remove all the malicious nodes from the networks which decrease the processing time and eliminate false positive detection ratio in packet overhead.

Chavan et al. [4] analysed the network performance of AODV and DSDV routing schemes with black hole attacks using different evaluation metrics scenarios. There is an improvement in packet delivery ratio, throughput, routing overhead in AODV routing protocol with a black hole attack is compared to DSDV routing protocol. There is a high end-to-end delay in AODV because of route discovery process when compared with DSDV protocol. So the author modified AODV routing protocol in the existence of single and collaborative outbreak with route reply and check verification process for efficient detection of black hole attack.

Dhende et al. [5] demonstrated SAODV protocol implemented for the detection and prevention of black hole and grey hole attack. There are two tables maintained to observe the activities of the node in the grid, namely neighbouring list table to maintain the record of all the neighbouring nodes with ID and another is opinion table is used to detect the black and grey hole kind of abuse. After getting reply from neighbouring node, source node sends opinion message to the entire node about the replied node within estimated time period. The source node confirms such that the neighbouring node is malicious one or otherwise. If the source node receives an ok message about the replied nodes, the node is legitimate node, otherwise, the node is malicious and correspondingly opinion tables are updated periodically. Alarm messages sent to the network about the new update by the source node.

Detecting attacks by intrusion detection systems [15–23] and securing in mobile ad hoc networks is a challenging because of no fixed infrastructure, dynamical topology and limited resources. From the literature review, it is observed that modified routing protocols have reduced packet delivery ratio, less throughput, high average end-to-end delay and high network overhead. In this paper, Software Agents based discovery and deterrence in Clustering Mobile Ad hoc Networks against Black Hole attacks (SABHDP-CAODV) are focused based on Ad hoc On-Demand Distance Vector Routing Protocol.

3 Proposed Methodology

In the existing papers, many researchers have implemented using different methods of black hole attack discovery and deterrence with various routing protocols. Even though using mobile agent affects network traffic congestion due to continuous reply to the agent in existing routing schemes. Still now security is a challenging task in mobile ad hoc networks. These problems can be overcome by deploying agents in clustering networks to free from congestion problems and also gives less energy consumption. In clustering networks, only agents can be communicated for data transmission by means of RREQ and RREP rather than individual node communication. Each agent controls the cluster members both intra-cluster and inter-cluster communication to avoid duplicate packet transmission. Agent can be elected based on received signal strength indicator (RSSI). In this figure, initially, agent sends hello message to the entire neighbour node for data transmission to share about individual unique node ID, location can be identified based on the distance from the source node, hop count details and wait for reply within predetermined time of 30 s about all information. Otherwise, it assumed as a malicious node and the same process will be continued for agent selection. Agent controls the data bundles in the clustering network. Three-zone agents are deployed for message packets transmission as shown in Fig. 2. If zone agent1 wants to transfer message packets to node B, it is an intra-cluster communication. In zone agent A1 has two nodes A&B, zone agent A2 has C, D & E and zone Agent A3 has F, G, H, & I. Suppose if zone agent A1 wants to transmit packets to node F, mainly it searches within the cluster zone A1, next it sends the requested copies to neighbouring cluster zone agents A2 and A3. If the node F is under agent 3, it sends RREP to agent 1. Other agent zone data packets are discarded.

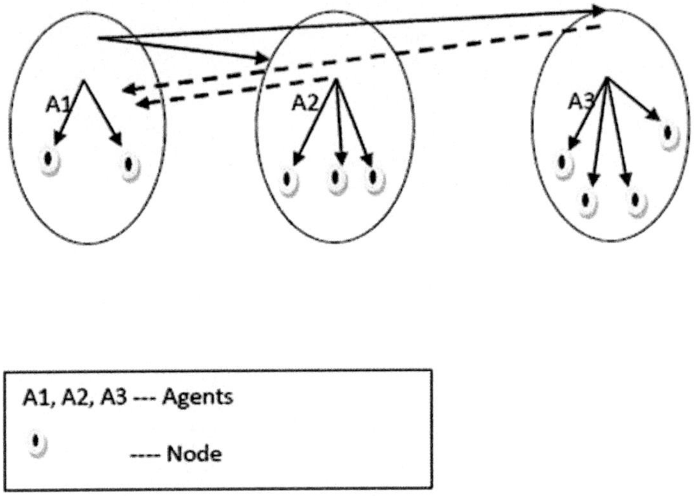

Fig. 2 Software agents based CAODV against black hole attack

It is called as agent-based inter-cluster communication networks. Routing table will be updated once in 50 s to avoid black hole node entry. Zone agents are mainly responsible for sharing data packets in both intra- and inter-cluster communication.

4 Results and Summary

4.1 Simulation Parameter

The proposed method SABHDP-CAODV was experimentally verified by network simulator tool (2.35 ns) with simulation metrics of throughput, network routing packet delivery ratio, overhead and avg. end-to-end delay. With this parameter, existing system of SABHDP-AODV basic AODV with black hole node (AODV-BH) and the proposed system has been analysed. In order to set simulation area of 1000 * 1000 m with a transmission range of 150 m with simulation time of 200 s, all the nodes have a mobility speed of (20–80) m/s, packet size of 512 bytes per second and pause time of 10 s. Black hole nodes are located in any of the path from source to destination nodes and drop the data packets for reducing the network performance. The simulation parameter is shown in Table 1.

4.1.1 Packet Delivery Ratio (PDR)

Figure 3 shows that as the numbers of black hole nodes are increased, hence a slight

Table 1 Simulation parameter of SABHDP-CAODV

Parameter	Value
Coverage area	1000 * 1000 m
Transmission range	150 m
Simulation time	200 s
Number of nodes	100
Routing protocol	ZRP
Number of black hole nodes	2, 4, 6, 8
Packet size	512 bytes
Mobility speed	(20–80) m/s
Pause time	10 s

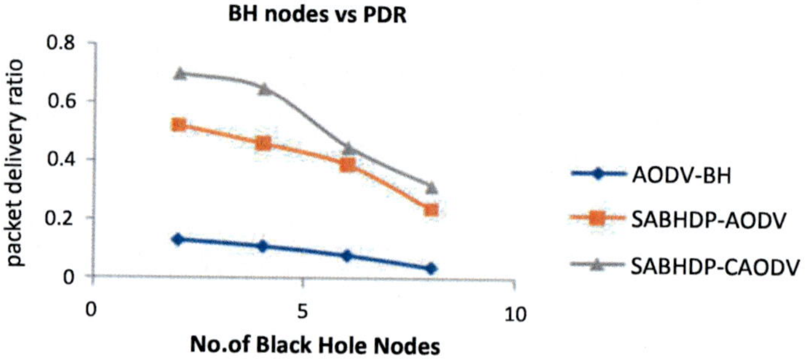

Fig. 3 Packet delivery ratio versus number of BH nodes

reduction in delivery of data packets due to black hole nodes drops some reply packets from unknown nodes in proposed SABHDP-CAODV when compared to the existing method of SABHDP-AODV and basic AODV protocol.

4.1.2 Throughput Ratio

Figure 4 depicts that the numbers of black hole nodes are increased, a reduction in throughput in proposed SABHDP-CAODV when compared to the existing method of SABHDP-AODV and basic AODV protocol.

4.1.3 Average End-to-End Delay

Figure 5 shows the numbers of black hole nodes are increased, there is a slight increase of end-to-end delay in SABHDP-CAODV when compared to the existing method of SABHDP-AODV and basic AODV protocol.

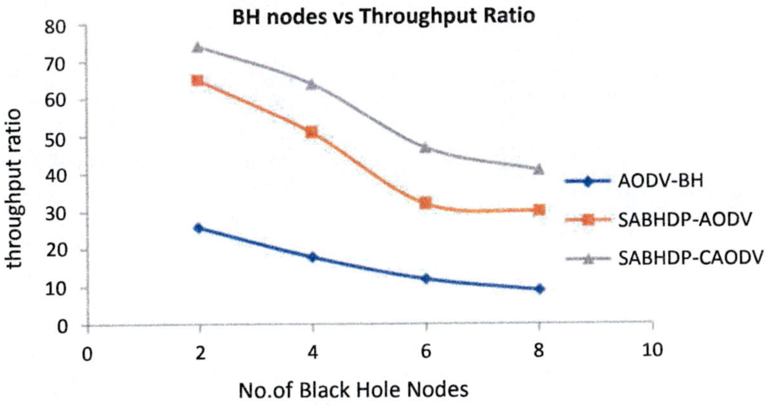

Fig. 4 Throughput versus number of BH nodes

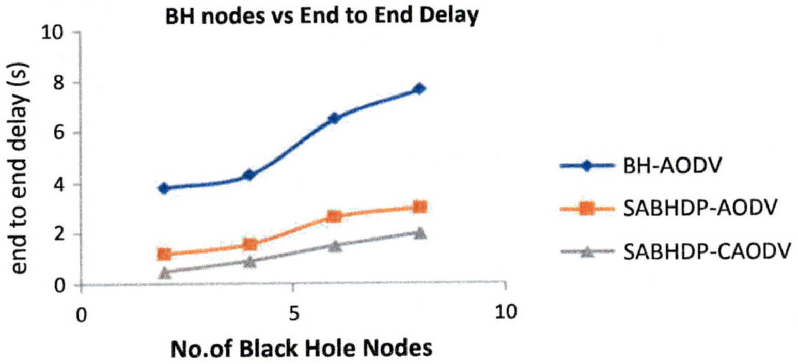

Fig. 5 Average end-to-end delay versus number of BH nodes

4.1.4 Network Routing Overhead

Figure 6 shows that an increase in black hole nodes, a slight increase of network overhead in found I the proposed SABHDP-CAODV when compared to the existing method of SABHDP-AODV and basic AODV protocol.

5 Summary

It is summarized that software agents based detection and prevention in clustering-based AODV protocol against multiple black hole attacks has been experimentally verified. In multiple black hole nodes, as the number of black hole nodes increases gradually, data packets are dropped gradually decreased. Simulation results show

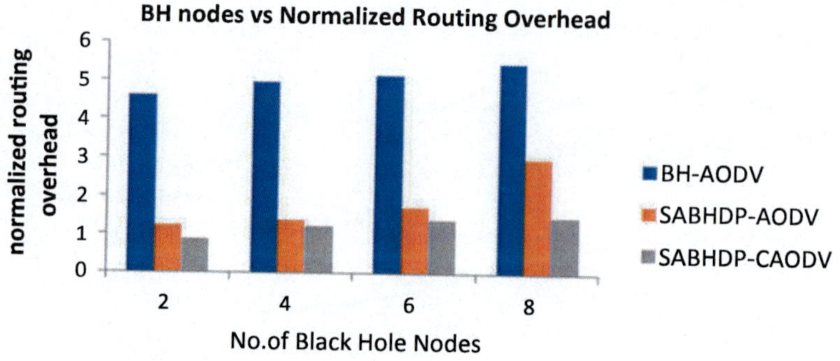

Fig. 6 Network routing overhead versus number of BH nodes

that Clustering-based Ad hoc On-Demand Distance Vector Routing (AODV) protocol performs better efficiency with respect to packet delivery ratio, throughput ratio, end-to-end delay and routing overhead compared to existing systems of SABHDP-AODV, basic AODV protocols against black hole attacks.

6 Conclusion and Future Work

Ad hoc On-Demand Distance Vector Routing mechanism is the efficient routing procedure of mobile ad hoc networks. It is widely used in military, disaster management, audio-video conferencing, etc. Therefore, in this paper, software agents based detection and prevention of black hole attacks have been successfully implemented in clustering mobile ad hoc networks. Simulation results achieved better performance in the proposed method in terms of simulation metrics are increased packet delivery ratio, high throughput, less end-to-end delay and minimum network overhead. The work can be extended to evaluate the performance analysis of reactive routing protocol, proactive routing protocol and multicast routing protocol with other routing protocols in clustering mobile ad hoc networks.

References

1. Kulbhushan, Singh, J.: Fuzzy logic based intrusion detection system against black hole attack on AODV in MANET. In: IJCA special on Security and Cryptography, pp. 28–35 (2011)
2. Liu, W., Nishiyama, H., Ansari, N., Yang, J., Kato, N.: Cluster based certificate revocation with vindication capability for mobile ad hoc networks. IEEE Trans Parallel Distrib. Syst. **24**(2), 239–249 (2013)
3. Dorri, A., Nikdel, H.: A new approach for detecting and eliminating cooperative black hole nodes in MANET, Department of Computer Engineering, Mashhad Branch, Islamic Azad University, Mashhad. In: IRAN IKT2015 7th International Conference on Information and Knowledge Technology, IEEE Xplore, ISBN 978-1-4673-7485-9, pp 1-6 (2015)

4. Chavan, A.A., Kurule, D.S, Dere, P.U.: Performance analysis of AODV and DSDV routing protocol in MANET & modification in AODV against black hole attack. In: 7th International Conference on Communication, Computing & Virtualization, Procedia Computer Science, pp. 835–844 (2016)
5. Dhende, S., Musale, S., Shirbahadurkar, S., Najan, A.: SAODV: Black hole and gray hole attack detection protocol in MANETs. In: International Conference on Wireless Communications, Signal Processing and Networking, pp. 2391–2394 (2017)
6. Nadeem, A., Howarth, M.P.: A survey of MANET intrusion detection & prevention approaches for network layer attacks. In: Proceedings of IEEE Communications Surveys & Tutorials, vol. 15(4), pp. 1–17 (2013)
7. Belding-Royer, E.M.: Hierarchical routing in ad hoc mobile networks. Wireless Commun. Mob. Comp. **2**(5), 515–532 (2002)
8. Pattanayak, B.K., Rath, M.: A mobile agent based intrusion detection system architecture for mobile ad hoc networks. J. Comput. Sci. **10**(6), 970–975 (2014). ISSN: 1549-3636
9. Kannhavong, B., Nakayama, H., Nemoto, Y., Kato, N.: A survey of routing attacks in mobile ad hoc networks, Tohoku University, Abbas Jamalipour, University of Sydney, Security in Wireless Ad hoc and Sensor Networks—IEEE Wireless Communications, pp. 85–91 (2007)
10. Chowdhury, C., Neogy, S.: Securing mobile agents in MANET against attacks using trust. Int. J. Network Secur. Its Appl. **3**(6), 259–279 (2011)
11. Hu C, Swami A, Chen RA (2011) Survey on trust management for mobile ad hoc networks. IEEE Commun. Surv. Tuts. **13**(4), 562–583 (2011)
12. Roy, D.B., Chaki, R.: BAIDS: detection of black hole attack in MANET by specialized mobile agent. Int. J. Comput. Appli. **40**(13):1–6 (2012)
13. Aranganathan, A., Suriyakala, C.D.: Software agents based trusted revocation for securing clustering MANETs. J. Adv. Res. Dynam. Control Syst. (JARCDS) **10**(special issue-02), 915–920 (2018). ISSN No: 1943-023X
14. Kukreja, D., Dhurandher, S.K., Reddy, B.V.R.: Power aware malicious nodes detection for securing MANETs against packet forwarding misbehavior attacks. **9**(4), 941–956 (2018)
15. Dutt, I., Borah, S., Moitra, I.K.: Immune system based intrusion detection system (IS-IDS): a proposed approach. IEEE Access **8**(1), 34929–34941 (2020). ISSN:2169-3536. https://doi.org/10.1109/access.2020.2973608
16. Dutt, I., Borah, S., Moitra, I.K.: Survey of bio-inspired techniques based on system resource usage in intrusion detection. Test Eng. Manage. **82**, 5734–5738 (2020). ISSN 0193-4120
17. Panigrahi, R., Borah, S.: A statistical analysis of lazy classifiers using Canadian Institute of Cybersecurity Dataset. In: Proceedings of ICDSM-2019, Bhubaneswar, India. Springer Lecture Notes on Data Engineering and Communication Technology, vol. 37, pp. 215–222 (2019). ISSN 2367-4512. ISBN 978-981-15-0978-0
18. Prasad, K.K., Borah, S.: Use of genetic algorithms in intrusion detection systems: an analysis. Int. J. Appl. Res. Stud. (IJARS) **2**(8), 01-07 (2013). ISSN: 2278-9480
19. Dutt, I., Borah, S.: Some studies in intrusion detection using data mining techniques. Int. J. Innov. Res. Sci. Eng. Technol. **4**(7), 5500–5511 (2015)
20. Dutt, I., Borah, S., Maitra, I.: Intrusion detection system using artificial immune system. Int. J. Comput. Appl. **144**(12), 19–22 (2016). ISSN: 0975-8887
21. Panigrahi, R., Borah, S.: Rank of normalizers through TOPSIS with the help of supervised classifiers. Int. J. Eng. Technol. **7**(3.24), 483–490 (2018)
22. Panigrahi, R., Borah, S.: A detailed analysis of CICIDS2017 dataset for designing intrusion detection systems. Int. J. Eng. Technol. **7**(3.24), 479–482 (2018)
23. Panigrahi, R., Borah, S.: Rank allocation to J48 group of decision tree classifiers using binary and multiclass intrusion detection datasets. Proc. Comput. Sci. **132**, 323–332 (2018)

Author Index

S. Borah et al. (eds.), *Soft Computing Techniques and Applications*,
Advances in Intelligent Systems and Computing 1248,
https://doi.org/10.1007/978-981-15-7394-1